European Navigation Conference ENC 2023

European Navigation Conference ENC 2023

Editors

Tom Willems
Okko Bleeker

MDPI

Basel • Beijing • Wuhan • Barcelona • Belgrade • Novi Sad • Cluj • Manchester

Editors

Tom Willems
Netherlands Institute of Navigation
Rotterdam
The Netherlands

Okko Bleeker
OFBConsult
Alphen aan den Rijn
The Netherlands

Editorial Office
MDPI
St. Alban-Anlage 66
4052 Basel, Switzerland

This is a reprint of articles from the Proceedings published online in the open access journal *Engineering Proceedings* (ISSN 2673-4591) (available at: https://www.mdpi.com/2673-4591/54/1).

For citation purposes, cite each article independently as indicated on the article page online and as indicated below:

Lastname, A.A.; Lastname, B.B. Article Title. *Journal Name* **Year**, *Volume Number*, Page Range.

ISBN 978-3-7258-0591-4 (Hbk)
ISBN 978-3-7258-0592-1 (PDF)
doi.org/10.3390/books978-3-7258-0592-1

Cover image courtesy of Pixabay contributor 13464740

Contents

About the Editors

Tom Willems

In 2006, Tom Willems obtained a PhD degree from Ghent University, Belgium, where he conducted research on regional augmentation systems. He is also an alumnus of the International Space University's Space Studies Program. During the summer of 2004, Willems was a visiting scholar at the Institute of Engineering Surveying and Space Geodesy of the University of Nottingham, thanks to a Marie Curie Fellowship. Between 2006 and 2021, he worked as an embedded software engineer in GNSS signal processing and as a system engineer at Septentrio and Antwerp Space. During his time at both companies, he was deeply involved in the development of the Galileo Test User Receivers for the European Space Agency. In 2018, he was elected as a Fellow of the Royal Institute of Navigation and was also a co-recipient of a GPS World Leadership Award. Currently, he works as an independent consultant. He is also affiliated with Department of Geography of Ghent University as a voluntary postdoctoral researcher. He serves as a board member of the Netherlands Institute of Navigation and the International Association of Institutes of Navigation, and as the Executive Secretary of the European Group of Institutes of Navigation.

Okko Bleeker

Okko Bleeker is the Technical Director for a long-range remotely controlled cargo aircraft (RCCA) project (CS-25 class 'Large Aeroplanes'). This includes RCCA business rationale, system functions, the ground control segment (GCS), and the networking aspect. The project has evolved to a key technology demonstration in order to clear TRL-7 in support of a proof of concept. He is involved in aircraft total system error (TSE) factor analysis, a part of RNP Approach/LPV procedure design, and related air traffic control practice/PBN compliance (ANSP projects). He is a designated expert for the European Commission. From 1995 to 2015, he was the Director of Research and Development for Europe with Rockwell Collins (now Collins Aerospace). His activities included developing numerous EU projects, commercial systems, as well as government/military and security projects. He conducted and initiated projects with the European Commission, SESAR JU, Eurocontrol, the European Space Agency (ESA), NATO/NIAG (NATO Industrial Advisory Group), and European airlines and Air Forces. These projects were generally guidance navigation and control (GNC) oriented. Between 1974 and 1995, he was employed by Fokker Aircraft. His final position was as a program manager for Corporate Research and Development Aircraft Systems. Before that, he was the Avionics Group Manager with responsibilities in design and certification for aircraft avionics and flight deck arrangements, specifically Fokker 100 passenger aircraft. Before that (1974–1980), he was a flight crew training instructor. Okko Bleeker graduated in 1971 with a BSc from the Leeuwarden Institute of Technology. His education was complemented by Cranfield University (aircraft design system safety assessment), the US Defense Systems Management College, and other post-graduate training courses.

engineering proceedings

MDPI

Editorial

Preface: The European Navigation Conference 2023 (ENC 2023) †

Tom Willems [1,*] and Okko Bleeker [2]

1 Netherlands Institute of Navigation, 3009 AD Rotterdam, The Netherlands
2 OFBConsult, 2402 PP Alphen Aan Den Rijn, The Netherlands; ofbleeker@wxs.nl
* Correspondence: tom.willems@navnin.nl

† All papers published in the volume were presented at the European Navigation Conference 2023 (ENC 2023), Noordwijk, The Netherlands, 31 May–2 June 2023.

1. Introduction

The European Navigation Conference 2023 (ENC 2023) took place at the European Space Research and Technology Centre (ESTEC) of the European Space Agency (ESA) in Noordwijk, The Netherlands, from 31 May until 2 June 2023.

ENC 2023 addressed Resilient Navigation, in its broadest sense, as the main theme. 'Navigation' comprises the art of monitoring and controlling a vehicle from its current position to an intended position along a route, with due recognition of constraints. Thus, the conference was structured to address resilience in those main navigation sub-functions.

Resilience in position determination concerns the robustness of individual components by considering redundancy, dissimilarity, specific algorithms, and data processing techniques. This applies to all elements of the Position Navigation and Timing (PNT) solution: the signal measurement domain, any infrastructural elements, service provision, and end-user implementation.

Over and above position determination, however, navigation resilience also requires vulnerabilities to be addressed in the various types of data used for navigation, including map information, various databases, routing/path construction, and guidance and control. Separate from those functional matters, application domains such as aviation, maritime, and surface navigation have widely different implementations. Examples of aspects that must also be considered in the context of overall navigation resilience are man–machine interfaces, function allocation between electronic elements and the human operator, crew complement, training and proficiency/skill base, developments in rules and regulations (ICAO/Aviation, COLREG/Maritime), and the quality and use of meteorological and oceanographic information and corresponding measures.

Citation: Willems, T.; Bleeker, O. Preface: The European Navigation Conference 2023 (ENC 2023). *Eng. Proc.* **2023**, *54*, 63. https://doi.org/10.3390/engproc2023054063

Published: 7 February 2024

2. Sessions

In response to submitted papers considering the topics described above, the conference was organized into five tracks, comprising a total of 134 presentations arranged in 18 sessions, including one poster session, to enable the audience to attend presentations consistent with their personal interest. The audience count was 350 from around the world, including about 50 students from both the Netherlands and international universities. The program consisted of the following tracks and parallel sessions:

- GNSS and Hybridized Navigation;
    - Interference and Jamming;
    - Spoofing and Meaconing;
    - Performance Prediction, Monitoring and Assessment;
    - Alternative and Complementary PNT;
    - Multi-Sensor and Augmented PNT;

- Advanced GNSS Techniques;
- Integrity Algorithms;
- Atmospheric Modelling and Sensing;
- High-Accuracy Techniques;

- Extra-Terrestrial Navigation;
  - Space Navigation;
  - Moon/Mars Navigation Services;
- AI, Machine Learning, Simulation and Testing;
  - Use of AI and Machine Learning;
  - Simulation, Testing, Analysis Tools and Results;
- Integral Navigation Systems;
  - Reference Trajectory Optimization;
  - End-to-End Navigation Systems;
- LEO PNT Developments;
  - LEO-GNSS Synergism;
  - LEO User Equipment Considerations.

3. Plenary Talks

A range of plenary talks was held, addressing contemporary requirements and developments in the field of navigation, the need for resilience and ways to achieve it.

Minister Mark Harbers' opening speech [1] addressed the subject quite elegantly, referencing astronomer Richard Carrington's observations of extreme sun activity back in September 1859 and, even in those days, the subsequent disruption of a range of services, including communication, navigation, and electrical power distribution. The example highlighted the need for resilience two centuries ago, but much more so in today's world, with the ever-increasing dependence on radio- and networked-computer-based navigation, communication, and other essential services.

Plenary talks included:

- Conference Opening Speech—Mark Harbers, Minister of Infrastructure and Water Management of the Netherlands;
- ESA ESTEC—Dietmar Pilz, ESA;
- Towards the Future of Satellite Navigation—Javier Benedicto, ESA;
- EUSPA—Pascal Claudel, EUSPA;
- Spaceopal—Marco Folino, Spaceopal;
- GPS Status Update—Robert Wray, US Space Force;
- Galileo Second Generation—Miguel Manteiga (ESA) and Heads of Industry;
- Lunar Navigation and Moonlight Programme Overview—Javier Ventura, ESA;
- Lunar Pathfinder Navigation Experiment—Pietro Giordano, ESA;
- U.S. President's National Space-Based PNT Advisory Board—Dana Goward (Resilient Navigation and Timing Foundation), Frank van Diggelen (Google), Terry Moore (Royal Institute of Navigation);
- LEO PNT Introduction—Sandra Verhagen, Delft University of Technology.
- FutureNAV LEO PNT In-Orbit Demonstration Program—Roberto Prieto, ESA;
- PULSAR LEO PNT GNSS Augmentations—Bryan Chan, Xona;
- Next Big Thing—Maarten Uijt de Haag, Technische Universität Berlin.

Of particular interest was an announcement regarding Galileo Second Generation (G2), including an on-stage ceremony. The ceremony involved contracts to Thales (Italy), Airbus Defence & Space (Germany), and Thales Six GTS (France) for system engineering and technical assistance in support of the full development phase of G2 [2].

Equally inspiring was the briefing by Lt. Col. Wray of the US Space Force (USSF). Mr. Wray oversees the team responsible for operating and maintaining the US Global Positioning System. He emphasized the international importance of the system, but also some key differences between GPS operation and that of other Global Navigation Satellite Systems worldwide, particularly GPS operation by the (military) USSF, with a view to the specific signals and tactics designed to support American, allied and partner military capabilities.

The plenary talks were completed by Prof. Maarten Uijt de Haag, with his stimulating and dynamic speech 'The Next Big Thing'. Prof. Uijt de Haag took the audience from the elementary notions of navigating in the real world, employing sensors, PNT algorithms, and state estimation, as well as comprehension and prediction, to the examples, such as urban swarm navigation. He was looking forward to future challenges in the field of PNT, where many use-cases require complementary PNT approaches, using extensive forms of integration to meet the required navigation performance (RNP). To enable a rapid and cost-effective development cycle, good tools and interface definitions will be needed including All Source Positioning and Navigation (ASPN) and the PNT Operating System (pntOS) that both have been developed to support the growing interest in the Modular Open Systems Approach (MOSA) for PNT. Such new concepts will reflect the elementary system, but will use an extensive networking of the various operational elements to achieve an improved system-level optimum.

4. Outstanding Paper Awards

During the conference, the audience was invited to express their opinions and appreciation regarding the presentations they attended. This public vote resulted in nine ENC 2023 Outstanding Paper Awards, which were handed out by the technical committee at the end of the final day of the conference.

In addition, on that same occasion, Ms. Valérie Renaudin, Editor-in-Chief of the *IEEE Journal of Indoor and Seamless Positioning and Navigation* (J-ISPIN) handed out the Best Early Stage Research Award, for a preliminary analysis of early datasets in 5G positioning.

5. Acknowledgements

ENC 2023 would like to take this opportunity to express our sincere gratitude to all involved in making the conference a success, including the Netherlands Institute of Navigation, the Local Organizing Committee, the Scientific–Technical Program Committee, the International Advisory Committee, the technical directors, session chairs and co-chairs, the paper reviewers, and the editorial team, who all and without exception voluntarily spent considerable time and effort over a period of more than one year.

In particular, we would like to thank our conference partners (Figure 1):

Figure 1. ENC 2023 conference partners.

- ESA;
- EUSPA;
- Ministry of Infrastructure and Water Management;
- Samenwerkende Maritieme Fondsen of the Netherlands.

We extend our special thanks to our gold sponsors (Figure 2):

- GMV;
- Netherlands Aerospace Centre;
- Spaceopal;
- S[&]T.

Figure 2. ENC 2023 gold sponsors.

In addition, we sincerely thank our silver sponsors (Figure 3):

Figure 3. ENC 2023 silver sponsors.

- Safran;
- Spirent.

We would also like to thank our media partners:

- Inside GNSS;
- European Journal of Navigation;
- IEEE Journal of Indoor and Seamless Positioning and Navigation (J-ISPIN);
- Delft Career Platform.

Finally, we sincerely thank all the authors who contributed to the publication of this volume. Thanks to their efforts, we believe that the scope and depth of these papers offers a great wealth of information, and we hope that you will enjoy reading this collection of peer-reviewed papers.

Conflicts of Interest: The authors declare no conflicts of interest.

References

1. Speech by Mark Harbers, Minister of Infrastructure and Water Management, at the European Navigation Conference 2023, Noordwijk. Available online: https://www.government.nl/documents/speeches/2023/05/31/speech-by-mark-harbers-minister-of-infrastructure-and-water-management-at-the-european-navigation-conference-2023-noordwijk (accessed on 1 July 2023).
2. Galileo Second Generation Enters Full Development Phase. Available online: https://www.esa.int/Applications/Navigation/Galileo_Second_Generation_enters_full_development_phase (accessed on 1 July 2023).

engineering proceedings

MDPI

Editorial

Statement of Peer Review †

Tom Willems [1,*] and Okko Bleeker [2]

1 The Netherlands Institute of Navigation, 3009 AD Rotterdam, The Netherlands
2 OFBConsult, 2402 PP Alphen Aan Den Rijn, The Netherlands; ofbleeker@wxs.nl
* Correspondence: tom.willems@navnin.nl
† All papers published in the volume were presented at the European Navigation Conference 2023 (ENC 2023), Noordwijk, The Netherlands, 31 May–2 June 2023.

In submitting conference proceedings to *Engineering Proceedings*, the volume editors of the proceedings certify to the publisher that all papers published in this volume have been subjected to peer review administered by the volume editors. Reviews were conducted by expert referees to the professional and scientific standards expected of a proceedings journal.

- Type of peer review: single-blind
- Conference submission management system: Sciforum and through e-mail.
- Number of submissions sent for review: 71 papers
- Number of submissions accepted: 58 papers
- Acceptance rate (number of submissions accepted/number of submissions received): 82%
- Average number of reviews per paper: 2
- Total number of reviewers involved: 30
- Any additional information on the review process: Only authors of accepted abstracts were invited to submit a paper for review.

Conflicts of Interest: The authors declare no conflicts of interest.

Citation: Willems, T.; Bleeker, O. Statement of Peer Review. *Eng. Proc.* **2023**, *54*, 62. https://doi.org/10.3390/engproc2023054062

Published: 7 February 2024

engineering proceedings

MDPI

Proceeding Paper

Quad-Band Multi-Constellation Global Navigation Satellite System Receiver Development Platform with System-on-Chip Architecture †

Muhammad Saad *,‡, Fabio Garzia ‡, Szu-Jung Wu ‡, Iñigo Cortés ‡, Frank Förster ‡, Matthias Overbeck, Santiago Urquijo and Wolfgang Felber

Satellite Based Positioning Systems Department, Fraunhofer IIS, Nordostpark 84, 90411 Nuremberg, Germany; fabio.garzia@iis.fraunhofer.de (F.G.); szu-jung.wu@iis.fraunhofer.de (S.-J.W.); inigo.cortes@iis.fraunhofer.de (I.C.); frank.foerster@iis.fraunhofer.de (F.F.); matthias.overbeck@iis.fraunhofer.de (M.O.); santiago.urquijo@iis.fraunhofer.de (S.U.); wolfgang.felber@iis.fraunhofer.de (W.F.)

* Correspondence: muhammad.saad@iis.fraunhofer.de

† Presented at the European Navigation Conference 2023, Noordwijk, The Netherlands, 31 May–2 June 2023.

‡ These authors contributed equally to this work.

Abstract: GNSS receivers with multi-system and multi-frequency capabilities allow more reliable positioning, especially in challenging environments. However, increased available systems and signals lead to hardware resource allocation problems. This issue escalates when ten correlators per tracked GNSS signal are used. Moreover, the communication interface between the PL and the PS affects the time complexity and challenges the capability of the maximum number of tracking channels. This paper presents the GOOSE v2, a MPSoC-based GNSS receiver, which is composed of a quad-band (L1, L2, L5/E5 and S band) RFFE and baseband digital signal processing. The SoC includes a quad-core 64-bit processor and an FPGA. The on-chip communication between the processors and the FPGA offers high bandwidth, significantly reducing the time complexity. The preliminary evaluation of this new receiver platform shows that, in addition to the legacy signals, it supports NavIC L5 and S-band signals. Furthermore, this receiver closes the tracking loops faster than its legacy version (GOOSE v1), opening the door to implementing more complex algorithms requiring higher time complexity.

Keywords: global navigation satellite system (GNSS); navigation with Indian constellation (NavIC); system-on-chip (SoC); field programmable gate array (FPGA)

Citation: Saad, M.; Garzia, F.; Wu, S.-J.; Cortés, I.; Förster, F.; Overbeck, M.; Urquijo, S.; Felber, W. Quad-Band Multi-Constellation Global Navigation Satellite System Receiver Development Platform with System-on-Chip Architecture. *Eng. Proc.* 2023, 54, 1. https://doi.org/10.3390/ENC2023-15439

Academic Editors: Tom Willems and Okko Bleeker

Published: 29 October 2023

1. Introduction

Global Navigation Satellite System (GNSS)-based positioning is growing at a rapid pace, and more devices (smartphones and upcoming application fields like autonomous driving or Internet of Things (IoT)) are equipped with GNSS receivers, enabling precise and reliable Position, Velocity and Time (PVT) solutions. According to the European Union Agency for the Space Programme (EUSPA) Earth Observation (EO) and GNSS market report for 2022 [1], it is forecasted that the annual shipments of GNSS receivers are to grow from 1.8 billion units to 2.5 billion units between 2021 and 2031. As a result, the global installed base of GNSS devices in use is expected to reach over 10 billion units by 2031. Consequently, the underlying satellite infrastructure system is also growing, and more countries tend to have independent satellite constellation systems. The same is true for Asia-Pacific and India. Instead of relying on other global systems (e.g., Global Positioning System (GPS), Galileo, BeiDou, or GLONASS), regional systems, such as Navigation with Indian Constellation (NavIC), were established to provide independent and tailored services to the users of a specific region.

More satellite systems mean more Satellite Vehicle (SV) availability, lower Dilution Of Precision (DOP) and an improved PVT solution. However, the increased availability of systems offers more challenges to the receiver manufacturers in terms of hardware and software resources, along with increased power consumption. Furthermore, GNSS receivers should maintain the synchronization of a higher number of SVs. GNSS receivers like the GNSS-Receiver with Open Software Interface (GOOSE)© platform [2] implement the tracking stage partly in hardware (correlators and Numerically Controlled Oscillator (NCO)) and partly in software (discriminators and loop filters). These receivers try to close the loop of all the tracking channels before a new correlation is performed. A low time complexity of the communication interface between the processor and the dedicated digital Hardware (HW) results in faster response and is essential to close the loop in time, avoiding synchronization failures. Also, the faster the response, the more tracking channels the GNSS receiver can manage. Moreover, a low time complexity in the communication interface enables the implementation of more complex and robust tracking architectures, such as Loop-Bandwidth Control Algorithm (LBCA)-based tracking architectures [3–5].

This work presents the GOOSE© v2 GNSS receiver, an enhanced version of the legacy GOOSE© [2], capable of receiving and processing four GNSS bands: L1, L2, L5/E5 and S-band. The latter could also be optionally replaced by E6. In addition to the existing legacy systems, the GOOSE© v2 also supports NavIC L5 and S-band signals. Figure 1 presents the GOOSE© v2 receiver platform.

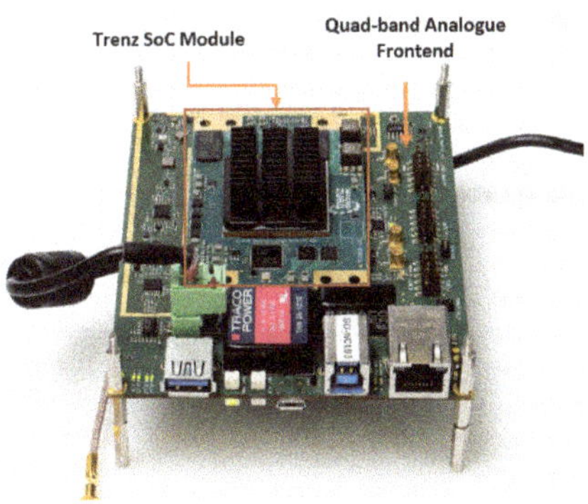

Figure 1. GOOSE© v2 receiver platform.

The receiver architecture is based on the Commercial-Off-The-Shelf (COTS) Multi-Processor System-on-Chip (MPSoC) module designed by Trenz Electronic [6]. The Programmable Logic (PL) part of this device contains a Fast Fourier Transform (FFT) core, a pseudo-random noise (PRN) generator module and 60 tracking channels. The Processing System (PS) includes a quad-core 64-bit processor and implements the code/carrier tracking loops, navigation message decoding and PVT algorithm. The System-on-Chip (SoC) feature allows the exchange between HW and Software (SW) with a very high bandwidth—thanks to the on-chip communication infrastructure.

This research performs a preliminary evaluation of the GOOSE© v2 by comparing it with its legacy version, evaluating the tracking of the NavIC signals and testing the capability of managed tracking channels.

The rest of the paper is organized as follows. Section 2 describes the proposed GNSS receiver named GOOSE© v2 and addresses the main updates compared to the legacy GOOSE© v1 receiver [7,8]. Section 3 presents the preliminary evaluation of the GOOSE© v2. Finally, Section 4 concludes and indicates future work.

2. Receiver Design

The GOOSE© v2 GNSS receiver is composed of an analog and a digital part. The digital part is characterized by the Xilinx MPSoC composed by the PS for the processing implemented in SW and the PL, a kind of integrated Field-Programmable Gate Array (FPGA), to accelerate digital processing functions. Figure 2 shows the main architecture of the proposed platform.

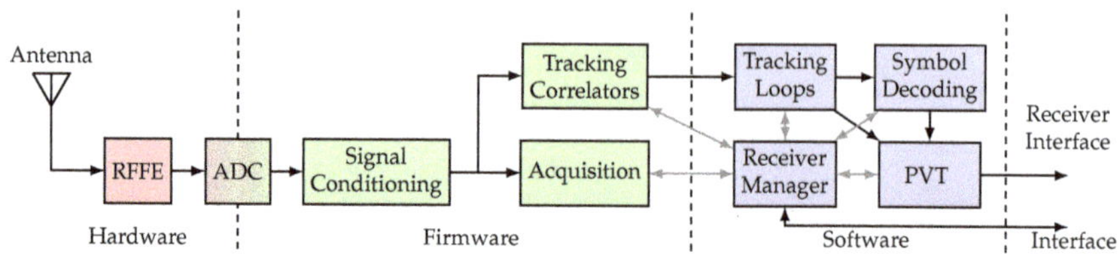

Figure 2. GOOSE© v2 architecture diagram.

2.1. Radio-Frequency Front-End

According to the targeted frequency bands, the frequency plan in Table 1 is proposed. The E5-band (including the NavIC L5) determines the maximum bandwidth of 54 MHz to enable the reception of the full Galileo E5 Alternative Binary Offset Carrier (AltBOC) signal.

Table 1. Proposed frequency plan.

Channel	Band	Center Frequency (MHz)	Bandwidth (MHz)	LO Frequency (MHz)	IF-Band (MHz)	Sampling Rate (MHz)
1	L5, E5	1192	54	1254	35–89	250
2	L2, G2	1235	30	1298	48–78	250
3	L1, E1, G1	1582	50	1520	37–87	250
4	S	2492	20	2430	52–72	250

Figure 3 depicts the four-channel analog front-end of the GOOSE© v2. The input stage separates the antenna signal into four channels (see Figure 3a). After the limiter, the signal is divided into the L-band and S-band. A second diplexer is included in the L-band to split it into an upper and lower band. The lower band is connected to a splitter that creates two identical GNSS signals, which are processed separately (RF-1 and RF-2).

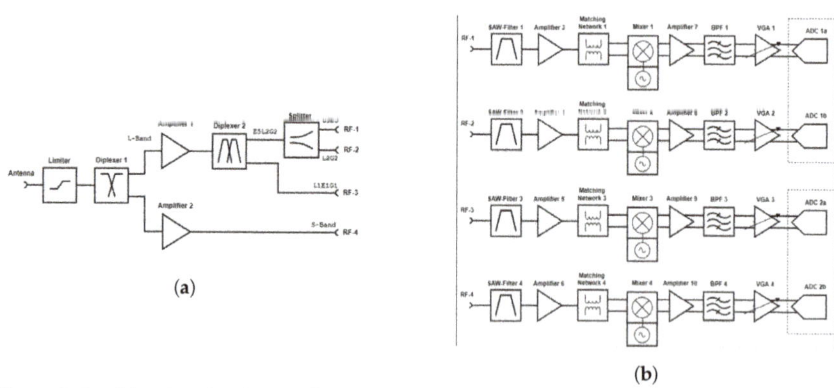

Figure 3. Quad-band analog front-end: (**a**) radio frequency (RF) input stage and (**b**) RF signal chains.

The subsequent RF chains are processed individually (see Figure 3b). As the receiver architecture consists of one mixer stage that provides an Intermediate Frequency (IF) signal with only the real part, the pre-selection of the RF signal is important. A second measure to reduce the non-desired side bands is that the LO frequency of the mixer is higher compared to a low IF architecture with a complex IF signal. The following analog processing is performed in four different paths. After band selection and amplification, the matching network converts the signal from single-ended to differential and is optimized for each band. The mixer stage shifts the signal according to the integrated local oscillator (see Table 1). The IF bandpass filter suppresses the out-of-band signals. The voltage gain amplifier (VGA) sets the appropriate level for the Analog-to-Digital Converter (ADC). Two dual 12-bit ADCs are selected to provide good performance in the digital domain. The sampling rate of the ADC is 250 MHz and enables the digital signal conditioning to use a simple $f_s/4$-mixer when shifting the signal to the baseband.

The schematic and the Printed Circuit Board (PCB) layout is optimized as follows:

- Same active components for all channels;
- Same footprint for all Surface Acoustic Wave (SAW) filters;
- Same structure for the matching network and the band pass filter, only the values of the components are changed.

The most significant benefit of this approach is that there is only one PCB layout design for all the channels, and the number of different active components is reduced to a minimum. This also guarantees fewer issues on the current component provisioning.

2.2. Digital Hardware Design

The digital HW board is based on a TE-0808 MPSoC module designed by Trenz Electronic [6]. This includes a Xilinx Zynq™ UltraScale+™, 2 gigabyte (GB) 64-Bit Double Data Rate fourth generation (DDR4) memory, 2× Serial Peripheral Interface (SPI) Boot Flash (dual parallel) and altogether 4 × 160 pins used for the connectivity to the analog front-end board.

The benefits of employing a system on module (SoM) are the easy development, maintenance efforts and migration. On the one hand, the functionality is guaranteed by the manufacturer and it is typically possible to upgrade to an MPSoC device with a larger PL. On the other hand, they avoid the risks associated with the development of a customized FPGA board, because some special components like Random Access Memorys (RAMs) or FLASH memories cannot be bought in low volumes. The integrated SoC device is the Zynq™ UltraScale+™ XCZU15EG-1FFVC900E manufactured by Xilinx. The PS is characterized by a quad-core advanced Reduced Instruction Set Computer (RISC) machine (ARM) Cortex–A53 processor and a PL with approx. 750 K logic cells. The PL allows the implementation of real-time digital signal processing tasks at higher throughput. The firmware (FW) design mapped on the PL is described in Section 2.3.

2.3. Firmware Design

The firmware has two main purposes, i.e., to provide a signal conditioning of the received samples and to implement the different correlation methods needed by the GNSS acquisition and tracking stage. The PS of the Zynq™ controls the digital HW modules implemented in the PL by reading and writing to the control and status registers. The communication interface integrated into the Zynq™ MPSoC between PS and PL is Advanced Extensible Interface (AXI).

Baseband Design

The baseband core performs the standard GNSS operations. It contains a time management module, FFT acquisition and correlation channels for tracking. The time management module is based on a 32-bit HW counter, which runs at the frequency of the digital samples, and provides a unique time reference to all the GNSS modules.

The FFT acquisition unit is composed of a fixed size FFT core and its control logic. The acquisition supports a fast, a medium and a sensitive mode, and its algorithm is based on a parallel code-phase search method. The FFT core consists of a Xilinx XFFT Intellectual Property (IP). The size is 16 K according to the resource available on FPGA used in the platform. The acquisition module is controlled by the processor which specifies which signal should be searched for according to its acquisition strategy. Once the acquisition performs a rough estimate of the main GNSS synchronization parameters, the tracking starts and refines the estimates. In total, there are 60 correlation channels in the presented GNSS platform.

Each correlation channel supports the correlation of two components (data and pilot) of a complex input signal. For each component, up to five delayed correlation points can be produced, which means that each channel is characterized in practice by ten complex correlators. The feedback control loops (Delay Locked Loop (DLL)/Frequency Locked Loop (FLL)/Phase Locked Loop (PLL)) are implemented in SW. They read out the complex integrate-and-dump data from the HW channels through a dedicated AXI interface. The loop steering values for the carrier and code NCOs are sent back through the channel control interface.

In addition to the control part, the correlation channels are characterized by the measurement interface, which collects information about all the internal counters and is triggered by the measurement interrupt generated by the time-management module. The measurements of all the channels are performed synchronously and allow for the calculation of the time of transmission of all the signals which are currently in tracking in SW. This is needed together with the time of arrival for the PVT calculation.

2.4. Software Design

The ogre_console is the main receiver program which manages the GNSS acquisition, tracking and PVT calculation processes. According to the signals selected (frequency and constellation), the receiver SW performs the acquisition in a continuous loop. For each acquired signal, a correlation channel is started. Different tracking options for each signal can be specified. As soon as enough signals are in tracking, a PVT can be calculated. The PVT is based on code measurements from the channels and the decoded Ephemeris.

3. Preliminary Receiver Evaluation

3.1. Resource Utilization on PL FPGA

Table 2 shows the resource utilization of the GOOSE© v2 receiver compared to v1. Since the same FFT size and amount of tracking channels are instantiated (as shown in Table 3), there is no significant difference in slice Lookup Tables (LUTs) between the two implementations. The 20,000 more LUTs are associated with the additional frequency band added in the digital signal conditioning. Each ADC frequency channel is first down-converted using the $f_s/4$-mixer and then down-sampled with a factor of 4. Therefore, an additional static Low-Pass Filter (LPF) is needed for anti-aliasing. This significantly increases the amount of Digital Signal Processor (DSP) blocks used, which is still only 10% of the blocks available on the device.

Table 2. Resource utilization.

Receiver	Xilinx Device	Slice LUTs	%	DSP Blocks	%
GOOSE© v1	Kintex7 410	221,783	78.25	71	4.61
GOOSE© v2	ZynqUS+ 15EG	239,506	70.18	370	10.49

Table 3. Firmware design components and properties.

Receiver	ADC Samples (bits)	ADC f_s (MHz)	Baseband Samples (bits)	Baseband f_s (MHz)	Static LPF	FFT Size	Tracking Channels
GOOSE© v1	8	81	4	81	-	16 K	60
GOOSE© v2	12	250	5	62.5	4	16 K	60

3.2. Performance Comparison between GOOSE v1 and GOOSE v2

In order to test the communication aspect of HW and SW, a parallel test with GOOSE© v1 and GOOSE© v2 is conducted in our Galileo lab using the Spirent Radio-Frequency Constellation Simulator (RFCS). The aim is to find out the number of tracking channels that the GNSS receiver can handle. The test setup is presented in Figure 4. Here, the RF output of the Spirent is connected to both receivers through a splitter.

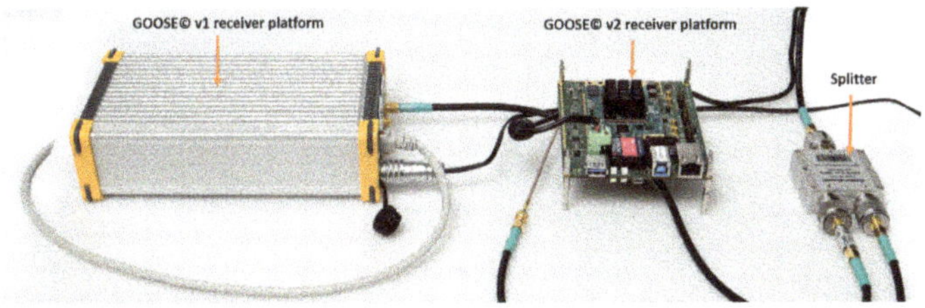

Figure 4. Test setup.

The selected test scenario includes GPS L1/L5 and Galileo E1/E5a signals. The HW integration time is set to 1 ms for both receiver platforms. Figure 5 shows the result of this test. At the start of the scenario, five SVs are enabled. Afterwards, one SV is added with at 1 min intervals. The number of tracked HW channels increases with respect to time. The trend is similar in both platforms, until E11 is added at the 480th second, where all of a sudden, the GOOSE© v1 stops while the other keeps tracking more SVs.

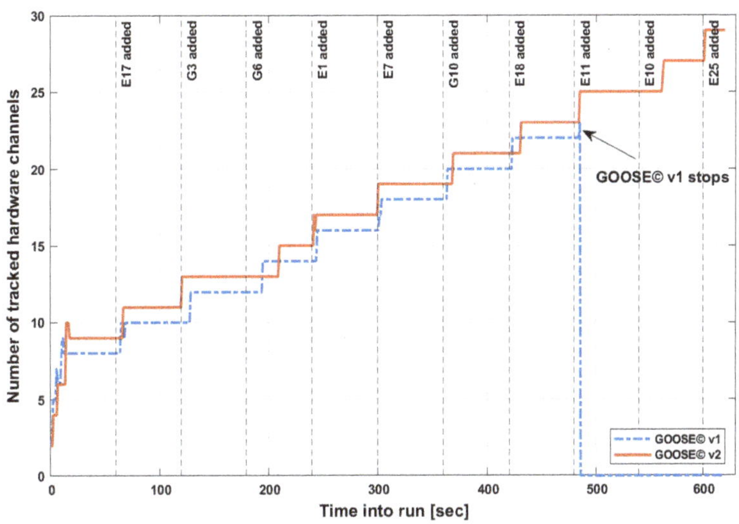

Figure 5. GOOSE© v1 and GOOSE© v2 performance result.

Figure 6 presents the computational load of both receivers at the break point of GOOSE© v1. Due to the high time complexity of the communication interface between PL and PS, the GOOSE© v1 cannot close the tracking channels in less than 1 ms. Consequently, the receiver SW application increases the computational load to a level until it terminates.

This result shows that GOOSE© v2 outperforms the GOOSE© v1 receiver when the HW integration time is reduced to 1 ms. This aspect plays a vital role for algorithms where the tracking loops need to be closed in the lowest possible time, such as in Vector Tracking (VT) [9], Deep Coupling (DC) and LBCA [5].

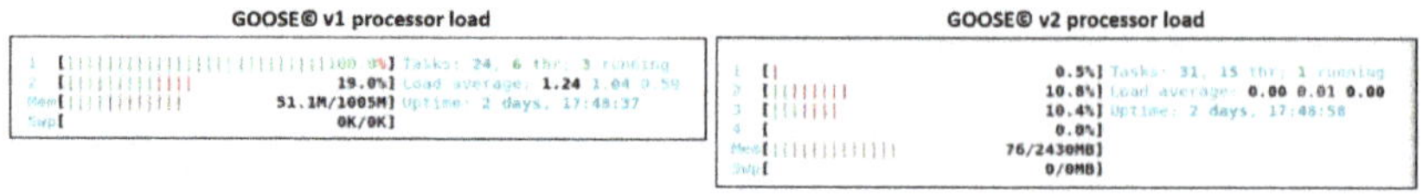

Figure 6. GOOSE© v1 and GOOSE© v2 processor load.

3.3. *NavIC Tests Using GOOSE v2*

The proposed platform was designed to support the NavIC processing. NavIC is a regional navigation satellite established by Indian Space Research Organization (ISRO). It is designed with a constellation of seven satellites and a network of ground stations. Currently, three satellites are placed in geostationary orbits and four satellites are placed in inclined geosynchronous orbits [10].

The first test demonstrates that the platform can acquire and track the L5 band of all the NavIC satellites visible from the test facility. The test setup is composed of GOOSE© v2 connected to the NavIC reference antenna. The NavIC reference antenna is mounted on the roof of the Fraunhofer IIS building in Nuremberg. Figure 7 shows that four L5 signals are acquired and are in stable tracking with a carrier-to-noise density ratio (C/N_0) of approx. 40 dBHz.

Table 4 shows a comparison of the C/N_0 between the proposed platform and a commercial receiver (Septentrio Mosaic-X5 [11]). The lower C/N_0 of the GOOSE© v2 might be due to the larger reception bandwidth, which also causes unwanted signals being processed. In order to improve the C/N_0, some additional filtering is needed in the S band channel or some interference mitigation capability.

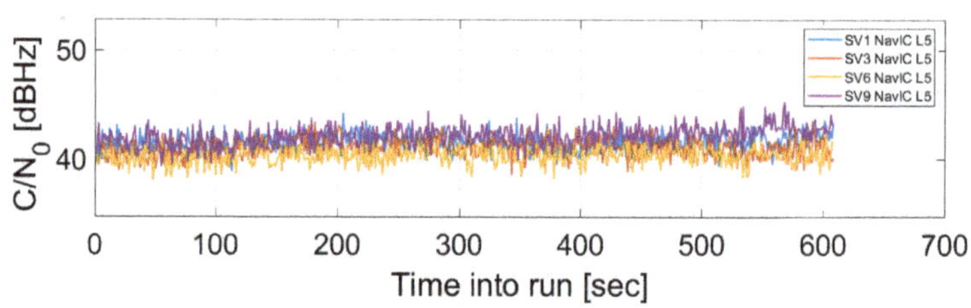

Figure 7. NavIC-L5 results using GOOSE© v2.

Table 4. NavIC-L5 C/N_0 mean values comparison between GOOSE© v2 and Mosaic-x5 receiver.

Receiver	SV1 (dBHz)	SV3 (dBHz)	SV6 (dBHz)	SV9 (dBHz)
GOOSE© v2	41.56	41.03	40.34	42.12
Septentrio Mosaic-X5	42.71	42.27	41.56	43.32

3.4. *L1–L5 Tests Using GOOSE v2*

To evaluate the more commonly used GPS and Galileo constellations, an additional test was performed. GPS L1CA and L5, Galileo E1BC and E5 AltBOC, and NavIC L5 and S were tracked using the NavIC reference antenna. In parallel, Septentrio PolaRx5 receivers also track signals of the same constellations. Figure 8a,b show the C/N_0 results of the GPS L1CA and Galileo E1BC using the GOOSE© v2. The C/N_0 results of the same signals using Septentrio PolaRx5 are provided in Figure 9. Here, we can see a slight difference in the amount of signals tracked due to the higher sensitivity of the commercial receiver.

Figure 8c,d show the C/N_0 results of the GPS L5 and Galileo E5 AltBOC on the GOOSE© v2. The C/N_0 results of the same signals using Septentrio PolaRx5 are provided in Figure 10. In this case, the C/N_0 of the two platforms is comparable and the same signals are tracked. No relevant difference can be observed for the L5-E5 band.

In addition, Figure 8e shows the C/N_0 for the NavIC L5- and S-band tracked satellites. The Septentrio receiver can track one low-C/N_0 signal more but cannot track signals in the S band.

In the test, we can see that both frequencies can be tracked for all the satellites except SVID1. The reason for this is not clear and should be further investigated. Additionally, L5 provides a higher C/N_0 of up to 3 to 5 dBHz. This could be due to the presence of interfering signals in the S band.

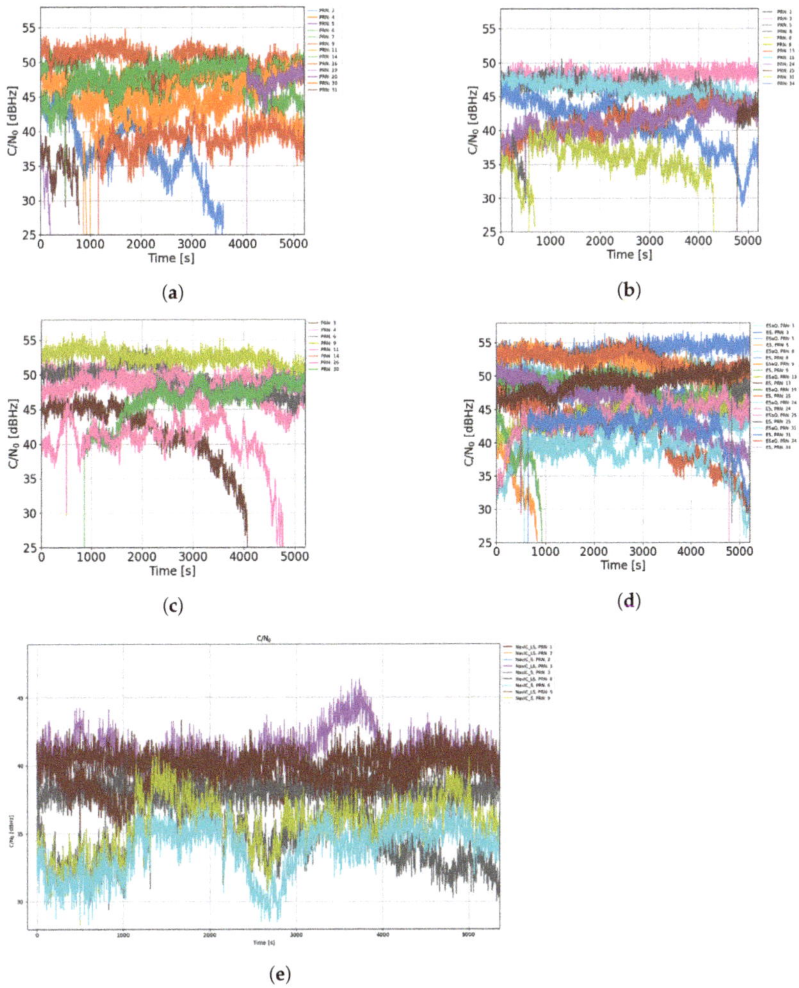

Figure 8. C/N_0 results with GOOSE© v2. (**a**) GPS L1, (**b**) Galileo E1, (**c**) GPS L5, (**d**) E5 AltBOC and E5A, and (**e**) NavIC L5 and S.

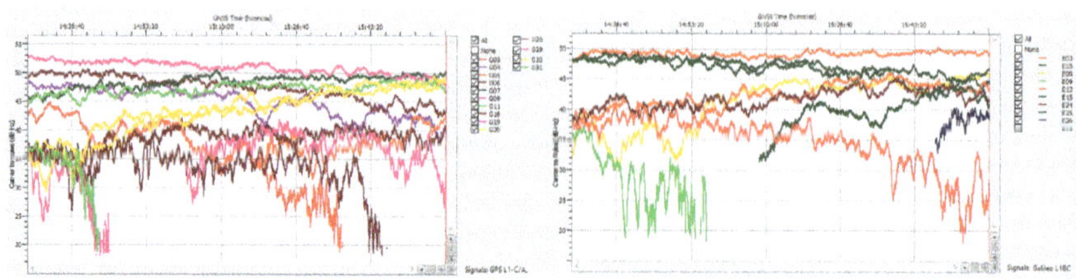

Figure 9. GPS L1 and Galileo E1 C/N_0 results using Septentrio PolaRx5.

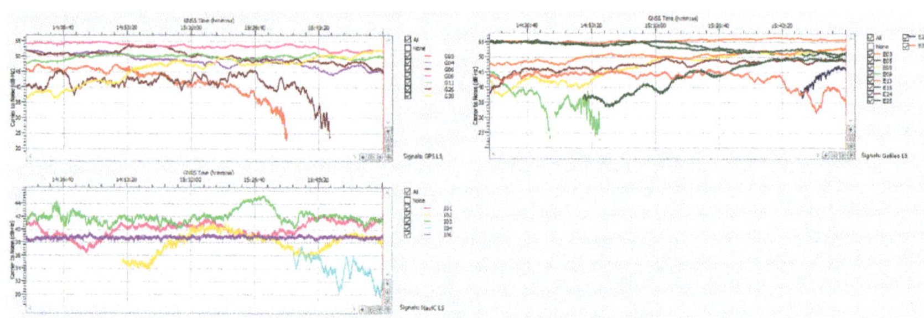

Figure 10. GPS L5 and Galileo E5 AltBOC C/N_0 results using Septentrio PolaRx5.

4. Conclusions

This paper presents an enhanced version of the legacy GOOSE©, the GOOSE© v2. This new platform is an MPSoC-based GNSS receiver that contains a quad-band (L1, L2, L5/E5 and S band) Radio-Frequency Front-End (RFFE) and baseband digital signal processing. The SoC feature allows at the same time the exchange between HW and SW with a very high bandwidth—thanks to the on-chip communication infrastructure. The results show that, in addition to the legacy signals, the new receiver platform supports the NavIC L5 and S-band signals along with the ability to close the tracking loops faster than its previous version. In addition, the proposed platform includes a Coral accelerator module suitable for exploiting potential GNSS Machine Learning (ML) algorithms [12].

As future work, the testing of the digital record and replay capability of the GOOSE© v2 using the four channels will be performed. In addition, the High-rate DFT-based Data Manipulator (HDDM) will be implemented in the PL of the platform [13]. Moreover, due to low time complexity of the PL-PS communication interface, more complex algorithms will be evaluated in SW, such as the VT and the LBCA.

Author Contributions: Conceptualization: F.G. and F.F.; Methodology: F.F. and S.U.; Software: S.-J.W. and M.S.; Validation: M.S., S.U. and F.G.; Formal analysis: F.F., I.C. and F.G.; Investigation: F.F. and M.S.; Writing—original draft preparation: M.S.; Writing—review and editing: M.S., I.C., S.-J.W., F.F. and M.O.; Visualization: S.-J.W. and F.G.; Project administration: M.S., M.O.; Funding acquisition: W.F. All authors have read and agreed to the published version of the manuscript.

Funding: Parts of the work presented in this paper are funded by the European Space Agency (ESA) Navigation Innovation and Support Program (NAVISP) Element 2 project "GOOSE-NavIC", which is managed by TeleOrbit GmbH (TOG) (ESA Contract No.: 4000137080/22/NL/MP/mk).

Institutional Review Board Statement: Not applicable.

Informed Consent Statement: Not applicable.

Data Availability Statement: The data presented in this study are available from the corresponding author on request. The data are not publicly available due to its data size and further explanations may be needed.

Acknowledgments: We acknowledge *ESA* and *TOG* for their valuable contributions, interest and time. We are also thankful to Gupta, Himanshu and Dietmayer, Katrin for the results analysis and review process. We also want to express our gratitude to Karim, Abrarul for the help in preparing the test setup.

Conflicts of Interest: The authors declare no conflict of interest. The funders had no role in the design of the study; in the collection, analyses, or interpretation of data; in the writing of the manuscript; or in the decision to publish the results.

References

1. European Union Agency for the Space Programme (EUSPA). *EUSPA EO and GNSS Market Report;* Market Report Issue 1; European Union Agency for the Space Programme (EUSPA): Luxembourg, 2022.
2. Overbeck, M.; Garzia, F.; Popugaev, A.; Kurz, O.; Forster, F.; Felber, W.; Ayaz, A.S.; Ko, S.; Eissfeller, B. GOOSE–GNSS Receiver with an Open Software Interface. In Proceedings of the 28th International Technical Meeting of The Satellite Division of the Institute of Navigation (ION GNSS 2015), Tampa, FL, USA, 14–18 September 2015.
3. Cortes, I.; Van der Merwe, J.R.; Rügamer, A.; Felber, W. Adaptive loop-bandwidth control algorithm for scalar tracking loops. In Proceedings of the IEEE/ION PLANS, Portland, OR, USA, 20–23 April 2020.
4. Cortés, I.; van der Merwe, J.R.; Nurmi, J.; Rügamer, A.; Felber, W. Evaluation of Adaptive Loop-Bandwidth Tracking Techniques in GNSS Receivers. *Sensors* **2021**, *21*, 502. [CrossRef] [PubMed]
5. Cortés, I.; van der Merwe, J.R.; Lohan, E.S.; Nurmi, J.; Felber, W. Evaluation of Low-Complexity Adaptive Full Direct-State Kalman Filter for Robust GNSS Tracking. *Sensors* **2023**, *23*, 3658. [CrossRef] [PubMed]
6. UltraSOM+ MPSoC-Modul mit Zynq™ UltraScale+™ XCZU15EG-1FFVC900E, 4 GB DDR4. Available online: https://shop.trenz-electronic.de/de/TE0808-05-BBE21-A-UltraSOM-MPSoC-Modul-mit-Zynq-UltraScale-XCZU15EG-1FFVC900E-4-GB-DDR4 (accessed on 5 May 2023).
7. Garzia, F.; Strobel, C.; Overbeck, M.; Kumari, N.; Joshi, S.; Förster, F.; Felber, W. A multi-frequency multi-constellation GNSS development platform with an open interface. In Proceedings of the 2016 European Navigation Conference (ENC), Helsinki, Finland, 30 May–2 June 2016; pp. 1–7. [CrossRef]
8. Seybold, D. GOOSE: Open GNSS Receiver Platform. Technical Report, TeleOrbit GmbH, 2020. Available Online: https://teleorbit.eu/en/satnav/goose/ (accessed on 14 May 2023).
9. Dietmayer, K.; Saad, M.; Strobel, C.; Garzia, F.; Overbeck, M.; Felber, W. Real time implementation of Vector Delay Lock Loop on a GNSS receiver hardware with an open software interface. In Proceedings of the 2019 European Navigation Conference (ENC), Warsaw, Poland, 9–12 April 2019; pp. 1–7. [CrossRef]
10. ISRO. *Indian Regional Navigation Satellite System: Signal in Space ICD for Standard Positioning Service (Version 1.1)*; Indian Space Research Organization: Bengaluru, India, 2017.
11. Septentrio Mosaic-X5 Development Kit. Available online: https://shop.septentrio.com/en/shop/mosaic-x5-devkit (accessed on 14 May 2023).
12. Accelerator Module: A Solderable Multi-Chip Module Including the Edge TPU. Available online: https://coral.ai/products/accelerator-module (accessed on 1 July 2022).
13. Garzia, F.; van der Merwe, J.R.; Rügamer, A.; Urquijo, S.; Felber, W. HDDM Hardware Evaluation for Robust Interference Mitigation. *Sensors* **2020**, *20*, 6492. [CrossRef] [PubMed]

engineering proceedings

Proceeding Paper

Multi-Constellation/Multi-Frequency GNSS Signal Degradation Due to Foliage and Reflective Environments †

Uttama Dutta [1,2,*] and Jan Johansson [1,2]

1 Department of Space, Earth and Environment, Chalmers University of Technology, 412 96 Gothenburg, Sweden; jan.johansson@chalmers.se
2 Department of Measurement Science and Technology, RISE Research Institutes of Sweden, Brinellgatan 4, 504 62 Borås, Sweden
* Correspondence: uttama.dutta@ri.se
† Presented at the European Navigation Conference 2023, Noordwijk, The Netherlands, 31 May–2 June 2023.

Abstract: Present day GNSS offers a variety of signals from different satellite constellations and at various frequencies. This paper is based on the work conducted in the project SiL 2.0 and will focus on the study of multi-constellation/multi-frequency GNSS signals as received on top of construction equipment as part of the SiL 2.0 dissemination solution. This paper aims to study the impact of foliage and reflective environments on the various signals of multi-constellation GNSS with a focus on GPS and Galileo. Signal strength indicators (SSI) have been used as a measure to understand the shadowing environment around a stationary GNSS antenna mounted on an excavator. It is quite clear from the analysis that traditional GPS L2 signals, C2W and L2W, are weaker in strength than the L2C and L1 C/A signals, and this is found to be consistent for all of the GPS satellites. The effect of signal degradation due to bending over sharp metallic edges is also discussed.

Keywords: GNSS; multi-constellation; signal bending or diffraction; reflective environment; foliage signal degradation

Citation: Dutta, U.; Johansson, J. Multi-Constellation/Multi-Frequency GNSS Signal Degradation Due to Foliage and Reflective Environments. *Eng. Proc.* **2023**, *54*, 2. https://doi.org/10.3390/ENC2023-15454

Academic Editors: Tom Willems and Okko Bleeker

Published: 29 October 2023

1. Introduction

Multi-constellation GNSS now offers a wide range of services, signals and frequencies to choose from. The L1 C/A signal remains the leading signal of the well-established GPS and is used in almost all GPS devices and technology. L2C, the new civilian-use GPS signal, is used to improve the accuracy of navigation, offers a signal that is very easy to track and provides signal redundancy for adverse situations. The improvement brought about by using L2C in GPS-based ionospheric research and monitoring has already been explored elsewhere [1,2]. In this study, we try to compare the positioning performance of L2C with traditional GPS signals and show that these signals have improved reliability in comparison to the traditional ones for field applications such as that on an excavator.

Andrei et al. [3,4] analysed the performance of Galileo L10 (tenth launch) in terms of Signals-in-Space (SiS) availability and accuracy and concluded with their numerical results that these satellites comply with their target values in terms of SiS performance. This demonstrates the reliability, robustness and consistency of the signal transmission for Galileo and is probably due to (apart from other technical improvements) the use of highly stable Passive Hydrogen Masers on board the satellites [5]. In this study, we also try to study Galileo signal positioning performance using signals transmitted on the E1 and E5b frequencies. This paper is based on the work conducted in the project SiL 2.0 and will focus on the study of multi-constellation/multi-frequency GNSS signals as received on top of construction equipment as part of the SiL 2.0 dissemination solution.

Section 2 briefly talks about the SiL 2.0 project; Section 3 gives the details of the data collection with the excavator. Section 4 shows the results. Section 4.1 deals with signal degradation due to foliage and compares different GNSS signals on the basis of degradation of their signal strength indicators. Section 4.2 deals with signal degradation due to reflective environments and discusses how a metallic edge could bring about such a degradation. Section 5 gives the conclusions.

2. SiL 2.0 Project

'Stomnät i Luften' 2.0 (SiL 2.0) is a research and innovation project with the overall goal to prepare the Swedish Transport Administration's construction operations for the challenges of the future regarding geodesy and machine control. This 3-year project financed by the Swedish Transport Administration uses GNSS-RTK position determination with the so-called project-adapted network RTK (PaNRTK) technology, which is used for geodetic measurements, machine control, etc., in large construction projects. In this study, a GNSS receiver mounted on the top of an excavator is used to study the positioning performance using GNSS PPP in an environment with foliage and reflective surfaces. The results from GipsyX version 2.3 [6,7] using PPP corrections as well as raw RINEX data are discussed.

3. Data Collection with the Excavator

Data collected during 3–11 September 2022 from a Septentrio PolarRx5 GNSS receiver, with its antenna (Leica AS10) mounted on an excavator, was used to study GNSS performance. The experimental set up demonstrating the antenna environment is as shown in Figure 1 below. From Figure 1c, it is clear that the antenna mounted on top of the excavator would experience signal degradation not only due to the metallic walls of the excavator but also from the building on the left (located at a distance of around 7 m), the hillock on the front and also from the vegetation on the right (located at a distance of around 10 m). It is to be noted for Figure 1c that the excavator is facing the NW direction, which means that the tree is located roughly on the NW side and the house on the SE direction.

(a)

Figure 1. *Cont.*

(b)

(c)

Figure 1. The experimental set up. (**a**) The GNSS antenna marked with a green circle on top of the excavator. (**b**) Top view of the excavator showing the GNSS antenna. (**c**) The excavator facing NW direction with the tree on the right and the house on the left.

4. Results

4.1. Signal Degradation Due to Foliage

In the following discussions, RINEX 3 data collected from the GNSS receiver on the excavator during 12:00 h on 3 September 2022 to around 10:00 h on 5 September 2022 were used for analysis. During this period, the excavator was stationary, and therefore, this is referred to as the 'stationary period' in multiple places in the document. The excavator at the beginning of the experiment on 3 September is made to face north, as mentioned before. Figure 2 shows the GNSS satellite visibility. During 40 percent of this time duration, 11 GPS satellites were visible, and 8 Galileo satellites were visible 50% of the time. A minimum of 5–7 satellites/constellations (GPS, Galileo and Glonass) and 22 satellites in total were visible almost all of the time during this period.

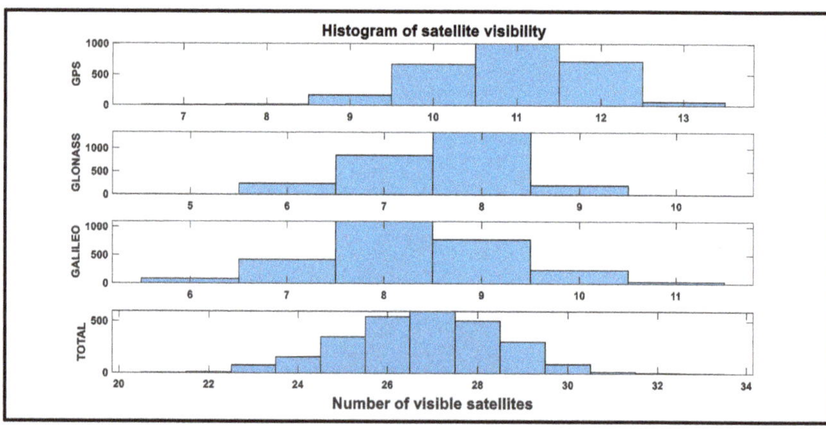

Figure 2. GNSS satellite visibility.

One-second sampled observation data from the excavator and IGS broadcasted navigation data were together used to calculate dilution of precision values and these are plotted along with the number of visible satellites in Figure 3. As can be seen from Figure 2, except a few occasions, the DOP value remains below 2.5.

Signal strength indicators (SSI) obtained from 1 s sampled RINEX 3 data files have been used here as a measure to understand the shadowing environment around the GNSS antenna mounted on the excavator. Figure 4 show plots of the SSI measurements on different frequencies and channels for GPS satellite G04 during the stationary period.

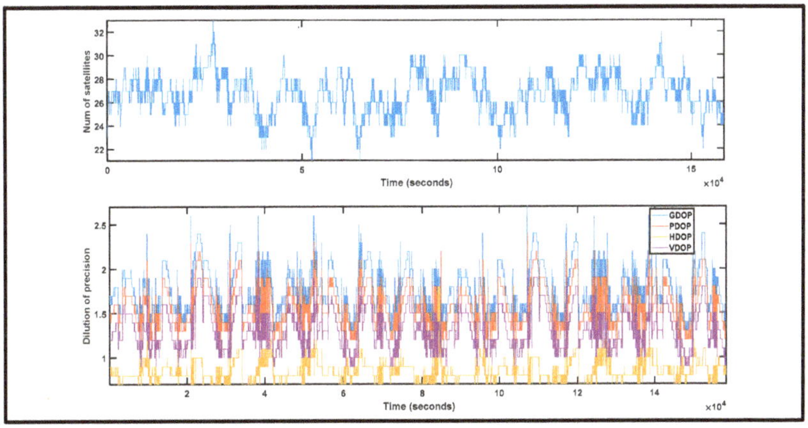

Figure 3. Plot of number of visible satellites and different dilutions of precision (GDOP, PDOP, HDOP and VDOP) as seen at the receiver on the excavator.

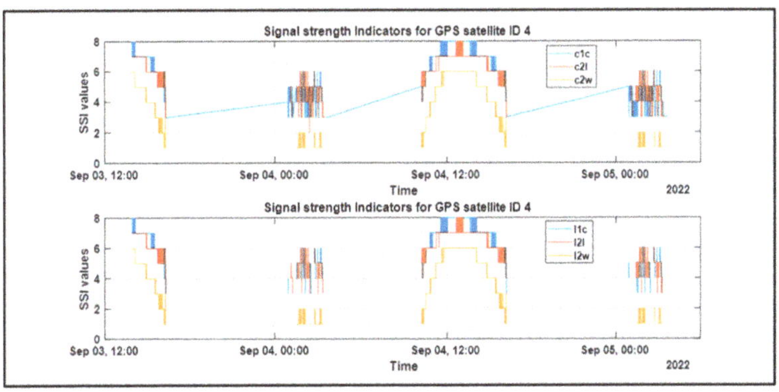

Figure 4. SSI for GPS satellite G04 during the stationary period (3–5 September 2022).

From Figure 4, it appears that C2W and L2W, which are the traditional GPS L2 signals, are weaker in strength than the L2C and L1 C/A signals. SSI values typically range between one and six, which means that the SNR for these signals range between a high of 36–41 dBHz to as low as less than 12 dBHz. Further analysis shows that this weakness of traditional L2 signals is found to be consistent for signals received from all of the GPS satellites. For some satellites, the SSI values for these signals have a maximum of eight or nine. But such cases are few and the time duration for such good signal strength is also very limited. Also from Figure 4, both C2W and L2W seem to be unavailable during certain periods between 2:00 and 3:00 h on 4 September and this is repeated also on 5 September. Also, the pseudorange signals on C1C and C2L seem to be less affected. The anomaly in the SSI measurements of G04 can be further confirmed with a skyplot showing the visibility of the satellites to the excavator antenna during 2:00–3:00 h on 4 and 5 September, and such a skyplot is shown in Figure 5.

As can be seen, the satellite is visible in the NW direction of the excavator during a certain period of time. It is possible that an obstruction (such as a tree) located in this direction blocks the line of sight from the satellite to the receiver antenna causing the drop in signal strength visible in Figure 4.

Figure 6 shows the plot of the SSI for the pseudorange measurements from G04 over varying azimuthal angles around the receiving antenna (during the stationary period).

One can notice anomalous C2W values already at 305° and a total loss of signal strength between 313 and 320° and again between 322 and 329°.

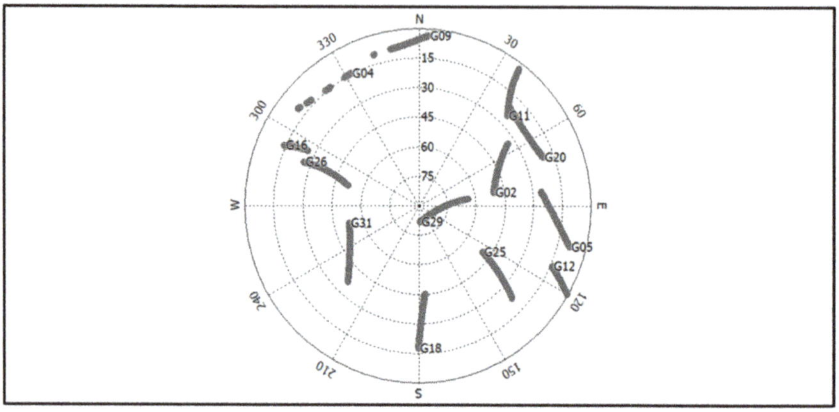

Figure 5. Skyplot showing the visibility of the satellites during 2:00–3:00 h on 4th and 5th September 2022. (The satellite numbers G02, G04 etc. in this case mean the GPS satellite numbers. The numbers on the skyplot indicate increasing values of azimuth in the clockwise direction and increasing values of elevation inwards in the plot).

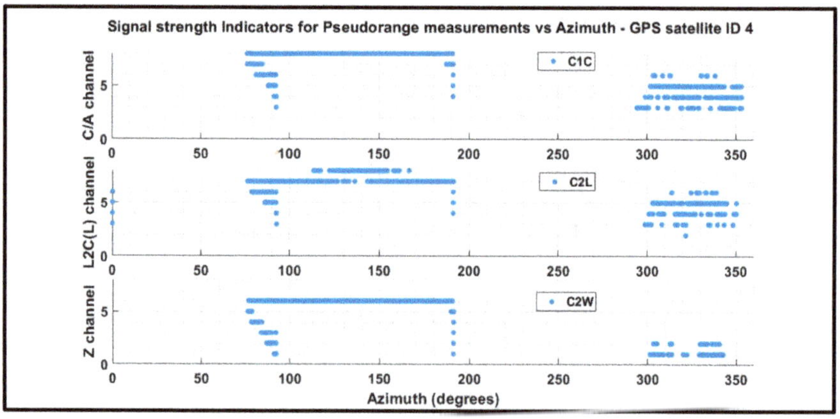

Figure 6. Plot of SSI for code measurements vs azimuth for GPS satellite G04 (during the stationary period).

Similar patterns of signal degradation at these azimuthal angles (around 300–350°) can be noticed for other GPS satellite signals such as those of G07, G09, G10, G14, G18, G23 and G29, as is clear from Figure 7a. Figure 7b clearly shows that C2L performance is much better than C2W.

In order to understand the effect of the obstruction on satellites/signals from other constellations, skyplots of signal strength indicators for C1C and C7Q signals of Galileo satellites E04, E07, E08, E11, E13, E15, E21, E24, E31, E34 and E36 are plotted, as shown in Figure 8 below.

The Galileo signals appear to be quite robust, and although the signal strength is lower at the azimuthal range between 300 and 330°, the signals are received throughout the period.

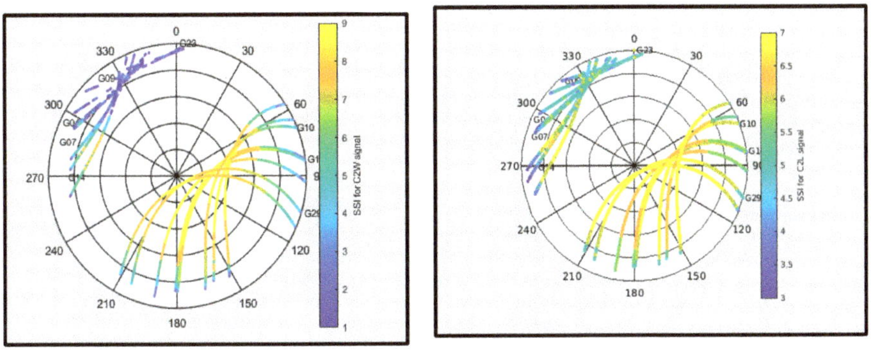

(a) (b)

Figure 7. Skyplots for different code signals of all affected GPS satellites (G04, G07, G09, G10, G14, G18, G23 and G29) during the stationary period: (**a**) C2W and (**b**) C2L.

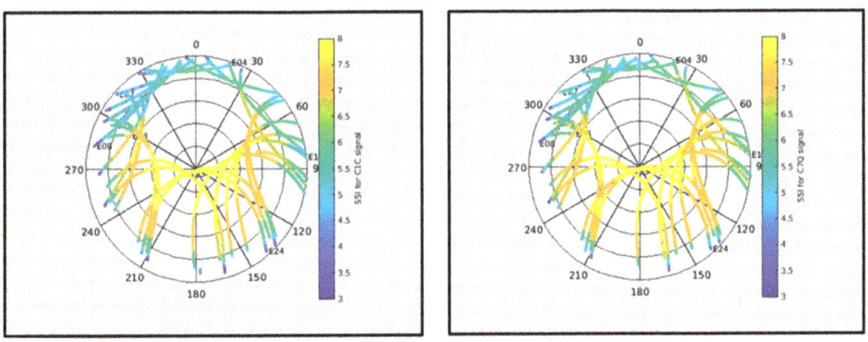

(a) (b)

Figure 8. Skyplots for different code signals of Galileo satellites during the stationary period: (**a**) C1C and (**b**) C7Q.

4.2. Signal Degradation Due to Reflective Environment

Data collected from the GNSS receiver on the excavator was post processed using GipsyX software version 2.3 to generate pseudorange residuals of the code signal using the Ionosphere-free combination of dual frequency GNSS during the stationary period.

Figure 9 shows the post processed code residuals for GPS using the traditional and the modern GPS signals. One can easily notice the improvement in the performance using the modern L2C GPS signal.

Due to reflective environments, GNSS signals become degraded due to multi-path. But signal degradation is also possible over sharp metallic edges such as that of the roof of the house to the left of the excavator shown in Figure 1c. In such cases, the GNSS signal diffracts or bends over the edge, thus traversing a longer path length than the direct line-of-sight path length and degrading the position computation of the receiver. Thus, the GNSS signal becomes available even in shadowed and non-line-of-sight regions. In order to demonstrate such an effect of signal bending due to reflective environments in the near vicinity of the antenna, controlled experiments were carried out both using the Leica AS10 and high-grade geodetic antenna, as shown in Figure 10a,b.

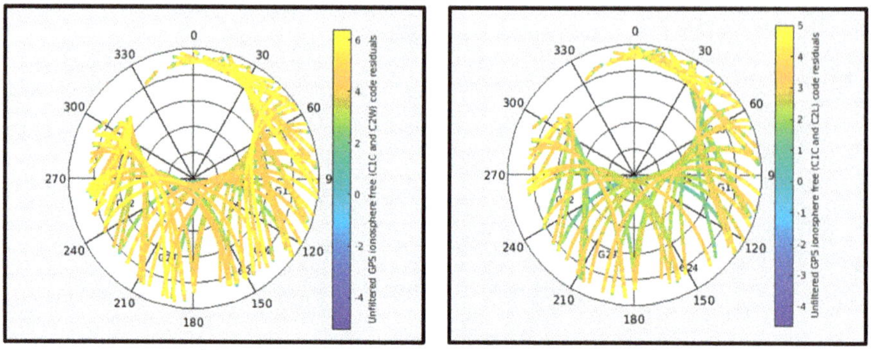

(**a**) (**b**)

Figure 9. GipsyX code residuals using the Ionosphere-free combination of dual frequency GPS. (**a**) Using C1C and C2W; (**b**) using C1C and C2L.

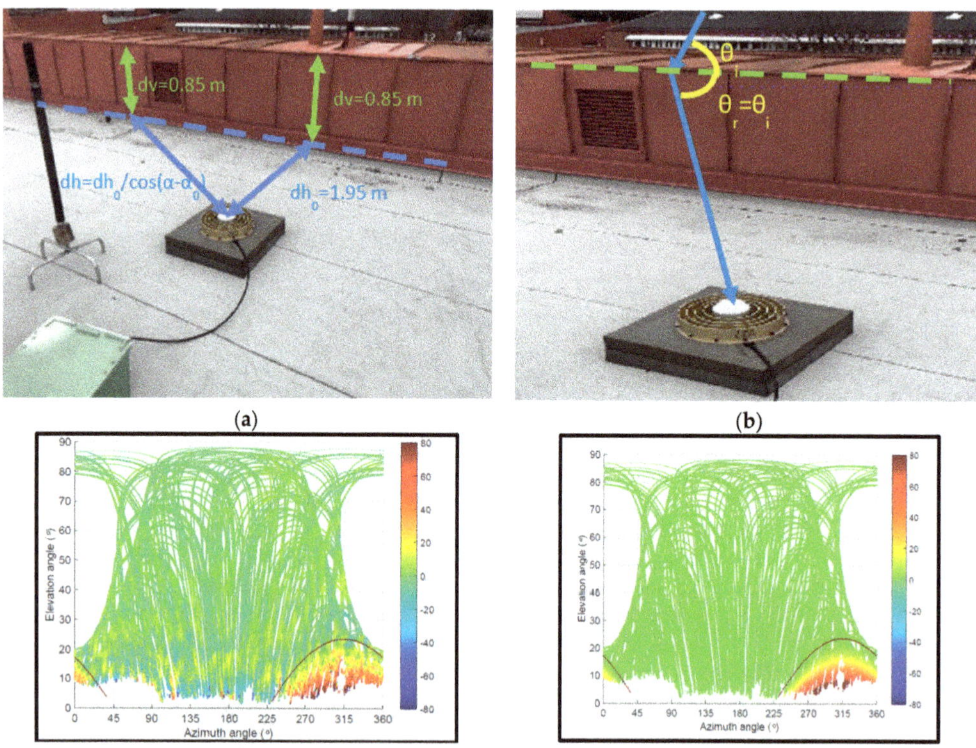

Figure 10. *Cont.*

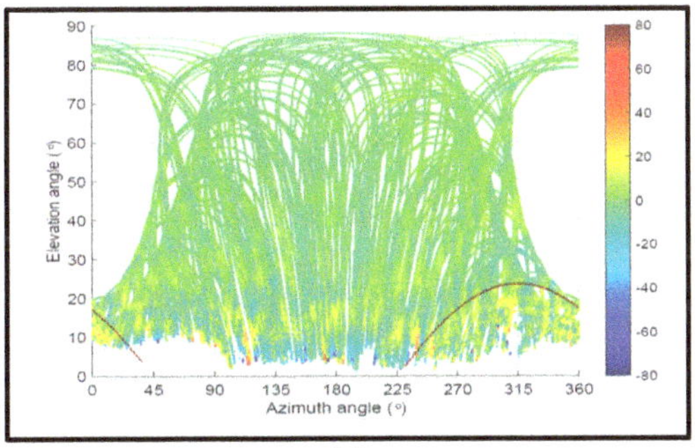

(**e**)

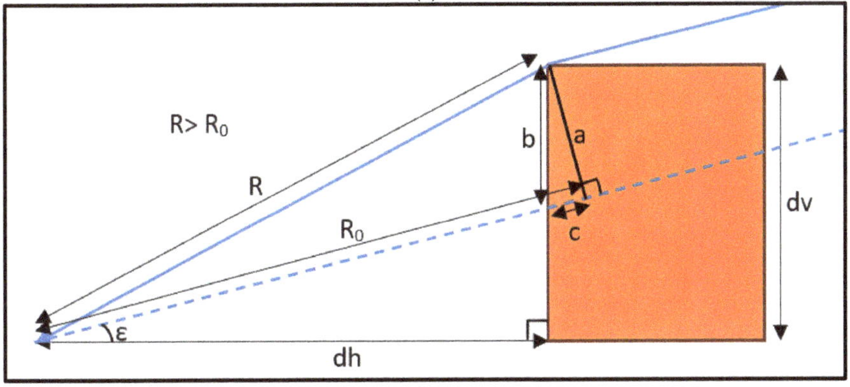

(**f**)

Figure 10. Demonstration of signal degradation due to reflective environments. (**a**) Antenna environment. (**b**) Signal diffraction over metallic edges. (**c**) GPS/GLONASS/Galileo L1/E1 and L2/E5a range residual plots. (**d**) Estimations using a diffraction model. (**e**) Corrected measurements using the model. (**f**) Signal diffraction model (plot dynamics ± 20 mm).

Figure 10c shows a plot of the pseudorange residuals using dual frequency GNSS measurements. The colour bars in Figure 10c–e indicate values of these residuals in mm. The enhanced values of residuals noted at low elevations and an azimuth of 225° to 360° is the effect of signal degradation due to bending over a steel edge, marked with the green dashed line in Figure 10b. The signal diffraction model of Figure 10f is used to estimate the residuals. From Figure 10f, the vertical height of the obstruction is dv and the horizontal distance to the obstruction from the receiver antenna is dh. Thus, for any elevation angle ε, at a given azimuth angle, the path length difference (R–R_0) can be estimated using the following equations.

$$b = dv - dh \cdot \tan(\varepsilon) \tag{1}$$

$$R = \sqrt{\left(dv^2 + dh^2\right)} \tag{2}$$

$$R_0 = \sqrt{\left((dh \cdot \tan(\varepsilon))^2 + dh^2\right)} + c \quad where \quad c = b \cdot \sin(\varepsilon) \tag{3}$$

Variations in dh for varying azimuth angle α can be computed as shown in Figure 10a. A plot of the modelled residuals based on these estimations is shown in Figure 10d. Figure 10e is the plot of the corrected residuals. GNSS signals can also be degraded due to metallic surfaces causing reflections. Such effects are expected to degrade the GNSS signal received on the antenna over the excavator, resulting in erroneous positioning performance.

The signal bending effect as well as signal degradation due to reflective environments will be studied as future work to this paper.

5. Conclusions

Skyplots obtained during the stationary period show that satellites in the NW direction of the excavator during a certain period of time face signal degradation due to a possible obstruction from a tree. Plots of SSI for the carrier phase and code measurements, respectively, from various GPS satellites over varying azimuthal angles around the receiver antenna show anomalous C2W (and L2W) values at around 300–330°. It can be therefore concluded that measurements using C2W (and L2W) may not prove to be very reliable and should be avoided. A similar analysis was conducted, and conclusions were drawn for signals from Galileo. It can also be concluded that Galileo signals are quite robust and reliable and suffer less due to degradation from foliage environments. GNSS signal degradation is also possible due to reflections and bending over sharp metallic edges. As future work, possible reflections from reflective surfaces will be inspected and analysed.

Author Contributions: The document is part of an industrial PhD and is based on the work carried out by U.D. under the supervision of J.J. All authors have read and agreed to the published version of the manuscript.

Funding: This research was funded by Trafikverket and the SiL project grant number TRV 2018/111087, 2019).

Institutional Review Board Statement: Not applicable.

Informed Consent Statement: Not applicable.

Data Availability Statement: The data presented in this study are available upon request from the authors.

Acknowledgments: The authors acknowledge NASA and Jet Propulsion Laboratory for the GipsyX Precise Point Positioning software package.

Conflicts of Interest: The authors declare no conflict of interest.

References

1. Dutta, U.; Jarlemark, P.; Rieck, C.; Johansson, J. Ionospheric Effects on GNSS RTK. In Proceedings of the 35th International Technical Meeting of the Satellite Division of The Institute of Navigation (ION GNSS+ 2022), Denver, CO, USA, 19–23 September 2022; pp. 1589–1598. [CrossRef]
2. McCaffrey, A.M.; Jayachandran, P.T.; Langley, R.B.; Sleewaegen, J.M. On the accuracy of the GPS L2 observable for ionospheric monitoring. *GPS Solut.* **2017**, *22*, 23. [CrossRef]
3. Andrei, C.-O.; Lahtinen, S.; Poutanen, M.; Koivula, H.; Johansson, J. Galileo L10 Satellites: Orbit, Clock and Signal-in-Space Performance Analysis. *Sensors* **2021**, *21*, 1695. [CrossRef] [PubMed]
4. Andrei, C.O.; Johansson, J.; Koivula, H.; Poutanen, M. Signal performance analysis of the latest quartet of Galileo satellites during the first operational year. In Proceedings of the 2020 International Conference on Localization and GNSS (ICL-GNSS), Tampere, Finland, 2–4 June 2020; pp. 1–6.
5. Montenbruck, O.; Steigenberger, P.; Hauschild, A. Comparing the 'Big 4'—A User's View on GNSS Performance. In Proceedings of the 2020 IEEE/ION Position, Location and Navigation Symposium (PLANS), Portland, OR, USA, 20–23 April 2020; pp. 407–418.
6. Bertiger, W.; Bar-Sever, Y.E.; Dorsey, A.; Haines, B.; Harvey, N.; Hemberger, D.; Heflin, M.; Lu, W.; Miller, M.; Moore, A.W.; et al. GipsyX/RTGx, a new tool set for space geodetic operations and research. *Adv. Space Res.* **2020**, *in press*. [CrossRef]
7. Zumberge, J.F.; Heflin, M.B.; Jefferson, D.C.; Watkins, M.M.; Webb, F.H. Precise point positioning for the efficient and robust analysis of GPS data from large networks. *J. Geophys. Res.* **1997**, *102*, 5005–5017. [CrossRef]

engineering proceedings

MDPI

Proceeding Paper

Design of a Signaling Scheme for Three Equal-Power Signals in a New Navigation Satellite System †

Hyoungsoo Lim

Satellite Navigation Research Section, Electronics and Telecommunications Research Institute, Daejeon 34129, Republic of Korea; lim@etri.re.kr; Tel.: +82-42-860-1608

† Presented at the European Navigation Conference 2023, Noordwijk, The Netherlands, 31 May–2 June 2023.

Abstract: Design methods are presented for chip pulses and a constant envelope multiplexing (CEM) scheme for a new navigation satellite system in the S band. If we assume three equal-power signals are needed to transmit, the chip pulses of at least two signals can have two or more sample magnitudes for a high CEM power efficiency. Considering a case with two dual-magnitude signals, we can find the magnitude values for a target efficiency and then design the magnitude and sign patterns. We analyze the interference for all possible chip pulse sets and sorted them according to selected priorities. We also consider spectral separation between two different signal groups for a flexible operation. The design methods and the results provided in this paper can be a good reference for the signal design of a new navigation satellite system.

Keywords: signal design; chip pulse; constant envelope multiplexing; navigation satellite system

Citation: Lim, H. Design of a Signaling Scheme for Three Equal-Power Signals in a New Navigation Satellite System. *Eng. Proc.* **2023**, *54*, 3. https://doi.org/10.3390/ENC2023-15469

Academic Editors: Tom Willems and Okko Bleeker

Published: 29 October 2023

1. Introduction

In navigation satellite systems, except for GLONASS (Global Navigation Satellite System) and FDMA (frequency-division multiple access) [1], several independent signals are transmitted syn-chronously in time at an identical carrier frequency. They use direct-sequence spread-spectrum signals but, unlike typical wireless communication systems, the signals can have different chip rates and chip pulses [2–6]. Although the signals use different spreading codes, the chip pulses are designed to guarantee far lower inter- and intra-system interference. Furthermore, the signals in a frequency band are properly multiplexed to have a constant envelope, in order to maximize the power efficiency of the high-power amplifiers.

Most legacy navigation satellite signals use single-magnitude bi-phase chip pulses, so the constant envelope multiplexing (CEM) scheme can then be designed independently of the chip pulse design. On the other hand, some special requirements can unite the two designs. In this paper, we assume that three equal-power signals are transmitted and present a set of design methods for chip pulses and CEM, with example design results.

2. Design Methodology

In most cases of legacy navigation satellite systems, the signals transmitted in a frequency band have different data rates and/or target QoS's (qualities of services), and thus have different transmit powers. In this paper, on the other hand, we assume that all the signals in a frequency band have an identical transmit power and only the data rates are determined by the target QoS's. It is similar to the case of Galileo E5, where four signals transmitted with an identical power. But, if the number of equal power signals is three, we need some modifications or constraints in the design of the constant envelope multiplexing scheme, as well as the chip pulse waveforms.

If we simply use bi-phase chip pulse waveforms, which have a single sample magnitude, for equal power assumption, the CEM power efficiency cannot exceed 75%. That is, as much as 25% of the total transmit power is consumed in order to tailor the baseband sample constellation to have identical magnitudes. However, if we allow some of the chip pulse waveforms to have two or more magnitudes with equal average powers, the instantaneous signal powers can be different, and thus we can achieve a high CEM power efficiency. In this case, the designs of the CEM scheme and the chip pulse waveforms are tightly coupled.

In this paper, we assume to have a pair of dual-magnitude signals and a single-magnitude signal to transmit in the S band centered at 2492.028 MHz with a bandwidth 16 MHz.

2.1. Constant Envelope Multiplexing Scheme

Let us denote the three baseband signals as $s_1(t)$, $s_2(t)$, and $s_3(t)$, where the chip pulse waveforms of $s_1(t)$ and $s_2(t)$ have two magnitudes, α and β; $0 < \alpha < \beta$ at an equal probability of 0.5 and $s_3(t)$ is a bi-phase signal with single magnitude, γ. The equal average power condition requires $\alpha^2 + \beta^2 = 2\gamma^2$. We further assume $|s_1(t)| = \alpha$ whenever $|s_2(t)| = \beta$, and vice versa, for simple design. Then, the magnitude values are determined as $\alpha = \sqrt{\frac{\eta - \sqrt{(4\eta-3)\eta}}{3}}, \beta = \sqrt{\frac{\eta + \sqrt{(4\eta-3)\eta}}{3}}, \gamma = \sqrt{\frac{\eta}{3}}$ for a given target CEM power efficiency, with η and the CEM results in $s_{CEM}(t) = s_1(t) + s_2(t) + j\left[s_3(t) - \frac{3}{\eta}s_1(t)s_2(t)s_3(t)\right]$ [7].

2.2. Chip Pulse Waveforms

Under the constraints from the previous Section 2.1, the magnitude and sign patterns of the chip pulse waveforms are designed. With some heuristic efforts, we can employ the exhaustive search method, analyzing the interference for all possible sets of chip pulses to find the best set with the minimum worst inter- and intra-system interference. For a fast and efficient preliminary design, we can use the spectral separation coefficient (SSC) as an asymptotic metric of the interference [8].

In this paper, we sort the analysis results, first by the worst inter-system interference from a candidate signal set to the legacy signals, and then by the maximum inter-system interference from the legacy signals to the candidate signal set, and intra-system interference among the candidate signal set. The list is then investigated one by one to find the best signal set with a low out-of-band emission and transmitter and receiver complexity. If necessary, we can also apply an additional requirement that one of the three signals has its spectrum near the center of the S band, whilst the spectra of the other two are located at the outer parts of the band. The spectral separation allows for flexible operation in the inner and outer parts of the frequency band, with marginal interference between them.

3. Design Results

We performed the exhaustive analyses described in Section 2.2 with the CEM scheme in Section 2.1, for an arbitrary CEM power efficiency of $\eta = 0.93$, and sorted them. The list was shortened by constraining each worst SSC to not exceed the minimum worst SSC observed for all possible chip pulse waveforms by 10 dB or more.

3.1. Design Results without Spectral Separation

First, we searched for proper chip pulse waveforms without the requirement of separated spectra. Figures 1 and 2 show the chip pulse waveforms and their spectra, designed for the case of two 1.023 Mcps signals and a 2.046 Mcps signal. The inter-system SSCs to the legacy signals are observed to be less than -74.86 dB/Hz and the inter-system SSCs from the legacy signals and the intra-system SSCs are less than -66.73 dB/Hz. For another case, with a 1.023 Mcps signal and two 2.046 Mcps signals, we found the chip pulse waveforms in Figure 3, whose spectra are shown in Figure 4. In this case, the inter-system SSCs to the legacy signals are observed to be less than -71.92 dB/Hz and the inter-system SSCs from the legacy signals and the intra-system SSCs are less than -68.27 dB/Hz.

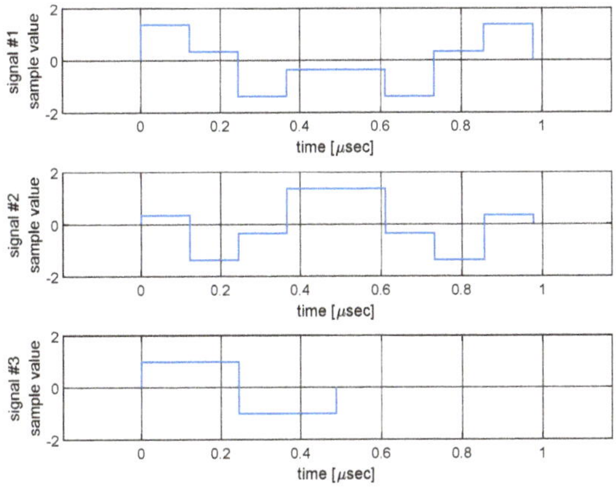

Figure 1. Chip pulses designed for two 1.023 Mcps signals and a 2.046 Mcps signal in the S band.

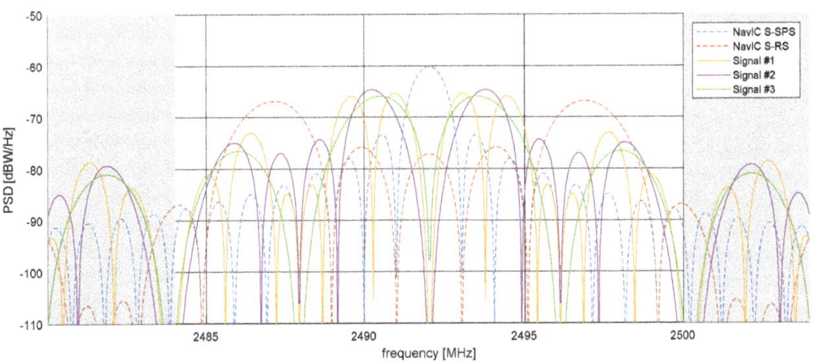

Figure 2. PSDs (power spectral densities) of the chip pulses designed for two 1.023 Mcps signals and a 2.046 Mcps signal in the S band.3.

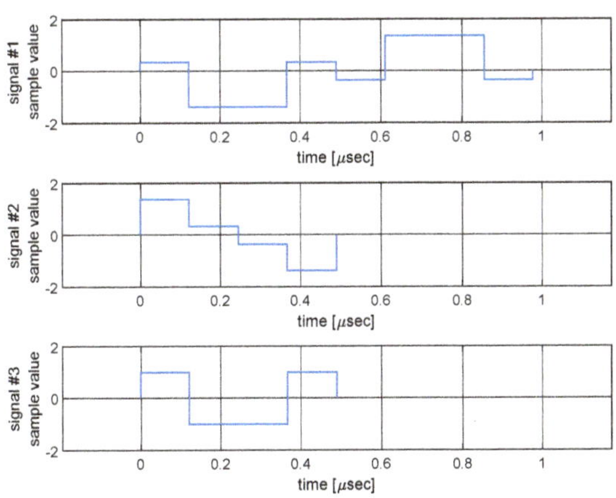

Figure 3. Chip pulses designed for a 1.023 Mcps signal and two 2.046 Mcps signals in the S band.

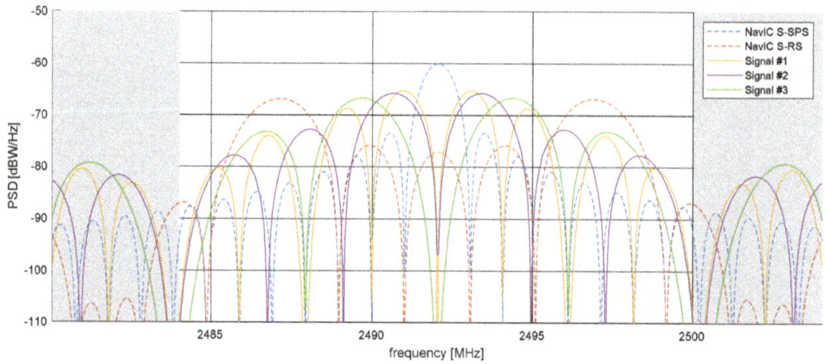

Figure 4. PSDs of the chip pulses designed for a 1.023 Mcps signal and two 2.046 Mcps signals in the S band.

3.2. Design Results with Separated Spectra

Let us assume that a signal has its spectrum in the inner band and the other two have their spectra in the outer band. Figures 5 and 6 show a proper design example for the case of two 1.023 Mcps signals and a 2.046 Mcps signal. In this case, the inter-system SSCs to the legacy signals are less than −63.50 dB/Hz, whilst the inter-system SSCs from legacy and the intra-system SSCs are less than −63.85 dB/Hz. It is obvious that the requirement of separated spectra significantly reduces the degree of freedom in the design, and we can see that the worst SSCs respectively increase by 11.36 dB and 2.88 dB, in comparison with those in Section 3.1.

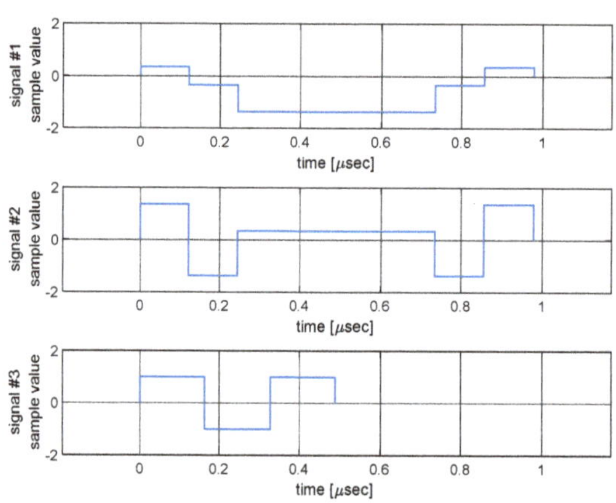

Figure 5. Chip pulses designed for two 1.023 Mcps signals and a 2.046 Mcps signal with spectral separation in the S band.

Unfortunately, we could not find any signal set for the case with a 1.023 Mcps signal and two 2.046 Mcps signals. That means that either the worst inter- or intra-system SSC is larger than the minimum for all possible chip pulse waveforms by 10 dB or more. In this case, we should consider loosening the worst interference requirement. Frequency division multiplexing (FDM) can be another solution. However, the FDM signals should be carefully designed to have a constant envelope, unless separate high power amplifiers are used for each signal, which is typically undesirable in practice.

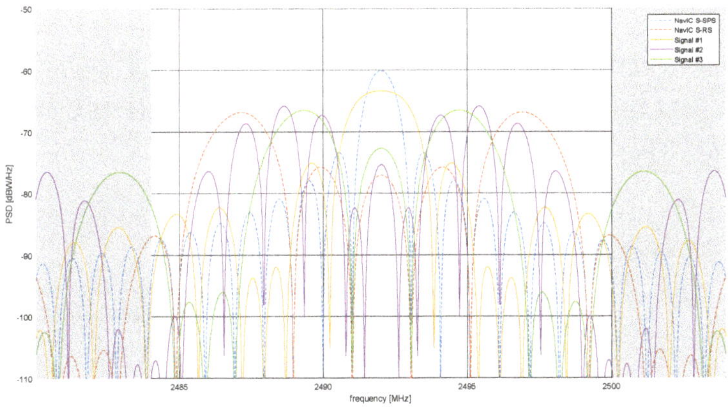

Figure 6. PSDs of the chip pulses designed for two 1.023 Mcps signals and a 2.046 Mcps signal with spectral separation in the S band.

4. Conclusions

A set of design methods for the chip pulses and CEM of three equal-power signals were presented. Under the condition of equal transmit powers, a pair of signals had dual magnitudes for a high CEM power efficiency. The proposed set of design methods had a high degree of freedom and thus novel sets of chip pulse waveforms could be obtained with relatively good interference performance. We designed the signals for the cases with and without the requirement of separated signal spectra and found that the worst interferences can significantly increase with spectral separation. The design methods presented in this paper and the design results can be a good reference for planning a new navigation satellite system, including the future Korea Positioning System.

Funding: This research was funded by Electronics and Telecommunications Research Institute (ETRI), grant number 23ZH1100 (Study on 3D communication technology for hyper-connectivity).

Institutional Review Board Statement: Not applicable.

Informed Consent Statement: Not applicable.

Data Availability Statement: Data are available in this manuscript.

Conflicts of Interest: The author declares no conflict of interest.

References

1. GLONASS Documents. Available online: http://www.glonass-svoevp.ru/index.php?option=com_content&view=article&id=146&Itemid=305&lang=en (accessed on 18 May 2023).
2. GPS Interface Control Documents. Available online: https://www.gps.gov/technical/icwg/ (accessed on 18 May 2023).
3. Galileo Programme Reference Documents. Available online: https://www.gsc-europa.eu/electronic-library/programme-reference-documents#Galileopub (accessed on 18 May 2023).
4. BeiDou Navigation Satellite System ICDs. Available online: http://en.beidou.gov.cn/SYSTEMS/ICD/ (accessed on 18 May 2023).
5. IRNSS Signal-in-Space Interface Control Document ICD Ver 1.1. Available online: https://www.isro.gov.in/IRNSSSignal.html (accessed on 18 May 2023).
6. Quasi-Zenith Satellite System (QZSS). Available online: https://qzss.go.jp/en/technical/ps-is-qzss/ps-is-qzss.html (accessed on 18 May 2023).
7. Lim, H.; Lee, S.U. Constant Envelope Multiplexing Scheme for Three Equal Power Signals. In Proceedings of the APCC, Jeju, Rpublic of Korea, 19–21 October 2022.
8. Betz, J.W. Effect of Partial-Band Interference on Receiver Estimation of C/N0: Theory. In Proceedings of the ION-NTM, Long Beach, CA, USA, 22–24 January 2001.

engineering proceedings

Proceeding Paper

Galileo Performance Improvements Employing Meta-Signals—Robustness Analysis against Payload and Receiver Distortions †

Florian C. Beck [1,2,*], Christoph Enneking [1], Steffen Thölert [1,2] and Michael Meurer [1,2]

1 Institute for Communications and Navigation, German Aerospace Center (DLR), Münchener Str. 20, 82234 Wessling, Germany; christoph.enneking@dlr.de (C.E.); steffen.thoelert@dlr.de (S.T.); michael.meurer@dlr.de (M.M.)
2 Chair of Navigation, RWTH Aachen University, Mies-van-der-Rohe-Str. 15, 52074 Aachen, Germany
* Correspondence: florian.beck@dlr.de
† Presented at the European Navigation Conference 2023, Noordwijk, The Netherlands, 31 May–2 June 2023.

Abstract: A concept that has been explored as a means to obtain decimeter-level positioning accuracy with global navigation satellite systems (GNSSs) is meta-signal processing (MSP), which treats several stand-alone GNSS signals as a single composite wideband signal. BeiDou Navigation Satellite System (BDS) III already offers with the B1I+B1C signal a meta-signal, while the forthcoming Galileo (GAL) E1D could be combined with E1B or E1C if the E1D signal is broadcast with a frequency offset to the L1/E1 carrier frequency. This would boost the ranging performance of GAL open service (OS) in the upper L-band through MSP. However, the cross-correlation function (CCF) of meta-signals contain numerous high side-maxima which can, when wrongly identified as the main peak, lead to significant pseudo-range errors of multiple meters. The probability of such a false lock is known to increase with decreasing signal-to-noise ratios but can significantly increase even further due to imperfections in the analog hardware components (e.g., linear and non-linear effects of a high-power amplifier (HPA), an output multiplexer (OMUX), a transmitter filter, and a front-end receiver), as these can distort the CCF. One remaining question is whether meta-signals are a well-suited approach to reliably increase ranging performance in the presence of payload and receiver distortions. This study presents the first systematic assessment of the robustness of several potential meta-signal options enabled by a forthcoming GAL E1D signal for different levels of distortion. The results show significant performance gains but also indicate constraints regarding the choice of signals when considering MSP under the influence of distortions.

Keywords: GNSS; Galileo; meta-signal processing; dual-band tracking

Citation: Beck, F.C.; Enneking, C.; Thölert, S.; Meurer, M. Galileo Performance Improvements Employing Meta-Signals —Robustness Analysis against Payload and Receiver Distortions. *Eng. Proc.* **2023**, *54*, 4. https://doi.org/10.3390/ENC2023-15472

Academic Editors: Tom Willems and Okko Bleeker

Published: 29 October 2023

1. Introduction

In recent decades, an evolving global navigation satellite system (GNSS) receiver landscape fueled the discussion of further introductions of new GNSS signals to meet emerging market needs. For instance, Galileo (GAL) E5-AltBOC and BeiDou Navigation Satellite System (BDS) B2-ACEBOC were added to the signal portfolio to offer a signal with increased bandwidth and therefore higher-ranging capabilities in the lower L-band. Options to broadcast new broadband signals in the upper L-band are limited due to spectrum restrictions as well as possible intra- and inter-system interference issues. In the upper L-band, a wideband open service (OS) ranging signal is missing, but to meet market demands, the introduction of a quasi-pilot signal for low-complexity receivers [1–3] with a potential offset to the L1 carrier was proposed by previous studies [4–6]. Adding this new signal with a frequency offset may enable meta-signal processing (MSP) of the GAL E1B/C with the E1D signal. The concept of MSP was first discussed in the context that a GNSS transmits a multitude of satellite navigation (satnav) signals across multiple frequencies [7,8]. MSP aims to track a combination of ranging signals which have different carrier frequencies and therefore have a larger Gabor bandwidth, which can in theory greatly increase tracking performance [9,10].

However, meta-signals have a severe ambiguity problem since their auto-correlation functions (ACFs) contain a large number of side-peaks which leads to larger probabilities of false locks in high-noise regimes [11]. To solve this issue, research has focused on developing tracking algorithms that optimize the benefits of meta-signals in particular for the BDS III signals in the upper L-band [10]. Notably, the extension of the double estimator/double phase estimator [12,13] and the wideband high-accuracy joint tracking technique [14] have been developed to improve the tracking performance of the BDS B1I-B1C meta-signal while reducing the number of required correlators [15].

The theoretical gains in terms of ranging performance and the mitigation of the ambiguity problem were the subject of previous work [9]. In [16], Nardin et al. conducted a first analysis of cross-correlation function (CCF) property changes in the presence of linear distortions. This study was limited to a meta-signal comprising a binary offset carrier (BOC)(1,1) and binary phase-shift keying (BPSK)(1)/BOC(1,1) with a frequency offset of 12×1.023 MHz. However, there is a lack of further findings regarding the influence of distortions on MSP performance dependent on the modulation and chosen frequency offset. In this paper, we present an extended analysis through a systematic assessment of the robustness of the time of arrival (TOA) estimation error standard deviation (stdev.) of several potential meta-signal options enabled by a forthcoming GAL E1D signal for different levels of distortions caused by a transmitter and a receiver. The results suggest that MSP may enable an improvement in the ranging performance depending on the exact shape and placement of the GAL E1D signal. Certain GAL E1D signal options were found to be sub-optimal in the context of MSP since the ambiguity problem is elevated in the presence of distortions.

2. System Model

2.1. Transmitter and Channel

The payload and channel of the system model consist of four building blocks, namely the input multiplexer (IMUX), high-power amplifier (HPA), output multiplexer (OMUX), and channel, as illustrated in Figure 1.

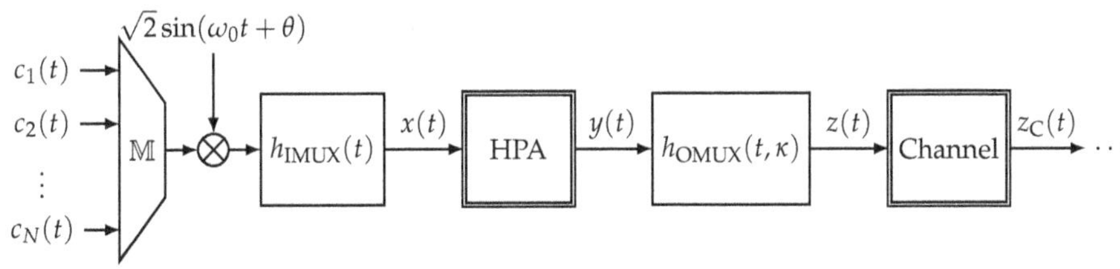

Figure 1. System model of transmitters and channels.

The IMUX $\mathbb{M}$ composes N analog code-division multiple access (CDMA) waveforms $c_n(t) = \sum_{m=-\infty}^{+\infty} c_n[m] d_n[m] p_n(t - mT_{c,n})$ to one composite signal

$$x(t) = h_{\mathrm{IMUX}}(t) * \left(\mathbb{M}\{c_1(t), c_2(t), \ldots, c_N(t) | \boldsymbol{p}, \boldsymbol{\phi}, \boldsymbol{f}^{\Delta}\} e^{j\omega_0 t}\right) \tag{1a}$$

$$= h_{\mathrm{IMUX}}(t) * \left(\left(\sum_{n=1}^{N} c_n(t)\sqrt{P_n} e^{j\phi_n} e^{j2\pi f_n^{\Delta} t} + x_{\mathrm{IM}}(t)\right) e^{j\omega_0 t}\right) \tag{1b}$$

modulated onto a joint carrier with the angular frequency $\omega_0 = 2\pi f_0$, where $c_n[m] \in \{+1, -1\}$ denotes the binary spreading code with a code rate of $f_{c,n} = 1/T_{c,n}$, $d_n[m] \in \{+1, -1\}$ denotes the navigation data, and $p_n(t)$ denotes the pulse shape of the n-th CDMA waveform. Furthermore, $f_n^{\Delta} = f_0 - f_n$ denotes the frequency offset of the

n-th CDMA waveform to the carrier frequency. In the case of the GAL E1 band, the first three CDMA waveforms are the BOC-modulated PRS and the composite binary offset carrier (CBOC)-modulated E1B and E1C signals. The fourth CDMA waveform was the potential E1D signal generated with a frequency offset. Possible imperfections in the digital-to-analog conversion may be captured by the filter $h_{\mathrm{IMUX}}(t)$. However, in the following, an all-pass IMUX was assumed. The multiplexed pass-band signal $x(t)$ was then fed to an HPA which amplifies the signal to a desired output power level. In this study, we constrain ourselves to an idealized, memoryless HPA model such that the resulting amplified signal is

$$y(t) = A(|x(t)|)e^{i(\Phi(|x(t)|)+\mathrm{atan2}(\mathrm{Im}\{x(t)\},\mathrm{Re}\{x(t)\}))} \tag{2}$$

where $A(|x(t)|)$ denotes the AM/AM mapping function and $\Phi(|x(t)|)$ denotes the AM/PM mapping function [17]. The amplified signal $y(t)$ is then passed to the OMUX which facilitates a band limitation, which can be modeled with a filter such that the signal radiated from the satellite can be expressed as

$$z(t) = h_{\mathrm{OMUX}}(t,\kappa) * y(t) \tag{3}$$

where $\kappa \in [0,1]$ denotes the payload distortion parameter, which controls the degree of filtering distortion introduced by the OMUX. Section 2.3 describes the implementation of a scalable distortion OMUX filter $h_{\mathrm{OMUX}}(t,\kappa)$ used in the numerical evaluations. After OMUX filtering, the satellite contribution

$$z_{\mathrm{C}}(t) = \gamma z(t-\tau_0) \tag{4}$$

is formed by the channel which delays the signal by τ_0 and attenuates the signals mainly due to path loss, which is captured by a constant complex scalar γ.

2.2. Receiver

A straight-forward receiver model may assume an introduction of additive white Gaussian noise (AWGN) $\xi(t)$ onto the pass-band signal $z_{\mathrm{C}}(t)$ on the ground before its demodulation with $\omega_{\mathrm{IF}} = 2\pi f_{\mathrm{IF}}$ and front-end filtering with a ideal brick-wall filter $h_{\mathrm{FE}}(t) = B\sin(\pi Bt)/(\pi Bt) \triangleq B\,\mathrm{sinc}(Bt)$, where B denotes the two-sided front-end filtering bandwidth, resulting in the analog receive signal $r(t)$. This receive signal is then digitized using a sampling rate of $1/T_s$, resulting in the receive signal vector:

$$\boldsymbol{r} = \begin{bmatrix} r(0) & r(T_s) & r(2T_s) & \cdots & r((N_s-1)T_s) \end{bmatrix}^{\mathrm{T}} \quad \in \mathbb{C}^{N_s\times 1} \tag{5}$$

where N_s denotes the number of samples accumulated for a specific integration time and $(.)^{\mathrm{T}}$ denotes the transpose operator. Applying the receive signal vector $\boldsymbol{r}$ to a correlator bank

$$\boldsymbol{C} = \begin{bmatrix} \bar{\boldsymbol{c}}^{(0)} & \bar{\boldsymbol{c}}^{(1)} & \cdots & \bar{\boldsymbol{c}}^{(n_c)} & \cdots & \bar{\boldsymbol{c}}^{(N_c-1)} \end{bmatrix} \quad \in \mathbb{C}^{N_s\times N_c}, \tag{6}$$

where $\bar{\boldsymbol{c}}^{(n_c)}$ denotes the vectorized signal replica $\bar{c}(t-\tau_{n_c})$ with the shift $\tau_{n_c} = \Delta((N_c-1)/2+n_c)$ using the uniform correlator spacing Δ, yields then a noisy, sampled CCF $\tilde{\boldsymbol{q}} = \boldsymbol{C}^{\mathrm{H}}\boldsymbol{r} \in \mathbb{C}^{N_c\times 1}$ for a specific noise realization of $\xi(t)$. Figure 2 shows the generation of the correlator output $\tilde{\boldsymbol{q}}$ for the straight-forward model.

For simulation purposes, it is convenient to replace the straight-forward receiver model with an equivalent alternative, shown in Figure 3, which adds colored noise $\tilde{\boldsymbol{n}}$ directly onto the noise-free version, a sampled CCF $\boldsymbol{q} = \boldsymbol{C}^{\mathrm{H}}\boldsymbol{s} \in \mathbb{C}^{N_c\times 1}$ resulting in a noisy CCF $\tilde{\boldsymbol{q}}$ with identical statistical properties as in the straight-forward model. For the simulation, a vector for different noise power levels has to be computed based on the carrier-to-noise-density ratio (C/N_0) that is to be investigated. The advantage of this approach is that the Monte-

Carlo simulation can be achieved with less computational effort since $N_s \gg N_c$. From a statistical viewpoint, the system models shown in Figures 2 and 3 are fully equivalent. This is established in the following theorem.

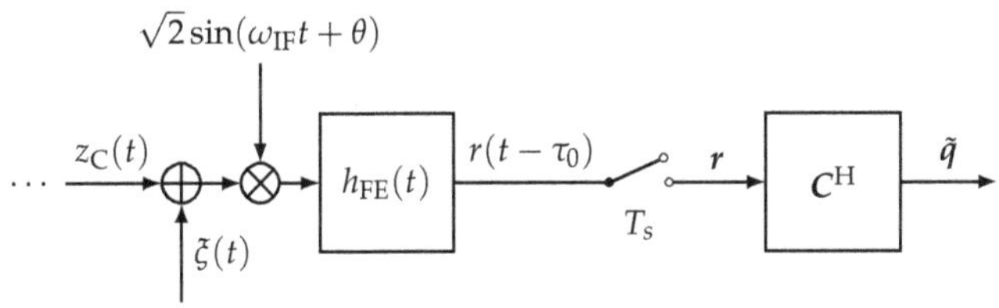

Figure 2. Straight-forward system model of the receiver with the addition of white noise onto a pass-band signal.

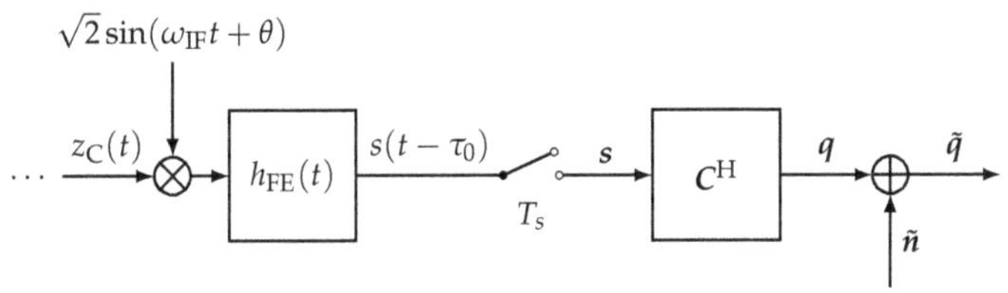

Figure 3. Alternative system model of the receiver with the addition of colored noise onto a noise-free correlator output.

Theorem 1. *The addition of the colored noise in the post-correlation domain is statistically equivalent to the addition of noise in the pre-correlation domain when generating the zero mean colored noise vector $\tilde{\boldsymbol{n}} = \mathbf{D}^{\text{H}}\boldsymbol{\eta}$ using a white noise vector $\boldsymbol{\eta} \sim \mathcal{CN}(0, \sigma^2\mathbf{I})$ with $\sigma^2 = \text{N}_0 B$ and a coloring matrix $\mathbf{D}$. The coloring matrix $\mathbf{D}$ is derived from the covariance matrix $\boldsymbol{\Sigma} = \text{E}[\tilde{\boldsymbol{q}}\tilde{\boldsymbol{q}}^{\text{H}}] = \mathbf{C}^{\text{H}}\boldsymbol{W}\boldsymbol{C}$ with*

$$\boldsymbol{e}_i^{\text{T}}\boldsymbol{W}\boldsymbol{e}_j = \text{sinc}(BT_s(i-j)) \quad \forall i,j \in \{0,\ldots,N_s-1\} \tag{7}$$

and by applying the eigendecomposition (ED) of

$$\boldsymbol{\Sigma} = \mathbf{Q}\boldsymbol{\Lambda}\mathbf{Q}^{-1} = \mathbf{Q}\boldsymbol{\Lambda}^{\frac{1}{2}}\left(\boldsymbol{\Lambda}^{\frac{1}{2}}\right)^{\text{H}}\mathbf{Q}^{\text{H}} \triangleq \mathbf{D}^{\text{H}}\mathbf{D} \tag{8}$$

exploiting $\mathbf{Q}^{-1} = \mathbf{Q}^{\text{H}}$, and defining the coloring matrix as $\mathbf{D} = \left(\mathbf{Q}\boldsymbol{\Lambda}^{\frac{1}{2}}\right)^{\text{H}}$.

Proof of Theorem 1. The vectorized receive signal can be expressed as

$$\boldsymbol{r} = \boldsymbol{s} + \boldsymbol{n} \quad \text{with} \quad \boldsymbol{n} \sim \mathcal{CN}(0, \sigma^2\boldsymbol{W}) \tag{9}$$

where $\sigma^2 = N_0 B$ and its covariance matrix $\boldsymbol{W} \in \mathbb{R}^{N_s \times N_s}$ with its matrix entries as defined in Equation (7) accounts for the band-limitation introduced by the front-end filter $h_{\text{FE}}(t)$ [18]. Hence, the received signal $\boldsymbol{r} \sim \mathcal{CN}(\boldsymbol{s}, \sigma^2\boldsymbol{W})$ follows a non-zero mean and non-Gaussian distribution. The noisy correlator output $\tilde{\boldsymbol{q}} = \mathbf{C}^{\text{H}}\boldsymbol{r}$ can be expressed as

$$\tilde{\boldsymbol{q}} \sim \mathcal{CN}(\mathbf{C}^{\text{H}}\boldsymbol{s}, \sigma^2\mathbf{C}^{\text{H}}\boldsymbol{W}\boldsymbol{C}) \tag{10}$$

such that the statistical properties of the colored noise $\tilde{\boldsymbol{n}}$ to be added into the post-correlation domain become evident and we can define

$$\tilde{\boldsymbol{q}} = \boldsymbol{q} + \tilde{\boldsymbol{n}} \quad \text{with} \quad \tilde{\boldsymbol{n}} \sim \mathcal{CN}(0, \sigma^2\boldsymbol{\Sigma}) \tag{11}$$

where $\mathbf{\Sigma} = \mathbf{C}^{\mathrm{H}}\mathbf{W}\mathbf{C}$ and $\mathbf{q} = \mathbf{C}^{\mathrm{H}}\mathbf{s}$. Since $\mathbf{\Sigma}$ is Hermitian and positive-definite, its ED can be expressed as $\mathbf{\Sigma} = \mathbf{Q}\mathbf{\Lambda}\mathbf{Q}^{\mathrm{H}}$, where $\mathbf{Q}$ denotes the matrix of column-wise collected eigenvectors and $\mathbf{\Lambda}$ denotes the diagonal matrix with the eigenvalues. This expression can be reformulated such that we have for the covariance matrix $\mathbf{\Sigma} = \mathbf{D}^{\mathrm{H}}\mathbf{D}$, as defined in Equarion (8). The, using the coloring matrix $\mathbf{D}$ and an AWGN noise vector $\boldsymbol{\eta} \sim \mathcal{CN}(0, \sigma^2\mathbf{I})$ to generate the colored noise vector $\tilde{\mathbf{n}} = \mathbf{D}^{\mathrm{H}}\boldsymbol{\eta}$ achieves the correct statistical properties of $\tilde{\mathbf{q}}$ as the covariance matrix is

$$\mathrm{cov}(\tilde{\mathbf{q}}, \tilde{\mathbf{q}}) = E[(\tilde{\mathbf{q}} - E[\tilde{\mathbf{q}}])(\tilde{\mathbf{q}} - E[\tilde{\mathbf{q}}])^{\mathrm{H}}] = E[(\tilde{\mathbf{q}} - \mathbf{q})(\tilde{\mathbf{q}} - \mathbf{q})^{\mathrm{H}}] \quad (12a)$$

$$= E[\tilde{\mathbf{n}}\tilde{\mathbf{n}}^{\mathrm{H}}] = E[\mathbf{D}^{\mathrm{H}}\boldsymbol{\eta}\boldsymbol{\eta}^{\mathrm{H}}\mathbf{D}] = \mathbf{D}^{\mathrm{H}}E[\boldsymbol{\eta}\boldsymbol{\eta}^{\mathrm{H}}]\mathbf{D} = \sigma^2\mathbf{D}^{\mathrm{H}}\mathbf{D} = \sigma^2\mathbf{\Sigma}. \quad (12b)$$

□

2.3. Scalable Filtering Distortion

The "morphing" between L states of payload distortion is achieved with a sinusoidal weight function $\boldsymbol{w}(\kappa, L) = [w_1(\kappa, L), ..., w_L(\kappa, L)]^{\mathrm{T}}$, where the l-th element is defined as

$$w_l(\kappa, L) = \begin{cases} \frac{1}{2}\cos(\Omega(L)(\kappa - m(l, L)) + \frac{1}{2} & \text{if } |\kappa - m(l, L)| \leq \frac{1}{L-1} \\ 0 & \text{else} \end{cases} \quad (13)$$

with $\Omega(L) = 2\pi(L-1)$ and $m(l, L) = (l-1)/(L-1)$. A filter with scalable distortions can be then described as

$$H(f, \kappa, \mathcal{H}) = \sum_{l=1}^{L} w_l(\kappa, L)H_l(f) = \boldsymbol{H}(f)\boldsymbol{w}(\kappa, L) \quad (14)$$

with $\kappa \in [0, 1]$ being the distortion scaling coefficient; $L = |\mathcal{H}|$; and the set of ordered filter functions $\mathcal{H} = \{H_1(f), ..., H_L(f)\}$, where $H_1(f)$ is the filter applying the least distortions (e.g., "all-pass" filter), $H_L(f)$ is the filter with maximal distortions, and the filters from $H_2(f)$ to $H_{L-1}(f)$ show an increase in filtering distortion in terms of band limitation and group delay. The scalable distortion OMUX filter in this study uses $L = 5$ filters. The first filter $H_1(f) = 1$ has an all-pass characteristic. The last filter $H_5(f)$ is generated through a polynomial fitting of a measured transfer characteristic of a GNSS satellite payload. $H_2(f)$ and $H_3(f)$ have a pre-distorted Butterworth characteristic of fourth and sixth order, respectively. Furthermore, $H_2(f)$ has two sides with 61.38 MHz and $H_3(f)$ has the same bandwidth as $H_5(f)$. Lastly, $H_4(f)$ is a pre-distorted version of $H_5(f)$, resulting in the same amplitude response but an all-zero phase response.

3. Signal Candidates

In the past, multiple potential signals were proposed to be transmitted as a future GAL E1D signal. In the early 2010s, E1D candidates with no offset to the already transmit signal were the subject of discussions [2,8]. In the second half of the decade, E1D signal candidates with an offset to the carrier of the E1B, E1C, and PRS signals were proposed and investigated [4–6,11,16]. Inspired by the aforementioned proposals, seven E1D signal candidates are investigated in the following, which are either BPSK-rectangular (R), CBOC, or BOC-modulated and have different chipping rates and carrier frequency placements within the upper L-band. The detailed parameters of the signal candidates can be found in Table 1. The spectral allocations of the main lobes and signal carrier placement of the E1D candidates, which where already suggested in earlier studies [5], are either slightly above or below the upper main lobe of the modernized authorized services provided by GAL and other GNSS to avoid inter- and intra-system interference. The signal candidates E1D-a/b/c have 4 × 1.023 MHz, E1D-g has 12 × 1.023 MHz, and E1D-d/e/f have 20 × 1.023 MHz center frequency offsets with respect to the L1/E1 carrier frequency. The most likely modulation for an E1D signal is BPSK-R due to its simplicity suited for reception by low-complexity receivers. The CBOC and BOC-modulated candidates were included due to interesting ACF properties when assuming a coherent signal generation. The shape of the meta-signal E1C+E1D ACFs is illustrated by Figure 4b. The complex E1C+E1D meta-signal replicas for the simulations are defined as

$$\bar{c}_n(t) = c_{\mathrm{E1C}}(t)e^{-j\pi f_n^{\Delta}t} + c_{\mathrm{E1D}}(t)e^{+j\pi f_n^{\Delta}t}. \tag{15}$$

Table 1. Parameters of investigated Galileo E1D signal candidates forming a meta-signal with the Galileo E1C signal.

Candidate	Modulation	$f_{c,\mathrm{E1D}}$/MHz	f^{Δ}/MHz	$f_0 + f^{\Delta}$/MHz	f_{IF}/MHz
E1D-a	BPSK-R	1.0230	4.092	1579.512	1577.466
E1D-b	BPSK-R	0.5115	4.092	1579.512	1577.466
E1D-c	CBOC(6,1,1/11,-)	1.0230	4.092	1579.512	1577.466
E1D-d	BPSK-R	1.0230	20.460	1595.880	1585.650
E1D-e	BPSK-R	0.5115	20.460	1595.880	1585.650
E1D-f	CBOC(6,1,1/11,-)	1.0230	20.460	1595.880	1585.650
E1D-g	BOC(8,1)	1.0230	12.276	1587.696	1581.558

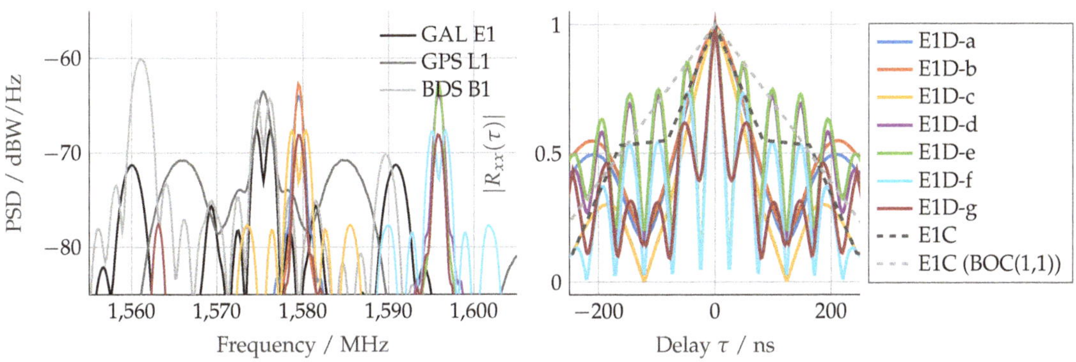

(**a**) (**b**)

Figure 4. Spectral allocation of individual signals and potential Galileo meta-signal ACFs options. (**a**) PSDs of broadcast GNSS L1/E1/B1 signals and potential E1D signal candidates in upper L-band. (**b**) Absolute part of E1C+E1D meta-signal ACFs assuming a centrally placed intermediate frequency.

4. Performance Measures

The first performance metric is the spectral separation coefficient (SSC) [19]

$$\mathrm{SSC}(G_1, G_2, B) = \frac{\int_{-B/2}^{+B/2} G_1(f)G_2(f)\mathrm{d}f}{\int_{-\infty}^{+\infty} G_1(f)\mathrm{d}f \int_{-\infty}^{+\infty} G_2(f)\mathrm{d}f}, \tag{16}$$

where $G_1(f)$ and $G_2(f)$ denote the power spectral densities (PSDs) of two signals to test. The second performance measure was the multipath error envelope assuming one line-of-sight (LOS) tap and one non-line-of-sight (NLOS) tap with a 3 dB attenuation for varying multipath tap delays [20]. The last investigated performance measure is the unsmoothed TOA estimation error stdev.

$$\sigma = \sqrt{\frac{1}{J}\sum_{j=1}^{J}(\hat{\tau}_0(j) - \mu)^2} \tag{17}$$

of the code phase, where $\hat{\tau}_0(j)$ denotes the j-th TOA estimate based on the j-th realization of $\tilde{\boldsymbol{q}}$ of, in total, J Monte-Carlo realizations, and $\mu = (1/J)\sum_{j=1}^{J}\hat{\tau}_0(j)$ denotes the sample mean of the TOA estimates, which are derived either by coherent early–late processing (CELP) or noncoherent early–late processing (NELP), as explored in [21].

5. Numerical Results

5.1. Simulation Parameters

The average power of the candidate meta-signals and current signals on the ground is set to −154.0 dBW by assuming an ideal HPA with an average output power of 15.0 dBW and a total propagation loss of 169.0 dB, accounting for path loss, implementation losses, and antenna gains. The power ratio between the E1C and E1D candidates was set to 1:1. The two-sided front-end filtering bandwidth B is 51.150 MHz. The integration time was 1.955 µs. The number of Monte-Carlo runs J for each data point is 200,000. No smoothing for the TOA estimation is assumed. The delay τ_0 was randomized in an interval $[-\frac{1}{3}T_{c,\text{BPSK-R}(1)}, +\frac{1}{3}T_{c,\text{BPSK-R}(1)}]$ to introduce scalloping losses [21]. Non-linear effects introduced by the HPA and an analog-to-digital converter are beyond the scope of this paper. The correlator banks have a spacing of $\Delta_n = T_{c,n}/250$ and $N_c = 501$. The TOA estimation of $\hat{\tau}_0$ uses a parabolic interpolation around the highest value of the absolute and real parts of the CCF vector $\tilde{\boldsymbol{q}}$ using an interpolation equation from [22].

5.2. Spectral Separation Analysis Using the Spectral Separation Coefficient

The self-SSCs of the already broadcast CDMA waveforms in the upper L-band range from −61.9 to −73.1 dB/Hz, assuming a 51.150 MHz two-sided reception bandwidth B. The SSCs between the civilian/OS signals range from −67.9 to −68.3 dB/Hz. The SSCs between the authorized signals range from −72.4 to −88.2 dB/Hz. Table 2 reports the expected SSCs for the currently present CDMA waveforms and the E1D signal candidates with the respective frequency offsets. The signal candidates E1D-a/b/c/g would interfere the most with the BPSK-R(10) waveform as their spectra have a significant overlap. In terms of the SSC, an introduction of the E1D-d/e/f would cause less interference. Among those, the CBOC-modulated E1D-f candidate performs the poorest. Among all investigated candidates, the E1D-e variant would cause the least interference with the already present CDMA waveforms.

Table 2. Spectral separation coefficients in dB/Hz of currently transmitted waveforms (with f_0 = 1575.42 MHz) with the investigated added E1D signal candidates within the upper L-Band placed off-center.

	E1D-a	E1D-b	E1D-c	E1D-d	E1D-e	E1D-f	E1D-g
BPSK-R(1)	−82.08	−85.09	−81.58	−96.06	−99.07	−94.91	−87.78
Sine BOC(1,1)	−82.08	−85.09	−75.31	−96.06	−99.07	−89.13	−87.78
MBOC(6,1,1/11)	−81.58	−84.59	−75.53	−94.91	−97.92	−89.19	−86.57
BPSK-R(10)	−72.59	−72.55	−72.85	−96.06	−99.07	−89.14	−76.59
Sine BOC(10,5)	−81.79	−82.12	−79.59	−96.06	−99.07	−90.30	−83.35
Cosine BOC(15,2.5)	−94.91	−96.49	−89.98	−87.13	−90.14	−80.70	−87.36
Sine BOC(14,2)	−90.07	−93.08	−84.60	−91.11	−94.12	−80.86	−87.65

5.3. Multipath Error Envelope

Figure 5 shows the multipath error envelopes of the meta-signals and current GAL signals when assuming an all-pass transmit characteristic, NELP with an early–late spacing $\Delta = 0.01 \times T_{c,\text{BPSK-R}(1)}$, and a 24.552 MHz brick-wall receive filter centrally placed at f_{IF}. The meta-signal using E1D-b/e with the longer chip duration has a poorer multipath rejection than the other signal candidates as the envelope is non-zero up to approximately 586 m in multipath tap length. The meta-signal combinations E1C+E1D-a/b/c/g perform poorer than using the E1C signal as a stand-alone signal in the region from 0 to 50 meters of NLOS tap length. The other meta-signals with E1D-d/e/f have a tighter multipath envelope but possess more ripples than the current signals' envelopes. According to these results, choosing a larger frequency offset would be beneficial for an E1D signal. Two-correlator NELP of a E1C+E1D-d/e/f signal can reduce the influence of multipath, but due to the sheer number of side peaks in the ACF, an oscillating multipath error can be observed up to the longer of the two chip durations of the stand-alone signals constituting the meta-signal. The minima of this oscillating multipath error of the E1C+E1D-a/b/c/d/e/f meta-signal are approximately located at $1/f^{\Delta}$.

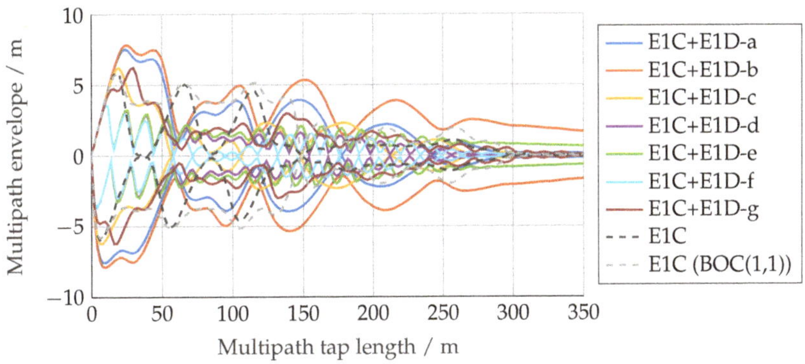

Figure 5. Multipath envelopes of investigated meta-signals assuming a very narrow NELP with an early–late spacing $\Delta = 0.01 \times T_{\mathrm{c,BPSK-R(1)}}$ and the application of a brick-wall filter with a two-sided bandwidth of 24.552 MHz.

5.4. Tracking Performance

Assessing the tracking performance in terms of unsmoothed TOA estimation error stdev. for a fixed C/N_0 is, in this study, the primary performance measure used to judge whether a potential E1D signal may also be a good candidate considering MSP. The carrier power $C = \mathbf{s}^{\mathrm{H}}\mathbf{s}/N_s$ is derived for each signal to individually test that a fair-ranging performance evaluation for each signal is achieved. However, when directly comparing the status quo tracking, for instance, the GAL E1C signal at a certain noise power N_0B with MSP of a potential E1C+E1D signal, one should keep in mind whether an E1D signal would be added with a 1:1 power ratio with respect to the E1C signal; the combined meta-signal's power would be 3 dB higher than the E1C signal alone, resulting in a boost of TOA estimation error stdev., where the degree depends on the reception situation in terms of noise regime but also the chosen E1D signal.

Figure 6 illustrates the code-phase ranging performance of each investigated signal for the two extreme cases, where either an ideal all-pass payload ($\kappa = 0$) or a realistic distortion of a measured filter is applied ($\kappa = 1$) during the signal formation. In the low-noise regime ($C/N_0 \geq 50$ dB-Hz), two performance tiers of signals can be identified. The meta-signals with the larger frequency offset (E1C+E1D-d/e/f/g) lead to a decreased TOA estimation error stdev. and therefore an increased ranging performance compared to E1C tracking at the same C/N_0. The E1C+E1D-g signal candidate accumulates deformations due its wider bandwidth and experiences the greatest ranging performance degradation for a realistic payload characteristic in comparison to an all-pass payload scenario. The E1C+E1D-f performs best above 45 dB-Hz and sees a performance difference between ideal and fully distorted payload in a transition region from 42 to 47 dB-Hz.

Tracking E1C+E1D-a/b meta-signals is worse than CBOC-tracking of the E1C signal in the all-pass case but even more so for the realistic payload case for a C/N_0 greater than 45 dB-Hz. In the fully distorted case, E1C+E1D-a/b-tracking is worse than BOC-tracking of an E1C signal above 50 dB-Hz C/N_0. Tracking E1C+E1D-a/b instead of E1C only would merely be improved due to the increased signal power but not due to MSP as the ACF main peak is less sharp than that of E1C using a CBOC replica.

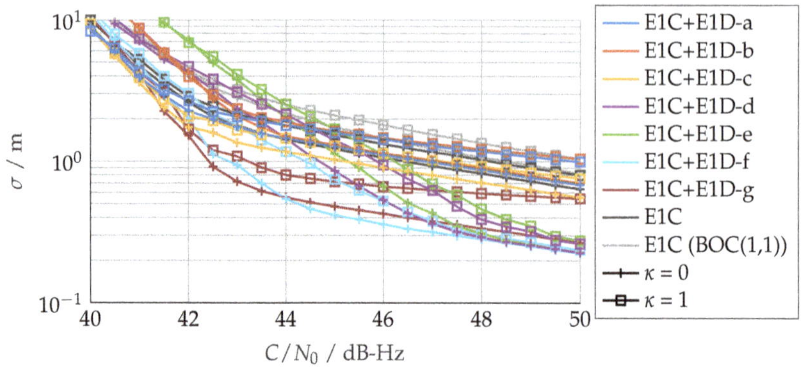

Figure 6. Unsmoothed TOA estimation error stdev. for different noise levels for the two extreme distortion cases for NELP.

Figure 7 shows the evolution of the stdev. and sample mean when varying the distortion parameter κ from 0 to 1 for a fixed C/N_0 of 45 dB-Hz.

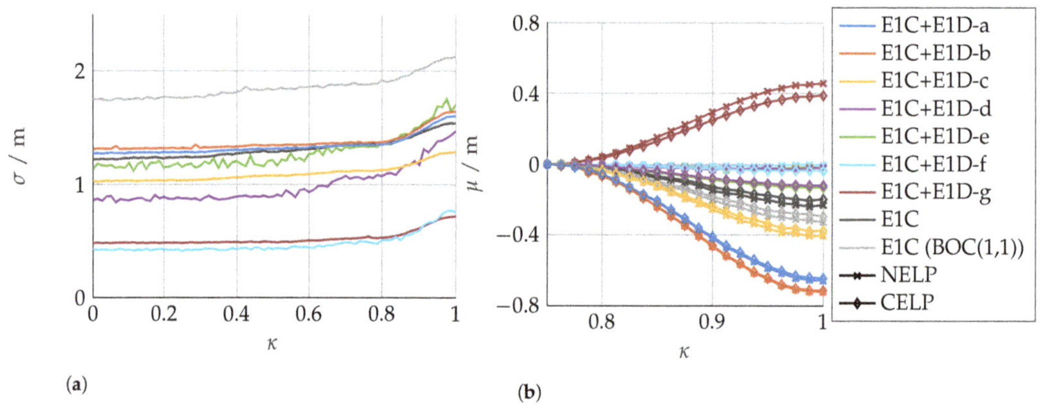

Figure 7. Evolution of first and second statistical moments for a fixed noise level of 45 dB-Hz. (**a**) NELP TOA estimation error sample stdev. (**b**) TOA estimation error sample mean.

The gradual increase in band limitation only slightly worsens the TOA estimation error stdev. For $\kappa > 0.75$, the influence of the measured phase response influences the mean and stdev., resulting in a range bias but also a worsening of ranging performance due to the introduced CCF asymmetry.

6. Conclusions

In the near future, MSP of the GAL E1B/C with an E1D signal could lead to significant performance gains for civil GNSS receivers in the upper L-band. The increase in ranging performance will depend on the exact shape and placement of the GAL E1D signal. However, the impact of payload and receiver distortions has to be closely monitored and expected theoretical performance gains will be reduced significantly for realistic levels of distortion. NELP TOA estimation with a single signal (e.g., E1C only) is more robust with respect to distortions and can even perform better than a solution based on MSP of signals with an insufficient frequency offset, if the distortions are too severe.

Author Contributions: Conceptualization, F.C.B. and M.M.; methodology, F.C.B. and C.E.; software, F.C.B.; validation, F.C.B., C.E., and M.M.; formal analysis, F.C.B. and C.E.; investigation, F.C.B.; resources, F.C.B.; data curation, F.C.B. and S.T.; writing—original draft preparation, F.C.B.; writing—review and editing, F.C.B. and C.E.; visualization, F.C.B.; supervision, M.M.; project administration, S.T.; funding acquisition, M.M. All authors have read and agreed to the published version of the manuscript.

Funding: This research received no external funding.

Institutional Review Board Statement: Not applicable.

Informed Consent Statement: Not applicable.

Data Availability Statement: The data presented in this study are available upon request from the corresponding author.

Conflicts of Interest: The authors declare no conflict of interest.

References

1. García-Molina, J.A.; Wallner, S.; Vazquez, C.; De Pasquale, G.; Del Peral-Rosado, J.A.; Da Broi, G.; Schmitt, T.; Melman, F.; Parro, J.M.; Soualle, F.; et al. Quasi-Pilot Signals: Acquisition and Fast Time Ambiguity Resolution. In Proceedings of the 2022 10th Workshop on Satellite Navigation Technology (NAVITEC), Noordwijk, The Netherlands, 5–7 April 2022.
2. Paonni, M.; Bavaro, M. On the Design of a GNSS Acquisition Aiding Signal. In Proceedings of the 26th International Technical Meeting of The Satellite Division of the Institute of Navigation (ION GNSS+ 2013), Nashville, TN, USA, 16–20 September 2013; pp. 1445–1456.
3. Wallner, S.; García-Molina, J.A.; Hahn, J.; Risueno, G.L.; Broi, G.d.; Floch, J.J.; Soualle, F.; Schmitt, T.; Ouedraogo, M.; Wörz, T.; et al. Quasi-Pilot Signal Design-Facilitating New Signal Processing Concepts. In Proceedings of the 34th International Technical Meeting of the Satellite Division of The Institute of Navigation (ION GNSS+ 2021), St. Louis, MO, USA, 20–24 September 2021; pp. 1859–1876. [CrossRef]
4. Anghileri, M.; Floch, J.J.; Margaria, D.; Motella, B.; Peña, A.G.; Julien, O.; Macabiau, C.; Chauvat, R.; Paonni, M. FUNTIMES–Future Navigation and Timing Evolved Signals. In Proceedings of the 31st International Technical Meeting of the Satellite Division of the Institute of Navigation (ION GNSS+ 2018), Miami, FL, USA, 24–28 September 2018; pp. 876–912. [CrossRef]
5. Sénant, E.; Gadat, B.; Charbonnieras, C.; Roche, S.; Aubault, M.; Marmet, F.X. Tentative New Signals and Services in Upper L1 and S Bands for Galileo Evolutions. In Proceedings of the 31st International Technical Meeting of The Satellite Division of the Institute of Navigation (ION GNSS+ 2018), Miami, FL, USA, 24–28 September 2018. [CrossRef]
6. Ortega Espluga, L. Signal Optimization for Galileo Evolution. Ph.D. Thesis, Institut National Polytechnique de Toulouse, Toulouse, France, 2019.
7. Issler, J.L.; Paonni, M.; Eissfeller, B. Toward centimetric positioning thanks to L-and S-Band GNSS and to meta-GNSS signals. In Proceedings of the 2010 5th ESA Workshop on Satellite Navigation Technologies and European Workshop on GNSS Signals and Signal Processing (NAVITEC), Noordwijk, The Netherlands, 8–10 December 2010; pp. 1–8. [CrossRef]
8. Paonni, M.; Curran, J.T.; Bavaro, M.; Fortuny-Guasch, J. GNSS Meta Signals: Coherently Composite Processing of Multiple GNSS Signals. In Proceedings of the 27th International Technical Meeting of the Satellite Division of the Institute of Navigation (ION GNSS+ 2014), Tampa, FL, USA, 8–12 September 2014; pp. 2592–2601.
9. Ortega, L.; Vilà-Valls, J.; Chaumette, E.; Vincent, F. On the Time-Delay Estimation Accuracy Limit of GNSS Meta-Signals. In Proceedings of the 2020 IEEE 23rd International Conference on Intelligent Transportation Systems (ITSC), Rhodes, Greece, 20–23 September 2020; pp. 1–6. [CrossRef]
10. Borio, D.; Gioia, C. Reconstructing GNSS Meta-signal Observations Using Side-Band Measurements. *NAVIGATION J. Inst. Navig.* **2023**, *70*, 1–7. [CrossRef]
11. Schwalm, C.; Enneking, C.; Thölert, S. Ziv-Zakai Bound and Multicorrelator Compression for a Galileo E1 Meta-Signal. In Proceedings of the 2020 European Navigation Conference (ENC), Dresden, Germany, 23–24 November 2020; pp. 1–9. [CrossRef]
12. Wang, C.; Cui, X.; Ma, T.; Zhao, S.; Lu, M. Asymmetric Dual-Band Tracking Technique for Optimal Joint Processing of BDS B1I and B1C Signals. *Sensors* **2017**, *17*, 2360. [CrossRef] [PubMed]
13. Borio, D.; Gioia, C. GNSS Meta-signals, Dual-Frequency Combinations and the Double Phase Estimator. In Proceedings of the 2022 International Technical Meeting of the Institute of Navigation, Long Beach, CA, USA, 25–27 January 2022; pp. 1015–1026. [CrossRef]
14. Zhang, W.; Yao, Z.; Lu, M. WHAT: Wideband High-Accuracy Joint Tracking Technique for BDS B1 Composite Signal. In Proceedings of the 2019 IEEE 9th International Conference on Electronics Information and Emergency Communication (ICEIEC), Beijing, China, 12–14 July 2019; pp. 178–182. [CrossRef]
15. Tian, Z.; Cui, X.; Liu, G.; Lu, M. LPRA-DBT: Low-Processing-Rate Asymmetrical Dual-Band Tracking Method for BDS-3 B1I and B1C Composite Signal. In Proceedings of the International Technical Meeting of the Institute of Navigation, Long Beach, CA, USA, 25–27 January 2022. [CrossRef]

16. Nardin, A.; Dovis, F.; Motella, B. Impact of non-idealities on GNSS meta-signals processing. In Proceedings of the 2020 European Navigation Conference (ENC), Dresden, Germany, 23–24 November 2020; pp. 1–8. [CrossRef]
17. Beck, F.C.; Enneking, C.; Thölert, S.; Antreich, F. Comparison of Constant and Non-constant Envelope Signals for Satellite Navigation. In Proceedings of the 2022 10th Workshop on Satellite Navigation Technology (NAVITEC), Noordwijk, The Netherlands, 5–7 April 2022; pp. 1–9. [CrossRef]
18. Enneking, C.; Beck, F.C.; Thölert, S.; Antreich, F. Transmit Pulse Shaping for 1-Bit GNSS Receivers. In Proceedings of the 2022 International Conference on Localization and GNSS (ICL-GNSS), Tampere, Finland, 7–9 June 2022; pp. 1–6. [CrossRef]
19. Betz, J.W.; Goldstein, D.B. Candidate designs for an additional civil signal in GPS spectral bands. In Proceedings of the 2002 National Technical Meeting of the Institute of Navigation, ION, San Diego, CA, USA, 28–30 January 2002; pp. 622–631.
20. Teunissen, P.J.G.; Montenbruck, O. *Springer Handbook of Global Navigation Satellite Systems*; Springer: Berlin/Heidelberg, Germany, 2017.
21. Betz, J.W. *Engineering Satellite-Based Navigation and Timing: Global Navigation Satellite Systems, Signals, and Receivers*; John Wiley & Sons, Inc.: Hoboken, NJ, USA, 2015. [CrossRef]
22. Smith, J.O. Spectral Audio Signal Processing. Availiable online: https://ccrma.stanford.edu/~jos/sasp/ (accessed on 25 April 2023).

engineering proceedings

MDPI

Proceeding Paper

Radar Absolute Positioning †

Chris Hargreaves *, Alan Grant and Lydia Hyde

GRAD (GLA {General Lighthouse Authorities of the UK & Ireland} Research & Development), Trinity House, Harwich CO12 3JW, UK; alan.grant@gla-rad.org (A.G.); lydia.hyde@gla-rad.org (L.H.)
* Correspondence: chris.hargreaves@gla-rad.org
† Presented at the European Navigation Conference 2023, Noordwijk, The Netherlands, 31 May–2 June 2023.

Abstract: The General Lighthouse Authorities (GLA) R&D Directorate has, for some time, advocated that alternate position navigation and timing (PNT) systems should be established to offset critical dependency on Global Navigation Satellite Systems (GNSS), both in the maritime world and across wider industry. The maritime world already has a number of viable systems that may provide a ship's navigator with position-fixing information independent of GNSS. A promising option is radar. Depending on size, merchant vessels are required to fit radar operating in the X-band (9 GHz) and, optionally S-band (3 GHz). The radar return from the shoreline provides a unique and identifiable fingerprint that depends on the location of the vessel. GRAD has developed a number of different processes that can convert a series of images from the ship's radar into a series of output position-fixes, independent of having access to satellite (GNSS) positioning. This report describes the methods we investigated and presents the results from a number of ship trials.

Keywords: navigation; resiliency; radar; SLAM; integration

Citation: Hargreaves, C.; Grant, A.; Hyde, L. Radar Absolute Positioning. *Eng. Proc.* **2023**, *54*, 5. https://doi.org/10.3390/ENC2023-15419

Academic Editors: Tom Willems and Okko Bleeker

Published: 29 October 2023

1. Introduction

Global Navigation Satellite Systems (GNSS) are in use worldwide and form the de-facto standard for positioning, navigation and timing systems. GNSS technology is accurate, reliable and available globally, free at the point of use. This has led to it becoming an integral part of most industries, adopted as a silent utility across almost all sectors, resulting in a large amount of industrial activity now being critically dependent on its functioning [1].

GNSS, in particular the American GPS, is used extensively in the maritime world, primarily for plotting a vessel's position on an electronic chart display (ECD) or ECDIS. It is also relied upon by other bridge equipment, such as for reporting a vessel's position via AIS, and for timing purposes, such as for AIS transmission slots and digital selective calling (DSC). The advent of e-Navigation, integrated bridge systems and greater automation within the maritime sector will only increase the sector's dependency on GNSS further.

GNSS signals are very weak, vulnerable to obstruction, local radio interference or deliberate jamming. The General Lighthouse Authorities (GLA) have, for a long time, advocated adopting alternate backup systems to mitigate the consequences of a potential loss of GNSS. The GLA Research & Development directorate (GRAD) has investigated a number of position-fixing technologies that operate independently of satellite systems, including making use of existing ship-borne equipment such as radar.

This paper provides a summary of our work on Radar Absolute Position Fixing and the results of our investigations so far.

2. Radar Absolute Position Fixing

SOLAS-class vessels are required to equip radar in the X-band (9 GHz) and, depending on size, also the S-band (3 GHz) [2]. Marine radar is a valuable tool for navigation; maintaining situational awareness; and collision avoidance, particularly in reduced-visibility conditions. There also exist a number of ways to employ a ship's radar to determine the vessel's position and to plot that position on a nautical chart [3]. A simple method is to pick several conspicuous landmarks such as headlands or radar-reflective beacons, to take range and bearing measurements off the radar screen, and plot these on the chart to determine a position. This is relatively easy to accomplish, but it requires the navigator's time to determine the targets, to take measurements and to transfer these to the chart. It is repetitive and laborious work to complete by hand. Easier, time-saving measures such as parallel indexing from the shoreline can help.

An automated process would relieve the navigator of much of this burden, but an automated system may struggle to unambiguously determine radar-conspicuous targets on the screen and relate these to charted landmarks. We have looked at a number of different methods to automatically employ radar data to determine the position of the vessel, these are described below.

2.1. E-RACON Positioning

A RACON is a radar beacon, a device that listens for the signature emission of a ship's radar and responds with a series of pulses on the same frequency. This "paints" a distinctive Morse code character on the interrogating ship's radar screen and can enhance the visibility of an Aid to Navigation (AtoN). An "e-RACON" is a device that operates in an almost identical way to a conventional RACON except it is able to send digitally encoded data in its Morse paint to a suitably equipped New Technology (NT) radar. This information includes the precise surveyed latitude and longitude of the beacon. The technology is transparent to conventional radars that do not decode the data but see only the expected Morse character that appears on the screen.

By measuring the range and bearing to one or more e-RACONs, the radar can then determine its own position. The accuracy with which a radar system can determine the range to a target is considerably better, usually a few meters [4], than its ability to determine the bearing angle. A bearing error of a quarter of one degree equates to a position uncertainty of about 40 m at five Nautical Miles from the beacon and 100 m at 12 Nautical Miles. Figure 1 shows the results of some GRAD trials of the technology.

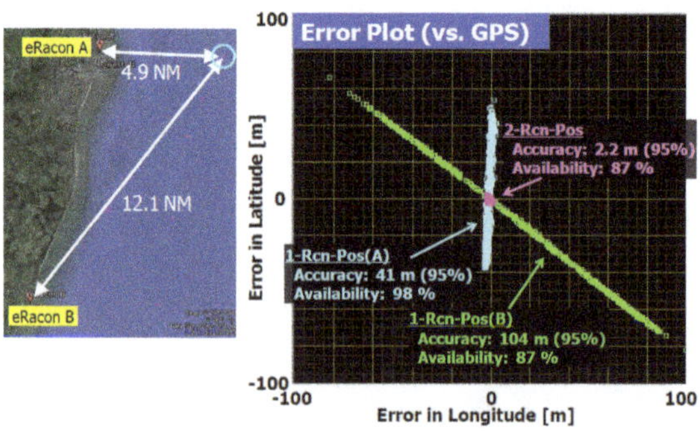

Figure 1. Position fixing using two e-RACONS. A single beacon can provide a position, but inaccuracies measuring the bearing result in large errors that increase with distance. Two beacons used simultaneously can deliver very high accuracy (~2 m).

This potentially limits the capability of the process, since the radar requires simultaneous measurements to at least two e-RACONs for the most accurate performance. The RACONs may only have a maximum visible range of about 10–15 NM, which requires deploying at least two e-RACONS for every 10 NM of coastline. The need to provide this hardware with associated maintenance and management costs led us to consider alternative approaches.

2.2. Image Matching

Image matching is the basic process of performing dead reckoning (DR) positioning by comparing the current image on the radar screen to one taken some time earlier at a known location. Indeed, this is not a new idea [5]. GRAD installed their own "black box" Voyage Data Recorder (VDR) on one of the GLA's vessels. This device captures an image from the vessel's radar screen at regular intervals. It also archives other navigation data such as NMEA 0183 messages output from the GNSS receiver and other on-board sensors. The DR process then works as follows:

- A screen grab is taken from the radar.
- Digital processing of this image extracts the targets that are currently visible to the radar. This image is associated to a GNSS-derived latitude and longitude position. This is shown in Figure 2.
- Later, another image is captured, and the same processing is performed. The two images are then overlaid, and a best-fit alignment between the two is found.
- Image correlation is used to determine the best alignment of both images: Pixels that represent targets in both images count as +1. Pixels that differ count as zero. A simple summation over all image pixels determines the correlation metric. Different pixel-offsets are tested, and the alignment with the strongest correlation is selected.
- The number of pixels offset between the two images is then translated into a movement of the vessel across the sea. A new position solution is derived based on the latitude and longitude of the old image and the distance travelled by the ship. For this reason, we call the process radar dead reckoning.

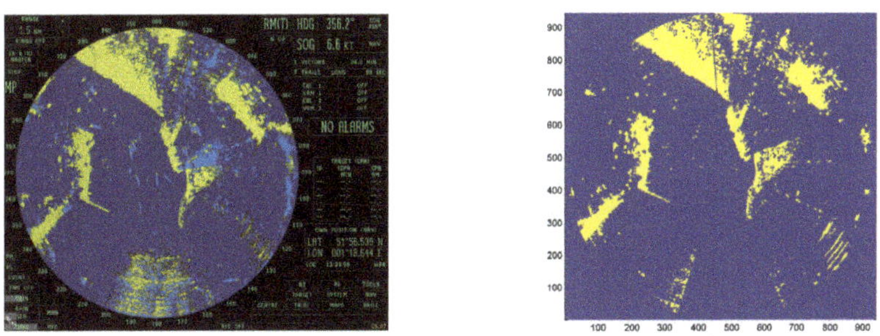

(a) (b)

Figure 2. Capturing the radar screen for navigation. (**a**) A screen grab taken from the ship's radar. (**b**) Target data is processed by eliminating artefacts such as AIS targets, trails and vectors. Noise from the ship's "blind sector" is also removed.

Two significant aspects have been observed by GRAD when using this process:

1. The correlation to the reference image falls very rapidly as the ship sails away from the location at which the image was taken.
2. The error in the position solution computed by the image-matching process increases as the image correlation drops.

This process can fix the vessel's position to within an accuracy of about 10–20 m, but only while the ship is within about half the radar range of the reference image. For example, with a radar range setting of 6 Miles, if the ship loses its GPS position, image matching can provide an accurate radar DR solution, but only for about the next 3 Miles of sailing. Depending on the speed of the vessel, this process may only be viable for a few minutes.

GRAD has investigated using more than one reference image for matching. Screen grabs can be taken regularly from the radar, and laid down as a "trail of breadcrumbs" to follow home. This is useful, especially if the vessel returns to an area it has previously been and allows accurate radar DR to operate for an extended time without GPS. However, it cannot help a vessel sailing into an area it has not previously visited. A solution to this is to pre-survey an area (e.g., a harbour approach) and publish a map of radar-conspicuous targets.

2.3. *Radar Conspicuity Mapping*

A logical extension to the trail of breadcrumbs process is to pre-survey the harbour, gathering a large number of radar images, and to process these into a single, authoritative radar map of the region. This map consists of a large grid of pixels covering the whole harbour area, with the value assigned to each pixel related to the strength and consistency of the radar returns seen at that location.

This process can be automated relatively easily: strong, reliable targets that appear across multiple images are given larger values in the map; weak or intermittent targets that are inconsistent across images are given lower values (Figure 3).

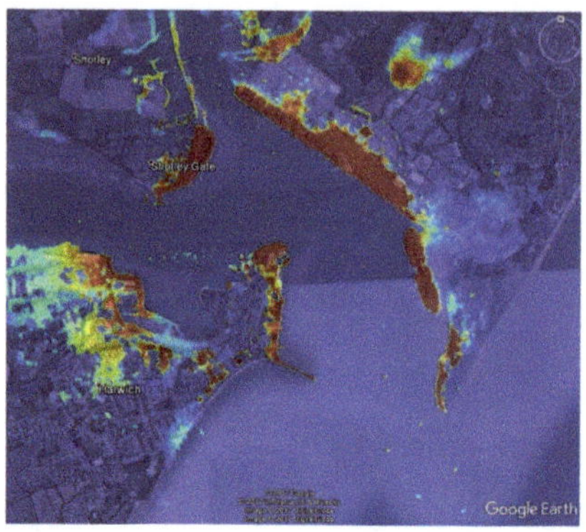

Figure 3. Radar conspicuity map of Harwich and Felixstowe, derived from the ship's radar and overlaid onto Google Earth. Hotter colours represent stronger, more reliable targets.

Wherever the vessel is in this area, it can then match the image on its radar screen to the conspicuity map of the area. The same image-correlation process described above is used, and trials have shown that this allows the vessel to fix its position reliably within about 30 m and achieve a 95% accuracy performance of about 20 m (Figure 4).

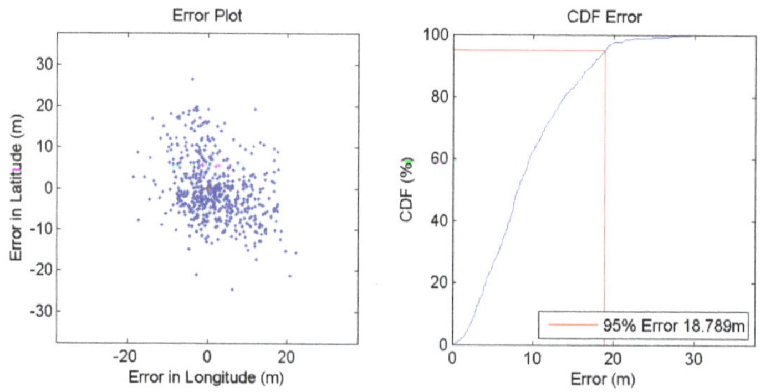

Figure 4. Error scatter and CDF for position-fixing using image matching to the above conspicuity map.

2.4. *Terrain Matching*

There exist concerns that a radar conspicuity map measured by one vessel may not be usable by other vessels. Issues such as radar configuration; sea and clutter settings; and the height of the antenna will mean different ships in the same area may see very different radar targets on their screens. It also takes a large amount of time and effort to survey a harbour, so GRAD has investigated using models of terrain elevation to automate the process of determining radar conspicuity maps.

We developed a process for generating simulated radar returns from the Digital Terrain Elevation Database (DTED), which is available for free at various resolutions down to a post-spacing of about 30 m. By employing a ray-tracing process and a simple 3D illumination model, GRAD can determine modelled radar conspicuity maps for any vessel configuration. An example of this process is shown in Figure 5.

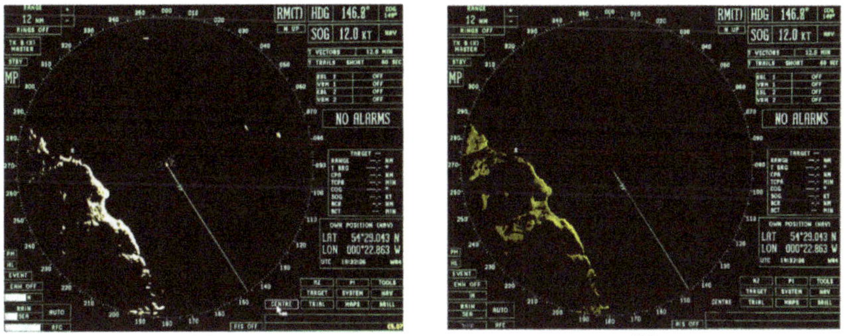

(a) (b)

Figure 5. Example radar conspicuity generated from a terrain database: (**a**) a real radar image captured from a vessel; (**b**) the same image with the radar returns replaced with simulated data.

To the human eye, these simulated images do look quite realistic, and the process does work reasonably reliably. However, the absolute accuracy performance so far has been poor (~100 m accuracy, 95%). This is thought to be due to issues such as the quality of the terrain data; the accuracy of the datum used; and the absence of radar returns from man-made structures in the simulated maps. GRAD has paused our research into terrain-matching pending the availability of more suitable terrain data.

2.5. *Simultaneous Location and Mapping (SLAM)*

If the ship is able to build its own conspicuity map as it sails, then this alleviates the need to perform extensive surveys and also solves the problem of whether the map is compatible between different ships' radar setups. However, this still leaves the problem of a vessel sailing into an area that it has not previously visited, and therefore does not have pre-existing conspicuity data to fall back on. The solution is Simultaneous Location and Mapping (SLAM) [6]. This allows the ship to continue building its conspicuity map, even if GNSS is lost, by using the radar-derived navigation solution as a ground-truth for mapping new radar-conspicuous targets.

Our trials have shown it is possible to perform extended runs on radar SLAM, even using only a single GNSS-fixed image to start the process. A number of significant issues with the process have been observed, most notably that errors in the heading data (from the ships gyrocompass) translate into rotational errors in the map datum, which then translate into significant position errors as the ship sails. The relationship between image pixels and the physical distance across the ground also needs to be carefully calibrated: for example, we have seen that with the radar set at a 1.5 nautical mile range, targets 1.6 NM from the vessel are visible on the screen. It is thought that the real-world implementation of radar SLAM on a vessel will require a substantial calibration process first to remove artefacts such as these from the system.

GRAD has also investigated using the ship's own speed log and gyrocompass sensors to perform "traditional" dead reckoning and has constructed a combined system-of-systems that integrates radar and traditional DR. This integration provides two useful pieces of information to the combined system:

1. Traditional DR can provide the initial Estimated Position (EP) of the vessel so the map-matching process has a relatively accurate position to begin with.

2. The inherent errors and biases in the speed log and gyro can be calibrated out by comparing the difference between the radar-derived navigation solution and the EP given by the DR process.

The result of our work is that the error in the solution can be reliably constrained to within less than about 25 m, even after sailing for several hours without GPS in previously un-surveyed areas. Figure 6 shows the results of a trial of traditional DR, both with and without radar aiding.

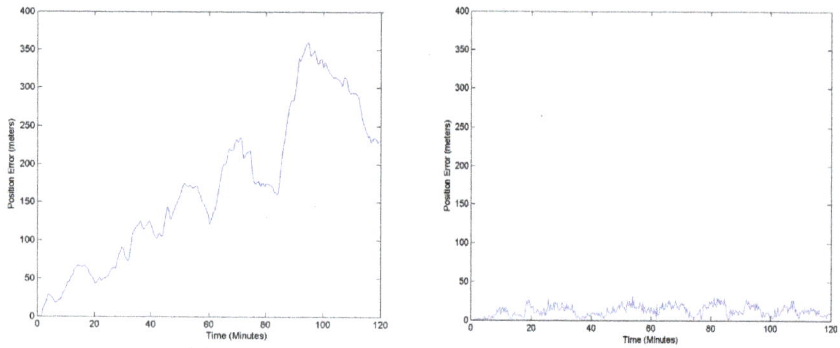

(**a**) (**b**)

Figure 6. Error growth over time: (**a**) using traditional log-and-gyro dead reckoning; (**b**) when radar SLAM is used to calibrate errors and biases in the DR sensors.

The combined system results in a 95% accuracy performance that has been shown to be reliably better than about 20 m. Figure 7 shows the scatter of position errors vs a DGPS ground-truth reference.

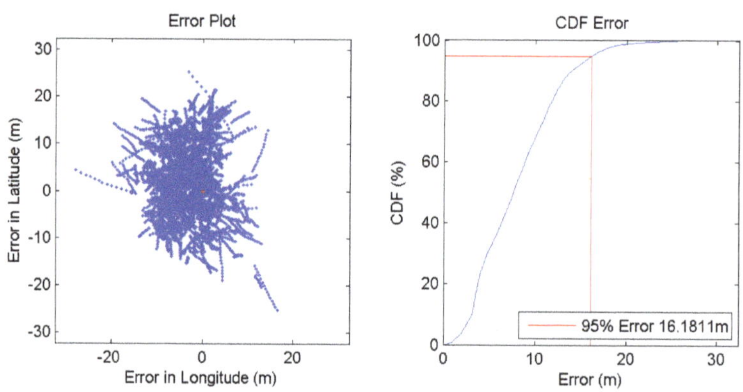

Figure 7. Error vs. GPS for a combined radar SLAM and DR solution.

So far, this is the most accurate and reliable radar positioning GRAD has been able to achieve. Work is currently underway to obtain access to the raw data output from the radar itself and to process this data directly in the radar SLAM algorithm. The use of a processed image captured from the radar screen is probably sub-optimal, as a number of artefacts have been added to the image to aid human interpretation, not least the overlay of other navigation information such as AIS targets and vectors. It is thought that an integrated system employing raw radar data, plus log-and-gyro DR should enable resilient position fixing in line with the 10 m (95%) accuracy requirement the IMO has set for electronic position-fixing systems [7]. This is what our future research into the technology will investigate.

3. Conclusions

GRAD has investigated a number of different radar absolute positioning techniques, these are summarised in Table 1, below.

Table 1. Summary of the radar position-fixing methods investigated by GRAD.

Method	Description	Accuracy Performance	Notes
e-RACON	NT radar obtains range and bearing data from one or more e-RACONs	2–100 m	Accuracy depends on how many RACONs are visible and the crossing angle between them. Potentially, a large number of e-RACONS will be needed.
Radar dead reckoning Single-image matching	The last radar image is associated with a GPS position and is used as a reference. Positioning is then achieved via image correlation.	~20 m	Accuracy falls sharply as the vessel sails away from the reference image and the correlation falls. Limited to about half the range of the radar.
Multiple-image matching	Several GNSS-referenced images are laid down and used for image matching.	<35 m	Good for sailing in areas where the ship has already been, and left itself a trail of breadcrumbs to follow.
Conspicuity mapping	A radar survey of an area is conducted to generate a single authoritative radar conspicuity map of the region. This is then used on the ship for image matching.	<20 m	Requires a significant amount of survey work. Conspicuity maps may not be compatible between different ships and may require frequent updates.
Terrain matching	Uses a terrain database (DTED) to build the conspicuity map using a ray-tracing process.	<100 m	Not currently viable for high-accuracy positioning. May be improved with better terrain data and 3D models of human-made structures.
DR + SLAM	Integrated speed-log and gyro dead reckoning with radar SLAM.	<20 m	It may be possible to improve this technique further by accessing raw data from the radar. We are aiming for <10 m accuracy.

It is envisaged that national maritime authorities may select a mixture of approaches depending on the most appropriate approach for their waters. The installation of physical hardware, such as e-RACONs may offer a better result for areas of low-lying coast with minimal radar features, while the SLAM and DR solution may perform better in areas with sufficient coastal radar returns.

Author Contributions: Conceptualization, C.H. and A.G.; methodology, C.H.; software, C.H.; validation, C.H. and A.G.; formal analysis, C.H.; investigation, C.H.; resources, GLA; data curation, C.H.; writing—original draft preparation, C.H.; writing—review and editing, C.H. and A.G.; visualization, C.H.; supervision, A.G.; project administration, C.H. and A.G.; funding acquisition, GRAD; presentation: L.H. All authors have read and agreed to the published version of the manuscript.

Funding: This research received no external funding.

Institutional Review Board Statement: Not applicable.

Informed Consent Statement: Not applicable.

Data Availability Statement: The data presented in this study are available upon request from the corresponding author.

Conflicts of Interest: The authors declare no conflict of interest.

References

1. Royal Academy of Engineering. *GNSS: Reliance and Vulnerabilities*; RAE: London, UK, 2011.
2. IMO. *International Convention for the Safety of Life at Sea (SOLAS)*; IMO: London, UK, 1974.
3. Bole, A.; Norris, A.; Wall, A. *Radar and ARPA Manual Radar*, 3rd ed.; Butterworth-Heinemann: Oxford, UK, 2014.
4. Ward, N.; Safar, J. Enhanced Radar Positioning. In Proceedings of the 2014 International Technical Meeting of the Institute of Navigation, San Diego, CA, USA, 27–29 January 2014.
5. Austin, G.L.; Bellon, A.; Riley, M.; Ballantyne, E. Navigation by Computer Processing of Marine Radar Images. *J. Navig.* **1985**, *38*, 3. [CrossRef]
6. Han, J.; Cho, Y.; Kim, J. Coastal SLAM With Marine Radar for USV Operation in GPS-Restricted Situations. *IEEE J. Ocean. Eng.* **2019**, *44*, 300–309. [CrossRef]
7. *IMO Resolution A.1046—World Wide Radio Navigation Service*; IMO: London, UK, 2011.

engineering proceedings

MDPI

Proceeding Paper

Preliminary Field Results of a Dedicated 5G Positioning Network for Enhanced Hybrid Positioning [†]

José A. del Peral-Rosado [1,*], Ali Y. Yildirim [1], Nils C. Klinger [1], Patric Nolle [1], Sara M. Razavi [2], Sagar Parsawar [3], Rakesh Mundlamuri [3], Florian Kaltenberger [3], Niilo Sirola [4], Stefano Garlaschi [5], Luca Canzian [5], Jukka Talvitie [6] and Detlef Flachs [1]

1 Airbus Defence and Space, 82024 Taufkirchen, Germany
2 Ericsson, 16483 Stockholm, Sweden
3 Eurecom, 06410 Sophia-Antipolis, France
4 Exafore, 33720 Tampere, Finland
5 Qascom, 36061 Bassano del Grappa, Italy
6 Electrical Engineering Unit, Tampere University, 33720 Tampere, Finland
* Correspondence: jose_antonio.del_peral_rosado@airbus.com
† Presented at the European Navigation Conference 2023, Noordwijk, The Netherlands, 31 May–2 June 2023.

Abstract: Dedicated fifth generation (5G) positioning networks are envisaged as a key technology to complement Global Navigation Satellite Systems (GNSS) in challenging situations, such as safety-critical automotive and aerial applications. These 5G networks are explicitly designed to fulfil the positioning needs in a specific environment, by using dedicated pilot signals and locating the base stations (BSs) in favorable conditions for user reception. To prove this concept, the Hybrid Overlay Positioning with 5G and GNSS (HOP-5G) testbed demonstrates the enhanced positioning capabilities resulting from the deployment of ground and aerial BSs. This paper presents a preliminary assessment of the first real-time ranging measurements obtained with a flying BS in a dedicated 5G positioning network. Under hovering and optimal conditions, field results obtained over an 80 MHz downlink bandwidth indicate a ranging precision below 1 m in 95% of cases.

Keywords: GNSS; 5G; positioning; flying base stations

Citation: del Peral-Rosado, J.A.; Yildirim, A.Y.; Klinger, N.C.; Nolle, P.; Razavi, S.M.; Parsawar, S.; Mundlamuri, R.; Kaltenberger, F.; Sirola, N.; Garlaschi, S.; et al. Preliminary Field Results of a Dedicated 5G Positioning Network for Enhanced Hybrid Positioning. *Eng. Proc.* **2023**, *54*, 6. https://doi.org/10.3390/ENC2023-15432

Academic Editors: Tom Willems and Okko Bleeker

Published: 29 October 2023

1. Introduction

There is nowadays an increasing demand for precise and reliable navigation in a wide range of applications, such as in Urban Air Mobility (UAM) or autonomous vehicles. Current solutions are based on Global Navigation Satellite Systems (GNSS) for absolute positioning, navigation, and timing (PNT), which are fused with relative positioning from multiple sensor technologies (e.g., inertial measurement units, cameras, radars, and LiDAR, among others) for further accuracy and robustness. However, complementary navigation technologies are still expected to be necessary to fulfil the stringent reliability requirements. In this sense, fifth generation (5G) cellular technologies are very promising thanks to their key positioning features and standard support [1], including 5G new radio (NR) terrestrial and non-terrestrial networks (NTN).

5G commercial and private networks are deployed over many different use cases, i.e., from mobile broadband connectivity to Internet of Things (IoT). Still, the exploitation of 5G positioning has been mainly focused on indoor industrial use cases, such as asset tracking and industrial automation. This has been reflected on the 3rd Generation Partnership Project (3GPP) standardization, where the studies on enhanced 5G positioning have been initially focused on indoor use cases [2]. As a result, very limited attention has been paid to the dedicated deployment of ground and aerial base stations (BSs) for enhanced 5G positioning service areas in outdoor applications. Indeed, aerial BSs (ABSs), unmanned aerial vehicle (UAV)-mounted BSs, or flying BSs have been mainly considered for communication purposes [3,4] but not for enabling enhanced-positioning services, which can be especially beneficial thanks to the increased diversity in the vertical dimension with respect to ground-fixed BSs. Thus, to prove this new opportunity, the Hybrid Overlay Positioning with 5G and GNSS (HOP-5G) testbed has been proposed as a first-of-a-kind demonstrator of dedicated aerial 5G positioning networks.

The design considerations of these dedicated 5G positioning networks have been studied in [5] and their expected positioning capabilities in a representative scenario in [6]. Experimental 5G positioning results have been presented in predecessor testbeds, such as in [7,8], as well as with signals of opportunity approaches in [9,10]. To the best of the authors' knowledge, the use of flying 5G BSs for positioning has not been demonstrated yet. Thus, this paper presents the first ranging results of a flying 5G BS deployed specifically for positioning purposes.

2. Dedicated 5G Positioning Network

2.1. HOP-5G High-Level Architecture

The HOP-5G testbed is based on a dedicated 5G positioning network. The key objective of this first-of-a-kind demonstrator is to assess the enhanced hybrid 5G and GNSS-positioning capabilities using dedicated ground and aerial BSs. Furthermore, the HOP-5G testbed operates in real-time and in post-processing, which allows for a wide range of experiments. The main building blocks of this testbed are:

- Ground support equipment (GSE): The testbed operations are performed within this building block. The GSE is used to control, command, collect data, monitor, and perform post-processing functionalities within the HOP-5G testbed.
- 5G BSs: The 5G BSs (or 5G gNBs as in the 3GPP nomenclature) can be deployed in ground or in the UAV payloads. The BSs are based on software-defined radios (SDRs) synchronized with GNSS-disciplined oscillators. For each BS, dedicated positioning reference signals (PRS) are transmitted in specific downlink slots.
- User Equipment (UE): The UE can be deployed in ground or in the UAV payload, and it is also based on SDR. In downlink operations, the UE exploits a specific upgrade of OpenAirInterface (OAI) to obtain 5G real-time ranging measurements. In addition, a COTS GNSS receiver is used to collect the raw observables and position estimates. Then, a positioning engine performs the UE position estimation also in real-time by combining either or both GNSS and 5G time of arrival (ToA) in a weighted least squares (WLS) or extended Kalman filter (EKF).

The baseline HOP-5G deployment is based on two ground-fixed BSs, two UAV BSs, and two UEs. Further details of the HOP-5G testbed architecture can be found in [6].

2.2. Testbed Equipment

The testbed equipment is based on commercial off-the-shelf (COTS) components while keeping a high degree of flexibility and modularity thanks to the use of SDR platforms and regular computers. Each node of the HOP-5G testbed shares the same high-level architecture, including both aerial and ground deployments. As shown in Figure 1, the nodes are based on four key components:

- Computer: The HOP-5G software modules are installed and executed on a regular computer with Ubuntu 20.04 operating system. When installed within the UAV payload, an Intel NUC with a Thunderbolt3-to-10GbE adapter is used to fulfil the size, weight, and power (SWaP) constraints. When fixed on ground, a workstation or tower computer with a 10GbE adapter is preferred to support very high computational performance. In both cases, a Wi-Fi module is used to establish a remote connection with the operator.
- SDR: The USRP X300 is considered the preferred SDR to transmit and receive 5G signals. This SDR is built with an UBX daughterboard to operate at frequency range

1 (FR1), i.e., below 6 GHz, and with a GPS-disciplined oscillator (GPSDO) to keep a tight synchronization among testbed nodes.

- GNSS receiver: The COTS GNSS receivers are the u-blox F9P for ground-fixed nodes and the u-blox F9R for moving aerial or ground nodes.
- Antennas: The two additional antennas of the node are a 5G monopole omnidirectional antenna and a GNSS helical antenna. An L-band splitter is used to connect the GNSS antenna output with the SDR and the GNSS receiver.

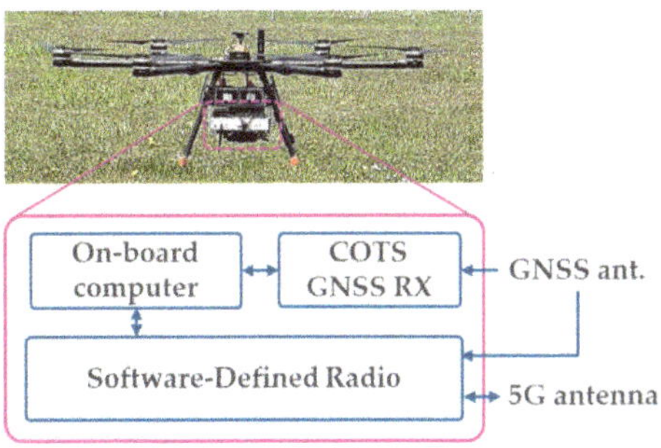

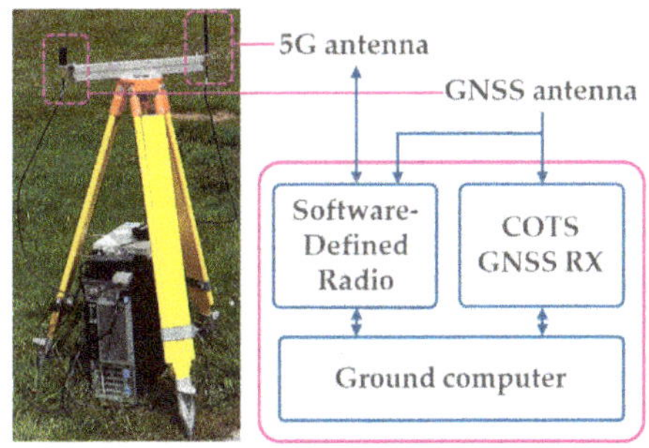

Figure 1. Example equipment of the HOP-5G testbed for UAV-mounted and ground deployments: (**a**) UAV-mounted BS or UE with the key building blocks of its payload; (**b**) Ground-fixed BS or static UE with its key building blocks.

Thanks to the hardware and software flexibility and modularity, a node can be used as a BS or as UE with only software configuration updates. All the nodes are remotely controlled and monitored with a regular laptop, e.g., by exploiting a Wi-Fi connection.

3. Experimental Methodology

3.1. Experimental Procedure

The HOP-5G testbed was deployed in the heliport at the Airbus premises, and it was used to assess the precision of the real-time downlink ranging measurements at a static UE from a flying 5G base station. The experimental procedure was based on hovering the UAV at several line-of-sight (LoS) points. As it is shown in Figure 2, each of those sequential points are classified based on the distance d to UE and the UAV height h from ground.

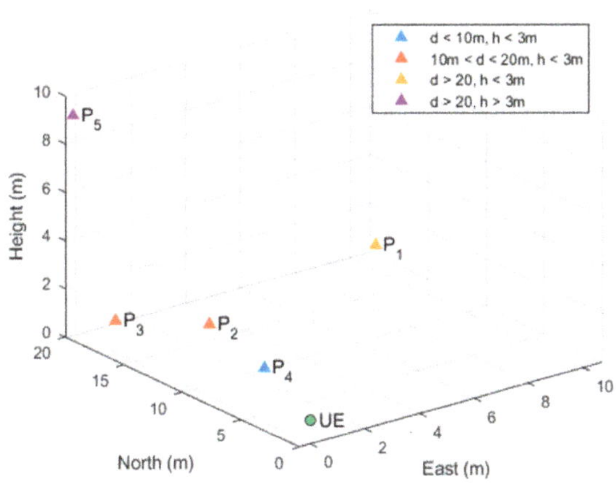

Figure 2. UAV hovering positions for the corresponding ranging measurements at the UE.

The position solution of the GNSS receiver within the UAV payload is logged and disseminated in real-time to the UE. The time of those UAV GNSS solutions are used to associate the 5G real-time measurements with the corresponding UAV hovering position. For each of those points, the 5G measurements are collected over 30s. As the 5G PRS transmissions are scheduled once every 10 ms, the 5G OAI UE produces a maximum of 5000 measurement messages over those 50 s for each UAV hovering point.

The downlink 5G PRS are transmitted over a bandwidth of approximately 80 MHz at 3.75 GHz and a subcarrier spacing of 30 kHz. This 80 MHz bandwidth is defined as result of the supported sampling rate of the USRP X300, which is set to 92.16 MSps. Each BS transmits its PRS in a different slot allocation to fully avoid inter-cell interference.

In addition to the flying BS, a ground-fixed BS is also deployed close to the UE, in order to fully stabilize the fine synchronization of the UE received radio frames. Thus, the measurement conditions are considered optimal.

3.2. Performance Metrics

The 5G-ranging performance assessment within this paper is based on the variability of the measurements over a period of 1 s. This assessment indicates the ranging precision of downlink UE measurements from the flying 5G BS.

Let us define the pseudorange measurement at the n-th time instant as follows:

$$\rho(n) = \|\mathbf{x}_{\mathrm{UAV}}(n) - \mathbf{x}_{\mathrm{UE}}\|^2 + c\cdot(\delta t_{\mathrm{UE}}(n) - \delta t_{\mathrm{UAV}}(n)) + \varepsilon(n), \tag{1}$$

where $\mathbf{x}_{\mathrm{UAV}}(n)$ is the three-dimensional (3D) position of the UAV, $\mathbf{x}_{\mathrm{UE}}$ is the 3D position of the static UE, $\delta t_{\mathrm{UE}}(n)$ is the receiver clock offset, $\delta t_{\mathrm{UAV}}(n)$ is the flying BS clock offset, and $\varepsilon(n)$ is the pseudorange error due to noise and multipath. As one can observe from this expression, the main sources of pseudorange variability are the slow-moving distance $d(n) = \mathbf{x}_{\mathrm{UAV}}(n) - \mathbf{x}_{\mathrm{UE}}{}^2$ due to UAV hovering movements, the BS and UE clock offsets, and the noise and multipath errors. The variation in these error sources can be calculated for each measurement as

$$\Delta\rho(n) = \rho(n) - \mu_\rho\,, \tag{2}$$

where μ_ρ is the average of all pseudoranges over the measurement period of 1 s. Note that the 50 s measurement set is split in subsets of 1 s, and μ_ρ is computed for each subset. Thus, in the absence of large pseudorange outliers, the pseudorange variation $\Delta\rho(n)$ is a suitable performance metric to assess the ranging precision of the 5G measurements.

As a complementary metric, the 5G downlink PRS reference signal received power (RSRP) in [11] is computed to further assess the results in terms of propagation conditions.

4. Field Results

The cumulative density function (CDF) of the pseudorange variation over 1 s and the CDF of the RSRP is computed for each of the UAV hovering points grouped per distance to the UE and UAV height from ground. As it is shown in Figure 3, there is a correlation between ranging precision and distance between flying BS and UE, due to the favorable LoS propagation. A remarkable ranging precision is obtained with the flying BS and the optimal receiver conditions. As it is summarized in Table 1, the static UE achieves a ranging variation below 1 m in 95% of cases, under the operational conditions of this experiment. Furthermore, the outlier rate (or amount of non-available measurements per UAV hovering point) is kept to around or below 1% depending on the experimentation point. Indeed, the first point (i.e., P_1) shows the worse degradation, even with respect to the fifth point (i.e., P_5). This degradation is assumed to be associated with the multipath errors due to ground reflections, although further experimentation and analyses are required as part of future work, e.g., by comparing ground and flying BSs ranging errors to fully characterize the UAV hovering impact. Future field results of the HOP-5G testbed are expected to further demonstrate the positioning capabilities of dedicated 5G networks.

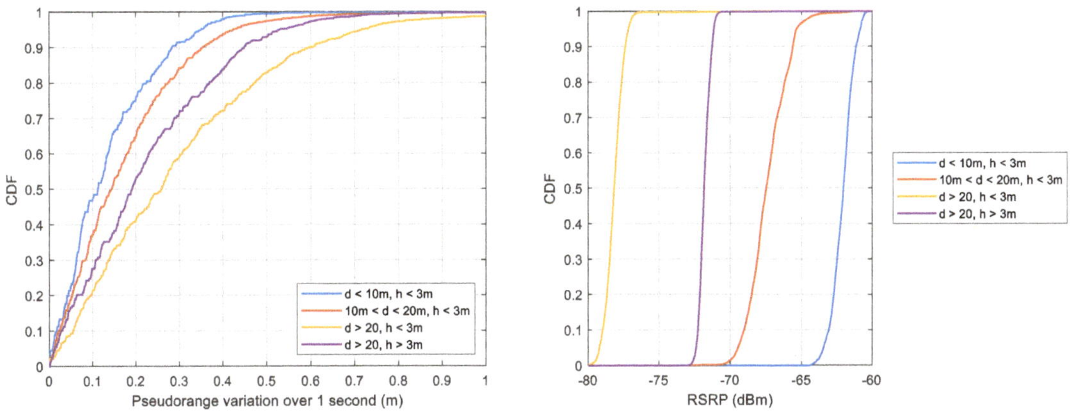

Figure 3. Preliminary field results of the downlink UE ranging measurements from a flying 5G BS: (**a**) CDF of the pseudorange variation over 1 s; (**b**) CDF of the RSRP.

Table 1. Ranging precision at 95% of cases and overall outlier rate of 5G ranging measurements for each group of UAV hovering points.

Point Group	Point	d	h	Ranging Precision @ 95%	Outlier Rate
$d < 10$ m, $h < 3$ m	P_4	7.1 m	1.10 m	0.35 m	0.18%
10 m $< d <$ 20 m, $h < 3$ m	P_2	10.8 m	2.23 m	0.42 m	0.28%
	P_3	16.3 m	1.43 m		
$d > 20$ m, $h < 3$ m	P_1	22.2 m	0.39 m	0.72 m	1.14%
$d > 20$ m, $h > 3$ m	P_5	21.1 m	9.35 m	0.53 m	0.32%

5. Conclusions

This paper presents the first field ranging measurements with a dedicated fifth generation (5G) flying base station (BS) for positioning purposes. This field experimentation is performed within the Hybrid Overlay Positioning with 5G and GNSS (HOP-5G) testbed, which exploits 80 MHz downlink positioning reference signals (PRS) to achieve real-time ranging measurements at the user equipment (UE). Considering hovering and optimal conditions for a static UE, a ranging precision is achieved below 1 m in 95% of cases, when considering the pseudorange variation over periods of 1 s within test points of 50 s. Future work is envisaged to further demonstrate the positioning capabilities of dedicated 5G positioning networks under static and dynamic UE conditions.

Author Contributions: Conceptualization, J.A.d.P.-R., A.Y.Y., N.C.K., P.N., S.M.R., S.P., R.M., F.K., N.S., S.G., L.C., J.T. and D.F.; experimentation, J.A.d.P.-R., A.Y.Y., P.N. and D.F.; data post-processing, J.A.d.P.-R., N.C.K. and P.N.; writing, J.A.d.P.-R.; project administration, J.A.d.P.-R. All authors have read and agreed to the published version of the manuscript.

Funding: This work from the Hybrid Overlay Positioning with 5G and GNSS (HOP-5G) project has been supported under NAVISP-EL1-050 activity by the European Space Agency Navigation Innovation and Support Programme (NAVISP), which aims to generate innovative concepts, technologies and system solutions in the wide-field of Positioning, Navigation and Time.

Institutional Review Board Statement: Not applicable.

Informed Consent Statement: Not applicable.

Data Availability Statement: Data sharing is not applicable to this article.

Conflicts of Interest: The authors declare no conflict of interest. The view expressed herein can in no way be taken to reflect the official opinion of the European Space Agency.

References

1. del Peral-Rosado, J.A.; Raulefs, R.; López-Salcedo, J.A.; Seco-Granados, G. Survey of Cellular Mobile Radio Localization Methods: From 1G to 5G. *IEEE Commun. Surv. Tutor.* **2018**, *20*, 1124–1148. [CrossRef]
2. 3GPP. *Study on NR Positioning Enhancements*; Technical Report TR 38.857, Rel. 17; 3GPP: Sophia Antipolis, France, 2021.
3. Li, B.; Fei, Z.; Zhang, Y. UAV Communications for 5G and Beyond: Recent Advances and Future Trends. *IEEE Internet Things J.* **2019**, *6*, 2241–2263. [CrossRef]
4. Zeng, Y.; Wu, Q.; Zhang, R. Accessing From the Sky: A Tutorial on UAV Communications for 5G and Beyond. *Proc. IEEE* **2019**, *107*, 2327–2375. [CrossRef]
5. del Peral-Rosado, J.A.; Nolle, P.; Razavi, S.M.; Lindmark, G.; Shrestha, D.; Gunnarsson, F.; Flachs, D. Design Considerations of Dedicated and Aerial 5G Networks for Enhanced Positioning Services. In Proceedings of the 2022 10th Workshop on Satellite Navigation Technology (NAVITEC), Noordwijk, Netherlands, 5–7 April 2022; pp. 1–12.
6. del Peral-Rosado, J.A.; Nolle, P.; Rothmaier, F.; Razavi, S.M.; Lindmark, G.; Jiang, X.; Flachs, D. Proof-of-Concept of Dedicated Aerial 5G and GNSS Testbed for Enhanced Hybrid Positioning. In Proceedings of the 35th International Technical Meeting of the Satellite Division of The Institute of Navigation (ION GNSS+ 2022), Denver, Colorado, 19–23 September 2022; pp. 1–15.
7. Maymo-Camps, R.; Vautherin, B.; Saloranta, J.; Crapart, R. 5G Positioning and Hybridization with GNSS Observations. In Proceedings of the International Technical Symposium on Navigation and Timing (ITSNT), Toulouse, France, 13–16 September 2018; pp. 1–7.
8. Mata, F.J.; Grec, F.; Azaola, M.; Blázquez, F.; Fernández, A.; Dominguez, E.; Cueto-Felgueroso, G.; Seco-Granados, G.; del Peral-Rosado, J.A.; Staudinger, E.; et al. Preliminary field trials and simulations results on performance of hybrid positioning based on GNSS and 5G signals. In Proceedings of the 33rd International Technical Meeting of the Satellite Division of The Institute of Navigation (ION GNSS+ 2020), Online, 22–25 September 2020; pp. 1–15.
9. Shamaei, K.; Kassas, Z.M. Receiver Design and Time of Arrival Estimation for Opportunistic Localization With 5G Signals. *IEEE Trans. Wireless Commun.* **2021**, *20*, 4716–4731. [CrossRef]
10. Abdallah, A.; Kassas, Z.M. UAV Navigation with 5G Carrier Phase Measurements. In Proceedings of the 34th International Technical Meeting of the Satellite Division of The Institute of Navigation (ION GNSS+ 2021), St. Louis, MO, USA, 20–24 September 2021; pp. 1–13.
11. 3GPP. *NR; Physical Layer Measurements*; Technical Specification TS 38.215, Rel. 17; 3GPP: Sophia Antipolis, France, 2023.

engineering proceedings

MDPI

Proceeding Paper

Comparison of NeQuick G and Klobuchar Model Performances at Single-Frequency User Level †

Ulrich Ngayap [1,*], Claudia Paparini [1], Marco Porretta [1], Peter Buist [1], Knut Stanley Jacobsen [2], Michael Dähnn [2], Natalia Hanna [3], Dzana Halilovic [3], Anna Świątek [4] and Paulina Gajdowska [4]

1 European Union Agency for Space Programme (EUSPA), Galileo Reference Centre (GRC), Zwarteweg 53, 2201 AA Noordwijk, The Netherlands; c.paparini@rheagroup.com (C.P.); marco.porretta@euspa.europa.eu (M.P.); peterjacob.buist@euspa.europa.eu (P.B.)
2 Geodetic Institute, Norwegian Mapping Authority (NMA), Kartverksveien 21, 3511 Hønefoss, Norway; knut.stanley.jacobsen@kartverket.no (K.S.J.); michael.daehnn@kartverket.no (M.D.)
3 Department of Geodesy and Geoinformation, TU Wien (TUW), Wiedner Hauptstraße 8-10, 1040 Wien, Austria; natalia.hanna@geo.tuwien.ac.at (N.H.); dzana.halilovic@bkg.bund.de (D.H.)
4 Centrum Badań Kosmicznych Polskiej Akademii Nauk (SRC), Bartycka 18A Str., 00-716 Warsaw, Poland; ana@cbk.waw.pl (A.Ś.); pwozniak@cbk.waw.pl (P.G.)
* Correspondence: ulrich.ngayap@euspa.europa.eu

† Presented at the European Navigation Conference 2023, Noordwijk, The Netherlands, 31 May–2 June 2023.

Abstract: In this study, the NeQuick G and Klobuchar models are evaluated by monitoring performance issues related to ionosphere activity for single-frequency users. The effects of radio frequency (RF) signal propagation through the ionosphere may have a significant impact on satellite communication and navigation systems because of geomagnetic field geometry near the magnetic equator and in the proximity to the high- and low-latitude zones. An ongoing challenge is determining how accurate the ionospheric models employed by existing Global Navigation Satellite Systems (GNSSs) are. This work investigates the patterns of total electron content (TEC) fluctuations over distinct zones from 1 January 2019 to 30 June 2022. Measurements are collected at station networks deployed worldwide. Firstly, monthly and seasonal variations of TECs are analysed. Secondly, the TEC 'availability' parameter, as the percentage of time when the TEC error is compliant with the specification of the Galileo Single-Frequency Ionosphere Algorithm ('NeQuick G' model), is introduced. The TEC error defines the difference between (a) the model TEC, obtained by either the NeQuick G or the Klobuchar model over a given station, and (b) the reference TEC, based on observations from networks of GNSS receivers. Finally, the position, velocity, and time (PVT), along with broadcast group delays (BGDs) are analysed and the PVT accuracy is compared between the NeQuick G and Klobuchar models. In 3.5 years, the seasonal behaviour of TEC shows maxima during the March and October equinox and minima during the June and December solstice. Moreover, an increase in the TEC values and the amount of TEC errors are visible as we are approaching the next solar maximum. Preliminary results show a larger associated positioning error using the Klobuchar than the NeQuick G model. However, the difference is zone-dependent, most evident in equatorial regions. This collaborative study of the GRC, NMA, TUW, and SRC was performed under the Framework Partnership Agreements (GSA/GRANT/04/2016).

Keywords: NeQuick G; Klobuchar; position accurracy

Citation: Ngayap, U.; Paparini, C.; Porretta, M.; Buist, P.; Jacobsen, K.S.; Dähnn, M.; Hanna, N.; Halilovic, D.; Świątek, A.; Gajdowska, P. Comparison of NeQuick G and Klobuchar Model Performances at Single-Frequency User Level. *Eng. Proc.* **2023**, *54*, 7. https://doi.org/10.3390/ENC2023-15475

Academic Editors: Tom Willems and Okko Bleeker

Published: 29 October 2023

1. Introduction

The collaborative study presented in this paper was performed under the European Union Agency for Space Programme (EUSPA) Framework Partnership Agreements (GSA/GRANT/04/2016) involving the entities Galileo Reference Centre (GRC) [1], Norwegian Mapping Authority (NMA), Technische Universität Wien (TUW), and Space Research Centre (SRC) of Polish Academy of Sciences, whose main objectives were to investigate the

GNSS performance at different levels. The ionosphere, the ionized part of the Earth's upper atmosphere, is highly dynamic and is influenced by various space weather phenomena. The presence of free electrons can significantly impact radio wave transmission, altering the electromagnetic radio frequency waves [2]. Satellite communication and navigation systems, such as the Galileo and the Global Positioning System (GPS), rely on radio signals transmitted through the ionosphere. However, the ionosphere can cause significant errors in these signals, which can lead to degraded performance and reduced accuracy, especially at the equatorial and high latitudes. To mitigate the effects of the ionosphere on satellite systems for single-frequency (SF) applications, ionospheric models have been developed [3]. Their final scope is to provide accurate estimates of the total electron content (TEC). Two widely used ionospheric models, enabling the ionospheric delay corrections computations for SF users, are the NeQuick G model, developed for Galileo, and the Klobuchar model, developed for GPS. NeQuick G [3] is a three-dimensional and time-dependent ionospheric electron density model based on an empirical climatological representation of the ionosphere, which predicts monthly mean electron density from analytical profiles, depending on solar-activity-related input values: sunspot number or solar flux, month, geographic latitude and longitude, height, and universal time (UT). On the other side, the Klobuchar model [2] is designed to minimise user computational complexity and user computer storage as well as to keep a minimum number of coefficients to transmit on satellite-user links. In this study, the comparison of NeQuick G and Klobuchar models at the SF user level is shown. Three parameters are used to present the results: the TEC accuracy (also called TEC error) defined as the difference between the model TEC obtained by either the NeQuick G or the Klobuchar model over a given station and the reference TEC based on observations from networks of GNSS receivers; the TEC availability characterization defined as the percentage of time when the TEC error is compliant with the specification of the Galileo Single-Frequency Ionosphere Algorithm (NeQuick G model); the Position, Velocity, and Time (PVT) accuracy characterization used to assess the impact of ionospheric disturbances on the accuracy of PVT estimates. The results are presented for a network of stations distributed worldwide. The computation method used for the TEC estimates is different for the networks of stations used by the involved entities GRC, NMA, TUW, and SRC. The results cover the period from 1 January 2019 to 30 June 2022. The comparison of the performance of the two models in the different networks is the key element of this work, providing a better understanding of the model performances. Moreover, this study provides valuable insights into the performance of the NeQuick G and Klobuchar models at the SF user level.

2. Data and Methods

This section explains the data and methods used to calculate the TEC accuracy, TEC availability, and SF position accuracy. The stations used by the different entities are spread worldwide as shown in Figure 1.

These stations are therefore divided into four groups according to the different geomagnetic latitude zones. Stations considered by individual entities are marked with different colours; in the case of TUW, in the mid-latitude region, 5 stations per quarter were considered. The station choice for each quarter was mainly dependent on the site location within the mid-latitude region and RINEX data availability (2019 Q1: AJAC, GRAS, JOZ2, LEOP, NICO; 2019 Q2–2019 Q4: AJAC, GRAS, LEOP, NICO, WROC; 2020 Q1–2021 Q4: EBRE, GRAS, GRAZ, NICO, WROC). Due to urgent maintenance activities at the Data Centre of the Bundesamt für Kartographie und Geodäsie (BKG) of the International Geodesy Service (IGS), and hence RINEX data unavailability, the majority of used validation sites had to be replaced in the 1st and 2nd quarter of 2022 (2022 Q1: DYNG, GRAS, MELI, NICO, WARN; 2022 Q2: DYNG, EBRE, GOP6, GRAS, WSRT).

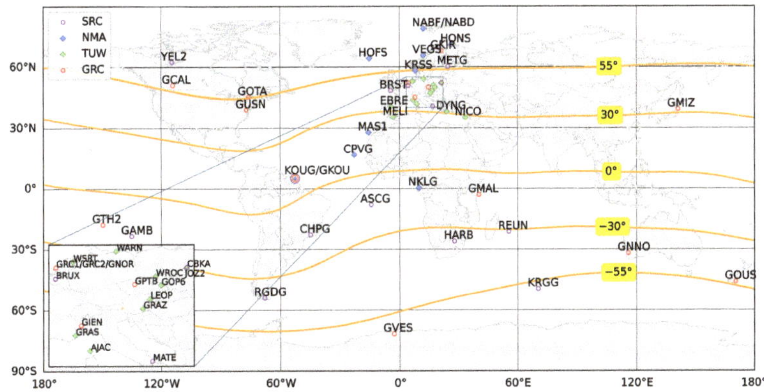

Figure 1. Locations of GNSS stations used in this study. The orange lines show the magnetic latitudes.

As described in the introduction, this paper presents two ionospheric models used for SF receivers, the NeQuick G and Klobuchar models. The NeQuick G model [4] is designed to reach a correction capability of at least 70% of the ionospheric code delay in terms of root mean square (RMS), with a lower Slant total electron content (STEC) residual error bound of 20 TEC Unit (TECU) for any location, time of day, season, and solar activity, excluding periods where the ionosphere is largely disturbed due to, for instance, geomagnetic storms. On the other hand, the Klobuchar model [2] procides for GPS SF users an accuracy of about 50% RMS of the ionospheric range delay, depending on the solar activity or the region. Hence, its accuracy is often unsatisfactory even for absolute positioning, especially in the case of heightened solar activity [5].

2.1. TEC Accuracy

To test the accuracy of the broadcast ionospheric models for both NeQuick G and Klobuchar, the observations derived from the selected networks of stations described above, are compared. For the investigation of the accuracy of the modelling we present results in a vertical total electron content (VTEC) format. This is directly comparable to common ionospheric models and avoids having the elevation angle as a dimension in the plots. We note this because for other parts of this paper STEC is used in the underlying calculations, as the NeQuick G requirement is defined using STEC. The VTEC accuracy, performed for the Galileo and the GPS constellations, is obtained following the three steps below:

1. The computed VTEC values are acquired using the navigation models. The models are different depending on the constellation, NeQuick G for Galileo and Klobuchar for GPS. The computation follows the algorithm as described in [3] and in [2] for NeQuick G and Klobuchar, respectively. The different institutions involved in the study, used different approaches, in particular, for NMA/TUW, the VTEC for NeQuick G and Klobuchar is calculated every 5 min using the received ionosphere parameters, for the set of satellites observed by the GNSS reference stations (Figure 1), using the actual observation geometry (receiver and satellite positions).
2. For the same set of points, the reference VTEC is computed. A description of the methodologies is provided, detailing the computation of the reference VTEC. For NMA, the observed VTEC values are calculated every 5 min based on dual-frequency measurements from the GNSS reference stations (Figure 1). For the high-latitude stations, differential code biases (DCBs) are retrieved from the NMA's ionosphere monitoring system as the receivers used here are part of the much larger network processed by that system. For the low-latitude stations, DCBs are retrieved from the International GNSS Service (IGS) archives. For TUW, the observed VTEC values are calculated every minute based on dual-frequency measurements from the GNSS reference stations (Figure 1). For the mid-latitude stations, DCBs are obtained by a

least squares adjustment based on P1 and P2 code observations, as well as satellite DCBs downloaded from the Astronomical Institute of the University of Bern (AIUB) CODE database. The standard Single Layer Model (SLM) is used to convert slant TEC to vertical TEC (ionosphere altitude 350 km).

3. The VTEC error is finally calculated as the difference between the model-derived VTEC values and the reference VTEC values obtained through the observations.

2.2. TEC Availability

To obtain the TEC availability, the STEC parameter is considered, instead of the VTEC one. The reason for this choice is to be in line with the NeQuick G requirements specification, described below. Based on the NeQuick G specification, described in Section 2, this section introduces the TEC availability parameter which assesses the capability to correctly estimate the ionospheric delay affecting the Signal in Space (SiS) measurement. The TEC availability is defined as the number of epochs (for a single station) where the STEC error is under the following Galileo NeQuick G requirements [3]:

- For an STEC $\leq$ 66.7 [TECU]: Threshold = 20 TECU;
- For an STEC > 66.7 [TECU]: Threshold = 30% of STEC.

From the GRC network of stations, the STEC is obtained for a grid of points using the navigation model and averaging the contribution of the different satellites; while the reference STEC is obtained with the tomographic ionosphere model (TOMION) software V1.6.4. [6] developed at the Universitat Politècnica de Catalunya (UPC). The STEC error is therefore defined as the difference between the STEC as computed from the navigation model and the reference STEC as computed using the TOMION software. To perform a fair and comprehensive performance assessment between the NeQuick G and Klobuchar ionospheric models, the Galileo SF ionospheric algorithm specification is used as a common benchmark for both models.

2.3. SF Position Accuracy

This section describes the position accuracy computation algorithm used by the different institutions. The algorithm consists of three main steps, which are detailed below.

1. The Galileo E1 SF solutions are calculated in post-processing mode based on RINEX files, whereby the settings are applied as shown in Table 1. In the first run, the NeQuick G model is applied for ionospheric corrections, while in the other one, the Klobuchar model is used.

Table 1. Settings applied by each institution to calculate SF position accuracy.

	GRC	NMA	SRC	TUW
Software	TOMION [6]	Where [7]	GALAT [1]	raPPPid [2] [8]
Sampling rate	30 s	30 s	30 s	30 s
Elevation cut-off mask	5 deg	5 deg	5 deg	5 deg
Troposphere delay	GPT2	GPT2w [9]	GPT [10]/GMF [11]	VMF3 [12]
Solid Earth tides	applied	applied	applied	applied
Elev.-dep. weighting	1/sin (elev.)	1/sin (elev.)	1/sin (elev.)	1/$\sin^2$ (elev.)
Estimation	Epochwise least square	Epochwise least square	Epochwise least square	Kalman filter
Estimated parameters	coordinates, receiver clock	coordinates, receiver clock	coordinates, receiver clock	coordinates, receiver clock, ZWD [3]
PDOP threshold	6	6	6	N/A

[1] SRC internal software; [2] TUW open-source software—https://github.com/TUW-VieVS/raPPPid, https://vievswiki.geo.tuwien.ac.at/raPPPid (accessed on 10 May 2023); [3] Zenith Wet Delays.

2. The reference site positions are based either on the official International Terrestrial Reference Frame 2014 (ITRF2014) SINEX file station position and velocity solution (epoch 1 January 2010) extrapolated to the selected epoch from the analysed quarter of

observation (TUW) or yearly calculated ITRF14 coordinates, which refer to an epoch on 1st January of each year (NMA, GRC).

3. Coordinates obtained from the two SF solutions are directly compared epoch-wise in a local topo-centric coordinate system (East, North, Up) to the reference values in ITRF 2014, whereby differences are computed for east, north, and upward coordinates. The horizontal positioning error (HPE) and vertical positioning error (VPE) can be derived based on the differences in east, north, and up ($\Delta E, \Delta N, \Delta U$) for each epoch i, as follows:

$$HPE_i = \sqrt{\Delta E_i^2 + \Delta N_i^2} \tag{1}$$

$$VPE_i = |\Delta U_i| \tag{2}$$

Then, the single-frequency position errors were calculated as a mean of the 95th percentile (HPE 95% and VPE 95%) for the reference sites on a daily basis. Hereby, epochs exceeding the position dilution of precision (PDOP) threshold are excluded. The results in Figure 2 show the monthly average of these daily HPE 95% and VPE 95% site values grouped into high-latitude, middle north, middle south, and low-latitude regions.

(**a**) HPE 95%

Figure 2. *Cont.*

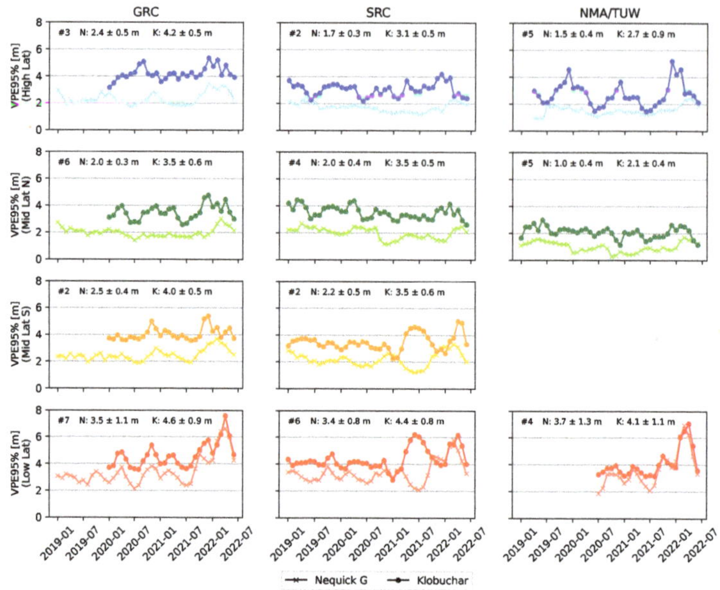

(**b**) VPE 95%

Figure 2. Monthly means of the daily (**a**) HPE 95% and (**b**) VPE 95% for the Nequick G and Klobuchar models. The top panels in (**a**,**b**) show results for high latitudes, the middle panels for middle latitudes divided into northern and southern hemispheres, and the bottom panel for low latitudes. The left panels present the GRC, the middle panels the SRC, and the right panels the NMA/TUW results. The text line at the top of each subplot includes information about the number of used stations (e.g., #5) and the average together with standard deviation over the complete period for the NeQuick G solution (e.g., N: 2.4 ± 0.5 m) and Klobuchar solution (e.g., K: 4.2 ± 0.5 m).

3. Results

3.1. TEC Accuracy

Figure 3 shows the monthly mean VTEC error for the high/middle/low latitudes and for the NeQuick G and Klobuchar model for the NMA and TUW network of stations as described above. Annual variations of the errors can be seen for the high and mid-latitudes. These variations are of opposite signs for the two models, i.e., when the Klobuchar-derived error decreases, the NeQuick G-derived error increases, and vice versa.

At high latitudes, the NeQuick G errors vary around zero, approximately between −2 and +2 TECU. The Klobuchar errors have an offset from zero of about 5–6 TECU, and slightly higher peak-to-peak values during a year. There may be a long-term decreasing trend in the Klobuchar error values. In the last quarter of 2021, the largest discrepancy (values comprises between 10 TECU and 15 TECU) in the time series error of the two models over the analysed period is observed. This may be related to the increasing ionization and space weather activity as the solar activity increases in the current solar cycle 25. At mid-latitudes, NeQuick G and Klobuchar have similar error magnitudes but with opposite offsets from zero. On average NeQuick G is slightly closer to zero, but this difference between Klobuchar and NeQuick G errors is smaller than the offsets of their individual time series and the inherent variations in the time series. At low latitudes, there is no obvious annual trend. Instead, both error graphs appear to have a long-term trend. This time series is not long enough to determine if this is related to the 11-year solar cycle, but this would be the most plausible explanation. It is known that the general ionization level varies with the solar cycle, and as the low latitudes have the largest TEC levels, they will also have the largest magnitude of long-term variation.

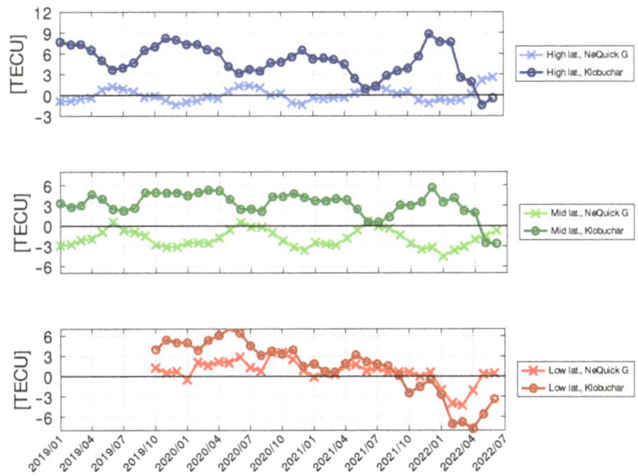

Figure 3. Monthly means of the VTEC error for the NeQuick G and Klobuchar model (NMA/TUW results). The top panel shows results for high latitudes, the middle panel for middle latitudes, and the bottom panel for low latitudes.

3.2. *TEC Availability*

This section describes the monthly mean TEC availability results for the high/middle north/middle south/low latitudes and for the NeQuick G and Klobuchar model for the GRC network of stations as described above. Numerical results are presented in Figure 4. As explained in Section 2.3, the Galileo SF ionospheric algorithm specification serves as a benchmark for both models. Being the specification not designed to fit the Klobuchar model, its worst performances with respect to the NeQuick G one are evident. However, the patterns of the Klobuchar follow as seeing for the VTEC accuracy a seasonal variation, showing the worst results during equinoctial months. On the contrary, NeQuick G performs better in the middle south, middle north, and high-latitude regions. In both models, this behaviour is more visible in low latitudes, in agreement with the disturbed ionospheric conditions at equatorial regions. In addition, during the last period analysed it is observed that the performances for both models are worst in the low-latitude regions. As described in Section 3.1, this time series is not long enough to conclude that it is driven by the 11-year solar cycle.

3.3. *SF Position Accuracy*

In Figure 2, SF monthly mean horizontal (HPE 95%) and vertical (VPE 95%) position accuracies are displayed, respectively, for the used stations for the high-/middle north/middle south/low-latitude regions. The position errors are estimated twice, once when the NeQuick G ionospheric model was applied to the positioning solution and another time when the Klobuchar model was applied. The monthly HPE 95% show, in general, smaller magnitudes compared to the monthly VPE 95%, which is mainly the case at high and both middle latitudes, especially for the Klobuchar model. Differences between the GRC, SRC, and NMA/TUW can be seen in the SF position accuracy solutions. This could be explained that each institution uses different kinds of station networks and that different software packages, models, and configurations are applied for the HPE 95% and VPE 95% calculation (see Table 1). At high latitudes (Figure 2) a more-or-less linear trend line of HPE 95% can be noted, which is gradually rising over time and reaching its maximum in April 2022 (beginning of Q2 2022), after which it declines. Before April 2022, the position error is on average 1 m for the NeQuick G model (maximum HPE 95% reached among all the results 2.35 m in 2022 Q2), while for the Klobuchar model values are slightly larger (maximum HPE 95% reached 2.52 m in 2022 Q2). Slightly—but not significantly—larger

HPE values for the Klobuchar model compared to the NeQuick G model can be observed at mid-latitudes. The low latitudes for both tested models have 2–4 times larger magnitudes compared to the two higher latitude regions. The largest errors are reached in March and April 2022 (end of Q1 2022, beginning of Q2 2022), being ≈5.5 m for the NeQuick G model and 6.93 m for the Klobuchar model. After those maxima are reached, a rapid decline in monthly HPE 95% at low latitudes can be noticed, approaching the high and middle latitude error values.

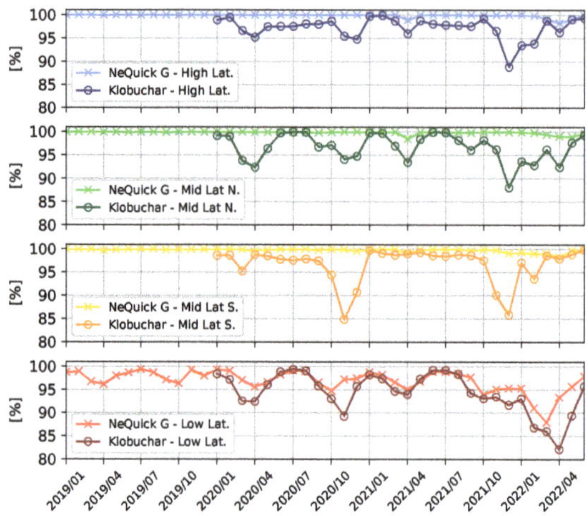

Figure 4. Monthly means of the TEC availability for the NeQuick G and Klobuchar models. The top panels show results for high latitudes, the middle panels for middle latitudes divided into the northern and southern hemispheres, and the bottom panel for low latitudes.

Significantly higher differences in positioning errors between the NeQuick G and the Klobuchar model are reflected in the monthly VPE 95% (Figure 2). Again, a slight linear trend can be recognized at high latitudes for the NeQuick G model (maximum VPE 95% reached 3.5 m in December 2021). On the other hand, the Klobuchar model does not indicate such a trend. The Klobuchar VPEs 95% are throughout the whole period larger compared to the NeQuick G values, with the largest difference being up to 5.36 m in December 2021. The difference between the NeQuick G and the Klobuchar model is also evident at middle latitudes. Moreover, in the case of SRC, for stations located below the 30° N parallel, in 2021, there is a significant difference between VPE 95% for Klobuchar and NeQuick G, reaching even 4 m in May 2021. There is also a visible difference between the two groups of mid-latitude stations, while the northern ones behave similarly to the high-latitude group, the southern ones have characteristics closer to equatorial stations. The vertical positioning errors from NeQuick G (maximum VPE 95% reached 3.7 m in February 2022) show, as well, for the whole period smaller magnitudes (up to 1 m smaller compared to Klobuchar). The VPE 95% values at low latitudes demonstrate a similar pattern as the horizontal positioning errors for both tested models. In fact, NeQuick G values are on a comparable level (difference | HPE 95%–VPE 95% | at low latitudes approximately up to 0.5 m) with the exception in April and May of 2022 (Q2 2022), where the difference between horizontal and vertical errors reached almost 1 m. During the analysed period, Klobuchar VPEs 95% were overall larger compared to the NeQuick G VPE 95% (one exception in March 2022).

4. Discussion

4.1. Network-Wise Performance Comparison

This study explores the benefits of using multiple networks of stations around the world to improve ionospheric models for SF users. NeQuick G and Klobuchar models are compared across a wide geographic range and under different ionospheric conditions to evaluate their performance. Data collection from multiple networks provides a more comprehensive evaluation of the models' ability to mitigate ionospheric effects. SF positioning accuracy is considered for each network of stations, but results may vary due to differences in ionospheric characteristics and data quality. Variations in the number and distribution of ground-based stations, data quality, and local ionospheric conditions could lead to differing results. Overall, this study highlights the benefits of using multiple networks to improve ionospheric models for SF users.

4.2. Model-Wise Performance Comparison

The NeQuick G and Klobuchar models are compared in their ability to mitigate ionospheric effects on GNSS signals for SF users. NeQuick G uses a physical-based approach and is more accurate than Klobuchar, especially at high latitudes and during intense solar activity, while the Klobuchar model uses an empirical approach. NeQuick G can model smaller ionospheric plasma structures and the 3D distribution of plasma density better than Klobuchar. The results, therefore, show better performances of NeQuick G compared to Klobuchar. TEC availability is more stable for NeQuick G at high middle north and middle south latitudes, and the effect on TEC availability is more evident at low-latitude regions with the best performances achieved by NeQuick G. Although, in terms of horizontal positioning, both models provide similar results at all analysed periods and stations, NeQuick G performs better than Klobuchar for vertical positioning.

4.3. Implications of the Results for SF Users

The implications of the results for SF users are significant, as these users typically have more limited access to multiple GNSS frequency signals and are, hence, more susceptible to the effects of ionospheric disturbances. The study founds that in terms of TEC accuracy, TEC availability, and SF position accuracy, the NeQuick G model outperformed the Klobuchar model. It suggests that the NeQuick G model is more effective for mitigating the effects of ionospheric disturbances on SF GNSS signals. For SF users, the improved TEC accuracy and availability offered by the NeQuick G model can result in more accurate position solutions and improved navigation performance. The improved SF position accuracy offered by the NeQuick G model may also lead to increased confidence in the accuracy of positioning solutions in challenging ionospheric conditions. Additionally, the results of this study may be useful in informing future developments of ionospheric models for SF users. Overall, the study highlights the importance of selecting an appropriate ionospheric model for mitigating ionospheric effects on GNSS signals, particularly for SF users who may have limited access to multiple-frequency signals.

Author Contributions: Methodology, U.N., C.P., K.S.J., M.D., N.H, D.H., A.Ś. and P.G.; software, U.N., K.S.J., M.D., D.H. and A.Ś.; validation, U.N., C.P., K.S.J., M.D., N.H. and A.Ś.; formal analysis, U.N., K.S.J., M.D., N.H., D.H., A.Ś. and P.G.; investigation, U.N., K.S.J., N.H., A.Ś. and P.G.; resources, U.N., C.P., A.Ś. and P.G.; data curation, U.N., K.S.J., M.D., A.Ś. and P.G.; writing—original draft preparation, U.N., C.P., K.S.J., M.D., N.H., D.H., A.Ś. and P.G.; writing—review and editing, U.N., C.P., K.S.J., M.D., N.H. and P.G.; visualization, U.N., C.P., K.S.J., M.D., N.H. and P.G.; supervision, M.P. and P.B.; project administration, M.P., P.B., U.N. and C.P.; funding acquisition, M.P. and P.B. All authors have read and agreed to the published version of the manuscript.

Funding: We acknowledge the European Union and the European Union Agency for the Space Programme (EUSPA) for supporting the cooperation of the Galileo Reference Centre (GRC) with Member States within the GRC-MS project and co-financed (Grant agreement nr. GSA/GRANT/04/2016), in support of an independent monitoring of the Galileo system performance. The SRC contribution was also co-funded by the Polish Ministry of Education and Science from funds for science in 2020 allocated to the implementation of a co-financed international project.

Institutional Review Board Statement: Not applicable.

Informed Consent Statement: Not applicable.

Data Availability Statement: Data are available in this manuscript.

Acknowledgments: We acknowledge the use of International GNSS Service (IGS) data and products.

Conflicts of Interest: At the time of writing, the authors C.P. and U.N. were employed by the RHEA Group, working as consultants for EUSPA at GRC. The remaining authors declare that the research was conducted in the absence of any commercial or financial relationships that could be construed as a potential conflict of interest.

References

1. Buist, P.; Mozo, A.; Tork, H. Overview of the Galileo Reference Centre: Mission, Architecture and Operational Concept. In Proceedings of the 30th International Technical Meeting of the Satellite Division of the Institute of Navigation (ION GNSS+ 2017), Portland, OR, USA, 25–29 September 2017; pp. 1485–1495.
2. Klobuchar, J.A. Ionospheric time-delay algorithm for single-frequency GPS users. *IEEE Trans. Aerosp. Electron. Syst.* **1987**, *AES-23*, 325–331. [CrossRef]
3. Commission, E. Ionospheric Correction Algorithm for Galileo Single-Frequency Users. *Eur. GNSS (Galileo) Open Serv.* **2016**, *1*, 1–96.
4. Hoque, M.M.; Jakowski, N.; Orús-Pérez, R. Fast ionospheric correction using Galileo Az coefficients and the NTCM model. *GPS Solut.* **2019**, *23*, 41. [CrossRef]
5. Böhm, J.; Schuh, H. *Atmospheric Effects in Space Geodesy*; Springer: Berlin/Heidelberg, Germany, 2013; Volume 5.
6. Juan, J.M.; Rius, A.; Hernández-Pajares, M.; Sanz, J. A two-layer model of the ionosphere using Global Positioning System data. *Geophys. Res. Lett.* **1997**, *24*, 393–396.
7. Kirkvik, A.S.; Hjelle, G.A.; Dähnn, M.; Fausk, I.; Mysen, E. Where-a new software for geodetic analysis. In Proceedings of the 23rd European VLBI Group for Geodesy and Astrometry Working Meeting, Gothenburg, Sweden, 15–19 May 2017; Volume 23, pp. 248–252.
8. Glaner, M.F. Towards Instantaneous PPP Convergence Using Multiple GNSS Signals. Ph.D. Thesis, TU Wien, Wien, Austria, 2022.
9. Böhm, J.; Möller, G.; Schindelegger, M.; Pain, G.; Weber, R. Development of an improved empirical model for slant delays in the troposphere (GPT2w). *GPS Solut.* **2015**, *19*, 433–441. [CrossRef]
10. Böhm, J.; Heinkelmann, R.; Schuh, H. Short note: A global model of pressure and temperature for geodetic applications. *J. Geod.* **2007**, *81*, 679–683.
11. Böhm, J.; Niell, A.; Tregoning, P.; Schuh, H. Global Mapping Function (GMF): A new empirical mapping function based on numerical weather model data. *Geophys. Res. Lett.* **2006**, 33. [CrossRef]
12. Landskron, D.; Böhm, J. VMF3/GPT3: Refined discrete and empirical troposphere mapping functions. *J. Geod.* **2018**, *92*, 349–360. [CrossRef] [PubMed]

engineering proceedings

Proceeding Paper

Assessing the High-Accuracy Service at High Latitudes †

Melania Susi [1,*,‡], Daniele Borio [1], Althaf Azeez [2], Jihye Park [2], Kinga Wezka [3] and Ignacio Fernandez-Hernandez [4]

1 Joint Research Centre (JRC), European Commission, 21027 Ispra, VA, Italy; daniele.borio@ec.europa.eu
2 School of Civil and Construction Engineering, Oregon State University, Corvallis, OR 97331, USA; althaf.azeez@oregonstate.edu (A.A.); jihye.park@oregonstate.edu (J.P.)
3 Faculty of Geodesy and Cartography, Department of Geodesy and Geodetic Astronomy, Warsaw University of Technology, 00-661 Warsaw, Poland; kinga.wezka@pw.edu.pl
4 Directorate General for Defence Industry and Space (DG DEFIS), European Commission, 1049 Brussel, Belgium; ignacio.fernandez-hernandez@ec.europa.eu
* Correspondence: susimelania1@gmail.com
† Presented at the European Navigation Conference 2023, Noordwijk, The Netherlands, 31 May–2 June 2023.
‡ Current address: Topcon Positioning System Inc., 41124 Modena, EMR, Italy.

Abstract: The Galileo High-Accuracy Service (HAS) was declared operational (initial service) in January 2023 after an intense testing phase aimed at assessing its service performance through the transmission of live HAS corrections. The HAS performance at high latitudes is evaluated by analyzing data collected at the Polish Polar Station located in Hornsund, in the Norwegian Svalbard archipelago. At such a location, the reception conditions can be affected by poor satellite geometries and ionospheric scintillation may degrade the signal quality. In this respect, data affected by scintillation events were identified and used for the performance analysis. The assessment is carried out by comparing the HAS corrections with the International GNSS Service (IGS) precise products. Moreover, the retrieved HAS corrections are used to assess the Precise Point Positioning (PPP) performance.

Keywords: Galileo High-Accuracy Service; HAS; Precise Point Positioning; PPP; high latitudes; ionospheric scintillation

Citation: Susi, M.; Borio, D.; Azeez, A.; Park, J.; Wezka, K.; Fernandez-Hernandez, I. Assessing the High-Accuracy Service at High Latitudes. *Eng. Proc.* **2023**, *54*, 8. https://doi.org/10.3390/ENC2023-15421

Academic Editors: Tom Willems and Okko Bleeker

Published: 29 October 2023

1. Introduction

The Galileo High-Accuracy Service (HAS) broadcasts Precise Point Positioning (PPP) corrections targeting the horizontal and vertical positioning accuracy, after convergence, below twenty and forty centimeters, respectively. In January 2023, the HAS initial service was declared operational after an intense testing phase aimed at assessing the service performance through the transmission of live corrections. During its testing phase and since the initial service declaration, HAS has attracted significant research interest given the potential offered by decimeter-level position accuracies.

In this respect, intensive tests have been conducted to assess the service's performance. For instance, an analysis was conducted in [1] to assess the accuracy achievable with HAS: two static receivers placed in Brussels, Belgium, and in Tres Cantos, Spain, were considered. For both receivers, it was shown that the HAS accuracy targets can already be achieved with the current service level, where ionospheric corrections are not broadcast. Similar results were obtained by [2], which considered a set of 33 International GNSS Service (IGS) stations and used their data to assess the HAS performance. Also, in this case, horizontal accuracies within the 20 cm target were observed. Similarly, the vertical accuracy achieved was within the 40 cm target.

The focus of the mentioned analyses is either limited to central Europe or aggregated at the level of several reference stations. In this work, we assess the HAS performance at high latitudes by exploiting data collected by the Polish Polar Station located in Hornsund, in the Norwegian Svalbard archipelago at a latitude around 77°. The station features a multi-constellation Septentrio PolaRx5S receiver connected to a choke-ring antenna. The binary files provided by the receiver are processed using an in-house developed decoder, denoted as Galileo HAS Parser (GHASP) [3,4], which is able to retrieve HAS corrections from the Galileo E6-B signal pages collected by the Septentrio PolaRx5S receiver. The corrections were then applied to the broadcast orbits and clocks to obtain precise satellite positions and clock offsets. Finally, the obtained precise orbits and clocks, and the decoded biases, were input to the Oregon State University Precise Point Positioning (ORPPP) software v 1.0, which was used to estimate the final position solution.

The assessment of the HAS performance at high latitudes is of particular interest because reception conditions can be affected by poor satellite geometries [5]. Moreover, at high latitudes, ionospheric scintillation may occur, degrading the signal quality.

Ionospheric scintillations are fast and random fluctuations in the signal amplitude and/or phase produced by strong ionospheric electron gradients [6]. This phenomenon is more frequent and severe at high latitudes and in equatorial areas [7].

However, since the ionospheric morphology is different in these two regions, the phenomenon presents different characteristics in these geographical areas. Specifically, in the equatorial regions, scintillation usually is characterized by correlated amplitude and phase signal variations, while at high latitudes, scintillation is mainly characterized by strong phase fluctuations and lower amplitude variations. Strong phase signal variations induced by scintillation can be particularly challenging for carrier-phase-based position techniques, such as PPP algorithms, which are exploited by the typical HAS users. In order to assess the impact of ionospheric scintillation on the HAS performance, data collected during a day affected by strong phase scintillation are considered. The HAS performance assessment is carried out by (1) comparing the quality of the HAS-corrected orbits and clocks with the precise products provided by the IGS (https://igs.org/products/ accessed on 31 January 2023), and (2) analyzing the HAS PPP solution in the position domain.

The remainder of this paper is organized as follows: Section 2 describes the methodology used to decode and apply the HAS corrections and obtain the final precise position. Section 3 illustrates the experimental set-up used to collect the data analyzed in this paper. Section 4 provides the assessment results through the statistics of the HAS orbit/clock and position errors. Finally, Section 5 draws some conclusions.

2. HAS Correction Decoding and Application

HAS corrections are broadcast through the Galileo E6-B Signal-in-Space (SiS) and disseminated using a high-parity encoding and dissemination scheme based on Reed–Solomon codes [8]. Several E6-B pages need to be combined in order to extract the actual HAS corrections. This operation is performed using the GHASP software v 1.0 [3,4] developed by the authors. GHASP supports several input formats and is able to convert messages collected using E6-B capable receivers into actual corrections. The data extracted using the GHASP software includes orbit and clock corrections and code and carrier biases. In the following, the quality of the orbit and clock corrections is analyzed by comparing them with the precise products obtained from the IGS. This process is schematically represented in Figure 1, which shows the different processing steps adopted for the analysis. HAS corrections were applied to the broadcast orbits and clocks to obtain precise satellite orbits and clocks. The latter were then formatted as standard Special Product 3 (.sp3) orbit and clock (.clk) files [9,10] using an in-house developed software so that they could be easily processed. The extracted HAS corrections in the standard orbit and clock file formats were tested using a PPP algorithm.

For this experiment, the authors used the ORPPP software, developed by the coauthors of this paper at the Oregon State University. The ORPPP software is capable of processing multi-constellation, multi-frequency GNSS signals.

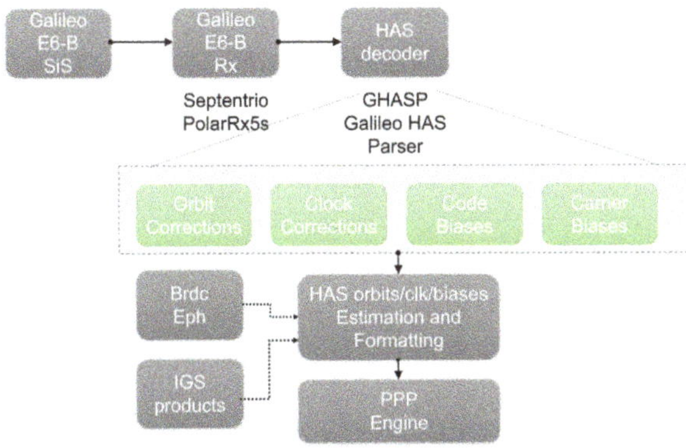

Figure 1. Representation of the process adopted for decoding and applying HAS corrections.

The primary observables used by the ORPPP software are ionosphere-free linear combinations, and the unknown parameters include the position vector, clock error, zenith wet delay, float ambiguity parameter, differential code biases, and intersystem bias. Tidal corrections (solid earth, ocean, pole tides), phase wind-up, and relativistic corrections are applied. The unknown parameters are estimated using an adaptive robust Kalman filtering method introduced by [11].

3. Experimental Set-Up

The data under analysis were collected by the WUTH reference station located in the Polish Polar Station of Hornsund in Spitsbergen/Norway (77.00° N, 15.54° E) and included in the IGS and the EUREF Permanent GNSS Network (EPN). Currently, this station is the second northernmost location of the IGS network. The station is equipped with a POLARX5S receiver connected to a SEPCHOKE_B3E6–SPKE choke-ring antenna from Septentrio, Leuven, Belgium. A view of the antenna used for the data collection is provided in Figure 2. The receiver is an Ionospheric Monitoring Scintillation Receiver (ISMR) embedded with an ultra-low phase noise oscillator, which allows the identification of the phase fluctuations due to ionospheric scintillation minimizing the clock noise contribution. The ISMR provides scintillation indices, such as S4 and Phi60, which are used to identify the level of amplitude and phase scintillation [12]. Moreover, the receiver was set to record the E6-B navigation message, which was exploited to retrieve the HAS corrections as described in Section 2.

Figure 2. View of the antenna of the WUTH Polish Polar Station in Hornsund (Svalbard, Norway).

4. HAS Assessment Results

While several days of data were considered, the assessment presented in this paper focuses on the data collected from the WUTH station on the 4th of September 2022 (DOY 247). In particular, the interval between 15:00 and 19:00 was analyzed, since it was characterized by very intense ionospheric activity.

Moving at high latitudes, the maximum reachable satellite elevation decreases and therefore the satellite visibility is reduced. Consequently, the occurrence of low-elevation satellites and intense ionospheric activity can render the high-latitude region a very challenging environment for high-accuracy applications.

During the observation period, the maximum satellite elevation was around 60 degrees for both the GPS and Galileo satellites, as shown in the skyplot provided in Figure 3. Despite this fact, a good satellite geometry can be observed in Figure 4, where the Position Dilution of Precision (PDOP), the Horizontal (HDOP) and Vertical (VDOP) are reported. The DOPs are always below 2.5 for the GPS only, Galileo only and GPS + Galileo cases. This result indicates good geometry conditions for the selected receiver location. In Figure 5, the phase scintillation index provided by the ISMR during the time interval considered for the analysis is also reported for all satellites in view. The Phi60 is an index used to quantify the level of phase scintillation and is obtained by computing the standard deviation of the detrended carrier phase, averaged over one minute of data [12]. The detrending operation is performed to eliminate the low-frequency variations in the phase measurements due to other noise sources, such as the receiver clock, multipath and the relative satellite user motion. In this way, only high-frequency components caused by ionospheric scintillation are retained.

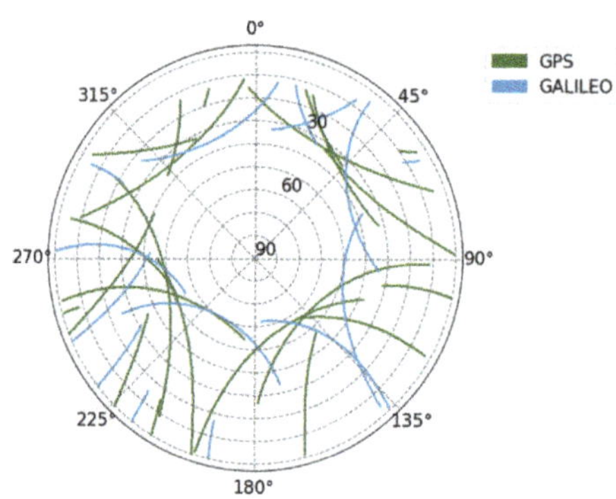

Figure 3. Skyplot with the track of the satellites in view over the WUTH station during the testing period.

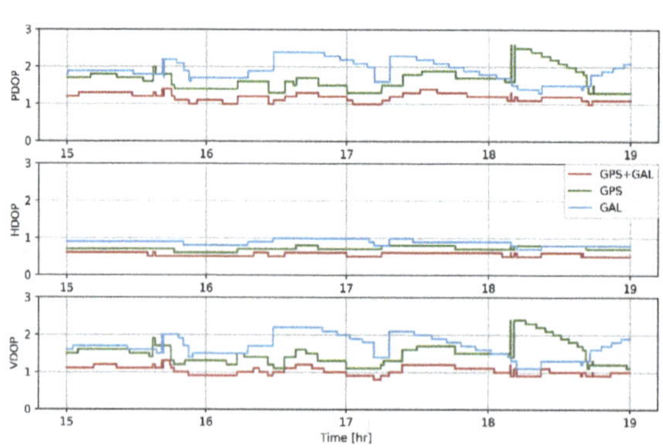

Figure 4. PDOP; HDOP and VDOP observed during the testing period.

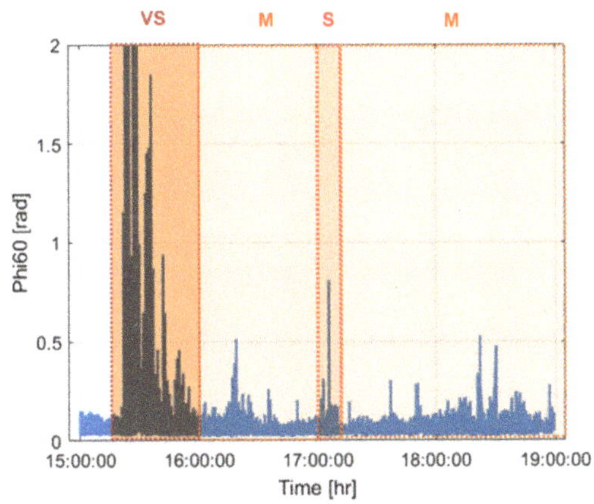

Figure 5. Phi60 index recorded at the WUTH station during the testing period and for all satellites.

Slightly different Phi60 thresholds are used in the literature to classify different scintillation levels. Adapting the values in [13,14], we classify the phase scintillation level as follows: values of Phi60 above 0.25 and below 0.4 radians indicate the occurrence of weak scintillation, values between 0.4 and 0.6 correspond to moderate (M) phase scintillation, values between 0.6 and 0.8 radians are associated with strong (S) phase scintillation events, while values above 0.8 radians indicate very strong (VS) scintillation. The different levels of phase scintillation are indicated on top of Figure 5.

Around 15:20, very intense ionosphere activity occurs, leading to the occurrence of very strong phase scintillation. Then, the activity decreases, but still some moderate/strong ionospheric activity is present, with Phi60 achieving values above 0.5 radians.

4.1. Orbit and Clock Errors

The quality of the HAS orbits and clocks was assessed using the Center for Orbit Determination in Europe (CODE) final products as reference. Details of the accuracy of these final products can be found in [15]. Given that the CODE satellite orbits are referred to the Centre of Mass (COM) while the HAS satellite orbits are referred to the Antenna Phase Centre (APC), satellite Phase Centre Offsets (PCOs) were applied to align the two products and allow a comparison. The PCOs were retrieved from the igs14.atx antenna model [16].

Moreover, the HAS clock corrections are referred to the Ionosphere Free (IF) combination of L1C/A/L2P measurements for GPS and E1/E5b for Galileo, while CODE clocks are referred to the IF combinations L1P/L2P and E1/E5a. Therefore, IGS satellite biases were used to align the two products to the same reference IF measurement combination and enable a fair comparison. After the alignment, the clock accuracy was evaluated by computing the difference between the two products and removing the mean error per constellation at each instant.

The Galileo and GPS 3D orbit errors are shown in Figures 6 and 7, respectively. The errors for all visible satellites are provided jointly in the figures. After computing the root mean squared (rms) 3D orbit error at each instant for all satellites, the 95th percentile values were computed and are reported in Table 1 for the Galileo and GPS cases. Specifically, the values are 10.5 and 11 cm for the Galileo and GPS constellations, respectively. These values are comparable with the ones found in [2] for the data collected on the same day but at a different geographical location. The clock errors are reported in Figures 8 and 9 for the Galileo and GPS constellations, respectively. In Table 1, the 95th percentile of the rms clock errors, computed for all satellites at each instant, are also reported. For the GPS case, this value was computed after removing satellite G02, which showed bigger error values, also in agreement with the findings in [2]. Moreover, in Figure 8, it can be observed that a higher clock error was estimated for one Galileo satellite, with values around 0.6 m. For this satellite (E12), similar error values were obtained in [2]. Note that the results were obtained during the HAS validation period, some months before the initial service operation. The HAS product accuracy at the moment is higher and is expected to gradually increase over time thanks to the algorithm tuning and the addition of stations.

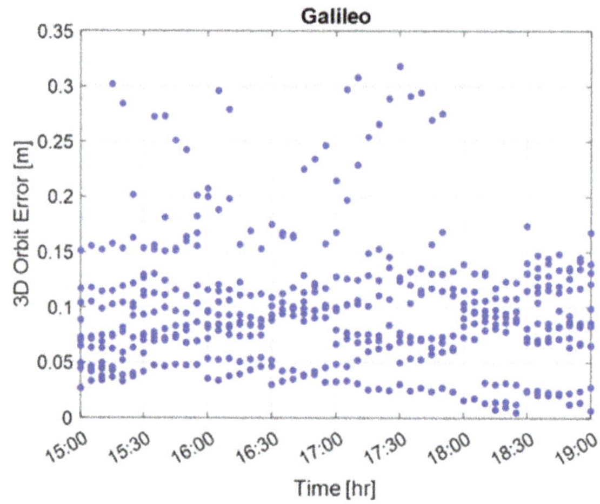

Figure 6. The 3D orbit error for all the Galileo satellites visible during the testing period.

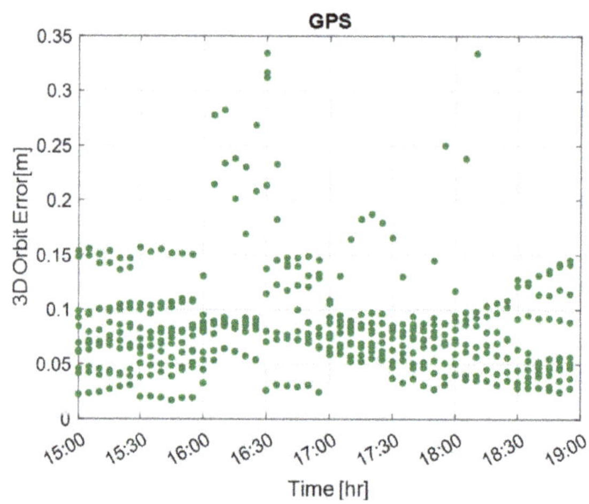

Figure 7. The 3D orbit error for all GPS satellites visible during the testing period.

Table 1. Orbit/Clock statistics (95th percentile).

Constellation	Gal (p95)	GPS (p95)
3D Orbital Error	10.5 cm	11 cm
Clock Error	15 cm	11 cm

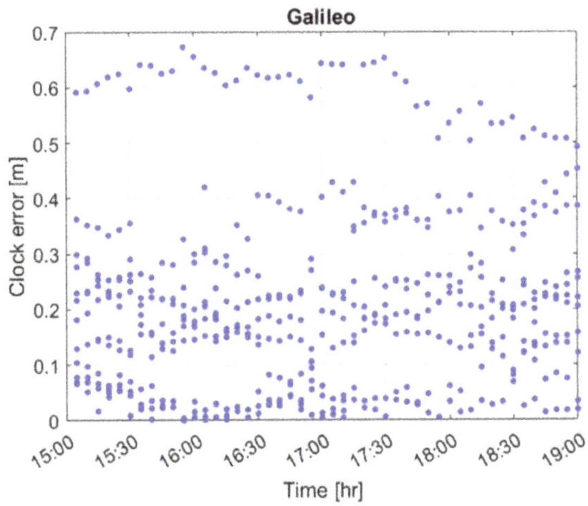

Figure 8. Galileo clock error during the testing period.

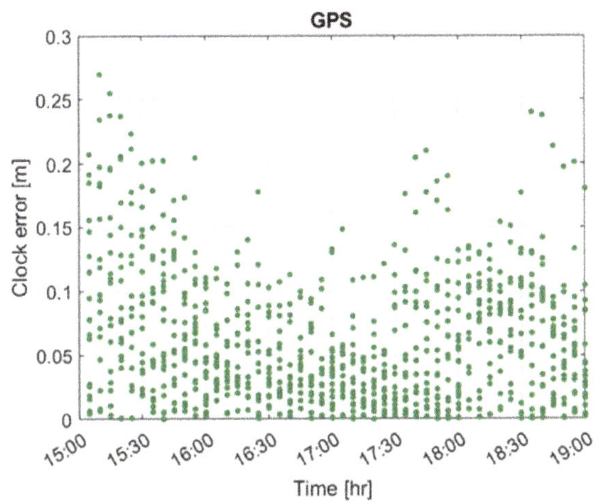

Figure 9. GPS clock error during the testing period.

4.2. Position Domain Results

The PPP engine was set to process Galileo E1/E5b and GPS L1C/A/L2P measurements. The processing started just before the occurrence of the strong phase scintillation event identified by the peak in Phi60 reported in Figure 5 to stress the PPP performance. The horizontal and vertical position errors are provided in the upper and bottom parts of Figure 10. The PPP filter converges in 15 min, reaching a horizontal and vertical position accuracy below 20 and 40 cm, respectively. No phase biases were used for the PPP processing (they were not available at the time of data collection) and no ambiguity resolution was performed to assist in reducing the convergence time. After convergence, the horizontal and vertical accuracies were found to be below 20 and 40 cm, respectively. This can be observed in Table 2, where the statistics (95th percentile and std) of the horizontal and vertical errors (HE and VE) computed after convergence are reported and the target values are indicated with red dashed lines. The solution remains below the target thresholds even if the ionospheric activity persists until the end of the data collection, as shown in Figure 5. For comparison purposes, in Figure 11, the HE and VE are reported for a period of quiet ionospheric conditions recorded during the same day. The associated statistics are also reported in Table 2. In this case, the convergence is reached after 230 s. However, after convergence, the positioning accuracy is comparable for the two cases.

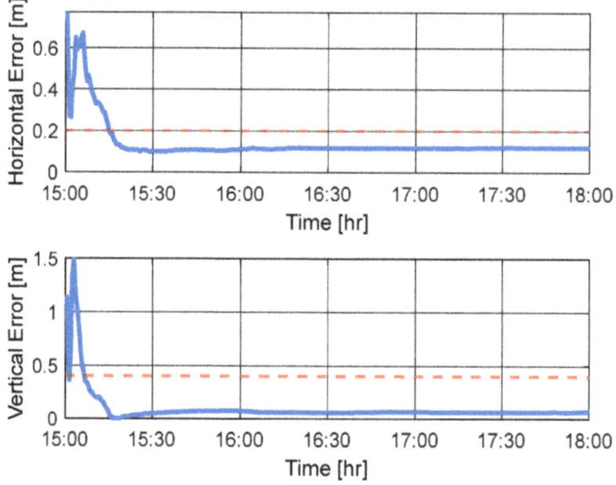

Figure 10. Horizontal/vertical accuracy under ionospheric scintillation.

Table 2. Positioning accuracy statistics after convergence and convergence time.

	HE (p95/std)	VE (p95/std)	Conv Time
Scintillation period	12/1.2 cm	7.9/1.5 cm	900 s
Quiet ionospheric period	11.5/1.7 cm	15.7/1.9 cm	230 s

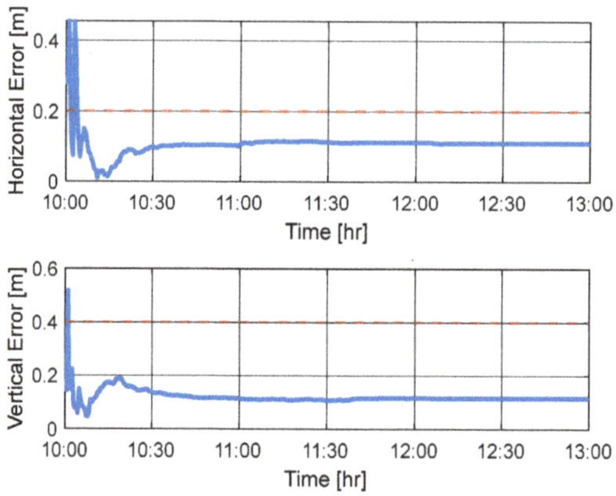

Figure 11. Horizontal/vertical accuracy under quiet ionospheric conditions.

5. Conclusions

At the beginning of 2023, the Galileo HAS was declared operational, providing free corrections through the Galileo E6-B data component worldwide. In this paper, we showcased the potential of HAS to achieve a high accuracy at high latitudes and under ionospheric scintillation. For this purpose, we exploited the data collected at the Polish Polar station located in Hornsund (Svalbard, Norway) at a 77° latitude during a day with very intense ionospheric activity. Despite the harsh conditions, the HAS-based PPP solution achieved horizontal and vertical accuracies, after convergence, below 20 and 40 cm, respectively. The results were obtained by analyzing the data collected during the HAS testing phase. Improved performance should be expected with the enhancement of the corrections transmitted after the HAS declaration. As future work, it will be of interest to analyze more data sets collected at the same station and affected by different ionospheric scintillation conditions. The analysis will be extended to an assessment under equatorial scintillation, which is characterized by amplitude and phase scintillation.

Author Contributions: All authors contributed to the preparation of this paper. Conceptualization and methodology, M.S., D.B., A.A., J.P., K.W. and I.F.-H.; software, M.S., D.B., A.A. and J.P.; validation, M.S., D.B., A.A., J.P., K.W. and I.F.-H.; data curation, K.W. and M.S.; writing—original draft preparation, M.S. and D.B.; writing—review and editing, M.S., D.B., A.A., J.P., K.W. and I.F.-H. All authors have read and agreed to the published version of the manuscript.

Funding: This research received no external funding.

Institutional Review Board Statement: Not applicable.

Informed Consent Statement: Not applicable.

Data Availability Statement: Data are available in this manuscript.

Acknowledgments: The data were provided by the Warsaw University of Technology EPN Analysis Center. The operation of the WUTH GNSS station is supported by: Polish National Science Center (NCN) Program Excellence Initiative: Research University (ID-UB).

Conflicts of Interest: The authors declare no conflict of interest.

References

1. Fernández-Hernández, I.; Chamorro-Moreno, A.; Cancela-Diaz, S.; Calle-Calle, J.D.; Zoccarato, P.; Blonski, D.; Senni, T.; Javier de Blas, F.; Hernández, C.; Simón, J.; et al. Galileo High Accuracy Service: Initial Definition and Performance. *GPS Solut.* **2022**, *26*, 65. [CrossRef]
2. Naciri, N.; Yi, D.; Bisnath, S.; Javier de Blas, F.; Capua, R. Assessment of Galileo High Accuracy Service (HAS) test signals and preliminary positioning performance. *GPS Solut.* **2023**, *27*, 73. [CrossRef] [PubMed]
3. Borio, D.; Susi, M.; Gioia, C. GHASP: A Galileo HAS Parser. *GPS Solut.* **2023**, *27*, 197. [CrossRef]
4. Gioia, C.; Borio, D.; Susi, M.; Fernández-Hernández, I. The Galileo High Accuracy Service (HAS): Decoding and Processing Live Corrections for Code-Based Positioning. In Proceedings of the International Technical Meeting of The Institute of Navigation, Long Beach, CA, USA, 25–27 January 2022; pp. 1065–1074. [CrossRef]
5. Susi, M.; Borio, D.; Gioia, C.; Brunes, M.T.; Dähnn, M.; Grinde, G.; Rost, C. Assessment of Galileo E6-B Data Demodulation Performance at High Latitudes: A Norwegian Vessel Case Study. *Remote Sens.* **2021**, *13*, 4669. [CrossRef]
6. Rino, C.L. A Power Law Phase Screen Model for Ionospheric Scintillation: 1. Weak Scatter. *Radio Sci.* **1979**, *14*, 1135–1145. [CrossRef]
7. Aarons, J. Global Morphology of Ionospheric Scintillations. *Proc. IEEE* **1982**, *70*, 360–378. [CrossRef]
8. Fernández-Hernández, I.; Senni, T.; Borio, D.; Vecchione, G. High-parity vertical Reed-Solomon codes for long GNSS high-accuracy messages. *Navigation* **2020**, *67*, 365–378. [CrossRef]
9. Hilla, S. The Extended Standard Product 3 Orbit Format (SP3-d). February 2016. Available online: https://epncb.eu/ftp/data/format/sp3d.pdf (accessed on 31 January 2023).
10. Ray, J.; Coleman, M. RINEX Extensions to Handle Clock Information. Version 3.04 8 July. 2017. Available online: https://files.igs.org/pub/data/format/rinex_clock304.t (accessed on 31 January 2023).
11. Guo, F.; Zhang, X. Adaptive robust Kalman filtering for precise point positioning. *Meas. Sci. Technol.* **2014**, *25*, 105011. [CrossRef]
12. Van Dierendonck, A.J. Eye on the Ionosphere: Measuring Ionospheric Scintillation Effects from GPS Signals. *GPS Solut.* **1999**, *2*, 60–63. [CrossRef]
13. Dubei, S.; Wahi, R.; Mingkhwan, E.; Gwal, A. Study of amplitude and phase scintillation at GPS frequency. *Indian J. Radio Space Phys.* **2005**, *34*, 402–407.
14. Tiwari, R.; Skone, S.; Tiwari, S.; Strangeways, H.J. WBMod assisted PLL GPS software receiver for mitigating scintillation effects in high latitude region. In Proceedings of the 2011 XXXth URSI General Assembly and Scientific Symposium, Istanbul, Turkey, 13–20 August 2011; pp. 1–4.
15. Prange, L.; Villiger, A.; Sidorov, D.; Schaer, S.; Beutler, G.; Dach, R.; Jäggi, A. Overview of CODE's MGEX solution with the focus on Galileo. *Adv. Space Res.* **2020**, *66*, 2786–2798. [CrossRef]
16. Rebischung, P.; Schmid, R. IGS14/igs14.atx: A New Framework for the IGS Products. In Proceedings of the AGU Fall Meeting, San Francisco, CA, USA, 12–16 December 2016.

engineering proceedings

MDPI

Proceeding Paper

Development of a Custom GNSS Software Receiver Supporting Supercorrelation †

Javier Gonzalo Garcia *, Johannes Rossouw van der Merwe, Paulo Esteves, Dana Jamal, Samir Benmendil, Chris Higgins, Rose Grey, Eugene Coetzee and Ramsey Faragher

Focal Point Positioning, Cambridge CB4 3NP, UK; j.r.vandermerwe@ieee.org (J.R.v.d.M.); paulo@fppnt.com (P.E.); dana.jamal@fppnt.com (D.J.); samir@fppnt.com (S.B.); chris@fppnt.com (C.H.); rose@fppnt.com (R.G.); eugene.coetzee@fppnt.com (E.C.); ramsey@fppnt.com (R.F.)

* Correspondence: javier.garcia@fppnt.com

† Presented at the European Navigation Conference 2023, Noordwijk, The Netherlands, 31 May–2 June 2023.

Abstract: Mitigating multipath interference is one of the biggest challenges in radio positioning. The Supercorrelation™ technology developed via Focal Point Positioning (FPP) suppresses multipath interference by performing long coherent integration while undergoing complex motion in order to isolate the Line-Of-Sight (LOS) signals from the unwanted multipath interference. This article presents the current status of a Supercorrelating Global Navigation Satellite System (GNSS) Software-Defined Radio (SDR) and a systematic testing framework. The SDR receiver is capable of real-time processing and facilitates independent testing and demonstrations. The testing framework uses synthetic signals with a Spirent Radio-Frequency Constellation Simulator (RFCS) with Sim3D to create controlled and repeatable scenarios. The initial results demonstrate the benefits of Supercorrelator Technology (S-GNSS) for navigation resilience.

Keywords: Supercorrelation; GNSS; multipath mitigation; Non-Line-Of-Sight (NLOS) suppression; SDR

Citation: Garcia, J.G.; van der Merwe, J.R.; Esteves, P.; Jamal, D.; Benmendil, S.; Higgins, C.; Grey, R.; Coetzee, E.; Faragher, R. Development of a Custom GNSS Software Receiver Supporting Supercorrelation. *Eng. Proc.* **2023**, *54*, 9. https://doi.org/10.3390/ENC2023-15423

Academic Editors: Tom Willems and Okko Bleeker

Published: 29 October 2023

1. Introduction

Multipath interference limits the accuracy of Global Navigation Satellite System (GNSS) receivers in urban and other cluttered environments. Mitigating it is one of the biggest challenges for reliable navigation, especially in dense urban scenarios. The Supercorrelation™ technology developed via Focal Point Positioning (FPP) suppresses multipath interference by enabling very long coherent integration times (many seconds) on low-cost platforms undergoing highly non-linear dynamics. Achieving this requires sub-wavelength scale corrections for receiver clock errors, the receiver motion, and the motion of the satellite during accumulation [1,2]. We refer to a GNSS receiver that contains this technology as an S-GNSS receiver. FPP is currently developing a real-time S-GNSS Software-Defined Radio (SDR) demonstrator platform with European Space Agency (ESA) support, and this paper provides information on the project progress and some interim performance testing and demonstrates the Non-Line-Of-Sight (NLOS) rejection and multipath mitigation capabilities of the technology.

The S-GNSS SDR demonstrator platform is a multi-band software receiver and can process the GPS and Galileo signals in the L1/E1 and L5/E5a bands. The evaluation includes controlled laboratory tests with synthesized multipath signals. A multi-band Spirent Radio-Frequency Constellation Simulator (RFCS) with Sim3D capabilities allows for accurate signal generation. Using controlled experiments, a systematic performance evaluation of the S-GNSS technology is conducted. The initial results show significant improvements in code phase and carrier frequency measurements compared to standard GNSS processing, which results in a more accurate and reliable Position, Velocity, and Time (PVT) solution.

The rest of the article is structured as follows: Section 2 presents the background of the project, and Section 3 briefly describes Supercorrelator Technology (S-GNSS). The SDR design approach and real-time performance are exhibited in Section 4. Section 5 defines the test setup, and Section 6 shows the results. Finally, the conclusions are drawn in Section 7.

2. Project Background

With the support of the ESA under the NAVISP Element 2 program (https://navisp.esa.int/element/competitiveness (accessed on 15 May 2023)).

FPP is designing, building, and deploying an S-GNSS SDR real-time receiver, which is a hardware and software platform, for the development and demonstration of the S-GNSS technology. The approach of this receiver is to use hardware Commercial-Off-The-Shelf (COTS) components for the Radio-Frequency Front-End (RFFE) and an SDR approach for the GNSS processing in the software. This approach allows for rapid development and maximal flexibility in design.

The existing demonstrators of the S-GNSS technology rely on recordings generated via partner and customer company GNSS receivers. Although this is a great approach to demonstrate the performance improvements that S-GNSS provides, the results and experiments are held under Non-Disclosure Agreements (NDAs) and cannot be publicized, and the receivers are generally not directly accessible to FPP for our own specific research and development. As such, this project allows FPP to demonstrate current and future capabilities to the public more freely and provides a mechanism for Original Equipment Manufacturers (OEMs) to trial S-GNSS directly without requiring support from the third-party chipset suppliers.

Figure 1 shows the main building blocks of the S-GNSS hardware: an NTLab® NT1065-USB3 RF IC EVALUATION KIT is the RFFE, a Linux-based Personal Computer (PC) is the processing unit of the receiver, and an FPP Inertial Sensor Data Logging and Streaming Unit (DLSU) is the inertial sensor data source. This paper focuses on the initial results of the GNSS SDR, but integration with the inertials from the DLSU is planned for future publications.

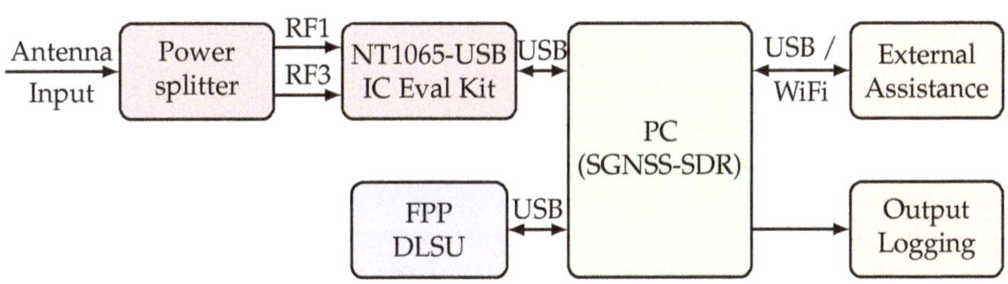

Figure 1. S-GNSS receiver building blocks.

The NTLab® NT1065-USB3 RFFE is configured for dual-band operation and receives the L1 and L5 frequency bands with a real-valued sample rate of approximately 48 MHz and uses 2-bit precision.

3. S-GNSS Background and Operation

S-GNSS suppresses multipath interference through extended correlation by correcting the receiver clock, the receiver motion, and the motion of the satellite during accumulation [2]. Longer correlation allows a separation between the correct Line-Of-Sight (LOS) signal and multipath interference using the carrier frequency offset. The technique is similar to synthetic aperture processing [3–5], although it is specifically designed to overcome the challenges [6–9] provided by the low-quality oscillators found in smartphones and smartwatches and the highly non-linear motion dynamics that those devices can undergo in typical use. S-GNSS is ideal for urban environments with challenging multipath interference [10] and has shown benefits when integrated with 3D mapping [11] and inertial navigation systems [12]. Furthermore, it has been adopted by mass-market receivers [1], has been tailored for easy integration into smartphones [13,14], and is being integrated into the automotive sector [15]. A benefit of S-GNSS is that it allows for spatial processing (i.e., synthetic aperture processing), which is ideal for GNSS (https://www.youtube.com/watch?v=9vj_yW0YJFg (access on 15 May 2023)) spoofing detection and correction [2].

The core of S-GNSS is the extended accumulation of the satellite signal based on the correlation outputs. It allows for processing after tracking, making it practical for existing receiver systems. However, improved performance is possible when it is tightly coupled to tracking. The Cross Ambiguity Function (CAF) after correlation is scaled inversely proportional to the integration time for the carrier frequency offset. This correlation property allows S-GNSS to discriminate between LOS and NLOS signal components according to their small variations in Doppler shifts.

Figure 2a shows the theoretical (i.e., noise-free) CAF for 20 ms integration for a multipath interference scenario. An LOS signal with zero code delay and frequency offset represents the correct peak. An NLOS signal with equal power, a code offset relating to 40 m, and a carrier frequency offset of 5 Hz represents a multipath interference. It is a challenging scenario as both components have the same power (e.g., such as for a reflection from a large metal surface). As a result, the Delay-Locked Loop (DLL) in tracking is expected to have a mean pseudorange bias of 20 m. It will significantly affect the PVT solution.

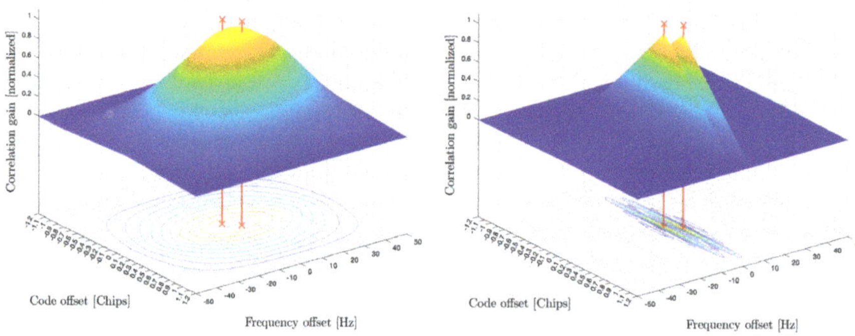

(a) (b)

Figure 2. Theoretical CAF with the LOS and NLOS peaks highlighted with the red points. (**a**) Normal correlation: 20 ms integration. (**b**) S-GNSS extended correlation: 500 ms.

Figure 2b shows the same scenario but extends the correlation to 0.5 s. Notice that the two peaks are separated, making the removal of the NLOS signals simple. It allows the pseudorange bias to be corrected by only accumulating correlation output values of several epochs. Therefore, this approach can be applied in post-processing to the correlator In-phase and Quadrature-phase (IQ) values of an existing receiver.

4. SDR Design and Performance

GNSS SDR receivers [16,17], also known as software receivers, are becoming more common due to the improvements in computing technology [18]. A typical approach is to use interpreted languages, such as Matlab® [19] or Python [20], as they facilitate rapid development and simple debugging but severely limit performance. Therefore, compiled languages, such as C++, which are significantly more processing-efficient but more challenging to design, are required for real-time performance. The SDR receiver in this project is developed with C++ and utilizes concurrency implemented through multi-threading for improved real-time processing capabilities.

Several open-source GNSS SDR projects are based on C++, such as GNSS-SDR (https://gnss-sdr.org/ (accessed on 8 May 2023)) and GNSS-SDRLIB (https://github.com/taroz/GNSS-SDRLIB (accessed on 8 May 2023)). However, these often have copyright limitations when it comes to commercial applications and code reuse. Therefore, it emphasizes the need for an in-house-developed but highly performant SDR for FPP. It allows development with fewer restrictions. Furthermore, developing an in-house SDR allows for the complete control and development of the entire GNSS processing chain. Therefore, future extensions of the Supercorrelation™ technology are possible.

The developed GNSS SDR interfaces to the NTLAB® NT1065-USB RFFE, de-interleaves the data stream, performs base-band signal conditioning, acquires L1 and E1 signals, and tracks L1-band and L5-band signals from GPS and Galileo. Currently, the S-GNSS processing and PVT calculation are conducted off-line with pre-existing software from FPP. However, these later stages will be integrated in the forthcoming months.

An Intel NUC computer with a 13th Gen Intel® Core™ i9-13900K processor is used for primary testing and development. It will also be the target device for later studies. The Central Processing Unit (CPU) has 24 cores supporting 32 threads and can be clocked up to 5.8 GHz. In the current version, it is capable of tracking at least 40 signals in real-time in live operation with three correlation taps each (i.e., complex early, prompt, and late). It simultaneously tracks ten signals each for GPS L1 C/A, GPS L5I, Galileo E1B, and Galileo E5aI. However, the development is still in the early optimization stages, and better performance and capabilities are expected in future development. Currently, S-GNSS is run in post-processing on the correlator IQ values from tracking, but the excess processing resources will facilitate the future integration of the S-GNSS for real-time processing. It is currently under development.

In the second performance test, the SDR is run on an Apple MacBook Pro with an M1 Pro processor and 32 GB Random Access Memory (RAM). In this test, a 100 s data recording is processed in 87 s using 24 tracking channels, each with 11 complex correlator taps. More taps accommodate superior signal reconstruction for S-GNSS. Both these cases showcase real-time processing with a significant number of signals and correlator taps.

5. Test Setup

Previous investigations with S-GNSS used real-world data [10,12–14]. Although these are great approaches to demonstrating the practical benefits of the technology, it does not allow for systematic performance analysis. Therefore, this paper uses a systematic approach using an RFCS. Figure 3 shows the test setup with a Spirent® GSS7000 (https://www.spirent.com/products/gnss-simulator-gss7000 (accessed on 13 May 2023)) RFCS to create the synthetic GNSS signals and the NTLAB® NT1065-USB (http://ntlab.com/section/sec:v:44979.htm (accessed on 13 May 2023)) RFFE to receive them.

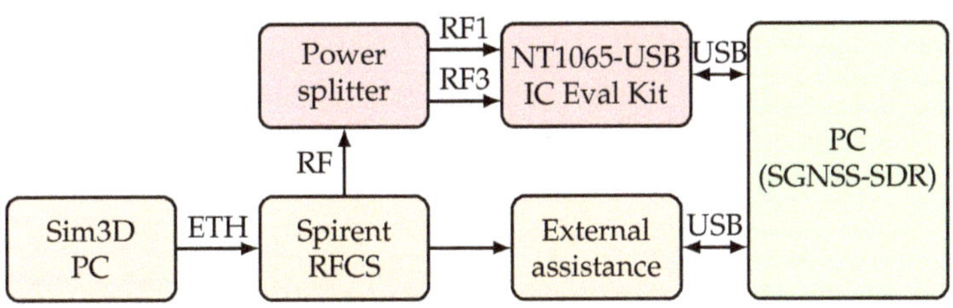

Figure 3. Hardware test setup for final tests.

A PC running Spirent® Sim3D™ (https://www.spirent.com/products/multipath-and-obscuration-simulation-sim3d (accessed on 13 May 2023)) Multipath Simulation Software, developed by Oktal-SE (https://www.oktal-se.fr/ (accessed on 13 May 2023)), generates a realistic multipath scenario. The Sim3D™ and GSS7000 combination allows for controllable and repeatable tests, which is ideal for systematic analysis and benchmarking. Furthermore, the same scenario can be run with and without multipath interference, allowing for a direct comparison with the same satellite constellation and trajectory to analyze the impact of multipath interference.

Figure 4 shows the generated scenario at the Trumpington Park and Ride near Cambridge, UK. This scenario is close to FPP's headquarters, allowing a future real-world recording to verify the scenario. The scenario starts in the car park and moves at a rate of 5 m/s toward a high building (a commercial delivery hub). The received NLOS intensity from the building is expected to increase as the receiver approaches the building. Once the receiver is near the building, it moves along briefly and then loops back to the starting point. This loop is repeated several times to allow for repeated comparisons of the multipath scenarios. Each loop takes just under a minute.

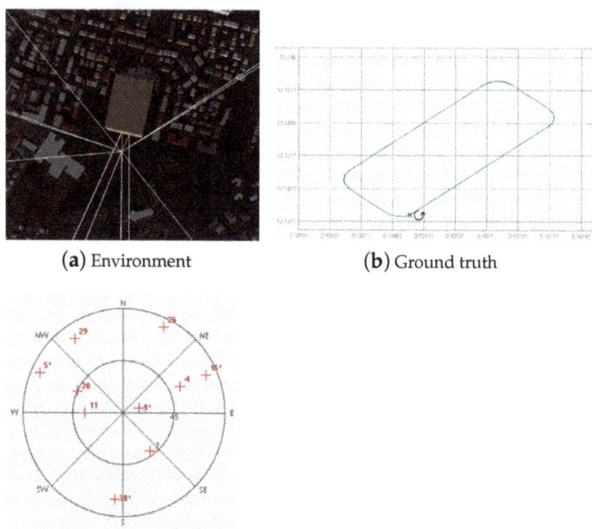

(**a**) Environment (**b**) Ground truth

(**c**) Skyplot

Figure 4. Generated scenario with SIM3D.

In the test setup, a scenario is recorded with NLOS signal components with the RFFE. GPS L1 CA and L5 signals and Galileo E1B/C and E5a signals are generated with the Spirent RFCS. The signals are processed with the GNSS SDR and evaluated in the next section.

6. Results

6.1. Scenario Tracking

A case study of two satellites demonstrates general tracking performance in this subsection and the improvement achievable with S-GNSS in the next one. The vehicular scenario described in Section 6 and in Figure 4a is used for this case study.

First, Figure 5 shows the tracking results of PRN07, which has a high elevation with high carrier-to-noise density ratio (C/N_0), resulting in generally good tracking performance. However, there are some disruptions (e.g., between 10 s and 20 s) caused by multipath interference when the receiver moves past the building to the west of the trajectory.

Figure 6 shows the tracking results of PRN30, which has a lower elevation and is opposite the large building, resulting in significant reflections from the building. Significant oscillations and noise are visible in tracking the observed multipath interference. It demonstrates how multipath interference degrades the receiver tracking performance. In particular, it shows that the multipath interference results in position-dependent distortions that exceed the effects of thermal noise in tracking. For example, compare the variance in the graphs between 25 s and 35 s (i.e., close to the building) with considerable multipath interference to the times between 45 s and 60 s (i.e., far from the building) with mostly thermal noise. As a result, the errors significantly influence later pseudorange determination, which will degrade the position solution. The purpose of Supercorrelation™ is to identify these multipath interference components and to suppress them.

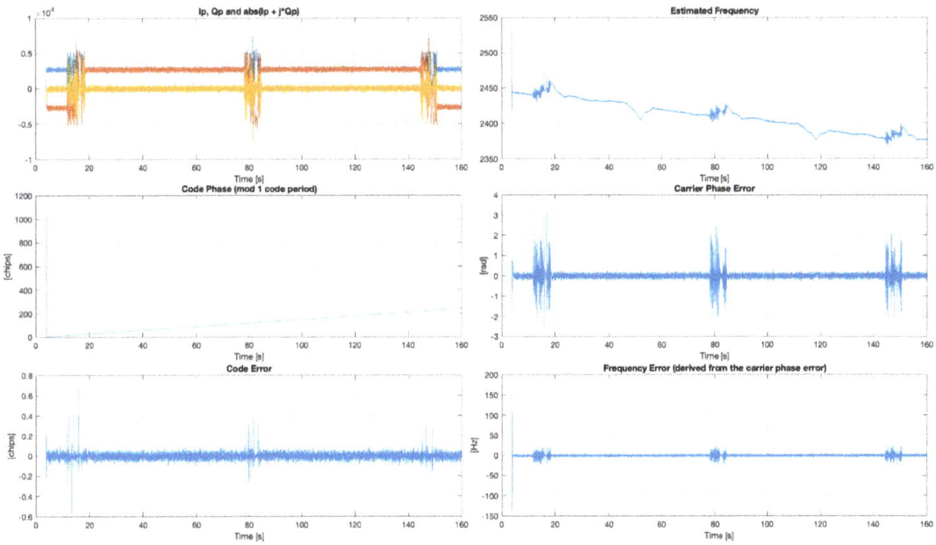

Figure 5. Tracking for PRN07.

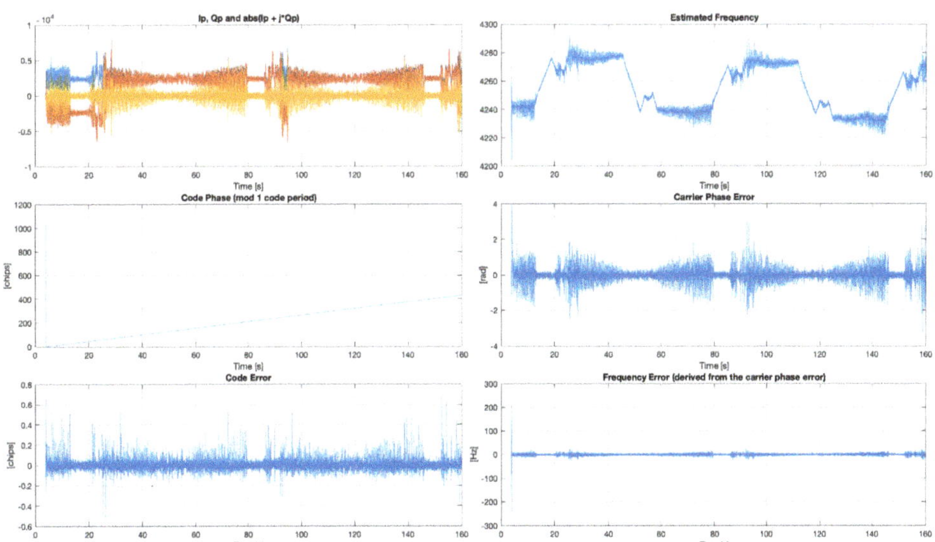

Figure 6. Tracking for PRN30.

6.2. S-GNSS Performance

S-GNSS processing is now applied to the scenario. Figure 7 shows the pseudorange errors over time along with the geometry of the satellites (highlighted in blue) and selected CAFs (blue is without S-GNSS correction, red is with) for PRN07. As this satellite is less frequently impacted by multipath interference, the CAF is not always improved by the S-GNSS processing (e.g., CAF at 87 s). However, there are still many cases—even with this high elevation satellite—where S-GNSS enhances the performance. It is emphasized in

Figure 8 where the error Cumulative Distribution Function (CDF) for the pseudorange and frequency measurements are displayed. The errors after S-GNSS processing are significantly reduced; for example, the 90% pseudorange error is reduced from 5 m to 2.4 m after applying S-GNSS.

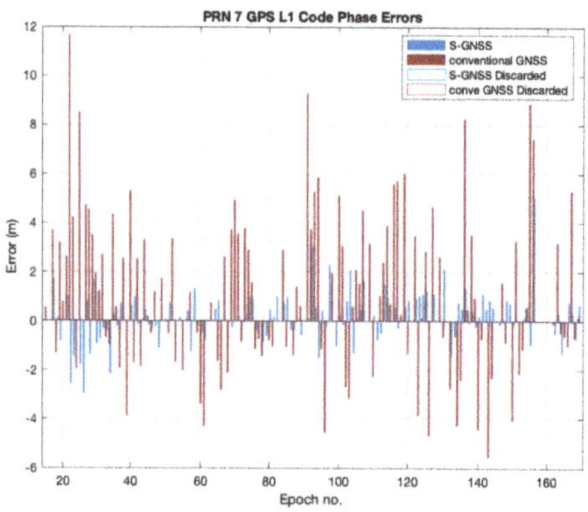

(**a**) Pseudorange error

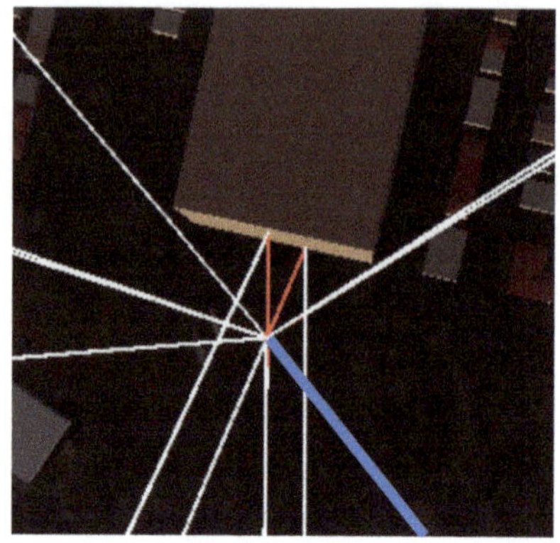

(**b**) Geometry

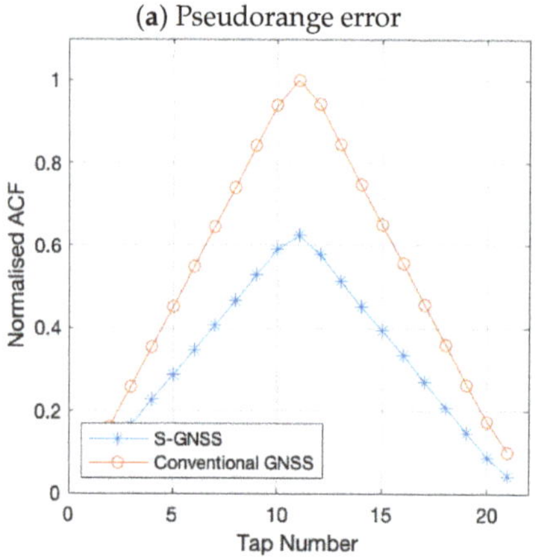

(**c**) CAF at 21 s

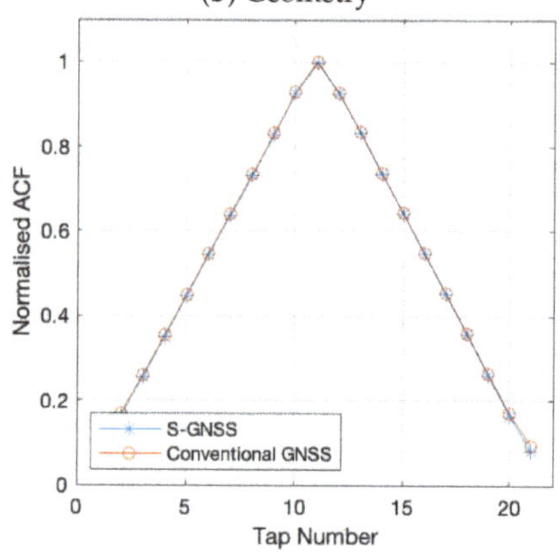

(**d**) CAF at 87 s

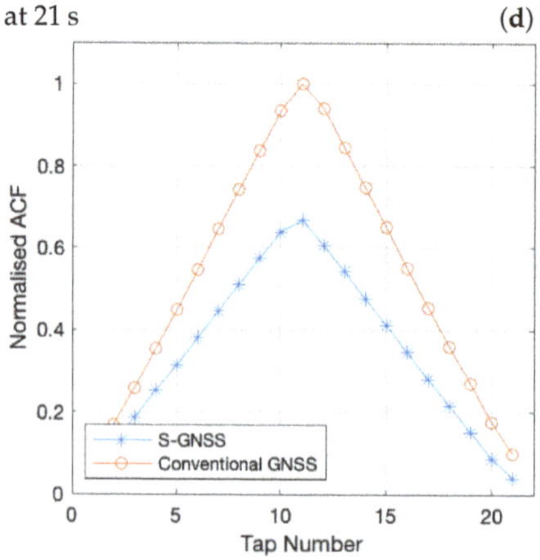

(**e**) CAF at 95 s

Figure 7. S-GNSS analysis for PRN07.

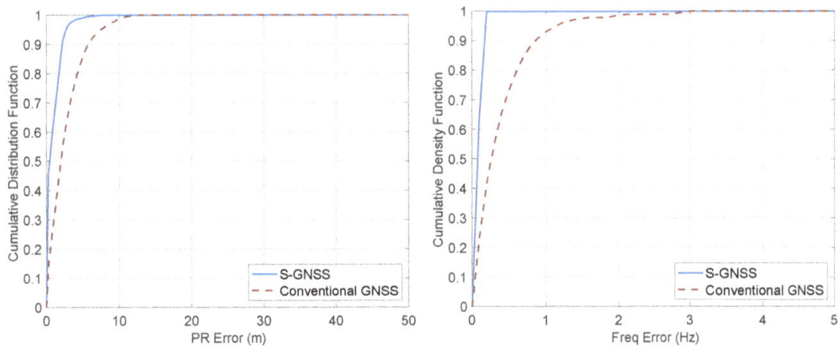

(a) Pseudorange (b) Frequency

Figure 8. S-GNSS error CDF for PRN07.

Figure 9 shows the processing results for PRN30, where the multipath interference is more present. In the CAFs, there are even multiple peaks visible. In this case, applying S-GNSS significantly improves the pseudorange and frequency measurements, as shown with the CAFs in Figure 10.

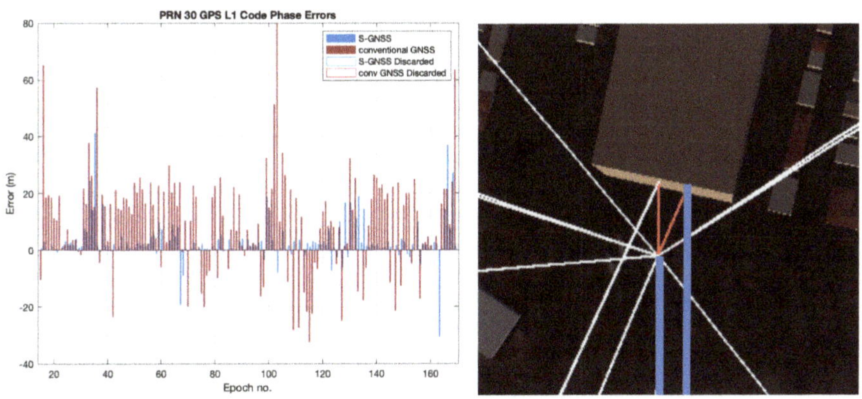

(a) Pseudorange error (b) Geometry

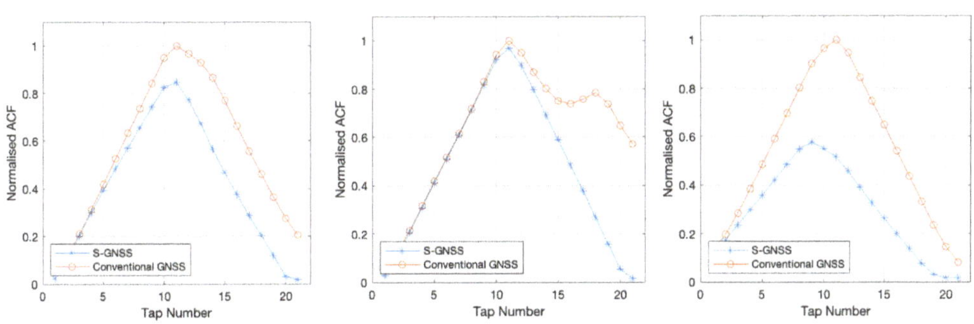

(c) CAF at 18 s (d) CAF at 62 s (e) CAF at 100 s

Figure 9. S-GNSS analysis for PRN30.

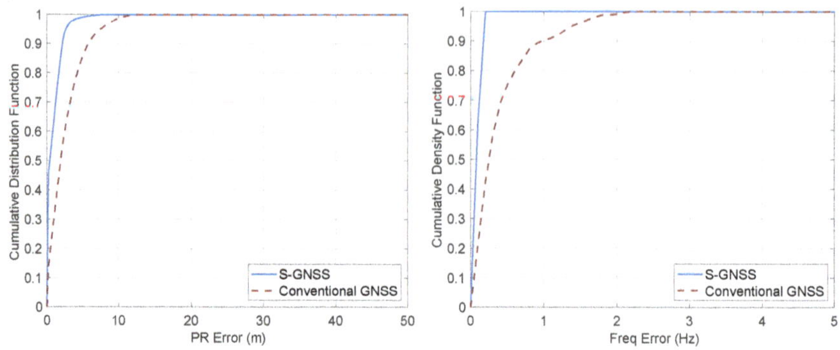

(**a**) Pseudorange (**b**) Frequency

Figure 10. S-GNSS error CDF for PRN30.

7. Conclusions

This paper showcases the current development of FPP's GNSS SDR software receiver capable of S-GNSS processing. Initial performance results are exhibited with a systematic testing methodology. This paper tests with controlled multipath interference scenarios and could compare with the ground truth, which is extremely challenging to do in real-world scenarios. Furthermore, Supercorrelation™ processing demonstrates improved position availability due to the multipath interference rejection. Future work includes a benchmarking comparison between Supercorrelation™ in multipath interference scenarios and nominal receiver performance in multipath-free scenarios.

The results show that the GNSS SDR software receiver is capable of multi-frequency multi-GNSS processing in real time. The S-GNSS performance showed that NLOS rejection is possible and improves PVT performance.

Future work includes further testing and evaluating the platform, as only the initial results are shown. Integration with the inertial measurements and deeper coupling with the S-GNSS tracking engine is also the next step on the project roadmap. Finally, some real-world evaluations are also planned.

Author Contributions: Conceptualization, J.G.G., J.R.v.d.M., P.E. and R.F.; methodology, J.G.G., P.E. and R.F.; software, J.G.G., D.J., S.B., C.H., R.G. and E.C.; validation, J.G.G., J.R.v.d.M. and P.E.; formal analysis, J.G.G., J.R.v.d.M. and P.E.; investigation, J.G.G.; resources, P.E. and R.F.; data curation, J.G.G., J.R.v.d.M. and P.E.; writing—original draft preparation, J.G.G. and J.R.v.d.M.; writing—review and editing, J.G.G., J.R.v.d.M., D.J., S.B., C.H., R.G., P.E., E.C. and R.F.; visualization, J.G.G., J.R.v.d.M. and P.E.; supervision, J.G.G. and P.E.; project administration, J.G.G., P.E. and R.F.; funding acquisition, P.E. and R.F. All authors have read and agreed to the published version of the manuscript.

Funding: This work was funded through the ESA project NAVISP-EL2-119.

Institutional Review Board Statement: Not applicable.

Informed Consent Statement: Not applicable.

Data Availability Statement: Data are available in this manuscript.

Conflicts of Interest: All the authors were employed by the company Focal Point Positioning Limited.

References

1. Inside GNSS. GNSS Supercorrelation to Increase Accuracy Is Basis of New Partnership of u-Blox, Focal Point. Inside GNSS. December 2020. Available online: https://insidegnss.com/gnss-supercorrelation-to-increase-accuracy-is-basis-of-new-partnership-of-u-blox-focal-point/ (accessed on 30 April 2023).
2. Focal Point Positioning. Supercorrelation Technical Overview White Paper. 2022. Available online: https://focalpointpositioning.com/ASSETS/IMG/header-img/FOCALPOINT_Supercorrelation_Technical_Overview.pdf (accessed on 13 May 2023).

3. Bochkati, M.; Dampf, J.; Pany, T. Receiver Clock Estimation for RTK-Grade Multi-GNSS Multi-Frequency Synthetic Aperture Processing. In Proceedings of the 2023 IEEE/ION Position, Location and Navigation Symposium (PLANS), Monterey, CA, USA, 24–27 April 2023.
4. Luo, Y.; Hsu, L.T.; El-Sheimy, N. A Baseband MLE for Snapshot GNSS Receiver Using Super-Long-Coherent Correlation in a Fractional Fourier Domain. *NAVIGATION J. Inst. Navig.* **2023**, *70*, 588. [CrossRef]
5. Abdallah, A.A.; Kassas, Z.M. Multipath Mitigation via Synthetic Aperture Beamforming for Indoor and Deep Urban Navigation. *IEEE Trans. Veh. Technol.* **2021**, *70*, 8838–8853. [CrossRef]
6. Faragher, R.; Couronneau, N.; Crockett, M.; Duffett-Smith, P. Method and System for Correcting the Frequency or Phase of a Local Signal Generated Using a Local Oscillator. CN111065937A, 4 April 2020; GB2564406B, 7 September 2023; US10816672B2, 27 October 2020; WO2019008327A1, 10 January 2019.
7. Faragher, R.; Couronneau, N.; Crockett, M.; Duffett-Smith, P. Method, Apparatus, Computer Program, Chip Set, or Data Structure for Correlating a Digital Signal and a Correlation Code. US11137500B2, 5 October 2021; US2017279486A1, 24 March 2016; US2019086553A1, 21 March 2007; US9780829B1, 3 October 2017.
8. Crockett, M.R.; Couronneau, N.; Faragher, R. A Method, Apparatus, Computer Program, Chip Set, or Data Structure for Correlating a Digital Signal and a Correlation Code. GB2548620A, 2 June 2021; GB2548620B, 1 January 2020.
9. Faragher, R.M.; Crockett, M.R.; Duffett-Smith, P. A Method, System and Computer Program Product for Performing Correlation in a Positioning System. CN115667997A, 25 January 2023; EP4121797A1, 25 January 2023; GB2593459A, 29 September 2021; KR20220156856A, 28 November 2022; US2023009945A1, 12 January 2023; WO2021186194A1, 31 January 2023.
10. Faragher, R. Robust Positioning for Autonomous Systems in Cities. In Proceedings of the 33rd International Technical Meeting of the Satellite Division of the Institute of Navigation (ION GNSS+ 2020), Online, 22–25 September 2019; pp. 2372–2388. [CrossRef]
11. Groves, P.D.; Zhong, Q.; Faragher, R.; Esteves, P. Combining Inertially-aided Extended Coherent Integration (Supercorrelation) with 3D-Mapping-Aided GNSS. In Proceedings of the 33rd International Technical Meeting of the Satellite Division of the Institute of Navigation (ION GNSS+ 2020), Online, 22–25 September 2022; pp. 2327–2346. [CrossRef]
12. Faragher, R. Ubiquitous Navigation using S-GPS and D-Tail. In Proceedings of the 30th International Technical Meeting of the Satellite Division of the Institute of Navigation (ION GNSS+ 2017), Portland, OR, USA, 25–29 September 2017; pp. 328–344. [CrossRef]
13. Faragher, R.; Powe, M.; Esteves, P.; Couronneau, N.; Crockett, M.; Martin, H.; Ziglioli, E.; Higgins, C. Supercorrelation as a Service: S-GNSS Upgrades for Smartdevices. In Proceedings of the 32nd International Technical Meeting of the Satellite Division of the Institute of Navigation (ION GNSS+ 2019), Miami, FL, USA, 16–20 September 2019; pp. 2372–2388. [CrossRef]
14. Faragher, R.; Couronneau, N.; Powe, M.; Esteves, P.; Crockett, M.; Martin, H.; Ziglioli, E.; Higgins, C.; Buckle, D. Supercorrelation: Enhancing the Accuracy and Sensitivity of Consumer GNSS Receivers with a DSP Upgrade. In Proceedings of the 31st International Technical Meeting of the Satellite Division of the Institute of Navigation (ION GNSS+ 2018), Miami, FL, USA, 24–28 September 2018; pp. 357–375. [CrossRef]
15. Focal Point Positioning. Announcing Strategic Investment from GM Ventures and Collaboration with General Motors on Integration of Next Generation GPS Technologies. 2023. Available online: https://focalpointpositioning.com/insights/focalpoint-collab-with-general-motors-on-integration-of-next-generation-gps-technologies (accessed on 5 January 2023).
16. Pany, T. *Navigation Signal Processing for GNSS Software Receivers*; GNSS Technology and Applications Series; Artech House: Boston, MA, USA; London, UK, 2010.
17. Akos, D.; Arribas, J.; Bhuiyan, M.Z.H.; Closas, P.; Dovis, F.; Fernandez-Hernandez, I.; Fernández-Prades, C.; Gunawardena, S.; Humphreys, T.; Kassas, Z.; et al. GNSS Software Defined Radio: History, Current Developments, and Standardization Efforts. In Proceedings of the 35th International Technical Meeting of the Satellite Division of the Institute of Navigation (ION GNSS+ 2022), Denver, CO, USA, 19–23 September 2022; pp. 3180–3209. [CrossRef]
18. Nichols, H.; Murrian, M.; Humphreys, T. Software-Defined GNSS is Ready for Launch. In Proceedings of the 35th International Technical Meeting of the Satellite Division of the Institute of Navigation (ION GNSS+ 2022), Denver, CO, USA, 19–23 September 2022; pp. 996–1013. [CrossRef]
19. Borre, K.; Fernández-Hernández, I.; López-Salcedo, J.A.; Bhuiyan, M.Z.H. (Eds.) *GNSS Software Receivers*; Cambridge University Press: Cambridge, UK, 2022. [CrossRef]
20. Gunawardena, S. A High Performance Easily Configurable Satnav SDR for Advanced Algorithm Development and Rapid Capability Deployment. In Proceedings of the 2021 International Technical Meeting of The Institute of Navigation, Online, 2–28 January 2021; pp. 539–554. [CrossRef]

engineering proceedings

MDPI

Proceeding Paper

Achieving Sub-Decimeter Accuracy with the Galileo High Accuracy Service: Results from GMV's HAS Positioning Engine †

Adrián Chamorro, Jorge Rocamora, Simón Cancela and Jesús David Calle *

GMV, 28760 Tres Cantos, Spain; achamorro@gmv.com (A.C.); jrocamora@gmv.com (J.R.); scancela@gmv.com (S.C.)
* Correspondence: jdcalle@gmv.com

† Presented at the European Navigation Conference 2023, Noordwijk, The Netherlands, 31 May–2 June 2023.

Abstract: High-accuracy (HA) positioning services allow users to achieve sub-decimeter-level positioning accuracy. Although these kinds of services are not new, the market is showing great interest in exploiting them for new applications within the mass-market domain. This growing interest is causing a change in the paradigm of HA services, moving from niche sectors to applications targeting billions of users. Considering this framework, the Galileo High Accuracy Service (HAS) provides an open-access service based on the provision of high-accuracy corrections transmitted through the Galileo E6-B signal (E6, data component). The data retrieved by the end-user, which includes orbit, clock, and bias corrections, is reconstructed to allow the computation of Precise Point Positioning (PPP) solutions. This paper is focused on the description and results of GMV's HAS Positioning Engine (HAS-PE) software library which, implements a PPP solution using the HAS corrections transmitted through Galileo Signal-in-Space. A high-level overview of the integration of the HAS in the Positioning Engine software is presented together with user performance assessments based on static and kinematic tests executed to process real data from GNSS receivers in real time. The static tests are performed using the GMV Global station network, which consists of geodetic grade receivers tracking the signal in open-sky locations around the globe. The kinematic tests are performed with a setup consisting of a mass-market receiver and a high-end receiver for obtaining the E6 pages. The PPP solutions are configured to process both Galileo and GPS corrections transmitted through the Galileo satellites. The assessment performed includes the computation of a set of performance indicators aimed at the analysis of high-accuracy positioning performances. The results of this assessment show that a PPP user taking advantage of this Galileo HAS initial service may easily achieve decimeter-level accuracy on the horizontal and vertical components.

Keywords: HAS; high accuracy; Galileo; IMU; GSHARP

Citation: Chamorro, A.; Rocamora, J.; Cancela, S.; Calle, J.D. Achieving Sub-Decimeter Accuracy with the Galileo High Accuracy Service: Results from GMV's HAS Positioning Engine. *Eng. Proc.* **2023**, *54*, 10. https://doi.org/10.3390/ENC2023-15427

Academic Editors: Tom Willems and Okko Bleeker

Published: 29 October 2023

1. Introduction

The declaration of the Initial Service phase for the High Accuracy Service (HAS) on 24 January 2023 marked a significant milestone in the democratization of high-precision GNSS capabilities for users worldwide [1]. This service, provided free of charge, is globally accessible via a signal in space (corrections are transmitted through the E6-B signal) and via an internet connection (corrections are also available through an NTRIP Caster available by registration). Due to its extensive coverage and cost-free nature, the stakeholders within the sector are likely to show considerable interest in implementing the use of this service to enhance the performance of their solutions. This encompasses receiver manufacturers, who may be interested in offering fully integrated solutions with HAS capabilities, as well as end-users, who may seek software positioning solutions compatible with the Galileo HAS service.

Considering this high potential interest of the users, GMV has developed a HAS Position Engine that is able to exploit all the potential of the HAS service. This paper presents the GMV's HAS Position Engine, including an overview of the software, together with results in real conditions under different scenarios. It ends with the conclusions of the product and the possibilities that are offered.

2. GMV'S HAS Position Engine Overview

The GMV's HAS Position Engine is a software library designed to integrate a Precise Point Positioning (PPP) solution together with the decoding and application of Galileo HAS corrections. This engine can operate in real time, utilizing GMV's Real Time wrapper, or can be integrated by users into their own platforms. This integration is facilitated through an API, which allows data loading, configuration of some parameters, and retrieval of high-accuracy positioning output.

The GMV's HAS Position Engine supports multiple formats for observations and navigation, including RTCM, SBF, and u-Blox. Additionally, it is equipped to retrieve raw navigation pages containing E6 data using the GalRawCNAV SBF message.

The PPP solution of the GMV's HAS Position Engine is based on the GMV GSHARP [2] PPP implementation, which also includes some advanced features such as IAR (Integer Ambiguity Resolution), compatibility with ionospheric corrections, and integrity features. In addition, the GMV's HAS Position Engine includes a tight-coupling integration with six-axis IMU measurements, which provides dead-reckoning features. The integration with IMU measurements is optional, so the users can benefit from this integration in setups where there is an IMU available, but they are not limited to setups with an IM. This complete implementation is designed for efficiency, being both single-threaded and lightweight in terms of CPU consumption and memory usage, capable of running at 10 Hz on low-end processors while keeping RAM consumption below 80 MB.

Figure 1 provides a high-level overview of the GMV's HAS Position Engine library's architecture. Components in red are the main components of the GMV's HAS Position Engine, which are the HAS decoder and the GSHARP Kalman Filter to compute the positioning. Components in grey are supporting components and white boxes provide a description of the information flow that is represented by each arrow. The users have several options to integrate the GMV's HAS Position Engine. They can use a standalone version, which also manages the communication channels for input acquisition (compatible with TCP, NTRIP, and Serial communication), or the client can integrate only the GMV's HAS Position Engine library and manage the communication channels by their own means for a higher level of integration.

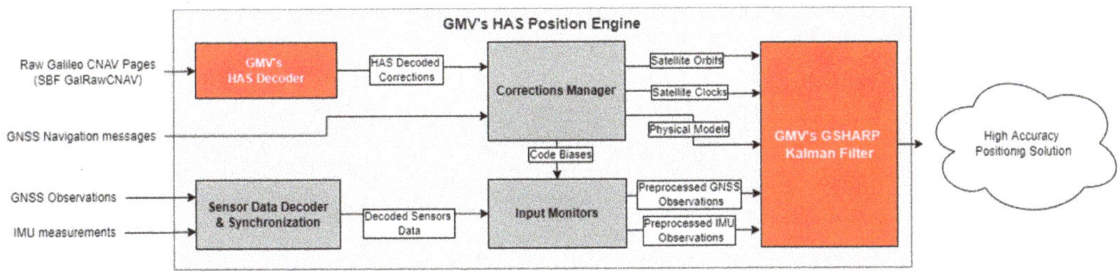

Figure 1. GMV's HAS Position Engine overview.

Once the inputs are loaded into the library, the processing begins. The Galileo CNAV pages are accumulated and decoded as soon as the necessary preconditions for applying the Reed–Solomon algorithm from the HAS are met [3]. In the first step of the process, GNSS observations and IMU measurements are decoded and synchronized. Subsequently, a set of input monitors preprocesses and checks the measurements for anomalies (high noise, jumps, cycle slips, etc.) before they are introduced into the GSHARP Kalman Filter. The Corrections Manager applies the HAS corrections to the navigation messages and manages all the physical models used to compute the residuals for the Kalman Filter. Ultimately, the Kalman Filter takes the preprocessed measurements from the IMU and the GNSS receiver, along with the information supplied by the Corrections Manager, to compute a High Accuracy Solution.

3. Performance Analysis

This section provides an assessment of the performance of GMV's HAS Position Engine under various scenarios. We will outline the specific test setups employed and present the results obtained. Furthermore, a comparative analysis with state-of-the-art positioning alternatives is conducted to highlight both the strengths and limitations of GMV's HAS Position Engine.

3.1. Performance Analysis in Static Scenarios

The starting point of the performance analysis will be the static scenarios. In these scenarios, the setup that has been used is based on the GMV's GGRN (Global GNSS Reference Network). The GMV's GGRN is a private network of GNSS reference stations that is composed of 40 redundant stations with multi-constellation and multifrequency support with a stable timing reference thanks to Atomic Frequency Standards [4]. The distribution of the stations around the globe and the typical setup of the antenna and the environment of the stations are provided in Figure 2.

(**a**)

(**b**)

Figure 2. (**a**) GMV's network of GNSS stations; (**b**) typical GMV GGRN station environment.

Observations, decoded navigation data, and raw E6 navigation pages were collected through an NTRIP caster to feed the GMV's HAS Position Engine in real time for the static tests. This information was obtained using the RTCM protocol for observations and decoded navigation messages (messages 1019, 1045, 1077, and 1097), while the raw navigation pages were acquired using the SBF format (GalRawCNAV message).

Previous analyses performed with a subset of GMV GGRN have shown the performance of the Galileo HAS using the GMV's HAS Position Engine during the HAS System Validation phase [5]. Nowadays, the HAS is currently operational in a remarkable way and the results and conclusions of that paper have been consistent during the Initial Service phase. Therefore, for the sake of simplicity, in this case, results from only two of the stations (one of them in Spain (station code: SPTR) and the other in the USA (station code: USNA)) of the network will be shown. The results presented consist of 48 h of continuous running and they have been extracted from longer runs to show the results after the convergence period because for a static user, this is the most interesting figure and including the convergence period can lead to misleading results in this use case. For reference, the typical convergence time for a static PPP scenario using HAS corrections is around 30 min. The horizontal and vertical positioning errors of these tests, together with the number of satellites that are used at every moment (blue line represents the number of satellites which are being used and green line represents the number of satellites used by the Ambiguity Fixing algorithm, which is only available for GMV correction scenario used as reference), are shown in Figures 3 and 4. The RMS and 95th percentile of the results have been collected in Tables 1 and 2 to summarize the results from the previous figures.

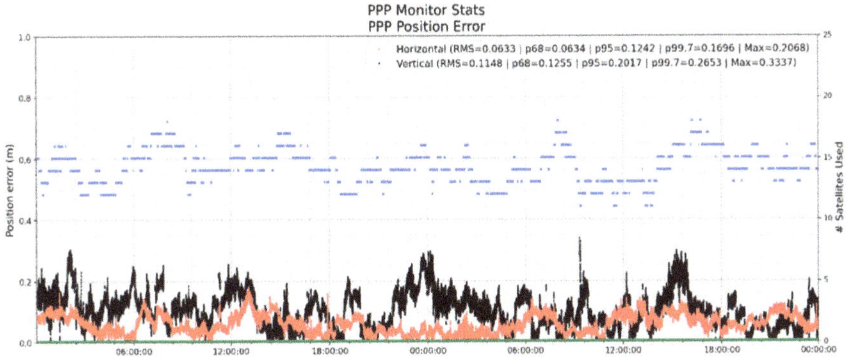

Figure 3. Results of a 48 h continuous run (9 May 2023 and 10 May 2023) with the GMV's HAS Position Engine in a GMV GGRN station in Madrid (Spain).

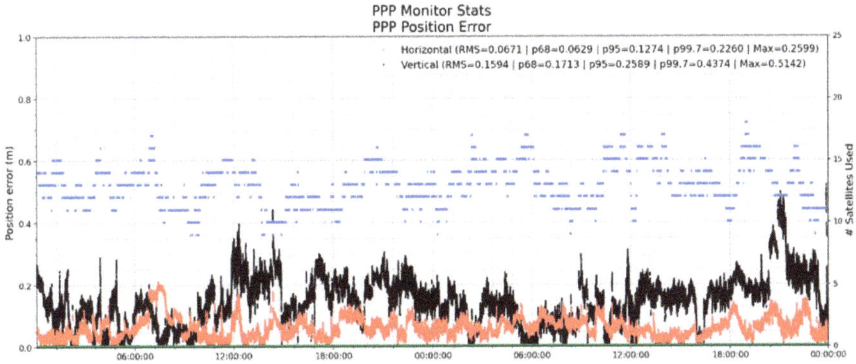

Figure 4. Results of a 48 h continuous run (9 May 2023 and 10 May 2023) with the GMV's HAS Position Engine in a GMV GGRN station in New Jersey (U.S.A.).

Table 1. RMS of positioning error for the static scenarios of the two stations considered (SPTR: Madrid, Spain; USNA: New Jersey, U.S.A.).

Errors RMS (cm)	Horizontal	Vertical
SPTR	6.3	11.5
USNA	6.7	15.9

Table 2. The 95th percentile of positioning error for the static scenarios of the two stations considered (SPTR: Madrid, Spain; USNA: New Jersey, U.S.A.).

Errors p95 (cm)	Horizontal	Vertical
SPTR	12.4	20.1
USNA	12.7	25.9

3.2. *Performance Analysis in Kinematic Scenarios*

The second scenario that will be shown is a kinematic scenario. The setup and the route that has been followed are described and then the results are presented.

The setup used is based on a Ublox F9P and an Asterx-4 Septentrio receiver. Other components of the setup are a RaspberryPi 4, where the GMV's HAS Position Engine is installed, an automotive grade IMU from ST Electronics (ASM330LHH), and a Novatel PwrPak7 for obtaining the ground truth. A schematic view of the setup can be found in Figure 5. Regarding communication, the u-blox F9P is connected via USB to the RaspberryPi and transmits RTCM messages, the IMU uses the I2C protocol and uses RTCM proprietary messages to send the IMU measurements to the RaspberryPi, and the Septentrio Asterx-4 is connected via ethernet to the RaspberryPi and transmits the raw E6 navigation pages using the GalRawCNAV SBF message. In this case, the rate of GNSS observations is 10 Hz, which is a more suitable rate for automotive or similar use cases.

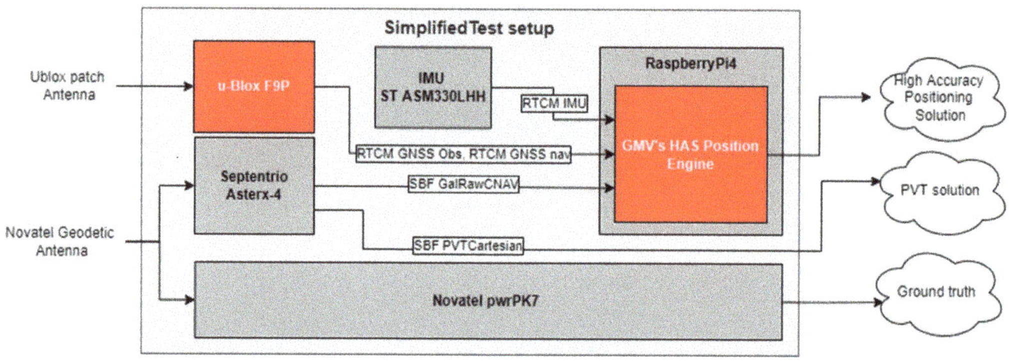

Figure 5. Test setup for the kinematic scenarios. Red boxes are the main components of the setup (mass-market receiver and GMV's software), grey boxes are the supporting components for the study (IMU for hybridization test, Septentrio Asterx as reference receiver and E6 pages recording, ground truth system and Raspberry Pi as processing unit for the GMV software) and white boxes represent the information flow details including its format.

The reason behind using two GNSS receivers is that the usage of a mass-market receiver together with a mass-market antenna has been preferred over a high-end setup in order to deepen the assessment of the possibilities of democratization of the high-accuracy solutions for the growth of this market that has been forecasted [6]. However, nowadays there are not many mass-market receivers that can track the E6 signal for obtaining the HAS corrections yet, and a high-end receiver has been used for this purpose only (another equivalent setup would be to use the mass-market receiver and obtain the HAS corrections from the HAS ground NTRIP Caster from GSC). Showing the results that are obtained with this kind of receiver together with HAS corrections could be an interesting input for the market to assess the possibilities of increasing the offer of mass-market-grade receivers with E6 capabilities.

The route, illustrated in Figure 6, commenced at GMV premises and consisted of a ~40 min journey from 07:33 UTC to 08:16 UTC along the M-607 highway and within the city of Tres Cantos (Madrid). The majority of the environment was under highway conditions and the last part of the scenario (from 08:06 UTC) to the end of the scenario can be considered mid-urban conditions. Figure 7 shows an example of the environmental conditions that are found in the mid-urban conditions.

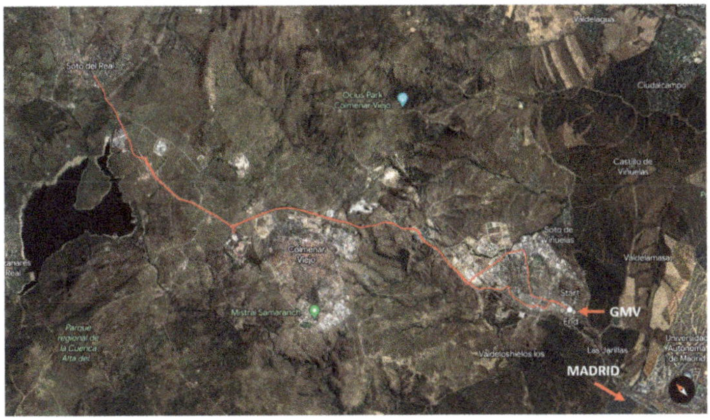

Figure 6. Highway and semi-urban kinematic trajectory.

Figure 7. Google Street View of one of the streets of the kinematic scenarios (Tres Cantos).

The positioning error of the GMV's Position Engine, operating with HAS corrections and a Ublox F9P receiver, is depicted in Figure 8. Performance degradation at ~07:36:30 is caused by a reconvergence due to a GNSS outage caused by a street with dense foliage.

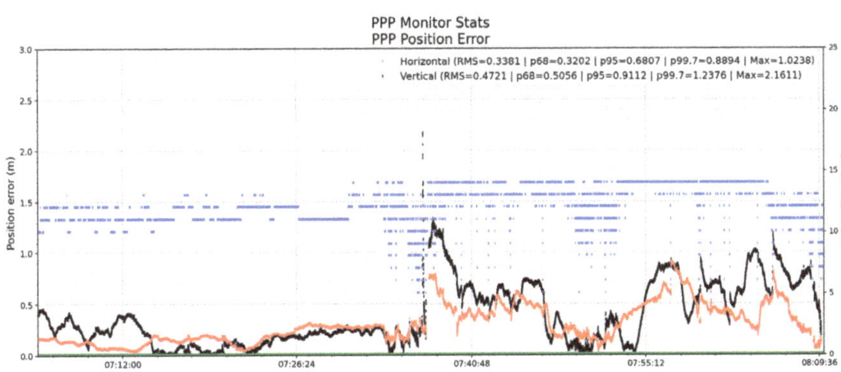

Figure 8. Positioning error of the GMV's HAS Position Engine using HAS corrections and observations from u-Blox F9P receiver (9 May 2023).

3.3. Performance in Kinematic Scenarios Using IMU Hybridization

In addition to GNSS-only scenarios, we also assessed the performance of GMV's HAS Position Engine using Inertial Measurement Unit (IMU) hybridization to enhance the positioning solution. The GMV's HAS Position Engine implements a tightly coupled IMU hybridization that provides dead-reckoning capabilities and also helps to mitigate some of the discontinuities that are observed in the GNSS-only solutions. As can be seen in Figure 9, the positioning error is smoother in this case and the performance is generally improved. As you can see, the Galileo HAS can be complemented with other sensors for obtaining synergies and better overall performance.

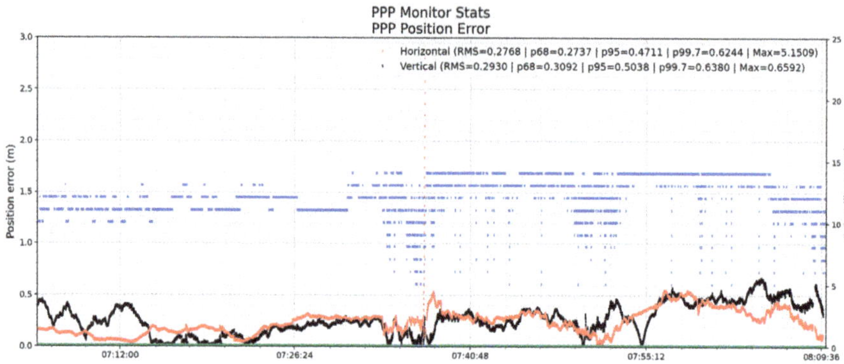

Figure 9. Positioning error of the GMV's HAS Position Engine using HAS corrections, observations from u-Blox F9P receiver, and IMU hybridization (9 May 2023).

3.4. Comparison with Other State-of-the-Art Solutions

The results from the previous scenarios can be complemented with a comparison with other state-of-the-art solutions. This comparison can help interpret and extract conclusions about the performance of the GMV HAS Position Engine. Two different solutions have been selected for this comparison:

- Asterx-4 PVT: This solution was selected for comparison with a state-of-the-art standalone PVT solution using a commercial receiver.
- GSHARP solution: This solution was selected for comparison with another high-accuracy solution. In this case, in addition to the orbit, clocks, and code biases, ionospheric corrections and phase biases are also available. This allows the implementation of additional techniques such as Integer Ambiguity Resolution to obtain better accuracy figures and to highly reduce the convergence time. However, an internet connection is needed to receive the corrections.

The results using these two solutions together are provided in Figures 10 and 11, respectively. In addition, a comparison collecting the results from all the kinematic scenarios has been added in Tables 3 and 4.

Table 3. RMS of positioning error for the kinematic scenarios comparing the different solutions presented.

Errors RMS (cm)	Horizontal	Vertical
GMV HAS PE	33.8	47.2
GMV HAS PE + IMU	27.7	29.3
State-of-the-art PVT	122.7	105.7
GMV GSHARP	8.9	11.2

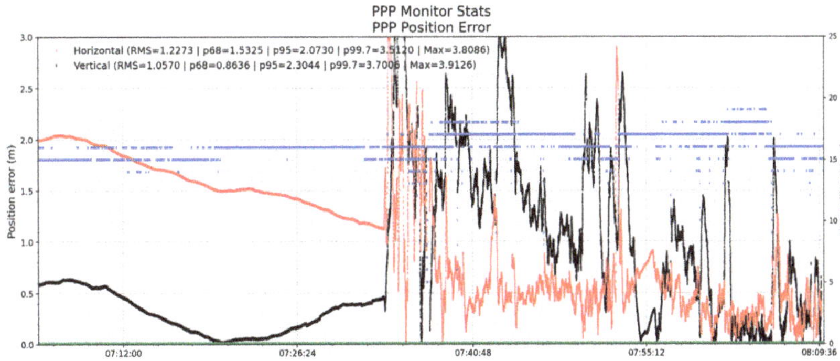

Figure 10. Positioning error of the Asterx-4 standalone PVT solution (9 May 2023).

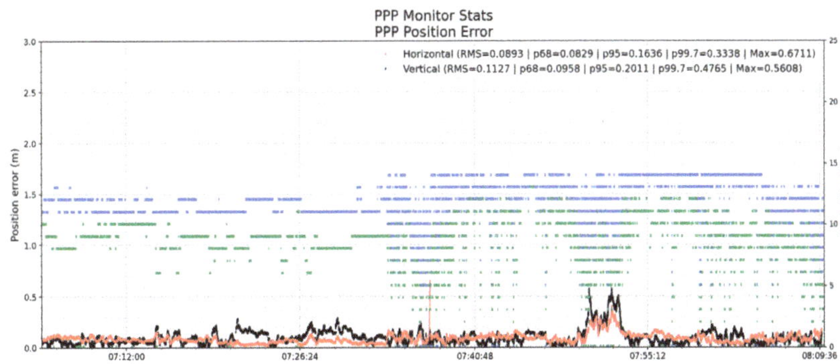

Figure 11. GMV's GSHARP Solution (9 May 2023). Green line indicates the number of fixed satellites by the Integer Ambiguity Resolution algorithm.

Table 4. The 95th percentile of positioning error for the kinematic scenarios comparing the different solutions presented.

Errors p95 (cm)	Horizontal	Vertical
GMV HAS PE	68.1	91.1
GMV HAS PE + IMU	47.1	50.4
State-of-the-art PVT	207.3	230.4
GMV GSHARP	16.4	20.1

4. Conclusions

This paper has introduced and evaluated the performance of GMV's HAS Position Engine under various static and kinematic scenarios. The findings highlight that the GMV's HAS Position Engine delivers high-accuracy solutions under these conditions.

From the user perspective, the benefits of using Galileo HAS in conjunction with applications like GMV's HAS Position Engine are remarkable. This is due to the improved accuracy it provides when compared with other standalone solutions that are commonplace in commercial applications. Moreover, the standalone nature of the Galileo HAS and GMV's HAS Position Engine is an added advantage as it liberates users from dependency on an internet connection, which can be a limited factor for certain applications or setups. On the other hand, it is important to acknowledge the current limitation that the availability of the E6 signal poses for a significant proportion of mass-market receivers, with wide availability currently confined to high-end receivers.

For applications with more demanding accuracy or convergence time requirements, advanced commercial solutions such as GMV's GSHARP, which include additional features like ionospheric corrections and Integer Ambiguity Resolution, are necessary to reach the desired performance level.

In conclusion, the Galileo HAS, together with the GMV's HAS Position Engine, can offer significant advantages to users for a number of applications. Nowadays, the users who can benefit from using E6-capable receivers can adopt this kind of solution in a standalone way, but mass-market applications can also benefit from using the HAS corrections via the NTRIP caster.

Author Contributions: Conceptualization, A.C., S.C. and J.D.C.; methodology, A.C. and S.C. and J.D.C.; software, A.C., S.C. and J.R.; validation, A.C.; formal analysis, A.C. and J.D.C.; investigation, A.C.; resources, J.D.C.; data curation, A.C.; writing—original draft preparation, A.C.; writing—review and editing, A.C. and J.D.C.; visualization, A.C.; supervision, J.D.C. and S.C.; project administration, J.D.C.; funding acquisition, J.D.C. All authors have read and agreed to the published version of the manuscript.

Funding: This research received no external funding.

Institutional Review Board Statement: Not applicable.

Informed Consent Statement: Not applicable.

Data Availability Statement: Data sharing is not applicable to this article.

Conflicts of Interest: All the authors were employed by the company GMV.

References

1. European GNSS Service Centre. Galileo High Accuracy Service. Available online: https://www.gsc-europa.eu/galileo/services/galileo-high-accuracy-service-has (accessed on 28 October 2023).
2. GMV GSHARP. GMV. Available online: https://www.gmv.com/en/products/space/gmv-gsharp (accessed on 28 October 2023).
3. European Union. *Galileo HAS SIS ICD*; Issue 1.0; European Union: Brussels, Belgium, 2022.
4. Global GNSS network data provision service. GMV. Available online: https://magicgnss.gmv.com/magicGNSS_Reference_Network.pdf (accessed on 28 October 2023).
5. Chamorro, A.; Cancela, S.; García, A.J.; Calle, D.; Fernandez-Hernandez, I.; de Blas, F.J.; Vazquez, J.; Hernández, C.; Fernandez-Hernandez, I. Early Demonstration of Galileo HAS User Performances. In Proceedings of the 35th International Technical Meeting of the Satellite Division of The Institute of Navigation (ION GNSS+ 2022), Denver, CO, USA, 19–23 September 2022.
6. EUSPA. EUSPA EO and GNSS Market Report. 2022. Available online: https://www.euspa.europa.eu/sites/default/files/uploads/euspa_market_report_2022.pdf (accessed on 28 October 2023).

engineering proceedings

Proceeding Paper

Addressing the Potential of L5/E5a Signals for Road ITS Applications in GNSS-Harsh Environments †

Amarildo Haxhi *, Manos Orfanos, Harris Perakis and Vassilis Gikas

School of Rural, Surveying and Geoinformatics Engineering, National Technical University of Athens, 9 Iroon Polytechneiou Str., 157 80 Zographos, Attic, Greece; orfanos_emmanouhl@mail.ntua.gr (M.O.); hperakis@central.ntua.gr (H.P.); vgikas@central.ntua.gr (V.G.)

* Correspondence: a_haxhi@mail.ntua.gr

† Presented at the European Navigation Conference 2023, Noordwijk, The Netherlands, 31 May–2 June 2023.

Abstract: This study explores the potential of satellite signals L5, E5a and B2a tracked by contemporary Android smartphones. Particularly, the objective is to investigate their performance capabilities and vulnerabilities concerned with L1, E1 and B1 bandwidths and a focus on land vehicle ITS (Intelligent Transportation Systems) applications aiming to address low to medium PVT (Positioning, Velocity and Timing) solutions. In this regard raw, kinematic GNSS measurements from two Android smartphones were collected (Xiaomi Mi 8 and One Plus Nord 2 5G) under GNSS-harsh environments. The Single Point Positioning (SPP) technique was adopted for processing the single-frequency, multi-constellation raw GNSS measurements through an Extended Kalman Filter (EKF). The results obtained indicate the benefits and difficulties of exploiting modernized GNSS signals for road ITS applications.

Keywords: GNSS raw measurements; Single Point Positioning; L5/E5a/B2a signals; multipath; Intelligent Transportation Systems

Citation: Haxhi, A.; Orfanos, M.; Perakis, H.; Gikas, V. Addressing the Potential of L5/E5a Signals for Road ITS Applications in GNSS-Harsh Environments. *Eng. Proc.* **2023**, *54*, 11. https://doi.org/10.3390/ENC2023-15430

Academic Editors: Tom Willems and Okko Bleeker

Published: 6 December 2023

1. Introduction

With the release of Android Nougat in August 2016, the raw GNSS measurements on L1 (1575.42 MHz) became available to users on an increasing number of low-cost devices. Since then, research from various centers worldwide have tested and analyzed the performance capability of their properties including the satellite elevation angle, the received C/N0 (Carrier-to-Noise density ratio), the measurement accuracy with respect to high grade receivers, the sensitivity to cycle slips and multipath [1]. Published results [1] have pointed out that GNSS measurements on the L1 frequency are C/N0 dependent, featuring values approx. 10 dBHz less than those from high-grade receivers, suggesting that the adoption of C/N0 weighting schemes being a more appropriate approach. Concerning the satellite signal multipath, it was demonstrated that GNSS measurements on L1 are affected more severely than those in high grade receivers, especially in GNSS-harsh environments [1]. However, the main challenge for the quality of the GNSS measurement is that the embedded GNSS antenna in the smartphones lead to L1 pseudorange accuracies of the order of 4–5 m [2].

Following these developments, an increasing number of smartphones featuring dual-frequency capabilities was developed. Concerning L5, E5a and B2a, all three signals can be tracked from the GPS, Galileo and BeiDou satellites, respectively. L5/E5a/B2a signals are superior to L1/E1/B1 ones as a result of their design specifications [1,3]. For example, the chipping rate of L5/E5a/B2a's is higher than that of the L1/E1/B1 leading to more precise measurements. The lower level of observation noise in the code observables and their resilience against the multipath for GPS/Galileo/BeiDou L5/E5a/B2a signals has been validated through field surveys using various smartphones [1]. Contrarily, fever satellites are being tracked on L5/E5a/B2a than on L1/E1/B1, whilst featuring lower C/N0 values. However, the rapid expansion of the European Galileo and BeiDou systems and the modernization of GPS raise the number of satellites (>60) transmitting on L5/E5a/B2a signals.

In this evolving landscape, this study aims to address the capabilities of L5/E5a/B2a signals in GNSS-harsh environments with a focus on land vehicle ITS applications requiring meter-level accuracy [4–6]. The structure of the paper is organized as follows. Firstly, the background and state of the art on L5/E5a/B2a signals is presented. Section 2 presents the methodology and fieldwork undertaken. The analysis of the raw GNSS measurements at the pre-processing and post-processing stage is presented in Section 3. Finally, Section 4 presents the key conclusions and discussion for future work.

2. Methods and Materials

2.1. Methods

Considering that the smartphone pseudorange measurements are characterized from high noise levels compared to high-grade geodetic GNSS receivers, at the pre-processing stage, they are smoothed using the Hatch filter [7,8].

$$\overline{p}_{r,k}^{s,S} = wp_{r,k}^{s,S} + (1-w)\left(\overline{p}_{r,k-1}^{s,S} + \varphi_{r,k}^{s,S} - \varphi_{r,k-1}^{s,S}\right)\overline{p}_{r,0}^{s,S} = p_{r,0}^{s,S} \quad (1)$$

where k, $k-1$ and 0 refer to the current, previous and initialization epochs accordingly, $p_{r,k}^{s}$ represents the pseudorange measurement, S indicates the GNSS system with $S \in (G: GPS,\ E: Galileo,\ C: BeiDou)$, s represents the satellite used, $\tilde{p}_r^{s}$ is the smoothed pseudorange, φ_r^s is the precise but ambiguous carrier phase measurement and w is a weight factor that is inversely proportional to the number of consecutive epochs. To obtain and adapt a weighting value for w, a similar approach to [8] was adopted.

For computing the GNSS PVT solution, the Single Point Positioning (SPP) technique is implemented for the case of single-frequency, multi-constellation measurements [7]. The linearized equations for the observed minus computed single-frequency, multi-constellation pseudoranges is given as follows (the time notation t is omitted for simplicity): [7]:

$$E\left(\begin{bmatrix} \Delta\tilde{\boldsymbol{p}}_{r,j}^{G} \\ \Delta\tilde{\boldsymbol{p}}_{r,j}^{E} \\ \Delta\tilde{\boldsymbol{p}}_{r,j}^{C} \end{bmatrix}\right) = \begin{bmatrix} \boldsymbol{G}_r^G & \boldsymbol{u}_{m_G} & 0 & 0 \\ \boldsymbol{G}_r^E & \boldsymbol{u}_{m_E} & \boldsymbol{u}_{m_E} & 0 \\ \boldsymbol{G}_r^C & \boldsymbol{u}_{m_C} & 0 & \boldsymbol{u}_{m_C} \end{bmatrix} \begin{bmatrix} \Delta\boldsymbol{r}_r \\ dt_{r,j}^G \\ ISB_{r,j}^{GE} \\ ISB_{r,j}^{GC} \end{bmatrix} \quad (2)$$

where $E(\cdot)$ denotes the expected operation, G, E and C are notations for the GPS, Galileo and BeiDou, respectively, r is the subscript for the receiver, j denotes the frequency used, m_S is the corresponding number of observations for each constellation and $\Delta\tilde{\boldsymbol{p}}_{r,j}^{S}$ denotes the observed minus computed pseudoranges for each constellation $S \in (G, E, C)$. Also, $\boldsymbol{G}_r^S$ and $\boldsymbol{u}_{m_S}$ are the matrix containing the LOS (Line of Sight) unit directions vectors and the unit vector, respectively, for each constellation S, $\Delta\boldsymbol{r}_r$ is the receiver position vector, $dt_{r,j}^G$ is the receiver clock error with respect to GPS lumped with the receiver hardware delays, and $ISB_{r,j}^{GE}$ and $ISB_{r,j}^{GC}$ are the GPS-Galileo and GPS-BeiDou inter-system biases (ISB), respectively. The notation $\tilde{\cdot}$ denotes that the observed pseudoranges are corrected for the orbit, clock, atmospheric delays, etc.

Another bias that should be handled appropriately refers to hardware delays that relate to the signals that the broadcasted satellite clock corrections are based on [7]. For example, the broadcasted satellite clocks of GPS satellites are based on the ionosphere-free (IF) linear combination (LC) of L1 + L2 frequencies, while for Galileo, they are based either on the IF LC of E1 + E5a or E1 + E5b frequencies. Instead, for the BeiDou system, the broadcasted satellite clock corrections refer to the B3 signal. Therefore, for the broadcasted satellite clocks ($dt^{s,S}$) to refer to the signals utilized in this study, the Timing Group Delay (TGD) and Broadcast Group Delay (BGD) should be applied for GPS/BeiDou and Galileo systems, respectively, as follows [7,9]:

$$dt^{s,G} - \frac{\mu_j^G}{\mu_1^G} TGD^s, \quad j \in (1,5) \tag{3}$$

$$dt^{s,E} - \frac{\mu_1^E}{\mu_j^E} BGD^s_{E5a-E1}, \quad j \in (1,5) \tag{4}$$

$$dt^{s,C} - TGD_1, \; for\ the\ B1\ signals \tag{5}$$

$$dt^{s,C} - TGD_{B2ap}, \; for\ the\ B2a\ signals \tag{6}$$

where μ_j^S is the frequency dependent coefficient, with $S \in (G : GPS, E : Galileo, C : BeiDou)$.

To estimate the PVT solution, the smartphone-based pseudorange and Doppler measurements are processed through an Extended Kalman Filter (EKF) [10]. The EKF algorithm was developed as an additional capability to the existing open-source software *PANGNAV 1.0.0* [10]. The EKF is described by

$$x_{k|k-1} = f\left(x_{k-1|k-1}\right) \tag{7}$$

$$\boldsymbol{P}_{k|k-1} = \boldsymbol{F}\boldsymbol{P}_{k-1|k-1}\boldsymbol{F}^T + \boldsymbol{Q} \tag{8}$$

$$\boldsymbol{K}_k = \boldsymbol{P}_{k|k-1}\boldsymbol{H}^T\left(\boldsymbol{H}\boldsymbol{P}_{k|k-1}\boldsymbol{H}^T + \boldsymbol{R}_k\right)^{-1} \tag{9}$$

$$\boldsymbol{x}_{k|k} = \boldsymbol{x}_{k|k-1} + \boldsymbol{K}_k\left(z_k - h\left(\boldsymbol{x}_{k|k-1}\right)\right) \tag{10}$$

$$\boldsymbol{P}_{k|k} = \boldsymbol{P}_{k|k-1} - \boldsymbol{K}_k\boldsymbol{H}\boldsymbol{P}_{k|k-1} \tag{11}$$

where the left indicator of $_{k|k-1}$ represents the current time instant whilst the right indicator represents the time instant value for the associated parameter being used, $\boldsymbol{x}_{k|k}$ is the state estimate at time k, $\boldsymbol{P}_{k|k}$ is the updated (a posteriori) estimate covariance at time k, $x_{k|k-1}$ is the prediction of the state at time k, $\boldsymbol{P}_{k|k-1}$ is the predicted (a priori) estimate covariance at time k, $\boldsymbol{K}_k$ is the Kalman Gain at epoch k, $\boldsymbol{R}_k$ is the covariance of the measurement error at epoch k, $\boldsymbol{Q}$ is covariance of the process noise, $\boldsymbol{H}$ is the geometric matrix of the measurements (pseudorange and Doppler), h is the state-to-measurement mapping function, $\boldsymbol{F}$ is the state transition matrix, f is the state transition mapping function and $\boldsymbol{Z}_k$ is the measurement vector at epoch k. For the EKF, a constant velocity dynamic model was adopted, whilst the ISBs are handled as white noise [11].

2.2. Materials

In order to evaluate the impact of L5, E5a and B2a signals in GNSS-harsh environments for ITS applications, a kinematic experiment was carried out on 21 May 2022, at the NTUA campus and the greater area (Figure 1). The kinematic scenario's test trajectory was constructed such that the vehicle runs through sub-urban, urban and deep-urban areas (Figure 1). Table 1 contains information on the duration and length of each sub-trajectory.

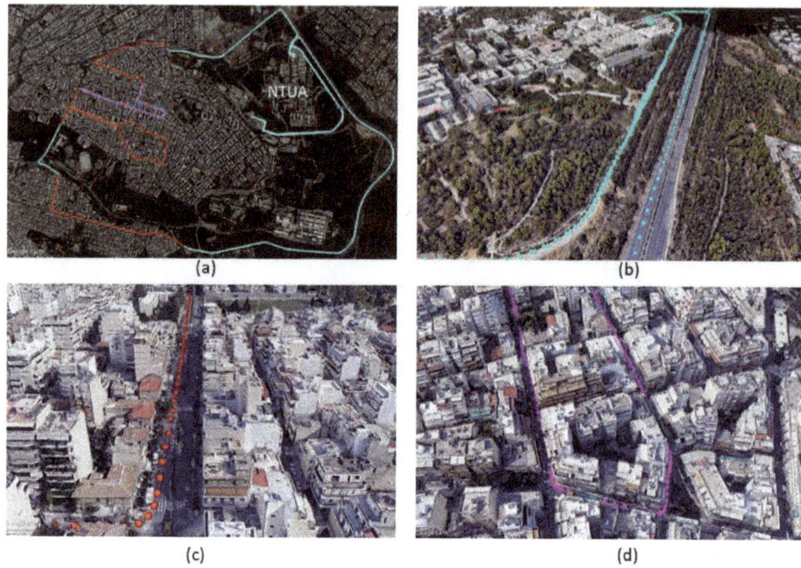

Figure 1. (**a**) Overall test trajectory which includes sub-urban (cyan), urban (red) and deep-urban (magenta) trajectories. (**b**) Sub-urban, (**c**) urban and (**d**) deep-urban environments.

Table 1. Characteristics of the total experimental trajectories and the underlying sub-trajectories.

Experimental Trajectory	Length (km)	Duration (min)
Sub-urban	9.62	21
Urban	5.35	17
Deep-urban	1.24	7
Total	16.21	45

Two separate datasets of spatio-temporal measurements were collected. The first one contains the raw GNSS measurements captured from two *Android* smartphones, the Xiaomi Mi 8 and the OnePlus Nord2 5G. Xiaomi Mi 8 is capable of tracking GPS, Galileo and BeiDou Signals on L1/L5, E1/E5a and B1, respectively. On the other hand, OnePlus Nord 2 5G can track L1/L5, E1/E5a and B1/B2a signals of GPS, Galileo and BeiDou, respectively. In order to realize an operational scenario, both smartphones were installed inside the vehicle, on the dashboard as shown in Figure 2. Finally, the GNSS raw measurements were retrieved from the smartphones using the *GNSSLogger* App [12].

The second dataset consists of the GNSS and IMU measurements collected from a tactical grade GNSS/IMU system (NovAtel® PwrPak7, iMAR IMU-FSAS) and a geodetic type GNSS antenna (Figure 2). The data from the tactical grade GNSS/IMU system were used to calculate the reference trajectory employing the PPK (Post Processed Kinematic) positioning technique. In this setup, the level arms among the various sensors mounted on the roof-top sensor platform and inside of the NTUA test vehicle are known thanks to the design specifications of the roof-top track. Furthermore, the level-arms among sensors are considered with respect to the IMU zero point.

Figure 2. (**a**) Roof-top sensor platform of the NTUA (National Technical University of Athens) test vehicle, (**b**) NovAtel® PwrPak7, (**c**) iMAR IMU-FSAS, (**d**) GNSS antenna, (**e**) Xiaomi Mi 8, (**f**) OnePlus Nord2 5G.

3. Analysis

3.1. Satellites' Visibility, C/N0 and Multipath

Figures 3 and 4 illustrate the performance of Xiaomi Mi 8 and OnePlus Nord 2 5G in terms of satellites being tracked and the received C/N0 values at specific frequency bands. For the Xiaomi Mi 8, 12, 9 and 11 satellites were being tracked for the GPS, Galileo and BeiDou systems on the L1, E1 and B1 bands, respectively. Regarding L5 and E5a, six GPS and nine Galileo satellites were tracked. OnePlus Nord 2 5G was able to track 12, 9 and 13 satellites for the GPS, Galileo and BeiDou systems on the L1, B1 and E1 bands, respectively. Finally, regarding L5, E5a and B2a, OnePlus Nord 2 5G was able to track eight GPS, nine Galileo and nine BeiDou satellites. The two smartphones have comparable tracking capabilities on L1/E1/B1, albeit OnePlus Nord 2 5G outperforms Xiaomi Mi 8 on the satellites being tracked on L5 for GPS.

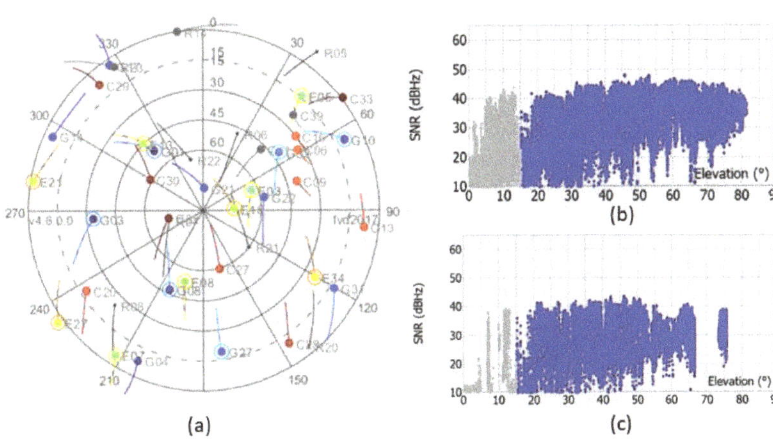

Figure 3. Xiaomi Mi 8: (**a**) Skyplot of the observed satellites on L1/E1/B1 (solid marks) and on L5/E5a (empty marks), (**b**) L1/E1/B1 and (**c**) L5/E5a/B2a C/N0 elevation-based (grey: elevation $\leq 15°$, blue: elevation > 15°) series (modified from *GNSSAnalysisApp* and *RTKLIB*).

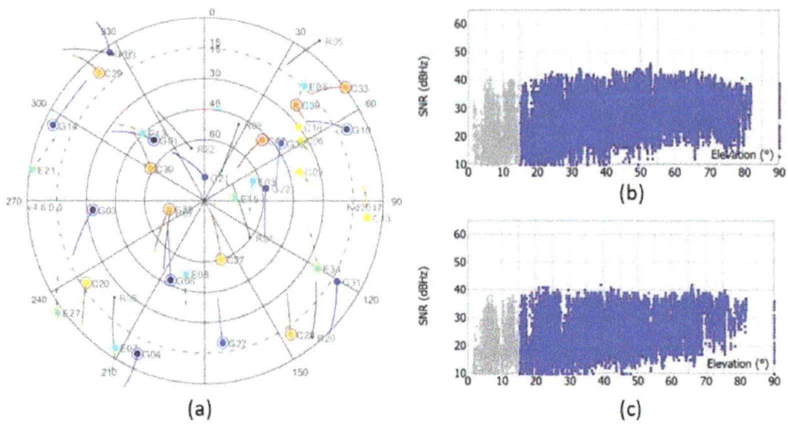

Figure 4. OnePlus Nord 2 5G: (**a**) Skyplot of the observed satellites on L1/E1/B1 (solid marks) and on L5/E5a (empty marks), (**b**) L1/E1/B1 and (**c**) L5/E5a/B2a C/N0 elevation-based (grey: elevation $\leq 15°$, blue: elevation $> 15°$) series (modified from *GNSSAnalysisApp* and *RTKLIB*).

Data analysis reveals two points regarding the elevation-based C/N0 series obtained for the Xiaomi Mi 8 and OnePlus Nord 2 5G (Figures 3 and 4). Firstly, the C/N0 values of the L1/E1/B1 signals both for the Xiaomi mi 8 and the OnePlus Nord 2 5G seem not to depend on the satellite elevation angle, as opposed for the L5/E5a/B2a signals. As a result, in this study, an elevation-dependent weighting scheme is used for L5/E5a/B2a signals and a C/N0-dependent weighting scheme [10] is implemented for L1/E1/B1 signals. Moreover, one should notice that the C/N0 values for L1/E1/B1 signals are higher than those of L5/E5a/B2a ones. On the other hand, the C/N0 values of L5/E5a/B2a signals seem to be less-susceptible to abrupt changes than those observed for the L1/E1/B1 signals.

In this study, in order to assess the multipath effect on L1/E1/B1 and L5/E5a/B2a, we use the closed form expression for the dual-frequency ionosphere corrected CMC (Code-Minus-Carrier) linear combination [7,13]. According to Figure 5, the multipath for Xiaomi Mi 8 ranges between −5 m and 5 m for L1/E1 signals, while for the L5/E5a signals, it lies in the range of −2 m to 2 m. This discrepancy supports the hypothesis that L5/E5a signals are more resilient to multipath by design, compared to L1/E1 ones. Also, multipath is seen to affect the GPS and Galileo measurements in the same way, as the multipath-RMS values for L1-E1 and L5-E5a are relatively close. Regarding the different sub-trajectories, it seems that multipath has more or less the same effect on the pseudorange measurements, irrespective of the environment type. Nevertheless, excessive multipath is observed both for the urban and deep-urban sub-trajectories.

For OnePlus Nord 2 5G, a lower impact of multipath is noticed for L5/E5a/B2a signals compared to those of L1/E1/B1 signals (Figure 6). In numbers, the RMS values observed due to multipath for the L1, E1 and B1 are 1.43 m, 0.96 m and 0.89 m, respectively, while for L5, E5a and B2a, they equal 0.25 m, 0.22 m and 0.17 m, respectively. Based on these results, one can also infer that E1 and B1 signals are more resilient to multipath than the L1, whilst for the L5, E5a and B2a, a more comparable performance is evident. Similar remarks could be made for the sub-trajectories for the OnePlus Nord 2 5G, as those made for Xiaomi Mi 8.

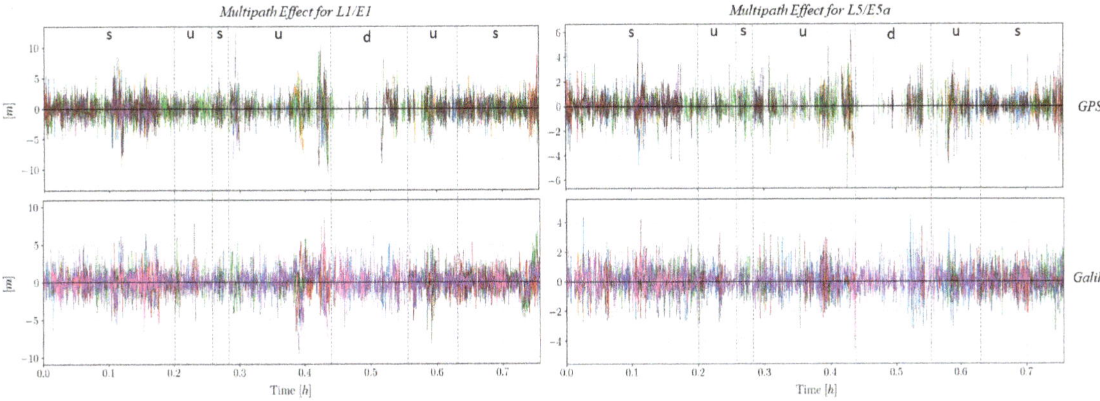

Figure 5. Multipath effect in meters (m) of L1/E1 and L5/E5a signals for Xiaomi Mi 8. The vertical dotted lines separate the intervals for each sub-trajectory, i.e., sub-urban (s), urban (u) and deep-urban (d). The different colors represent the GPS and Galileo satellites.

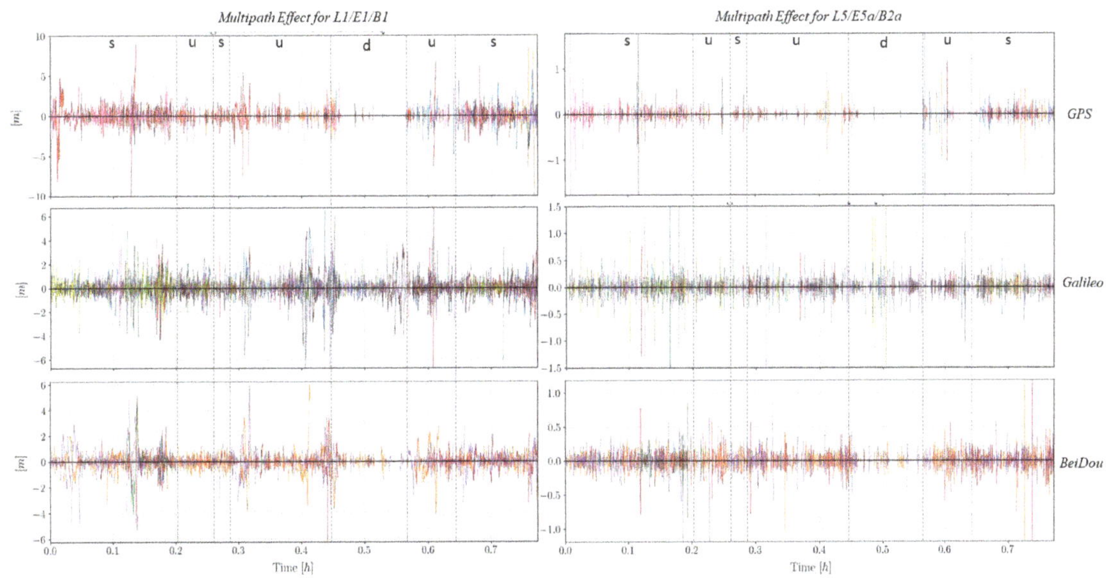

Figure 6. Multipath effect in meters (m) of L1/E1/B1 and L5/E5a/B2a signals for One Plus Nord 2 5G. The vertical dotted lines separate the intervals for each sub-trajectory, i.e., sub-urban (s), urban (u) and deep-urban (d). The different colors represent the GPS, Galileo and BeiDou satellites.

3.2. *Accuracy of the PVT solution*

The assessment of the PVT quality resides on measures of trueness (i.e., deviation from reference trajectory) and precision (i.e., repeatability). Table 2 summarizes the deviation (Root Mean Square, RMS) of the position solution for Xiaomi Mi 8 for the two sets of E1/L1/B1 and L5/E5a multi-constellation raw GNSS data.

Table 2. Horizontal position and velocity RMS errors of Xiaomi Mi 8 with respect to the reference trajectory.

Trajectories	Horizontal Position RMS (m)		Horizontal Velocity RMS (m/s)	
	L1/E1/B1	L5/E5a	L1/E1/B1	L5/E5a
Sub-Urban	2.9	3.5	1.4	1.7
Urban	5.3	6.8	1.5	1.8
Deep-Urban	6.6	7	1.6	2.0
Total	4.7	5.6	1.5	1.8

Clearly, concerning Xiaomi Mi 8, Table 2 indicates a better performance for the L1/B1/E1 position solution (RMS: 2.9 m) compared to the L5/E5a (RMS: 3.5 m). Furthermore, a horizontal RMS value of 2.9, 5.3 and 6.6 m is achieved for the sub-urban, urban and deep-urban trajectories, respectively. These results indicate the superiority of the estimated solution when multi-constellation GNSS measurements are used.

Two points are clearly evident concerning the L5/E5a solution no matter whether or not the accuracy obtained is significantly lower than that obtained for L1/E1/B1. Firstly, the Xiaomi MI 8 does not receive signals on B2a; therefore, more satellites are being monitored on the first frequency than on the second one. Secondly, this outcome is heavily affected by the poor observation geometry due to smartphone placement (dashboard).

In contrast, opposite conclusions result for the horizontal position errors obtained for the OnePlus Nord 2 5G (Table 3). For the case of L5/E5a/B2a solutions, the horizontal RMS errors improve by 21%, 17.1%, 14.7% and 22.5% for the total, sub-urban, urban and deep-urban trajectories, respectively, with respect to the L1/E1/B1 solution. This improved performance is due to the superior precision of the GNSS measurements on L5/E5a/B2a bands when compared to the L1/E1/B1 counterparts. Furthermore, as already stated, the One Plus Nord 2 5G is capable of receiving an extra signal (i.e., B2a) compared to Xiaomi Mi 8, which allows for tracking more satellites on the second frequency. Finally, it is worth noting that the tuning process of OnePlus Nord 2 was not as straightforward as that for Xiaomi Mi 8, indicating that better or more advanced filter tuning could potentially lead to superior results for the OnePlus Nord 2 5G.

Table 3. Horizontal position and velocity RMS errors of OnePlus Nord 2 with respect to the reference.

Trajectories	Horizontal Position RMS (m)		Horizontal Velocity RMS (m/s)	
	L1/E1/B1	L5/E5a/B2a	L1/E1/B1	L5/E5a/B2a
Sub-Urban	6.0	4.7	2.5	2.0
Urban	6.2	4.8	1.8	1.5
Deep-Urban	6.8	5.8	1.9	1.8
Total	6.2	4.9	2.1	1.7

In terms of horizontal velocity errors, both smartphones perform in a precision range of a few m/s (Tables 2 and 3). This performance could be attributed to the low precision of the Doppler-shift measurements acquired from *Android* devices [14]. The Xiaomi Mi 8 performs best when L1/E1/B1 measurements are employed, resulting in an accuracy of 1.5 m/s over the whole experimental trajectory. Regarding OnePlus Nord 2, comparing the L5/E5a/B2a signal velocities to the L1/E1/B1 ones, an improvement of 19%, 20%, 16.7% and 5% is noted for the complete, sub-urban, urban and deep-urban trajectories.

4. Conclusions

In this study, the performance capabilities of the raw GNSS measurements on L5/E5a/B2a signals from *Android* devices is investigated for road ITS applications. Specifically, a representative dataset was created by collecting GNSS raw measurements from two *Android* smartphones, namely, the Xiaomi Mi 8 and OnePlus Nord 2 5G, under GNSS-harsh environments. The raw GNSS pseudoranges obtained on L1/E1/B1 and L5/E5a/B2a were smoothed using a Hatch-Filter and processed through an EKF filter algorithm for computing the estimated PVT solution. In addition, the Doppler-shift measurements derived from the *Android* devices were integrated into the EKF filter. The PVT solutions for L1/E1/B1 and L5/E5a/B2a were compared against a reference PPK trajectory produced from high-grade navigation equipment and software.

Data analysis revealed that both smartphones used in this study track more satellites on L1/E1/B1 signals than on L5/E5a/B2a. Additionally, a slight improvement in the tracking capabilities has been observed for the most recently released device (i.e., OnePlus Nord 2 5G) for the L5/E5a/B2a signals compared to Xiaomi Mi 8. Furthermore, data collection on the L1/E1/B1 seems to exhibit higher C/N0 values, compared to those for L5/E5a/B2a; however, abrupt changes occur more frequently irrespective of the satellite's elevation. Contrariwise, measurements on L5/E5a/B2a tend to exhibit lower C/N0 values, but with fewer abrupt changes, which however relate to an increase in the satellite view angle. In terms of multipath, the investigation shows that pseudorange measurements on L5/E5a/B2a signals are more robust than those obtained on L1/E1/B1.

Regarding the PVT performance, the results obtained for both devices collecting data on the L5/E5a/B2a signals are promising. In particular, for Xiaomi Mi 8, even though the PVT solution refers to only L5/E5a measurements, it achieves a comparable performance with that of L1/E1/B1, in terms of horizontal position and velocity errors. On the contrary, for the case of L5/E5a/B2a signals, the PVT solution for the OnePlus Nord 2 5G shows a significant improvement when compared to the PVT solution for L1/E1/B1 signals.

In order to further improve the PVT solution, it is proposed that future work will focus on the single differences between satellites as the basic functional model. In this approach, it is foreseen that receiver specific biases should be mitigated to a great extent, and therefore, a better performance of the Hatch-Filter should be realized [8]. Also, a range of additional, contemporary smartphones shall be tested, as in the experiment conducted in this paper, a Record and Replay system (Racelogic's LabSat 3 Wideband) was utilized.

Author Contributions: Conceptualization, A.H., H.P. and V.G.; methodology, A.H.; software, A.H.; validation, A.H. and V.G.; formal analysis, A.H.; investigation, A.H., M.O., H.P. and V.G.; resources, A.H.; data curation, A.H.; writing—original draft preparation, A.H.; writing—review and editing, A.H. and V.G.; visualization, A.H.; supervision, V.G. All authors have read and agreed to the published version of the manuscript.

Funding: This research has been co-financed by the European Regional Development Fund of the European Union and Greek national funds through the Operational Program Competitiveness, Entrepreneurship and Innovation, under the call RESEARCH–CREATE–INNOVATE (project code:T2EDK-03966).

Institutional Review Board Statement: Not applicable.

Informed Consent Statement: Not applicable.

Data Availability Statement: Data supporting the reported results can be accessed by contacting the authors at a_haxhi@mail.ntua.gr.

Conflicts of Interest: The authors declare no conflict of interest.

References

1. Zangenehnejad, F.; Gao, Y. GNSS smartphones positioning: Advances, challenges, opportunities, and future perspectives. *Satell. Navig.* **2021**, *2*, 24. [CrossRef] [PubMed]
2. Bahadur, B. A study on the real-time code-based GNSS positioning with Android smartphones. *Measurement* **2022**, *194*, 111078. [CrossRef]
3. Miao, W.; Li, B.; Gao, Y. The Superiority of Multi-GNSS L5/E5a/B2a Frequency Signals in Smartphones: Stochastic Modeling, Ambiguity Resolution, and RTK Positioning. *IEEE Internet Things J.* **2023**, *10*, 7315–7326. [CrossRef]
4. GSA. Report on Road User Needs and Requirements Outcome of the European GNSS' User Consultation Platform. Jan. 2019. Available online: https://www.euspa.europa.eu/gnss-applications/user-needs-and-requirements (accessed on 5 June 2023).
5. Clausen, P.; Gilliéron, P.; Perakis, H.; Gikas, V.; Spyropoulou, I. Assessment of Positioning Accuracy of Vehicle Trajectories for Different Road Applications. *IET Intell. Transp. Syst.* **2017**, *11*, 113–125. [CrossRef]

6. Haxhi, A.; Perakis, H.; Gikas, V.; Fafoutellis, P.; Mantouka, E.; Vlahogianni, E.I.; Pnevmatikou, A.; Fortsakis, P. Towards Smartphone-based Enhanced GNSS Positioning for Eco Driving ITS Services. In Proceedings of the 10th International Congress on Transportation Research, Rhodes, Greece, 2–3 September 2021.
7. Teunissen, P.J.G.; Montenbruck, O. (Eds.) *Springer Handbook of Global Navigation Satellite Systems*; Springer International Publishing: Cham, Switzerland, 2017. [CrossRef]
8. Geng, J.; Jiang, E.; Li, G.; Xin, S.; Wei, N. An Improved Hatch Filter Algorithm towards Sub-Meter Positioning Using only Android Raw GNSS Measurements without External Augmentation Corrections. *Remote Sens.* **2019**, *11*, 1679. [CrossRef]
9. Romero, I. *The Receiver Independent Exchange Format Version 4.00*; IGS: Pasadena, CA, USA, 2021.
10. Haxhi, A.; Perakis, H.; Mpimis, T.; Gikas, V. Testing of a Combined Hatch Filter/RAIM Algorithm for SPP Smartphone Kinematic Positioning in GNSS Harsh Environments. In Proceedings of the 2nd Symposium of IAG Commission 4 "Positioning and Applications, Potsdam, Germany, 5–8 September 2022. [CrossRef]
11. Petovello, M. *GNSS Solutions: Calculating Time Offsets*; InsideGNSS: Red Bank, NJ, USA, 2013.
12. GnssLogger App—Apps on Google Play. Available online: https://play.google.com/store/apps/details?id=com.google.android.apps.location.gps.gnsslogger&hl=en&gl=US (accessed on 23 October 2022).
13. Aarnes, P.H. GNSS Multipath Analysis. 5 May 2023. Available online: https://github.com/paarnes/GNSS_Multipath_Analysis_Software (accessed on 16 May 2023).
14. Ramesh, A.; Renaudin, V.; Ragoin, C.; Ortiz, M. External-Connection Free, Accurate Standalone GNSS Positioning on Smartphones. In Proceedings of the Positioning and Navigation for Intelligent Transport Systems POSNAV, Berlin, Germany, 3–4 November 2022.

engineering proceedings

MDPI

Proceeding Paper

A Review of GNSS Carrier Phase Ambiguity Resolution and Conceptual AI-Driven Approaches [†]

Amarildo Haxhi * and Vassilis Gikas

School of Rural, Surveying and Geoinformatics Engineering, National Technical University of Athens, 9 Iroon Polytechneiou Str., 157 80 Zographos, Attica, Greece; vgikas@central.ntua.gr

* Correspondence: a_haxhi@mail.ntua.gr

† Presented at the European Navigation Conference 2023, Noordwijk, The Netherlands, 31 May–2 June 2023.

Abstract: Precise Point Positioning (PPP) with Integer Ambiguity Resolution (IAR) is an effective way of improving the overall performance of the PPP technique both in terms of convergence time and accuracy. Even though PPP-IAR has seen a tremendous development in the last two decades, there is still room for improvement. Specifically, the search stage of candidate carrier phase ambiguities is characterized as an NP-hard problem that requires a long processing time, leading to limitation in the reliability of the identified optimal solution. A field that would have an impact on the search process of carrier phase ambiguities, and is addressed conceptually in this paper, is Artificial Intelligence (AI).

Keywords: Precise Point Positioning; Integer Ambiguity Resolution; Artificial Intelligence; GNSS

Citation: Haxhi, A.; Gikas, V. A Review of GNSS Carrier Phase Ambiguity Resolution and Conceptual AI-Driven Approaches. *Eng. Proc.* **2023**, *54*, 12. https://doi.org/10.3390/ENC2023-15431

Academic Editors: Tom Willems and Okko Bleeker

Published: 29 October 2023

1. Introduction

Precise Point Positioning (PPP) is a GNSS technique employed for obtaining a high-accuracy positioning solution using minimal infrastructure and data corrections [1]. Nevertheless, a major drawback of the method is the long convergence time that for the single constellation case is of an order of 30 min [2]. There are many ways to reduce this time, albeit upon circumstances most of them are deemed cost-inefficient [2]. Among them, a prominent approach resides on Integer Ambiguity Resolution (IAR), which is proven to reduce the convergence time and improve the accuracy of the estimated solution [3]. If the ambiguities are resolved correctly to their integer counterparts, then the carrier phase measurements can be used as precise pseudoranges.

A variety of applications including road ITS (Intelligent Transportation Systems) could benefit from PPP-IAR even if they do not ask for cm level accuracy. Specifically, applications for which the PPP-IAR is exercised, in general can attain robustness and reliability in the estimated solution, leading to a high integrity performance [3]. Moreover, in [4], the impact of PPP-IAR was investigated in urban- and deep-urban environments where satellite obstruction occurs more frequently. The analysis determined that employing dual frequency PPP-IAR can yield better solutions than PPP float solutions for which more satellites and/or constellations are enabled. This suggests that processing an ever-increasing number of satellites might not be the answer in order to substantially improve the PPP performance.

Considering the importance of PPP-IAR, a great number of variations in the technique have been proposed and tested since early 2000. Representations of a PPP-IAR range from the Single-Constellation, Dual-Frequency models of Linear Combinations to the recently developed Multi-Constellation, Multi-Frequency ones that rely on "Undifferenced" and Uncombined GNSS measurements. These developments have decreased the convergence time of PPP-IAR but still, instantaneous convergence while assuring correct IAR is difficult to achieve [5].

A field that might have an impact on fast and reliable PPP with IAR is Artificial Intelligence (AI). Notwithstanding the fact that the mathematical models describing the

GNSS measurements are well defined, AI, and particularly its subset Machine Learning (ML), has been an area of increased research interest for GNSS [6]. According to published work in GNSS research, the most typical applications that increasingly make use of ML include satellite signal acquisition exploiting Multi-Layer Perception (MLP), the Convolutional Neural Network (CNN), NLOS (Non-Line of Sight)/multipath/evil waveform detection and classification via Logistic Regression (LR), Support Vector Machines (SVMs), CNN, Recurrent Neural Networks (RNN), etc., included Earth observation/monitoring harnessing the Artificial Neural Network (ANN); GNSS position error estimation utilizing the Long Short-Term Memory (LSTM) network [7]; the detection of falsely resolved integer ambiguities using ANN or CNN [8]; the prediction of ionospheric corrections; the detection of ionospheric scintillation, the improvement of the satellite clock, orbit prediction accuracy, and parameter prediction in missing GNSS corrections [9]; the calibration of the PVT algorithm's filter parameters (e.g., tuning of the Extended Kalman Filter (EKF)) [10]; and the enhancement of the accuracy of outputted PVT information from the positioning module [11] by deploying Reinforcement Learning (RL).

However, it is realized that current ML approaches have not been used for the IAR problem. Even though it is not explicitly reported in the literature, possible reasons for that could be that most ML techniques require a large amount of data and excessive computational resources for training the algorithms. Moreover, the models developed relying on ML might not perform sufficiently in real-time scenarios due to the lack of representative data, considering that most training sets comprise simulated data. However, instead, another AI approach that belongs to the family of computational intelligence [12] has been tested for tackling the IAR problem. Specifically, computational intelligence algorithms, or metaheuristics [12,13], have been used successfully, but not exhaustively, for resolving the ambiguities to their integer counterparts in the ambiguity domain for relative precise positioning GNSS.

Consequently, accounting for the importance of IAR for a fast-converging, high-integrity solution in challenging GNSS environments, this paper offers a brief state-of-the-art review on PPP-IAR. Next, the role of Artificial Intelligence (AI) in GNSS is explored, focusing on metaheuristic algorithms. Finally, after a summary on the limitations of PPP-IAR is provided, the possible integration schemes of AI-IAR are discussed, including their pros and cons and the generic methodology that shall be adopted in future implementations.

2. Precise Point Positioning with Integer Ambiguity Resolution

In PPP, by exploiting dual-frequency GNSS measurements, it is feasible to solve the ambiguities to their integer counterparts, subject to additional state space representation (SSR) corrections, precise orbits, and clocks made available to the user via a network of (global or regional) permanent stations [1]. Currently, six PPP-IAR (or else known as PPP-RTK) models are available, namely the Common Clocks model 1 (CC-1), the Distinct Clocks model (DC), the Common Clocks model 2 (CC-2), the Integer Recovery Clock model (IRC), the Decoupled Clock Model (DCM), and the Uncalibrated Phase Delays (UPD)/Fractional Cycle Biases (FCB) model [1].

The main differences among the aforementioned models lie (i) in the choice of the S-basis, (ii) in the choice of reparameterization, and (iii) in the choice of handling the ionospheric delay [1]. CC-1, DC, and CC-2 employ an Uncombined (UC) representation of the measurements, while IRC, DCM, and UPD/FCB adopt an Ionosphere-Free (IF) representation of the observations. From the literature, it is well known that adopting linear combinations for the representation of the measurements can (i) eliminate the ionospheric delay (first order) and (ii) amplify the corresponding wavelengths, meaning that the ambiguities could be resolved easier. Analytically, for the IRC, DCM, and UPD/FCB techniques, the procedure for IAR in dual-frequency PPP consists of two steps resolving first the Hatch–Melbourne–Wübbena Wide-Lane (WL) ambiguities, which then enables the resolving of the Narrow-Lane (NL) ambiguities. Step one is mandatory, considering that the IF ambiguities exhibit a relatively small wavelength (of an order of 0.6 cm), whereas when decomposed to WL and NL (which practically is equivalent to an L1 ambiguity [14]), the corresponding wavelengths are 86 cm and 11 cm long, respectively.

In the sequel, as the evolution of GNSS at space, network, and user side became more pervasive, a more elegant deterministic model started to become more frequently adopted, which relied on UC GNSS measurements. In the UC case, the ionospheric delay is handled as an unknown parameter, which enables the utilization of ionospheric constraints. If the ionospheric constraints exhibit better accuracy than the pseudoranges, then a better performance is expected in terms of accuracy and convergence than the LC models; otherwise, they are treated as equivalent [14]. For example, the authors in [15–17] adjusted the UPD/FCB model for UC representation of the GNSS measurements to perform IAR. Furthermore, the authors in [17] stress out the ability of the UC model to achieve rapid re-convergences by exploiting the estimated ionospheric delays.

In addition to leveraging precise ionospheric corrections, another approach for reducing the convergence time of PPP is to use multi-frequency signals. By introducing one additional step to the two-step procedure of dual-frequency PPP-IAR for triple-frequency PPP-IAR, the convergence time can be reduced significantly. Analytically, in this case, first the Extra-Wide-Lane (EWL) ambiguities are resolved to their integer counterparts, followed by the settlement of WL and afterwards the NL ambiguities. This additional step reduces the convergence time of triple-frequency PPP-IAR as a consequence of the long wavelength of the EWL ambiguities (e.g., 5.86 m, 9.77 m, and 4.88 m for GPS, Galileo, and BeiDou, respectively [18]), despite their excessive noise. For instance, the authors in [2] formulated the UPD/FCB model for UC observations, without ionospheric corrections, for triple-frequency, multi-constellation PPP with IAR for instantaneous positioning. It was shown that the concentricity of the N1 residuals within 1 cycle is 90.2% due to the extremely fast ambiguity resolution of the EWL, and subsequently, the instantaneous convergence of the WL ambiguities with a confidence level of 95%, for horizontal and vertical thresholds of 5 cm and 10 cm, respectively.

In addition, triple-frequency, additional frequencies (four and even five frequencies) can be exploited for PPP-IAR as they can provide more choices for EWL and WL combinations [19]. However, since earlier research demonstrated that triple-frequency and quadruple-frequency PPP-IAR perform similarly in accuracy and convergence time [19], some authors aimed to enhance the performance of dual- and triple-frequency PPP-IAR. Specifically, in [18], a near-optimal linear combination for dual- and triple-frequency PPP-AR was derived utilizing the equivalence of mathematical forms between the UC and LC models, considering that LC provide slightly less computational complexity compared to UC functional models, while a functional model with rank 4 or 3 (i.e., a set of linear independent observables) instead of 6 (uncombined observations) can further reduce the computational burden. The results indicate an average of 10 min convergence time with a success rate of 95% in kinematic mode. Finally, in [20], the authors proposed a PPP-IAR-based model that resides on the extended DCM model of simultaneously processing dual- and triple-frequency UC measurements using all available constellations. Their study concludes that quasi-instantaneous centimeter accuracy-level positioning is possible with a moving receiver.

3. Artificial Intelligence for Integer Ambiguity Resolution

Most of the aforementioned implementations of PPP-IAR algorithms, after revealing the integer nature of the corresponding ambiguities, search for the integer set in the ambiguity domain that would result in a highly accurate solution. Notably, this is due to the remaining unmodeled biases, especially when the raw observables contain high noise [21]. Most of the dual- and/or triple-frequency PPP-IAR implementations mentioned in Section 2 implement the LAMBDA (Least-Squares Ambiguity Decorrelation Adjustment) algorithm [22].

The PPP-IAR approach generally employs the LAMBDA algorithm for resolving the NL ambiguities. This is due to the high correlation degree observed between the

NL ambiguities and their accelerated resolution, which contributes significantly in the reduction in the initialization period [15]. Furthermore, for the case that a full ambiguity resolution (FAR) is not feasible, a subset of the ambiguities is fixed, leading to Partial Ambiguity Resolution (PAR). Nevertheless, despite the profound effectiveness of the LAMBDA method [5], it is still computationally cumbersome because all integer candidates need to be checked. This results in increased search times, particularly as the number of ambiguities and the search space increases [23], whereas the major problem with PAR is the decision on which a set of ambiguities should finally be chosen [24].

The carrier phase ambiguity problem is an NP-hard (Non-Polynomial) problem [25], meaning that no explicit algorithm exists to find the exact solution for the integer ambiguities in polynomial time. The PPP-IAR problem, assuming that "Undifferenced" or Single-Differenced, Uncombined, or Combined deterministic functional models are used, in its linearized form reads as follows [26]:

$$y = Aa + Bb + e \tag{1}$$

where $y \in R^q$ contains the observed minus computed code and carrier-phase measurements, $a \in Z^n$ represents the integer carrier phase ambiguities with the design matrix A, $b \in R^p$ contains the real unknowns with the design matrix B, and $\mathbf{e}$ represents the additive noise component. Applying the least-squares criterion to Equation (1) for estimating the unknown parameters yields

$$min(\mathbf{y} - Aa - Bb)^T Q_y^{-1} (y - Aa - Bb) \tag{2}$$

where Q_y is the variance–covariance matrix of the GNSS observations. The minimization problem in Equation (2), referred to as the Mixed Integer Non-Linear Programming (MINLP) problem, is equivalent to solving the following standard Integer Least-Squares problem (ILS) or the Integer Quadratic Problem (IQP) [27]:

$$\check{a} = min(\hat{a} - a)^T Q_{\hat{a}}^{-1} (\hat{a} - a) \tag{3}$$

where $\hat{a}$ is the float solution of Equation (2) by neglecting the integer property of the ambiguities, $Q_{\hat{a}}$ is the variance–covariance matrix of the estimated float ambiguities, and $\check{a}$ is the integer solution of Equation (3). In the recent past, AI methods originating from the intelligence computation field have been exploited for solving the ILS problem in Equation (3), which in principle is a closest lattice point problem [27]. More specifically, this problem could be solved by employing principles of metaheuristics [13,27]. Metaheuristics are iterative optimization techniques for solving combinatorial problems, i.e., finding an optimal solution among a finite large solution space, such as the IAR problem.

Most of the well-known metaheuristic algorithms that have been implemented to IAR are the Genetic Algorithm (GA), Differential Evolution (DE), Ant Colony Optimization (ACO), and Artificial Fish Swarm Algorithm (AFSA) [27–33]. GA and DE draw their inspiration from the process of natural selection, which adopts the idea of the survival of the fittest [28,33]. Both GA and DE belong to Evolutionary Algorithms (EA) [12]. ACO is based on biological ants' foraging behavior, which uses pheromones as a communication tool [27], while AFSA simulates a variety of ecological behaviors of fish schooling in the water [31,32]. Both ACO and AFSA apply to Swarm Intelligence. GA, DE, ACO, and AFSA all belong to population-based optimization methods, meaning that they operate on a set of solutions (candidates), often called population.

In the context of the IAR problem, GA represents the candidate integers (or decision variables) as chromosomes encoded in particular bit strings [28–30,34], where each chromosome corresponds to an individual and all the individuals together are the population. The most common encodings for candidate integer ambiguities in GA are binary- [34], real- [28], and grey-coding [29]. Real coding does not have the drawbacks of binary-coding, i.e., longer code lengths, larger solution space, and longer time costing [28,30], while gray-coding can effectively avoid the hamming cliff (i.e., a situation in which a small alteration in a solution's binary representation can result in a big change in its fitness value) [29]. The DE algorithm represents the decision variables as individuals of integer vectors [33]. In AFSA, the state of each artificial fish is represented by the dimension of the decision variable, i.e., ambiguities [31,32]. In ACO, a candidate solution (artificial ant) is also characterized by the dimension of the decision variable [27].

According to [13], the fundamental idea behind any metaheuristic applied to the IAR problem consists of four stages (Figure 1). Firstly, the ambiguities should be appropriately represented using a model from which an a priori search space $Q(N)$ and initial solution (population) shall be constructed (stage 1). The initial population is mainly generated at random [34] or initiated either via random numbers generated from a normal distribution $X \sim N(\hat{a},\ Q_{\hat{a}})$ [32] or by using a priori information from the baseline vector (in the case of double-difference ambiguities) as a constraint for the generated individuals of candidate integer ambiguities [28]. Depending on how the initial solution and constraints are formed for the optimization problem, it is indicated to use some form of decorrelation of the ambiguities, which enhances the convergence properties of the metaheuristic algorithm to the global minima [28,31–33].

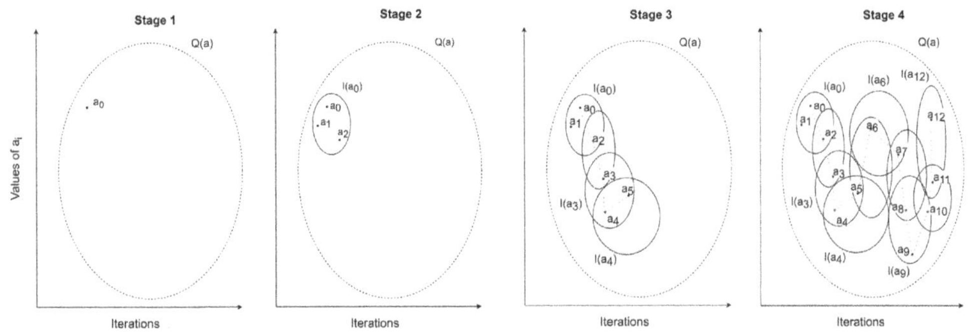

Figure 1. Schematic representation of GNSS ambiguity resolution stages based on metaheuristic principles, modified from [13].

At the following stage (stage 2), a set of neighbor values $I(N_i)$ adjacent to the initial solution is generated by choosing a suitable mechanism. GA and DE utilize crossover, a genetic operation that involves combining genetic information from two parent solutions to produce one or more offspring, and mutation, a genetic operation that introduces small random changes into an individual solution to promote diversity within the population [28–30,33,34]. Despite the fact that both GA and DE use crossover and mutation, in GA, crossover is applied to fixed positions in the encoded string, representing a candidate solution [28–30,34], while on the contrary, DE combines three or more individuals from the population, usually via vector arithmetic [33]. On the other hand, AFSA utilizes the random, preying, swarm, and following behavior of fish [31,32]. Praying corresponds to the behavior of a fish to relocate itself to a location with a higher consecration of food, swarming epitomizes the behavior of fish to avoid danger and over-crowded areas, and following emphasizes the following behaviors of artificial fish when another fish finds a location with a higher concentration of food. Conversely, ACO exploits pheromone levels parametrized using a weighted sum of several one dimensional Gaussian functions, which enables the reconstruction of new solutions [27].

To avoid premature convergence and stagnation at the local minima, some sub-routines of the aforementioned mechanisms undertake explorative behaviors (a series of downhill moves), while others undertake explorative behaviors (a series of uphill moves). In the case of GE and DE, the role of exploitation is adopted via the crossover operation, while exploration is performed via the mutation operation. In AFSA, the random and swarming behaviors of artificial fish give exploration properties to the algorithm, while preying and following act as exploitation mechanisms. ACO implements exploitation by following trails with high-levels of pheromones laid by artificial ants and assures exploration by allowing artificial ants to choose paths with a lower pheromone level as well [27].

Next, the best integer ambiguity candidates are searched and selected based on an acceptable strategy in the vicinity of $I(N_i)$ (stage 3). In GA, DE, and AFSA, searching and selection are applied via the appropriate transformation of the objective function, Equation (2) or Equation (3) [28–33], to a fitness function. In most cases, the transformation involves incorporating a positive scaling factor and applying a logarithmic operation to the original objective function. This transformation occurs in order to avoid premature convergence and trapping in the local optimum. On the contrary, ACO directly uses the objective function of Equation (3) [27]. Finally, the searching process stops when no further improvement is observed (i.e., the population of individuals has converged) [34] or a pre-defined number of iterations is reached (stage 4) [13,27,28,32,33].

The inductions from the several metaheuristic algorithms implemented in the IAR for double-difference ambiguity resolution encompass multiple aspects, including feasibility, a reduction in search space, comparative analysis among different metaheuristic algorithms, and comparison with the LAMBDA algorithm in terms of convergence speed and success rate. Analytically, implementations of GA [28–30,34], DE [33], ACO [27], and AFSA [31,32] demonstrate the feasibility of achieving convergence to the global optimum for the IAR problem. As a case in point, in [27], ACO was used, exploiting a sufficiently large dataset of simulated double-differenced float ambiguities with high precision (inferred via the Ambiguity Delusion of Precision, or ADoP [27]). In these conditions, ACO was able to converge to the global optima for a variety of ambiguity dimensionalities, e.g., 3–20, 21–30, and 31–40, with increasing convergence time [27].

Regarding the reduction in search space, an early implementation of the binary-coded GA implemented in [34] achieved the global optimum for one unknown ambiguity in just over 2 s of CPU processing while exploring something less than 0.7 of the total search space. In [28], a real-coded GA was introduced with adaptive probabilities for crossover and mutation based on individual fitness (Adaptive Genetic Algorithm, AGA). With the same experimental conditions (population number, maximum iterations, etc.), real-coded AGA outperformed the binary-coded GA in terms of iterations needed for convergence, and subsequently, the total search space required. Furthermore, in [31], the total space required for searching for the correct integer ambiguities with AFSA accounts for only 2.18% of the total problem space (41^3).

The authors in [32] used simulated double-difference ambiguities to compare AFSA and AGA. Ensuring that the experimental conditions were the same for both algorithms, they found that AFSA converged faster (6.37 s) compared to AGA (12.92 s). Later, the authors in [31] improved AFSA by adding an attenuation factor to the visual field and an adaptive step size for the artificial fish for better global and local exploration. A simple comparison with GA showed that AFSA can converge to the optimum solution in 2.5 s, while GA requires 5.8 s. Lastly, a recent implementation of an adaptive DE (ADE), where mutation and crossover gained a more adaptive behavior leading to robustness and reliability, outperformed other metaheuristic algorithms like DE and GA [33].

As for comparing metaheuristic algorithms with LAMBDA [5], a few attempts have been realized. Specifically, in [28], the real-coded AGA was compared to LAMBDA in terms of required convergence time. Utilizing a snapshot of low-dimensional simulated double-differenced ambiguities, it was shown that LAMBDA converges under 91.5 s and real-coded AGA under 14.7 s. An evaluation this time of the success rate of LAMBDA and ADE was implemented in [33], using a more representative simulated dataset of 14-dimensional float ambiguities of 100 epochs in succession. The success rate achieved by ADE was 94%, while for LAMBDA, it was 81%. For both AGA [28] and ADE [30] algorithms, the fitness function was subjected to the baseline constraint.

4. Discussion/Proposed Methodology

Undoubtedly, the development of PPP models and algorithms for IAR has seen tremendous progress in the last two decades. Today, it is possible to achieve a highly accurate and robust position solution within a few minutes or quasi-instantaneously when multi-constellation and triple-frequency GNSS measurements are used. Furthermore, a significant improvement in the convergence and robustness of dual-frequency, multi-constellation PPP-IAR is observed. However, there is still room for further improvement in the search process for resolving the correct ambiguities after their integer property is retrieved, in the case of PPP-IAR.

Today, the standard, state-of-the-art methodology for searching integer ambiguities after their integer property is secured is the LAMBDA method [5]. Although it is very effective, it is still computationally complex for the reasons explained in Section 3. In this regard, an interesting alternative for IAR in the ambiguity domain relies on AI, and to this effect, metaheuristics have great potential due to their inherent properties (e.g., parallel search, global optimization, and robustness). Early implementations suggest comparable results to well-established approaches and adaptation, to some extent, to the properties of the specific problem at hand, which is how an optimization algorithm can achieve a better performance than others according to the no-free-lunch theorem [35]. However, all of the existing implementations are designed for double-differenced ambiguities; the results so far are based on simulated instances of them inferred from strong models, and in most cases, the baseline constraint is used to subject the fitness function. Therefore, further investigation is required, utilizing appropriate datasets, undergoing experiments, and exploiting the advancements that have occurred in metaheuristics for solving NP-hard problems (calibration, initialization, noise addition, hybrid metaheuristic algorithms, etc. [12,36,37]).

In this direction, it is envisioned that future research should focus on metaheuristics by studying their impact on searching for integer ambiguities in the ambiguity domain after their integer property is assured. Regarding the latter one, a PPP-IAR method will be selected, considering its efficiency in terms of position quality capabilities (i.e., convergence time and accuracy) and computational resources; specifically, achieving equivalent results by efficiently reducing the dimensionality of the problem (e.g., estimable parameters). The suggested AI approach for PPP-IAR will be assessed via tests involving a set of known elements that are expected to affect its performance (e.g., dimensionality of the unknown ambiguities) and a comprehensive comparison with the LAMBDA algorithm.

Author Contributions: Conceptualization, A.H.; methodology, A.H.; investigation, A.H.; writing—original draft preparation, A.H.; writing—review and editing, A.H. and V.G.; visualization, A.H.; supervision, V.G.; All authors have read and agreed to the published version of the manuscript.

Funding: This research has been co-financed by the European Regional Development Fund of the European Union and Greek national funds through the Operational Program Competitiveness, Entrepreneurship and Innovation, under the call RESEARCH–CREATE–INNOVATE (project code: T2EDK-03966).

Institutional Review Board Statement: Not applicable.

Informed Consent Statement: Not applicable.

Data Availability Statement: Data sharing not applicable.

Conflicts of Interest: The authors declare no conflict of interest.

References

1. Teunissen, P.J.G.; Khodabandeh, A. Review and principles of PPP-RTK methods. *J. Geod.* **2015**, *89*, 217–240. [CrossRef]
2. Tao, J.; Chen, G.; Guo, J.; Zhang, Q.; Liu, S.; Zhao, Q. Toward BDS/Galileo/GPS/QZSS triple-frequency PPP instantaneous integer ambiguity resolutions without atmosphere corrections. *GPS Solut.* **2022**, *26*, 127. [CrossRef]
3. Seepersad, G.; Aggrey, J.; Bisnath, S. Do We Need Ambiguity Resolution in Multi-GNSS PPP for Accuracy or Integrity? In Proceedings of the 30th International Technical Meeting of The Satellite Division of the Institute of Navigation (ION GNSS+ 2017), Portland, OR, USA, 25–29 September 2017; pp. 2204–2218.

4. Naciri, N.; Bisnath, S. Multi-GNSS Ambiguity Resolution as a Substitute to Obstructed Satellites in Precise Point Positioning Processing. In Proceedings of the 33rd International Technical Meeting of the Satellite Division of The Institute of Navigation (ION GNSS+ 2020), Online, 21–25 September 2020; pp. 2960–2971.
5. Atiz, O.F.; Ogutcu, S.; Alcay, S.; Li, P.; Bugdayci, I. Performance investigation of LAMBDA and bootstrapping methods for PPP narrow-lane ambiguity resolution. *Geo-Spat. Inf. Sci.* **2021**, *24*, 604–614. [CrossRef]
6. Siemuri, A.; Kuusniemi, H.; Elmusrati, M.S.; Valisuo, P.; Shamsuzzoha, A. Machine Learning Utilization in GNSS—Use Cases, Challenges and Future Applications. In Proceedings of the 2021 International Conference on Localization and GNSS (ICL-GNSS), Tampere, Finland, 1–3 June 2021; pp. 1–6.
7. Lan, T.; Dodinoiu, A.; Geffert, A.; Becker, U. 3.1 Machine-Learning-Based Position Error Estimation for Satellite-Based Localization Systems. In Proceedings of the Proceedings-ettc2020, Virtual Conference, 23–25 June 2020; AMA Service GmbH: Wunstorf, Germany; pp. 89–97.
8. Dabove, P.; Manzino, A.M. Artificial neural network for detecting incorrectly fixed phase ambiguities for L1 mass-market receivers. *GPS Solut.* **2017**, *21*, 1213–1219. [CrossRef]
9. Messineo, R.; Dovis, F.; Leuzzi, C.; Nardin, A. navisp Element 1 Final Presentation EL1-035 BIS "Machine Learning Techniques to Model GNSS". 15 December 2022. Available online: https://navisp.esa.int/project/details/144/show (accessed on 15 August 2023.).
10. Gao, X.; Luo, H.; Ning, B.; Zhao, F.; Bao, L.; Gong, Y.; Xiao, Y.; Jiang, J. RL-AKF: An Adaptive Kalman Filter Navigation Algorithm Based on Reinforcement Learning for Ground Vehicles. *Remote Sens.* **2020**, *12*, 1704. [CrossRef]
11. Zhang, E.; Masoud, N. Increasing GPS Localization Accuracy With Reinforcement Learning. *IEEE Trans. Intell. Transport. Syst.* **2021**, *22*, 2615–2626. [CrossRef]
12. Simon, D. *Evolutionary Optimization Algorithms: Biologically-Inspired and Population-Based Approaches to Computer Intelligence*; Wiley: Hoboken, NJ, USA, 2013; ISBN 978-0-470-93741-9.
13. Saleh, H.A. Artificial Intelligence for Optimizing the GNSS Carrier Phase-based Positioning. In Proceedings of the 2003 National Technical Meeting of The Institute of Navigation, Anaheim, CA, USA, 22–24 January 2003; pp. 407–416.
14. Seepersad, G. Improving Reliability and Assessing Performance of Global Navigation Satellite System Precise Point Positioning Ambiguity Resolution. Ph.D Thesis, York University, Toronto, ON, Canada, 2018.
15. Geng, J.; Shi, C. Rapid initialization of real-time PPP by resolving undifferenced GPS and GLONASS ambiguities simultaneously. *J. Geod.* **2017**, *91*, 361–374. [CrossRef]
16. Zhou, P.; Wang, J. Evaluating PPP Ambiguity Resolution Methods with Ionosphere-Free and Raw GPS Observation Models. In *China Satellite Navigation Conference (CSNC) 2016 Proceedings: Volume III*; Sun, J., Liu, J., Fan, S., Wang, F., Eds.; Lecture Notes in Electrical Engineering; Springer: Singapore, 2016; Volume 390, pp. 529–539. ISBN 978-981-10-0939-6.
17. Li, X.; Ge, M.; Zhang, H.; Wickert, J. A method for improving uncalibrated phase delay estimation and ambiguity-fixing in real-time precise point positioning. *J. Geod.* **2013**, *87*, 405–416. [CrossRef]
18. Duong, V.; Harima, K.; Choy, S.; Laurichesse, D.; Rizos, C. An optimal linear combination model to accelerate PPP convergence using multi-frequency multi-GNSS measurements. *GPS Solut.* **2019**, *23*, 49. [CrossRef]
19. Instantaneous Centimeter-Level Multi-Frequency. *GPS World*. 4 July 2018. Available online: https://www.gpsworld.com/innovation-instantaneous-centimeter-level-multi-frequency-precise-point-positioning/ (accessed on 2 March 2023).
20. Naciri, N.; Bisnath, S. Approaching Global Instantaneous Precise Positioning with the Dual- and Triple-Frequency Multi-GNSS Decoupled Clock Model. *Remote Sens.* **2021**, *13*, 3768. [CrossRef]
21. Wen, Q.; Geng, J.; Li, G.; Guo, J. Precise point positioning with ambiguity resolution using an external survey-grade antenna enhanced dual-frequency android GNSS data. *Measurement* **2020**, *157*, 107634. [CrossRef]
22. Teunissen, P.J.; Kleusberg, A. (Eds.) *GPS for Geodesy*; Springer: Berlin/Heidelberg, Germany, 2012; ISBN 978-3-642-72013-0.
23. Verhagen, S.; Tiberius, C.; Li, B.; Teunissen, P.J.G. Challenges in ambiguity resolution: Biases, weak models, and dimensional curse. In Proceedings of the 2012 6th ESA Workshop on Satellite Navigation Technologies (Navitec 2012) & European Workshop on GNSS Signals and Signal Processing, Noordwijk, The Netherlands, 5–7 December 2012; IEEE: Piscataway, NJ, USA, 2013; pp. 1–8. [CrossRef]
24. Teunissen, P.J.G.; Montenbruck, O. (Eds.) *Springer Handbook of Global Navigation Satellite Systems*; Springer International Publishing: Cham, Switzerland, 2017; ISBN 978-3-319-42926-7.
25. Xu, P.; Shi, C.; Liu, J. Integer estimation methods for GPS ambiguity resolution: An applications oriented review and improvement. *Surv. Rev.* **2012**, *44*, 59–71. [CrossRef]
26. Brack, A. *Partial Carrier-Phase Integer Ambiguity Resolution for High Accuracy GNSS Positioning; Thesis Commons*; Verlag Dr. Hut: Munich, Germany, 2019.
27. Jazaeri, S.; Amiri-Simkooei, A.R.; Sharifi, M.A. Fast GNSS ambiguity resolution by ant colony optimisation. *Surv. Rev.* **2013**, *45*, 190–196. [CrossRef]
28. Yi, Q.; Yi, X.; Shi, M. Fast Integer Ambiguity Resolution Based on Real-Coded Adaptive Genetic Algorithm. *Int. J. Res. Eng. Sci. (IJRES)* **2017**, *4*, 67–72.
29. Xu, D.; Liu, M.; Zhu, L. Single frequency GNSS integer ambiguity resolution with adaptive genetic algorithm. In Proceedings of the 2013 IEEE Third International Conference on Information Science and Technology (ICIST), Yangzhou, China, 23–25 March 2013; IEEE: Piscataway, NJ, USA, 2013; pp. 1049–1051.

30. Liu, Z.; Liu, J.; Jiang, W.; Li, T.; Liu, H. Ambiguity resolution of double-difference GPS short baseline using genetic algorithm. In Proceedings of the International Conference on Space Information Technology, Wuhan, China, 19–20 November 2005; Wang, C., Zhong, S., Hu, X., Eds.; p. 598529.
31. Wang, L.; Wang, Y.-G.; Teng, H.-L. An Integer Artificial Fish Swarm Algorithm and Its Application on Ambiguity Resolution. *DEStech* **2017**. [CrossRef] [PubMed]
32. Liu, S.; Zhang, Y. Integer ambiguity solution based on artificial swarm algorithm. In Proceedings of the 2016 5th International Conference on Measurement, Instrumentation and Automation (ICMIA 2016), Shenzhen, China, 17–18 September 2016; Atlantis Press: Amsterdam, The Netherlands, 2016.
33. Dou, Z.; Wu, Y. Integer ambiguity search algorithm based on adaptive differential evolution algorithm. In Proceedings of the Eleventh International Conference on Graphics and Image Processing (ICGIP 2019), Hangzhou, China, 12–14 October 2019; Pan, Z., Wang, X., Eds.; SPIE: Bellingham, WA, USA, 2020; p. 107.
34. Hassan, A.; Mezera, D. Method for Determining a Value for the Phase Integer Ambiguity and a Computerized Device and System Using Such a Method 2000. US6052082A, 18 April 2020. Available online: https://patents.google.com/patent/US6052082/en (accessed on 17 August 2023).
35. Ho, C.; Pepyne, D.L. Simple Explanation of the No Free Lunch Theorem of Optimization. *Cybern. Syst. Anal.* **2002**, *38*, 292–298. [CrossRef]
36. Choi, K.P.; Kam, E.H.H.; Tong, X.T.; Wong, W.K. Appropriate noise addition to metaheuristic algorithms can enhance their performance. *Sci. Rep.* **2023**, *13*, 5291. [CrossRef] [PubMed]
37. Sarhani, M.; Voß, S.; Jovanovic, R. Initialization of metaheuristics: Comprehensive review, critical analysis, and research directions. *Int. Trans. Operational. Res.* **2023**, *30*, 3361–3397. [CrossRef]

engineering proceedings

MDPI

Proceeding Paper

Cycle Slip Detection of Single-Frequency Measurements in Drone Platforms †

Chan-hee Lee [1] and Euiho Kim [2,*]

[1] Department of Mechanical Engineering, Hongik University, 94, Wausan-ro, Mapo-gu, Seoul 04066, Republic of Korea; cjessh0824@gmail.com
[2] Department of Mechanical & System Design Engineering, Hongik University, 94, Wausan-ro, Mapo-gu, Seoul 04066, Republic of Korea
* Correspondence: euihokim@hongik.ac.kr
† Presented at the European Navigation Conference 2023, Noordwijk, The Netherlands, 31 May–2 June 2023.

Abstract: For the precise operation of Unmanned Aerial Vehicles (UAV), Real-Time Kinematic (RTK) techniques of Global Navigation Satellite System (GNSS) have been used as a positioning source. In a typical drone platform with a low-cost multi-frequency receiver, there are several dual-frequency measurements and a small set of single-frequency measurements at one time. In this paper, considering the measurement characteristics, we introduce a novel method that aims to detect a cycle slip of a small set of single-frequency measurements using dual-frequency measurements of other satellites. The performance of the proposed single-frequency cycle slip detection methods is compared with a conventional Doppler-based approach using flight test data of a drone platform.

Keywords: cycle slip detection; GNSS; drone flight test

Citation: Lee, C.-h.; Kim, E. Cycle Slip Detection of Single-Frequency Measurements in Drone Platforms. *Eng. Proc.* **2023**, *54*, 13. https://doi.org/10.3390/ENC2023-15436

Academic Editor: Tom Willems and Okko Bleeker

Published: 29 October 2023

1. Introduction

The use of a carrier phase is mandatory for precise positioning using GNSS, such as real-time kinematic (RTK) positioning or precise point positioning (PPP) [1–4]. To implement a carrier phase-based positioning, an integer ambiguity must be resolved, and its continuity should be reliably monitored. If a discontinuity of the carrier phase called a cycle slip occurs, a new integer ambiguity of the carrier phase should be found [5,6], or a cycle slip should be repaired [7,8].

In the case of single-frequency measurements, it is difficult to eliminate nuisance terms in measurements like dual-frequency measurements. Here, a typical approach is to compare a range increment from a Doppler frequency and one from carrier phase measurements, referred to as Doppler-aided cycle slip detection (DACSD) [9,10]. While Doppler measurements are immune to cycle slips, they have a high noise level that makes it difficult to detect small cycle slips. Instead of using Doppler measurements, some other approaches used an inertial measurement unit, which would make a cycle slip detection problem more complicated [11,12].

Nowadays, dual-frequency GNSS receivers are typically installed in a small UAV platform. However, a lower-grade GNSS antenna and receiver on a small UAV platform and a dynamic motion with vibrations during flights often lead to unexpected cycle slips and poor detection performance. Also, a fair amount of single-frequency measurements, along with some dual-frequency measurements, are typically observed from a dual-frequency receiver on a small UAV platform.

In the case of mixed single- and dual-frequency measurements, rather than applying a Doppler-aided cycle slip detection method like the prior works, this study proposes a novel single-frequency cycle slip detection method based on UAV velocity and receiver clock drift estimates from dual-frequency carrier measurements, providing a lower noise level in cycle slip detection test statistics.

The structure of the paper is as follows. Section 2 presents the prior DACSD and introduces a novel cycle slip detection algorithm for the mixed single- and dual-frequency measurements. Section 3 presents flight test results, followed by the conclusions.

2. Cycle Slip Detection Methods and Proposed Approach for Single-Frequency Measurements

In this section, we briefly review DACSD as a prior representative method and suggest a cycle slip detection algorithm for single-frequency measurement that utilizes only the carrier phase measurement.

2.1. Doppler-Aided Cycle Slip Detection Method for Single-Frequency Measurements

GNSS measurement models of a receiver are

$$\begin{aligned} \lambda_i\phi_i &= r - I_i + T + (c\delta t_u - c\delta t^s) + \lambda_i N_i + \varepsilon_\phi, \\ d_i &= \frac{d}{dt}(\phi_i) = -\frac{1}{\lambda_i}\left[\dot{r} + (c\delta\dot{t}_u - c\delta\dot{t}^s) + \varepsilon_d\right] \end{aligned} \tag{1}$$

where ϕ represents a carrier measurement, r is the actual distance between a satellite and a user, I is an ionospheric delay error, and T is a tropospheric delay error. GNSS receiver and satellite clock errors are δt_u and δt^s, respectively. c is the speed of light. λ is the wavelength of a carrier, and N is an integer ambiguity in a carrier phase measurement. ε is the noise of each measurement. d is a Doppler measurement, which is equal to the time derivative of the carrier measurement. The subscript i denotes the frequency of each measurement.

Doppler measurements have been typically used to detect a cycle slip for single-frequency measurement. A Doppler is an apparent frequency change in a wave and is caused by the relative motion between a satellite and a user. An integrated Doppler frequency shift over some time period is equal to a carrier measurement change such that [13]

$$\Delta\phi_i = \phi_i(k) - \phi_i(k-1) = \int_{k-1}^{k} f_d dt, \tag{2}$$

where k is the current epoch, and f_d is an instantaneous Doppler frequency between a receiver and a satellite. If the time interval between epochs is short, (2) can be expressed as

$$\int_{k-1}^{k} f_d dt \approx \frac{d_i(k) + d_i(k-1)}{2} \cdot \Delta t + \varepsilon_{\Delta\phi_i}, \tag{3}$$

where d_i is a coarse Doppler shift measured by a receiver, which is immune to a cycle slip. Δt represents the time interval between successive epochs. $\varepsilon_{\Delta\phi_i}$ includes nuisance errors like carrier phase measurements and noise. The shorter the time interval between successive epochs, the smaller the error caused by the Doppler measurement [9].

If there is no cycle slip between two epochs, the difference of carrier phase measurement should be equal to the integrated Doppler measurements. However, these two values will have a significant difference when there is a cycle slip because the integer ambiguity in the carrier phase measurement at the two epochs is no longer the same. Note that the Doppler measurements, d, are not impacted by a carrier phase cycle slip.

Therefore, a cycle slip can be inferred from the following

$$T_{MDACSD} = \Delta\phi_i^s - \left\{\frac{d_i^s(k) + d_i^s(k-1)}{2} \cdot \Delta t\right\}, \tag{4}$$

where the superscript s refers to the satellite from which the signal is transmitted.

The DACSD method has the advantage of being able to detect cycle slips even in a single-frequency environment. Additionally, unlike the cycle slip detection method that detects through a differential method between frequencies or receivers, we can identify a specific frequency or measurement at which a cycle slip occurs. However, the noise level of the test statistics is high, making it difficult to detect a small cycle slip.

2.2. Proposed Single-Frequency Cycle Slip Detection for Small UAV Platforms

In this section, we propose a time-difference carrier phase (TDCP) combination method that better detects a cycle slip of single-frequency carrier measurements with a lower noise level than Doppler measurements. The TDCP combination method uses the differential value of the carrier phase measurement and can be expressed as

$$\begin{aligned} \lambda_i \cdot \frac{\phi_i^s(k) - \phi_i^s(k-1)}{\Delta t} &\approx \lambda_i \dot{\phi}_i^s \\ &= (\mathbf{v}^s - \mathbf{v}) \cdot \mathbf{1}^s + \dot{b} + \varepsilon_{\dot{\phi}_i^s}, \end{aligned} \quad (5)$$

where $\mathbf{v}^s$ is the satellite velocity vector, $\mathbf{v}$ is the user velocity vector, and $\dot{b}$ is change rate in a receiver clock. The rates of ionospheric and tropospheric delay are assumed to be negligible. $\varepsilon_{\dot{\phi}_i^s}$ is the combined noise term during measurement interval, and $\mathbf{1}^s$ is a line-of-sight vector of the satellite s. (5) can be rearranged as below

$$\lambda_i \cdot \frac{\phi_i^s(k) - \phi_i^s(k-1)}{\Delta t} - \mathbf{v}^s \cdot \mathbf{1}^s = -\mathbf{v} \cdot \mathbf{1}^s + \dot{b} + \varepsilon_{\dot{\phi}_i^s}, \quad (6)$$

Here, the left-hand side of (6) represents a value that we can calculate or obtain through measurements. The receiver velocity vector and receiver time drift value on the right-hand side are unknown and need to be estimated with a set of dual frequency measurements. When (6) is expressed in a matrix form with two frequencies, it can be represented as

$$\mathbf{Y}_{\text{dual}} = \mathbf{G}\mathbf{X} + \varepsilon_{\dot{\phi}^s}, \quad (7)$$

where $\mathbf{Y}_{\text{dual}} = \begin{bmatrix} \lambda_i \cdot \frac{\phi_1^1(k) - \phi_1^1(k-1)}{\Delta t} - \mathbf{v}^1 \cdot \mathbf{1}^1 \\ \vdots \\ \lambda_i \cdot \frac{\phi_2^m(k) - \phi_2^m(k-1)}{\Delta t} - \mathbf{v}^m \cdot \mathbf{1}^m \end{bmatrix}$, $\mathbf{G} = \begin{bmatrix} -\mathbf{1}^1 & 1 \\ \vdots & \vdots \\ -\mathbf{1}^m & 1 \end{bmatrix}$, and $\mathbf{X} = \begin{bmatrix} \mathbf{v} \\ \dot{b} \end{bmatrix}$. Here, $\mathbf{X}$ is estimated by pseudo-inverse using matrix $\mathbf{Y}$ and matrix $\mathbf{G}$

$$\hat{\mathbf{X}} = (\mathbf{G}^T\mathbf{G})^{-1}\mathbf{G}^T\mathbf{Y}_{\text{dual}}. \quad (8)$$

In this method, the measurements used for the user velocity and receiver clock error rate estimation should be free of any cycle slips. Therefore, we only used m dual-frequency measurements that have passed the dual-frequency carrier phase (DFCP) combination, which is often referred as a geometry-free method [14]. This is because the data that have passed through DFCP are highly likely to be free of cycle slips. Then, the $\hat{\mathbf{X}}$ is applied for the cycle slip detection of s single-frequency carrier phase measurement as follows:

$$T_{TDCP} = \lambda_i \cdot \frac{\phi_i^s(k) - \phi_i^s(k-1)}{\Delta t} - \mathbf{v}^s \cdot \mathbf{1}^s - \begin{bmatrix} -\mathbf{1}^s & 1 \end{bmatrix} \cdot \hat{\mathbf{X}}. \quad (9)$$

When there are no cycle slips, T_{TDCP} includes noise in carrier phase as well as estimated errors in $\hat{\mathbf{X}}$. An experimental distribution of T_{TDCP} is addressed in later sections.

The TDCP combination method has the advantage of being able to detect a small cycle slip even in single-frequency measurements. In contrast to the DFCP method, it is possible to specify the measurement in which cycle slip has occurred. Compared to DACSD, the noise level of the proposed TCDP is reduced by using the differential value of the carrier measurement. For this reason, the TDCP approach detects a smaller cycle slip and its false alarm is relatively lower than the one of DACSD.

3. Results

We collected data via flight tests and set a threshold for each cycle slip method with the flight test data. The flight tests were conducted at Hongik University's 4th Industrial Revolution Campus located in Hwaseong, South Korea. During flight, dual-frequency measurements of GPS, Galileo, and Beidou constellations were collected at 1Hz rate. Each flight took approximately 10 min to complete, and a total of 45 flight tests were conducted. Our testbed of a small UAS was based on a S550 frame shown in Figure 1a. The receiver and antenna onboard the testbed were Ublox ZED-F9P and Trimble AV17, respectively. The speed of the UAV ranged from 1 to 6 m/s.

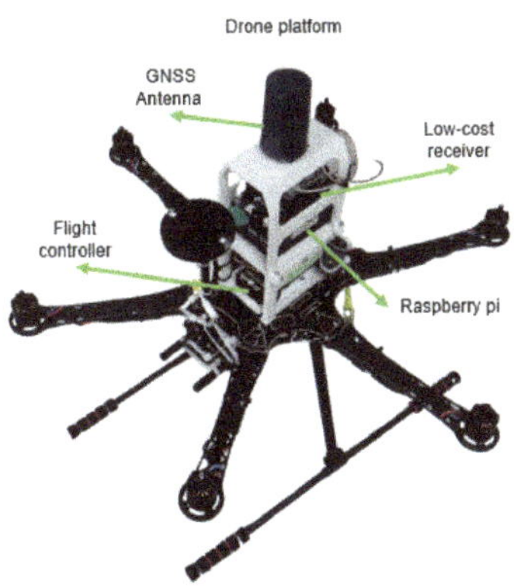

(a)

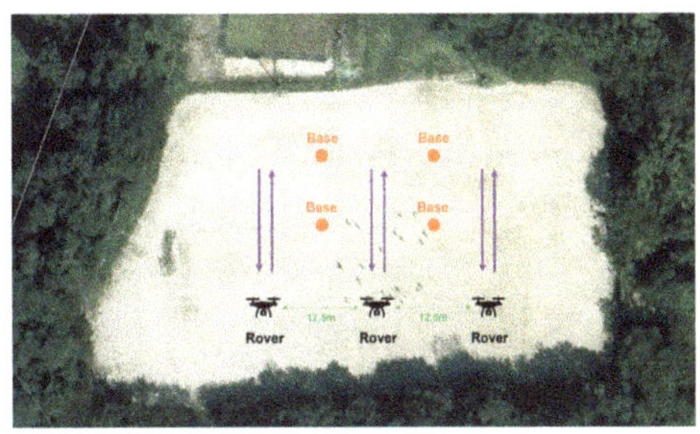

(b)

Figure 1. Hexacopter drone platform and flight test set-up. (**a**) Hexacopter drone platform with GNSS Antenna, flight controller, Ublox F9P receiver, and raspberry-pi board. (**b**) Trajectory of three drones during flight test.

Figure 2 shows the test statistics of DACSD and TDCP based on the flight test data. To establish a criterion for determining the occurrence of actual cycle slips, we adopted the results of the DFCP method by using dual-frequency carrier phase measurements only. Table 1 lists the number of false alarms and missed detection of cycle slips resulted from the DACSD and TDCP methods. There were no false alarms in the TDCP combination. On the other hand, in the case of DACSD, it can be observed that false alarms have occurred in many cases. Table 2 shows the number of false alarms and missed detection of simulated cycle slips injected to flight test data. For the smaller cycle slip of one wavelength, TDCP performs significantly better than DACSD.

Table 1. Performance comparison of MDACSD and DFCP in the aspects of false alarm and missed detection of flight data.

Method	False Alarm Rate	Missed Detection Rate	Data
DFCP	0%	0%	738,612
DACSD	22.4913%	0.058669%	738,612
TDCP	0%	0.025206%	738,612

Table 2. Performance comparison of DACSD and DFCP in the aspects of missed detection through injected cycle slips.

Dataset	Injected Cycle Slip of 1 Wavelength		Injected Cycle Slip of 2 Wavelength		No. of Simulation
	TDCP	DACSD	TDCP	DACSD	
1	952	0	952	291	959
2	711	0	711	302	711
3	1190	0	1190	288	1190
4	1080	0	1080	318	1080

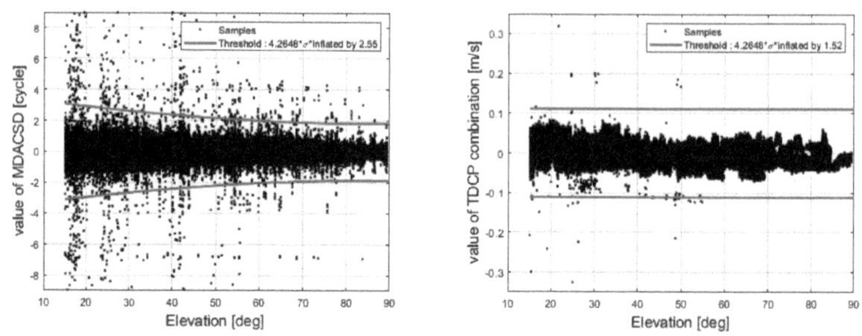

(a) (b)

Figure 2. Test statistics from the flight test data. The outliers much beyond the threshold were removed in the process of determining the thresholds. (**a**) The DACSD combination. (**b**) The TDCP combination.

4. Conclusions

We introduced a novel cycle slip detection method for single-frequency measurements using mixed dual- and single-frequency measurements. This method could result in more precise cycle slip test statistics compared to previous Doppler-based approaches. Our test results based on flight test data confirmed that the proposed TDCP approach provides lower false alarms and missed detection rates than Dopper-based approaches. Therefore, using this approach is advantageous in the RTK process as it allows a tighter cycle slip check on the set of single-frequency measurements obtained along with dual-frequency measurements.

Author Contributions: C.-h.L. developed algorithm and analyzed flight test data. E.K. provided advices throughout this research. All authors have read and agreed to the published version of the manuscript.

Funding: This research was supported by the Unmanned Vehicles Core Technology Research and Development Program through the National Research Foundation of Korea (NRF) and the Unmanned Vehicle Advanced Research Center (UVARC) funded by the Ministry of Science and ICT, Republic of Korea (No. 2020M3C1C1A01086407). This work was also supported by Future Space Navigation & Satellite Research Center through the National Research Foundation funded by the Ministry of Science and ICT, Republic of Korea (2022M1A3C2074404).

Institutional Review Board Statement: Not applicable.

Informed Consent Statement: Not applicable.

Data Availability Statement: Data sharing is not applicable to this article.

Conflicts of Interest: The authors declare no conflict of interest. The funders had no role in the design of the study; in the collection, analyses, or interpretation of data; in the writing of the manuscript; or in the decision to publish the results.

References

1. Vollath, U.; Landau, H.; Chen, X.; Doucet, K.; Pagels, C. Network RTK versus single base RTK-understanding the error characteristics. In Proceedings of the 15th International Technical Meeting of the Satellite Division of the Institute of Navigation (ION GPS 2002), Portland, OR, USA, 24–27 September 2002.
2. Rizos, C. Alternatives to current GPS-RTK services and some implications for CORS infrastructure and operations. *GPS Solut.* **2007**, *11*, 151–158. [CrossRef]
3. Bisnath, S.; Gao, Y. Current state of precise point positioning and future prospects and limitations. *Obs. Our Chang. Earth* **2009**, *133*, 615–623.
4. Zumberge, J.F.; Heflin, M.B.; Jefferson, D.C.; Watkins, M.M.; Webb, F.H. Precise point positioning for the efficient and robust analysis of GPS data from large networks. *J. Geophys. Res. Solid Earth* **1997**, *102*, 5005–5017. [CrossRef]
5. Xu, G.; Xu, Y.; Xu, G.; Xu, Y. *GPS: Theory, Algorithms and Applications*, 3rd ed.; Springer: Berlin/Heidelberg, Germany, 2016; pp. 229–261.
6. Yoon, Y.; Lee, B.; Heo, M. Multiple Cycle Slip Detection Algorithm for a Single Frequency Receiver. *Sensors* **2022**, *22*, 2525. [CrossRef] [PubMed]
7. Cai, C.; Liu, Z.; Xia, P.; Dai, W. Cycle slip detection and repair for undifferenced GPS observations under high ionospheric activity. *GPS Solut.* **2013**, *17*, 247–260. [CrossRef]
8. Li, T.; Melachroinos, S. An enhanced cycle slip repair algorithm for real-time multi-GNSS, multi-frequency data processing. *GPS Solut.* **2019**, *23*, 1–11. [CrossRef]
9. Zhao, J.; Hernández-Pajares, M.; Li, Z.; Wang, L.; Yuan, H. High-rate Doppler-aided cycle slip detection and repair method for low-cost single-frequency receivers. *GPS Solut.* **2020**, *24*, 1–13. [CrossRef]
10. Ren, Z.; Li, L.; Zhong, J.; Zhao, M.; Shen, Y. A real-time cycle-slip detection and repair method for single frequency GPS receiver. *Int. Proc. Comput. Sci. Inf. Technol.* **2011**, *17*, 224–230.
11. Colombo, O.L.; Bhapkar, U.V.; Evans, A.G. Inertial-aided cycle-slip detection/correction for precise, long-baseline kinematic GPS. In Proceedings of the 12th International Technical Meeting of the Satellite Division of The Institute of Navigation (ION GPS 1999), Nashville, TN, USA, 14–17 September 1999; pp. 1915–1922.
12. Takasu, T.; Yasuda, A. Cycle slip detection and fixing by MEMS-IMU/GPS integration for mobile environment RTK-GPS. In Proceedings of the 21st international technical meeting of the satellite division of the Institute of Navigation (ION GNSS 2008), Savannah, GA, USA, 16–19 September 2008; pp. 64–71.
13. Hernández-Pajares, M.; Juan, J.M.; Sanz, J.; Aragón-Àngel, À.; García-Rigo, A.; Salazar, D.; Escudero, M. The ionosphere: Effects, GPS modeling and the benefits for space geodetic techniques. *J. Geod.* **2011**, *85*, 887–907. [CrossRef]
14. Dai, Z. MATLAB software for GPS cycle-slip processing. *GPS Solut.* **2012**, *16*, 267–272. [CrossRef]

engineering proceedings

MDPI

Proceeding Paper

The Galileo High-Accuracy Service: Evaluating the Quality of the Corrections and Initial PPP Performance †

Camille Parra [1,2,*], Andreas Schütz [3], Urs Hugentobler [1], Thomas Pany [3] and Stefan Baumann [2]

1 Technical University of Munich, 80333 Munich, Germany
2 Industrieanlagen-Betriebsgesellschaft mbh (IABG), 85521 Ottobrunn, Germany
3 Universität der Bundeswehr München, 85579 Neubiberg, Germany
* Correspondence: camille.parra@tum.de

† Presented at the European Navigation Conference 2023, Noordwijk, The Netherlands, 31 May–2 June 2023.

Abstract: The Galileo High-Accuracy Service (HAS) is providing free Precise Point Positioning (PPP) corrections for Galileo and GPS satellites through the E6b signal and the internet. Currently, HAS should provide a horizontal and vertical 95% accuracy below 20 centimetres and 40 centimetres, respectively. To assess the accuracy of the current HAS corrections, software has been developed to receive, decode and use HAS corrections. HAS corrections were acquired in the Munich area (Germany) during two sessions in November 2022 and April 2023. The decoder was validated by comparing recorded corrections to data from the Galileo High Accuracy Reference Algorithm and User Terminal (HAUT). Then, the availability and quality of HAS corrections were analysed. The use of orbit and clock corrections significantly improves the broadcast product accuracy; for instance, the Galileo orbit RMS error decreases up to 43% and the GPS RMS up to 80%. The code bias accuracy is at sub-nanosecond level. Finally, to validate the use of HAS corrections, PPP positioning has been achieved. For the used station network, we reached a 95% horizontal and vertical accuracy of 19 cm and 34 cm, respectively, for the kinematic and bi-constellation positioning matching with targeted HAS performances.

Keywords: Galileo; high-accuracy service; precise point positioning

Citation: Parra, C.; Schütz, A.; Hugentobler, U.; Pany, T.; Baumann, S. The Galileo High-Accuracy Service: Evaluating the Quality of the Corrections and Initial PPP Performance. *Eng. Proc.* **2023**, *54*, 14. https://doi.org/10.3390/ENC2023-15450

Academic Editors: Tom Willems and Okko Bleeker

Published: 29 October 2023

1. Introduction

Since the 1990s, Precise Point Positioning (PPP) has become an increasingly popular alternative to the Real-Time Kinematic (RTK). While the RTK, or Network RTK (NRTK), is based on differential positioning, which uses one reference station or a local network to eliminate the measurement errors, the PPP technique relies on precise estimations of satellite orbit and clock products [1]. Moreover, nowadays, with the State Space Representation (SSR) of corrections, the PPP achieves similar performances to RTK with centimetre-level accuracy in real time [2] or post-processing and for multiple static [3] or kinematic applications [4].

However, with the evolution of our needs, centimetre accuracy is not essential for every application, so Quasi–Zenith Satellite System (QZSS), BeiDou and Galileo constellations have created free-of-charge services to meet decimetric needs. The Japanese Centimetre Level Augmentation Services (CLAS) [5] service is the first operational service of its kind, and perhaps the most comprehensive service available today, because it is the only satellite centimetre-level augmentation service to provide atmospheric corrections. The CLAS service has the same area coverage as QZSS and should achieve positioning with a 95% accuracy, better than 12 cm in horizontal and 24 cm in vertical. PPP-B2b is a Chinese regional PPP service that provides PPP corrections [6]. These BeiDou and GPS corrections are transmitted, through the B2b signal of the BeiDou's GEO satellites, for a user located in China or surrounding regions. The expected performances for 95% accuracy are 20 cm in the horizontal direction and 40 cm in the vertical direction. The targeted performances of each service are summarize in Table 1.

Like their Asian counterparts, in January 2023, Galileo initiated a similar service: the Galileo High-Accuracy Service (HAS). Indeed, since the beginning of the year, the Galileo HAS has officially provided corrections for PPP free of charge [7]. These corrections are openly available through the internet and the E6b signal. Nonetheless, thanks to their Medium Earth Orbit (MEO) satellites, this is the first globally available service and it utilizes a scheme called High-Parity Vertical Reed–Solomon (HPVRS). This scheme improves the reception performance of common corrections broadcast by different satellites [8]. The HAS is divided into two service levels: the first one (SL1), currently available worldwide, is broadcasting orbit and clock corrections and code biases. The second service level (SL2), which will be available in 2024 for the European Coverage Area (ECA), will send atmospheric corrections. For the two service levels, the Galileo HAS expected horizontal and vertical 95% accuracies were originally below 20 cm and 40 cm, respectively. However, with atmospheric corrections, the convergence time should be shorter than 100 s; in other words, three times faster than with SL1.

Table 1. PPP services provided by the constellations.

Service Name	Constellations with Corrections	95% Horizontal Accuracy [cm]	95% Vertical Accuracy [cm]	Coverage
Centimetre Level Augmentation Services (CLAS)	QZSS, GPS, Galileo	12	24	Regional
PPP-B2P	Beidou, GPS	20	40	Regional
Galileo High-Accuracy Service (HAS)	GPS, Galileo	20	40	Global (SL1), Regional (SL2)

Hauschild et al. [9] focused on the use of Galileo HAS for the orbit determination of satellites in Low Earth Orbit (LEO) and highlighted satisfactory results with Galileo HAS. Their results showed a Signal-In-Space Ranging Error (SISRE) Root Mean Square (RMS) of 7.4 cm for Galileo and 12.2 cm for GPS. A PPP solution for an International GNSS Service (IGS) permanent station has been computed too. They achieved a 95th percentile error of 11.7 cm, 14.5 cm and 29.7 cm for the North, East and Up component, respectively.

Despite these promising performances, with the publication of the Service Definition Document (SDD) in January 2023 [10], the European Union Agency for the Space Program (EUSPA) has relaxed the HAS performances and reduced the service area. Presently, the target horizontal and vertical accuracies are 15 cm and 20 cm respectively, but they are defined for the 68th percentile, for a static user and outside of the Pacific area. Users in this excluded area could still use the HAS corrections, but they may have downgraded results.

After a short description of the HAS data, the present study aims to confirm the targeted performances of HAS with respect to the expected performances published by the EUSPA. For this purpose, the HAS corrections availability and accuracy were analysed, and the HAS satellite code biases assessed with respect to IGS products. Then, once the quality of the corrections was proven, PPP was performed and compared to the expected accuracy. The article finishes with a discussion and conclusions for future work using the preliminary results of this study.

2. HAS Signal-in-Space

HAS provides free PPP corrections for high-accuracy global positioning applications [11]. These corrections are composed of orbit and clock corrections, as well as code and phase biases for four frequencies of Galileo (E1, E5a, E5b, E6) and the three frequencies of GPS (L1, L2C and L5). All these corrections are available in real time and are broadcast through the

E6b signal and the internet. Currently, because the network includes 5 uplink stations with 4 antennas each, a maximum of 20 satellites could transmit the corrections simultaneously.

The reference time of HAS is the Galileo System Time (GST) and the reference frame is Galileo Terrestrial Reference Frame (GTRF) [12]. The HAS corrections refer to the satellite Antenna Phase Center (APC). The APC is defined as the centre of the phase of the satellite antenna for the dual-frequency ionosphere-free signals combination for GPS and Galileo.

The orbit and clock corrections have to be applied to the navigation messages. The orbit corrections are provided in the Satellite Coordinate System NTW composed of the radial (N), tangential (T) and normal (W) components; thus, rotation must be performed to obtain the corrections in the Earth-Centred Earth-Fixed (ECEF) reference frame of the broadcast ephemeris. The clock corrections and the biases are added directly to the satellite clock error and to the measurements, respectively.

3. Data

For the first part of our work, we validated our Galileo HAS decoder, which was integrated in the software receiver Multi-Sensor Navigation Analysis Tool (MuSNAT) [13]. We compare our received corrections to the data decoded using the Galileo High-Accuracy Reference Algorithm and User Terminal (HAUT) belonging to EUSPA and developed by the SpaceOpal company [14]. Both receivers were connected to the same antenna (Trimble Zephyr 2) and received the same data feed. The data were recorded for two Days of Year (DOY)—days 307 and 308—in 2022.

To obtain more recent and longer-term data from after the service was notified as operational, a one-week dataset was recorded in April 2023, from DOY 091 to 098. As stated by EUSPA, no phase biases were broadcast at this time. It is important to note that the study was performed at the beginning of the service and the results presented in this paper may have evolved in the coming months.

The statistical results are similar for the two sessions, and except for the validation with the HAUT receiver, only the results from the second period will be shown.

As mentioned before, we compare our decoded corrections to the data from the HAUT receiver. We also conduct comparisons between the navigation messages that have been corrected by Galileo HAS and the precise products from an analysis center. To limit the potential causes of problems, and evaluate only Galileo HAS corrections, the broadcast messages used are the combined products from IGS. The precise products used are the final ephemeris and clocks products from the Groupe de Recherche de Géodésie Spatiale (GRG) and the Differential Code Bias (DCB) from the Chinese Academy of Sciences (CAS). These products could be used as a reference because they are among the most precise products available nowadays.

4. Assessment of HAS Data

4.1. Comparison between HAUT and MuSNAT

To verify the message readout, the correction parameters provided by MuSNAT were compared with an external reference, which was provided by the HAUT receiver, in a dedicated measurement campaign in November 2022. For this purpose, a zero-baseline test was performed by connecting both receivers to the same GNSS antenna using a splitter. This ensured that the same input signal was used for both receivers.

This measurement campaign successfully proved the correct functioning of the entire chain of our decoder. Figure 1 shows the difference between the orbit correction data received by MuSNAT and the reference. The disparity is plotted, in meters, for each component during the first 7 h of the DOY 307. Each color represents a different Galileo satellite.

The differences in the millimetre range that could occur during the day are due to differences in the time stamping with the reception time. The reception time is determined by means of code measurements in a Single Point Positioning (SPP) algorithm, which works independently on both receivers. That is the reason why these disparities can be observed at the same epoch for all three components. The results are similar for GPS and for the second day.

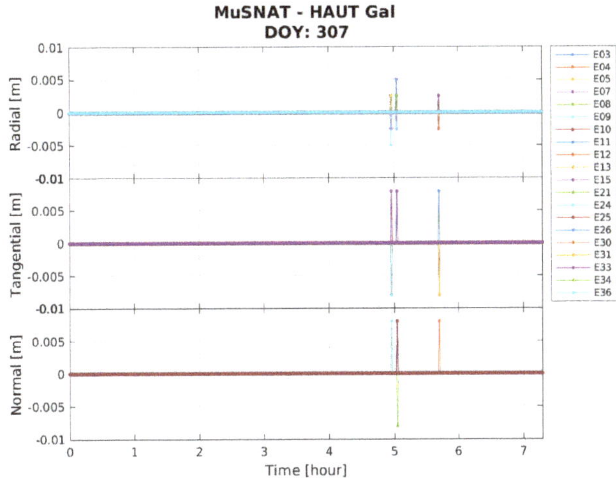

Figure 1. HAS orbit correction differences between Multi-Sensor Navigation Analysis Tool (MuSNAT) and High-Accuracy Reference Algorithm and User Terminal (HAUT) for DOY 307 of year 2022.

4.2. Availability Analysis

In this section, the availability of Galileo HAS corrections is analysed from DOY 091 to 098. During this period, the HAS messages contained orbit and clock corrections and code biases. As the code biases were always available during our assessment, the analysis only concerns orbits and clocks.

The availability of HAS corrections was checked every 30 s over the considered week. A filter was applied to flag every correction exceeding the interval of valid values defined in the ICD [12] and summarized in the Table 2. Indeed, if the corrections were not included in their interval, they were considered bad and not used. The code biases had a validity interval of 300 s, but they were relatively constant over the day.

Table 2. Range and validity interval of the HAS orbit and clock corrections and code biases.

Correction	Values Interval [m]	Validity Interval [s]
Orbit-radial	±10.2375	300
Orbit-along-track	±16.376	300
Orbit-cross-track	±16.376	300
Delta clock	±10.2375	60
Code bias	±20.46	300

Figure 2 shows the mean availability per satellite of usable HAS orbit corrections over the studied time interval. The HAS clock corrections availability is similar to the orbit corrections. Satellites E14, E18, E20 and E22 were not analysed because they were flagged as "not usable" by the Galileo operator. Satellites E24 and G02 were not usable, according to their constellation information, for DOY 091 and from DOY 096 to 098, respectively.

For the evaluated week, almost all the evaluated satellites had an availability above 90% (Figure 2), with an average value of 97% for Galileo and 95% for GPS. The lower availability of GPS was partly due to some unusable corrections. Indeed, only the GPS orbit corrections had some out-of-range values, which were the corrections exceeding the validity interval (Table 2), most of the time less than 1% for the time interval. However, spikes could be noticed for some satellites and for some days of up to 13%. Only the satellite

G02 had a mean availability below 90%, mainly because more than 5% of its corrections were set as unusable during its four days of availability. This could be because this satellite was set as unusable during 3 days and the corrections needed a period to converge.

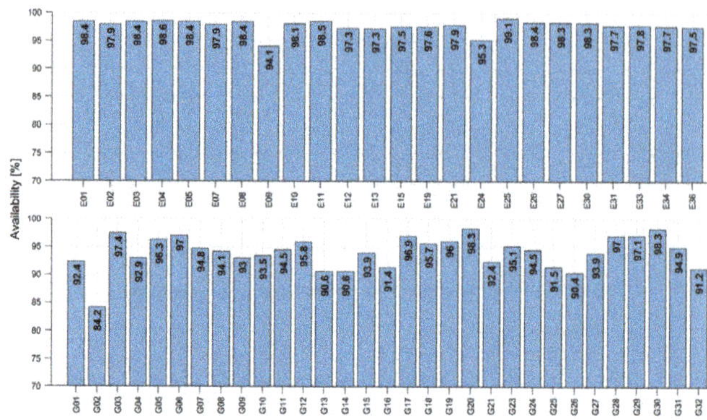

Figure 2. Mean availability of orbit corrections for Galileo (**top**) and GPS (**bottom**) from DOY 091 to 098.

To conclude the availability analysis for the evaluated week, the mean availability was 97% for Galileo and 95% for GPS. The expected availability of the Galileo HAS service was 99%, so this figure has yet to be achieved for both constellations, but these results are promising for further work. For the purpose of comparison, according to [15], the availability of GPS corrections with the RTS products of the IGS was similar to that obtained using Galileo HAS.

4.3. Products Comparison

To evaluate the HAS correction accuracy, corrected navigation messages were assessed over seven days with respect to precise products from the GRG for ephemeris and clocks, and from the CAS for code biases.

The orbit comparison was based on the difference between the broadcast and the GRG ephemeris coordinates, both in Earth-Centred Earth-Fixed frame (ECEF). As the GRG products refer to the centre of mass, while the HAS ephemeris refer to the ionosphere-free antenna phase centre, the satellite antenna corrections from IGS had to be applied using the latest ANTEX file [16]. Then, the conversion from ECEF to NTW was performed. In the clock comparison procedure, a mean value per constellation was estimated based on the difference between the broadcast clocks correcting with HAS and the GRG products, and subsequently eliminated. This value was due to the different reference clock used for the clock computation and could be removed because the receiver clock would absorb this offset during the user positioning and would thus not impact it. An alignment had to be performed, too, because the GRG clocks refer to the ionosphere-free combination of C1C/C5Q for Galileo and C1W/C2W for GPS [17], while HAS corrections refer to the ionosphere-free combination of the used navigation message, i.e., INAV (C1C/C7Q) for Galileo and LNAV (C1C/C2P) for GPS. All the signals were identified according to the RINEX convention [18]. This alignment was performed using the code biases from CAS.

Furthermore, to complete this comparison of the products, the Signal-In-Space Ranging Error (SISRE) was evaluated. The SISRE reflects the combined error of broadcast ephemeris and clock offset and thus shows the performance of the space and ground segment. In [19], Montenbruck et al. highlighted that SISRE could be a key indicator for the GNSS performance monitoring due to its dependence on the space segment (clock stability and predictability) and on the ground segment (orbit and clock determination performance). Table 3 summarizes the accuracy of the broadcast products with or without HAS corrections for the whole week. To detail the error distribution, Figure 3 represents the RMS errors for the radial, along-track and cross-track components, the clock error and the SISRE, for each satellite of Galileo and GPS, for the analysed time interval. Satellite G02 has been removed of the statistical analysis because its corrections were not broadcast for several days and it has a relatively high percentage of unusable corrections during the other days.

Table 3. RMS errors for radial, along-track and cross-track orbit components, clock error and SISRE RMS for broadcast ephemerides of GPS and Galileo corrected with HAS. The values to the left refer to the non-corrected broadcast.

Constellation	Radial [cm]	Along-Track [cm]	Cross-Track [cm]	Clock [cm]	Signal-In-Space Ranging Error (SISRE) [cm]
GPS	51/18	75/15	65/17	35/17	60/23
GAL	17/14	23/13	20/14	14/8	19/16

Without the HAS corrections, the Galileo ephemeris and clocks seem better than the GPS products. This is mainly because of the update rate. Indeed, Galileo navigation messages are updated every 10 min, while the GPS navigation messages are refreshed every two hours. So, with the higher update rate of HAS, the improvement was more significant for GPS. Thus, the GPS orbit errors decreased by 65%, 80% and 74% for the radial, tangential and normal components, respectively. For Galileo, the improvement was less significant, but remained above 17% for each component. Regarding the clock errors, the accuracy was improved with the use of HAS (51% and 43% for GPS and Galileo accuracies). For the SISRE, the analysis is similar. For GPS, the major error seems to be the imprecision of the clock, whereas for Galileo, the SISRE is dominated by the orbit errors.

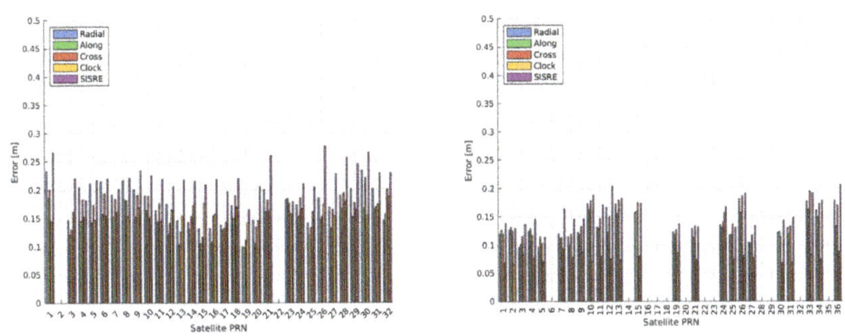

Figure 3. Orbit, clock and Signal-In-Space Ranging Error (SISRE) RMS errors for GPS (**left**) and Galileo (**right**) with respect to Groupe de Recherche de Géodésie Spatiale (GRG) final products, from DOY 091 to 098.

To conclude the HAS product evaluation, the biases were compared to the CAS DCB for the entire week. Contrary to the DCB, which is a difference between two code biases, the HAS code biases are given as observable-specific signal biases (OSB), thus non-differenced code bias. So, they have been differenced to obtain DCB. With the biases available, three Galileo combinations were possible: C1C-C5Q, C1C-C7Q, C1C-C6C; and one GPS combination: C1C-C2L. A constellation average value of −1.74 ns was removed from all the GPS bias differences. During the user positioning, this constellation-dependent mean is absorbed by the receiver clock estimation; thus, it has no impact on the positioning result. To conform to the CAS DCB convention, the results of this section are presented in nanoseconds and not in meters.

Figure 4 shows the median of absolute values and the standard deviation of the difference between HAS code biases and CAS DCB. For the week, the medians of the four couples are below 0.2 ns, with 0.12 ns for C1C-C5Q, 0.16 ns for C1C-C7Q, 0.11 ns for C1C-C6C and 0.14 ns for C1C-C2L. Furthermore, STDs are around 0.1 ns. This shows good consistency and stability of the DCB over the week. Similar results can be found in the literature [20,21].

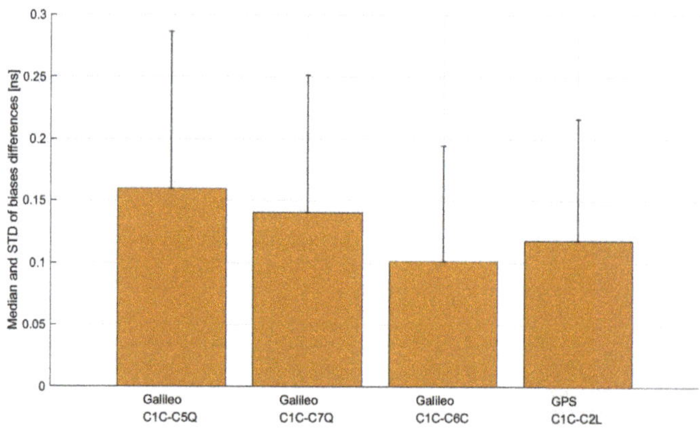

Figure 4. Median and STD of HAS code biases minus the Differential Code Bias (DCB) from the Chinese Academy of Sciences (CAS), from DOY 091 to 098 of year 2023 (except DOY 097).

5. PPP Application

5.1. PPP Configuration

To complete our analysis of HAS products, PPP is performed. The PPP engine is a custom-made software, which is currently under development as a part of the MuSNAT project [13]. The software is still quite new; nevertheless, it has already provided good results. The PPP is performed on a network of six European permanent stations (Figure 5) for the first 10 h of DOY 096 in the year 2023, with a sampling of 1 second. All the stations belong to the IGS network. The processing is performed in kinematic mode using an Extended Kalman Filter, with float ambiguities. The observations are processed in dual-frequency mode (L1C/L2W for GPS and E1C/E5Q for Galileo), with a between-satellite single-difference linear combination (BSSD) of un-combined measurements. BSSD combination should remove some receiver-related biases, like receiver clock error. Regarding the products comparison, the satellite and receiver antenna are corrected using the latest ANTEX file, and the combined broadcast file is used to compute the ephemeris. The accuracy is computed using the GNSS station position products file of the processed DOY. All three files are obtained from IGS. To remove the low elevation satellites, a cut-off of 10° is applied.

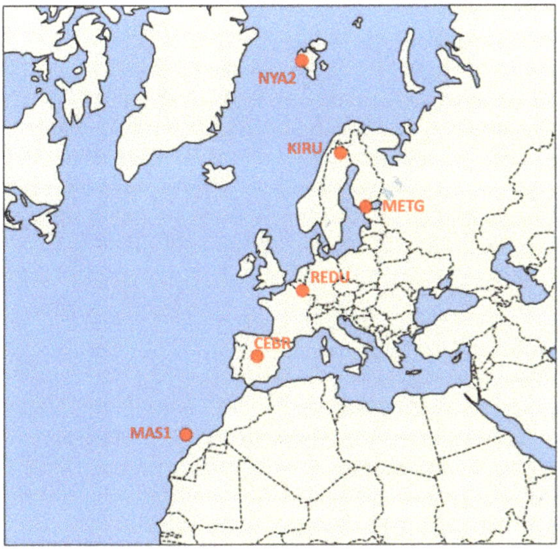

Figure 5. Permanent stations used for the PPP performance evaluation.

5.2. PPP Results

Figure 6 summarizes the accuracy obtained with the PPP processing. The accuracy is given as RMS of the 95% accuracies of the six stations during the 10 h of kinematic processing. It has been computed for joint Galileo GPS processing, but also separately.

All processes achieve a horizontal and vertical accuracy of a few decimetres. The GPS-Galileo and Galileo-only solutions have similar results, with horizontal and vertical accuracies of 18.9 cm and 34 cm for GPS-Galileo, and 15.8 cm and 37.9 cm for Galileo only. Both have accuracies better than the expected performance (20 cm and 40 cm according to [7]). The GPS-only solution has a slightly degraded result, with horizontal and vertical accuracies of 21.5 cm and 49.7 cm, respectively. This degradation is consistent with what we can expect. Indeed, the GPS-only position is degraded not only due to the lower number of satellites, resulting in reduced redundancy and poorer geometry, but also, as we observed during the HAS corrections evaluation, because the GPS SISRE is worse than that of Galileo.

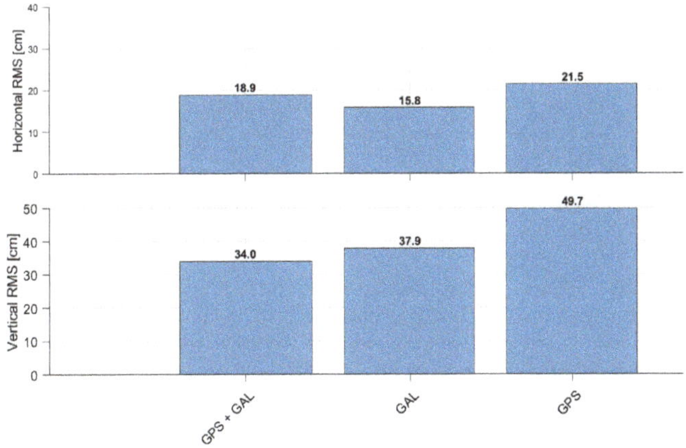

Figure 6. RMS of 95% accuracies for the Galileo + GPS, Galileo-only and GPS-only, in kinematic mode.

6. Discussion and Conclusions

With HAS, Galileo is providing a global PPP service for free and directly through the signal E6b. It is the first constellation to present such globally available service. The goal of this paper is to show an evaluation of the Galileo HAS corrections with a quality assessment of corrected products and with PPP positioning.

First, it is worth noticing that the HAS availability is quite good, with 97% and 95% availability for Galileo and GPS corrections over a one-week period. This is slightly below the expectations of 99% of availability, but this is promising for the future and similar to the availability of the RTS products of IGS. The GPS corrections are a little bit less available, mostly because some of them are flagged as "unusable". The accuracy of the corrected navigation messages with Galileo HAS obtained through comparison with IGS final products is high. The SISRE is 23 cm for GPS and 16 cm for Galileo, over the analysed week; this is dominated by the clock errors for GPS and by the ephemeris errors for Galileo. Furthermore, code biases have been evaluated with respect to the CAS DCB. They reveal a sub-nanosecond-level accuracy, with a median of 0.12 ns for C1C-C5Q, 0.16 ns for C1C-C7Q, 0.11 ns for C1C-C6C and 0.14 ns for C1C-C2L. These figures show good consistency over a week of the HAS code bias.

PPP positioning was achieved for a European network of six IGS stations. The results show consistency with the expected values (20 cm and 40 cm for the horizontal and vertical accuracy), because the Galileo and GPS solution achieves a horizontal and vertical accuracy of 18.9 cm and 34 cm, respectively.

In light of this paper, future work will involve the acquisition of new data as soon as phase biases are broadcasted, which should also allow ambiguity resolution. Moreover, a new Galileo Ground Sensor Station (GSS) is installed in Wallis, so an analysis of its contribution is planned. Finally, because Galileo HAS target markets are mainly moving users (as rail, road, maritime or space applications [11]), a test conducted in real kinematic conditions will be carried out on a train or a car.

Author Contributions: Methodology, C.P. and A.S.; software, C.P. and A.S.; visualization, C.P.; writing—original draft, C.P.; writing—review and editing, C.P., A.S., U.H., T.P. and S.B.; supervision, U.H., T.P. and S.B. All authors have read and agreed to the published version of the manuscript.

Funding: This research received no external funding.

Institutional Review Board Statement: Not applicable.

Informed Consent Statement: Not applicable.

Data Availability Statement: All data could be find in IGS archive center https://cddis.nasa.gov/archive, except the HAS data which are live stream recording.

Acknowledgments: The authors acknowledge the IGS for the provision of GNSS products used in this study. The authors would like to thank Munich Aerospace for the grant that made this study possible. The authors gratefully acknowledge SpaceOpal and EUSPA for the loan of the HAUT receiver.

Conflicts of Interest: The authors and IABG declare no conflict of interest.

References

1. Zumberge, J.F.; Heflin, M.B.; Jefferson, D.C.; Watkins, M.M.; Webb, F.H. Precise point positioning for the efficient and robust analysis of GPS data from large networks. *J. Geophys. Res.* **1997**, *102*, 5005–5017. [CrossRef]
2. Laurichesse, D.; Blot, A. Fast PPP Convergence Using Multi-constellation and Triple-frequency Ambiguity Resolution. In Proceedings of the 29th International Technical Meeting of the Satellite Division of The Institute of Navigation (ION GNSS+ 2016), Portland, OR, USA, 12–16 September 2016; pp. 2082–2088.
3. Petit, G.; Kanj, A.; Loyer, S.; Delporte, J.; Mercier, F.; Perosanz, F. 1×10^{-16} frequency transfer by GPS PPP with integer ambiguity resolution. *Metrologia* **2015**, *52*, 301–309. [CrossRef]
4. Carlin, L.; Hauschild, A.; Montenbruck, O. Precise point positioning with GPS and Galileo broadcast ephemerides. *GPS Solut.* **2021**, *25*, 77. [CrossRef]

5. Chiu, P.C.; Yeh, S.J.; Jan S.S. Performance Analysis of QZSS Centimeter Level Augmentation Services (CLAS). In Proceedings of the Stanford Position, Navigation and Time, Stanford, CA, USA, 9–30 October 2019.
6. Chen, L.; Zhou, G.; Chen, G.; Sun, W.; Pan, L. Signal-in-space and positioning performance of BDS open augmentation service. *Math. Probl. Eng.* **2022**, *2022*, 1112646. [CrossRef]
7. European GNSS Agency Galileo High Accuracy Service (HAS) Info Note. Available online: https://www.gsc-europa.eu/sites/default/files/sites/all/files/Galileo_HAS_Info_Note.pdf (accessed on 26 April 2023).
8. Fernández-Hernández, I.; Senni, T.; Borio, D.; Vecchione, G. High-parity vertical Reed-Solomon codes for long GNSS high-accuracy messages. *Navigation* **2020**, *67*, 365–378. [CrossRef]
9. Hauschild, A.; Montenbruck, O.; Steigenberger, P.; Martini, I.; Fernandez-Hernandez, I. Orbit determination of Sentinel-6A using the Galileo high accuracy service test signal. *GPS Solut.* **2022**, *26*, 120. [CrossRef]
10. European GNSS Agency Galileo High Accuracy Service Service Definition Document (HAS SDD). Available online: https://www.gsc-europa.eu/sites/default/files/sites/all/files/Galileo-HAS-SDD_v1.0.pdf (accessed on 26 April 2023).
11. European GNSS Agency Galileo High Accuracy Service (HAS). Available online: https://www.gsc-europa.eu/galileo/services/galileo-high-accuracy-service-has (accessed on 26 April 2023).
12. European GNSS Agency Galileo High Accuracy Service Signal-In-Space Interface Control Document (HAS SIS ICD). Available online: https://www.gsc-europa.eu/sites/default/files/sites/all/files/Galileo_HAS_SIS_ICD_v1.0.pdf (accessed on 26 April 2023).
13. Arizabaleta, M.; Ernest, H.; Dampf, J.; Kraus, T.; Sanchez-Morales, D.; Dötterböck, D.; Schütz, A.; Pany, T. Recent Enhancements of the Multi-Sensor Navigation Analysis Tool (MuSNAT). In Proceedings of the 34th International Technical Meeting of the Satellite Division of The Institute of Navigation (ION GNSS+ 2021), St. Louis, MI, USA, 20–24 September 2021; pp. 2733–2753.
14. Pintor, P.; González, E.; Senado, A.; Bohlig, P.; Sperl, A.; Henkel, P.; Simón, J.; Hernández, C.; de Blas, J.; Vázquez, J. Galileo High Accuracy Service (HAS) Algorithm and Receiver Development and Testing. In Proceedings of the 35th International Technical Meeting of the Satellite Division of The Institute of Navigation (ION GNSS+ 2022), Denver, CO, USA, 19–23 September 2022; pp. 836–851.
15. Hadas, T.; Bosy, J. IGS RTS precise orbits and clocks verification and quality degradation over time. *GPS Solut.* **2015**, *19*, 93–105. [CrossRef]
16. Rebischung, P.; Schmid, R. IGS14/igs14.atx: A new framework forthe IGS products. In Proceedings of the AGU Fall Meeting 2016, San Francisco, CA, USA, 12–16 December 2016.
17. Montenbruck, O.; Steigenberger, P.; Prange, L.; Deng, Z.; Zhao, Q.; Perosanz, F.; Romero, I.; Noll, C.; Stürze, A.; Weber, G.; et al. The multi-GNSS experiment (MGEX) of the international GNSS service (IGS)—Achievements, prospects and challenges. *Adv. Space Res.* **2017**, *59*, 1671–1697. [CrossRef]
18. RINEX the Receiver Independent Exchange Format Version 4.00. Available online: https://files.igs.org/pub/data/format/rinex_4.00.pdf (accessed on 27 November 2023).
19. Montenbruck, O.; Steigenberger, P.; Hauschild, A. Broadcast versus precise ephemerides: A multi-GNSS perspective. *GPS Solut.* **2015**, *19*, 321–333. [CrossRef]
20. Montenbruck, O.; Hauschild, A.; Steigenberger, P. Differential code bias estimation using multi-GNSS observations and globalionosphere maps. *Navig. J. Inst. Navig.* **2014**, *61*, 191–201. [CrossRef]
21. Villiger, A.; Schaer, S.; Dach, R.; Prange, L.; Sušnik, A.; Jäggi, A. Determination of GNSS pseudo-absolute code biases and their long-term combination. *J. Geod.* **2019**, *93*, 1487–1500. [CrossRef]

engineering proceedings

Proceeding Paper

Estimating Satellite Navigation Broadcast Ephemeris via Inter-Satellite and Ground-to-Satellite Ranging [†]

Andrea Testa [1], Grzegorz Michalak [1], Manuele Dassié [1,2,*], Karl Hans Neumayer [3] and Gabriele Giorgi [1]

1 German Aerospace Center (DLR), Münchener Strasse 20, 82234 Wessling, Germany; gabriele.giorgi@dlr.de (G.G.)
2 Faculty VI Planning Building Environment, Technische Universität Berlin (TUB), Strasse des 17. Juni 135, 10623 Berlin, Germany
3 German Research Centre for Geosciences (GFZ), Telegrafenberg, 14473 Potsdam, Germany
* Correspondence: manuele.dassie@dlr.de; Tel.: +49-8153-28-4963
† Presented at the European Navigation Conference 2023, Noordwijk, The Netherlands, 31 May–2 June 2023.

Abstract: This paper investigates the potential of performing orbit determination directly in the Earth-fixed frame based on Inter-Satellite Ranging (ISR) measurements as primary observables, combined with Ground-to-Satellite Ranging (GSR) measurements from a small regional ground network. Current Global Navigation Satellite Systems (GNSSs) use L-band pseudo-range and carrier phase measurements from global or regional ground station networks to perform dynamic Orbit Determination and Time Synchronization (ODTS), whereas sparse Satellite Laser Ranging measurements are mainly used for validation. Future GNSSs may be equipped with inter-satellite links (ISLs) to enable inter-satellite clock offset estimation, ranging and data relay. These capabilities carry the potential to significantly improve ODTS procedures. In this work, we assume a fully connected constellation via pair-wise ISLs, with measurement topology assigned by appropriate link schedulers. The satellite orbits are parametrized with the standard 15 Galileo broadcast perturbed Keplerian elements, estimated by using ISR and GSR measurements. This processing strategy eliminates the complex modeling of gravitational and non-gravitational forces, making it particularly suitable for on-board applications and offering an alternative to classical GNSS orbit determination processing architectures. The proposed orbit determination scheme can be used in case of a ground segment failure as a back-up procedure to estimate the orbits of the GNSS satellites onboard of each satellite and guaranteeing a continuous navigation message generation for the system users. The performance of the proposed method depends on a number of factors, such as the length of the data fitting interval, the measurement quality (precision and accuracy), the scheduling and geometry of ISR and GSR measurements, the number and distribution of ground stations, and the accuracy of the ground station coordinates. Preliminary results show that an orbit-only Signal-in-Space Range Error (SiSRE) in the order of 7–9 cm can be obtained by processing 2 to 3 h data with a limited set of supporting ground stations. In this study, the orbit determination scheme proposed is tested on different scenarios, providing a first assessment of attainable performance.

Keywords: on-board orbit determination, inter-satellite links, laser ranging, ground-to-satellite ranging, perturbed Keplerian elements, signal-in-space range error

Citation: Testa, A.; Michalak, G.; Dassié, M.; Neumayer, K.H.; Giorgi, G. Estimating Satellite Navigation Broadcast Ephemeris via Inter-Satellite and Ground-to-Satellite Ranging. *Eng. Proc.* **2023**, *54*, 15. https://doi.org/10.3390/ENC2023-15463

Academic Editors: Tom Willems and Okko Bleeker

Published: 29 October 2023

1. Introduction

Global Navigation Satellite Systems (GNSSs) transmit satellite positions and clock offsets in a navigation message that enables real-time navigation on Earth and in space. In order to generate the message, a dynamic precise orbit determination and time synchronization (ODTS) is performed, mainly based on L-band observables—pseudoranges and carrier phases—collected by ground sensor stations deployed globally (in the case of Galileo and GPS) or regionally (Beidou and GLONASS). The ODTS process estimates a large number of nuisance parameters necessary for accurate modeling of satellite dynamics (e.g., solar radiation pressure), observation biases (e.g., antenna offsets), atmospheric delays, satellite and station clock offsets, and Earth rotation parameters for the transformation of orbits from inertial to an Earth-fixed frame [1]. This entails a computationally extensive processing of long data batches performed by the GNSS ground segment. Once the orbit determination is completed, orbital arcs are predicted, parametrized with a set of broadcast parameters, and encoded into the navigation message, which is then uploaded to the satellites. The accuracy of the broadcast ephemeris is currently at the level of decimeters to meters [1].

Optical Inter-Satellite Links (OISLs) are being considered in GNSS to provide inter-satellite ranging (ISR) and time-transfer capabilities [2–4]. By exploiting active ISR, capable of cm-level accuracy, classical Precise Orbit Determination (POD) can be significantly enhanced. It is shown in [5] that orbits can be estimated with sub-cm accuracy, with an improvement of one order of magnitude with respect to the capabilities of current GNSSs. It is of interest to investigate the potential of performing autonomous orbit determination using ISR as a primary observable, without exploitation of an extended network of ground sensor stations logging navigation data for subsequent processing.

Direct estimation of the satellite Cartesian coordinates in an Earth-fixed frame with only ISR measurements would not be feasible in general due to rank deficiency of the system and rotational and translational ambiguities. Any orbit solution solely based on ISR would be ambiguous by a rotation and a translation of the entire constellation. This can be resolved by including Ground-to-Satellite Ranging (GSR) observations from ground reference stations, by parametrizing the orbit via a reduced number of parameters (here, we used the 15 Galileo F/NAV broadcast ephemeris, i.e., the classical six Keplerian elements complemented by nine correction parameters) and by batch-processing data collected over longer time intervals. The GSR measurements can be passive, e.g., Satellite Laser Ranging, or active, e.g., optical satellite-to-ground links.

2. Functional and Stochastic Models for the Ranging Observations

For the purpose of this study, let us consider the following variables. The three-element vector $\mathbf{r}_i(t) = (x_i(t), y_i(t), z_i(t))^{\mathrm{T}}$ denotes the 3D coordinates identifying the position of object i (satellite's center of mass, or ground station reference point) at time t in the Earth-Centered Earth-Fixed (ECEF) reference frame. The ranging observable, denoted with $\rho_{ij}(t)$, is the measured range between objects i and j at time t,

$$\rho_{ij}(t) = \|\mathbf{r}_i(t) - \mathbf{r}_j(t - \Delta_{ij})\| + \varepsilon_{ij}(t), \quad \varepsilon_{ij}(t) \sim \mathcal{N}(b_{ij}(t), \sigma_{ij}^2), \tag{1}$$

where Δ_{ij} is the signal flight time and $\varepsilon_{ij}(t)$ is the measurement error (σ_{ij}^2 denotes the measurement variance, and $b_{ij}(t)$ is the measurement bias, generally time dependent). The latter is associated to a number of unmodeled effects: range biases due to residual clock offsets (we are here assuming one-way ranging based on the assumption of synchronized clocks at the two objects i and j), uncorrected relativistic effects, hardware delay calibration errors, satellite attitude modeling errors, as well as ground station coordinates uncertainties in GSR measurements.Functional Model (1) links the ranging observables with the unknown satellite coordinates. As a purely kinematic solution relying only on ISR and sparse GSR measurements is not feasible, satellite orbit is parametrized by a set of 15 parameters of a Galileo standard F/NAV navigation message [6]. We then denote the vector of unknown orbital parameters $\mathbf{p}$ at time t_{oe} (time of ephemeris) for unknown m satellites as

$$\mathbf{p}(t_{oe}) \equiv \left[\mathbf{p}_1^{\mathrm{T}}(t_{oe}), \mathbf{p}_2^{\mathrm{T}}(t_{oe}), \ldots, \mathbf{p}_m^{\mathrm{T}}(t_{oe})\right]^{\mathrm{T}}, \tag{2}$$

where $\mathbf{p}_s(t_{oe})$ is a 15-element vector containing the perturbed Keplerian parameters given in Table 1 (a), referred to satellite s at time of ephemeris t_{oe}.

Satellite Cartesian ECEF coordinates $\mathbf{r}(t)$ are computed from the 15 F/NAV orbital parameters following the user algorithm (non-linear functions) summarized in Table 1 (b) [6]. The advantage of using an orbit model based on the 15 perturbed Keplerian parameters lies in its simplicity, which removes the need for a complex modeling of all forces acting on the satellite and their integration over time. Moreover, the data relay capabilities of the ISLs enable distributing ISR and GSR measurements across the whole constellation, potentially enabling autonomous orbit determination onboard each satellite. In this approach, the performance of orbit determination is determined by the accuracy of (a) the ranging observables, (b) the quality of the parametric model used to represent the actual satellite orbit and (c) on the length of the fitting interval (the longer the interval, the larger the errors, as it is shown in Section 5.1). It should be noted that such parametrized orbit has a very limited validity when extrapolating outside the fitting interval.

Table 1. The 15 F/NAV orbital parameters and the user algorithm.

(a) Ephemeris data definition.

Parameter	Description
M_0	Mean anomaly
Δn	Mean motion correction
e	Eccentricity
$\sqrt{a}$	Square root of semi-major axis
Ω_0	Longitude of ascending node
$\dot{\Omega}$	Rate of change in longitude of ascending node
i_0	Inclination angle
$\dot{i}$	Rate of change in inclination angle
ω	Argument of perigee
C_{uc}	Amplitude of cosine correction to the argument of latitude
C_{us}	Amplitude of sine correction to the argument of latitude
C_{rc}	Amplitude of cosine correction to the orbit radius
C_{rs}	Amplitude of sine correction to the orbit radius
C_{ic}	Amplitude of cosine correction to the inclination
C_{is}	Amplitude of sine correction to the inclination

(b) The user algorithm.

Computation

$$a = (\sqrt{a})^2$$
$$n = \sqrt{\frac{\mu}{a^3}} + \Delta n$$
$$\Delta t = t - t_{0e}$$
$$M = M_0 + n\Delta t$$
$$M = E - e \sin E$$
$$\sin \nu = \frac{\sqrt{1-e^2} \sin E}{1 - e \cos E}$$
$$\cos \nu = \frac{\cos E - e}{1 - e \cos E}$$
$$\phi = \nu + \omega$$
$$\Omega = \Omega_0 + (\dot{\Omega} - \omega_{\mathrm{E}})\Delta t - \omega_{\mathrm{E}} t_{\mathrm{oe}}$$
$$u = \phi + C_{us} \sin(2\phi) + C_{uc} \cos(2\phi)$$
$$r_s = a \cdot (1 - e \cos E) + C_{rs} \sin(2\phi) + C_{rc} \cos(2\phi)$$
$$i = i_0 + \dot{i} \cdot \Delta t + C_{is} \sin(2\phi) + C_{ic} \cos(2\phi)$$
$$\begin{Bmatrix} x_s = r_s \cos u \cos \Omega - r_s \sin u \cos i \sin \Omega \\ y_s = r_s \cos u \sin \Omega + r_s \sin u \cos i \cos \Omega \\ z_s = r_s \sin u \sin i \end{Bmatrix}$$

$\mu = 3.986004418 \times 10^{14}\ \mathrm{m}^3/\mathrm{s}^2$
$\omega_{\mathrm{E}} = 7.2921151467 \times 10^{-5} \mathrm{rad/s}$
$c = 299792458\ \mathrm{m/s}$

3. Nonlinear Estimation of the Perturbed Keplerian Parameters

Let us consider a scenario in which m satellites in MEO are all equipped with two terminals for ISLs. The constellation is maintained fully connected via a link scheduling that accounts for the satellite positions, orientations and terminal field of regards; see details in [7,8]. Moreover, GSR measurements are obtained from n ground stations. We assume that each ground station provides GSR observables to just one MEO satellite at a time, and that every satellite is only observed by one ground station at a time.

In closed-loop link topology, all satellites are assumed linked in a 'closed-ring' manner, with $(m+n)$ range observations (ISR and GSR) available at each epoch, and $15m$ unknowns. The ephemeris state vector $\mathbf{p}$ can be solved by using a non-linear least squares adjustment to minimize objective function

$$\sum_{k=1}^{K}\left\{\left(\mathbf{y}_{t_k}-\mathbf{f}_{t_k}(\mathbf{p}(t_{oe}))\right)^{\mathrm{T}}\mathbf{Q}_{\mathbf{y}_{t_k}}^{-1}\left(\mathbf{y}_{t_k}-\mathbf{f}_{t_k}(\mathbf{p}(t_{oe}))\right)\right\}, \tag{3}$$

where K is the number of epochs used, $\mathbf{y}_{t_k}$ is the vector containing the $(m+n)$ observed ranges at time t_k, $\mathbf{f}_{t_k}(\mathbf{p}(t_{oe}))$ is the vector of nonlinear functions mapping the $15m$ broadcast parameters to the range at time t_k (given in Table 2b), and $\mathbf{Q}_{\mathbf{y}_{t_k}}^{-1}$ is the inverse of the covariance matrix of the observables.

The minimization problem in (3) is linearized by taking $\mathbf{f}(\mathbf{p}(t_{oe}))$ at each epoch t_k about an orbital set obtained with a first guess, denoted with $\hat{\mathbf{p}}^{(0)}(t_{oe})$ such that

$$\mathbf{f}(t_k,\mathbf{p}(t_{oe})) \approx \mathbf{f}(t_k,\widehat{\mathbf{p}}^{(0)}(t_{oe})) + \frac{\partial\mathbf{f}(t_k,\mathbf{p})}{\partial\mathbf{p}}\Big|_{\mathbf{p}=\hat{\mathbf{p}}^{(0)}(t_{oe})}\cdot\left(\mathbf{p}(t_{oe})-\hat{\mathbf{p}}^{(0)}(t_{oe})\right), \tag{4}$$

where $\left.\frac{\partial\mathbf{f}(t_k,\mathbf{p})}{\partial\mathbf{p}}\right|_{\mathbf{p}=\hat{\mathbf{p}}^{(0)}(t_{oe})}$ is a $((m+n)\times 15m)$ matrix containing the partial derivatives of the vector function $\mathbf{f}()$ at time t_k with respect to the 15 perturbed Keplerian orbital elements of all m satellites, computed in $\hat{\mathbf{p}}^{(0)}(t_{oe})$. Expression (3) can then be linearized and rewritten in matrix form as

$$(\Delta\rho-\mathbf{J}_{\mathbf{f}}\Delta\mathbf{p}(t_{oe}))^{\mathrm{T}}\mathbf{Q}_{\mathbf{y}}^{-1}(\Delta\rho-\mathbf{J}_{\mathbf{f}}\Delta\mathbf{p}(t_{oe})), \tag{5}$$

where $\mathbf{J}_{\mathbf{f}}=\left[\mathbf{J}_{\mathbf{f},t_1}(\widehat{\mathbf{p}}^{(0)}(t_{oe})),\ldots,\mathbf{J}_{\mathbf{f},t_K}(\widehat{\mathbf{p}}^{(0)}(t_{oe}))\right]^T$ is the Jacobian matrix introduced in (4), containing all $((m+n)K\times 15m)$ partial derivatives of the observations over the K epochs; $\Delta\rho$ is the residual range error, i.e., the difference between the observed ranges and the ranges computed from initial values of the orbital parameters at the linearization point; $\Delta\mathbf{p}(t_{oe})$ is the correction to the initial state vector minimizing observation residuals; and $\mathbf{Q}_{\mathbf{y}}$ is the block-diagonal covariance matrix of all observations, with diagonal components obtained from matrices $\mathbf{Q}_{\mathbf{y}_{t_k}}$.

The vector of orbital parameters is computed iteratively, starting from the initial guess:

$$\hat{\mathbf{p}}^{(i+1)}(t_{oe})=\hat{\mathbf{p}}^{(i)}(t_{oe})+\Delta\hat{\mathbf{p}}^{(i)}(t_{oe}),\quad \Delta\hat{\mathbf{p}}^{(i)}(t_{oe})=\left(\mathbf{J}^{(i)}{}_{\mathbf{f}}^{\mathrm{T}}\mathbf{Q}_{\mathbf{y}}^{-1}\mathbf{J}^{(i)}{}_{\mathbf{f}}\right)^{-1}\mathbf{J}^{(i)}{}_{\mathbf{f}}\mathbf{Q}_{\mathbf{y}}^{-1}\Delta\rho. \tag{6}$$

The iterations are stopped when the norm of the increment, $\Delta\hat{\mathbf{p}}^{(i)}(t_{oe})$, falls below a given threshold. Convergence of the iterative method to a global minimum depends, among other factors, on the choice of the initial conditions. It has been observed that convergence is achieved even with an initial orbital error of a few hundred metres.

4. Data Simulation

To test the performance of the algorithm presented in Section 3, we resorted to simulation. Realistic Medium Earth Orbits (MEOs) were simulated via the software Earth Parameter and Orbit System—Orbit Computation [9], developed and provided by the German Research Center for Geosciences. The software applies up-to-date, complex dynamical models to generate realistic trajectories of Galileo satellites. The orbits were simulated with a 30 s time step for a period of 10 days. A scheduler was implemented to generate links between satellites and ground stations using available orbital data. The ISR and GSR observations are generated every 30 s. The scheduling used to define which links occur at a given epoch is planned such that satellite-to-satellite links have a persistence of at least 1 min, whereas the ground-to-satellite links have a persistence of at least 15 min. This implies that approximately every minute, the links between satellites change, and every 15 min, a ground station links to a different satellite. Further details about the link scheduling are provided in [7,8]. The true range measurements were generated by computing the Euclidean distance (the norm term of (1)) between two satellites and between a satellite and a ground station. Gaussian noise and biases were then added to the true ranges to create observations according to stochastic Model (1): $\sigma_{\text{ISR}} = 1$ mm and $\sigma_{\text{GSR}} = 50$ mm. To emulate systematic error components, constant biases, randomly distributed in the interval 1–10 cm, were added to both ISR and GSR observations for the entire duration of the links.

5. A Note on Orbit Fitting and Orbit Extrapolation Accuracies

Before analyzing the performance of orbit determination from ISR and GSR measurements, this section focuses first on assessing the intrinsic accuracy of the broadcast orbit fitted to simulated dynamic orbits as well as on orbit accuracy when extrapolating outside the fit interval.

5.1. Accuracy of Orbit Fitting

The set of 15 F/NAV broadcast parameters (Section 2), including six Keplerian and nine perturbation corrections, has limited ability to reproduce true dynamic orbits. The orbit error is expected to be small for a short fitting interval and increase with interval length. Figure 1a shows the average accuracy over the entire constellation for a fit interval of 10 min. The orbit error is decomposed in the radial, along-track and cross-track components. Over such short intervals, the broadcast orbit model is very accurate, with fitting errors below the mm-level.

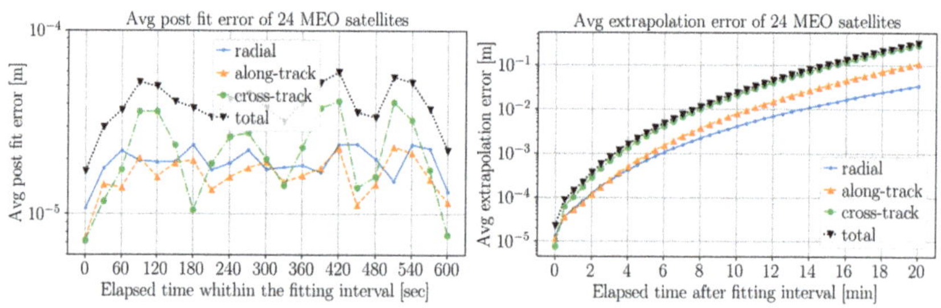

(**a**) (**b**)

Figure 1. Average absolute post fit error and extrapolation error for the 24 MEO satellites when fitting 15 perturbed Keplerian parameters to dynamic orbits. The fitting interval is set to 10 min. (**a**) Post fit error. (**b**) Extrapolation error.

Figure 2a shows the maximum recorded Root-Mean-Square Error (RMSE) over the entire constellation for fitting intervals of 10–120 min, while Figure 2b presents results for fitting intervals of 2–6 h.

The maximum 3D (total) fitting error grows from a cm-level for two-hour intervals, a dm-level for a typical broadcast fit interval of 4 h, to an m-level when using 5 to 6 h intervals. These plots show that a cm-level accuracy is only guaranteed within 1 to 2 h fitting intervals. Therefore, any orbit determination method based on this orbit parametrization is always affected by an error equal to or larger than the orbit fitting error.

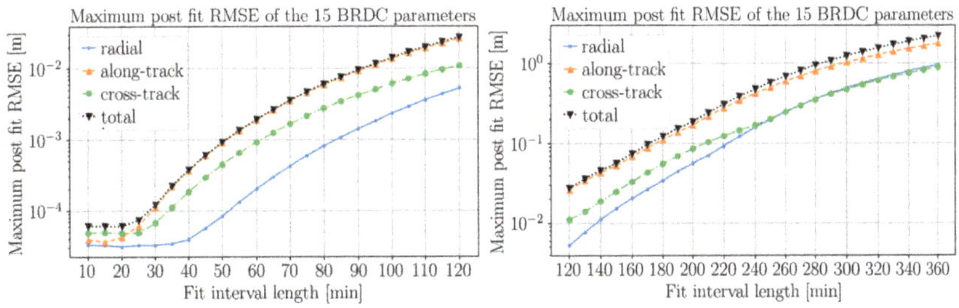

(**a**) (**b**)

Figure 2. Maximum post fit RMSE for the 24 MEO satellites when fitting 15 perturbed Keplerian parameters to dynamic orbits, as function of the fitting interval length. (**a**) Fitting interval: 10 to 120 min. (**b**) Fitting interval: 2 to 6 h.

5.2. Accuracy of Orbit Extrapolation

The estimated broadcast parameters can be used to extrapolate satellite trajectories with the user algorithm (Table 1b) outside of the fitting interval. Figure 1b shows the average orbit extrapolation error for an interval of up to 20 min using broadcast parameters obtained from a 10 min long fit interval (cf. Figure 1a). It can be seen that the extrapolation error hits 1 cm after 8 min (elapsed time from the end of the fitting interval), and 1 decimeter after just 14 min. This rapid error growth suggests that the period of validity of the ephemeris produced when fitting to 10 min of a dynamic orbit arc should not exceed 10 to 20 min when an orbit prediction accuracy better than dm-level is targeted. This, in turn, means that in order to employ the proposed orbit determination scheme for real applications, new orbital parameters should be produced every few minutes, e.g., by implementing sliding batch processing or a Kalman filter to continuously update the estimation of the ephemeris.

6. Analysis of Orbit Determination Performance

In this section, we present simulation results of the orbit determination method detailed in Section 3 for a Galileo-like constellation of 24 MEO satellites. The 15 × 24 broadcast parameters are fitted to the ISR and GSR measurements collected over a time interval varying from 1 to 3 h. We analyze two different scenarios, with different numbers and distribution of ground stations. The first three stations, GS1-GS3 (Tromsø, Nemea and Tenerife), are located within the European geographical area. A fourth station GS4 (Papeete, French Polynesia) is added to analyze the impact of placing a station on the opposite side of the Earth.

Two scenarios are considered: (1) a regional European network of three ground stations GS1–GS3 and (2) a global network of stations GS1–GS4. Two figures of merit are used to assess the performance of the OD method. The first is the RMSE of the estimated orbit in radial, along- and cross-track directions with respect to the simulated true orbit (Section 4). The second is the orbit-only Signal-in-Space Range Error ($\text{SiSRE}_{\text{orb}}$), which provides a metric to express the range error introduced by the space segment. The $\text{SiSRE}_{\text{orb}}$ combines errors in all three spatial components [10]:

$$\text{SiSRE}_{\text{orb}} = \sqrt{(W_{\text{R}}\Delta R)^2 + W_{\text{AC}}^2(\Delta A^2 + \Delta C^2)}, \tag{7}$$

where ΔR, ΔA, ΔC are the average absolute radial, along- and cross-track errors, respectively, over all 24 satellites in a given time interval. Weights W_{R} and W_{AC} are chosen as for the Galileo system ($W_{\text{R}} = 0.98$ and $W_{\text{AC}}^2 = 1/61$) [10].

6.1. Noise-Free Analysis

First, in order to analyze the noise floor of the OD scheme devised, we estimated the orbital parameters from a set of error-free observations. To this aim, we set $\varepsilon_{ij}(t) = 0$ in (1) when simulating the ISR and GSR observables, and $\mathbf{Q_y} = \mathbf{I}$ in (6).

Table 2 (a) gathers the average OD RMSE and $SiSRE_{orb}$ of the 24 MEO satellites in the scenario with three European ground stations GS1-GS3. It can be observed that the longer the arcs, the larger the orbit estimation errors. The $SiSRE_{orb}$ measures less than 1.5 cm for a 1 h fit interval and grows to 9.42 cm for a 3 h fit interval.

Table 2. Average OD RMSE for 24 MEO satellites linked to a network of three or four ground stations. No noise or biases are added to the observations.

Orbital arc Length	Radial RMSE (cm)	Along-tr. RMSE (cm)	Cross-tr. RMSE (cm)	$SiSRE_{orb}$ (cm)
		(a) Three Ground Stations		
1 h	1.23	2.49	3.46	1.33
2 h	2.89	19.31	23.80	4.84
3 h	7.13	35.00	39.92	9.42
		(b) Four Ground Stations		
1 h	0.98	3.27	4.74	1.21
2 h	2.11	16.95	20.52	3.99
3 h	5.26	25.17	28.81	7.11

The same analysis was performed for a scenario with four ground stations, and the corresponding results are given in Table 2 (b). The addition of a fourth ground station improves the OD accuracy, mainly due to a better observation geometry, and the larger the orbital arc considered, the bigger the improvement, from 9% for 1 h intervals to 25% for 3 h intervals. It should be noted that OD accuracy is significantly lower than the accuracy of the F/NAV fit to the simulated orbits (Section 5.1) due to geometry and a much lower number of observations (ranges only).

6.2. Analysis with Measurement Errors—Regional Network

In this Section, we analyze the OD performance with simulated range measurements affected by errors and biases, as described in Section 4.

Table 3 gathers the corresponding OD RMSE and $SiSRE_{orb}$ values for arc lengths set at 1 h, 2 h and 3 h, and different combinations of simulated noise and biases. In general, for the three intervals considered, a longer observation period provides better OD results, from an m-level on 1 h arcs to a dm-level on 3 h arcs, in absence of observation biases. When adding observation biases, SiSRE increases by 40–80% with respect to a bias-free scenario, highlighting the importance of proper calibration of the terminals performing range measurements. In contrast to the analysis without noise (Section 6.1), we note that the use of a longer fitting interval has the effect of averaging the observation noise and bias, thus reducing the OD error.

Figure 3a shows the average error inside the estimated orbital arc for the 2 h scenario affected by random biases. It can be seen that the errors increase on the extremes of the fitting interval leading to a reduced accuracy of the extrapolated orbit outside the interval. This is demonstrated in Figure 3b, where total orbit extrapolation error reaches 40 m for a 1 h prediction.

Table 3. Average OD RMSE for a MEO constellation linked to a regionl network of three ground stations. ISR and GSR measurements are affected by noise and biases.

Orbital arc Length	σ_{GSR} (mm)	σ_{ISR} (mm)	b_{ij} (cm)	Radial RMSE (cm)	Along-tr. RMSE (cm)	Cross-tr. RMSE (cm)	$SiSRE_{orb}$ (cm)
1 h	50	1	None	62.83	441.24	335.83	93.98
2 h	50	1	None	7.56	44.73	45.65	11.04
3 h	50	1	None	6.32	36.46	36.23	9.04
1 h	50	1	[1, 10]	82.13	488.99	630.62	130.07
2 h	50	1	[1, 10]	12.80	90.01	75.75	19.60
3 h	50	1	[1, 10]	10.71	57.86	58.25	14.86

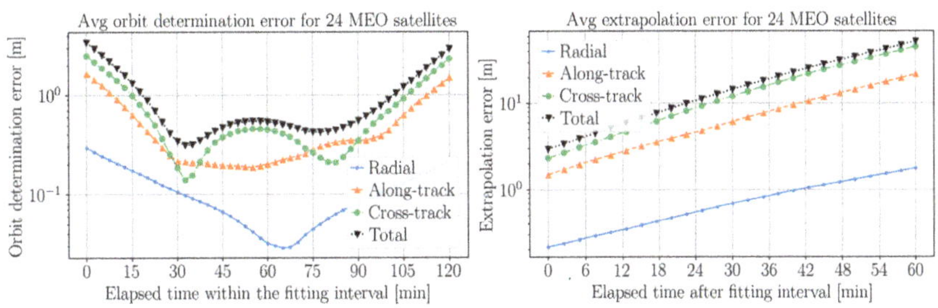

(**a**) (**b**)

Figure 3. Average absolute post fit error and extrapolation error for the 24 MEO satellites obtained for a regional network of three ground stations, with ISR and GSR observations affected by noise and biases. The orbit parameters are estimated over a 2 h long arc. (**a**) Post fit error. (**b**) Extrapolation error.

6.3. Analysis with Measurement Errors—Global Network

When adding a fourth ground station, observation geometry improves. Orbit determination results based on this global network are shown in Table 4. It can be observed that the RMSE and the $SiSRE_{orb}$ significantly improve by 30–45% with respect to the scenario with three ground stations, especially for shorter orbital arcs. Again, it can be seen that increasing the measurement interval reduces the negative impact of noise and biases.

The behavior of the OD error inside the estimated 2 h orbital arc is provided in Figure 4a. Similarly to the three ground station scenarios, OD errors tend to grow when approaching the extremes of the estimation interval. The average extrapolation error obtained with the parameters estimated over the 2 h fitting interval (Figure 4a) is shown in Figure 4b, where the three radial, across-track and along-track components are also provided. It can be seen that the prediction error at 1 h, approx. 25 m, is smaller than in the three regional ground station scenarios (40 m). It should be noted that the radial error is not significantly affected and remains at a 1.5 m level. Both OD and prediction performance show the benefits of including an additional ground station, located such as to significantly improve observation geometry.

Table 4. Average OD RMSE for a MEO constellation linked to a global network of four ground stations. ISR and GSR measurements are affected by noise and biases.

Orbital arc Length	σ_{GSR} (mm)	σ_{ISR} (mm)	b_{ij} (cm)	Radial RMSE (cm)	Along-tr. RMSE (cm)	Cross-tr. RMSE (cm)	$SiSRE_{orb}$ (cm)
1 h	50	1	None	39.76	293.31	321.88	68.02
2 h	50	1	None	4.44	20.83	27.25	6.18
3 h	50	1	None	5.46	24.55	29.50	7.27
1 h	50	1	[1, 10]	49.29	334.88	457.35	87.17
2 h	50	1	[1, 10]	12.18	54.38	54.12	15.46
3 h	50	1	[1, 10]	9.78	54.47	61.64	14.24

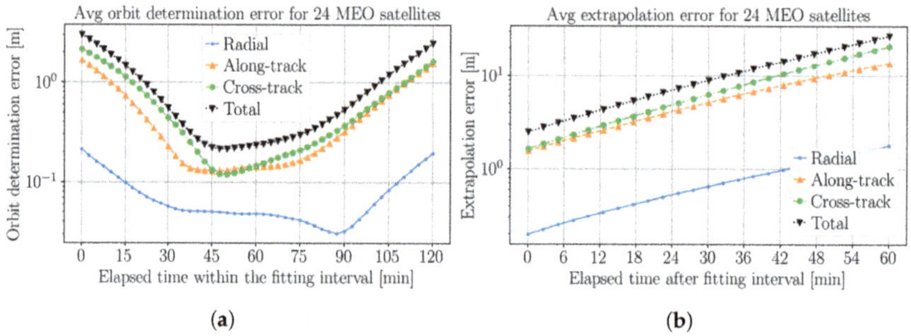

Figure 4. Average absolute post fit error and extrapolation error for the 24 MEO satellites obtained in the scenario with a global network of four ground stations, with ISR and GSR observations affected by noise and biases. The orbit parameters are estimated over a 2 h long arc. (**a**) Post fit error. (**b**) Extrapolation error.

7. Conclusions and Future Work

This work addressed the potential of performing orbit determination using inter-satellite ranging as a primary observable, without the intervention of an extended network of ground sensor stations collecting navigation data. The orbits of a GNSS MEO constellation of 24 satellites were estimated over observation intervals spanning 1, 2, or 3 h orbit arcs, aided by ground-to-satellite ranges obtained from three or four ground stations. Errors and biases were injected into the range observations to emulate realistic measurement data sets. It was shown that dm- to m-level orbit determination accuracy is attainable with a limited ground network extension. When focusing on GNSS applications, the radial orbital error component has the largest weight in defining GNSS navigation performance. To this aim, we also inspected the orbit-only signal-in-space range error $SiSRE_{orb}$ in all the simulated scenarios, showing that $SiSRE_{orb}$ below a decimeter can be reached when orbital arcs of sufficient length are processed, between two and three hours. However, this method cannot be further improved by extending the observed orbital arcs, since the Galileo broadcast F/NAV orbit parametrization introduces larger fitting errors.

The proposed orbit determination scheme could be used as a back-up procedure to estimate the orbits of GNSS satellites, guaranteeing a continuous navigation message generation in case of ground segment failures, as long as sparse GSR measurement can still be provided. However, we also noted a rapidly deteriorating accuracy of the predicted (extrapolated) orbit based on the estimated F/NAV parameters, which limits the validity interval of the ephemeris.

As a next step, the proposed method will be extended to process real-time observations with a Kalman filter to guarantee fast updates of the estimated orbital parameters. Additionally, in order to further improve orbit determination performance with the proposed method, it will be assessed whether low-dynamic biases affecting ISR and GSR measurements can be co-estimated, mitigating the impact of systematic biases with an identifiable signature.

Author Contributions: Conceptualization, A.T. and G.G.; methodology, A.T. and G.G.; software, A.T., G.M., M.D. and K.H.N.; validation, A.T.; writing, A.T., G.G., G.M. and M.D.; supervision, G.G. All authors have read and agreed to the published version of the manuscript.

Funding: This research received no external funding.

Institutional Review Board Statement: Not applicable.

Informed Consent Statement: Not applicable.

Data Availability Statement: Simulated data are available for internal use only.

Conflicts of Interest: The authors declare no conflict of interest. The funders had no role in the design of the study; in the collection, analyses, or interpretation of data; in the writing of the manuscript; or in the decision to publish the results.

Abbreviations

The following abbreviations are used in this manuscript:

GNSS	Global Navigation Satellite Systems
GSR	Ground-to-Satellite Ranging
ISR	Inter-Satellite Ranging
MEO	Medium Earth Orbit
OISL	Optical Inter-Satellite Link
ODTS	Orbit Determination and Time Synchronization
POD	Precise Orbit Determination
RMSE	Root-Mean-Square Error
SiSRE	Signal-in-Space Range Error

References

1. Teunissen, P.J.G.; Montenbruck, O. (Eds.) *Springer Handbook of Global Navigation Satellite Systems*; Springer Handbooks; Springer International Publishing: Berlin/Heidelberg, Germany, 2017. [CrossRef]
2. Giorgi, G.; Schmidt, T.D.; Trainotti, C.; Mata-Calvo, R.; Fuchs, C.; Hoque, M.M.; Berdermann, J.; Furthner, J.; Günther, C.; Schuldt, T.D.; et al. Advanced technologies for satellite navigation and geodesy. *Adv. Space Res.* **2019**, *64*, 1256–1273. [CrossRef]
3. Calvo, R.M.; Poliak, J.; Surof, J.; Wolf, R. Evaluation of optical ranging and frequency transfer for the Kepler system: Preliminary laboratory tests. In Proceedings of the 2020 European Navigation Conference (ENC), Dresden, Germany, 23–24 November 2020. [CrossRef]
4. Surof, J.; Poliak, J.; Wolf, R.; Macri, L.; Calvo, R.M.; Blümel, L.; Dominguez, P.; Giorgi, G.; Agazzi, L.; Furthner, J.; et al. Validation of Kepler Time and Frequency Transfer on a Terrestrial range of 10.45 km. In Proceedings of the ION GNSS+ 2022, Denver, CO, USA, 19–23 September 2022.
5. Michalak, G.; Glaser, S.; Neumayer, K.; König, R. Precise orbit and Earth parameter determination supported by LEO satellites, inter-satellite links and synchronized clocks of a future GNSS. *Adv. Space Res.* **2021**, *68*, 4753–4782. [CrossRef]
6. Galileo–Open Service. *Signal in Space Interface Control Document (OS-SIS-ICD-V1.3)*; Technical Report; European Union: Maastricht, The Netherlands, 2016.
7. Giorgi, G.; Kroese, B.; Michalak, G. Future GNSS constellations with optical inter-satellite links. Preliminary space segment analyses. In Proceedings of the 2019 IEEE Aerospace Conference, Big Sky, MT, USA, 2–9 March 2019. [CrossRef]
8. Deprez, C.; Giorgi, G. Operational Envelope and Link Scheduling for Inter-Satellite Links in Next-Generation GNSSs. In Proceedings of the 2021 IEEE Aerospace Conference (50100), Big Sky, MT, USA, 6–13 March 2021. [CrossRef]
9. Zhu, S.; Reigber, C.; König, R. Integrated adjustment of CHAMP, GRACE, and GPS data. *J. Geod.* **2004**, *78*, 103–108. [CrossRef]
10. Montenbruck, O.; Steigenberger, P.; Hauschild, A. Broadcast versus precise ephemerides: a multi-GNSS perspective. *GPS Solut.* **2015**, *19*, 321–333. [CrossRef]

engineering proceedings

Proceeding Paper

Breaking the One-Meter Accuracy Level with Smartphone GNSS Data †

Marcus Franz Glaner * and Robert Weber

Research Division Higher Geodesy, Department of Geodesy and Geoinformation, TU Wien, 1040 Vienna, Austria; robert.weber@geo.tuwien.ac.at
* Correspondence: marcus.glaner@geo.tuwien.ac.at

† Presented at the European Navigation Conference 2023, Noordwijk, The Netherlands, 31 May–2 June 2023.

Abstract: Smartphones are usually equipped with simple, cost-effective GNSS chips and antennas. They provide mainly single-frequency, low-quality, and challenging GNSS measurements. We demonstrate the difficulties of processing raw GNSS data from Android devices and introduce solutions to break the one-meter accuracy level with smartphones and Precise Point Positioning. With the logged data of a Samsung Galaxy S23+ and a Google Pixel 7 smartphone, a horizontal position accuracy of around one meter and a few decimeters was accomplished, respectively. These results were achieved after about two minutes of convergence time with our open-source software raPPPid in quasi-real-time settings. Furthermore, the corrections provided by the Galileo High-Accuracy Service proved to be sufficient to achieve sub-meter accuracy with smartphones.

Keywords: GNSS; PPP; Android; smartphone; raPPPid; Galileo HAS; low cost; uncombined; ionospheric constraint

Citation: Glaner, M.F.; Weber, R. Breaking the One-Meter Accuracy Level with Smartphone GNSS Data. *Eng. Proc.* **2023**, *54*, 16. https://doi.org/10.3390/ENC2023-15465

Academic Editors: Tom Willems and Okko Bleeker

Published: 29 October 2023

1. Introduction

Before Android 7.0 was officially released in 2016, it was only possible to read out the position solution of a smartphone's internal algorithm [1,2]. Nowadays, everyone can access raw GNSS measurements tracked by Android smartphones and directly use the smartphone's GNSS observations to estimate the user's position [3–5]. Therefore, developing specialized algorithms and applying correction data can enhance positioning performance. However, smartphones are typically equipped with simple, cost-effective chips and antennas and, consequently, provide low-quality, single-frequency measurements with various challenges [6–8]. Altogether, experience has shown that the GNSS measurements of different smartphone devices behave diversely and unpredictably. To successfully process raw GNSS measurements from smartphones, we must study and understand their characteristics and differences from data recorded with geodetic equipment.

Precise Point Positioning (PPP) is characterized by applying precise satellite products (e.g., satellite orbits, clocks, and biases) and complex models to estimate the user's position [9]. PPP is a well-established processing technique for multi-frequency GNSS data from geodetic receivers and allows reaching accuracies at the centimeter level, or even below, after a notable convergence period [10]. Reducing or eliminating this convergence period is a major topic in scientific research. Usually, high-quality GNSS observations on two frequencies are used for PPP to build the ionosphere-free linear combination (IF LC), and researchers have found ways to integer fix the phase ambiguities [11,12]. Several adjustments of the PPP technique are necessary to process the challenging GNSS measurements from smartphones in an adapted PPP mode [13,14]. The approaches presented in the following are implemented in our open-source PPP software raPPPid [15,16], available at https://github.com/TUW-VieVS/raPPPid (accessed on 24 November 2023). Additionally, the raPPPid wiki (https://vievswiki.geo.tuwien.ac.at/en/raPPPid (accessed on 24 November 2023) provides documentation and processing examples (e.g., with smartphone GNSS data).

The following section introduces the difficulties presented by using raw GNSS data from smartphones and an adapted PPP approach. Afterwards, Section 3 presents processing examples and Section 4 discusses the results. The final section provides a summary and outlook.

2. Challenges

Android devices significantly differ in their capabilities regarding raw GNSS measurements [17,18]. Considerable differences exist between smartphone devices of the same model, for example, sold in different regions. Access to the raw GNSS measurements might be limited (e.g., phase measurement) even on the most recent smartphone models [19]. Furthermore, a specific smartphone might not provide dual-frequency (DF) measurements due to, most likely, software restrictions, although it has a DF chip installed. Apart from that, DF capability means that the smartphone tracks signals on the L1 and L5 bands, and this frequency is currently not covered by all GNSS satellites (e.g., GPS modernization).

Naturally, Android repeatedly powers the GNSS hardware on and off to save battery energy, called duty-cycling [1]. This behavior causes non-continuous signal tracking and, for example, destroys the constant property of the phase ambiguity. Since Android 9, the user can force full GNSS measurements to deactivate duty-cycling and track all GNSS and frequencies in the developer settings [20]. On Android 12 or higher, this key configuration for precise positioning is available as an in-app setting. Figure 1 shows the triple-time-differenced GPS C1C code measurements of four consecutive epochs of a Samsung Galaxy S20 FE without deactivating duty-cycling. Building the triple-time difference with GNSS measurements recorded at a one-second observation interval typically nearly eliminates all effects, and the residuals represent the measurement noise. Consequently, the resulting code and phase residuals of geodetic GNSS equipment are usually the size of a few decimeters and a few millimeters, respectively. In Figure 1, the code residuals are significantly larger, and discontinuities occur due to non-deactivated duty cycling around epochs 82 and 116. These jumps also appear for other GNSS (same epochs, different magnitude) and complicate the processing of multiple GNSS in the PPP solution. However, they disappear when deactivating duty-cycling.

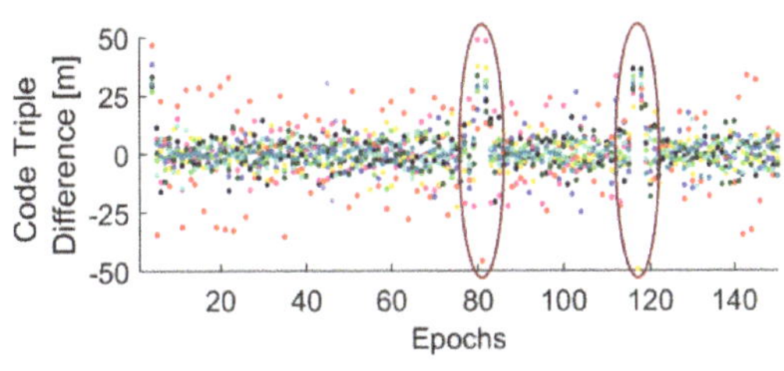

Figure 1. Triple-time-differenced code observations (GPS C1C, color-coded by satellite) for a Samsung Galaxy S20 FE when duty-cycling is not deactivated. Discontinuities due to duty-cycling are marked in dark ellipses.

The IF LC is commonly used for PPP, requiring observations on two frequencies to eliminate the ionospheric delay from the observation model. However, smartphones typically only provide single-frequency observations or a limited number of DF observations. Therefore, the uncombined PPP model with ionospheric constraint was chosen to connect the measurements and unknown parameters. This flexible PPP approach effectively handles the variable nature of smartphone GNSS observations by managing any number of frequencies, maintaining the raw observation noise, and including ionospheric pseudo-observations. For example, the uncombined model can consider different numbers of frequencies for each satellite since missing observations are easily excluded from the parameter estimation process.

$$C_1 = \rho + c(dt_r + \delta t^g) + dTrop^{wet} + dIono_1 + \varepsilon \quad (1)$$

$$L_1 = \rho + c(dt_r + \delta t^g) + dTrop^{wet} - dIono_1 + \lambda_1 N_1 + \varepsilon \quad (2)$$

In the case of a single-frequency smartphone, Equations (1) and (2) present the observation equations of the uncombined model for the code observation C and the phase observation L, respectively. The geometric distance between the satellite and the smartphone is denoted as ρ and includes the unknown user position. The GPS receiver clock error dt_r and receiver clock offset δt^g for GNSS g are multiplied by the speed of light c. Furthermore, $dTrop^{wet}$ denotes the wet part of the tropospheric delay, and $dIono_1$ corresponds to the ionospheric delay. Equation (2) includes the float ambiguity N, which absorbs satellite and receiver phase biases and is multiplied by the corresponding wavelength λ. For readability reasons, these observation equations do not have an index to differentiate between GNSS or discriminate the term ε, containing noise and other errors. These observation Equations (1) and (2) comprise the unknown parameters, estimated with a Kalman filter in the PPP solution. Other error sources are modeled, like solid Earth tides, relativistic effects, and satellite phase center offsets.

$$C_2 = \rho + c(dt_r + \delta t^g) - DCB_{12} + dTrop^{wet} + f_1^2/f_2^2\, dIono_2 + \varepsilon \quad (3)$$

$$L_2 = \rho + c(dt_r + \delta t^g) - DCB_{12} + dTrop^{wet} - f_1^2/f_2^2\, dIono_2 + \lambda_2 N_2 + \varepsilon \quad (4)$$

In the case of a DF smartphone, Equations (3) and (4) are added to the parameter estimation for the corresponding satellites. DCB_{12} denotes the smartphone's differential code bias between the first and second processed frequency (e.g., GPS L1 and L5). The squared ratio of the signal frequencies f_1^2/f_2^2 converts the ionospheric delay to the first frequency.

$$dIono_{pseudo} = dIono_1 + \varepsilon \quad (5)$$

Furthermore, the uncombined model can integrate ionosphere models into the parameter estimation by adding the modeled ionospheric delays to the first frequency as pseudo-observations for each satellite ($dIono_{pseudo}$, Equation (5)). In this way, the inevitable imperfections of the ionosphere model can be adequately considered. Therefore, this so-called ionospheric constraint adds valuable information to the PPP solution, strengthens parameter estimation, and shortens the convergence time.

The phase measurements of smartphones are usually affected by ambiguity jumps and cycle slips. Detecting all these events is essential for PPP but complex, especially in single-frequency data, and not always feasible. Moreover, smartphones often do not provide access to the phase measurement. In such cases, a code-only solution has to be calculated by omitting Equations (2) and (4). Even such a code-only PPP approach allows a horizontal position accuracy around the one-meter level, as shown in the results section. However, by deactivating duty-cycling and considering the inconsistencies of some logger applications [21], the phase measurements of state-of-the-art smartphones can significantly contribute to breaking the one-meter accuracy level. Thereby, reliable cycle-slip detection is crucial, and the following approach based on Doppler observations proved to be effective.

$$L_{now} = L_{old} + dt \cdot sqrt(D_{now} \cdot D_{old}) \quad (6)$$

The phase observation of the previous epoch L_{old} is used with the Doppler observations of the previous and current epoch (D_{old} and D_{now}, respectively) to predict the phase observation of the current epoch L_{now}. For this method, the observation interval dt has to be sufficiently small, which is usually the case for smartphone data (1 s). The predicted phase observation is compared with the actual phase measurement and a suitable threshold (e.g., 0.5 or 1 cycle).

The quality of GNSS observations from smartphones is generally worse than that of geodetic receivers. The carrier-to-noise density (C/N0) is typically around 10 dB.Hz lower and more variable (Figure 2). Furthermore, the code and phase observation noise is 10 to 100 times larger than for geodetic equipment (Figure 3). Anomalies like multipath or cycle-slips are usually much more pronounced due to the incorporated ultra-low-cost equipment. Therefore, the PPP processing should use a cutoff angle higher than usual (e.g., 10°) and exclude observations below a certain C/N0 threshold (e.g., 20 dB.Hz). Furthermore, the software should carefully check the measurements to eliminate suspicious observations. For example, the triple-time-differenced code measurements and the difference between the code measurement and observation model can be compared with reasonable thresholds (e.g., 50 and 25 m, respectively).

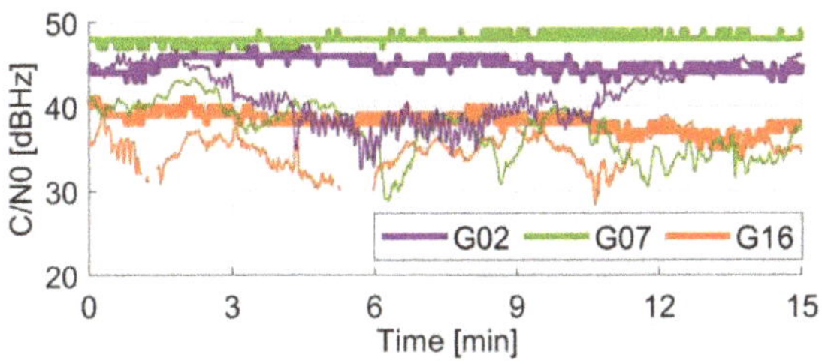

Figure 2. The carrier-to-noise density of three simultaneously observed GPS satellites tracked with a Trimble Spectra SP80 (thicker lines) and a Google Pixel 5 (thinner lines).

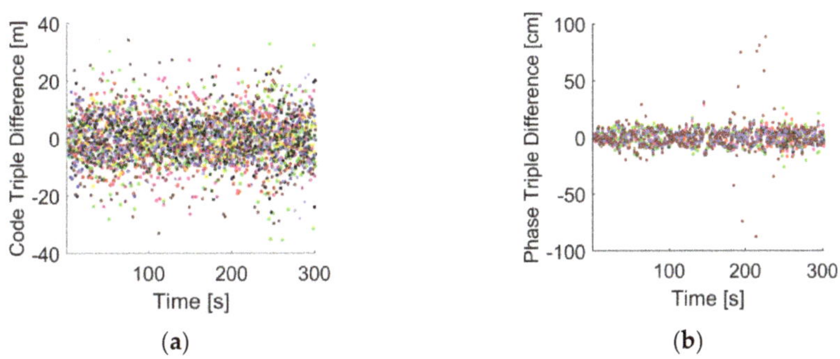

Figure 3. Triple-time differenced code (**a**) and phase (**b**) observations of a Google Pixel 7 (GPS C1C and L1C, color-coded by satellite). The corresponding standard deviations are 8.45 m and 12.9 cm.

Generally, the inaccurate weighting of the observations extends the convergence period or introduces biases to the parameters estimated with PPP. An elevation-based weighting approach (e.g., $\sin^2(e)$) is typically used in PPP since the observations' residuals are correlated with the elevation in the case of geodetic GNSS equipment. However, no such relation usually exists for smartphones, but it can be found between the observations' residuals and the C/N0 (Figure 4). Therefore, weighting the smartphones' measurements based on the C/N0 usually significantly improves the performance and should be adjusted to the specific smartphone device. A simple CN/0 weighting function is used in the following (Table 1), leading to decent results in our experience.

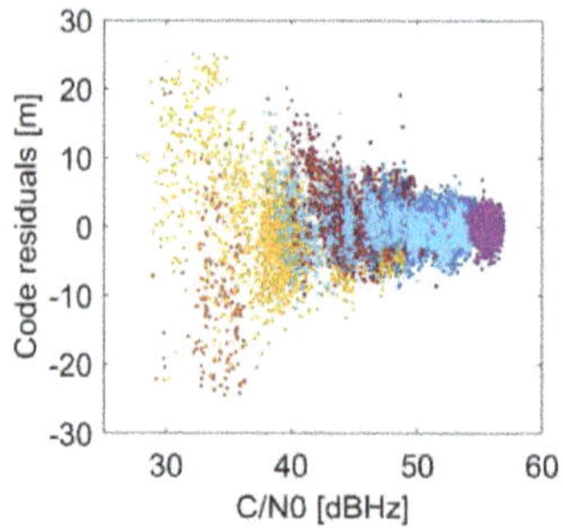

Figure 4. Code residuals (color-coded by satellite) over carrier-to-noise density for a Samsung Galaxy S20 FE.

Table 1. Quasi-real-time processing settings of raPPPid for processing raw GNSS measurements of smartphones. Options listed twice differentiate for the Galaxy S23+ and Pixel 7, respectively.

Smartphones	Samsung Galaxy S23+, Google Pixel 7
Date and period	4 and 19 April 2023; 60 min in total
Observation interval	1 s, reset of the solution every 6 min
Processed measurements	GPS (C1C + C5Q), GLONASS (C1C), Galileo (C1C + C5X), BeiDou (C2I)
Satellite orbits, clocks, and biases	Real-time corrections stream from CNES [22] or the Galileo HAS [23]
Processing mode	Static, quasi-real-time
Observation model	Uncombined model with ionospheric constraint
Raw observation noise	Code: 1 m/7 m, phase: NaN/0.01 m, ionosphere: 3 m
Observation weighting	$10^{-\frac{\max(0\ ;\ 55-SNR)}{a}}$, with a = 20 or 10
Ionosphere model	GIOMO predicted [24] as ionospheric constraint, constant/released after 1 min
Troposphere model	GPT3 [25,26], ZWD not estimated/estimated
Correction models	Solid Earth tides, relativistic effects
Satellite exclusion criteria	C/N0 < 20 dB.Hz or elevation < 10°
Data quality checks (thresholds)	Observed minus computed (code: 25 m, phase 5 m), code triple-time difference (1 m), cycle-slip detection with Doppler (0.7 cy)

3. Results

This section presents position coordinate results obtained with two state-of-the-art smartphones (Google Pixel 7 and Samsung Galaxy S23+) and our open-source PPP software raPPPid [15,16]. The smartphones were placed on geodetic reference points on the rooftop of TU Wien, ensuring suitable reference coordinates. Considering the imperfections introduced by various logger applications to the code and phase observations [21], the raw output of all sensors was logged with the Google GnssLogger application for 30 min on two different days. The GNSS measurements of the resulting text files were converted to the RINEX format using UofC CSV2RINEX, available at https://github.com/FarzanehZangeneh/csv2rinex (accessed on 24 November 2023).

The PPP processing was performed in quasi-real-time (Table 1), applying real-time capable atmosphere models and satellite products (recorded real-time correction stream data from CNES [22] or the Galileo High-Accuracy Service (HAS) [23]). The PPP solution was reset every 6 min during the PPP processing. Such a reset corresponds to a restart of the PPP calculations, and the resulting ten convergence periods can be used to assess the convergence behavior and accuracy of the coordinates.

Generally, GPS, GLONASS, Galileo, and BeiDou (GREC) observations are processed for both devices. The variance of GLONASS and BeiDou observations is increased by a factor of 1.25 because their real-time satellite products are usually less accurate. Since the Galileo HAS provides real-time corrections for GPS and Galileo, a GPS and Galileo (GE) solution was calculated with these satellite orbits, clocks, and code biases. Unfortunately, the Samsung Galaxy S23+ does not provide access to the phase measurements. Therefore, a code-only solution is calculated for this device. On the other hand, code and phase measurements are processed in the case of the Google Pixel 7. Due to the different characteristics of code-only and code + phase PPP processing, some settings must be modified (listed twice in Table 1). For example, the ZWD is estimated only for the Google Pixel 7 because the phase observations substantially improve the accuracy of the solution.

Figure 5 shows the convergence behavior of the horizontal position error for the Samsung Galaxy S23+ and the Google Pixel 7, applying real-time corrections from CNES. Table 2 presents the corresponding statistics. Convergence is defined as the period until the horizontal position difference is below 1 m for the whole remaining period. The introduced PPP approach achieves a 2D position accuracy around the one-meter level after a convergence time of 2–4 min using only code observations from the Galaxy S23+. In the case of the Google Pixel 7, the 2D position accuracy is clearly below one meter, thanks to the phase observations, and the convergence time is typically 1–2 min. After 6 min, even the 3D position error is at the one-meter level for this device (Table 2).

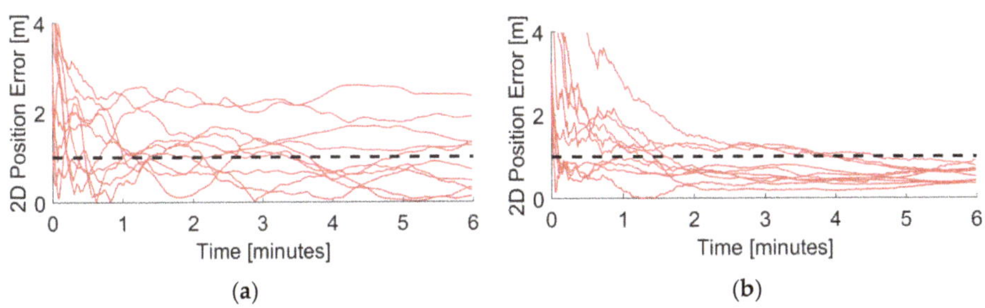

Figure 5. Horizontal position difference applying real-time corrections from CNES in a GREC processing for (**a**) Samsung Galaxy S23+ and (**b**) Google Pixel 7.

Table 2. Accuracy and convergence time applying real-time corrections from CNES in a GREC processing.

Smartphone	Median 2D Position Error [cm]	Mean 2D Position Error after 6 min [cm]	Mean 3D Position Error after 6 min [cm]	Mean Convergence [min]	Converged [%]
Samsung Galaxy S23+	112.5	101.4	258.8	2.33	50
Google Pixel 7	66.5	58.7	108.8	1.70	100

Furthermore, Figure 6 shows the convergence behavior using Galileo HAS corrections in a GPS and Galileo processing, and Table 3 presents the corresponding statistics. Naturally, these results based only on GPS and Galileo observations are worse than the GREC results (Table 2) due to significantly fewer satellites and observations. Furthermore, when comparing Table 2 with Table 4, the real-time clock corrections provided by the Galileo HAS seem slightly worse than the corresponding real-time corrections from CNES. Nevertheless, the Galileo HAS is sufficient to achieve sub-meter accuracy with smartphones.

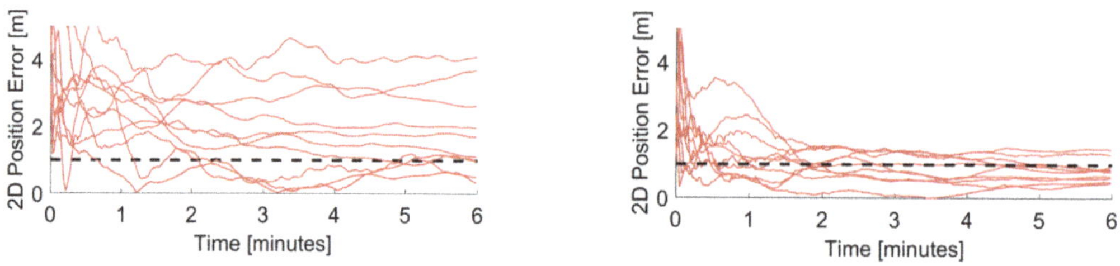

(**a**) (**b**)

Figure 6. Horizontal position difference applying corrections from the Galileo HAS in a GE processing for (**a**) Samsung Galaxy S23+ and (**b**) Google Pixel 7.

Table 3. Accuracy and convergence time applying corrections from the Galileo HAS in a GE processing.

Smartphone	Median 2D Position Error [cm]	Mean 2D Position Error after 6 min [cm]	Mean 3D Position Error after 6 min [cm]	Mean Convergence [min]	Converged [%]
Samsung Galaxy S23+	190.3	180.9	384.3	3.96	30
Google Pixel 7	99.9	84.0	261.3	2.63	80

Table 4. Accuracy and convergence time applying real-time corrections from CNES in a GE processing.

Smartphone	Median 2D Position Error [cm]	Mean 2D Position Error after 6 min [cm]	Mean 3D Position Error after 6 min [cm]	Mean Convergence [min]	Converged [%]
Samsung Galaxy S23+	182.6	133.9	245.6	3.05	60
Google Pixel 7	93.9	79.0	208.4	2.25	70

4. Discussion

The presented results were accomplished under good conditions since the smartphones were statically placed on geodetic reference points, ensuring an open sky. Notably, they were achieved using quasi-real-time processing, and the smartphones could perform these PPP calculations directly. Therefore, sub-meter accuracy is possible with smartphones and PPP after a convergence time of 2–3 min, even when only GPS and Galileo observations are processed with corrections from the Galileo HAS. More complex modeling approaches, like a high-quality ionosphere model or more sophisticated observation weighting, might improve the results and are the subject of future investigations. Furthermore, smartphone phase center offsets are not considered due to a lack of calibrations, introducing an uncertainty of several centimeters. If possible, including smartphones' phase observations significantly improves the PPP performance. They proved the key to achieving a 2D position accuracy at the decimeter level. Unfortunately, not all Android devices currently provide access to phase measurements. Furthermore, several key issues must be considered to ensure consistency between the code and phase observations.

5. Summary

This article describes the key challenges of processing the GNSS measurements from Android smartphones and introduces solutions to achieve decimeter-level accuracy with PPP. The presented approaches are implemented in our open-source software raPPPid, available at https://github.com/TUW-VieVS/raPPPid (accessed on 24 November 2023). With the GNSS data from two state-of-the-art smartphones recorded under good conditions, raPPPid achieved a 2D position accuracy around the one-meter level (code-only, Galaxy S23+) and at the few-decimeter level (code + phase, Pixel 7) in quasi-real-time processing after a convergence time of around two minutes. Also, the real-time corrections from the Galileo HAS enable sub-meter accuracy with smartphones after a similar convergence period.

Due to the variable characteristic of GNSS measurements from smartphones, more adaptations and refinements of the observation model are subject to further studies. Future work will focus on kinematic test cases and the fusion of PPP with other sensors (e.g., accelerometer). Currently, we are developing an Android application performing these PPP calculations directly on the smartphone.

Author Contributions: Conceptualization, M.F.G. and R.W.; methodology, M.F.G.; software, M.F.G.; validation, M.F.G. and R.W.; formal analysis, M.F.G.; investigation, M.F.G.; resources, M.F.G.; data curation, M.F.G.; writing—original draft preparation, M.F.G.; writing—review and editing, M.F.G. and R.W.; visualization, M.F.G.; supervision, R.W.; project administration, M.F.G. and R.W.; funding acquisition, R.W. All authors have read and agreed to the published version of the manuscript.

Funding: This research was funded by the Austrian Research Promotion Agency (FFG) within the research project APPP (Advanced Precise Point Positioning for mass market GNSS receivers on Android devices).

Institutional Review Board Statement: Not applicable.

Informed Consent Statement: Not applicable.

Data Availability Statement: The data used in this study are available on request from the corresponding author. The open-source PPP software raPPPid is available on GitHub (https://github.com/TUW-VieVS/raPPPid, accessed on 24 November 2023), and the raPPPid wiki provides documentation and processing examples (https://vievswiki.geo.tuwien.ac.at/en/raPPPid, accessed on 24 November 2023).

Acknowledgments: We want to thank all organizations which are operationally computing and disseminating satellite products enabling PPP.

Conflicts of Interest: The authors declare no conflict of interest.

References

1. Zangenehnejad, F.; Gao, Y. GNSS Smartphones Positioning: Advances, Challenges, Opportunities, and Future Perspectives. *Satell. Navig.* **2021**, *2*, 24. [CrossRef] [PubMed]
2. European GNSS Supervisory Authority. *Using GNSS Raw Measurements on Android Devices: White Paper*; Publications Office: Luxembourg, 2017.
3. Wang, L.; Li, Z.; Wang, N.; Wang, Z. Real-Time GNSS Precise Point Positioning for Low-Cost Smart Devices. *GPS Solut.* **2021**, *25*, 69. [CrossRef]
4. Everett, T.; Taylor, T.; Lee, D.-K.; Akos, D.M. Optimizing the Use of RTKLIB for Smartphone-Based GNSS Measurements. *Sensors* **2022**, *22*, 3825. [CrossRef] [PubMed]
5. Suzuki, T. Precise Position Estimation Using Smartphone Raw GNSS Data Based on Two-Step Optimization. *Sensors* **2023**, *23*, 1205. [CrossRef] [PubMed]
6. Zhang, X.; Tao, X.; Zhu, F.; Shi, X.; Wang, F. Quality Assessment of GNSS Observations from an Android N Smartphone and Positioning Performance Analysis Using Time-Differenced Filtering Approach. *GPS Solut.* **2018**, *22*, 70. [CrossRef]
7. Liu, W.; Shi, X.; Zhu, F.; Tao, X.; Wang, F. Quality Analysis of Multi-GNSS Raw Observations and a Velocity-Aided Positioning Approach Based on Smartphones. *Adv. Space Res.* **2019**, *63*, 2358–2377. [CrossRef]
8. Wanninger, L.; Heßelbarth, A. GNSS Code and Carrier Phase Observations of a Huawei P30 Smartphone: Quality Assessment and Centimeter-Accurate Positioning. *GPS Solut.* **2020**, *24*, 64. [CrossRef]
9. Kouba, J.; Lahaye, F.; Tétreault, P. Precise Point Positioning. In *Springer Handbook of Global Navigation Satellite Systems*; Teunissen, P.J.G., Montenbruck, O., Eds.; Springer International Publishing: Cham, Switzerland, 2017; pp. 723–751. ISBN 978-3-319-42928-1.
10. Choy, S.; Bisnath, S.; Rizos, C. Uncovering Common Misconceptions in GNSS Precise Point Positioning and Its Future Prospect. *GPS Solut.* **2017**, *21*, 13–22. [CrossRef]
11. Ge, M.; Gendt, G.; Rothacher, M.; Shi, C.; Liu, J. Resolution of GPS Carrier-Phase Ambiguities in Precise Point Positioning (PPP) with Daily Observations. *J. Geod.* **2008**, *82*, 389–399. [CrossRef]
12. Glaner, M.; Weber, R. PPP with Integer Ambiguity Resolution for GPS and Galileo Using Satellite Products from Different Analysis Centers. *GPS Solut.* **2021**, *25*, 102. [CrossRef]

13. Shinghal, G.; Bisnath, S. Conditioning and PPP Processing of Smartphone GNSS Measurements in Realistic Environments. *Satell. Navig.* **2021**, *2*, 10. [CrossRef]
14. Li, Z.; Wang, L.; Wang, N.; Li, R.; Liu, A. Real-Time GNSS Precise Point Positioning with Smartphones for Vehicle Navigation. *Satell. Navig.* **2022**, *3*, 19. [CrossRef]
15. Glaner, M.F.; Weber, R. An Open-Source Software Package for Precise Point Positioning: raPPPid. *GPS Solut.* **2023**, *27*, 174. [CrossRef]
16. Glaner, M.F. Towards Instantaneous PPP Convergence Using Multiple GNSS Signals. Ph.D. Thesis, TU Wien, Vienna, Austria, 2022.
17. Crowdsourcing GNSS features of Android Devices. Available online: https://barbeau.medium.com/crowdsourcing-gnss-capabilities-of-android-devices-d4228645cf25 (accessed on 4 April 2023).
18. GPSTest Database. Available online: https://docs.google.com/spreadsheets/d/1jXtRCoEnnFNWj6_oFlVWflsf-b0jkfZpyhN-BXsv7uo/edit#gid=0 (accessed on 4 April 2023).
19. Raw GNSS Measurements. Available online: https://developer.android.com/guide/topics/sensors/gnss (accessed on 4 April 2023).
20. GNSS, Interrupted: The Hidden Android Setting You Need to Know. Available online: https://barbeau.medium.com/gnss-interrupted-the-hidden-android-setting-you-need-to-know-d812d28a3821 (accessed on 4 April 2023).
21. Zangenehnejad, F.; Jiang, Y.; Gao, Y. GNSS Observation Generation from Smartphone Android Location API: Performance of Existing Apps, Issues and Improvement. *Sensors* **2023**, *23*, 777. [CrossRef] [PubMed]
22. PPP-WIZARD: Daily Products. Available online: http://www.ppp-wizard.net/daily.html (accessed on 11 April 2023).
23. Galileo High Accuracy Service (HAS). Available online: https://www.gsc-europa.eu/galileo/services/galileo-high-accuracy-service-has (accessed on 4 April 2023).
24. Magnet, N. Giomo a Robust Modelling Approach of Ionospheric Delays for GNSS Real-Time Positioning Applications. Ph.D. Thesis, TU Wien, Vienna, Austria, 2019.
25. Landskron, D.; Böhm, J. VMF3/GPT3: Refined Discrete and Empirical Troposphere Mapping Functions. *J. Geod.* **2018**, *92*, 349–360. [CrossRef] [PubMed]
26. Landskron, D.; Böhm, J. Refined Discrete and Empirical Horizontal Gradients in VLBI Analysis. *J. Geod.* **2018**, *92*, 1387–1399. [CrossRef] [PubMed]

engineering proceedings

MDPI

Proceeding Paper

Testing the Galileo High Accuracy Service User Terminal (HAUT) in Static Scenarios †

Emilio González [1,*], Pedro Pintor [1], Ana Senado [1], Narayan Dhital [1], Javier Ostolaza [2], Carmelo Hernández [2], Juan Vázquez [2], Javier de Blas [2] and Stefano Lagrasta [2,*]

[1] Spaceopal GmbH, 80335 Munich, Germany; pedro.pintor.ext@spaceopal.com (P.P.); ana.senado.ext@spaceopal.com (A.S.); narayan.dhital@dlr-gfr.de (N.D.)

[2] EU Agency for the Space Programme (EUSPA), 170 00 Prague, Czech Republic; javier.ostolaza@euspa.europa.eu (J.O.); carmelo.hernandez@euspa.europa.eu (C.H.); juan.vazquez@euspa.europa.eu (J.V.); franciscojavier.deblasfernandez@euspa.europa.eu (J.d.B.)

* Correspondence: emilio.gonzalez@ext.spaceopal.com (E.G.); stefano.lagrasta@euspa.europa.eu (S.L.)

† Presented at the European Navigation Conference 2023, Noordwijk, The Netherlands, 31 May–2 June 2023.

Abstract: In just one year, Spaceopal and its partners developed the Galileo HAS Performance Characterization User Algorithm for the EU Agency for the Space Programme (EUSPA). The Galileo HAS User Terminal (HAUT) hosts the Galileo HAS Performance Characterization User Algorithm. The Galileo HAS User Terminal is a portable, configurable and autonomous device powered by a triple-frequency Galileo and GPS receiver and calculates a single- (Galileo) or multi-constellation (Galileo + GPS) Galileo HAS and Open Service (OS) positioning, velocity and time (PVT) solution. The User Terminal can be configured to retrieve Galileo HAS corrections either from Galileo Signal-in-Space (SIS) over E6-B or Internet Data Distribution (IDD) over NTRIP in an RTCM3 format and works in different frequency combinations that can be configured by the user. The User Terminal is a robust device (IP64) with multiple communication and logging capabilities. The Galileo HAS Initial Service was declared on 24 January by the European Commission, and provides free-of-charge, high-accuracy Precise Point Positioning (PPP) corrections (orbits, clocks) and code biases for Galileo and GPS to achieve real-time improved user positioning performance. The Galileo HAS Service Definition Document (SDD) and the HAS SIS Interface Control Document (HAS SIS ICD) are freely available to users on the web portal of the European GNSS Service Centre and HAS Internet Data Distribution Interface Control Documents (HAS IDD ICD) are available after registration. Using the Galileo HAS User Terminal, this article presents the results of Galileo HAS User Terminal's performance, configuring the User Algorithm to assume static dynamics. It is to be noted that this configuration provides a significant performance benefit with respect to a configuration compatible with kinematic operation. Preliminary results indicate the Galileo HAS User Terminal achieves excellent accuracy.

Keywords: Galileo HAS; high accuracy; PPP

Citation: González, E.; Pintor, P.; Senado, A.; Dhital, N.; Ostolaza, J.; Hernández, C.; Vázquez, J.; de Blas, J.; Lagrasta, S. Testing the Galileo High Accuracy Service User Terminal (HAUT) in Static Scenarios. *Eng. Proc.* **2023**, *54*, 17. https://doi.org/10.3390/ENC2023-15471

Academic Editors: Tom Willems and Okko Bleeker

Published: 29 October 2023

1. Introduction

The Galileo High Accuracy Service (HAS), in its initial stage, was declared in January 2023. In order to support the validation of the service and test its performance in different environments, EUSPA launched a procurement for an algorithm and user terminal, which was awarded to an industrial team led by Spaceopal GmbH (Munich, Germany).

This paper introduces the abovementioned algorithm and user terminal and provides performance results from the weeks following the launch of the service.

2. The Galileo High Accuracy Service

As one of the services offered by the European Global Navigation Satellite System, Galileo, the Galileo High Accuracy Service is an open access and free-of-charge service based on the provision of precise corrections transmitted by the Galileo E6 signal from the Galileo space segment as well as via the Internet.

3. The HAUT Project

As indicated in the introduction, EUSPA procured the implementation of the Galileo HAS User Algorithm for the Initial Service within a User Terminal (UT) equipment that was later used in the service validation. This project (HAUT) was launched under the Fundamental Elements (FE) program.

The project was led by Spaceopal and conducted during one year (throughout the COVID-19 pandemic) together with its industrial partners: ANavS (responsible for algorithm definition and overall hardware design and assembly), DLR Institute of Communication and Navigation (in charge of the production of synthetic data and the provision of their GNSS testing infrastructure), IABG (responsible for housing and physical and environmental tests) and Iguassu Software Systems (in charge of the human–machine interface on the computer). Spaceopal conducted the project management and overall verification and validation tasks.

The requirements for this work targeted a wide range of applications, configurations, dynamics and environmental constraints. Five receivers were delivered to the customer for service validation and the algorithm was also provided as an offline tool, useful for experimentation and scenario replay. In the following sections, details about the UA and UT will be provided.

4. The Galileo HAS User Algorithm for HAS Initial Service

The Galileo HAS User Algorithm is a real-time precise positioning algorithm that processes the corrections provided by the Galileo HAS Initial Service (orbit and clock corrections and code biases) and calculates position and velocity referred to the GTRF (Galileo Terrestrial Reference Frame), time-stamped with respect to the GST (Galileo System Time) using Galileo L-Band observables and, in the case of a multi-constellation configuration, GPS L-Band observables as well. The User Algorithm was designed to work in all of the following dual- and triple-frequency modes:

- Dual-frequency modes:
  - Galileo only: E1-E5a/E1-E5b/E1-E5AltBOC/E1-E6B.
  - Galileo and GPS: E1-E5a/E1-E5b/E1-E5AltBOC/E1-E6B + L1-L2P/L1-L2C/L1-L5.
- Triple-frequency modes:
  - Galileo only: E1-E5a-E6B.
  - Galileo and GPS: E1-E5a-E6B + L1-L2C-L5.

The Galileo HAS User Algorithm used to characterize the typical positioning performance of the Galileo HAS described in the HAS SDD was configured for kinematic operation.

4.1. User Algorithm Description

The selected position estimation technique for the User Algorithm was an extended Kalman filter (EKF) with state predictions and state updates to improve convergence and accuracy, as it achieves, in the opinion of the authors, an optimum trade-off between the measurement model and the state space model by minimizing the variance of the updated states. The User Algorithm processing follows an uncombined, satellite-to-satellite single-differenced (SD) processing.

Compared to an ionosphere-free approach, the proposed uncombined processing avoids the amplification of residual satellite orbit, clock, and phase bias errors, avoids noise amplification and brings no loss of information, which occurs when using measurement combinations. Conversely, ionosphere-free combinations do not require the integration of ionospheric corrections or models. Moreover, the significant bias amplification achieved by the wide-lane combination is considered more detrimental than the benefit of a larger wavelength.

When comparing the SD approach of the User Algorithm to an undifferenced setup, the elimination of receiver-specific errors and noise amplification by single differences are on a comparable performance level but only if the process noise of the clock and other receiver-specific unknowns can be modeled sufficiently well. Thus, the elimination of these unknowns by SD is the preferred option.

The algorithm has preprocessing steps, including outlier detection and elimination of single-satellite failure for each GNSS constellation with RAIM/DIA and detection and elimination of complete GNSS constellation outages. Preprocessing also addresses cycle slip detection and correction by using a time-differenced Melbourne–Wübbena combination and ionosphere-free phase combination. With this set of techniques, the proposed UA provides robustness to different fault behaviors on the site of satellites, the level of the correction service and the receiver and its environment.

4.2. Main Blocks of User Algorithm

The algorithm consists of two main blocks (Figure 1), including their input and output interfaces. The first main block is preprocessing of observation, navigation and correction data (PPO), which includes, e.g., the determination of the standard PVT solution and the forming of satellite–satellite single differences. The second main block includes the actual High Accuracy Service PVT solution (HAS) with an extended Kalman filter and partial integer ambiguity resolution.

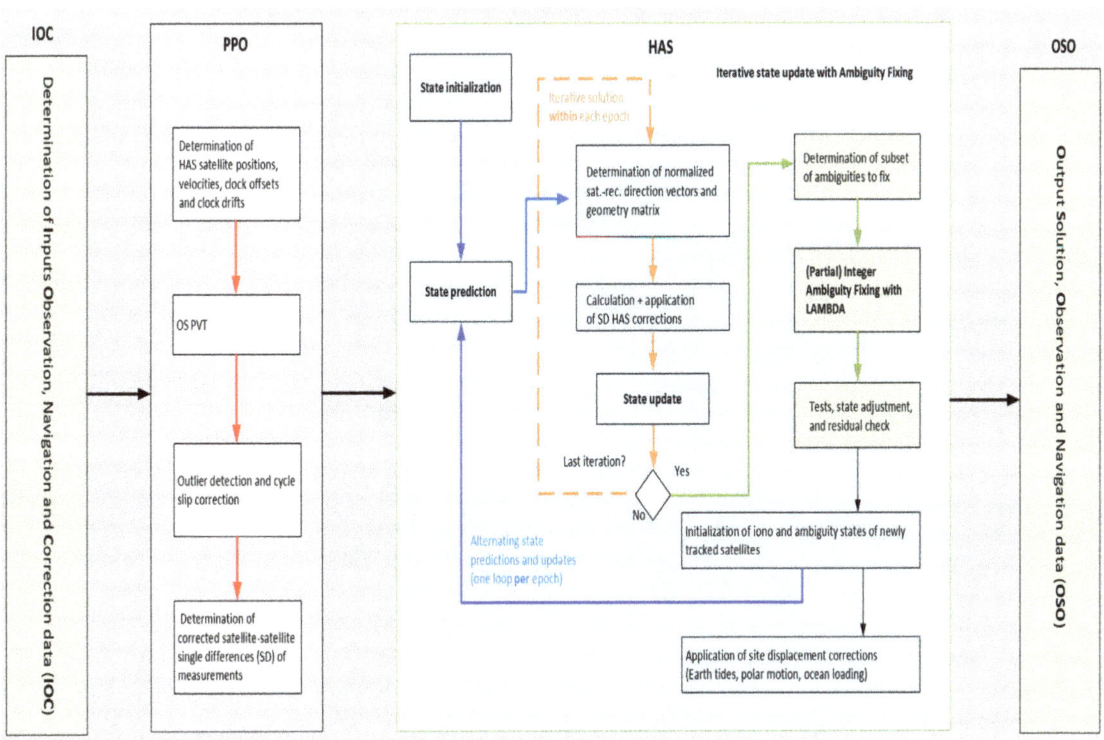

Figure 1. Main blocks of user algorithm.

Additional details on this algorithm are provided in a previous paper [1].

5. The Galileo HAS User Terminal

The Galileo HAS User Terminal provides a precise positioning, velocity and time (PVT) solution based on the Galileo HAS User Algorithm for SL1, guaranteeing access to corrections from either the SIS or the Internet and offering a flexible approach to the configuration of the algorithm, communication means and logging capabilities.

The User Terminal (Figure 2) is a small (8 cm, 18 cm, 25 cm), light (2.1 kg), configurable, portable and autonomous device that includes a professional multi-constellation, multi-frequency receiver and calculates a single- or multi-constellation (Galileo + GPS) Galileo HAS and Open Service (OS) PVT solution with an update rate ranging from 1 Hz to 100 Hz. The User Terminal can be configured to retrieve Galileo HAS corrections either from Galileo E6-B or RTCM 3 corrections over NTRIP and works in any of the User Algorithm dual- and triple-frequency modes.

Figure 2. HAUT user terminal.

The User Terminal has RF interference detection capabilities in the E1, E5 and E6 bands. Its ingress protection is IP64 and its operating temperature range is −40°–65 °C. Its ruggedized aluminum housing was designed and tested against vibration, shock, temperature, humidity and corrosion to make the unit a resilient piece of equipment for future testing campaigns. It can be interfaced with an HMI application running on a PC for configuration and real-time monitoring purposes.

The UT has a wide logging capability: real time HAS/OS solution and monitoring streams, GNSS observations in RINEX and binary files (up to 100 Hz) and HAS corrections in ASCII and binary format. It has 1 TB internal memory and USB, WiFi and Bluetooth interfacing capabilities.

The UT can be powered through USB or Ethernet (PoE I/F). It has internal batteries allowing autonomy for more than 3 h off the grid. Portability is also considered (transport box, external batteries).

The current HAS specifications are described in references [2–4], which include the service definition document and signal and ground specifications. Additional details on the development and verification of the User Terminal are provided in a previous paper [1]. Previous work on HAS user performance from other developments and using different algorithms can be found in [5–7].

6. Positioning Performance by Galileo HAS User Terminal Using Galileo HAS Corrections

The positioning performance provided by the HAUT using Galileo HAS corrections and configuring the User Algorithm to assume no dynamics (as opposed to the configuration used for the Galileo HAS performance characterization presented in the HAS SDD, which considers kinematic conditions) was evaluated for a period of 31 days from 25 February to 27 March 2023. The Galileo HAS User Terminal, located in Munich (Germany), was switched on on 24 February to converge and left untouched until 8 March, when the campaign had to be interrupted for a few hours to perform maintenance of the antenna installation. The day on which the Galileo HAS User Terminal was switched on again is excluded from the analysis. The positioning accuracy provided by the HAUT for 30 days in 24 h periods from 0 h UTC is presented in Table 1.

Table 1. Real time Galileo HAS User Terminal horizontal and vertical HAS positioning accuracy (68%) over 24 h periods using Galileo E1-E5a, GPS L1-L2C and Galileo HAS Initial Service corrections over E6-B.

Date	Vertical Accuracy 68% (cm)	Horizontal Accuracy 68% (cm)	Date	Vertical Accuracy 68% (cm)	Horizontal Accuracy 68% (cm)
02.25	2	2	03.13	2	3
02.26	2	2	03.14	1	3
02.27	2	1	03.15	1	3
02.28	4	1	03.16	1	3
03.01	6	2	03.17	1	3
03.02	4	2	03.18	1	2
03.03	3	2	03.19	1	2
03.04	3	2	03.20	1	2
03.05	2	2	03.21	1	2
03.06	2	2	03.22	1	2
03.07	2	1	03.23	1	2
03.09	3	3	03.24	1	2
03.10	5	4	03.25	1	2
03.11	3	4	03.26	1	2
03.12	2	3	03.27	1	2

It is to be noted that the positioning accuracy results from Table 1 are only considered representative of static, open-sky conditions after full convergence time of the EKF. Therefore, they are not representative of the performance in kinematic conditions, which are the conditions assumed for the HAS performance characterization presented in the HAS SDD [2]. This explains the different level of performance between Table 1 results and the SDD performance levels.

7. Informative Convergence Assessment by Galileo HAS User Terminal Using Galileo HAS Corrections

A key matter for high-accuracy applications is convergence time. In this study, we have defined the convergence time as the time from warm start until the Galileo HAS User Terminal provides the first solution with the required instantaneous positioning accuracy below the specified threshold.

A warm start assumes that the whole set of input broadcast data is already available at the receiver and in this way, the time to demodulate navigation data from SIS is excluded from the computation.

To consider the converged position, a horizontal and vertical threshold must be defined. In this study, we have chosen two sets, the first being 15 cm and 20 cm, respectively. The second set is double these values, so 30 cm horizontal and 40 cm vertical.

The set-up for the first campaign was repeated, but the Galileo HAS User Terminal was configured to execute 1 h automatic warm restarts. One-hour restarts allowed us to evaluate 24 restarts every day. The User Terminal convergence time using HAS Initial Service corrections was evaluated for a period of 30 days from 29 March to 27 April 2023. The overall statistics for convergence time are presented in Table 2.

Table 2. Real-time Galileo HAS User Terminal convergence time 50th, 68th and 95th percentiles for 30 days with 1 h automatic warm restarts using Galileo E1-E5a, GPS L1-L2C and Galileo HAS Initial Service corrections over E6-B for two different thresholds.

	Convergence Time 15 cm H and 20 cm V	Convergence Time 30 cm H and 40 cm V
50th percentile	345 s	80 s
68th percentile	638 s	170 s
95th percentile	1836 s	586 s

8. Conclusions

In a record one-year time, Spaceopal and its partners developed the Galileo High Accuracy Algorithm and the Galileo HAS User Terminal (HAUT), which hosts the algorithm for the EU Agency for the Space Programme (EUSPA).

The Galileo HAS User Terminal is a portable, configurable and autonomous device powered by a triple-frequency Galileo and GPS receiver and calculates a single- (Galileo) or multi-constellation (Galileo + GPS) Galileo HAS and Open Service (OS) positioning, velocity and time (PVT) solution. The User Terminal can be configured to retrieve Galileo HAS corrections either from Galileo E6-B or Internet link corrections over NTRIP in an RTCM3 format and works in all dual- and triple-frequency modes of the algorithm. The User Terminal is a robust device (IP64) with multiple communication and logging capabilities.

This article has presented the results of the positioning performance provided by the Galileo HAS User Terminal, using Galileo HAS corrections and configuring the User Algorithm to assume static dynamics (as opposed to the configuration used for the Galileo HAS performance characterization presented in the HAS SDD, which considers kinematic conditions) from February to April 2023. The results of the data campaign reach centimeter level at the 68th percentile.

Author Contributions: Writing: E.G. (Sections 1–5) and P.P. (Sections 6–8); resources: A.S.; validation: P.P. and N.D.; supervision, J.O., C.H., J.V., S.L. and J.d.B. All authors have read and agreed to the published version of the manuscript.

Funding: The HAUT project was funded by EUSPA through the Fundamental Elements program, contract GSA-OP-25-19.

Institutional Review Board Statement: Not applicable.

Informed Consent Statement: Not applicable.

Data Availability Statement: Galileo HAS corrections are publicly available. The data presented in this study are not publicly available as they belong to the administrators of the HAUT project.

Acknowledgments: We deeply acknowledge the contribution of the industrial team to the development and validation of the HAS algorithm and User Terminal, as well as the technical supervision of the work by the relevant EUSPA officers and partners.

Conflicts of Interest: The authors and Spaceopal GmbH declare no conflict of interest.

References

1. Pintor, P.; González, E.; Bohlig, P.; Sperl, A.; Henkel, P.; Simón, J.; Hernández, C.; de Blas, J. Galileo High Accuracy Service (HAS) Algorithm and Receiver Development and Testing. In Proceedings of the 35th International Technical Meeting of the Satellite Division of The Institute of Navigation (ION GNSS+ 2022), Denver, CO, USA, 19–23 September 2022; pp. 836–851.
2. European Union. Galileo High Accuracy Service-Service Definition Document (HAS SDD), Issue 1.0. Available online: https://www.gsc-europa.eu/sites/default/files/sites/all/files/Galileo-HAS-SDD_v1.0.pdf (accessed on 1 April 2023).
3. European Union. Galileo High Accuracy Service Signal-In-Space Interface Control Document (HAS SIS ICD), Issue 1.0. Available online: https://www.gsc-europa.eu/sites/default/files/sites/all/files/Galileo_HAS_SIS_ICD_v1.0.pdf (accessed on 1 April 2023).

4. European Union. Galileo High Accuracy Service-Internet Data Distribution Interface Control Document (HAS IDD ICD), Issue 1.0. Available online: https://www.gsc-europa.eu/galileo/services/galileo-high-accuracy-service-has/internet-data-distribution-registration-form (accessed on 1 April 2023).
5. Fernandez-Hernandez, I.; Chamorro-Moreno, A.; Cancela-Diaz, S.; Calle-Calle, J.D.; Zoccarato, P.; Blonski, D.; Senni, T.; de Blas, F.J.; Hernández, C.; Simón, J.; et al. Galileo high accuracy service: Initial definition and performance. *Gps Solut.* **2022**, *26*, 65. [CrossRef]
6. Chamorro, A.; Cancela, S.; García, A.J.; Calle, D.; Fernández-Hernández, I.; de Blas, F.J.; Vazquez, C.; Hernández, C. Early Demonstration of Galileo HAS User Performances. In Proceedings of the 35th International Technical Meeting of the Satellite Division of The Institute of Navigation (ION GNSS+ 2022), Denver, CO, USA, 19–23 September 2022; pp. 828–835.
7. Naciri, N.; Yi, D.; Bisnath, S.; de Blas, F.J.; Capua, R. Assessment of Galileo High Accuracy Service (HAS) test signals and preliminary positioning performance. *GPS Solut.* **2023**, *27*, 73. [CrossRef] [PubMed]

engineering proceedings

MDPI

Proceeding Paper

GNSS Radio Frequency Interference Mitigation in Collins Commercial Airborne Receivers †

Angelo Joseph, Patrick Bartolone, Joseph Griggs, Bernard Schnaufer, Huan Phan and Vikram Malhotra *

Navigation Systems Engineering, Collins Aerospace, 795 W NASA Blvd, Building 312, Melbourne, FL 32901, USA; patrick.bartolone@collins.com (P.B.)
* Correspondence: vikram.malhotra@collins.com

† Presented at the European Navigation Conference 2023, Noordwijk, The Netherlands, 31 May–2 June 2023.

Abstract: Nowadays, commercial aeronautical Global Navigation Satellite Systems (GNSS) receivers are more and more exposed to Radio Frequency Interference (RFI) threats from GNSS jammers and spoofers. On commercial aircraft GNSS, receiver outputs, in general, are integrated or cross-monitored with other navigation sensors such as IRS and DME, etc., and, in many cases, the GNSS receiver outputs are used directly by on-board aircraft systems. The advent of modernized dual-frequency and multi-constellation signals will improve the availability and integrity of GNSS receivers in the presence of RFI. To be further resilient to the various types of RFI threats, the airborne GNSS receiver will need to perform additional receiver-based detection/mitigation techniques and should be able to determine position integrity in the presence of spoofers. This paper focuses specifically on two techniques under development that will be incorporated via a field loadable software update to the GLU-2100. The first method, Receiver Autonomous Signal Authentication (RASA), and a second type of technique, Staggered Examination of Non-Trusted Receiver Information (SENTRI). The paper will provide a brief description of the RASA and SENTRI algorithms, followed by results from both simulation and real-world tests. Finally, the limitations of the algorithms will also be provided.

Keywords: GNSS; RFI; anti-spoofing

Citation: Joseph, A.; Bartolone, P.; Griggs, J.; Schnaufer, B.; Phan, H.; Malhotra, V. GNSS Radio Frequency Interference Mitigation in Collins Commercial Airborne Receivers. *Eng. Proc.* **2023**, *54*, 18. https://doi.org/10.3390/ENC2023-15420

Academic Editors: Tom Willems and Okko Bleeker

Published: 29 October 2023

1. Introduction

Nowadays, commercial aeronautical GNSS receivers are more and more exposed to RFI threats from GNSS jammers and spoofers. Worldwide reports from air navigation service providers and Collins customers have confirmed that commercial airborne GNSS receiver products have been exposed to both jamming and spoofing. The following three publications, Collins Aerospace RFI Assessment [1], SIB EASA's Safety information bulletin [2], and the 2022 ITU RRB circular [3] particularly show the increasing RFI threats and vulnerabilities of commercial airborne GNSS receivers.

1.1. EASA Related RFI Events

Through its safety information bulletin released in February 2023 [2], the European Union Aviation Safety Agency (EASA) reported a significant increase in jamming and or possible spoofing of GNSS since the beginning of 2022. The bulletin lists the areas affected by these events and provides a list of issues such as an inability to use GNSS for waypoint navigation, a loss of RNAV and RNP operations, inconsistent aircraft positions, a loss of ADS-B, wind shear and surface functionalities, the failure or degradation of ATM/ANS/CNS and aircraft systems which use GNSS as a time reference, potential airspace infringements, and/or route deviations that a degradation of GNSS signal could generate. EASA has recommended a series of mitigating measures to address issues related to GNSS jamming and spoofing. These measures focus on being prepared with alternate procedures when GNSS is unavailable due to jamming and spoofing.

1.2. Collins Aerospace RFI FlightAware Data Analysis

Collins developed a method using FlightAware's Firehose service to identify airborne GNSS RFI by detecting ADS-B Out drops within a given time interval from co-located aircraft equipped with ADS-B Out transponders [1]. This analysis shows that, while less than 1% of all flights over the CONUS during the time period of analysis were impacted by RFI, sustained RFI-related outages were observed in specific regions (Figure 1). Validation of the results showed that the method was successful at detecting RFI when and where intentional RFI was expected via aviation NOTAMs.

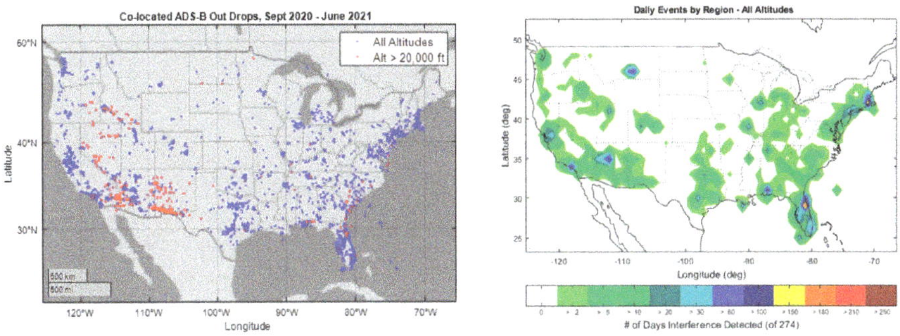

(a) (b)

Figure 1. Map of GNSS RFI threats with respect to ADS-B FlightAware data analysis.

1.3. Circular from the ITU the Radiocommunication Bureau

The ITU Radio Regulations Board (RRB) has advanced a circular letter [3] to member states with recommendations for mitigating harmful interference to RNSS receivers using measures such as reinforcing navigation systems' resilience to interference, increasing collaboration between regulatory and enforcement authorities, and retaining conventional navigation infrastructure for contingency support in case of RNSS outages.

2. The Collins Aerospace GLU-2100 Multi-Mode Receiver (MMR)

The Collins Aerospace GLU-2100 MMR features a hardware, software, and firmware baseline that has been certified on Airbus and Boeing platforms. It is an ARINC 755-4 compliant unit whose GNSS navigation is currently based on GPS and SBAS L1 signals. The baseline GLU-2100 software, hardware, and firmware are designed to Level A DO-178C [4] and DO-254 [5] standards. Further, a dual thread processing architecture is used to meet the safety requirements of a unit whose failure could lead to catastrophic conditions. The hardware baseline architecture allows the GLU-2100 to be upgraded via a software/firmware field load to support GPS, SBAS, Galileo, and Beidou dual frequency signals.

The baseline GNSS engine within the GLU-2100 supports a Universal Channel Architecture [6] that can be configured at run-time to track any of the supported signal types. The DFMC capability of the GLU-2100 was demonstrated as part of the Boeing ecoDemonstrator [7] program. Further, as part of the European Union MUGG project, the universal channel architecture was leveraged to implement the core GNSS acquisition, tracking, and demodulation functions as defined in the DFMC MOPS ED-259A [8]. The navigation software was updated to support future navigation modes such as DFMC SBAS and H-ARAIM.

3. Overview of the RFI Technology in Collins Aerospace GNSS Solutions

Collins Aerospace has the following three-pronged strategy to address GNSS RFI threats, as illustrated in Figure 2:

- RFI Detect—generic flag to warn the user of the presence of RFI.

- RFI Mitigate—identify more precisely the magnitude of induced position errors, jammer strengths, or the specific satellite signals that are being spoofed.
- RFI Playthrough—play through jammer and spoofer attacks by continuing to output authentic position solutions.

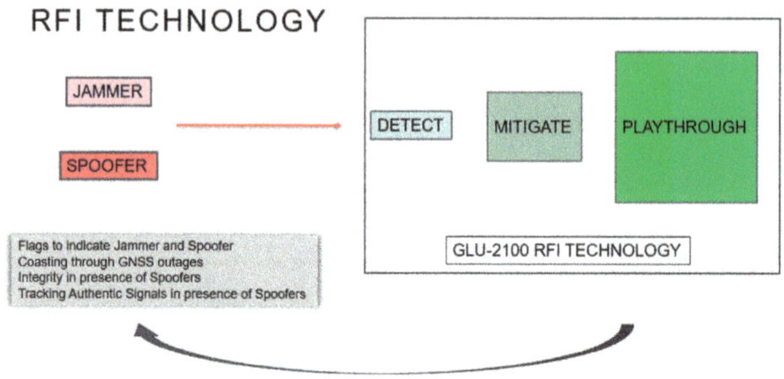

Figure 2. Collins RFI Response Strategy.

The next generation of Collins Aerospace GNSS receivers will introduce multiple RFI resilience techniques via field loadable software updates to satisfy the goals of the RFI mitigation strategy. The remainder of this paper will focus on two techniques known as RASA and SENTRI that will be used for spoofer detection.

4. Receiver Autonomous Signal Authentication (RASA)

RASA [9,10] is a Collins patented technique that uses the effect of the spoofer that is observed on the receiver's clock bias estimate to detect the presence of spoofers. When spoofed GNSS signals are received from a single transmitter, the path lengths between the transmitter and the aircraft are common to all the spoofed signals and so affect the estimated receiver clock bias (unsynchronized spoofing, Figure 3a). Even if the path length is estimated and accounted for by the spoofer (synchronized spoofing, Figure 3b), the clock bias estimate will be impacted by the error in this estimate. If the baseline no-fault behavior of the clock is known under worst-case environmental conditions such as vibration and temperature ramps, the change in the clock behavior due to spoofing can be used to detect a spoofer at a fixed P_{FA}. The clock behavior is assessed by computing the Allan variance of the clock bias, with the performance varying based on the τ (time between clock samples) and averaging period over which the Allan variance is computed.

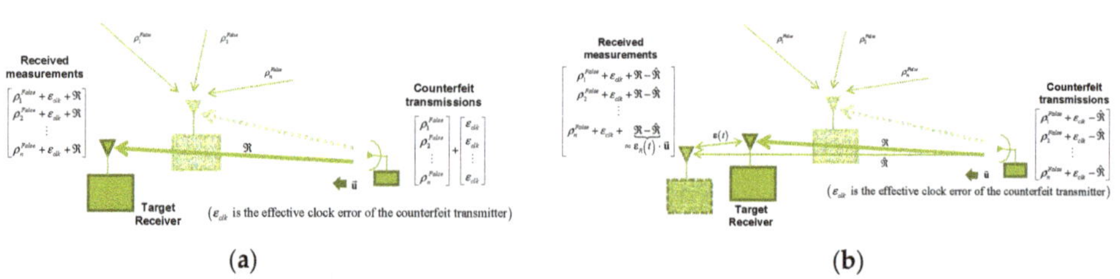

Figure 3. GNSS receiver operation in the presence of unsynchronized (**a**) and synchronized (**b**) spoofing.

An RASA detection performance analysis was performed with live-sky authentic measurement logs captured with multiple GLU-2100 units during flight trials and other dynamic scenarios, as well as scenarios generated on a GNSS RF simulator. The RASA statistic was computed offline over the full duration of each scenario for each simulated spoofer. The simulation is somewhat conservative, as it assumes that the spoofer can track and rebroadcast the spoofed measurements without any errors. Figure 4 shows the performance of different combinations of τ and Allan variance averaging periods in terms of the relative LoS acceleration with the worst-case fixed P_{fa} threshold to achieve a detection rate of 99%, as well as the detection performance of the chosen τ of 1 s and averaging period of 24 s.

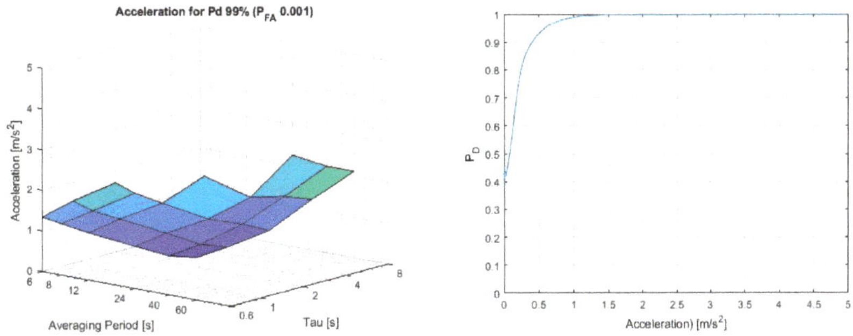

(a) (b)

Figure 4. LoS Acceleration for a fixed detection performance (**a**) and performance (probability of detection P_D) of the fixed threshold over acceleration for fixed $P_{FA} = 10^{-3}$, $\tau = 1$ s and 24 s averaging period (**b**).

4.1. RASA Simulations: Real World Commercial Flight Dynamics

Twenty real world flight trajectories of different lengths and aircraft types were analyzed to produce statistics of the LoS acceleration between static ground-based spoofers and aircraft position throughout the duration of flight (Figure 5a).

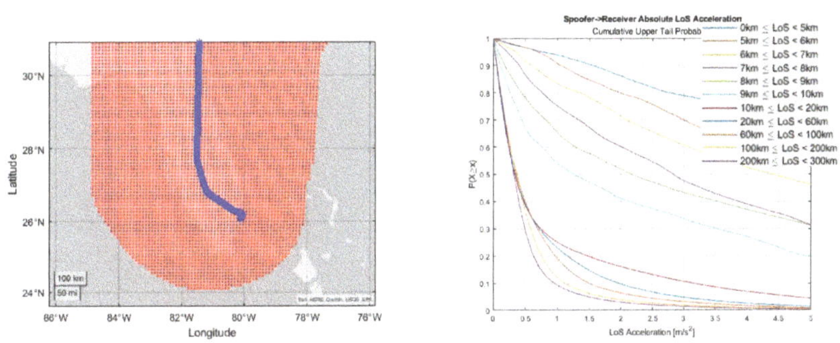

(a) (b)

Figure 5. Example simulated spoofer locations (red dots) versus aircraft trajectory (blue line) as far as LoS distance to horizon per altitude (**a**) and real-world LoS acceleration distribution (**b**).

Figure 5b shows the cumulative upper tail probability function of the aggregate LoS acceleration. With the tunable parameters already discussed and a 99% probability of detection at an LoS acceleration of 1 m/s^2, RASA could be expected to achieve a total P_D of around 95% (per sample) for an unsynchronized spoofer within 5 km of the aircraft, decreasing to around 45% for a spoofer between 9 and 10 km.

4.2. RASA Simulations: Mixed Sets

The following simulation results show the effectiveness of RASA when a combination of authentic LoS and spoofed signals are tracked. The ratio of authentic to counterfeit signals used in RASA offline processing, as well as the specific set of spoofed measurements within the same ratios, were varied. The results shown in Figure 6 indicate that RASA detection performance was not significantly reduced if the LoS acceleration was greater than ~1.1 m/s^2, even when only one of the measurements used in the navigation solution was counterfeit. As the LoS acceleration was reduced below this value, the performance diminished.

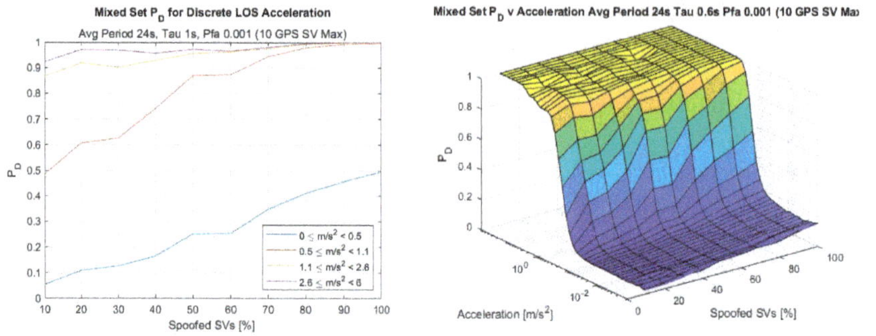

Figure 6. Mixed set detection performance for discrete acceleration bins (**a**) and as a function of acceleration (**b**).

4.3. *RASA Simulations: Synchronized Spoofers*

Residual estimation error from an in imperfectly synchronized spoofer was simulated with a first-order Gauss Markov model. This spoofer model generalizes the estimation error to a time variation of error magnitude along the LoS vector. The ASA detection performance in the presence of this model is shown in Figure 7. A standard deviation greater than 6 cm is detectable at 100% per sample.

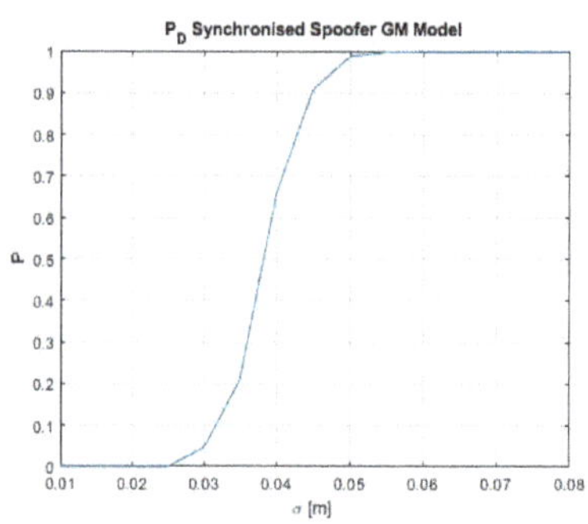

Figure 7. Synchronized spoofer detection performance.

5. Staggered Examination of Non-Trusted Receiver Information (SENTRI)

SENTRI (Figure 8) is a Collins patented [11] technique that uses multiple inertial coasted position solutions as trusted sources to protect a master GNSS position solution. Initially, the master solution will be the L1 SBAS solution or the L1 GPS-only solution, but in the future, the master solution can be a DFMC SBAS or H-ARAIM solution. In this architecture, the error estimation Kalman Filter maintains a fully blended GNSS-Inertial position solution. Three staggered coasting solutions (1, 2, and 3) are derived from the blended solution by propagating their solutions on pure inertial data without any GNSS measurements. The premise behind the SENTRI algorithm is that, when the GNSS measurements are being spoofed, the master solution will deviate from the coasting solutions. Depending on the stability of the IRS, the coasting solutions will gradually degrade in position accuracy, but they will not be tainted by the spoofed measurements. When considering a single coasted solution, the optimal performance of the of the SENTRI detector is achieved when the start of coasting occurs just before the GNSS measurements are spoofed. In this case, the coasting solution is untainted by the spoofed measurements and the master solution deviates maximally from the coasted solution. In the case where the start of a coasting solution occurs just after the GNSS measurements are spoofed, the master solution deviates minimally from the coasted solution. Since the start time of the spoofer is unknown, the coasted solutions are staggered in time such that the spoofer can be detected regardless of the relationship between the coasting start and spoofer start times. In the GLU-2100, SENTRI is implemented with three staggered coasted solutions, each with a coasting time of 30 s and staggered with an offset of 10 s between them.

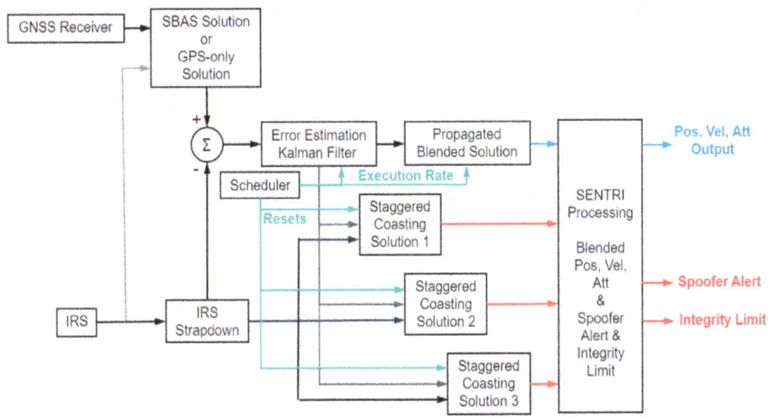

Figure 8. SENTRI Architecture.

SENTRI uses parity space techniques to determine a protection bound of the spoofing-induced master solution position error. The fault detection and protection level computation are mechanized by considering the measurement vector of the least square solution to be composed of the coasting solution coordinate (in either x, y, or z direction of the local level frame) and the corresponding master solution coordinate. The parity vector is dependent on the difference between the master and coasted solution coordinates. Standard parity space methods are used to derive fault detection and protection level statistics.

5.1. SENTRI Simulations and Performance

The RTCA MOPS for GNSS-Aided Inertial Systems DO-384 [12] Appendix Q defines the simulation approach to generating performance data to populate spoofer claim tables. Three types of spoofers are considered, position step, velocity step, and acceleration step. The scenario includes initialization time, exposure time, and recovery time. In general, the detection performance of the SENTRI algorithm in the presence of different spoofer types is dependent on the aircraft dynamics and the accuracy of the master solution. For this analysis, the aircraft dynamics were divided into normal and abnormal dynamics, as defined in Section 2.1.2.5 of DO-229E [13]. The GPS navigation mode and SBAS PA (Precision Approach) navigation mode accuracy values were used to model the master solution accuracy.

As seen in Table 1, the SENTRI algorithm performs better under normal dynamics and when the master solution is a more accurate solution such as the SBAS master solution. The above results are preliminary in nature and were determined via simulations that used the same model that is used to generate the code that is executed on the GLU-2100. For purposes of illustration, one run of the SENTRI simulation is shown below to demonstrate the spoofer detection capability when a velocity ramp of 1 m/s is introduced in SBAS mode. In Figure 9, the 1 m/s velocity ramp spoofing begins at 800 s into the run. The SENTRI alert occurs at 820 s and the SENTRI HPL bounds the master solution error.

Table 1. SENTRI Detection for GPS L1 and SBAS L1 Master.

	Normal Dynamics		Abnormal Dynamics	
	GPS L1	**SBAS L1**	**GPS L1**	**SBAS L1**
Min Position Step (m)	40	30	45	35
Min Velocity Step (m/s)	1.5	1.0	2.5	2.0
Min Acceleration Step (m/s^2)	0.24	0.15	0.28	0.20

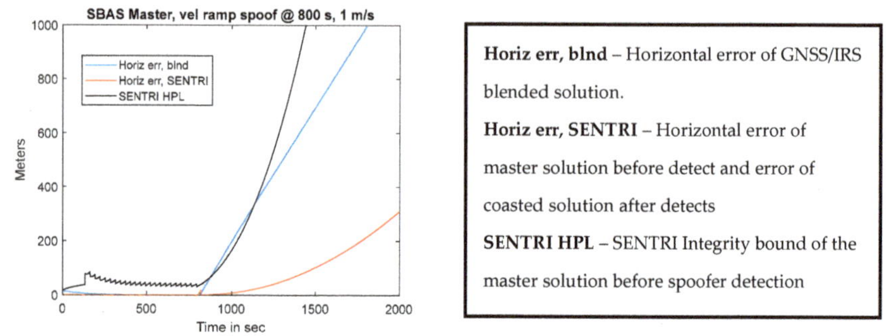

Figure 9. SBAS master velocity ramp (1 m/s).

5.2. Real-Time Demonstration of SENTRI Spoofer Detection Capability

In this test, the performance of the Real-Time SENTRI software (version 2.0) was tested in the lab with hardware in the loop by connecting a GLU-2100 unit loaded with the SENTRI software to a GNSS signal simulator. The NAV grade inertial data were simulated by generating the attitude and inertial position data from the simulator scenario and then adding the IMU noise to the simulated data by using the model of an NAV grade inertial system. The simulated IRS ARINC 429 labels were then sent to the GLU-2100 on the IRS input bus, thus mimicking the aircraft installation conditions. For purposes of illustration, the results from a scenario that mimics a true flight test on a Collins 604 Challenger Flight Test aircraft are discussed below. In this scenario, the spoofer was generated to gradually drift the GNSS signal simulated position from the true flight path.

In Figure 10, it is seen that the spoofer attempts to gradually drift the aircraft position from the true position. The SENTRI algorithm raises a spoofer alert before the position error exceeds the protection level.

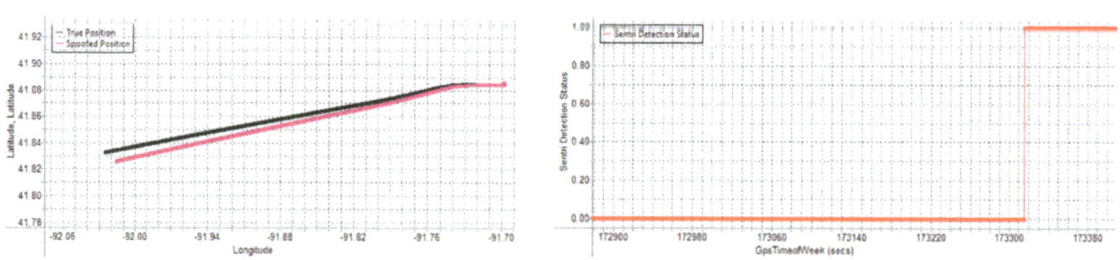

(a) (b)

Figure 10. Challenger 604 simulation trajectory (**a**) and SENTRI detection flag (**b**).

6. Integration of RASA and SENTRI in the GLU-2100

The previous sections demonstrated the spoofer detection performance of the RASA and SENTRI algorithms. The limitations of the algorithms were also presented. The RASA algorithm was shown to have a poor detectability when the line-of-sight acceleration between the aircraft and spoofer was low. Further a protection bound cannot be calculated for the RASA detector. RASA also has a longer time-to-detect and it generally requires the probability of false alarm to be high to reliably detect spoofers. The SENTRI algorithm provides a protection bound around the spoofer detection. SENTRI requires that it is initialized with authentic signals. In the GLU-2100, the complementary benefits of RASA and SENTRI are used together to improve the resilience of the GNSS to spoofing. RASA (with a higher probability of false alarm) is initially used to confirm that all the signals being tracked are authentic signals. Once this authentication is confirmed, the GNSS measurements are passed on to SENTRI to initialize the blending filters. After SENTRI initialization, RASA continues to execute in the background and its detector outputs are only used for maintenance logging. SENTRI is used to detect the presence of spoofers and, if spoofers are detected, the GLU-2100 may enter a pure inertial coasting mode. Once a spoofer is detected, the SENTRI processing is terminated, and RASA is once again activated to monitor for authentic signals. Upon the successful acquisition of authentic signals, SENTRI is once again re-initialized and the RASA-SENTRI workflow repeats again.

The SENTRI protection level is expected to be used only when inertial coasting is triggered because of spoofer detection. The use of the SENTRI protection level has the potential to reduce the availability of GNSS integrity. Until a standardized availability analysis for spoofer detection is accepted by the avionics community, the GLU-2100 will not output the SENTRI protection level as part of its integrity levels under normal operation when spoofing has not been detected.

In summary, while the combination of RASA and SENTRI techniques has proven to be highly effective in detecting and mitigating GPS spoofing attacks, it is important to be aware of their limitations and to implement appropriate measures to address these challenges. By doing so, we can ensure that our GPS systems remain secure and reliable, even in the face of sophisticated spoofing attacks.

7. Conclusions

GNSS RFI threats have become increasingly common and are being reported regularly on a worldwide basis. GNSS receivers that are used in all applications, including commercial aviation, will need to consider implementing techniques to address the threats of GNSS spoofing and jamming. Collins Aerospace plans to introduce features via a software field loadable update to the GLU-2100 product to improve its resiliency to GNSS spoofing and jamming. This update introduces two techniques, RASA and SENTRI, that can be used together in a complementary fashion to reliably detect the presence of spoofers. Additionally, this update will provide improved robustness to data spoofing attacks that induce errors in ephemeris, almanacs, and GPS time jumps, etc. Further, this update will enable the GLU-2100 to coast through GNSS outages that are induced due to spoofing or jamming. Future technologies will leverage antenna techniques, signal analysis, DFMC signals, and APNT to increase the robustness to new and evolving threats. The goal of this RFI mitigation roadmap is to continue to ensure that GNSS can be used in a safe and reliable manner in civil aviation.

Author Contributions: Conceptualization, All; methodology, H.P., A.J. and B.S.; software, H.P., J.G. and V.M.; validation, J.G. and P.B.; formal analysis, A.J., J.G. and P.B.; investigation, J.G. and P.B.; resources, A.J., J.G. and P.B.; data curation, J.G. and P.B.; writing—original draft preparation, All; writing—review and editing, A.J., J.G., P.B. and H.P.; visualization, B.S.; supervision, A.J.; project administration, A.J.; funding acquisition, A.J. All authors have read and agreed to the published version of the manuscript.

Funding: This research received no external funding.

Institutional Review Board Statement: Not applicable.

Informed Consent Statement: Not applicable.

Data Availability Statement: No new data were created or analyzed in this study. Data sharing is not applicable to this article.

Conflicts of Interest: The authors declare no conflict of interest.

References

1. Kazmierczak, J.; Joseph, A.; Cook, G. Aviation GNSS Interference Analysis Based on ADS-B Out Data. In Proceedings of the 34th International Technical Meeting of the Satellite Division of The Institute of Navigation (ION GNSS+ 2021), St. Louis, MO, USA, 20–24 September 2021; pp. 1108–1121.
2. Operations–ATM/ANS, EASA Safety Information Bulletin. Global Navigation Satellite System Outage Leading to Navigation/Surveillance Degradation. Available online: https://www.easa.europa.eu/en/newsroom-and-events/news/easa-updates-sib-gnss-outage-and-alterations (accessed on 1 February 2023).
3. ITU Radiocommunication Bureau (BR) Circular Letter CR/488, 8 July 2022. Available online: https://www.itu.int/dms_pub/itu-r/md/00/cr/cir/R00-CR-CIR-0488!!PDF-E.pdf (accessed on 1 February 2023).
4. *DO-178C:2011*; Software Considerations in Airborne Systems and Equipment Certification. RTCA: Washington, DC, USA, 2011.
5. *DO-254:2000*; Design Assurance Guidance for Airborne Electronic Hardware. RTCA: Washington, DC, USA, 2011.
6. Joseph, A.J.; Wahab, S.R. Universal Channel for Location Tracking System. U.S. Patent 9,702,979, 11 July 2017.
7. Joseph, A.; Kazmierczak, J.; Harris, M.; Schlais, P.; Murphy, T. Global Flight Test Results for a DFMC Primary Navigator on a Civil Air Transport Aircraft. In Proceedings of the 33rd International Technical Meeting of the Satellite Division of The Institute of Navigation (ION GNSS+ 2020), Online, 21–25 September 2020; pp. 140–160.
8. *ED-259A*; Minimum Operational Performance Standard For GALILEO/Global Positioning System/Satellite-Based Augmentation System Airborne Equipment. EUROCAE: Lucerne, Switzerland, 2021.
9. Hwang, P.Y.; McGraw, G.A. Receiver Autonomous Signal Authentication (RASA) based on clock stability analysis. In Proceedings of the 2014 IEEE/ION Position, Location and Navigation Symposium-PLANS 2014, Monterey, CA, USA, 5–8 May 2014; pp. 270–281.
10. Hwang, P.Y. GNSS Receiver Autonomous Signal Authentication Using Signal Stability Analysis System and Related Method GNSS Receiver Autonomous Signal Authentication Using Signal Stability Analysis System and Related Method (RASA). U.S. Patent US9094392B1, 28 July 2015.
11. Doty, J.H.; Anderson, D.A.; Hwang, P.Y. Staggered Examination of Non-Trusted Receiver Information (SENTRI). U.S. Patent US11337034B1, 17 May 2022.
12. *DO-384:2020*; Minimum Operational Performance Standards (MOPS) for GNSS Aided Inertial Systems. RTCA: Washington, DC, USA, 2020.
13. *DO-229E:2016*; Minimum Operational Performance Standards for Global Positioning System/Satellite-Based Augmentation System Airborne Equipment. RTCA: Washington, DC, USA, 2016.

engineering proceedings

MDPI

Proceeding Paper

On the Quantification of the GNSS Signals' Quality for RFI Assessment †

Naveed Ahmed

Department of Engineering Cybernetics, Norwegian University of Science and Technology, 7034 Trondheim, Norway; naveed.ahmed@ntnu.no

† Presented at the European Navigation Conference 2023, Noordwijk, The Netherlands, 31 May–2 June 2023.

Abstract: The performance of radio-frequency interference (RFI) detection and localization could be improved if they are applied on good quality inputs. RFI assessment is an important aspect of evaluating overall signal quality. The worthiness of a Global Navigation Satellite System (GNSS) receiver output can be generally quantified by a number of parameters readily available from a receiver. However, such discrete parameters do not give a detailed picture of the quality of the received GNSS signals. Statistical treatment of the received signals both in the absence and presence of interference gives some interesting insight about the data. In this paper, we study if the baseband data from the front-end Analog to Digital Converter (ADC) are normally distributed and if the presence of interference affects the statistical behavior of the distribution, often characterized by its probability density function (PDF) and other related parameters, such as skewness and kurtosis. In the second part of the paper, we study the feasibility of the Shapiro–Wilk (SW) test as a method to study the effect of interference on the GNSS signal while also serving as a potential approach to assess RFI. Skewness and kurtosis are statistical measures used to examine the shape of the distribution of a set of data. The implementation of the Shapiro–Wilk test is also studied, which is a normality test used to check whether a set of data follows a normal distribution. The above approaches have been evaluated using an experiment, where an RTLSDR is used as a reference GNSS receiver and simulated noise is added in the real signals. The data have been logged both in the presence and absence of wideband interference. The obtained results show the potential of the techniques presented for both the quantification of GNSS signal quality and the RFI assessment, alike.

Keywords: RFI assessment; goodness of fit; Shapiro–Wilk (SW) test; skewness; kurtosis

Citation: Ahmed, N. On the Quantification of the GNSS Signals' Quality for RFI Assessment. *Eng. Proc.* **2023**, *54*, 19. https://doi.org/10.3390/ENC2023-15440

Academic Editors: Tom Willems and Okko Bleeker

Published: 29 October 2023

1. Introduction

A compromise in the safety of the GNSS dependent applications has been widely reported in recent years [1–3]. This leads to an escalation in the demand for more robust, resilient, and secure navigation systems for such navigation-reliant applications. The impact of RFI on the receiver operations has been observed at both pre-correlation and post-correlation stages of the GNSS receiver [4].

The subject of RFI detection has been widely covered in the literature, where both statistical and non-statistical approaches have been proposed. The non-statistical approaches often encompass the monitoring of observables in the received signal such as C/N_0, AGC, and spectrum in terms of power spectral density (PSD), pseudorange, etc., and aim to observe the abnormalities in the behavior of these observables. In [5], interference is detected using day to day C/N_0 monitoring, where the observation of distinct patterns declares the presence of RFI. The Wavelet-Hough transform approach is proposed in [6], which is a combination of wavelet transform, which breaks down a signal into different frequency components extracting local spectral and temporal information in the signal, and Hough transform, which is a popular feature extraction technique that is widely used in image processing; here, it is used for a template-matching purpose. A joint Time-Frequency (TF) analysis has been proven to be an instrumental technique [7]. These techniques work as the interference exhibits difference behavior in the time and frequency domain, and even a very short duration interference that might not be visible in the time domain could be identified in the frequency domain. The statistical approaches work around analyzing signals in the time domain or the pre-correlation domain and exploit the established statistical concepts and principles for RFI detection purposes [8]. In [9], a Chi-square-based goodness-of-fit method is proposed. The in-phase (I) and quadrature (Q) IF samples from the RF front end are investigated if they follow the normal distribution. From the literature, there is no attempt to apply normality tests for detecting distortions in the GNSS signals. It is important to clarify the difference between the normality testing and goodness-of-fit (GoF) methodologies. Apart from the obvious mathematical differences, the normality testing methods take a sample from the population and check whether its population follows certain normal distribution, whereas the GoF methods aim to fit the available dataset in order to follow the normal distribution.

The rest of the paper is organized as follows: Section 2 explains the statistical techniques used in the paper for RFI detection, Section 3 presents the SW-based distribution analysis technique, Section 4 presents the scenario considered for validating the proposed methods, Section 5 discusses the obtained results, and finally, Section 6 concludes the paper.

2. Statistical Characteristics of Data Distribution

In our analysis, we underscore the importance of the distribution modeled by the selected data both in the absence and presence of noise. In the current case, we are interested in two important parameters, skewness and kurtosis, that potentially give a good picture of how the distribution might look like.

2.1. Skewness

Skewness is a measure of the asymmetry of a distribution. It describes how much the shape of the distribution deviates from the symmetry around its mean.

If the distribution is symmetric, then the skewness is zero. If the distribution has a longer tail on the right side (i.e., the positive side of the x-axis), then it is said to be positively skewed, and the skewness value will be positive. Conversely, if the distribution has a longer tail on the left side (i.e., the negative side of the x-axis), then it is said to be negatively skewed, and the skewness value will be negative. Mathematically speaking, it is computed as

$$skewness(x) = \frac{3\overline{x} - M_d}{s} \tag{1}$$

where $\overline{x}$ is the arithmetic mean, M_d is the median of the distribution, and s is the standard deviation of the distribution.

The skewness is generally interpreted as:

1. If the skewness is between −0.5 and 0.5, the data are nearly symmetrically distributed around the mean.
2. If the skewness is between −1 and −0.5 (negative skewed) or between 0.5 and 1 (positive skewed), the sample data are representative of a slightly skewed distribution.
3. If the skewness is lower than −1 (negative skewed) or greater than 1 (positive skewed), the sample data would represent an extremely skewed distribution.

2.2. Kurtosis

The kurtosis is a numerical quantification of the tailedness of a distribution. The standard normal distributions have a kurtosis of around 3, where most of the data are concentrated around its mean. If the kurtosis is greater than 3 or, in other words, the excess kurtosis is positive, it means most of the data are concentrated around its mean. Such types of distributions are termed as a leptokurtic distribution. On the contrary, if the kurtosis is less than 3 or, in other words, the excess kurtosis is negative, it means that most of the data are concentrated around its ends/tails. Such types of distributions are termed as platykurtic. Mathematically, kurtosis is defined as

$$kurtosis(x) = \frac{\frac{1}{n}\sum_{i=1}^{n}(x_i - \overline{x})^4}{\left[\frac{1}{n}\sum_{i=1}^{n}(x_i - \overline{x})^2\right]^2} \quad (2)$$

where n is the number of samples, x_i is the ith value, and $\overline{x}$ is the sample mean.

3. Distribution Analysis for RFI Assessment Techniques

The literature covers many normality testing and goodness-of-fit methods that are suitable for certain applications [10]. Some of the methods are applied on the empirical distribution formed by the dataset [11,12], whereas others are applied on the subset of the whole dataset that is assumed to be the representative of the whole dataset. In the latter category, two popular methods for normality testing are the Shapiro–Wilk (SW) [13] and Shapiro–Francia (SF) [14] methods. Generally, both methods exhibit equally good performance; however, in some cases, the SW method shows some sensitivity to the sample size and the SF method tends to be powerful in alternate hypothesis testing [15]. For the sake of generality, we do not discriminate the two methods in terms of performance.

In this section, we discuss the normality testing using the SW method and present them in context of the GNSS RFI assessment. We initiate the analysis by formulating the hypotheses test for the presence of interference in the system.

$$H_0(RFI\ absent)\text{: the sample is taken from a normal distribution} \quad (3)$$

$$H_1(RFI\ present)\text{: the sample is not taken from a normal distribution} \quad (4)$$

Mathematical Formulation of the Shapiro–Wilk Test

Let us suppose x to be a sample of n GNSS observations derived from a population with an unknown distribution, where the sample is characterized by its mean ($\overline{x}$) and standard deviation (σ).

$$x = \{x_1, x_2, x_3, \ldots, x_n\} \quad (5)$$

$$\overline{x} = \frac{1}{n}\sum_{i=1}^{n} x_i \quad (6)$$

$$\sigma^2 = \frac{1}{n-1}\sum_{i=1}^{n}(x_i - \overline{x})^2 \quad (7)$$

We need to reorder the sample values in an ascending order, i.e., from smallest to largest.

$$x(1) \leq x(2) \leq \cdots \leq x(n) \quad (8)$$

The mean or the expected values of the sorted sample values under normality can be computed and the standardized values for the SW test can be calculated.

$$m(i) = E[x(i)] = \frac{i - 0.375}{n + 0.25},\ \forall i = 1, 2, \ldots, n \quad (9)$$

$$z(i) = (x(i) - \overline{x})/s,\ \forall i = 1, 2, \ldots, n \quad (10)$$

Finally, we calculate the coefficients a of the normal distribution, the test statistic W, and the p-value.

$$a(i) = \frac{m(i) - \overline{m}}{s_m},\ \forall i = 1, 2, \ldots, n \quad (11)$$

where

$$\overline{m} = \frac{1}{n}\sum_{i=1}^{n} m(i) \quad (12)$$

$$s_m^2 = \frac{1}{n-1}\sum_{i=1}^{n}(m(i) - \overline{m})^2 \quad (13)$$

$$W = \frac{(\sum_{i=1}^{n} a(i)z(i))^2}{\sum_{i=1}^{n}(x(i) - \overline{x})^2} \quad (14)$$

$$p = P(W > \alpha) \quad (15)$$

where α is the significance level, that is set to be 0.05; W is the observed value of the test statistic; and P is the probability function of the null hypothesis, which states that the sample comes from a normally distributed population.

4. Experimental Setup

A static test is performed using a multiple purpose and low-cost RTL-SDR that is mainly used as a TV tuner [16] (See Figure 1). However, due to a wide-range demodulator, the reception of GPS L1 signals is also possible .

Figure 1. RTL-SDR (sixth generation) USB Dongle showing antenna connector (left) and USB port to connect with PC (right).

The RTL-SDR design is based on the RTL2832U demodulator chip and the raw I/Q data can be accessed directly, which allows the DVB-T TV tuner to be used as a wide-band software defined radio via a custom software driver. Some important technical specifications, and the settings used to record the dataset, are given in Table 1.

Table 1. RTL-SDR Technical Specifications [9].

Parameter	Value
Tune Low (MHz)	24
Tune Max (MHz)	1766
RX Bandwidth (MHz)	3.2/2.56 (Stable)
ADC Resolution (Bits)	8
ADC Sampling Frequency (MHz)	2.048
Intermediate Frequency (Hz)	0

5. Results and Discussion

The frequency spectrum of the GPS L1 signals at the baseband level is shown in Figure 2. The effect of varying noise levels on the spectrum is evident in the figures. The wideband noise or white noise has a constant power spectral density across all frequencies, where the noise energy is spread out across the entire spectrum of the signal, effectively masking or obscuring any underlying signal components. The same observation can be seen that the noise adds many unwanted frequencies to the original signal spectrum, making it more difficult to identify and analyze the desired signal.

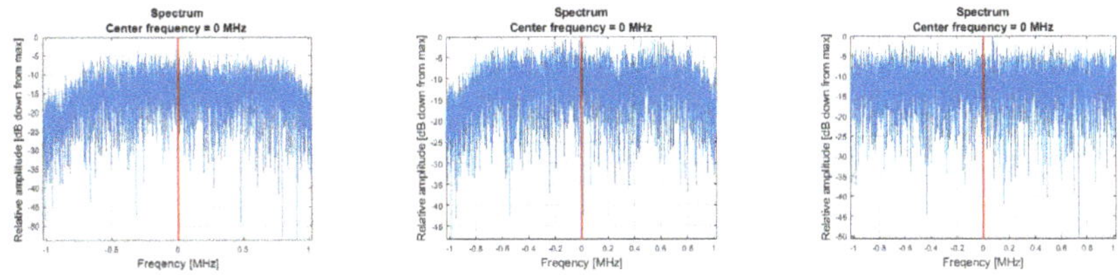

Figure 2. Power Spectrum Density (PSD) of clean (**left**) and noisy GNSS signals at −20 dB (**center**) and −60 dB (**right**).

5.1. Effect of Noise on the Statistical Properties

The test results obtained using different noise characteristics have been summarized in Table 2.

Table 2. Results showing the impact of noise on parameters of PDF (Sample Size = 4096).

GNSS Signal and Noise	Sample Size	Skewness	Kurtosis
L1 with no noise	1024	1.43	5.66
L1 with white noise @ −20 dB	1024	1.30	5.29
L1 with white noise @ −40 dB	1024	0.56	3.25
L1 with white noise @ −60 dB	1024	0.72	3.48
L1 with white noise @ −80 dB	1024	0.58	3.00
L1 with no noise	2048	1.52	6.67
L1 with white noise @ −20 dB	2048	1.39	6.15
L1 with white noise @ −40 dB	2048	0.67	3.25
L1 with white noise @ −60 dB	2048	0.61	3.27
L1 with white noise @ −80 dB	2048	0.6	3.00
L1 with no noise	4096	1.26	5.17
L1 with white noise @ −20 dB	4096	1.16	4.96
L1 with white noise @ −40 dB	4096	0.62	3.21
L1 with white noise @ −60 dB	4096	0.60	3.11
L1 with white noise @ −80 dB	4096	0.70	3.48
L1 with no noise	8192	1.39	6.00
L1 with white noise @ −20 dB	8192	1.25	5.52
L1 with white noise @ −40 dB	8192	0.62	3.14
L1 with white noise @ −60 dB	8192	0.62	3.26
L1 with white noise @ −80 dB	8192	0.66	3.32

We observe some very interesting trends from the above results. In comparison with the reference parameters of the PDF computed using clean signals, we observe that the distributions tend to become less skewed and so does the kurtosis approach to limiting the kurtosis value to 3.0. A slight increase for the added noise of intensity −60 dB can be ignored here as it does not have a substantial impact on the data distribution. This means the higher noise tends to overcome the real GNSS signals, leaving to follow an even greater corruption, whereas the lower intensity noise tends to normalize the signals to follow a moral and normal distribution-like behavior.

As far as the impact of sample size on the distribution shape is concerned, the values in the table indicate that there is no drastic change in the shape of the bell curve characterized by the corresponding skewness and kurtosis values, as shown in Figure 3.

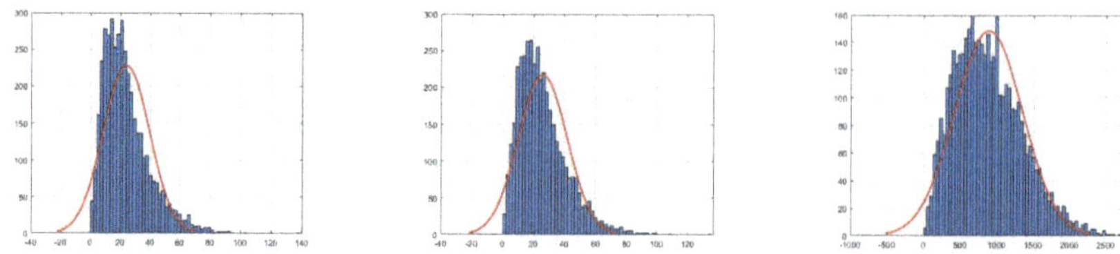

Figure 3. Histogram and distribution fitting for the samples taken from the clean (**left**) and noisy GNSS signals at −20 dB (**center**) and −60 dB (**right**), where the sample size is 4096.

5.2. Shapiro–Wilk Test Results

Typically, the test results are interpreted based on the statistics value and *p*-value that are compared against the significance level, which is considered to be 0.05. If the *p*-value is greater than the significance level (usually 0.05), then there is not enough evidence to reject the null hypothesis that the data are normally distributed. This means that the sample is likely to be normally distributed. If the *p*-value is less than or equal to the significance level, then there is evidence to reject the null hypothesis and conclude that the data are not normally distributed. This means that the sample is unlikely to be normally distributed.

Table 3 tabulates the SW test results for various test cases. It can be seen that the null hypothesis is not rejected in any case, meaning that the injection of noise does disturb the normal behavior of the distribution. However, the parameters pertaining to the SW test paint an interesting picture. Firstly, the decrease in the noise level in the signal decreases the value of the test statistic, and this trend is visible for each sample size in the table. Secondly, the impact of sample size is also evident, as having a bigger sample size increases the value of the test statistic, leading to the conclusion that the SW, in the current case, is sensitive to the sample size.

Table 3. Shapiro–Wilk Test Results performed for various noise levels.

GNSS Signal	Sample Size	Test Statistics Value (w)	*p*-Value	SK-Test Inference
L1 with white noise @ −20 dB	1024	0.04	1	Null Hypothesis not rejected
L1 with white noise @ −40 dB	1024	0.004	1	Null Hypothesis not rejected
L1 with white noise @ −60 dB	1024	4.4×10^{-5}	1	Null Hypothesis not rejected
L1 with white noise @ −80 dB	1024	4.4×10^{-7}	1	Null Hypothesis not rejected
L1 with white noise @ −20 dB	2048	0.09	1	Null Hypothesis not rejected
L1 with white noise @ −40 dB	2048	0.009	1	Null Hypothesis not rejected
L1 with white noise @ −60 dB	2048	8.8×10^{-5}	1	Null Hypothesis not rejected
L1 with white noise @ −80 dB	2048	8.5×10^{-7}	1	Null Hypothesis not rejected
L1 with white noise @ −20 dB	4096	0.13	1	Null Hypothesis not rejected
L1 with white noise @ −40 dB	4096	0.016	1	Null Hypothesis not rejected
L1 with white noise @ −60 dB	4096	1.79×10^{-4}	1	Null Hypothesis not rejected
L1 with white noise @ −80 dB	4096	1.78×10^{-6}	1	Null Hypothesis not rejected
L1 with white noise @ −20 dB	8192	0.32	1	Null Hypothesis not rejected
L1 with white noise @ −40 dB	8192	0.03	1	Null Hypothesis not rejected
L1 with white noise @ −60 dB	8192	3.48×10^{-4}	1	Null Hypothesis not rejected
L1 with white noise @ −80 dB	8192	3.5×10^{-6}	1	Null Hypothesis not rejected

6. Conclusions

The statistical methods of monitoring the signal quality and assessing the interference in the GNSS signals have been presented in the paper. The assessment covers the subject from two different loosely related methods. The first method is based on the deriving the statistical properties of the distribution formed by the dataset and comparing it with the same using the clean dataset. The second method is based on the SW-based normality testing method that identifies if the selected sample from the complete dataset is representative of the complete dataset by checking if the sample exhibits the properties of a normal distribution. In both cases, we studied the effect of the noise level and sample size on the desired observables. Such assessment could be instrumental in an offsite assessment of the dataset. Even though the preliminary results help us explain the phenomenon, the implementation of the presented techniques should be implemented on a wider range of scenarios with different receiver types, signals, and noise sources.

Funding: This research received no external funding.

Institutional Review Board Statement: Not applicable.

Informed Consent Statement: Not applicable.

Data Availability Statement: Data sharing is not applicable to this article

Acknowledgments: The author would like to acknowledge the support of Imtiaz Nabi at the Institute of Space Technology, Islamabad, Pakistan, for assisting with real data collection using the RTL-SDR.

Conflicts of Interest: The author declares no conflict of interest.

References

1. U.S. Coast Guard Navigation Center. GPS Navigation Problems. Available online: https://www.navcen.uscg.gov/?pageName=GPSProblems (accessed on 20 March 2023).
2. Nandra, R.; Sridevan, S.; Hancke, R. GPS Spoofing: The Next Generation. *IEEE Commun. Mag.* **2018**, *56*, 180–186.
3. Humphreys, T.E.; Ledvina, B.M. Assessing the Spoofing Threat: Development of a Portable GPS Civilian Spoofer. *J. Field Robot.* **2013**, *30*, 531–548.
4. Dovis, F. *GNSS Interference Threats and Countermeasures*; Artech House: Norwood, MA, USA, 2015.
5. Thompson, R.J.; Wu, J.; Balaei, A.T.; Dempster, A.G. Detection of RF interference to GPS using day-to-day C/N_0 differences. In Proceedings of the International Symposium on GNSS, Taipei, Taiwan, 26–28 October 2010.
6. Sun, K.; Zhang, T. A New GNSS Interference Detection Method Based on Rearranged Wavelet–Hough Transform. *Sensors* **2021**, *21*, 1714. [CrossRef] [PubMed]
7. Sun, K.; Jin, T.; Yang, D. An Improved Time-Frequency Analysis Method in Interference Detection for GNSS Receivers. *Sensors* **2015**, *15*, 9404–9426. [CrossRef] [PubMed]
8. Balaei, A.T.; Dempster, A.G. A statistical inference technique for GPS interference detection. *IEEE Trans. Aerosp. Electron. Syst.* **2009**, *45*, 1499–1511. [CrossRef]
9. Motella, B.; Pini, M.; Presti, L.L. GNSS interference detector based on Chi-square Goodness-of-fit test. In Proceedings of the 2012 6th ESA Workshop on Satellite Navigation Technologies (Navitec 2012) and European Workshop on GNSS Signals and Signal Processing, Noordwijk, The Netherlands, 5–7 December 2012; pp. 1–6. [CrossRef]
10. Specht, M. Statistical Distribution Analysis of Navigation Positioning System Errors—Issue of the Empirical Sample Size. *Sensors* **2020**, *20*, 7144. [CrossRef] [PubMed]
11. Lilliefors, H.W. On the Kolmogorov-Smirnov Test for Normality with Mean and Variance Unknown. *J. Am. Stat. Assoc.* **1967**, *62*, 399–402. [CrossRef]
12. Smirnov, N. Table for Estimating the Goodness of Fit of Empirical Distributions. *Ann. Math. Stat.* **1948**, *19*, 279–281. [CrossRef]
13. Shapiro, S.S.; Francia, R. An Approximate Analysis of Variance Test for Normality. *J. Am. Stat. Assoc.* **1972**, *67*, 215–216. [CrossRef]
14. Shapiro, S.S.; Wilk, M.B. An Analysis of Variance Test for Normality (Complete Samples). *Biometrika* **1965**, *52*, 591–611. [CrossRef]
15. Ahmad, F.; Khan, R.A. A power comparison of various normality tests. *Pak. J. Stat. Oper. Res.* **2015**, *11*, 331–345. [CrossRef]
16. RTL-SDR. Available online: https://www.rtl-sdr.com/about-rtl-sdr/ (accessed on 20 March 2023).

engineering proceedings

MDPI

Proceeding Paper

Continuous Localization-Assisted Collaborative RFI Detection Using the COTS GNSS Receivers †

Naveed Ahmed

Department of Engineering Cybernetics, Norwegian University of Science and Technology, 7034 Trondheim, Norway; naveed.ahmed@ntnu.no

† Presented at the European Navigation Conference 2023, Noordwijk, The Netherlands, 31 May–2 June 2023.

Abstract: Radiofrequency Interference (RFI) is a growing concern for many navigation-reliant applications. The dual benefits of RFI localization are considered: first, it can help with situational awareness by estimating the location of the interference source, and secondly, the results can be used to verify the detection of significant interference. The paper exploits the latter by proposing detection techniques making use of the localization results. The performance of the algorithms is evaluated using an experiment in a controlled lab environment where a wideband interference source is emulated in a UAV-based scenario. The detection results are validated using a reference detector operating in a non-position domain.

Keywords: RFI detection; carrier-to-noise density ratio (C/N_0); automatic gain control (AGC); minimum to second minimum error (MSME) ratio; majority voting

Citation: Ahmed, N. Continuous Localization-Assisted Collaborative RFI Detection Using the COTS GNSS Receivers. *Eng. Proc.* **2023**, *54*, 20. https://doi.org/10.3390/ENC2023-15441

Academic Editors: Tom Willems and Okko Bleeker

Published: 29 October 2023

1. Introduction

The vulnerability of Global Navigation Satellite Subsystem (GNSS) signals due to their extremely low power that is often in the order of femtowatts has been widely known, and various disruptive events impacting the proper functioning of the receivers have been identified. The RFI can affect a receiver at different stages of its operation. An interference signal may attempt to target the pre-correlation stage, post-correlation stage or final position, velocity and time (PVT) computation of the receiver [1]. Therefore, it is important to identify where in the receiver chain the effect of the interference is observed. There has been a lot of work already conducted in the RFI detection domain, where research is more focused on the pre-correlation, post-correlation or hardware-based detection. The novelty in the work is introduced when the detection is studied starting from the fact that the localization, which often assumes the jamming has been pre-detected, could be used to determine whether the RFI source is actually presented. Hence, the localization results help to detect the presence of an interference source in an operating environment. AGC is a proven RFI detection tool that monitors the changes in the power level of the received signals [2,3]. Modern front-ends, which are usually equipped with a multi-bit analog-to-digital converter (ADC), can detect the interference by observing the histograms of the received samples. The drawbacks of such techniques are the sensitive nature of AGC itself and the need for careful received calibration. Pre-correlation detection techniques are implemented on the RF front-end output. The statistical assessment is performed on the received signals aiming to study the deviation in behavior indicating the presence of an interference [4]. The Finite Fourier Transform (FFT) properties are exploited to see the effect of interference on parameters like the FFT point and the test statistics mean and variance. In [5], detection is performed based on various parameters computed using the RF front-end output samples. The parameters computed in the paper include time domain energy, power spectral density (PSD), minimum detectable bias and the mean value estimator. A non-parametrical approach to detect interference is presented in [6], where the authors studied the received signals using the Welch windowed periodogram.

The technique is tested for both GPS L1 C/A and Galileo binary offset carrier (BOC(1,1)) signals. Such techniques are easily implemented on digital signal processing (DSP) or field-programmable gate array (FPGA) devices. The post-correlation family of detection techniques includes techniques that use the acquisition, tracking and navigation results for detection purposes. The pseudorange and C/N_0 are the parameters that appear to be most sensitive to the RFI [7]. A detection method using the position output monitoring is presented in [8]. It is observed that there is a deviation in the parameters pertaining to position accuracy such as C/N_0 and horizontal dilution of precision (HDOP), and large pseudorange errors were also noticed in the presence of interference.

In our work, we present two different RFI detection methods—the first uses the raw C/N_0 and AGC measurements obtained from nodes operating in a collaborative navigation scenario, whereas the second one uses the RFI localization results obtained using the method given in [9]. The first technique works on a principle of majority voting, where each participating node locally determines its detection state, and the final detection decision is made based on the collective detection flags. The second technique depends on the results obtained used a centralized approach for RFI localization. This approach defines a new detection metric that is mathematically represented as a minimum to second minimum position error (MSME) ratio. An experiment is conducted in a lab environment where an unmanned aerial vehicle (UAV)-based scenario is emulated and the UAVs follow a predefined trajectory in the presence of an interference source. The detection outcomes are compared with a reference detector designed using the receivers' observables in the non-position domain that, when considered together, indicate the overall jamming status.

The rest of the paper is organized as follows: Section 2 introduces different detection approaches in detail. Section 3 gives an insight of the experiments used to evaluate the detection algorithms. Section 4 discusses the results, and Section 5 concludes the paper.

2. RFI Detection Techniques

In this paper, we present post-correlation detection techniques and cover various detection approaches based on the jammer localization results discussed in [9]. The outcomes of the detection schemes are compared to a reference detector that is based on the receivers' parameters pertaining to their jamming detection capability.

2.1. Detection Using MSME Ratio Metric

It is crucial to have better localization results to utilize them for detection purposes. A better localization technique should have the least possible errors between estimated and real jammer positions. However, in certain marginal situations when the receivers become closer to entering or exiting the jamming zone of the jammer, this metric is not always a true indicator of jammer detection. A new metric is introduced that exploits the variability in localization results. In the current case, we are interested in observing the accuracy of the estimated jammer position. The metric is based on the ratio of the minimum position error to the second minimum position error. Considering the assumption of a ground-based jammer in our case, the MSME can be empirically defined for two-dimensional position errors. Based on relative MSME values, this metric can be used to identify both weak and strong interference zones. The MSME, γ_d, can be mathematically defined as

$$\gamma_d = \frac{\min(\Delta d)}{second\ \min(\Delta d)} \tag{1}$$

where Δd is the estimated position error. The γ_d is a dimensionless quantity and ranges from 0 to 1, with a lower value indicating the more erroneous estimation results that could either be due to presence of strong interference or some other reason such as poor geometry formed by nodes.

2.2. *Majority Voting Detection Technique*

The technique is based on the collective impact of the interference on the receivers' observables. A majority voting scheme is implemented on the instantaneous change in the C/N_0 and AGC. A jammer is potentially detected when the majority of the votes indicate the presence of jammer. Figure 1 shows the flow chart of how the scheme works for a system of N nodes. Based on the comparison of the instantaneous change in the C/N_0 with a defined threshold, the detection state is determined for each node individually. A majority voting is then implemented using the detection states obtained for all the nodes. The cycle then repeats for the observations from the next epoch.

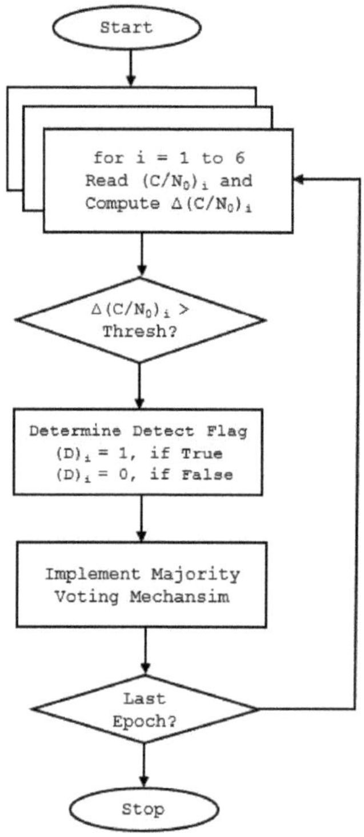

Figure 1. Majority voting scheme flow chart.

The threshold for comparison has been obtained based on the overall observation of C/N_0 degradation in the presence of interference.

Based on the observed jammer value in comparison to the threshold, an alert window is defined. Based on the length of the alert window, it is investigated how quickly the reference detection method, which is introduced in next section, is triggered. And considering the same C/N_0 degradation tolerance level, the robustness of the proposed detection method is determined.

2.3. *Reference Detector Based on Receivers' Jamming Indicators*

In order to evaluate the performance of the above two approaches, the results are compared with a reference detection approach. In our case, it is based on collective interpretation of the receivers' parameters pertaining to jamming detection. Table 1 shows some important parameters in the `UBX-MON-HW` message as defined in the u-Blox protocol specification document [10]. According to the document, the `jamInd` field only works for narrowband interference (NBI) and continuous wave interference (CWI). Other values in the table should be studied together with the variation in C/N_0 or AGC.

Under normal situations, the AGC works on the thermal noise and hence the indicating factor, `agcCnt`, remains constant. Sometimes, it is mapped to the AGC voltages (0 to 3.3 V). In the case of RFI, the AGC reduces the RF gain to keep the signal within ADC dynamic range. Hence, the `agcCnt` value or AGC voltage decreases depending on the received signal strength of interference signals. The `noiseperMS` parameter gives the noise measured by the GPS core in dBm. `jamInd` is a CWI detection parameter, and its value lies on a scale from 0 to 255 and indicates the severity of the interference effect on the receiver. The `jammingstate` bits in the flag give information about the general performance of the receiver in the presence of interreference. A collective investigation of all such parameters helps in efficient detection.

Table 1. Jamming-related parameters in UBX-MON-HW receiver message.

Field Name	Field Description
agcCnt	AGC monitor (range 0 to 8191)
noisePerMS	Noise level as measured by the GPS core
jamInd	CW jamming indicator, scaled (0 = no CW jamming, 255 = strong CW jamming)
flags/JammingState	Output from jamming/interference monitor 0 = unknown or feature disabled; 1 = ok—no significant jamming; 2 = warning—interference visible but fix OK; 3 = critical—interference visible and no fix

3. Experiment Scenario Description

A network of six identical UAVs is deployed to fly over an open field of dimensions 2 km × 2 km. In the current context, an open field is considered that guarantees an open sky with clear reception of GNSS signals and offering no signal attenuation and blocking (multipath reflection due to urban canyon, foliage and so on), except due to interference. To ensure better observability during the scenario, the UAVs are programmed to fly at different altitudes following the predefined trajectory. For the sake of simplicity, no wind, jerk or other effects experienced by the UAVs have been considered in the article. Figure 2 shows UAVs and jammer trajectories in the scenario with a color bar representing the UAVs' ground speeds at various instants in the scenario that is also shown separately in Figure 3.

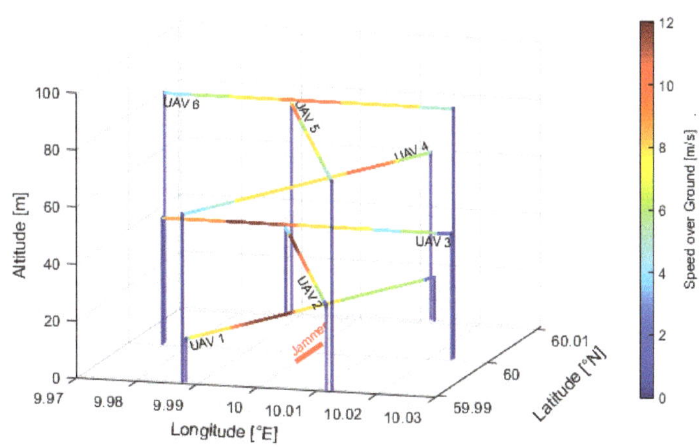

Figure 2. UAVs' and jammer's trajectories during the experiment in a 2 km × 2 km field.

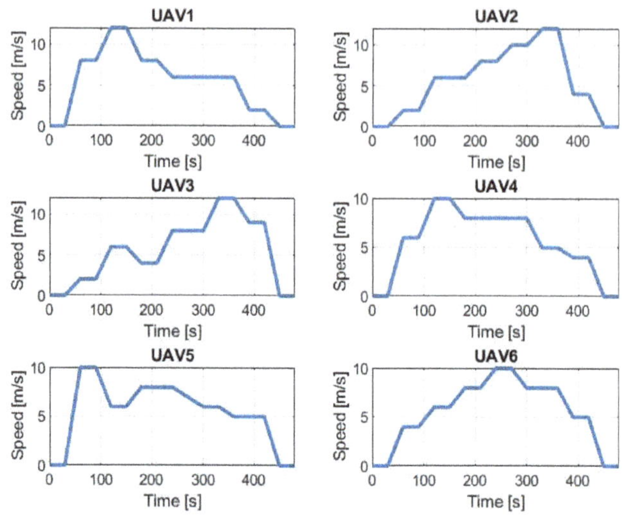

Figure 3. Simulated dynamics of the UAVs in the scenario.

3.1. Jammer Signals' Characteristics

We have considered wideband signal as our jamming signal of interest to study the impact of these signals on the C/N_0 and AGC of the receiver. Table 2 provides some important details about the simulated platform dynamics and other characteristics of the wideband interference signals. In the table, T_0 denotes the time at the start of simulation.

Table 2. Jammer's signal characteristics and dynamics simulated in the scenario.

Property	Value
Frequency (MHz)	−1575.42
Signal Power (dBm)	−60
Signal Bandwidth (MHz)	24
Speed (m/s)	2.5
Course over ground (deg)	0
ON time	(T_0 + 3 min) − (T_0 + 7 min)

3.2. Scenario Implementation

The aforementioned scenario is implemented on a Spirent HW GNSS simulator, and the interference signal is generated using the Spirent's Interference Signal Generator. The two signals are combined using an RF combiner, and the resulting results are given at the antenna input of the u-Box NEO-M8T GNSS receiver.

3.3. Brief Overview of RFI Localization

A centralized localization approach is followed, where it is assumed that all the cooperating nodes are capable of transmitting the distance from the jammer, calculated based on (2), to the central processing unit/node.

$$d_i = antilog_{10}\left(\frac{1}{10\alpha}\left(\frac{C_a}{N_0}\bigg|_{eff}^{i} - \bar{\beta}\right)\right) \tag{2}$$

where d_i is the distance of the ith UAV from the interference source, α is the path loss constant, C_a/N_0 is the effective average C/N_0 computed for the ith UAV and $\bar{\beta}$ is the receiver specific constant that is obtained after calibration. Since we have six nodes (or UAVs in our case) in the system, we can formulate a system of six nonlinear equations that is solved using the iterative least squares estimation method. The unknown jammer coordinates can be given in the form of a state vector.

$$\mathbf{x} = \begin{bmatrix} x_j \\ y_j \\ z_j \end{bmatrix} \tag{3}$$

with a range measurement vector given as

$$\mathbf{y} = \begin{bmatrix} d_1 \\ d_2 \\ \cdots \\ d_6 \end{bmatrix} \tag{4}$$

The relation between the change in the range and the change in the estimated state can be obtained by subtracting the predicted node to jammer measurement from the ranges obtained using (2) and applying first-order Taylor expansion.

$$\mathbf{\Delta y} = H\mathbf{\Delta x} \tag{5}$$

where H is the measurement or geometry matrix. The final solution can be obtained using the following:

$$\mathbf{\Delta\hat{x}} = (H^T H)^{-1} H^T \mathbf{\Delta y} \tag{6}$$

4. Results and Discussion

Figure 4 shows the estimated jammer positions obtained after collaborative RFI localization using C/N_0 and AGC [9]. The green boundary shows the region where the UAVs maneuver during the experiment. The quasistatic jammer's trajectory is represented by a red line. For the sake of analysis, the results within the 200 m × 200 m boundary are considered, shown with a magenta boundary.

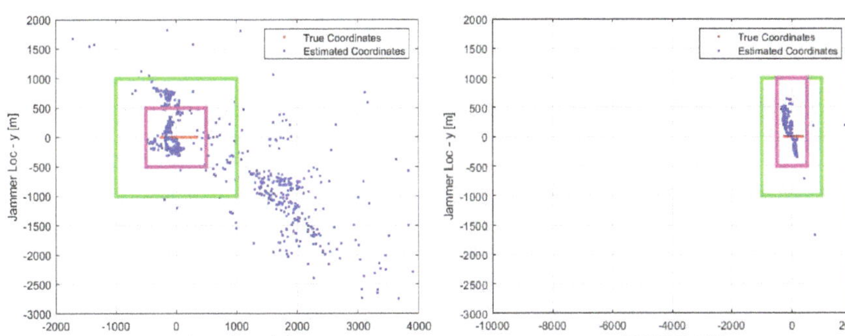

Figure 4. RFI localization results using C/N_0 (**top**) and AGC (**bottom**); area of UAVs' deployments zone is shown in green, whereas jammer detection region is shown in magenta—empirically selected based on the localization results.

4.1. MSME Results

Figure 5 shows the results of the MSME metric applied to the estimated position results. The figures show the variability in the localization results in terms of the ratio between the minimum to second minimum position error. The C/N_0 seems to be more sensitive compared to all other cases since the metric value goes to the least value, i.e., 0.955, which corresponds to a 4.5% variation in localization results, in contrast to 0.997 in the case of AGC, which corresponds to 0.3%. Comparing the trend of metric values computed with C/N_0 with what was obtained using the AGC case, it seems that the former is more consistent at indicating the presence of interference since the variations happen in regular intervals during the jamming intervals and have the maximum effect when the interference source is turned ON and turned OFF, respectively.

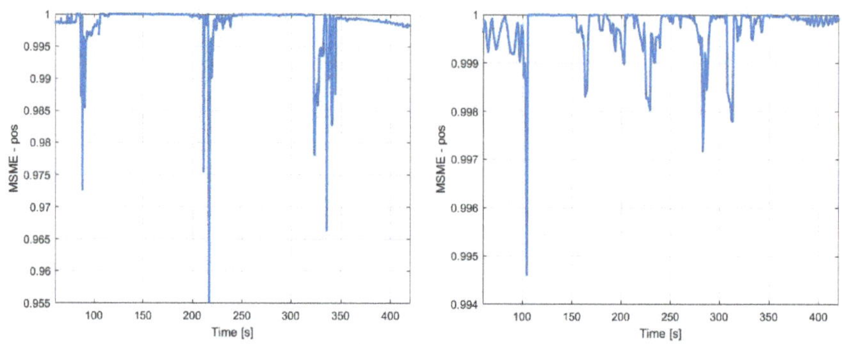

Figure 5. MSME results obtained using the localization results based on C/N_0 and AGC measurements.

4.2. Majority Voting Results

The majority voting approach has been implemented for a selected threshold using the method discussed in preceding sections. In the current case, the threshold is considered as 15 dB-Hz. Table 3 shows the final results for the detection technique based on the C/N_0 measurements. The technique is not fully adaptable for the case of AGC owing to its behavior and challenges in determining the threshold. The results from the table show that the RFI is collectively detected at the intervals between t = 210 s and t = 310 s. These are the instances when most or all the UAVs appear to be in closer proximity to the source. The level of contribution of each UAV in determining the overall decision can be determined through the state of each UAV, as given in different columns.

Table 3. Jammer signal characteristics and dynamics simulated in the scenario.

t[s]	Detection State—UAV1	Detection State—UAV2	Detection State—UAV3	Detection State—UAV4	Detection State—UAV5	Detection State—UAV6	Final Decision
200	1	1	0	1	1	0	1
205	1	0	0	1	1	0	0
210	1	1	0	1	1	0	1
215	1	1	0	1	1	1	1
220	1	1	0	1	1	1	1
225	1	1	0	1	1	1	1
230	1	1	0	1	1	1	1
235	1	1	1	1	1	1	1
240	1	1	1	1	1	1	1

Table 3. *Cont.*

t[s]	Detection State—UAV1	Detection State—UAV2	Detection State—UAV3	Detection State—UAV4	Detection State—UAV5	Detection State—UAV6	Final Decision
245	1	1	1	1	1	1	1
250	1	1	1	1	1	1	1
255	1	1	1	1	1	1	1
260	1	1	1	1	1	1	1
265	1	1	1	1	1	1	1
270	1	1	1	1	1	1	1
275	1	1	1	1	1	1	1
280	1	1	1	1	1	1	1
285	1	1	1	1	1	1	1
290	0	1	1	0	1	1	1
295	0	1	1	0	1	1	1
300	0	1	1	0	1	1	1
305	0	1	1	0	1	1	1
310	0	1	1	0	1	1	1
315	0	1	1	0	1	1	1
320	0	0	1	0	1	1	0
325	0	0	1	0	1	1	0
330	0	0	1	0	1	1	0
335	0	0	1	0	1	1	0
340	0	0	0	0	1	1	0
345	0	0	0	0	0	0	0
350	0	0	0	0	0	0	0
355	0	0	0	0	0	0	0
360	0	0	0	0	0	0	0
365	0	0	0	0	0	0	0
370	0	0	0	0	0	0	0
375	0	0	0	0	0	0	0
380	0	0	0	0	0	0	0
385	0	0	0	0	0	0	0
390	0	0	0	0	0	0	0
395	0	0	0	0	0	0	0
400	0	0	0	0	0	0	0

4.3. Reference Detection Approach Results

Figure 6 shows the plots of the parameters introduced in Table 1 obtained for the two UAVs closest to the jammer (UAV1 and UAV2) in the scenario (Figure 7 shows results for all the UAVs). The trends observed from the plots and the peculiar trend of variations indicate the presence of interference. However, the jamming severity could be declared as LOW based on the lower values of these jamming indication parameters. However, the important point is to obtain the temporal information of the interference injected into the scenario. This also serves as reference for comparing the results of other two techniques, as we notice the detection results obtained using MSME metric and majority voting comply with the reference detection technique.

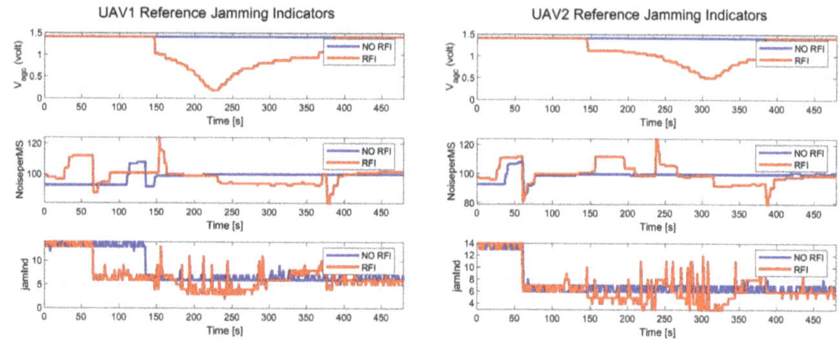

Figure 6. Reference jamming indicating parameters along with AGC voltage shown for UAV1 and UAV2, both in absence and presence of interference.

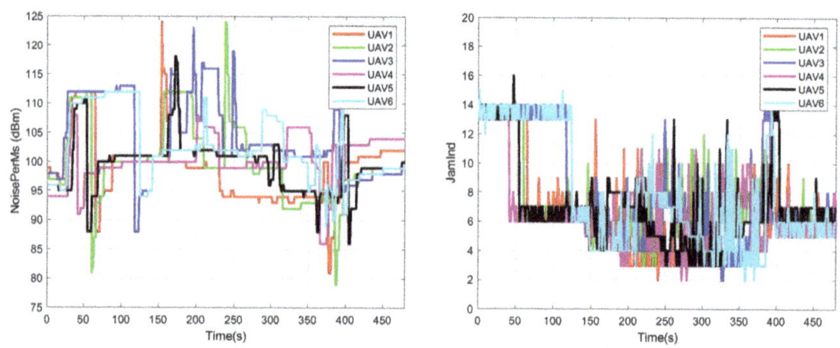

Figure 7. Comparison of NoisePerMs and JamInd values for different UAVs in the presence of interference.

5. Conclusions

The non-traditional methods of detecting interference have been presented in this paper. The variability in the localization results has been quantified using the MSME metric that proves to be instrumental in understanding detection. The majority voting technique detects interference based on the overall impact of the jammer on each node. The results are studied and presented in comparison with the reference techniques. The simulation results and observations show not only the potential of the discussed techniques to detect an RFI source but also to ensure that the localization results could possibly be used for detection. This is useful for situational awareness in practical situations and implementing a contingency plan.

Funding: This research received no external funding.

Institutional Review Board Statement: Not applicable.

Informed Consent Statement: Not applicable.

Data Availability Statement: Data sharing is not applicable to this article.

Conflicts of Interest: The author declares no conflict of interest.

References

1. Fabio, D. *GNSS Interference Threats and Countermeasures*; Artech House: Norwood, MA, USA, 2015.

2. Bastide, F.; Akos, D.; Macabiau, C.; Roturier, B. Automatic gain control (AGC) as an interference assessment tool. In Proceedings of the 16th International Technical Meeting of the Satellite Division of The Institute of Navigation (ION GPS/GNSS 2003), Portland, OR, USA, 9–12 September 2003; pp. 2042–2053.
3. Ndili, A.; Enge, P. GPS receiver autonomous interference detection. In Proceedings of the 1998 Position Location and Navigation Symposium, Palm Springs, CA, USA, 20–23 April 1996; pp. 123–130
4. Balaei, A.T.; Dempster, A.G. A statistical inference technique for GPS interference detection. *IEEE Trans. Aerosp. Electron. Syst.* **2009**, *45*, 1499–1511. [CrossRef]
5. Marti, L.; van Graas, F. Interference Detection by Means of the Software Defined Radio. In Proceedings of the 17th International Technical Meeting of the Satellite Division of The Institute of Navigation (ION GNSS 2004), Long Beach, CA, USA, 21–24 September 2004; pp. 99–109.
6. Tani, A.; Fantacci, R. Performance Evaluation of a Precorrelation Interference Detection Algorithm for the GNSS Based on Nonparametrical Spectral Estimation. *IEEE Syst. J.* **2008**, 2, 20–26. [CrossRef]
7. Groves, P.D. GPS Signal to Noise Measurement in Weak Signal and High Interference Environments. In Proceedings of the 18th International Technical Meeting of the Satellite Division of the Institute of Navigation (ION GNSS 2005), Long Beach, CA, USA, 13–16 September 2005; pp. 643–658.
8. Balaei, A.T.; Motella, B.; Dempster, A.G.; Rizos, C. Mutual Effects of Satellite Quality and Satellite Geometry on Positioning Quality. In Proceedings of the ION GNSS 2007 Conference, Fort Worth, TX, USA, 25–28 September 2007.
9. Ahmed, N.; Winter, A.; Sokolova, N. Low Cost Collaborative Jammer Localization Using a Network of UAVs. In Proceedings of the 2021 IEEE Aerospace Conference (50100), Big Sky, MT, USA, 6–13 March 2021; pp. 1–8.[CrossRef]
10. u-blox 8/u-blox M8 Recevier Description Including Protocol Specification. Available online: https://content.u-blox.com/sites/default/files/products/documents/u-blox8-M8_ReceiverDescrProtSpec_UBX-13003221.pdf (accessed on 23 March 2023).

engineering proceedings

Proceeding Paper

Robustness Levels of Critical Infrastructures Against Global Navigation Satellite System Signal Disturbances †

André Bos [1,2,*], Merle Snijders [2,3], Alexandra Zevenbergen [2,4], Kirsten Drost [1,2], Hein Zelle [2,3] and Bas van der Hoeven [2,4]

1 S[&]T BV, 2616 LR Delft, The Netherlands
2 EGNSS Centre of Excellence, 2201 DK Noordwijk, The Netherlands; merle.snijders@nlr.nl (M.S.); alexandra.zevenbergen@cgi.nl (A.Z.); hein.zelle@nlr.nl (H.Z.); bas.van.der.hoeven@cgi.nl (B.v.d.H.)
3 Netherlands Aerospace Center (NLR), 1059 CM Amsterdam, The Netherlands
4 CGI, 3068 AX Rotterdam, The Netherlands
* Correspondence: andre.bos@stcorp.nl

† Presented at the European Navigation Conference 2023, Noordwijk, The Netherlands, 31 May–2 June 2023.

Abstract: Resilience against signal disturbances is an important characteristic of GNSS-based PNT solutions. In particular, for critical infrastructures, failure to provide correct PNT information in these domains may have a major societal impact. The Resilience Framework by the Department of Homeland Security (DHS) provides a set of requirements and guidelines to design a PNT solution of a certain level of resilience. Over the lifetime of the applications, it will be of prime importance to assess the resilience of the PNT solutions on a regular basis. Given how often GNSS-based solutions are being applied, partly automating the assessment process will be needed to make this task feasible. To automate the generative process, a machine-readable structure with well-established meaning is required. In this work, the use of fault trees as a formal system to encode the resilience framework is investigated.

Keywords: resilience; PNT solutions; jamming; spoofing; assessment; questionnaire generation

Citation: Bos, A.; Snijders, M.; Zevenbergen, A.; Drost, K.; Zelle, H.; van der Hoeven, B. Robustness Levels of Critical Infrastructures Against Global Navigation Satellite System Signal Disturbances. *Eng. Proc.* **2023**, *54*, 21. https://doi.org/10.3390/ENC2023-15443

Academic Editors: Tom Willems and Okko Bleeker

Published: 29 October 2023

1. Introduction

Global navigation satellite systems (GNSS) have become immensely popular for providing positing, navigation, and timing (PNT) information to a wide range of applications. The popularity has a downside as well. It has been well documented (see, e.g., [1]) that GNSS-based PNT information can be easily compromised, either intentionally or accidentally. However, this kind of vulnerability knowledge of GNSS-based PNT information is not very well known outside the community of GNSS specialists. Especially for applications that control critical infrastructures, this lack of knowledge may lead to less effective operation and maintenance of the infrastructure and eventually to the potential loss of essential services and resources.

In order to better inform designers and operators of critical infrastructures, the knowledge of robustness should be better disseminated. This study describes the first results of an attempt to make this information more readily available: The resilience of applications that use PNT information for their correct operation is investigated. The assessment will be based on a resilience framework pioneered by the US Department of Homeland Security (DHS) [2]. Various ways of assessing resilience are foreseen, from questionnaires to automated test approaches. The proposed approach in this study is to largely generate the assessment methods automatically using a computer program that uses a computer-readable and formalized description of the resilience framework. Resilience is obtained by the robustness of the GNSS receiver and the additional equipment to determine the PNT information. Various signal disturbances are considered, including natural radio frequency interference (RFI), jamming, and spoofing. There are many different formalisms for describing the (lack of) resilience knowledge. This study considers the "fault tree" formalism to capture the resilience knowledge.

2. GNSS Basics, Disturbances, and the Potential Effect on Critical Infrastructures

2.1. GNSS Basics

A GNSS receiver tracks the signals of all the GNSS satellites in view. One of the indicators of the quality of the tracking process is the carrier-to-noise density (C/N0), which is expressed in decibel-Hertz (dB-Hz). Too low a value of C/N0 (typically < 25 dB-Hz) usually renders the receiver to lose its lock on the signal. The higher the values of the C/N0, the better the receiver estimates the pseudo range.

2.2. Disturbances

Unintentional interference mainly comprises natural, inter- and intra-, out-of-band, and in-band interference. Such interference can come from faulty machinery. Also, a multipath is one of the major contributors of error to satellite positioning, especially in urban canyons.

Jamming: Jamming is the process of purposefully disturbing the reception of GNSS signals by receivers in a targeted geographic—the jammed—region. In case of jamming, a powerful signal is applied to the target receiver's antenna, such that the front end will need to handle the high-power signal, thereby reducing the capability to handle the low-power authentic GNSS signal. Usually, the front end will adjust the automatic gain control (AGC) so that the processor can handle the powerful jamming signal. The effects of jamming are [3]: potential loss of tracking, decrease in measured signal strength (C/N0), adjustment of AGC values, increase in noise on pseudo range measurements (code phase and carrier phase) and position, and increase in cycle slips.

A rather new jamming technique is called systematic jamming and uses measurements of the signal-to-be-distorted to jam the signal in an intelligent way. Systematic jamming relies on a standard receiver that determines at which points in time jamming is most effective and only jams the signal briefly, e.g., to deny reception of the time of week indicator.

Spoofing: Spoofing is the transmission of forged GNSS-like signals, with the purpose of producing a false position or time at the victim's receiver without seemingly disrupting GNSS operations and thus effectively taking control of the receiver. With the advent of software-defined radio (SDR) technology, it has become a practical method to implement a spoofing system and thus could pose a real threat to the trustworthiness of GNSS-based PNT information.

2.3. Example of Critical Infrastructure Applications and Electricity Networks

Applications within critical infrastructures often require reliable and accurate PNT information. Just relying on a simple GNSS receiving system for this PNT information may not be enough to satisfy the availability and safety requirements that are imposed on the applications running in the critical infrastructures domain.

Some example applications from the electrical energy sector domain include [4]:

- Sequence of Event (SoE) recording is a data logging system that stores timestamped event information. The exact order of events is essential in quickly finding the real cause of the problem in electricity systems. A fault in one system may result in a whole avalanche of events at the subsystems connected to the source of failure. These events happen so fast that it almost seems like one instantaneous event with too coarse a time resolution for the SoE recorder. The timing accuracy requirement is < 1 ms;
- A phasor measurement unit (PMU) is a device used to estimate the magnitude and phase angle of an electrical phasor quantity in the electricity grid. The timing accuracy requirement is < 1 μs;
- Wire fault localization using the traveling wave method. Such methods calculate fault locations by measuring the arrival times of the naturally occurring traveling waves caused by a transmission line fault. The timing accuracy requirement is < 100 ns.

GNSS-disciplined clocks are used in such applications to obtain these strict time constraints. The basic principle of these clocks is that the GNSS time information controls a local oscillator. The basic principles actually being used are often not disclosed, but in general, the frequency of the local oscillator is kept within strict bounds by one or more servo loops, such as a Phased Locked Loop (PLL). After a GNSS disturbance has occurred, the clock will still continue to operate, but the clock stability parameters will degrade over time. In case the disturbance occurs for a longer period, the clock might not meet the application's requirements. It may take minutes to hours before the clock is degraded such that it fails to meet the accuracy requirements, depending on the quality of the local oscillator.

3. Resilience Framework

In the previous section, the various ways to disturb GNSS signals were introduced, as well as how these can affect various applications within the electricity-critical infrastructure. There is great interest in the development and analysis of resilient implementations. One of the approaches is the Resilience Framework by the US Department of Homeland Security (DHS) [2]. This framework could be seen as a set of very useful guidelines to achieve a certain level of resilience during the implementation of PNT sources for applications. The framework defines the following concepts:

PNT System: The components, processes, and parameters that collectively produce the final PNT solution for the consumer. Note that this is not necessarily restricted to GNSS components.

PNT Source: A PNT system component that is used to produce a PNT solution. Examples include GNSS receivers, networked and local (stable) clocks, inertial navigation systems (INS), and/or timing services provided over a wired or wireless connection.

PNT Solution: The full solution provided by a PNT system or source, including time, position, and velocity. A PNT system or source may provide a full PNT solution or a part of it. For example, a GNSS receiver with a clock may provide a full PNT solution, while a local clock only provides a timing/frequency solution.

Component: A part or element of a larger PNT system with well-defined inputs and outputs and a specific function. Examples may include individual PNT sources or subsystems of PNT sources, discrete software functions that implement resilient PNT processing algorithms, hardware modules providing a supporting function internal to the PNT system, antennas, firewalls (between antenna and receiver), and external detectors such as those based on SDR technology.

Recover from atypical errors to return to a proper working state and defined performance.

Then, the framework identifies four levels of resilience (see Table 1).

The framework is not necessarily geared towards GNSS-PNT in particular but is agnostic with respect to a particular technology that is being used. Furthermore, although the framework provides guidelines to assess the resilience of PNT solutions, in reality, the assessment will not be straightforward. For example, the effectiveness of identifying compromised PNT sources (point 4 in Table 1) will vary from one implementation to another. By further detailing the conditions for the resilience levels (as can be part of the formalization process), one can better identify why a certain solution fails to qualify for a certain resilience level.

Table 1. Resilience level requirements overview (from [2]).

Level	Minimum Requirements
Level 1 Ensures recoverability after removal of the threat.	1. Must verify that stored data from external inputs adheres to values and formats of established standards; 2. Must support full system recovery by manual means, making all memory clearable or resettable, enabling a return to a proper working state, and returning the system to the defined performance after removal of the threat; 3. Must include the ability to securely reload or update firmware.
*Level 2 ** Provides a solution (possibly with unbounded *** degradation) during a threat.*	Includes capabilities enumerated in Level 1 plus: 4. Must identify compromised PNT sources and prevent them from contributing to erroneous PNT solutions; 5. Must support automatic recovery of individual PNT sources and systems without disrupting system PNT output.
Level 3 Provides a solution (with bounded degradation) during a threat.	Includes capabilities enumerated in Levels 1 and 2 plus: 6. Must ensure that corrupted data from one PNT source cannot corrupt data from another PNT source; 7. Must cross-verify between PNT solutions from all PNT sources.
Level 4 Provides a solution without degradation during a threat.	Includes capabilities enumerated in Levels 1, 2, and 3 plus: 8. Must have a diversity of PNT source technology to mitigate common-mode threats.

Notes: Level 0 indicates a source or system that does not meet the criteria in Level 1 and thus is considered a non-resilient system or source. ** Critical infrastructure applications will likely require Level 2 resilience at a minimum. *** The output can deviate within a manufacturer-defined envelope.

Reference Architectures

The framework [2] also describes a reference architecture for acquiring Level 1 and Level 2 resilience. For Levels 3 and 4, no reference architecture is given yet. A PNT solution with resilience Level 3 may experience a bounded level of degradation when operating in a threat environment, whereas a Level 4 solution should not experience that degradation. The framework also provides guidelines for assessing the resilience level of a particular solution and defining testing procedures compliance for a certain level. The objective of the study presented in this paper is to largely automate these tasks, as described in the next section.

4. Formal Description of PNT Resilience

The following section will focus on the formalization of the framework. Such a formalization will ease the automatic processing by a computer as such a formalization can be used to generate computer programs performing a certain analysis task. One of the goals is to construct a computer-assisted resilience assessment and an automatic testing system that is able to generate test data to test resilience.

There are various types of formalisms that describe events (failures and exceptional conditions) and their impact on system behavior. One of these formalisms (and, maybe, conceptually, one of the simplest to comprehend) is the modeling of system behavior by fault trees [5]. This section briefly describes the fault tree (FT) formalism and then how it can be used to capture the resilience knowledge.

4.1. Fault Trees

One of the possible formalisms that can capture the failed operation of system components and, to a certain level, the sequence of occurrence of those failures is fault trees. Fault tree analysis (FTA) is a type of failure analysis in which an undesired state of a system is examined. This analysis method is mainly used in safety engineering and reliability engineering to understand how systems can fail, identify the best ways to reduce risk, and determine (or obtain a feeling for) event rates of a safety accident or a particular system-level (functional) failure. The main elements of a fault tree are a TOP event, which is the description of the critical system event, such as the tracking loss lock; Basic events are the lowest level of identified causes, such as excessive RFI or atmospheric scintillations; and Logic gates, such as OR or AND gates, which give the logical relationship between the TOP event and the basic events.

The AND gate is used to model redundancy in a system design or alternative test methods. A redundant system's function will only not be realized if all the alternatives fail. The OR gate is used to model serial dependencies in a system design or test method. If one of the components fails, then the overall system functionality fails to be realized. Given system components 1, 2, and 3 for a hypothetical system, then layer (i) of Figure 1 depicts the OR gate of these components, layer (ii) the AND gate of these components, and layer (iii) a combination of the OR and AND gates.

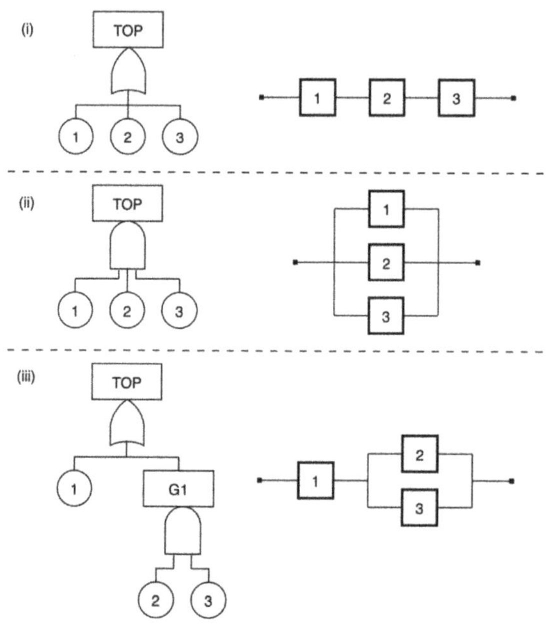

Figure 1. Construction of a fault tree.

There are several analysis methods that can be applied to the FT structure. One such method is the dependency analysis, or the estimation of the minimal set (minimal: deletion of one element from the minimal set invalidates the defining property) of Basic Events, causing the Top Event to be True. Such a minimal set gives the combination of events that renders the resilience to be invalid. Thus, it will be important to check that this combination will not happen during the lifetime of the PNT system. A well-known graph algorithm, the minimal cut set algorithm, exactly corresponds to solving this dependency analysis problem. Another analysis is the computation of the Probability of a Failure (PoF). If the probability of the Basic events can be determined and independence of events is assumed, the probability of the Top Event can be determined by propagating the probabilities in a bottom-up fashion.

4.2. Formalisation of the Resilience Framework

The fault tree method will not be applied to a particular system design (as in the usual case of applying fault trees) but will be used to encode the requirements of the resilience framework. One could say that the fault tree will basically model the robustness behavior of a reference architecture for a certain resilience level. Each of the resilience levels of the framework has attached a number of requirements that must be satisfied in order to qualify the PNT system to the corresponding resilience level. Furthermore, the resilience framework contains various high-level requirements for which the formalization may need to be made more specific to allow the automatic and effective generation of assessment applications (questionnaire and test program generation). Different fault trees can be defined for these requirements. For example, the framework specifies the rather high-level Requirement 4 of Level 2: "Must identify compromised PNT sources and prevent them from contributing to erroneous PNT solutions." For this the concept of "compromised PNT sources" and "prevent them from contributing" must be specified. Compromised PNT can mean various things, for example, Compromised by Jamming, Compromised by Spoofing, and Compromised by Antenna position. Concentrating further on the case of "Compromised by Jamming", the basic events would include, for example, "lock lost" and "increased code and carrier noise". However, these events can also come from other than jamming events. Jamming detection can be performed by inspecting the RF samples or monitoring the receiver observations, such as the carrier-to-noise density (C/N0). Once jamming has been detected, the effects of jamming can be mitigated using, e.g., signal processing techniques (such as the adaptive notch filter (ANF)) or spatial filtering (such as the controlled reception pattern antenna (CRPA)). The "Loss of Lock" fault tree for the Jamming Case could look like the one given in Figure 2.

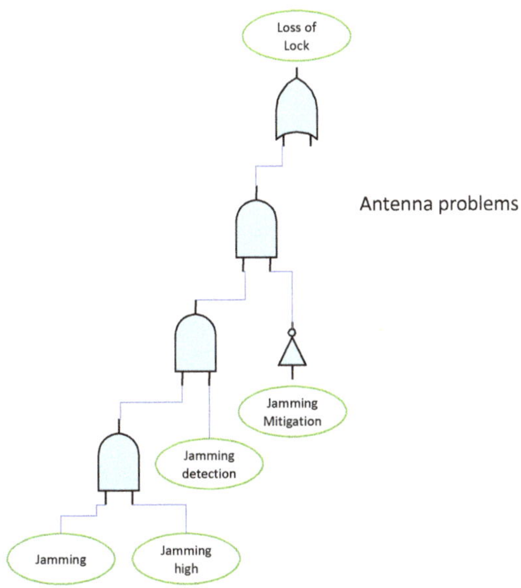

Figure 2. Fault tree for jammed signal.

"Loss of lock" can also be caused by the antenna position, due to the fact that the line of sight to the satellite is obscured. This can be caused by buildings that block the view. A similar fault tree can be constructed. For Compromised by Spoofing, the basic events would include "tracking of encrypted signals" (such as Galileo PRS [6]) and cryptographic checking of the navigation data (such as in OSNMA [7]). By correctly combining these basic events, one can encode the satisfaction conditions for this requirement.

In general, there can be several detection methods to recognize jamming, spoofing, or other disturbances. These methods will have, in general, different detection characteristics and will, therefore, trigger alarms under different conditions.

4.3. Generating a Resilience Questionnaire

The formalization of resilience knowledge can be used for various analysis techniques, such as the computer-assisted evaluation of resilience levels. The FT will be further attributed with evaluation questions. That is, for each of the "basic events", a question about the presence of the events can be assigned such that the truth value of these basic events can be established, as is depicted in Figure 3.

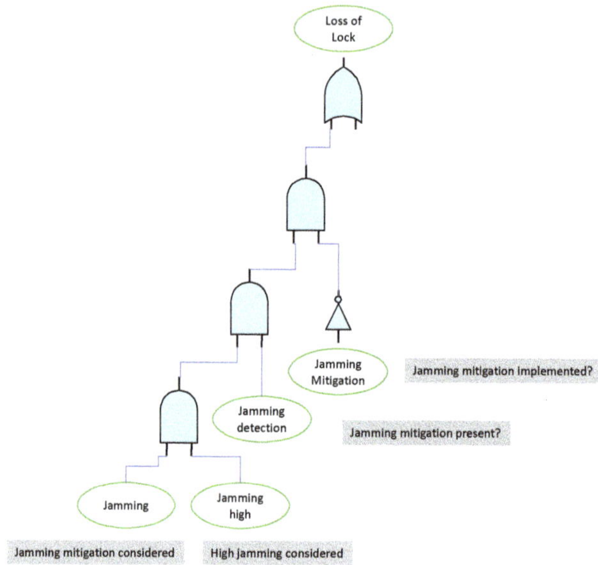

Figure 3. Fault tree augmentation with questions.

Now, as according to Figure 4, the FT can be evaluated as either "bottom up" or "top down". Top down starts at the TOP event and evaluates the value of the tree by descending the nodes of the tree and keeping track of the truth value of the basic events and the sub-trees.

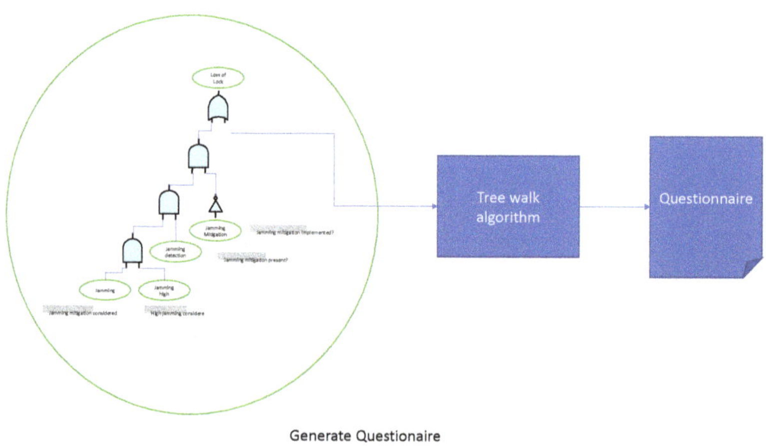

Figure 4. Automatic assessment method generation.

The formalization process is rather labor intensive, but once the knowledge is formalized, the generation of questionnaires is rather straightforward. Having a formal model allows various types of analyses, including the generation of test data.

5. Concluding Remarks

As the usage of GNSS-based PNT information for applications within critical infrastructures is present, it becomes of vital importance to keep an eye on the resilience of these systems. In this study, the partly automatic generation of questionnaires, and eventually test programs, is researched. This approach requires the cumbersome formalization of a resilience framework into a form that a computer can process. The formalization work ideally has to be performed once. Still, this formalization task is laborious and prone to errors. Alternative methods will be investigated in further research. The current popularity of generative Artificial Intelligence (AI) [8] seems to suggest that this technique would provide an alternative to questionnaire generation. Much more research is needed in this direction as well.

Author Contributions: Conceptualization, A.B.; methodology, writing—review and editing, A.B., M.S., A.Z., K.D., H.Z. and B.v.d.H. All authors have read and agreed to the published version of the manuscript.

Funding: This research received no external funding.

Institutional Review Board Statement: Not applicable.

Informed Consent Statement: Not applicable.

Data Availability Statement: Not applicable.

Acknowledgments: The authors thank the reviewers for their valuable comments.

Conflicts of Interest: The authors declare no conflict of interest.

References

1. Ioannides, R.T.; Pany, T.; Gibbons, G. Known Vulnerabilities of Global Navigation Satellite Systems, Status, and Potential Mitigation Techniques. *Proc. IEEE* **2016**, *104*, 1174–1194. [CrossRef]
2. DHS, Resilient Positioning, Navigation, and Timing (PNT) Reference Architecture. V2.0, The Department of Homeland Security (DHS) Science and Technology Directorate (S&T). 2022. Available online: https://www.dhs.gov/sites/default/files/2022-06/22_0609_st_resilient_pnt_ra.pdf (accessed on 13 December 2023).
3. Jansen, P. The Impact of Jamming and Spoofing on GNSS Signals. Master's Thesis, Delft University of Technology, Mekelweg, The Netherlands, 2022.
4. Dagle, J.; O'Neil, L.R.; Tuffner, F.; Bonebrake, C.; Engels, M.; Dindlebeck, G.B. *Energy Sector Position, Navigation, and Time Profile*; Technical Report, PNNL-30780; Pacific Northwest National Laboratory: Richland, WA, USA, 2021.
5. Fovino, I.; Masera, M.; De Cian, A. Integrating cyber attacks within fault trees. *Rel. Eng. Syst. Saf.* **2009**, *94*, 1394–1402. [CrossRef]
6. PRS. Available online: https://www.euspa.europa.eu/european-space/galileo/services/prs (accessed on 12 May 2023).
7. Galileo Open Service Navigation Message Authentication (OSNMA). Available online: https://www.gsc-europa.eu/galileo/services/galileo-open-service-navigation-message-authentication-osnma (accessed on 12 May 2023).
8. The AI writing on the wall. *Nat. Mach. Intell.* **2023**, *5*, 1. [CrossRef]

engineering proceedings

MDPI

Proceeding Paper

Jammertest 2022: Jamming and Spoofing Lessons Learned †

Aiden Morrison [1,*], Nadezda Sokolova [1], Anders Solberg [2], Nicolai Gerrard [3], Anders Rødningsby [4], Harald Hauglin [5], Thomas Rødningen [5] and Tor Dahlø [2]

1 SINTEF, 7034 Trondheim, Norway; nadia.sokolova@sintef.no
2 The Norwegian Mapping Authority, 3511 Hønefoss, Norway; anders.martin.solberg@kartverket.no (A.S.); tor-ole.dahlo@kartverket.no (T.D.)
3 Norwegian Communications Authority, 4790 Lillesand, Norway; nicolai.gerrard@nkom.no
4 Norwegian Defense Research Establishment, 2007 Kjeller, Norway; anders.rodningsby@ffi.no
5 Norwegian Metrology Service, 7075 Tiller, Norway; hha@justervesenet.no (H.H.); thr@justervesenet.no (T.R.)
* Correspondence: aiden.morrison@sintef.no

† Presented at the European Navigation Conference 2023, Noordwijk, The Netherlands, 31 May–2 June 2023.

Abstract: Jammertest 2022 was a week-long series of satellite navigation and timing signal jamming and spoofing exercises carried out on the Norwegian island of Andøya in September of 2022. Organized via a collaboration between the Norwegian spectrum management authority, defense research establishment, public roads administration, metrology service, and others, the result was the largest known GNSS jamming and spoofing event open to international collaboration and provided an open-access data and publication policy for participants. This paper reviews the event's organization, scheduled tests, noteworthy jamming observations, noteworthy spoofing observations, and the unexpected observations found during the event and also presents information on what data are publicly available to interested parties, along with the contact information needed to obtain these data.

Keywords: GNSS; jamming; spoofing; RFI; PPD

Citation: Morrison, A.; Sokolova, N.; Solberg, A.; Gerrard, N.; Rødningsby, A.; Hauglin, H.; Rødningen, T.; Dahlø, T. Jammertest 2022: Jamming and Spoofing Lessons Learned. *Eng. Proc.* **2023**, *54*, 22. https://doi.org/10.3390/ENC2023-15445

Academic Editors: Tom Willems and Okko Bleeker

Published: 29 October 2023

1. Introduction—What Was Jammertest?

Jammertest 2022 was an event held from the 19th through to the 23 September 2022 on the island of Andøya on the Norwegian coast, and was an exceptional opportunity for the international navigation community due to each of the locations selected, the data access policies enacted by the organizers, and the selection of the tests conducted.

The location, adjacent to the community of Bleik, was ideal for carrying out jamming and spoofing tests for a number of reasons. The first and primary beneficial feature was the high ridgeline which surrounded the test location on three sides. With peaks exceeding 400 m above sea mean sea level and lying close to the transmission site, this ridge line not only protected the mainland from the emitted signals but also prevented the signal from propagating into much of the airspace surrounding the island. Signals could only freely propagate within the test area and to the northeast out to the northern Atlantic.

The Jammertest data access policies were also ideal in that attendees were allowed to freely discuss and publish their observations and results. Typically, jamming and spoofing exercises, along with the data collected through these exercises, are considered to be sensitive, possibly due to the perceived potential for misuse. Given that both jamming and spoofing are widely discussed in publicly accessible papers in both technical and general media publications, the decision was made that the utility of allowing relevant industrial, governmental, research, and academic institutions to freely discuss the observed phenomena outweighed remaining concerns around publication. The specific tests carried out at Jammertest were numerous and are detailed in Section 3.

2. Test Site, Setup, and Selection

Jammertest made use of several locations on the island of Andøya for different purposes, including the hosting of the participants, static emission sites, static reception sites, dynamic testing areas, and ad hoc evaluation points (all shown in Figure 1). The map in Figure 1 is labeled with five zones that were used for different purposes during the week.

Figure 1. Modified map showing test and lodging locations with numbered locations one through five as described below. Original map from norgeskart.no, accessed on 1 April 2023.

Location 1 at the northernmost extent of the island was the local host community where the majority of test participants were housed.

Location 2 to the North of the community of Bleik and adjacent to the local cemetery was used as an equipment deployment and emissions site for the high-power jamming sources, as well as for the testing of helicopter navigation systems.

Location 3 at the Bleik community center was the primary meeting base camp, setup and equipment preparation point, and deployment area for static receiver testing. It was also used as the first low-power Personal Privacy Device (PPD) test area. A collection of PPDs used by the Norwegian Communication Authority (Nkom) during the test activities is shown in Figure 2.

Figure 2. Twenty-one assorted jamming devices of various designs used throughout the PPD tests.

Location 4 at Bauktuva was a roadside parking area on a low hilltop used for making safe turns for single vehicles and platoons of vehicles (participating in tests).

Location 5 at Grunnvatnet was protected from emissions by the high-power jamming sources by local topography and was used as a second low-power/PPD test area.

The testing carried out during the event was designed to expose participants to as many forms of signal interference as possible and in as many ways as possible to help evaluate navigation system behavior in conditions of interest to as many different user segments as practical. Since the most common form of jamming most navigation users will encounter is emitted either unintentionally or from PPDs, it was natural to include both an array of PPDs as well as custom-generated signals. Since many sources of unintentional Radio Frequency Interference (RFI) are narrowband in nature [1], it was decided that several of the sub-tests should include Continuous Wave (CW) signals.

In addition to the PPD and 'unintentional' signal-like CW sources, a number of other modulations and power profiles were used to help simulate other scenarios or to try to stimulate failure modes in receiving equipment which may not normally arise. One example of this approach includes the progressive jamming of multiple Global Navigation Satellite System (GNSS) signal bands and modulations where a new modulation was periodically added incrementally to the list of jammed signals before being removed in the reverse order. This so-called modulation 'pyramid' was intended to test the potential fallback behavior of modern multi-constellation multi-frequency receivers. Similarly, some tests were conducted with ramping power levels to better evaluate the sensitivity thresholds of various systems or, in other cases, persistent high-power jamming on one or more bands to provide both static and dynamic testing opportunities.

3. Scheduled Tests

In order to accommodate as many use cases as possible, the testing was split into multiple morning and afternoon sessions over the five days of testing at a primary site, in addition to periodic tests at a secondary site and the convoy testing of vehicles on the local roads. While it is not possible to list all of the tests carried out by all of the various participants here, a summary in the form of the transmission schedule used by the Norwegian Communications Authority, the Norwegian Defense Research Establishment, and the Norwegian Metrology Service for the primary test site has been reproduced for the purposes of this paper and is shown in Table 1.

Table 1. Primary site tests.

Time	Transmission Plan	Comments
Monday 14:00–14:45	Single-band PPD jammers	Sixteen one-minute tests of an assortment of single-band PPD jammers, including four with multiple individual jammers of the same model. These tests were first conducted with the jammers located inside a car, then repeated with the jammers moved to the roof of the vehicle.
Monday 15:00–15:30	Multi-band PPD jammers	Six one-minute tests of multi-band PPD jammers (repeated once with the devices placed inside a static test vehicle, then repeated with the devices on the roof of the vehicle). The dual-band jammers had two models, with two units of each model type, while the jammers with more than two bands of coverage had only single units of two different types.
Monday 15:30–17:45	High-power jammers	Signals emitted at 20 W Equivalent Isotropically Radiated Power (EIRP) from a directional antenna placed at location 2 directed towards location 3 with 15 min of test time per modulation. The modulations used were CW and PRN in combinations of L1 only, L1 + G1, L1 + G1 + L2 and L1 + G1 + L2 + L5.
Tuesday 9:00–12:45	Power ramp RFI	Signals generated commensurate with a 20 W EIRP transmission passed through a variable attenuator starting from 100 dB attenuation (2 nW EIRP) and stepping to 0 dB, then back to 100 dB attenuation in steps of 2 dB every 10 s. Tests were repeated for modulations CW and PRN on carrier combinations of L1, L1 + G1 + L2 + L5 and L2 + L5 + G2 + E5b, plus a test of only CW (no PRN) modulation conducted for the carrier combination of L1 + L5 + E5b.

Table 1. Cont.

Time	Transmission Plan	Comments
Tuesday 14:00–15:30	Sustained high-power jamming	Twenty-minute transmissions of 20 W EIRP jamming signals, first with CW modulation and then PRN modulation on the L1 only band, followed by the L1 + G1 + L2 + L5 band combination.
Tuesday 16:30–18:00	Modulation pyramid.	With 10 W of power and PRN modulation on each identified band, one modulation per step was added first, and this step was repeated at intervals of 3 min; then, one modulation per step was removed. The order of the bands added in steps was as follows: E5b, L5, G2, L2, B1I, G1, L1. These were then removed in reverse order.
Wednesday 9:00–10:30	Sustained high-power jamming II	A repetition of the sustained high-power jamming to allow vehicles to execute driving tests in the same environment.
Thursday 9:00–13:00	Simple spoofing	Numerous spoofing activities, some preceded by jamming, some including persistent jamming on non-spoofed signals. Various signals with stale ephemeris, synchronization at the tens of seconds level and large transmitted antenna position errors, each of which should have been easily detectable, were adopted.
Thursday 14:00–18:00	Advanced spoofing	Numerous spoofing activities, some preceded by jamming, some including persistent jamming on non-spoofed signals. Current ephemerides with synchronization at the ten nanoseconds level and accurate spoofed positions preceding dynamic spoofed positions.
Friday 9:00–13:00	Ad hoc testing	Various [1].

[1] The ad hoc testing on Friday was an opportunity to repeat or modify tests that attendees requested based on their observations during the previous days of testing (in this case, focusing on personal privacy device re-testing and mobile user evaluations).

4. Jamming Observations

The behavior of the receivers observed by the authors ranged from completely expected to entirely unexpected. Outcomes that were in line with the expectations of the authors included the relatively good resistance of mass-market receivers to CW interference, as several of these models are advertised to contain CW detection and or mitigation support in the receiver firmware. While the 20 W EIRP transmissions were still enough to cause loss of lock to these receivers, their resilience to one of the most commonly observed forms of RFI based on long term international monitoring [2] is encouraging.

A neutral observation from the jamming tests related to the relative behavior of survey receivers compared to mass-market devices when subjected to persistent strong jamming signals. When the high-end receivers are subjected to 'moderate' levels of RFI on all their tracked signal bands, their firmware will selectively halt the output of observables, with some models appearing to have a cut out at 36 dB-Hz. In comparison, mass-market receivers that come from product families formerly marketed as being 'high sensitivity' tend to continue outputting observables and position solutions at these and lower signal levels. This latter case led to multiple observations of a low-cost receiver continuing to generate a position solution even as the position error increased to hundreds of meters, as is shown in Figure 3. Appropriately, the receiver generated increased uncertainty estimates and appeared to consistently bound the actual position error, meaning that the generated information is not hazardously misleading if the reported uncertainty is used by the target application.

An unexpected observation that is reasonable in retrospect is that multi-band multi-constellation receivers can manifest problems with timing synchronization when a subset of their signal bands are jammed. In the context of positioning, these systems appear to fall back almost universally to remaining available signals when a subset of their tracked signals are denied by persistent jamming and continue to generate observables and position plus velocity information to the user; however, timing synchronization can encounter unexpected behavior. When the signals from the constellation selected for time scale synchronization are denied for an extended period, some receiver models appear to fail to maintain their synchronization to the selected timing reference and begin to rapidly drift in time while maintaining other outputs. While not expected at the time of the tests, this behavior does pose an interesting design question for receiver firmware as it is not immediately obvious what the 'correct' answer to this situation is. One potential answer is that the receiver should rely on the data transmissions from other constellations, which include relative time offset information to indirectly track the selected time scale even when signals from the reference constellation are unavailable. Another potential approach is to simply halt outputting a solution as it is no longer possible to directly observe the selected time scale. Based on observed real-world behavior, it seems that some models of receivers have undefined or at least undocumented behavior with respect to their time scale synchronization when exposed to the jamming of the reference constellation, which could be more concerning to some users than a lack of outputs.

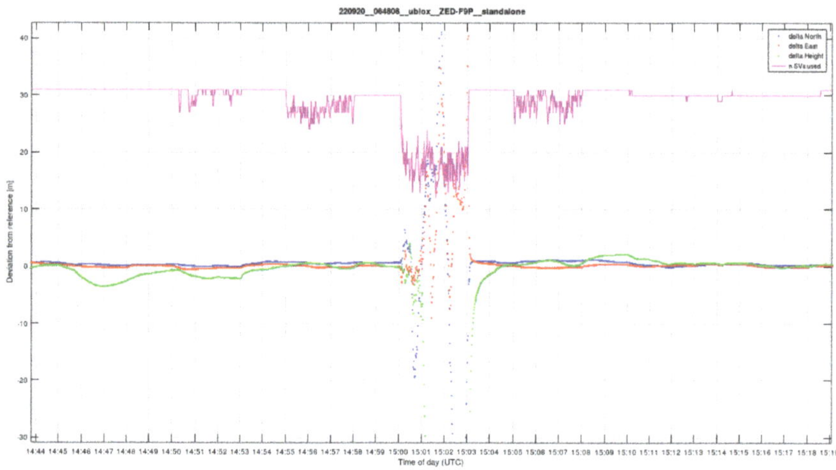

Figure 3. Number of tracked satellites and reported positional deviation of a static Ublox F9P receiver under jamming.

5. Spoofing Observations

The spoofing tests were physically arranged to allow as many participants as feasible to place their equipment under test within the line of sight of the transmitting antenna at a distance of less than approximately 20 m where possible. Based on this combination of criteria and the layout of the site in terms of where the spoofing signal generation equipment could be set up, the transmission antenna was placed in the middle of the front parking lot of the Bleik community center. One of the motivating factors for allowing users to operate as close to the transmission location as possible was the desire to minimize the amount of transmitted power required while also allowing the spoofed position information to be correct. The testing plan for the spoofing signal generation tests is included in Annex 1 for tests categorized as simple and Annex 2 for tests categorized as advanced.

Generating GNSS navigation signals that successfully masquerade as legitimately generated signals is difficult, with several publications indicating specific signal generation defects that can be used to aid in the detection of spoofed signals [3,4], including but not limited to time variation, correlation variation, position discontinuities, inconsistent observable data, incorrect ephemeris, or timing errors. In the Jammertest 2022 spoofing program, the distinction between the 'simple' spoofing tests and the 'complex' spoofing tests is the use of correct current ephemerides provided by the Norwegian Mapping Authority (Kartverket), together with the precise time keeping at the level of tens of nanoseconds for the latter. While the so-called simple tests could still entail careful pre-jamming, stable and high-fidelity signal generation, correct simulated ephemerides, and accurate initial position generation, the inclusion of precise synchronization enabled the evaluation of network timing receivers in addition to navigation-focused GNSS equipment. Attention to the generation of a representative and stable spoofing signal is particularly important given that some methods of potential spoofing investigation and detection might rely on the interaction of the produced signal with intentional antenna motion [5,6] or other isolation techniques which may be complicated by an unstable signal source. This presents a challenge where the given signal would ideally be useful to everyone, which can be partly addressed through providing both 'simple' and 'complex' spoofing scenarios such as those provided at Jammertest 2022.

Some of the observations of note from the author's experiences during the test include expected results such as entry-level receivers (such as those integrated into the phones of the participants) being very quickly fooled by the spoofed signal. In one case, one of the participants took advantage of the spoofing tests to activate a fitness training app which happily believed the owner ran a distance of six kilometers (climbing over 450 m vertically) within six minutes and allowed them to post it to social media as a new 'personal best', indicating that the phone has no inertial sanity checking and that the app has no filtering for superhuman feats (Figure 4).

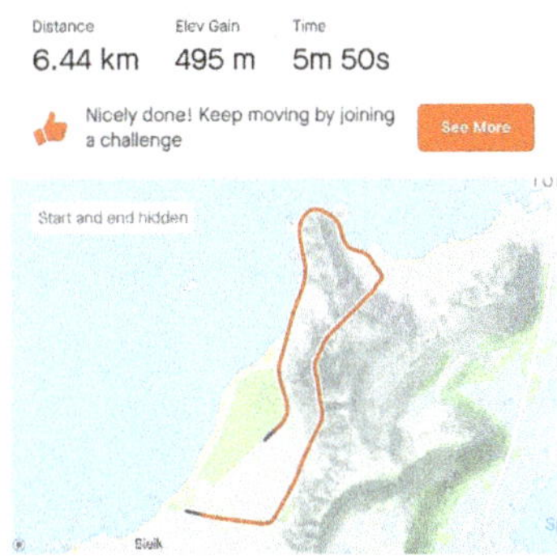

Figure 4. Screenshot from the Strava app showing a spoofed smartphone's reported superhuman run. Map data from Mapbox OpenStreet.

An unfortunate observation was that some high-end receiver systems supporting quadruple constellation triple frequency operation could be driven into apparently unrecoverable states via the single-frequency spoofing and multi-frequency jamming combinations of even the simple spoofing tests. Despite having an internal oscillator that is relatively stable, the receiver transitioned from tracking real signals to tracking the spoofed signals after only a short period of jamming but then would not recover after the removal of the spoofing and jamming. Instead, more than 40 min after the removal of the perturbing signals, the receiver reported that it was tracking only several GLONASS satellites and one QZSS satellite, stating that the receiver altitude was 1500 m higher than in reality and that it was slowly floating away ever higher.

A beneficial outcome of the spoofing tests was that even when the spoofing signals were generated with a power level approximately 20 dB (too weak to impact the assembled receivers), systems with in-band power level monitoring capabilities still noted significant and obvious changes in the level of their respective automatic gain control feedback states or measured in-band power levels. This is encouraging as it implies that even relatively low-cost receivers that generate Automatic Gain Control (AGC) level outputs have an additional avenue available to them for the detection and reporting of potential spoofing signals.

6. Unexpected Observations

The first of the notable unexpected observations was that on the morning of Tuesday 20th September, the scheduled power ramp test was apparently interrupted by multiple events of unintentional RFI. This test presented an opportunity for evaluating the detection thresholds of various pieces of equipment, including the Advanced RFI Detection Analysis and Alerting System (ARFIDAAS) monitoring system [2], which includes direct band power measurement instruments as part of its normal detection process. Monitoring using the ARFIDAAS system started approximately three minutes before the designed test start time, and the system almost immediately began detecting and logging events. Over the three-minute period between the start of logging and the start of the test, three separate RFI events were captured. Since the test was scheduled to start with an extremely low EIRP of approximately 2 nW, it was initially thought that these detections implied that either an incorrect attenuator setting had been selected at the transmission site or that an unrelated project participant had mistakenly activated one of the test PPDs without permission; however, upon a closer inspection of the three events, none of them matched either the expected test signals or the known signals from the PPD selection. Upon further review, three distinct RFI signals of unknown origins had been captured coincidentally in the time prior to the start of the test and despite the remote nature of the test site. The first of these was a 'chirping' signal thought to be formed by self-resonant feedback in car-borne navigation or entertainment systems impacting the G1 band; the second was a pulsed CW source near the Beidou B6 center frequency, while the third was a continuously active 1200 MHz narrowband signal. A spectrogram of the unintentionally generated 'chirping' signal is presented in Figure 5. Here, it should be noted that while the modulation appears to be a form of chirp signal, the center frequency and limited bandwidth of the signal indicate that it would be a relatively ineffective jammer to use against GPS, Galileo, and Beidou signals, though might present difficulties for GLONASS reception.

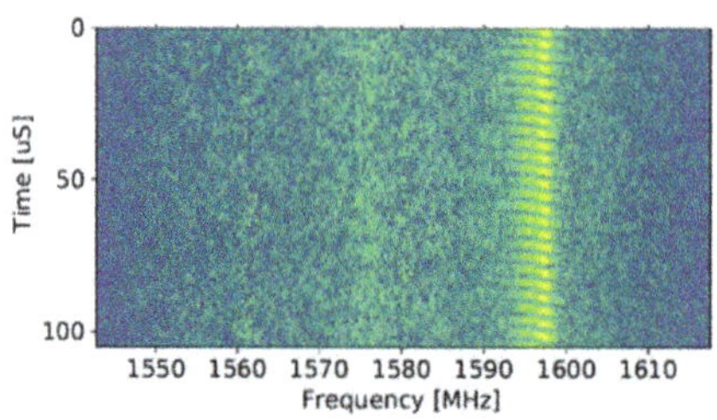

Figure 5. A time–frequency plot of a suspected unintentionally generated signal emitted from a car passing the test range just prior to the start of the low-power ramp test (detected using ARFIDAAS system hardware). Color encodes relative signal strength with lighter yellows indicating concentrated signal power.

A second unexpected observation was that some of the receivers used by the participants to evaluate Network RTK service performance during jamming and spoofing were disrupted by the RFI in a way that was not predicted to be a point of failure before the testing. In at least one receiver model, the loss of RTK performance was not caused by the degradation of the GNSS observables but instead by the disruption of the 2.4 GHz Wi-Fi-based data link between the receiver and the controlling computer acting as a correction relay to the receiver. It is unclear how much energy the jamming devices in question were leaking at or near the 2.4 GHz band used by the receiver's Wi-Fi link, but observations in the wild have shown the opposite situation, whereby malfunctioning Wi-Fi devices act as GNSS jammers by emitting at half of their expected operating frequency. This shows an additional danger of cheap jammer devices, in addition to their intended purpose the poor production value can cause additional interference in other, not-intended bands.

A third surprise came in the form of the difficulty encountered in calibrating the power level employed in spoofing the collected GNSS receivers. Before shipping the signal generation equipment to the venue, the selected scenarios were generated and tested in the Norwegian metrology service's (Justervesenet) lab to tune parameters such as the transmitted power level. However, after setting up on site and initial attempts were made to generate the spoofing signals for the test scenarios, none of the user receivers were initially captured by the generated spoofing signal. It was eventually necessary to increase the transmitted spoofing power by 20 dB relative to the starting point. Some of this difference was eventually explained by the use of a different cable assembly between the signal generator and the transmitting antenna, but this is not thought to explain the entirety of the change power level requirement. Despite careful preparation, generating a realistic minimum power spoofing signal was demonstrated to be challenging once more.

A final unexpected observation was that some of the highest-end navigation systems evaluated had more significant problems as a result of the generated signals than their much lower-end counterparts. For example, one of the mass-market automotive platforms appeared to be unaffected in terms of reported position solution by the generated spoofing due to integration with a set of low-cost Micro Electro-Mechanical Systems (MEMS) inertial sensors, while one of the high-end navigation systems, which integrated a ring laser gyroscope-based Inertial Navigation System (INS), accumulated a kilometer-level navigation error during jamming tests and could not recover until power cycled. The assumption is that the automotive receiver was successfully leveraging Zero-velocity Update (ZUPT) detection while the high-end platform intended to operate in an airborne platform may not have supported this state detection, as this effect was not expected to be observed in normal operation.

7. Conclusions and Data Availability

One of the standout observations from the tests was that various pieces of receiver equipment manifested behaviors that could be considered naive or counterintuitive when faced with jamming and spoofing signals, which indicate that the device firmware does not appear to be designed to monitor for or guard against the impacts of these threats. For example, integrated systems such as smartphones or network-connected receivers make no observable attempt to mitigate spoofing attacks that utilize obviously incorrect timing information. While the timing and positioning gap is not as egregious as that observed during the spoofing incident at ION 2017 [7], even a modest cost quartz oscillator should not be fooled by jumps of several seconds after a short period of signal denial, but this was observed during Jammertest. Hopefully, in the coming years, manufacturers will continue to deploy anti-jamming and anti-spoofing technologies within their receivers that can leverage even simple AGC state behavior to mitigate the impacts of these threats, implement firmware that can exploit the diversity of timing sources such as network timing protocol servers to prevent gross errors, or, at least, adjust their firmware parameters to allow receivers to gracefully recover within an hour of spoofing termination if they insist on allowing their clock to jump by tens of seconds in the first place.

The activities at Jammertest 2022 were an invaluable opportunity for commercial, governmental, research, and academic organizations to observe the behavior of GNSS-dependent navigation and timing equipment under jamming and spoofing conditions. While other events of this nature have been held before, the policy of open data access sets this event apart and provides a chance for others to learn from the experiences of the attendees not just through abstracted publication but also through the direct sharing of captured data.

The authors of this paper are pleased to be able to make available selected subsets of the testing data captured at Jammertest 2022 from the primary site tests. These data include raw IF spectrum captures of the L1/E1/G1 band captured with 40 MHz complex sampling during the entirety of the spoofing tests, as well as other data captured by the ARFIDAAS monitoring station via SINTEF, ublox F9P receiver data captured during various jamming and spoofing tests throughout the course of the test plan (captured via Kartverket and a centimetric network), and RTK receiver data captured throughout the tests via Kartverket. For access to these data, please contact the corresponding authors.

Author Contributions: A.M.: primary document drafting. N.S.: primary document drafting and proofing. A.S.: contributed receiver position deviation results. N.G.: documentation of test setup and descriptions. A.R.: setup of jamming and power level descriptions. H.H.: description of spoofing test setup and considerations. T.R.: description of spoofing test setup and document proofing. T.D.: contribution of receiver spoofing position domain observations and discussions. All authors have read and agreed to the published version of the manuscript.

Funding: The authors would like to thank the European Space Agency NAVISP program for funding the development of the ARFIDAAS system (NAVISP3-EL3-015), as well as the Norwegian Council of Research (project number 305051) for funding the participation of the primary authors and data analysis activities. Authors H.H. and T.R. would like to thank the Norwegian Space Agency for funding through project 74GA2203.

Institutional Review Board Statement: Not applicable.

Informed Consent Statement: Not applicable.

Data Availability Statement: For data availability please contact the authors via email.

Conflicts of Interest: The authors declare no conflict of interest.

References

1. Rødningsby, A.; Morrison, A.; Sokolova, N.; Gerrard, N.; Rost, C. RFI Monitoring of GNSS Signals on Norwegian Highways. In Proceedings of the 33rd International Technical Meeting of the Satellite Division of the Institute of Navigation (ION GNSS+ 2020), Virtual, 21–25 September 2020.
2. Morrison, A.; Sokolova, N.; Swinden, R.; Musumeci, L.; Caparra, G. Advanced RFI Detection, Alert and Analysis System Design and Monitoring Campaign Results. In Proceedings of the NAVITEC 2022, Virtual, 5–7 April 2022.
3. Kujur, B.; Khanafseh, S.; Pervan, B. Detecting GNSS Spoofing using Temporal Behavior of Spoofed Signals. In Proceedings of the 34th International Technical Meeting of the Satellite Division of The Institute of Navigation (ION GNSS+ 2021), St. Louis, MO, USA, 20–24 September 2021; pp. 3668–3682.
4. Turner, M.; Wimbush, S.; Enneking, C.; Konovaltsev, A. Spoofing Detection by Distortion of the Correlation Function. In Proceedings of the 2020 IEEE/ION Position, Location and Navigation Symposium (PLANS), Portland, OR, USA, 20–23 April 2020; pp. 566–574.
5. Psiaki, M.L.; Powell, S.P.; O'Hanlon, B.W. GNSS Spoofing Detection using High-Frequency Antenna Motion and Carrier-Phase Data. In Proceedings of the 26th International Technical Meeting of the Satellite Division of The Institute of Navigation (ION GNSS+ 2013), Nashville, TN, USA, 16–20 September 2013; pp. 2949–2991.
6. Humphreys, T.E.; Ledvina, B.M.; Psiaki, M.L.; O'Hanlon, B.W.; Kintner, P.M., Jr. Assessing the Spoofing Threat: Development of a Portable GPS Civilian Spoofer. In Proceedings of the 21st International Technical Meeting of the Satellite Division of the Insitute of Navigation (ION GNSS 2008), Savannah, GA, USA, 16–19 September 2008; pp. 2314–2325.
7. Scott, L. Spoofing Incident Report: An Illustration of Cascading Security Failure. Inside GNSS Magazine—Global Navigation Satellite System Engineering Policy, and Design. Available online: insidegnss.com/spoofing-incident-report-an-illustration-of-cascading-security-failure/ (accessed on 28 April 2023).

engineering proceedings

MDPI

Proceeding Paper

LO and Calibration Signal Distribution in a Multi-Antenna Satellite Navigation Receiver †

Uwe Stehr [1,*], Syed N. Hasnain [1], Björn Bieske [2], Marius Brachvogel [3], Michael Meurer [3,4] and Matthias A. Hein [1]

1 Thuringian Center of Innovation in Mobility, RF & Microwave Research Group, Technische Universität Ilmenau, 98693 Ilmenau, Germany; syed-naser.hasnain@tu-ilmenau.de (S.N.H.); matthias.hein@tu-ilmenau.de (M.A.H.)
2 IMMS Institut für Mikroelektronik- und Mechatronik-Systeme Gemeinnützige GmbH (IMMS GmbH), 98693 Ilmenau, Germany; bjoern.bieske@imms.de
3 Chair of Navigation, RWTH Aachen University, 52074 Aachen, Germany; marius.brachvogel@nav.rwth-aachen.de (M.B.); michael.meurer@nav.rwth-aachen.de (M.M.)
4 Institute of Communications and Navigation, German Aerospace Center (DLR), 82234 Weßling, Germany
* Correspondence: uwe.stehr@tu-ilmenau.de
† Presented at the European Navigation Conference 2023, Noordwijk, The Netherlands, 31 May–2 June 2023.

Abstract: Due to the low signal power of global navigation satellite signals, the receivers are prone to radio frequency interference. Employing multi-antenna arrays is one method to mitigate such effects, by incorporating spatial processing techniques. The large size of the uniform rectangular arrays prevents their use in applications where installation space is limited. Therefore, we proposed a new approach, namely to split one full array into a number of smaller, spatially distributed, sub-arrays to reduce their size and exploit available installation spaces. This concept challenges the distribution of the local oscillator and calibration signals to the respective sub-arrays. This paper compares qualitatively different design concepts for a satellite navigation receiver with two two-element sub-arrays, installed multiple wavelengths apart from each other, in support of establishing an optimal choice for our intended applications in the automotive sector in terms of electrical performance and required hardware and software efforts. In general, weighing the pros and cons of the different concepts, as discussed in the paper, will assist in optimizing the system design approach for a specific application.

Keywords: compact distributed array; GNSS receiver frontend; LO signal distribution; system calibration; satellite navigation; interference mitigation

Citation: Stehr, U.; Hasnain, S.N.; Bieske, B.; Brachvogel, M.; Meurer, M.; Hein, M.A. LO and Calibration Signal Distribution in a Multi-Antenna Satellite Navigation Receiver. *Eng. Proc.* **2023**, *54*, 23. https://doi.org/10.3390/ENC2023-15447

Academic Editors: Tom Willems and Okko Bleeker

Published: 29 October 2023

1. Introduction

Global navigation satellite systems (GNSS) are key to safety-critical applications like automated and connected or tele-operated driving or unmanned flight maneuvering. These applications depend on the reliable availability and functionality of GNSS signaling. Due to their positioning on medium earth orbits (MEO) and the available link budgets, the respective received signal power may be as low as -130 dBm (corresponding to 10^{-16} W) at ground level [1]. This low signal strength makes the receivers prone to deliberate or unintended RF interference (RFI). Sources of unintentional distortions are, e.g., harmonic tones of local oscillators or strong signals transmitted by other wireless services in adjacent frequency bands.

Among the deliberately caused disturbances, signal jamming and spoofing attacks are of major concern [2]. The term jamming refers to the transmission of a strong RF signal in or close to the GNSS frequency bands. The signal saturates the analog frontend of the GNSS receiver, thus blocking the reception of the navigation signals. As is widely known though legally prohibited in many countries, truck drivers use so-called personal privacy devices (PPD) to hide their location from third parties [3]. Due to the low GNSS signal power, jamming is easily achieved with an appreciable areal coverage. The detrimental impact of such jammer devices on safety-critical infrastructure or processes is frequently under-appreciated. Spoofing describes the process of deceiving a GNSS receiver, leading to false position, velocity, and time (PVT) data. Spoofing attacks are more challenging to identify, but are also more complex to implement compared to jamming.

Several methods have been proposed to achieve the immunity of a GNSS receiver against intentional or unintentional interference [4]. One possibility is to enhance the robustness of the analog receiver frontend. As an example, a receiver chain architecture able to handle lower output signal levels was less susceptible to saturation [5,6]. Another approach is the use of multi-antenna receivers in array configuration, enabling beamforming and null steering [7–9]. Dedicated digital array signal processing helps to suppress unwanted signals by directing radiation nulls towards the interfering source, or estimating the directions-of-arrival (DoA) to verify the authenticity and validity of incoming signals [10].

We previously introduced a 2×2-element uniform rectangular array (URA) connected to a dual-frequency receiver frontend [11]. This receiver system provides both a high positioning accuracy, due to the dual-band design, and high robustness against interference, due to the array performance. Despite the compact dimensions of the array with the radiating elements separated by a quarter wavelength, these features come at the expense of a large module size of 10 cm $\times$ 10 cm $\times$ 10 cm. Such dimensions exceed typical installation spaces, e.g., of passenger cars, and thus prevent their penetration into a mass market. Therefore, our goal was to realize a more compact GNSS receiver system. We proposed to split the full number of four antenna elements into two distributed two-element sub-arrays [12]. This approach effectively increases the number of possible mounting spaces by enabling the installation of sub-arrays in distant parts of a car, such as side mirrors or other suitable car body appliances. A drawback of this approach, however, is the creation of grating lobes due to the electrically large element separation that challenges array signal processing like DoA estimation. This effect can be mitigated by optimizing the arrangement of the sub-arrays [13].

On the RF frontend side, the concept of spatially separated sub-arrays necessitates new concepts for the design and calibration of the receiver hardware. The local oscillator (LO) signal required for the down conversion of the received signals, as well as the calibration signal required for the precise characterization of the multiple receiver chains have to be connected to the distant sub-arrays. This paper compares different options for the distribution of the LO and calibration signals and contrasts their respective advantages and disadvantages for the example of a four-element array consisting of two identical sub-arrays mounted in the side-mirrors of a passenger car. For this installation scenario, design parameters like mechanical dimensions and the cost of the solution are the most important, along with electrical performance. If the achievable electrical performance dominates the optimization goal, the system design can lead to different conclusions.

2. GNSS Receiver with a 2×2 Antenna Array

In our dual-band 2×2 URA receiver, each patch antenna element covers the L1 band with a center frequency of f_1 = 1575.42 MHz and E5a band with f_5 = 1176.45 MHz. The dual-polarized patches enable the reception of right-hand circular polarized (RHCP) as well as left-hand circular polarized (LHCP) signals [14]. A simplified block diagram of the receiver is shown in Figure 1. The antenna array provides a total number of 16 RF channels, namely $4\times$ RHCP and $4\times$ LHCP at f_1 and $4\times$ RHCP and $4\times$ LHCP at f_5. Each channel is amplified, filtered and down-converted. The down conversion in the heterodyne receivers must preserve the magnitude and phase relationships between all received antenna signals to preserve phase coherence and thus synchronicity between the channels. A calibration signal (CAL) is provided at each channel, to determine differences in the transmission characteristics between the receiver chains. The CAL signal consists of an artificial GNSS signal which is generated in the digital baseband and up-converted to the RF domain by the LO-signal. It passes through the entire signal processing chain of each frontend channel, is processed in the digital baseband together with the actual received GNSS signals, and is eventually compared to the initial calibration signal. This allows for a correction of gain and phase differences caused by imbalances between the channels.

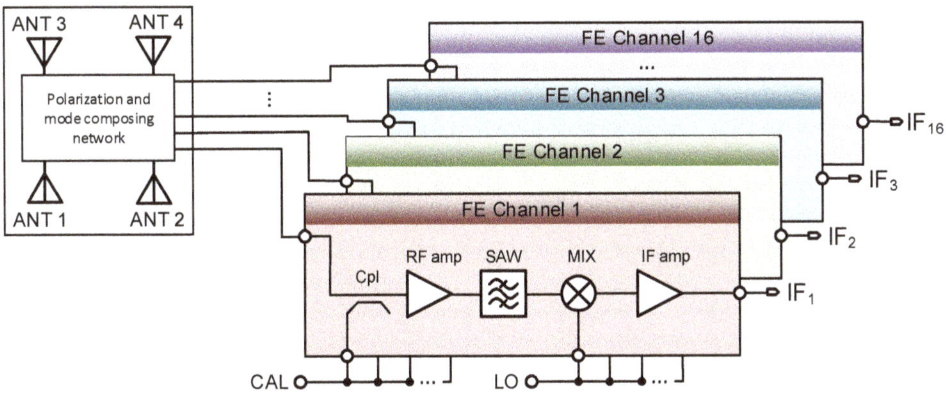

Figure 1. Simplified block diagram of the 2 × 2 multi-antenna dual-band dual-polarized GNSS array receiver with calibration (CAL) and local oscillator (LO) signals, leading to a total number of 16 channel outputs.

A prerequisite for the calibration algorithm is that the CAL signal be applied with identical amplitudes and phases to each receiver channel. In a compact antenna array, this condition can be met straightforwardly because all antenna elements are arranged on a common substrate; consequently, the calibration signal can be multiplexed using passive components like power splitters and distributed to the frontend channels by length-matched transmission lines. A similar conclusion holds for the distribution of the LO signal to the mixers of all receiver channels.

For the case of a distributed antenna array receiver, the above-mentioned conditions are much more challenging to meet because of the required cabling in the calibration and LO distribution paths between the two sub-arrays. As a consequence, the signals depend on the cable characteristics in terms of mechanical length differences, thermal variations and so on, leading to unforeseeable phase variations.

3. Distributed GNSS Receiver Concept with Two 2 × 1 Antenna Sub-Arrays

3.1. Receiver Design

Our approach of splitting one 2 × 2 antenna array into two sub-arrays resulted in 50% reduction of the array footprint per sub-array. Additionally, we reduced the complexity of the front-end design by processing the RHCP signals only at both frequencies, f_1 and f_5. This led to four channels per sub-array, namely two per frequency band. A dedicated application-specific integrated circuit (ASIC) was designed for the receiver in 180 nm United Microelectronics Corporation (UMC) technology. The ASIC contains four complete receiver chains, each with two RF low-noise amplifiers (LNA), a down-converting mixer, and an IF amplifier. An external bandpass filter can be connected between the LNAs for the pre-selection of the chosen frequency band. The ASIC also contains two phase-locked loop (PLL) synthesizers to generate the LO signals at the two operational frequencies. Alternatively, external LO signals can be applied to the ASIC.

Each sub-array module incorporates one ASIC and provides four receiver channels, two for each frequency, fed by two commercially available dual-band ceramic patch antennas APAKM2507 [15], as shown in Figure 2.The complete GNSS receiver consists of two sub-arrays. The common LO and calibration signals for both frequency bands, provided by the central base-band module, have to be split and fed to the sub-arrays by cable connections, as illustrated in Figure 3. Due to such a distributed arrangement of the two sub-arrays, the provision and distribution of identical and phase-coherent CAL and LO signals is challenging and requires dedicated considerations of the various trade-offs involved.

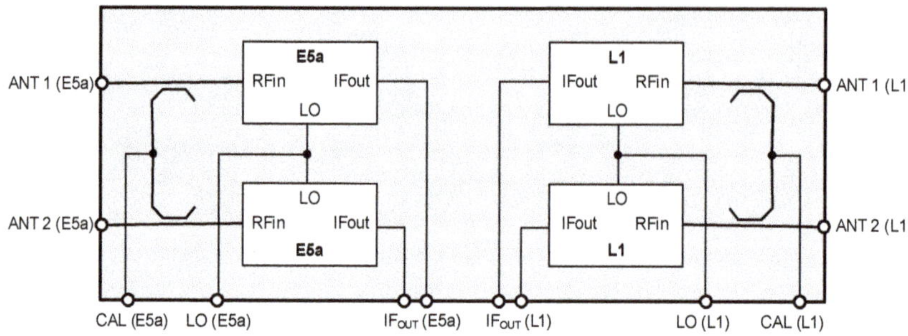

Figure 2. Principal block diagram of a dual-band sub-array of the distributed GNSS receiver. The four receiver channels are shown with their respective RF and LO inputs and IF outputs. Each sub-array contains a CAL and LO signal distribution between the two receiver chains at each frequency band.

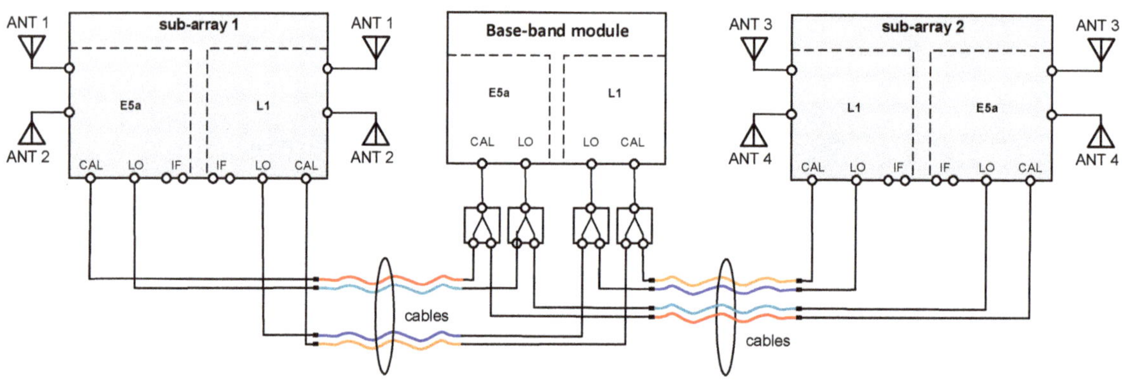

Figure 3. Block diagram of the complete GNSS receiver consisting of two sub-arrays and a central base-band module. The LO (REF) and calibration signal for both frequency bands are split and distributed to the sub-arrays by cable connections.

3.2. System Calibration

3.2.1. LO Signal Distribution

There are two options to provide the LO signal; the first and most obvious one is to generate the LO signal in a common PLL-based synthesizer using a single crystal-based reference oscillator, to split the signal in two identical signals by means of a power divider and to distribute it to both sub-arrays with low-loss coaxial cables. A second option is to distribute the reference signal (REF) instead, and employ two PLLs at each sub-array to synthesize the LO signal locally. Since both signals are derived from the same reference signal, the output signals of both PLLs remain phase-coherent.

The schematic in Figure 4. illustrates the two options. In Figure 4a, one common PLL module synthesizes the LO signal using the 10 MHz REF signal. The LO signal is split into two identical signals and distributed to the sub-arrays, where they are directly available for down-converting the input signals. In Figure 4b, a common reference signal at a frequency of f_{ref} = 10 MHz is split into two identical copies and distributed to the sub-arrays. Each sub-array incorporates its own PLL-based LO signal generator. In both cases, cables of significant lengths are required to feed the sub-arrays with the REF and, respectively, LO signal. The phase coherence required for the array signal processing is challenged by mechanical and thermal tolerances and variations, as well as other environmental effects. As one example, one side of the car can be subject to sunshine, the other side being shaded. The resulting temperature differences may lead to significant changes in the electrical length. The thermal phase sensitivity depends strongly on the type of cable. For teflon-insulated coaxial cables, maximum phase changes above 2500 ppm were reported [16]. For such a cable of 2 m length, as used in our study, this would result in phase changes of up to 17.5° at the LO frequency of 1648 MHz, causing errors in the DoA estimation algorithms. Phase-stable cables would come at the expense of increased costs.

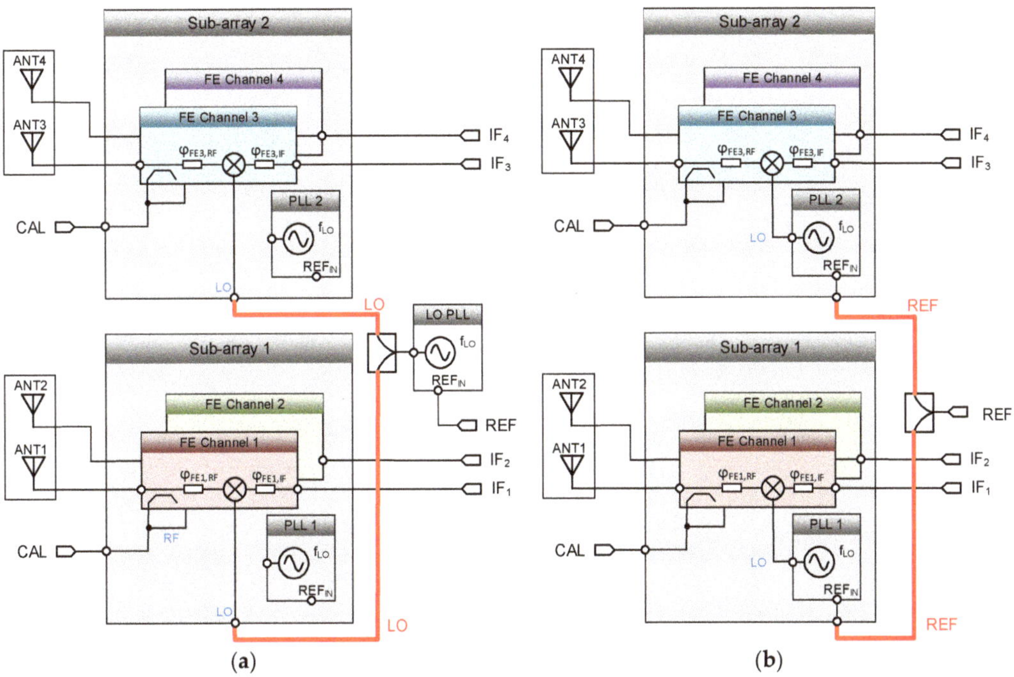

Figure 4. Options for LO signal distribution: (**a**) distribution of a common REF signal and separate LO generation at each sub-array; (**b**) common LO generation in a single PLL module and distribution of the LO signals to both sub-arrays.

For the REF signal distribution, despite the lower phase imbalance at both sub-arrays due to the lower frequency, the multiplication by the frequency synthesizer leads to the same total phase imbalance as for the direct LO signal distribution. Therefore, the electrical properties in terms of phase imbalance of both concepts are similar.

In the case of the REF signal distribution, the internal PLL of the ASIC is used to synthesize the LO signal. Using the internal PLL, the phase noise of the synthesized LO signal amounts to about −82 dBc/Hz @ 10 kHz offset and −110 dBc/Hz @ 1 MHz offset from the carrier frequency. In the case of the LO distribution, by using a dedicated PLL synthesizer IC (e.g., LTC6948 from Analog Devices, Wilmington, DE, USA), lower phase noise values of about −100 dBc/Hz @ 10 kHz and −130 dBc/Hz @ 1 MHz are achievable. The LO phase noise limits the carrier-to-noise ratio (CNR) of the output signal, narrowing the carrier tracking loop bandwidth and limiting the effective integration interval [17]. Therefore, the importance of this parameter for the system performance is an additional criterium to decide between both approaches.

A number of additional criteria have been weighted against each other to decide between the two options for the LO signal generation, including hardware and software efforts and costs; the low frequency of the REF signal allows twisted-pair cables to be employed instead of coaxial cables, which leads to a reduction in costs and weight. Another advantage of the REF signal distribution is that only one cable is required in contrast to the LO signal distribution, which needs to cover two frequency bands and thus two PLL modules, which have to be set up and monitored by additional software. Table 1 summarizes the pros and cons of both concepts. Eventually, the lower hardware and software efforts at similar RF characteristics led us to the decision to use the REF signal distribution for the generation of the LO signal at each sub-array.

Table 1. Comparison between REF-signal and LO-signal distribution.

	REF-Signal Distribution	**LO-Signal Distribution**
Hardware and software effort	(+) internal ASIC PLLs (+) programming via ASIC control software	(−) two additional PLL modules required (−) additional control software and lock-detect monitoring
Number of cables	(+) 1x twisted pair	(−) 2x coaxial cable
Phase imbalance due to cable length	(o) similar to LO-signal distribution	(o) depends on cables
Cable loss	(+) lower	(−) higher
LO phase noise	(−) higher, due to usage of the internal frontend ASIC PLL	(+) lower, due to usage of an advanced external PLL IC

3.2.2. CAL Signal Distribution

For calibration, an artificial GNSS signal is generated in the digital baseband and applied to all receiver chains. The calibration signal is generated in the IF domain at $f_{IF} \approx 70$ MHz and must be up-converted to the RF frequency bands with the synthesized LO signals. This guarantees phase coherence between the real GNSS receiver signals and the artificial GNSS calibration signal.

Upon down-conversion, all signals share the same IF frequency band. Similar to the LO signal distribution, two options also exist for the CAL signal distribution. Either the calibration signal is distributed to the subarrays and up-converted there (CAL signal distribution at IF), as illustrated in Figure 5a, or it is up-converted centrally and then distributed to the subarrays (CAL signal distribution at RF), as shown in Figure 5b.

Actually, the hardware effort on a sub-array level is larger for the CAL signal distribution at IF because an additional up-converter is required for each frequency band at each sub-array; in a future re-design, the up-converters could be integrated into the ASIC. In contrast, for the CAL signal distribution at RF, two LO signals have to be generated and distributed, with similar implications for the hardware and software efforts, as discussed, for the LO signal distribution.

For a comparison of the phase imbalance between the CAL signals at both sub-arrays, similar considerations are applicable as for the LO signal distribution, as the calibration signal is connected to the sub-arrays by cables in both cases. Therefore, the same constraints on length differences and temperature changes apply. The lower frequency of the CAL signal at $f_{IF} = 70$ MHz in comparison to the more than 20-fold higher RF signal in the L-band results in a correspondingly lower phase imbalance between the sub-arrays for a given change in the electrical length of the cables. This advantage, however, is lost upon up-conversion with the LO signal at each subarray. Additionally, it turns out that the calibration concept originally applied to the 2×2 URA cannot be applied to the distributed array in the same way. Rather, calibration with an artificial GNSS signal can be achieved at sub-array level only. There, the phase balance between the calibration signals for each receiver channel can be guaranteed by the use of a passive signal distribution network.

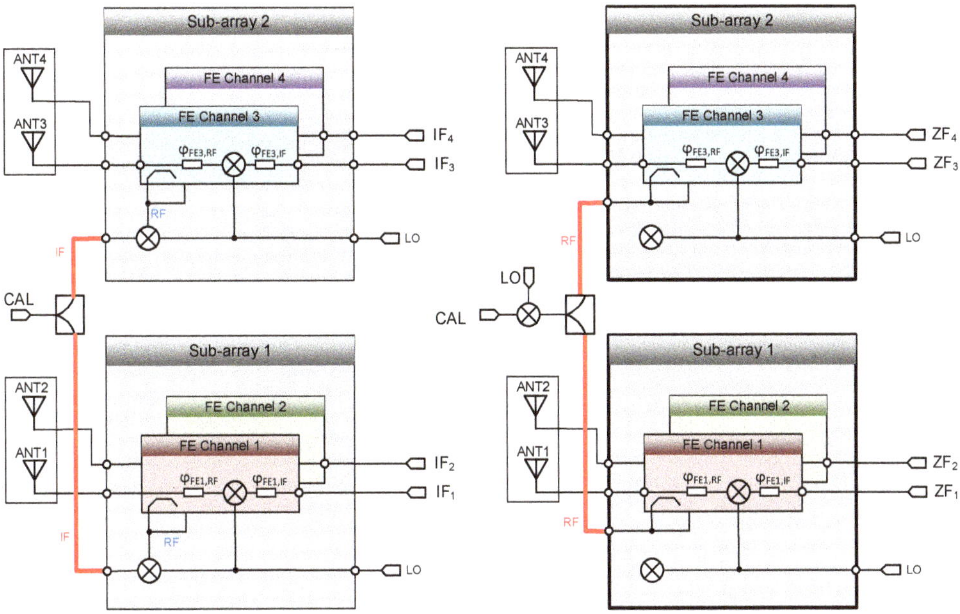

Figure 5. Calibration signal distribution options: (**a**) distribution of a common CAL signal to both sub-arrays, where it is up-converted using the local LO signals; (**b**): the CAL signal is generated in a common module and the radio frequency (RF) signal is distributed to each sub-array.

Table 2 compares the different properties of both CAL signal distribution concepts. Because of the lower effort and implementation costs, we selected the CAL distribution at IF for the compact distributed GNNS receiver.

Table 2. Comparison between CAL-signal distribution at IF and RF.

CAL Signal Distribution	... at IF	... at RF
Hard- and software effort	(o) usage of internal ASIC PLLs (+) programming via ASIC control software (o) up-converter required at sub-arrays	(−) 2 additional PLL modules required inside the car (−) additional control software and lock-detect monitoring (o) up-converter required in PLL module
Number of cables	(+) 2x twisted Pair	(−) 2x coaxial cable
Phase imbalance due to cable length	(o) similar to distribution at RF due to LO phase imbalance	(o) directly dependent on length difference between the RF cables towards each sub-array

The full calibration of the distributed array must be conducted in the digital domain, resulting in a blind or hybrid calibration concept [18]. In [19], we proposed a solution to determine the accurate position and attitude of a multi-antenna system without the explicit use of a dedicated calibration signal. The capability to mitigate interference sources with the distributed array receiver was presented in [3,20].

A photograph of one sub-array is shown in Figure 6.The sub-array module consists of two stacked printed circuit boards, one containing the antenna network and the other the

RF frontend including the ASIC. The size measures about 10 cm × 5 cm × 2 cm. With these dimensions, an integration into typical automotive installation spaces appears feasible.

(a) (b)

Figure 6. (**a**) Receiver sub-array consisting of antenna and frontend board (the module height is 2 cm); (**b**) antenna board (**top**) and frontend board with dedicated GNSS receiver ASIC (**bottom**) with CAL and LO signal input connectors.

4. Conclusions

In this paper, different concepts for the distribution of the calibration and local oscillator signals to distributed sub-arrays of a dual-band multi-antenna GNSS receiver were compared in terms of the resulting system performance and practical implementation issues. RF performance as well as hardware and software efforts and costs were reflected in our design considerations. The distributed arrangement of sub-arrays necessitates layout cable connections, which challenge the phase coherence required for the array signal processing. Despite the similar performance in terms of phase imbalance between the signals at both sub-arrays, the REF signal distribution and the CAL signal distribution at IF were eventually selected. This decision is based on the most important specification parameters of the intended application, installation space and cost, while considering the targeted system performance. For other applications, the required application parameters could lead to a different conclusion. Accordingly, the criteria compared in Tables 1 and 2 provide guidance for the selection process. The distributed system approach facilitates the integration of robust GNSS receivers into the limited installation spaces available in series cars and thus opens the market for an improved level of navigation safety.

Author Contributions: Conceptualization, U.S., B.B., M.B., S.N.H. and M.A.H.; methodology, U.S.; software, M.B.; investigation, U.S. and B.B.; writing—original draft preparation, U.S.; writing—review and editing, S.N.H., B.B., M.B. and M.A.H.; validation, supervision, project administration, funding acquisition, M.A.H. and M.M. All authors have read and agreed to the published version of the manuscript.

Funding: This work was supported by the German Aerospace Center (DLR) Space Administration on behalf of the German Federal Ministry for Economic Affairs and Climate Action under the grant numbers 50NA1901 and 50NA1902.

Institutional Review Board Statement: Not applicable.

Informed Consent Statement: Not applicable.

Data Availability Statement: The data that support the findings of this study are available upon reasonable request from the authors.

Acknowledgments: The authors wish to thank all colleagues of the ROSANNA project teams at TU Ilmenau, RWTH Aachen University and DLR Institute of Communication and Navigation, and IMMS GmbH for their valuable contributions and useful discussions.

Conflicts of Interest: Author B.B. is with IMMS GmbH, a non-profit limited liability company. The authors declare no conflict of interest.

References

1. Kaplan, E.; Hegerty, C. *Understanding GPS/GNSS: Principles and Applications*, 3rd ed.; Artech: London, UK, 2017.
2. Ruegamer, A.; Kowalewski, D. Jamming and Spoofing of GNSS Signals—An Underestimated Risk?! In Proceedings of the FIG Working Week 2015, Sofia, Bulgaria, 17–21 May 2015.
3. Brachvogel, M.; Niestroj, M.; Zorn, S.; Meurer, M.; Hasnain, S.N.; Stephan, R.; Hein, M.A. Jamming an uncalibrated GNSS array receiver of distributed antenna elements for concealed installation in passenger cars. In Proceedings of the 2020 European Navigation Conference (ENC), Dresden, Germany, 23–24 November 2020.
4. Kurz, L.; Taşdemir, E.; Bornkessel, D.; Noll, T.G.; Kappen, G.; Antreich, F.; Sgammini, M.; Meurer, M. An architecture for an embedded antenna-array digital GNSS receiver using subspace-based methods for spatial filtering. In Proceedings of the 2012 6th ESA Workshop on Satellite Navigation Technologies (Navitec 2012) & European Workshop on GNSS Signals and Signal Processing, Noordwijk, Netherlands, 5–7 December 2012; pp. 1–8. [CrossRef]
5. Amoroso, F. Adaptive A/D Converter to Suppress CW Interference in DSPN Spread-Spectrum Communications. *IEEE Trans. Commun.* **1983**, *31*, 1117–1123. [CrossRef]
6. Grabowski, J.; Hegarty, C. Characterization of L5 Receiver Performance Using Digital Pulse Blanking. In Proceedings of the 15th International Technical Meeting of the Satellite Division of The Institute of Navigation (ION GPS 2002), Portland, OR, USA, 24–27 September 2002; pp. 1630–1635.
7. Dabak, Ö.C.; Erdem, F.; Sönmez, T.; Alatan, L.; Koç, S.S. Interference suppression in a GPS receiver with 4 element array design and implementation of beamforming algorithms. In Proceedings of the IEEE/ION Position, Location and Navigation Symposium (PLANS), Savannah, GA, USA, 11–14 April 2016.
8. Curran, J.T.; Bavaro, M.; Fortuny-Guasch, J. Null-Steering Antennas. *GPS Word. GPS World Mag.* **2016**, *27*, 62–68.
9. Irteza, S. Compact Adaptive Planar Antenna Arrays for Robust Satellite Navigation Systems. Doctoral Thesis, TU Ilmenau, Ilmenau, Germany, 2016.
10. Meurer, M.; Konovaltsev, A.; Appel, M.; Cuntz, M. Direction-of-Arrival Assisted Sequential Spoofing Detection and Mitigation. In Proceedings of the 2016 International Technical Meeting of The Institute of Navigation, Monterey, CA, USA, 25–28 January 2016; pp. 181–192. [CrossRef]
11. Ibraheam, M. Design and Testing of Compact Dual-band Dual-polarized Robust Satellite Navigation Antenna Arrays. Doctoral Thesis, TU Ilmenau, Ilmenau, Germany, 5 June 2019.
12. Brachvogel, M.; Niestroj, M.; Zorn, S.; Meurer, M.; Hasnain, S.N.; Stephan, R.; Hein, M.A. A new array concept using spatially distributed subarrays for unambiguous GNSS interference mitigation in automotive applications. *Navigation* **2020**, *67*, 23–41. [CrossRef]
13. Hasnain, S.N.; Stephan, R.; Brachvogel, M.; Meurer, M.; Hein, M.A. Performance testing of a distributed automotive satellite navigation array with jammer in virtual environment. In Proceedings of the 15th European Conference on Antennas and Propagation (EuCAP), Dusseldorf, Germany, 22–26 March 2021; pp. 1–5. [CrossRef]
14. Ibraheam, M.; Bieske, B.; Blau, K.; Schäfer, E.; Jäger, A.; Butt, S.I.; Stephan, R.; Hein, M.A. Feasibility of dual-polarized antenna arrays for GNSS receivers at low elevations. In Proceedings of the 11th European Conference on Antennas and Propagation (EUCAP), Paris, France, 19–24 March 2017; pp. 857–861. [CrossRef]
15. Abracon, Datasheet Dual-band Antenna APAKM25075. Available online: https://abracon.com/datasheets/APAKM2507S-SGL5.pdf (accessed on 14 April 2023).
16. Times Microwave Systems, Current Innovation in Phase Stable Coaxial Cable Design. Microwave Product Digest. Available online: https://timesmicrowave.com/current-innovations-in-phase-stable-coaxial-cable-design/ (accessed on 11 April 2023).
17. Serna, E.P.; Thombre, S.; Valkama, M.; Lohan, S.; Syrjälä, V.; Hurskainen, H.; Nurmi, J. Local Oscillator Phase Noise Effects on GNSS Code Tracking. *InsideGNSS* **2010**, *5*, 52–62.
18. Brachvogel, M.; Niestroy, M.; Meurer, M. Full Wideband Calibration for an Array of Spatially Distributed Subarrays. In Proceedings of the IEEE/ION Position, Location and Navigation Symposium (PLANS), Monterey, CA, USA, 24–27 April 2023.

19. Zorn, S.; Siebert, C.; Niestroj, M.; Brachvogel, M.; Meurer, M. Accurate Position and Attitude Determination in a Severe Multipath Environment Using an Uncalibrated Multi-Antenna-System. In Proceedings of the 2020 IEEE/ION Position, Location and Navigation Symposium (PLANS), Portland, OR, USA, 20–23 April 2020; pp. 1247–1255. [CrossRef]
20. Brachvogel, M.; Niestroj, M.; Schoenbrod, S.; Meurer, M.; Hasnain, S.N.; Stephan, R.; Hein, M.A. Interference and Jammer Suppression of High Bandwidth Signals using an Array of Spatially Distributed Subarrays. In Proceedings of the 34th International Technical Meeting of the Satellite Division of The Institute of Navigation (ION GNSS+ 2021), St. Louis, MO, USA, 20–24 September 2021; pp. 3812–3823. [CrossRef]

engineering proceedings

MDPI

Proceeding Paper

Interference Detection, Localization, and Mitigation Capabilities of Controlled Reception Pattern Antenna for Aviation †

Annemarie van Zwol *, Jan-Joris van Es, Daniel Kappelle, Hein Zelle, Fennanda Doctor and Yuri Konter

Netherlands Aerospace Centre, Anthony Fokkerweg 2, 1059 CM Amsterdam, The Netherlands
* Correspondence: annemarie.van.zwol@nlr.nl
† Presented at the European Navigation Conference 2023, Noordwijk, The Netherlands, 31 May–2 June 2023.

Abstract: Global Navigation Satellite System (GNSS) interference poses an increasing threat for civil aviation, and the detection and mitigation of interferences can help to make the sector more robust. This paper focuses on the detection and mitigation capabilities of a software-based Controlled Reception Pattern Antenna (CRPA) as part of a wider study in which different detection and mitigation methods are tested and compared. The proposed CRPA uses eigenvalue decomposition to determine the weight vector and is combined with MUltiple SIgnal Classification (MUSIC) for detection purposes. Simulations are used to test the software CRPA for its robustness against different types of interference in static and dynamic scenarios. The test method and processing pipeline are described. Initial results show the CRPA algorithm under test is capable of detecting and mitigating different types of interferences, and mitigation can help a receiver to maintain a position velocity time (PVT) solution for higher levels of interference power.

Keywords: GNSS; CRPA; MUSIC; jamming; spoofing; detection; mitigation; aviation

Citation: van Zwol, A.; van Es, J.-J.; Kappelle, D.; Zelle, H.; Doctor, F.; Konter, Y. Interference Detection, Localization, and Mitigation Capabilities of Controlled Reception Pattern Antenna for Aviation. *Eng. Proc.* **2023**, *54*, 24. https://doi.org/10.3390/ENC2023-15452

Academic Editors: Tom Willems and Okko Bleeker

Published: 29 October 2023

1. Introduction

Civil aviation shows an increasing dependence on Global Navigation Satellite Systems (GNSS) for navigation, both en route as well as in critical operations such as takeoff and landing. The use of GNSS can improve the accuracy, integrity, and continuity of positioning, communication, and surveillance services. However, due to the low power of the signals, GNSS is also vulnerable to radio frequency interference (RFI). In our current society, the interference threat is of increasing relevance. In 2022, the European Union Aviation Safety Agency (EASA) issued a warning in which authorities and air operators were warned for the intensified occurrence of GNSS jamming and spoofing in proximity of the Ukraine conflict zone [1] and the International Telecommunication Union (ITU) issued a circular letter in which the jamming and spoofing threat for receivers onboard aircraft was addressed [2]. In March 2023, the International Federation of Air Line Pilots (IFALPA) issued a statement warning pilots in the Pacific for radio interference and GNSS jamming [3]. Although these indications are directly related to a conflict zone, it is to be expected that occurrences of harmful GNSS interference to civil aviation will increase in the future.

To increase aviation resilience against the threat of jamming and spoofing, the European Commission funded AIRING, a study on aviation resilience to GNSS frequency jamming and cyber threats [4]. As part of this study, multiple onboard interference detection and mitigation techniques were tested. Controlled Reception Pattern Antennas (CRPA) can actively adjust their gain pattern, thereby suppressing interferences. CRPAs were originally developed for use in the military domain, but have found their way to civil applications. They can also be used to increase GNSS interference robustness in civil aviation. In this paper, a variation of the MUltiple SIgnal Classification (MUSIC) algorithm is used to develop a virtual software-based 4-element CRPA antenna. The tests to assess the RFI detection, Angle of Arrival (AoA) estimation, and mitigation capabilities are described. The aim is to not only discuss the results obtained for this software-based CRPA configuration, but to also provide insight into the processing steps of this simulation-based test method and the possible applicability on a civil aircraft.

2. MUSIC

The antenna elements of a CRPA are connected to a central processing unit where the signals from the different antenna elements are weighted and linearly combined to form a single output signal. Effectively, the reception pattern of the combined antenna is modified by the weighing, creating flexible nulls and peaks in the reception pattern. The MUSIC algorithm is a subspace-based algorithm that can be used for direction finding of signals in a narrow frequency band [5,6]. For this project, a variation of the MUSIC algorithm is adapted to be the central processing unit of the CRPA under test.

MUSIC uses eigenvalue decomposition to determine the Angle of Arrival and the power of the interference signals. The CRPA additionally determines a weight vector orthogonal to the interference signals. To mitigate the interference, the obtained weight vector is multiplied with the signals received by the different antenna elements, thereby steering nulls with the appropriate depths in the combined radiation pattern in the direction of the interference.

The signal received on the different antenna elements is written to m by the n matrix X, where m is the number of antenna elements and n is the number of complex In-phase Quadrature (IQ) samples to be processed. The sample correlation matrix $\widehat{R_x}$ is defined as

$$\widehat{R_x} = \frac{1}{n}\mathbf{X}\mathbf{X}^* \tag{1}$$

where the * operator indicates the complex conjugate transpose of a matrix. Performing a singular value decomposition of the sample correlation matrix results in a diagonal matrix Σ with the eigenvalues on the diagonal, an m by m unitary matrix U of which the column vectors span the orthonormal basis of the signal space and the noise space, and the complex conjugate transpose V^* of the m by m unitary matrix V.

$$\widehat{R_x} = U\Sigma V^* \tag{2}$$

The eigenvalues in Σ and its respective eigenvectors in U are sorted based on decreasing eigenvalues. For p interference signals, the signal space U_s that contains the interference signals is defined by the first p eigenvectors and the noise space U_n is defined by eigenvectors $p+1$ to m. To determine whether interference is present, the eigenvalues of the signal space are compared to a detection threshold. For GNSS, the satellite signals can have any AoA above the horizon; therefore, the desired hemispherical antenna radiation pattern with upward focus is wide. To maintain an optimum gain pattern in the case of no interference, the complex gain vector $\mathbf{e}$ is chosen to only use the gain of a single central antenna element and therefore reduces to a unit vector with a single non-zero value (in a generalized form, $\mathbf{e}$ is defined as $[1, 0, \ldots, 0]^T$). The weight vector $\mathbf{w}$ is determined by projecting the complex gain vector to the noise space and scaling the projection to maintain unitary amplitude in the original direction of $\mathbf{e}$.

$$\mathbf{w} = \frac{U_n U_n^* \mathbf{e}}{\mathbf{e}^* U_n U_n^* \mathbf{e}} \tag{3}$$

By using this method for the weight vector determination, the gain of the central element remains constant and the complex gain of the other antenna elements is adjusted to place a null in the direction of the incoming interference signal. Apart from the wide antenna gain pattern that is maintained in this configuration, an extra advantage is that no inverse matrices are required. Instead, the noise space derived in AoA detection is directly used to update the CRPA weights.

The steering vector **a** for a specified Angle of Arrival can be calculated from the spatial distribution of the antenna elements. By comparison of the response of the cost function

$$P = \frac{\mathbf{a}^* U_n U_n^* \mathbf{a}}{\mathbf{a}^* \mathbf{a}} \tag{4}$$

for different steering vectors, and for determining for which steering vector the cost function has its minimum, both the azimuth and elevation of the incoming interference signal can be determined. Instead, the noise space derived in AoA detection is directly used to update the CRPA weights.

Implementation

In the MUSIC implementation developed in this project, the number of potential interference sources is fixed to one. Only one interference signal can be detected and mitigated. To improve the computational efficiency of the algorithm, the AoA estimation is split into two steps. First, a coarse estimate is obtained by comparing the steering vectors of 50 sectors. The sector that contains the interference signal is then subjected to a refinement over 10 sectors. This means that the final resolution in azimuth is 0.72° and the resolution in elevation is 0.36°. The use of a coarse step and a refinement step reduces the number of evaluation steps from 450 to 60 per AoA evaluation and therefore significantly improves the calculation time to reach the same resolution. The MUSIC algorithm is configured to use 10^5 samples for every iteration. The layout of the antenna elements is given in Table 1.

Table 1. CRPA layout with respect to CRPA center.

Antenna Element	X [mm]	Y [mm]	Z [mm]
1	0	0	25
2	80	0	0
3	−40	69.282	0
4	−40	−69.282	0

3. Test Architecture

To test the software CRPA antenna for the set of scenarios, a processing pipeline is developed. The processing pipeline consists of five steps.

1. IQ data simulation
2. IQ data combination
3. CRPA processing via MUSIC implementation
4. GNSS-SDR processing
5. Combine output KPI of MUSIC and GNSS-SDR

Since the test of this software CRPA antenna is part of a larger test campaign concerning multiple consortium partners that each test a different detection and/or mitigation technique, the simulation of the input IQ files (step 1) is centralized. The input data are generated at a baseband frequency with a sample rate of 25 Msamples/s. The IQ files consist of interleaved I shorts and Q shorts. For each frequency band (upper L-band and lower L-band), separate clean signal files and interference files are simulated and distributed to the respective consortium partners. The CRPA under test consists of four antenna elements, to account for the spatial distribution of the antenna elements, the input IQ files are simulated for each antenna element individually.

The block diagram of the processing pipeline is displayed in Figure 1. This pipeline accounts for the processing steps after data distribution (steps 2 to 5). To run the processing pipeline, the settings for the different steps are summarized in a *.csv* file. A script is written to automatically guide the files through the processing pipeline.

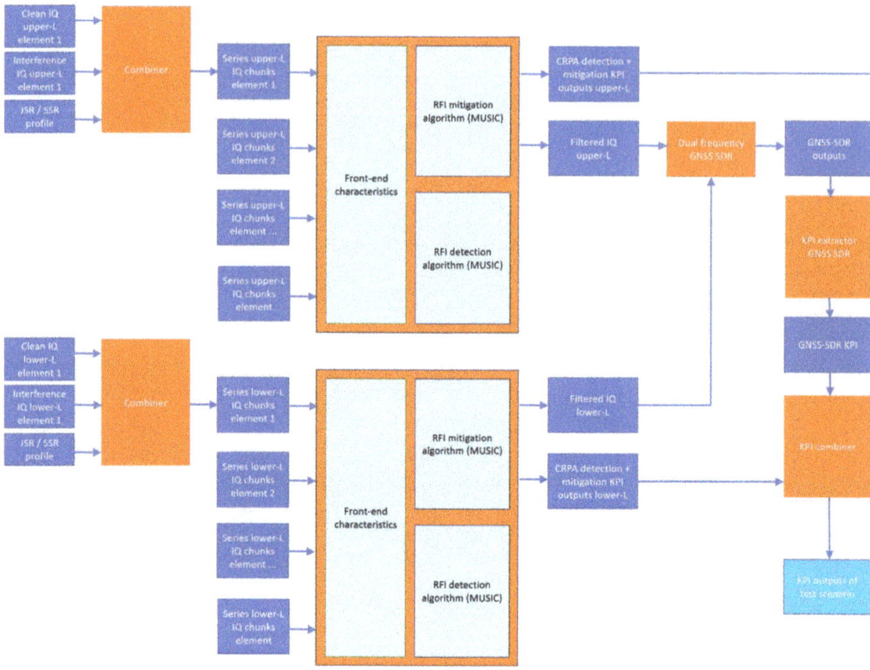

Figure 1. In this block diagram, the processing pipeline is visualized. The dark blue boxes represent a data file, the orange boxes represent a processing step, and the bright blue box represents the final output of the test.

The clean nominal signal, the interference signal, and a predefined jamming to the signal ratio (JSR) or spoofing to the signal ratio (SSR) profile over time are combined by each individual partner to a series of smaller-sized chunks. In addition, a white Gaussian noise with a 0 dB signal-to-noise ratio (SNR) in a 4 MHz bandwidth is added to the signal.

The application of mitigation is indicated in the general configuration file used by the script defining the test scenarios. Using the input IQ chunk series of the four antenna elements as input, the CRPA implementation determines whether interference was present. When interference is detected, the AoA estimation and mitigation are performed.

As the output of the CRPA is a Radio Frequency (RF) signal, the output still needs to be processed via a GNSS receiver. In this project, we have used the GNSS-SDR software receiver. GNSS-SDR is an open-source GNSS software-defined receiver platform capable of processing multi-frequency multi-constellation input signals. GNSS-SDR is configured to operate in multi-frequency, multi-constellation mode. The configuration includes 10 channel for each of the compatible signal types: GPS L1 C/A, Galileo E1 B/C, GPS L5, and Galileo E5a. Although the goal of this test is not to design or optimize a GNSS receiver architecture, the receiver settings must be agreed on between the different AIRING partners to be able to compare the detection and mitigation techniques under test.

4. Test Scenarios

In the AIRING project, a number of test scenarios is defined with the aim to cover a range of different types of attacks. The CRPA implementation is tested for a subset of these test scenarios, including scenarios with different types of dynamics and different types of interferences. Each scenario runs for 10 min. In the static scenarios, the true azimuth of the interference is 348.2° and the true elevation of the interference is 0.0°. The dynamic scenarios are designed to simulate an in-flight situation where the receiver has a constant velocity of approximately 800 km/h. The true azimuth and elevation of the static interference source with respect to the dynamic receiver are plotted as a function of time in Figure 2. To be able to determine the maximum JSR or SSR for which the detection and/or mitigation technique is effective, the respective JSR and SSR are gradually increased over time. In Table 2, an overview of the test scenarios is given. The JSR profiles for the static and dynamic jamming scenarios are displayed in Figure 3.

To run one scenario consisting of both upper L-band and lower L-band signals, four sets of two clean and two interference IQ files are needed, with each file respectively being 58 Gb. The processing of a single scenario on a four-core machine takes approximately 75 h. The file size and scenario duration put a considerable strain on the available memory, storage capacity, and efficiency of the processing pipeline.

Table 2. Summary of the test scenarios.

Dynamics	Interference Type	Interference
Static	Clean	
Static	Jammer	Continuous Wave, Chirp, Pulsed, Noise-Like
Dynamic	Jammer	Chirp
Dynamic	Spoofer	Reradiator, Untargeted (1 GPS L1 C/A signal)

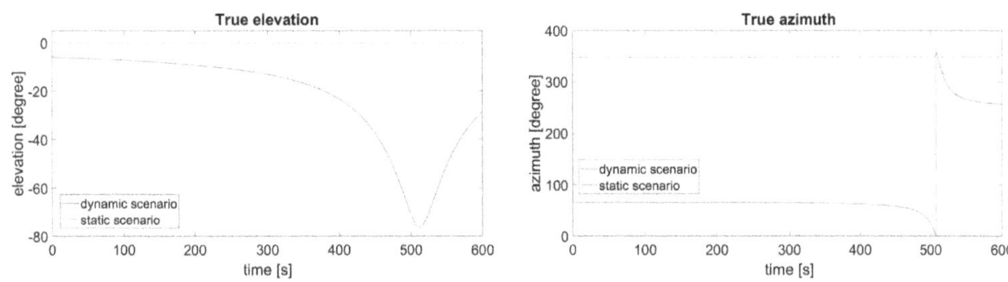

Figure 2. The true (**left**) elevation and (**right**) azimuth of the interference source with respect to the receiver.

5. Test Results

For the static jamming scenarios, the JSR gradually increases from 5 to 45 dB. An inspection of Figure 3 shows that for the static scenario, saturation seems to occur at a JSR of 35.77 dB and higher. This means that after a simulation time of 390 s, the input signals start to become affected by saturation effects of the I and Q values. For the static jamming scenarios including interference (the dirty run and the mitigated run), they can become unreliable upon saturation.

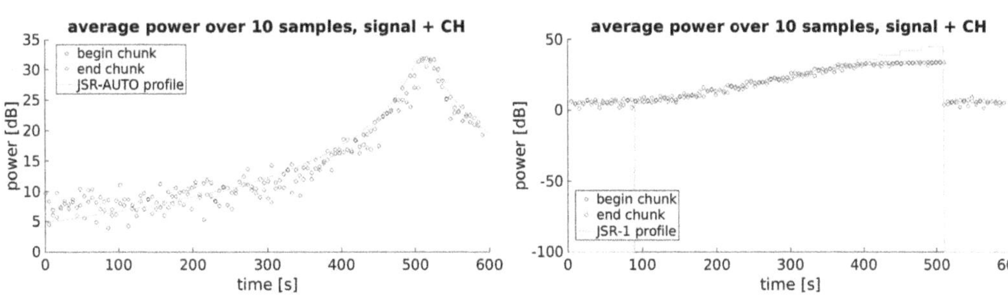

Figure 3. The theoretical JSR profile and the power in the IQ data as a function of time for the (**left**) dynamic jamming scenarios and (**right**) static jamming scenarios.

The chirped jamming signal, continuous wave jamming signal, and noise-like jamming signal were detected via the CRPA implementation. The three jamming signal types were all first detected at the moment the RFI power surpassed the detection threshold, which was at 11.15 dB JSR. The pulsed interference (12.5 µs pulse duration, 1% duty cycle) was not detected. For all static cases, the first interference detection in the lower band and the upper band occurred simultaneously and mitigation was applied directly when the detection flag was raised. To illustrate the results obtained from the performed tests, the results for the static chirp are taken as an example. The AoA estimates and power estimate for the interference signal obtained for the static chirp scenario are displayed in Figure 4. The azimuth estimate is accurate and the estimate obtained from the lower L-band and the upper-L band are in agreement. The elevation estimate shows larger variations over time, and a larger difference between the estimate obtained from the lower L-band signals and upper L-band signals. It is expected that the difference in accuracy between the azimuth and elevation estimates can be explained using the geometry of the CRPA with respect to the test scenario.

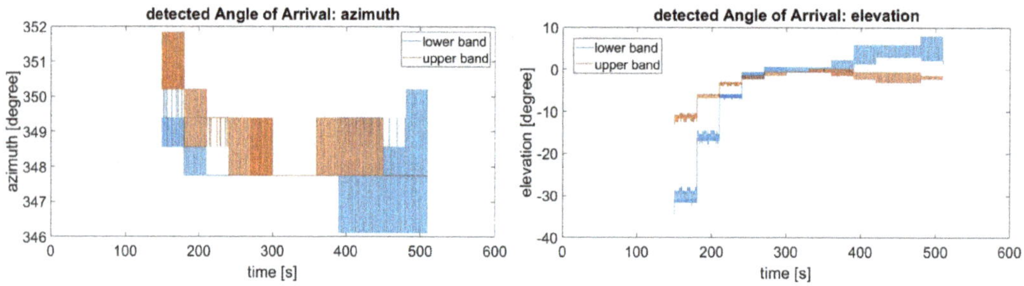

Figure 4. The AoA estimates over time for (**left**) the azimuth and (**right**) the elevation for the static chirp scenario.

Comparison of the GNSS-SDR output of the dirty run on the interfered signal, and the mitigated run provides insight into the effectiveness of the applied mitigation. The difference in carrier to noise (CN0) between the dirty run and the mitigated run shows that in general, the mitigation helps to keep the tracks locked to a signal up to a higher jamming power. As an example, the carrier-to-noise plots of the static chirp scenario are displayed in Figure 5. The CN0 of some of the tracked satellites in the mitigated signal shows an increase with respect to the CN0 of the tracked satellites in the clean signal. To apply mitigation, the CRPA gain pattern is adjusted. This can result in an increased gain in certain directions and can therefore increase the CN0 of satellite signals in certain directions. The effectiveness of the mitigation does depend on the type of interference. Mitigation seems to be most beneficial in the case of continuous wave jamming and chirped jamming. Pulsed jamming was not effective in disturbing the signal and was neither detected nor mitigated. Noise-like interference does not seem to be suppressed by mitigation.

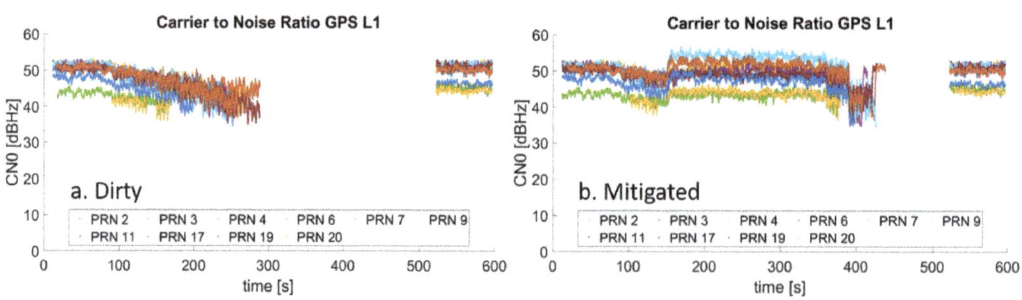

Figure 5. Carrier-to-noise ratios of the GPS L1 C/A satellite signals in track as a function of time for the static chirp scenario for (**a**) the "dirty" run, including interference, and (**b**) the run where mitigation was applied.

During the test phase, it was found that there is a significant offset (order of 30 m) between the position solution calculated from the clean lower L-band signals and the true position. This offset is not present for the upper L-band signals. The expected cause is a synchronization issue in the signal simulation process. To be able to interpret the PVT output of the multi-band receiver, the deviations of the PVT solution with respect to the true position of the clean signal is provided in Figure 6a. The deviations of the PVT solution with respect to the true position obtained from the interfered signal and the mitigated signal are displayed in Figure 6b,c. The prolonged period for which a PVT solution could be obtained in Figure 6c indicates that the mitigation for this scenario was effective.

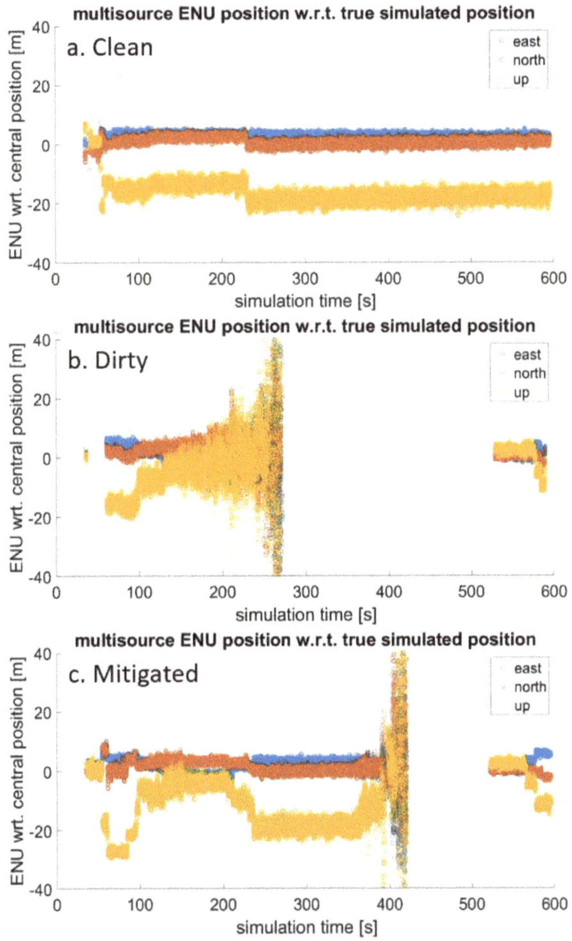

Figure 6. The deviation of the obtained PVT solution with respect to the true location in the east north and up direction for the static chirp scenario for (**a**) the clean run without interference, (**b**) the "dirty" run including interference, and (**c**) the run where mitigation was applied.

5.1. Dynamic Receiver

Apart from the static jamming scenarios that have been tested, the effect of dynamics on the effectiveness of this CRPA implementation can be studied using a comparison between the static chirp scenario and the dynamic chirp scenario in which the receiver had a constant velocity of 800 km/h with respect to the transmitter. It was observed that the CRPA detected and mitigated the interference on the lower L-band at a JSR of 9.62 dB and on the upper L-band at a JSR of 10.13 dB and applied mitigation when interference was detected. Mitigation of the interference allowed the satellite signals to continued to be tracked, but the robustness of the position solution for this particular scenario did not improve with respect to the interfered run. The mitigation does seem to be somewhat effective, but the effect of the high dynamics of this scenario should be studied in further detail.

5.2. Spoofing Scenarios

The original set of test scenarios included a repeater (meaconing) and a single satellite unsynchronized spoofing attack. In the test phase, it was observed that for the "dirty" signal, these scenarios were unable to spoof GNSS-SDR. Instead, the spoofing signal had a jamming-like effect. Therefore, no conclusions on the mitigation capability in the case of a successful spoofing attack can be made. It was observed that the MUSIC detected and mitigated the spoofing interference at an SSR of approximately 10.26 dB and applied mitigation when interference was detected. After mitigation was applied, the satellite signals continued to be tracked and the position solution was maintained for the entire duration of the scenario. Although a sophisticated spoofing attack could potentially spoof a receiver by using low SSR, the performed tests show that for higher-power spoofing attacks, the current CRPA implementation shows promising results. As CRPA is a precorrelator detection and mitigation technique, the detection and mitigation capabilities are expected to be independent of the effectiveness of the spoofing attack.

6. Discussion

Promising results confirm that the use of a CRPA antenna is an effective detection and mitigation method against most types of GNSS interference. There are, however, some topics that require extra attention via further study or adjustments in the test strategy.

Some of the tests were performed on a saturated input signal. Therefore, no reliable results on the maximum power that can effectively be mitigated are obtained. To determine the maximum jamming power for which the implementation under test is effective, the tests should be repeated with an unsaturated input signal.

A second issue encountered in the test campaign is the inconsistency between the used lower L-band and upper L-band signals. It is recommended to solve this issue for future tests, or to study the mitigation capabilities of the mitigation technique under test separately for the lower L-band and upper L-band signals (use a single-frequency receiver implementation for analysis).

The dynamic scenarios are particularly interesting for the aviation use case in which a CRPA is used as the aircraft GNSS antenna. The current implementation shows promising results where the satellite signals continued to be in track, but more research efforts should be given to the effect of high velocity (order of 800 km/h) on the mitigation performance.

In the selection of test scenarios for which the CRPA was tested, spoofing scenarios were included. Effective spoofing of a highly dynamic receiver is complicated and the spoofing scenarios were unsuccessful in spoofing the receiver. Although the results are promising as the CRPA is capable of mitigating the interference, tests with successful spoofing attacks should be performed in order to confirm the capability of the CRPA to mitigate such an attack.

Finally, the efficiency of the processing pipeline is limited by the file sizes of the required input data and tests take a significant amount of time. For future test campaigns, it is advised to be critical during the test scenario definition and to minimize the sample rate and scenario duration.

7. Conclusions

In this paper, the effectiveness of a software-based CRPA based on signal space decomposition for the detection and mitigation of different types of GNSS interference for civil aviation has been investigated. The algorithm under test is a modified version of the MUSIC algorithm, in which the gain of the central antenna element is kept constant, while the complex gains of the remaining three antenna elements are adjusted to place a null in the direction of the incoming interference signal. The proposed CRPA algorithm shows good results but has yet to be compared to alternative null-steering CRPA techniques, such as minimum variance distortionless response (MVDR) or linearly constrained minimum variance (LCMV) beamforming.

Initial results of the interference detection and mitigation tests show that this implementation is capable of detecting and mitigating jamming. Further study is required in order to make the detection and mitigation capabilities of the CRPA under test more robust. Although there is no definitive proof for the effectiveness of spoofing detection and mitigation, it is expected that the CRPA has a beneficial effect in the case of spoofing, as it agnostically detects and attempts to suppress the spoofing signal just like a jamming interference source. Further test with more complex spoofing scenarios should be performed in order to confirm this.

This study was conducted, specifically aiming at the detection and mitigation of GNSS interference in aviation. CRPAs are currently not used in the civil aviation sector, primarily due to their complexity, form factor, and export control limitations on certain specific technologies. There seems to be potential for further developments of CRPA for civil aviation, either as a standalone solution or in combination with other detection and mitigation measures investigated as part of the AIRING study [7].

The software and simulation defined test method used in this project has proven to be a valuable platform for interference tests. It offers a highly controllable test environment, which can be used to systematically perform repetitive tests. Furthermore, it offers the opportunity to test interference detection and mitigation techniques for interferences not available for tests with Signal in Space (SiS). The flexibility in-test scenarios and CRPA setup offer an accessible research tool. However, it has to be kept in mind that the data volumes and processing times can form a considerable constraint on large test campaigns.

Author Contributions: Research, A.v.Z., J.-J.v.E., D.K., H.Z., F.D. and Y.K.; Paper writing draft, A.v.Z.; Review, H.Z. and Y.K. All authors have read and agreed to the published version of the manuscript.

Funding: This research was funded by the European Commission under contract DEFIS/2020/OP0006.

Institutional Review Board Statement: Not applicable.

Informed Consent Statement: Not applicable.

Data Availability Statement: Data are contained within the article. For access to the complete datasets please contact the authors.

Acknowledgments: The authors thank the project partners GMV, NLS, GMV-NSL, ENAIRE, and AXENTEK for their technical support and contributions in the AIRING project.

Conflicts of Interest: The authors declare no conflict of interest. The funder had no role in the design of the study; in the collection, analyses, or interpretation of data; in the writing of the manuscript; or in the decision to publish the results.

References

1. European Union Aviation Safety Agency. Global Navigation Satellite System Outage Leading to Navigation/Surveillance Degradation. *Saf. Inf. Bull.* **2022**, *2*, 2022-02R1.
2. International Telecommunication Union. Prevention of harmful interference to Radio Navigation Satellite Service Receivers in the 1559–1610 MHz frequency band. *Circ. Lett.* **2022**, *488*, CR/488.
3. International Federation of Air Line Pilots. Communication Interference by Military Warships in the Pacific Region. 2023. Available online: https://www.ifalpa.org/media/3905/23sab03-communication-interference-by-military-warships-in-the-pacific-region.pdf (accessed on 1 October 2023).
4. European Commission. Aviation Resilience to GNSS Frequency Jamming and Cyber Threats. In *Invitation to Tender (Ares 4341987)*; DEFIS/2020/OP/0006; European Commission: Brussel, Belgium, 2020.
5. Schmidt, R. Multiple emitter location and signal parameter estimation. *IEEE Trans. Antennas Propag.* **1986**, *34*, 276–280. [CrossRef]

6. Zhang, J.; Cui, X.; Xu, H.; Lu, M. A two-stage interference suppression scheme based on antenna array for GNSS jamming and spoofing. *Sensors* **2019**, *19*, 3870. [CrossRef] [PubMed]
7. GMV; NLR; NLS; GMV-NSL; ENAIRE; AXENTEK. *Aviation Resilience to GNSS Frequency Jamming and Cyber Threats, Final Report*; EC Project Report AIRING-GMV-FR; European Union: Brussel, Belgium, 2023.

engineering proceedings

MDPI

Proceeding Paper

GNSS Interference Monitoring and Detection (GIMAD) System †

Enric Obiols-Bernaus [1,*], Lorena Tortajada-Ropero [1], Àngel Creus-Blanch [1], Antonio González-Novell [1], Fran Fabra [2,3] and Gonzalo Seco-Granados [2,3]

1 Satellite Navigation Department, INDRA Sistemas, 08018 Barcelona, Spain; ltortajada@indra.es (L.T.-R.); acreus@indra.es (À.C.-B.); agnovell@indra.es (A.G.-N.)
2 Institut d'Estudis Espacial de Catalunya (IEEC), 08034 Barcelona, Spain; fabra@ieec.cat (F.F.); gonzalo.seco@uab.cat (G.S.-G.)
3 Department of Telecommunications and Systems Engineering, IEEC-CERES, Universitat Autònoma de Barcelona (UAB), Bellaterra, 08193 Barcelona, Spain
* Correspondence: eobiols@indra.es
† Presented at the European Navigation Conference 2023, Noordwijk, The Netherlands, 31 May–2 June 2023.

Abstract: In the last few years, gradually more systems have relied on the Global Navigation Satellite System (GNSS) for their correct functioning. These systems include safety-critical applications such as airports or emergency services. Given the wide number of GNSS applications and the current availability of affordable and easily configurable Software-Defined Radio (SDR) devices, GNSS has become the target of numerous Radio Frequency Interference (RFI) attacks. Thus, RFI has become a real threat for GNSS and, hence, for those systems relying on it. With the purpose of detecting, characterizing, and localizing RFI not only in GNSS frequency bands but also in other daily-used frequency bands, a GNSS Interference Monitoring and Detection (GIMAD) prototype has been developed, with special emphasis on its deployment in safety-critical environments such as airports. GIMAD contemplates European Geostationary Navigation Overlay Service (EGNOS) V3 and International Civil Aviation Organization (ICAO) Ground Based Augmentation System (GBAS) RFI masks covering both in-band and out-band RFIs. In addition, GIMAD was tested in a real-field scenario.

Keywords: GNSS; GPS; Galileo; RFI; Jamming; RFI detection; RFI characterization; AoA; ICAO; EGNOS; GBAS

Citation: Obiols-Bernaus, E.; Tortajada-Ropero, L.; Creus-Blanch, À.; González-Novell, A.; Fabra, F.; Seco-Granados, G. GNSS Interference Monitoring and Detection (GIMAD) System. *Eng. Proc.* **2023**, *54*, 25. https://doi.org/10.3390/ENC2023-15460

Academic Editors: Tom Willems and Okko Bleeker

Published: 29 October 2023

1. Introduction

Radio Frequency Interference (RFI) jeopardizes satellite-based navigation applications. This is especially relevant in safety-critical activities where aspects such as availability, integrity, and/or continuity are of outmost importance. In order to mitigate and face an RFI in critical infrastructures/areas, the design of GNSS Interference Monitoring and Detection (GIMAD) was already presented in [1]. Now, this design has been implemented into a prototype.

The use case from which the system requirements are derived is presented in Section 2. Considering the above-mentioned requirements, a GIMAD prototype description is presented in Section 3, including the different parts of the architecture. Algorithms implemented in the design stage have been verified, and the key verification results are shown in Section 4. Finally, Section 5 contains the main conclusions.

2. Use Case and System Requirements

Airports are critical infrastructures where satellite-based navigation applications are needed to be particularly protected. RFI detection and monitoring campaigns have already been conducted around several airports around the world. For example, at Newark's airport in the USA, the Ground Based Augmentation System (GBAS) was affected due to the presence of jammers onboard vehicles passing through the highway near the airport [2]. The jammer device made the GBAS reference receiver lose track of GPS satellites, thus impacting GBAS operations. As a result, the Spanish Air Navigation Service Provider (ENAIRE) has brought out the risks that such interferences may cause and the real need for a monitoring system for early detection of RFIs.

In order to protect GNSS availability and thus airport operations, International Civil Aviation Organization (ICAO) RFI masks have been devised in order to provide the maximum recommended interference levels [3]. For a similar purpose, there are also European Geostationary Navigation Overlay Service (EGNOS) v3 masks. However, RFI monitoring campaigns have demonstrated that these masks are not always fulfilled. For example, infractions of such RFI masks have been observed at Frankfurt airport in Germany [4] including violations of the ICAO SARPs CW interference thresholds.

For that reason, both ICAO and EGNOS RFI masks have been adopted in the design of the GIMAD system, covering the frequency range from 800 to 1900 MHz.

Considering the outcomes of RFI detection campaigns at international airports, an operational use case has been devised by the Spanish ANSP (ENAIRE), as illustrated in Figure 1 shows that when a jammer is detected by any of the monitoring stations in the monitoring network, the station shall send an alarm message to the central processor. In the case in which the power of the transmitter is high enough to be detected by more than one monitoring station, the central processor shall provide the accurate position of the interference source. In parallel, the central processor shall also trigger an alarm inside the ATC HMI (Air Traffic Control Human Machine Interface) warning about the presence of an undesired interference. In some cases, the Air Traffic Control Officers (ATCOs) can send warnings to the pilots reporting the unreliability of GNSS signals. As a consequence of alarms, the ANSP as well as national authorities are informed of the events with the objective of localizing and dismantling the interference source.

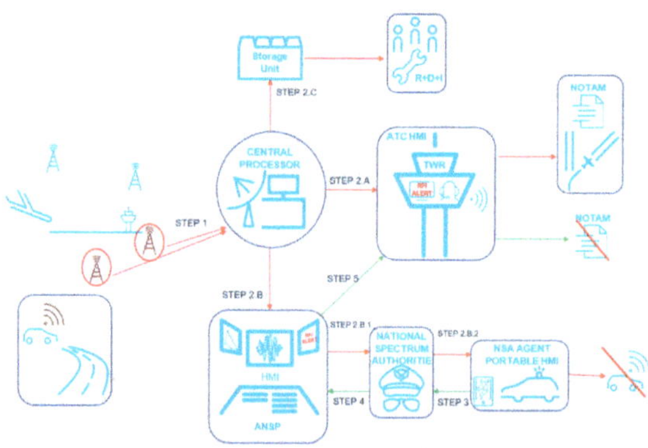

Figure 1. Operation use case (ENAIRE).

3. GIMAD Prototype Description

Following the GIMAD design introduced in [1], a GIMAD prototype has been implemented and tested.

The main components of the architecture are represented in Figure 2 and consist of the following.

1. **Antenna set**, integrating three different antenna types for different purposes:
    (a) Two GNSS antennas receiving in-band GNSS signals. One antenna for RFI detection purposes and another one for GNSS clock steering with a narrow-band pass filter.
    (b) Wide-band Fixed Reception Pattern Antenna (FRPA), whose main mission is detecting out-of-band interferences.

(c) Two Controlled Reception Pattern Antennas (CRPAs), used to determine the Angle of Arrival (AoA) of the RFI. One antenna set is used for in-band and the other for out-band RFI direction of arrival (DoA) determination. Each CRPA is made up of four individual receiving elements, allowing the possibility of determining the direction of up to three interferences.

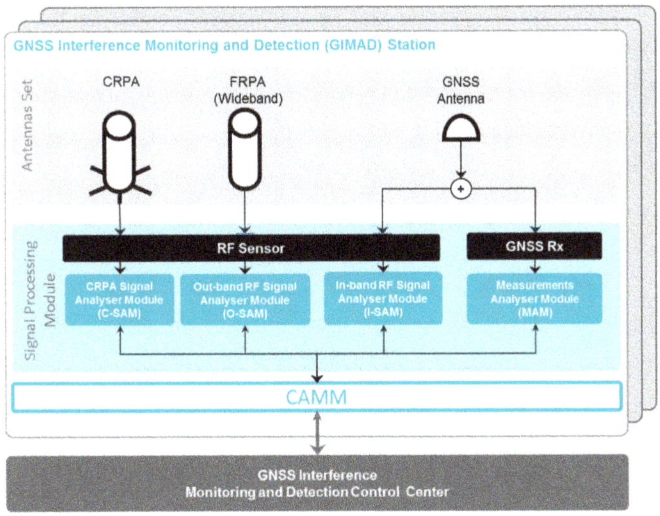

Figure 2. GIMAD Station architecture.

The GIMAD antenna set has been deployed at INDRA's Barcelona 22@ roof headquarters, as shown in Figure 3.

Figure 3. GIMAD antenna set installed at the INDRA's Barcelona 22@ building.

2. **Signal Processing Module** (SPM), with specific techniques for identification, characterization, and localization of RFI. It is composed of:
    (a) Measurement Analyzer Module (MAM). Post-correlator block that analyzes the observables of the GNSS receiver. It provides extra information about constellations/bands/satellite status.

   (b) In-band RF Signal Analyzer Module (I-SAM). The RF output of the GNSS antenna is redirected to this module, which digitalizes the received signal and applies pre-correlation algorithms for RFI detection, characterization, and localization to the baseband IQ samples.
   (c) Out-band RF Signal Analyzer Module (O-SAM). It is analogous to I-SAM but focuses on out-band processing.
   (d) CRPA Signal Analyzer Module (C-SAM). It provides a preliminary determination of the AoA RFI source.

3. **Control, Alert, and Monitoring Module (CAMM)**, which consolidates the SPM outputs, saves data for further analysis, generates alerts, and performs monitoring and control of the stations interfacing with the control center.
4. **GIMAD Control Center (GIMAD CC)**, which aims to monitor the status of the different GIMAD stations, manage the network, and provide a more accurate localization of the RFI source. An example of the GUI interface of the GIMAD CC is shown in Figure 4.

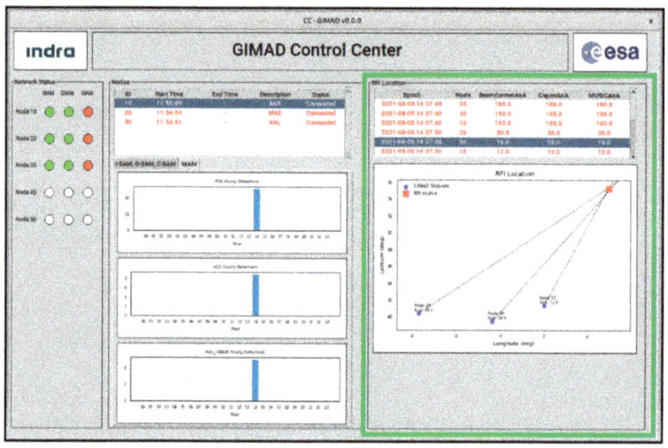

Figure 4. GIMAD CC example.

3.1. Signal Processing Module (SPM)

Three main operational blocks run under the SPM, which are named after the three main functionalities that they pursue: Detection, Characterization and Localization. Figure 5 summarizes the SPM modules in a block diagram.

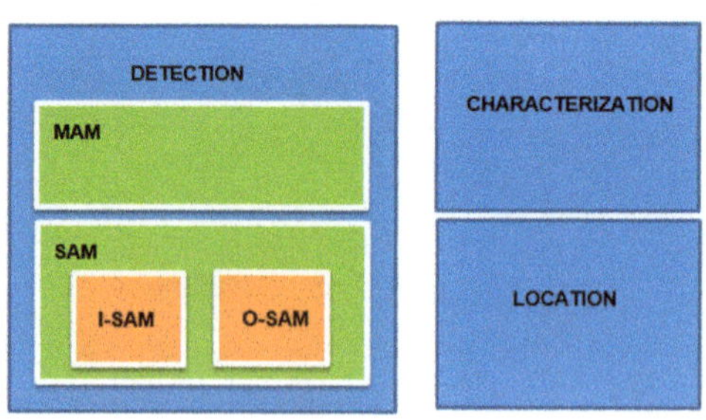

Figure 5. SPM modules.

The selected algorithms running behind them are described next.

3.1.1. Detection Modules/Algorithms

I-SAM and O-SAM modules from the Detection block in Figure 5 will assess the presence or not of RFI signals, performing a time-frequency characterization when a detection is confirmed.

The selected algorithms for interference detection in I-SAM/O-SAM are:

- Excess Kurtosis;
- Normalized sixth cumulant;
- Power level;
- Excess Kurtosis Fast Fourier Transform (FFT);
- Sixth cumulant FFT;
- Power Spectral Density (PSD) Peak Power.

If an interference alert is provided by any of the previous algorithms, the time-frequency characterization of the collected signal is performed.

Further details on algorithms and decision keys for selection can be found in [1].

Complementary to previous detection modules, the MAM module will be in charge of monitoring the GNSS receiver post-correlation observables for both detecting the presence of interference in the GNSS band and assessing the impact of the interference signals in the GNSS receiver. This is done by constantly comparing the current time evolution of each monitoring sudden changes in the measured observables.

The selected parameters for RFI detection and impact assessment in MAM are:

- Automatic Gain Control (AGC) level variations;
- Carrier-to-Noise Ratio (C/N0) variations;
- Pseudorange deviations;
- Carrier phase deviations;
- Position accuracy.

3.1.2. Characterization Modules/Algorithms

The characterization of the time–frequency of the monitored signal can be undertaken by means of (i) a spectrogram, (ii) the Wigner–Ville distribution, and (iii) a cyclostationary spectrum.

Further details on algorithms and decision keys for selection can be found in [1].

3.1.3. Localization Modules/Algorithms

C-SAM will assess the RFI AoA by performing a statistical analysis of the incoming signal at the CRPA antenna. The C-SAM concept is based on antenna array processing; thus, it combines the phase offsets of the received signal for each CRPA radiating element in order to estimate the AoA.

In order to accurately measure these phase differences and consequently enhance the AoA accuracy, the CRPA needs to be properly calibrated.

The selected algorithms for performing AoA estimations are (i) Spatial Filtering (Beamformer), (ii) MUltiple SIgnal Classification (MUSIC), and (iii) CAPON beamformer.

MUSIC is the default algorithm, while Beamformer and CAPON are kept as secondary methods, mainly because of the typically higher angle resolution of MUSIC with respect to the other two algorithms, although MUSIC is more computationally demanding.

4. Results

The GIMAD system was tested by capturing real GNSS signals and combining them with a generated RFI with an HP Agile signal generator. Testing included both injected (I-SAM, O-SAM, and MAM) and radiated (C-SAM only) RFI for a GIMAD station. The full capabilities of the system will be available after deploying more stations. In the following subsections, the key results for I-SAM, O-SAM, MAM, and C-SAM modules are presented hereafter.

4.1. I-SAM Test Results

A real GNSS signal was combined with an RFI signal generated by a signal generator. Then, this signal was injected into a National Instruments USRP-2945 where the signal was digitalized. IQ samples were sent to the SPM, where detection algorithms run. As expected, the probability of detecting RFI is higher when the Interference-to-Noise Ratio (INR) is higher. The GIMAD system is capable of detecting a pulsed interference RFI of −75 dBm with a duty cycle of 14%, a pulse duration of 1 ms, and a pulse repetition of 7 ms, as shown in Figure 6.

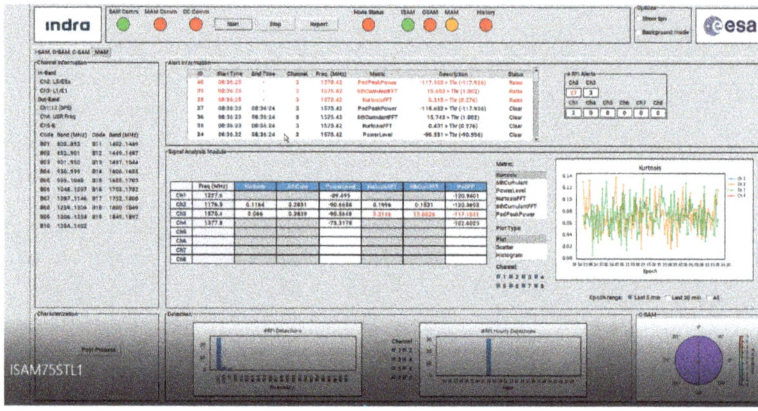

Figure 6. GIMAD CAMM (ISAM) detecting a pulsed RFI on L1.

Figure 6 shows the techniques Excess Kurtosis FFT, 6th Cumulant FFT, and PSD Peak Power in red, which trigger an alarm. Other techniques (such as Power Level) cannot detect the RFI. This result is in line with the Receiver Operating Characteristic (ROC) curve analysis previously simulated, where algorithms working with signals in the frequency domain achieve a higher probability of detection (pd) at the same pfa (probability of a false alarm).

4.2. MAM Test Results

A real GNSS signal was combined with an RFI signal generated by a signal generator. Then, the combined signal was injected into a GNSS receiver. GNSS observables extracted from the receiver were sent to the SPM, where algorithms based on detecting RFI from GNSS observables were applied. The GIMAD system can detect single-tone interference centered at L1 (1575.42 MHz) with a −40 dBm.

The GNSS data received from a satellite varies depending on its trajectory. Taking this into account, the MAM algorithm is conceived as an adaptive algorithm in the sense that it adapts to the evolutionary changes of the C/N_0, phase, and pseudorange of the satellites. Figure 7 below shows the RFI detection through the abnormal deviations of the AGC and the phase. Similar results for L5 have been obtained.

4.3. O-SAM Test Results

A competitive result in O-SAM where a challenging pulsed interference signal with a 1% duty cycle and pulse duration of 125 µs detected by GIMAD is shown in Figure 8 below.

Again, the pulsed RFI combined with the received real signal from the wide-band antenna in the B02 band (covering 50 MHz, from 850 to 900 MHz) has been injected using a combiner into the USRP NI-2945.

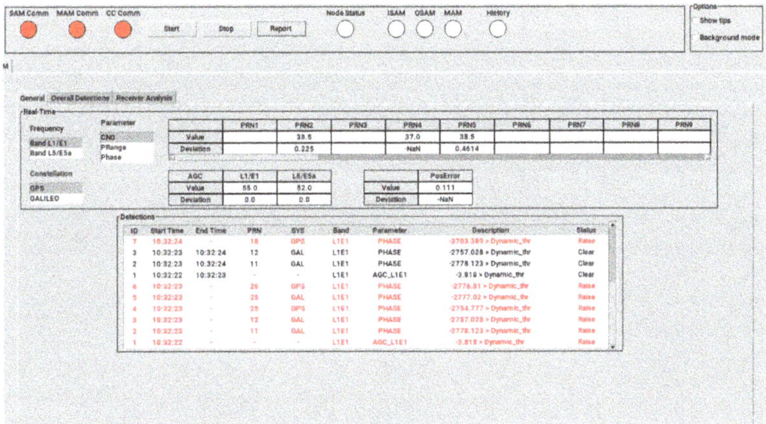

Figure 7. GIMAD CAMM (MAM) real-time detection on L1.

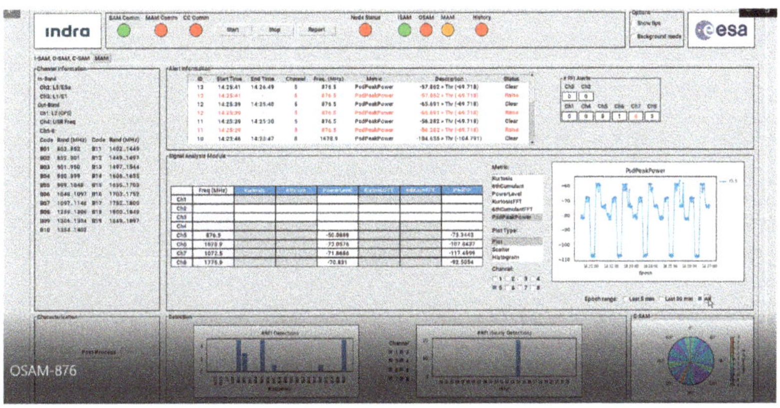

Figure 8. GIMAD CAMM (O-SAM) real-time detection at 876 MHz.

4.4. C-SAM Test Results

The out-band CRPA and the in-band CRPA of the GIMAD antenna set underwent calibration in an anechoic chamber, as is illustrated in Figure 9. During this calibration process, angular calibration parameters (relative amplitude and relative phase) were obtained for each carrier frequency and certain array orientations. The obtained calibration corrections had to be applied for the estimation of the AoA at the carrier frequency under analysis.

Figure 9. AOA calibration in an anechoic chamber.

In order to obtain the calibration parameters, the CRPAs were rotated with a moving platform rather than a fixed transmitting antenna.

In a second round, once the calibration parameters were already computed, the CSAM algorithms were tested first in the anechoic chamber.

Two cases are presented: the Out-of-Band (OOB) RFI AoA estimation and the In-Band RFI AoA estimation.

In order to address the OOB AoA estimation, an RFI of 827.5 MHz is broadcast inside the anechoic chamber. This OOB RFI is active from snapshot 200 to 700, as shown in Figure 10. The estimated AoA error is given for the different AoA techniques. Beamformer is depicted in blue, Capon in orange, and MUSIC in yellow. We observe in Figure 11 that the maximum AoA error is for Beamformer, which is about 10°. For Capon and MUSIC, the AoA error is below 4°.

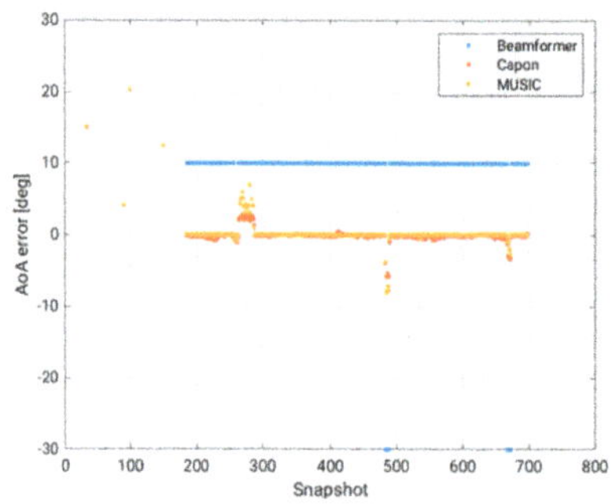

Figure 10. AoA estimation for 827.5 MHz RFI in the anechoic chamber.

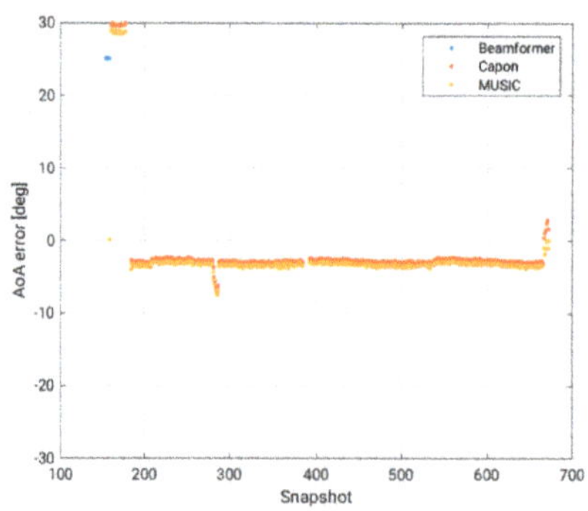

Figure 11. AoA estimation for 1575.42 MHz RFI in the anechoic chamber.

The technique that provides better AoA estimation without filtering and is considered the primary localization technique is MUSIC. In order to address the in-band AoA estimation, an RFI of 1575.42 MHz (L1) is broadcast inside the anechoic chamber. It shows an accurate estimation of the angle of arrival of the L1 RFI using MUSIC and Capon techniques.

5. Conclusions

GIMAD has been developed and verified. The verification results have shown that GIMAD is capable of detecting, characterizing, and localizing RFI both in-band and out-band. The GIMAD prototype uses algorithms based not only on GNSS observables but also on IQ samples. The combination of the selected algorithms maximizes the detection probability, as they emphasize different metrics. The capability of detecting very short interference signals (125 µs) has been verified, in line with EGNOS v3/ICAO GBAS RFI masks. Thus, it targets not only the aeronautical segment but also safety-critical environments.

Author Contributions: Conceptualization, software, validation: E.O.-B., L.T.-R., À.C.-B., A.G.-N., F.F. and G.S.-G. All authors have read and agreed to the published version of the manuscript.

Funding: This research was funded by CDTI and it was articulated through the program NAVISP EL2-034 (ESA contract 4000128898/19/NL/MP).

Institutional Review Board Statement: Not applicable.

Informed Consent Statement: Not applicable.

Data Availability Statement: Data are contained within the article.

Acknowledgments: The GIMAD project has been carried out in the frame of the NAVISP EL2-034 Program with the invaluable collaboration of CDTI, ESA, and ENAIRE.

Conflicts of Interest: Authors Enric Obiols-Bernaus, Lorena Tortajada-Ropero, Àngel Creus-Blanch, and Antonio González-Novell was employed by INDRA Sistemas. The remaining authors declare that the research was conducted in the absence of any commercial or financial relationships that could be construed as a potential conflict of interest.

References

1. Fabra, F.; Seco-Granados, G.; López-Salcedo, J.A.; Obiols, E.; Creus, À.; González-Novell, A. Detection, Characterization, and Localization of Interference Signals in Wideband GNSS Environments: The GIMAD System. In Proceedings of the IEEE Xplore: 2022 10th Workshop on Satellite Navigation Technology (NAVITEC), Noordwijk, The Netherlands, 5–7 April 2022.
2. Hill, K. Jamming GPS Signals Is Illegal, Dangerous, Cheap, and Easy. Gizmodo (12 August 2017). Available online: https://gizmodo.com/jamming-gps%20signals-is-illegal-dangerous-cheap-and-e-1796778955 (accessed on 5 October 2022).
3. *Annex 10*, 7th ed.; Aeronautical Telecommunications, Volume I: Radio Navigation Aids; ICAO: Montreal, QC, Canada, July 2018.
4. Dunkel, W.; Butsch, F. FGNSS Monitoring and Information Systems at Frankfurt Airport. In Proceedings of the 13th International Technical Meeting of the Satellite Division of the Institute of Navigation (ION GPS 2000), Salt Lake City, UT, USA, 19–22 September 2000; pp. 773–783.

engineering proceedings

MDPI

Proceeding Paper

The Vulnerability of Inland Waterway AIS to GNSS Radio Frequency Interference†

Jakub Steiner [1,2,*], Jakub Havlíček [1], Tomáš Duša [1] and Günter Heinrichs [3]

[1] GNSS Centre of Excellence, 14200 Prague, Czech Republic
[2] Department of Air Transport, Faculty of Transportation Sciences, Czech Technical University in Prague, 12800 Prague, Czech Republic
[3] Spirent Communications, Crawley RH10 1BD, UK; guenter.heinrichs@spirent.com
* Correspondence: steinja8@fd.cvut.cz

† Presented at the European Navigation Conference 2023, Noordwijk, The Netherlands, 31 May–2 June 2023.

Abstract: GNSS is an indispensable source of positioning, navigation and timing for many sectors, including inland waterway transport. Unfortunately, GNSS is also vulnerable to interference, including intentional jamming and spoofing. This paper evaluates the vulnerability of one of the key inland waterway systems—the automatic identification system (AIS)—to GNSS jamming and spoofing. The vulnerability is explored via a series of tests conducted in both laboratory and live-sky environments. The results clearly show the negative impact of both types of interference on AIS. The impact included denial of service and reporting of false position. Additionally, the effects on subsequent systems like river information services or nearby vessels are also showcased. The results presented provide valuable insight into the vulnerability of inland waterway transport. The need for understanding the system limitations and vulnerability rises with the increase in the implementation of autonomous systems into the inland waterway sector, as well as other critical infrastructure sectors.

Keywords: GNSS; GPS; interference; jamming; spoofing; automatic identification system (AIS); vulnerability; inland waterways; critical infrastructure

Citation: Steiner, J.; Havlíček, J.; Duša, T.; Heinrichs, G. The Vulnerability of Inland Waterway AIS to GNSS Radio Frequency Interference. *Eng. Proc.* **2023**, *54*, 26. https://doi.org/10.3390/ENC2023-15461

Academic Editors: Tom Willems and Okko Bleeker

Published: 29 October 2023

1. Introduction

Over the past decades, Global Navigation Satellite Systems (GNSS) have become an indispensable utility for many sectors of critical infrastructure (CI) as a reliable and accurate source of positioning, navigation and timing (PNT) [1]. In the CI sector of inland waterway transport, GNSS is predominantly used for vessel position determination and vessel tracking where GNSS is the primary and often the sole PNT source [1,2].

The specific system used by inland vessels for position determination is called the inland automatic identification system (AIS). The AIS also broadcasts the vessel's position and its identification, course, speed and other key parameters to other AIS units via a VHF transceiver. The AIS, in combination with the Electronic Chart Display and Information System (ECDIS), allows the captain to view the vessel's location on a map, as well as the positions of other nearby vessels. The use of inland AISs is mandatory in many European countries [3]. The AIS is also standardized for maritime transport [4]. The information from AIS units is also received by the National River Information Services (RISs), which are used by the national inland waterway authority as a surveillance tool to manage inland waterway transport.

Though GNSS reaches very high levels of availability and continuity, due to the low power of the signal when received on Earth, GNSS is vulnerable to unintentional and intentional terrestrial sources of radio frequency interference (RFI). Intentional interference may be in the form of jamming, spoofing or beaconing signals. This AIS's vulnerability to GNSS RFI is a known issue. There have been numerous documented cases of GNSS RFI impacting AISs. These cases include jamming in the Black Sea and around Crimea, in the sea near Syria and around Cyprus, and even cases of spoofing [5–7].

Even though these cases of GNSS RFI impacting AISs are known, the full extent of the vulnerability is not fully understood. Although a few studies have already been conducted [8–11], they were more focused on the feasibility of RFI attacks and included only one type of interference and environment. Furthermore, the impact of an either jammed or spoofed AIS unit on subsequent systems is completely unknown.

This paper presents a comprehensive vulnerability testing of the inland waterway AIS unit including jamming and spoofing tests in laboratory and live-sky environments. The paper provides performance and resilience parameters of AIS under RFI. The impact of multiple jammers differing in their jamming signal characteristics is compared. Additionally, the effect of the jammed and spoofed AIS unit on the RIS server and other vessels is described.

2. Materials and Methods

The materials and methodology used for the comprehensive vulnerability testing are divided based on the test conditions and type of interference into 3 parts: laboratory testing, jamming under live-sky and spoofing under live-sky

2.1. Laboratory Testing

The aim of the laboratory testing was to precisely measure the jamming signal power which impacts the AIS unit performance in a controlled environment. However, the impact may differ based on the jamming signal used. Therefore, two different jammers were used and compared. The AIS unit under test was a standalone unit receiving only GPS L1. The specific model number and manufacturer are intentionally not included.

GNSS simulator model GSS6700 from Spirent was used to generate an authentic GNSS signal. As a source of interference, two different types of GNSS jammers were used. The first was a commercial GNSS jammer, and the second was a jamming signal created by a DVB-T modulator DTU-315 from DekTec (Hilversum, The Netherlands) and vector generator SMA100A from Rohde&Schwarz (Muenchen, Germany). The different characteristics of the two jamming signals can be seen in Figure 1, which shows waterfall diagrams of a 160-microsecond-long sample of the signal at the 60 MHz bandwidth, centred at the 1575.42 MHz frequency.

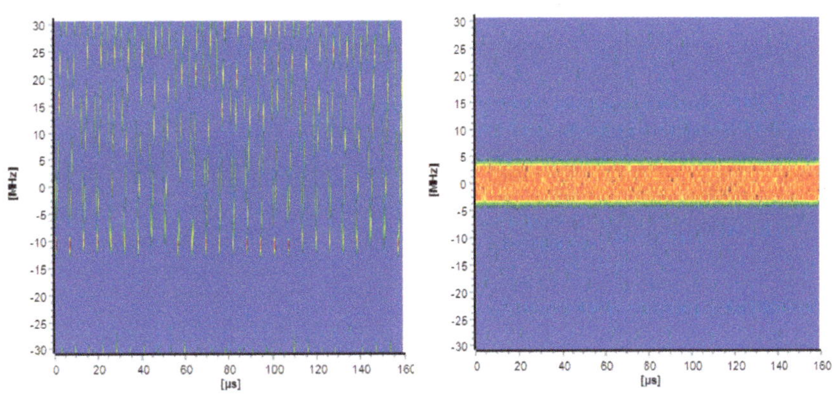

(a) (b)

Figure 1. The waterfall diagram of the commercial jammer (**a**) and the waterfall diagram of the jamming signal created by the DVB-T modulator (**b**).

The jamming signal power was controlled by a Hewlett Packard (Palo Alto, California, United States) attenuator, model 8494B. The authentic and the jamming signal were coupled using a signal coupler from Hewlett Packard, model 778D-012. During both scenarios, the AIS unit was connected via RF cable to the coupled signal and the unit was subjected to a slowly increasing jamming signal power. The power was increased by 1 dBm increments. The coupled signal was simultaneously monitored and analysed via a Rohde&Schwarz FSH13 spectrum analyser.

2.2. Jamming Testing under Live-Sky

The second part of the vulnerability testing took place on a 16 km long section of the river Elbe during normal operation. The AIS unit under test was on board a passenger ship called PORTA BOHEMICA (ENI: 32201626) providing a regular transport service on the river. The AIS unit under test receives only the GPS L1 frequency signal and the Differential GNSS corrections. The aim was to observe the impact of jamming on AIS, as well as how other AIS units and the RIS server will react in the presence of a jammed AIS unit.

The testing consisted of two static scenarios when a moored vessel was jammed from the shore and its deck and two dynamic scenarios when a sailing vessel was jammed from the shore and its deck. GNSS jammer model TG5CA was used for the live-sky tests. The jammer transmits at the GPS L1 frequency with 20 MHz bandwidth, has a 4.0 watts power output and its jamming signal can be characterized as a Sawtooth signal.

2.3. Spoofing Testing under Live-Sky

The third part of the vulnerability testing was very similar to the second one. The same AIS unit was tested, and the experiments were carried out at the same location. The aim was identical as well, only instead of jamming, a spoofing signal was used. The testing consisted of static and dynamic spoofing both from shore and the deck. In total, six scenarios were carried out. In two of the six scenarios, spoofing was preceded by jamming of the AIS unit. The difference between the true position of the vessel and the generated spoof position was between 100 and 1500 m, and the time difference was between a few minutes to several hours.

As a source of the spoofing signal, the portable GNSS simulator from Spirent was used. The portable GNSS simulator supports the simulation of multi-constellation and multi-frequency scenarios with up to 36 channels and a maximum RF output power level of −45 dBm. The spoofing signal generated was based on the GPS constellation transmitting at the GPS L1 frequency by using a passive transmit antenna with a 4.0 dBi gain. The RF output power level in the simulation scenarios was chosen in such a way that the target power level received by the AIS unit was −120 dBm.

2.4. Captured Data and Impact Evaluation

During all the testing, NMEA messages of the AIS unit under test were recorded together with the alerts issued by the unit. Additionally, for the live-sky testing, NMEA messages from nearby vessels and the recording of the RIS server screen were captured. To evaluate the vulnerability of the AIS unit, the following metrics were used:

- Horizontal position error (HPE);
- The number of satellites in use;
- Whether an integrity alert was issued or not;
- Time to alert;
- The ability to re-acquisition the authentic GPS signal after interference is gone;
- Time to re-acquisition the authentic GPS signal after interference is gone;
- General system behaviour and the ability to fulfil its function.

The power of the interference signal impacting the AIS unit was directly measured on a spectrum analyser in the laboratory testing and computed using the Free-space path loss formula in all of the live-sky tests. During performance testing, the HPE value did not exceed 2 m. Therefore, HPE over 2 m was considered a performance impacted by RFI.

3. Results

Similarly, to Section 2, the results of the vulnerability testing are structured into three parts based on the type of interference and test environment.

3.1. Laboratory Testing

The results of the laboratory testing are split into two sub-sections according to the jammer used.

3.1.1. Impact of the Commercial Jammer

Figures 2 and 3 show the effect of a slowly increasing jamming signal power generated by a commercial jammer on the HPE and the number of satellites in use parameters.

In Figure 2, we can observe that up to a jamming signal power of −64 dBm, the number of satellites in use corresponds to the unjammed state of the unit. From the power level of −61 dBm onwards, an increased fluctuation in the number of satellites in use and their gradual decrease is observed. From −56 dBm, the AIS unit is no longer able to receive the authentic signal in this experiment. The AIS unit displayed an alert, warning about the unavailability of the position information at a power of −57 dBm.

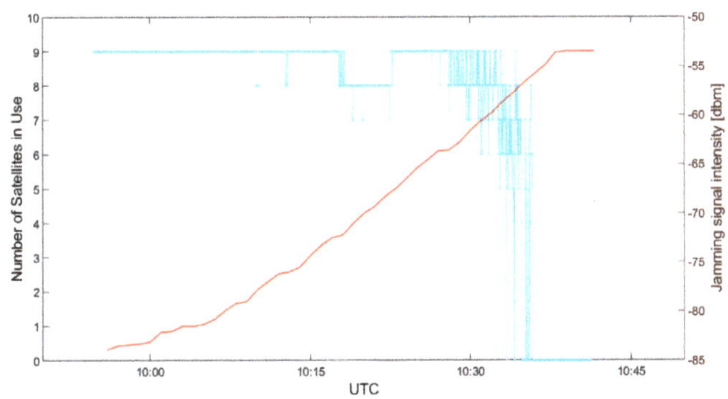

Figure 2. The number of satellites in use to calculate the position from GPS (in blue) while increasing the power of the jamming signal generated by the commercial jammer (in red).

In Figure 3, the impact of GNSS jamming on AIS performance expressed through the HPE parameter can be observed already at the power level of −70 dBm, when the HPE rises above 2 m for the first time. Up to this point, the HPE value has been kept below 1 m. From a power level of −62 dBm, the HPE value goes above 5 m. The HPE did not exceed 10 m for the duration of the experiment.

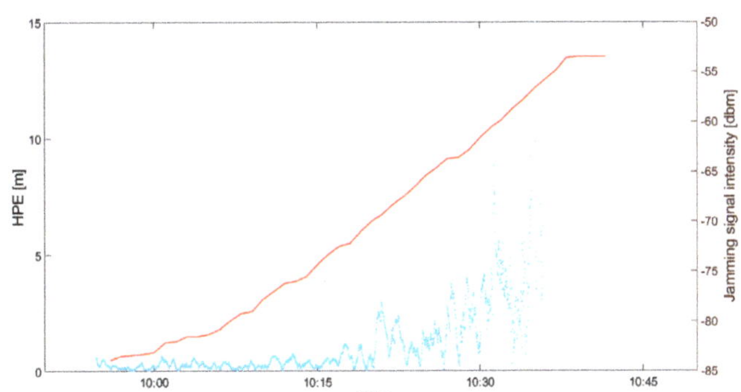

Figure 3. The horizontal position error parameter (in blue) while increasing the power of the jamming signal generated by the commercial jammer (in red).

3.1.2. Impact of the Jamming Signal Generated by the DVB-T Modulator

Figures 4 and 5 show the effect of a slowly increasing the jamming signal power generated by the DVB-T modulator on the HPE parameter and the number of satellites in use.

In Figure 4, at a jamming signal power of −71 dBm, the first decrease in the number of satellites in use by two can be observed. In the interval from −71 dBm to −68.5 dBm, the number of tracked satellites fluctuates between 6 and 11. From −68.3 dBm, the AIS unit is no longer able to track the authentic signal. At the same power, the AIS unit issued an alert, indicating a loss of position.

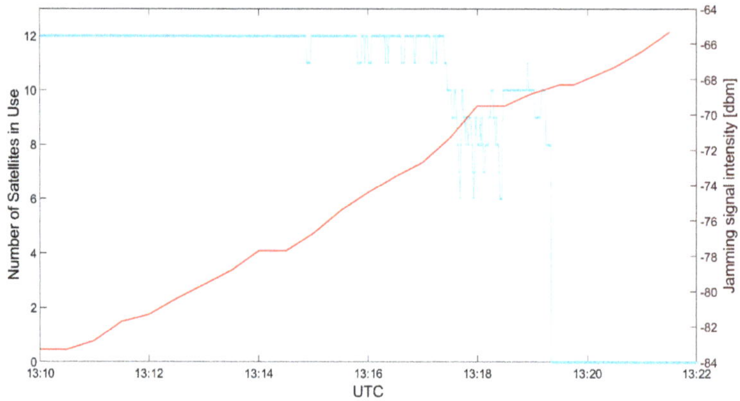

Figure 4. The number of satellites in use to calculate the position from GPS (in blue) while increasing the power of the jamming signal generated by the DVB-T modulator (in red).

In Figure 5, the interference generated by the DVB-T modulator caused an HPE greater than 2 m at −79 dBm. From −75.5 dBm, the HPE increases exponentially up to a maximum of 35 m. The HPE then decreases to the range of 10 to 20 m, which corresponds with the increase in the number of satellites in use in Figure 4. After −68.3 dBm, the HPE could no longer be computed since the unit stops reporting its position.

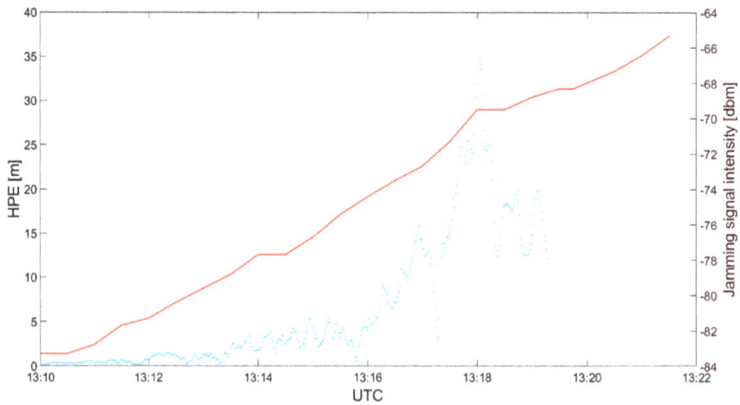

Figure 5. The horizontal position error parameter (in blue) while increasing the power of the jamming signal generated by the DVB-T modulator (in red).

3.2. *Jamming Testing under Live-Sky*

The AIS unit was successfully jammed during all four scenarios. The user of the unit was informed about the GPS outage in all cases. The time to alert was measured in the interval from 5 to 9 s, with a mean of 6.25 s. The unit was also able to re-acquisition the authentic GPS signal once the jamming was turned off. The time to re-acquisition was measured between 2 and 3 s. Nevertheless, the position was displayed on the AIS screen much later, specifically 7 to 8 s after the jamming ceased. The jamming signal power at which the AIS unit loses track of the authentic signal and later regains it was evaluated during one of the dynamic scenarios. Figure 6 shows the number of satellites in use while the vessel was approaching the jammer, passing it and moving away from it.

In Figure 6, with the gradual increase in the jamming signal power, the number of satellites in use starts to fluctuate and decrease. The loss of position occurs when the number of satellites in use drops below four, which corresponds with the signal power of −49.5 dBm. Once the vessel passed the jammer, the signal power gradually decreased. The first re-acquisition occurred at −50.5 dBm. However, a stable position reading in NMEA messages was reached later, when the power was reduced to −56.2 dBm.

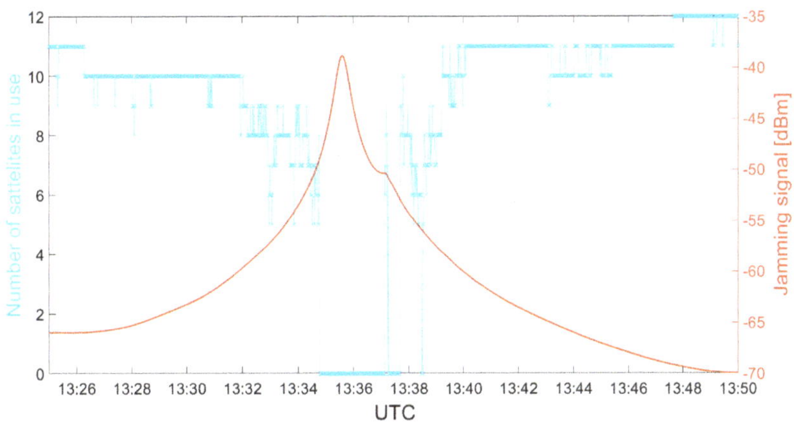

Figure 6. The number of satellites in use during a dynamic scenario when the vessel was approaching and moving away from a jammer on the shore of the river.

3.2.1. Impact on the RIS Server

Despite the absence of a position reading from the jammed AIS unit, the jammed vessel did not disappear from the RIS server screen. The jammed vessel remained visible at its last reported location. The vessel remained motionless for over 15 min until the unit was no longer jammed. Once no longer jammed, the vessel marker "jumped" to the newly reported location. No alert or warning was provided by the RIS server during the entire experiment.

The time between the first NMEA message containing the vessel position after the re-acquisition of the authentic GPS signal and the update of the vessel position on the RIS server screen were also measured. The time to update the vessel position was between 8 and 26 s with a mean of 14.5 s.

3.2.2. Impact on the Nearby Vessels

Once the AIS unit under test was jammed and unable to compute its position, its marker simply disappeared from the nearby vessels' screens. The captains of nearby vessels were not informed about the disappearance in any way.

3.3. *Spoofing Testing under Live-Sky*

None of the scenarios, when the spoofer was located at the shore, were successful. Conversely, all spoofing scenarios from the deck were successful.

3.3.1. Impact of Spoofing-Only Attack

In Figure 7, the impact of spoofing is evaluated by comparing the vessel trajectory from three different sources: the RIS server, the AIS unit NMEA messages and the spoofed trajectory. As Figure 7 demonstrates, shortly after the spoofer was turned on, in the city of Libotenice, the AIS unit started tracking the spoofed signal rather than the authentic one and began to report the counterfeit location. However, this spoofed location did not translate into the RIS surveillance situation.

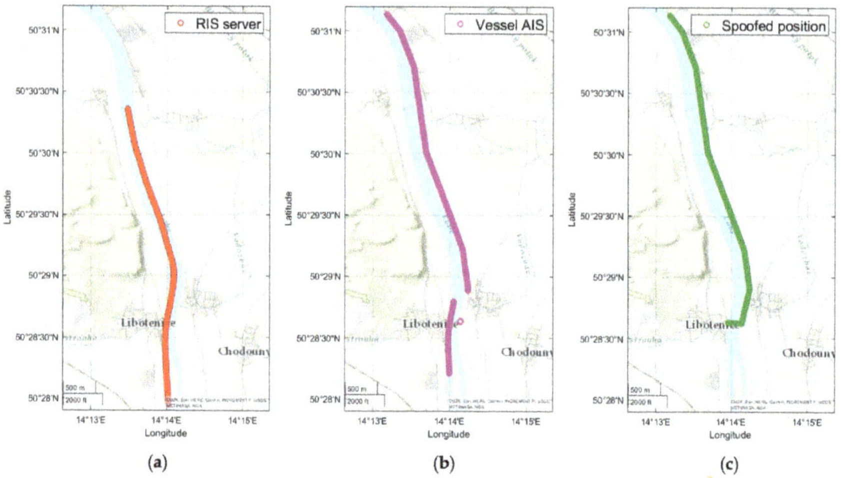

Figure 7. Comparison of vessel trajectories from three different sources: the RIS server (**a**), the AIS unit under test (**b**) and the spoofer (**c**) during the spoofing-only attack.

Despite the AIS unit transmitting the spoofed location, the RIS server displayed the real position of the vessel. During this scenario, the AIS user was alerted to the spoofing by a time synchronization error. The difference between the authentic UTC time and the spoofed time was in the order of hours. After the spoofer was turned off, the unit under test was not able to recover and required a manual restart to start tracking the authentic GPS signal.

3.3.2. Impact of an Attack Combining Jamming and Spoofing

In this scenario, the time difference between the authentic UTC time and the spoofed time was reduced to a few minutes, and the spoofing was preceded by strong jamming. The impact of this combined attack on the RIS server and the AIS under test is showcased in Figure 8. The combined attack led to the AIS unit tracing the spoofed signal. The RIS server displayed partly the authentic location which drifted into the land once the vessel approached the location generated by a spoofer. After the spoofer was turned off, the unit under test was not able to recover and required a manual restart.

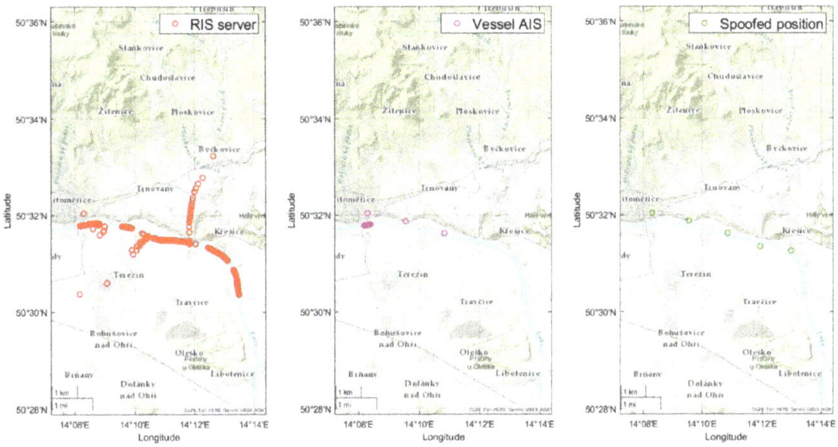

(**a**) (**b**) (**c**)

Figure 8. Comparison of vessel trajectories from three different sources: the RIS server (**a**), the AIS unit under test (**b**) and the spoofer (**c**) during the attack combining jamming and spoofing.

4. Discussion

The results presented in Section 3 clearly demonstrate the AIS's vulnerability to GPS jamming and spoofing and describe it in great detail. The impact on performance was first detected in the range of the jamming signal power from −79 dBm to −70 dBm depending on the type of jamming. The jamming signal power when the AIS unit loses complete track of the GPS satellites was measured within the interval from −68.3 dBm to −49.5 dBm based on the jammer used. Overall, it was observed that the degree of vulnerability and the impact on performance is greatly dependent on the character of the jamming signal.

An HPE of up to 35 m was measured during the experiments. Unfortunately, the user was not alerted about any decrease in performance. On the other hand, the user was informed about the complete loss of the GPS signal within 9 s at the latest after the loss of the signal occurred.

The spoofing attacks in multiple cases manipulated the vessel output which shows that spoofing is a real threat to AISs. The AIS unit also required a manual restart to recover from the spoofing attacks. Unless the spoofing was preceded by jamming and the time difference was in the order of minutes the unit did display a time synchronization warning message.

The RIS operator was not informed about the absence of position information from the jammed vessel in any way. The jammed vessel stayed "frozen" in the last reported location. Such misinformation may cause issues in inland waterway transport management. On the other hand, the vessel under spoofing-only attack had no effect on the RIS server which kept displaying the true position of the spoofed vessel. This result was surprising since, according to the research team's knowledge confirmed by the operator of the system, the AIS units are the only source of position for the RIS server. However, once the spoofing was preceded by jamming, even the RIS server was affected, resulting in the vessel position drifting in seemingly random directions. At one moment, the vessel position was displayed over 100 km away from its true location.

In regards to the impact of jammed and spoofed vessel on other AIS units, when subjected to jamming, the vessel simply disappeared from the screens of other AIS units. Captains of unjammed vessels were not alerted in any way about this disappearance. This may create a hazardous situation in conditions of reduced visibility, like foggy weather. When the vessel under test was subjected to successful spoofing, nearby AIS units displayed the spoofed location of the spoofed vessel.

5. Conclusions

This paper presents the results of a comprehensive GNSS RFI vulnerability testing of the AIS system. The results also describe the effect of the jammed or spoofed AIS unit on subsequent systems. The results provide valuable insight into the vulnerability of inland waterway transport. The paper provides a basic scheme which can be followed and/or adapted for future testing of the vulnerability of other CI systems.

The results presented are limited by the available hardware. In light of the presented results, vulnerability tests shall be conducted using a pre-determined set of jammers, each representing a common jamming signal characteristic. Additionally, having a spectrum analyser present at the spoofing testing might answer why some of the spoofing scenarios failed and bring more insights into the system's resilience.

The presented vulnerability testing was limited to a stand-alone AIS unit and to a vessel with an AIS unit as the only positioning system. However, more sophisticated vessel systems can combine AIS unit information with other technologies such as radar. Unfortunately, at the time of testing, no vessel with such equipment was available. Therefore, vulnerability testing of these vessels is recommended for future research.

Having in mind the trend of maritime and inland waterway transport of implementing more autonomous systems, where GNSS is a key source of PNT, vulnerability testing similar to the one presented in this paper shall be conducted to understand the vulnerability to GNSS RFI and its effects.

Author Contributions: Conceptualization, T.D.; methodology, J.S., J.H. and T.D.; software, J.S.; formal analysis, J.S. and J.H.; investigation, J.S., J.H., T.D. and G.H.; resources, T.D. and G.H.; data curation, J.S.; writing—original draft preparation, J.S. and J.H.; writing—review and editing, J.H., T.D. and G.H.; visualization, J.S.; supervision, T.D.; project administration, T.D.; funding acquisition, T.D. All authors have read and agreed to the published version of the manuscript.

Funding: This research was funded by the European Space Agency (ESA), specifically the NAVISP program element 3, project name "GNSS Vulnerability and Mitigation in Czech Republic".

Institutional Review Board Statement: Not applicable.

Informed Consent Statement: Not applicable.

Data Availability Statement: The data presented in this study are available on request from the corresponding author.

Acknowledgments: The laboratory testing was performed in collaboration with the Research Institute of Posts and Telecommunications in Slovakia. The Inland Waterway Authority of the Czech Republic provided the team with the standalone AIS test unit and participated in the live-sky testing.

Conflicts of Interest: Author Günter Heinrichs was employed by the company Spirent Communications, Crawley, UK. The remaining authors declare that the research was conducted in the absence of any commercial or financial relationships that could be construed as a potential conflict of interest.

Abbreviations

The following abbreviations are used in this manuscript:

AIS	Automatic Identification System
CI	Critical Infrastructure
ECDIS	Electronic Chart Display and Information System
ESA	European Space Agency
GNSS	Global Navigation Satellite Systems
HPE	Horizontal Position Error
PNT	Positioning, Navigation and Timing
RFI	Radio Frequency Interference
RIS	River Information Services

References

1. *EUSPA EO and GNSS*; Publications Office of the European Union: Luxembourg, 2022; Volume 2022. [CrossRef]
2. Whitty, C.; Walport, M. *Satellite-Derived Time and Position: A Study of Critical Dependencies*; Government Office for Science: London, UK, 2018.
3. *Commission Implementing Regulation (EU) 2019/838 of 20 February 2019 on Technical Specifications for Vessel Tracking and Tracing Systems and Repealing REGULATION (EC) No 415/2007*; European Commission: Brussels, Belgium, 2019.
4. *SOLAS—International Convention for the Safety of Life at Sea: Chapter V-Safety of Navigation, Regulation 19*; International Maritime Organization: London, UK, 2002.
5. *Above Us Only Stars: Exposing GPS Spoofing and Jamming in Russia and Syria*; C4ADS: Washington, DC, USA, 2019.
6. *Resolution MSC.428(98)*; Maritime Cyber Risk Management in Safety Management Systems. International Maritime Organization: London, UK, 2017.
7. *GPS World: Moves ship Locations Thousands of Miles*; GPS World: Cleveland, HI, USA, 2020.
8. Androjna, A.; Perkovič, M.; Pavic, I.; Mišković, J. AIS Data Vulnerability Indicated by a Spoofing Case-Study. *Appl. Sci.* **2021**, *11*, 5015. [CrossRef]
9. Caprolu, M.; Pietro, R.D.; Raponi, S.; Sciancalepore, S.; Tedeschi, P. Vessels Cybersecurity: Issues, Challenges, and the Road Ahead. *IEEE Commun. Mag.* **2020**, *58*, 90–96. [CrossRef]
10. Grant, A.; Williams, P.; Ward, N.; Basker, S. GPS Jamming and the Impact on Maritime Navigation. *J. Navig.* **2009**, *62*, 173–187. [CrossRef]
11. Medina, D.; Lass, C.; Marcos, E.P.; Ziebold, R.; Closas, P.; Garcia, J. On GNSS Jamming Threat from the Maritime Navigation Perspective. In Proceedings of the 2019 22th International Conference on Information Fusion (FUSION), Ottawa, ON, Canada, 2–5 July 2019. [CrossRef]

engineering proceedings

MDPI

Proceeding Paper

Beamforming Techniques for Resilient Navigation with Small Antenna Arrays †

Lucía Pallarés-Rodríguez *, Sergi Locubiche-Serra, Guillem Foreman-Campins, Gonzalo Seco-Granados and José A. López-Salcedo

Department of Telecommunications and Systems Engineering, IEEC-CERES, Universitat Autònoma de Barcelona (UAB), Bellaterra, 08193 Barcelona, Spain; jose.salcedo@uab.cat (J.A.L.-S.)

* Correspondence: lucia.pallares@autonoma.cat

† Presented at the European Navigation Conference 2023, Noordwijk, The Netherlands, 31 May–2 June 2023.

Abstract: This paper presents a comparative benchmarking of different beamforming techniques for robust multipath and spoofing mitigation in handheld receivers equipped with a small number of antenna elements. In the field of Global Navigation Satellite Systems (GNSSs), multipath and spoofing have become major sources of signal degradation, whereby the presence of closely spaced replicas of the signal of interest hampers the receiver's performance. Spatial filtering through the use of antenna arrays is one appealing approach to combat these effects. However, the upward trend in device miniaturization poses a serious concern for spatial filtering in handheld devices such as tablets or smartphones, particularly for the case of decimeter-wave signals such as in GNSSs. The terminal size constraints lead to deploying arrays with very few antennas, thus degrading the mitigation performance of the above impairments. This work explores the feasibility of performing robust multipath and spoofing mitigation in GNSS receivers with antenna arrays of very small sizes. Simulation results on four-antenna rectangular and two-antenna linear array distributions are provided, showing that the reduction in array size can be counteracted by the use of advanced beamformers.

Keywords: array processing; beamforming; handheld; multipath mitigation; spatial filtering; spoofing mitigation

Citation: Pallarés-Rodríguez, L.; Locubiche-Serra, S.; Foreman-Campins, G.; Seco-Granados, G.; López-Salcedo, J.A. Beamforming Techniques for Resilient Navigation with Small Antenna Arrays. *Eng. Proc.* **2023**, *54*, 27. https://doi.org/10.3390/ENC2023-15466

Academic Editors: Tom Willems and Okko Bleeker

Published: 29 October 2023

1. Introduction

Wireless communications and navigation applications are experiencing a massive proliferation in urban environments thanks to the increasing widespread deployment of enabling technologies such as 5G, the Internet-of-Things (IoT) and Global Navigation Satellite Systems (GNSSs). This has unveiled the need to cope with signal propagation impairments such as interference, multipath and spoofing that are abundant in such urban environments [1,2]. Multipath is certainly a major concern, and it appears whenever the line-of-sight signal (LOSS) of interest is received from multiple paths and different directions of arrival (DoA) due to reflection in nearby objects. The signal replicas exhibit an additional delay and carrier phase offset with respect to the LOSS, thus distorting the latter and hampering the receiver's performance. Of particular concern is the case of coherent multipath, the worst yet most frequent form of multipath in dense urban cities. It appears that the signal reflections are highly correlated to the LOSS due to the fact that they are received very closely spaced in time, coming from reflecting objects very close to the user's receiver. The need to mitigate coherent signal replicas also appears when the receiver is subject to a spoofing attack, a threat of increasing concern in positioning and navigation applications whereby a malevolent party forges the receiver position after gaining illegitimate control over the latter [3].

The use of array-processing techniques, also known as spatial filtering or beamforming, is known to be the most powerful approach to combat the above impairments [4]. The goal is to steer the array beam toward the LOSS for signal-to-noise ratio (SNR) maximization, and to place nulls at the DoA of the undesired signals. Some examples of conventional beamformers are the phased array, the Capon [5] and the linear minimum mean square error (LMMSE) [6] techniques. However, more advanced beamformers have recently been proposed in the literature, such as the LMMSE–Capon hybrid [7] and the power-based Capon [8] techniques, which have been specially designed to deal with coherent reflections, thus becoming of particular interest for multipath and spoofing mitigation in challenging environments.

An unexplored problem appears, though, when moving to the arena of handheld devices such as tablets or smartphones, which are driven by constraints on miniaturization, low power consumption and low-cost components. The advent of 5G cellular networks has definitely paved the way for the deployment of antenna arrays in such devices as a de-facto standard in the context of multiple-input, multiple-output (MIMO) systems [9–11]. This is motivated by the use of millimeter-wave signals, allowing the implementation of arrays in an area with tiny dimensions. Some examples can be found in [12,13] for 5G applications. However, this is not the case for other technologies such as GNSS, where the terminal dimensions are still and will be, in the long-term, comparable to the signal carrier wavelength. This poses a serious concern for the practical use of spatial filtering in these devices. While extensive research on antenna arrays has been performed for professional GNSS receivers [14,15], where external arrays with many antennas are used, the case of mass-market handheld receivers where just a small number of embedded antennas can be used still remains, to our best knowledge, an open problem.

Hence, the objective of this paper is to explore the feasibility of performing robust multipath and spoofing mitigation in handheld GNSS devices with a small number of antenna elements. Additionally, the paper evaluates the suitability of new beamformers such as the LMMSE–Capon hybrid and the power-based Capon for compensating the limitations and technological challenges posed by such small arrays. To these ends, a benchmarking of the above-mentioned conventional and new beamforming techniques is carried out in two representative scenarios, namely urban and suburban. Two representative antenna array distributions are considered: four-antenna uniform rectangular and two-antenna uniform linear array distributions, henceforth denoted as URA4 and ULA2, respectively.

2. Fundamentals of Array Processing

2.1. Signal Model

Let us consider an array with L antenna elements that perceives the superposition of a LOSS of interest and a set of M undesired replicas. The signal as perceived by the different antenna elements can be arranged into an $L \times 1$ vector $\mathbf{x}(n)$, thus leading to the following complex baseband signal model at the array output,

$$\mathbf{x}(n) = \alpha_0 \mathbf{a}(\theta_0) s(\tau_0, \varphi_0) + \sum_{m=1}^{M} \alpha_m \mathbf{a}(\theta_m) s(\tau_m, \varphi_m) + \mathbf{e}(n) \tag{1}$$

with $s(\tau, \varphi) \doteq c(n-\tau)e^{j\varphi}$ being the signal bearing the useful information $c(n)$, τ and φ being the propagation delay and carrier phase, and the subscript $_0$ referring to the LOSS. In (1), α is a complex amplitude encompassing information about the signal power, and $\mathbf{e}(n)$ embeds the noise introduced by each antenna channel. The term $\mathbf{a}(\theta) \in \mathbb{C}^{L\times 1}$ is the spatial signature of the signal coming from direction θ as perceived by the different antennas in the array, and it depends on the antenna inter-separation and array distribution. In the remainder of the paper, we will refer to $\mathbf{a}(\theta_0)$ as simply $\mathbf{a}_0$.

The samples in (1) collected during an observation interval of N samples can be arranged into the following $L \times N$ matrix

$$\mathbf{X} \doteq \begin{bmatrix} \mathbf{x}(0) & \mathbf{x}(1) & \dots & \mathbf{x}(N-1) \end{bmatrix} \tag{2}$$

and we also define the following $N \times 1$ signal vector,

$$\mathbf{s}(\tau,\varphi) \doteq \left[s(-\tau,\varphi) \quad s(1-\tau,\varphi) \quad \ldots \quad s(N-1-\tau,\varphi)\right]^T. \tag{3}$$

The definitions in (2) and (3) will be used later in Section 3 in the description of the beamforming techniques.

2.2. The Beamforming Principle

Beamforming is nothing but the linear combination of the signal samples coming from each of the antennas in the receiver array, with the aim of concentrating the array beam toward a specific DoA and filtering out any other contribution coming from a different DoA. This is performed by applying a set of spatial filtering coefficients or *weights* to the signal in (1) perceived by the array. This is

$$y(n) = \mathbf{w}^H \mathbf{x}(n) \tag{4}$$

where $\mathbf{w} \in \mathbb{C}^{L\times 1}$ is the weight vector, and $y(n)$ is the signal at the beamformer output. Note that $y(n)$ is a scalar magnitude, and it can thus be understood as the output of an equivalent *smart* single antenna. The problem, then, boils down to determining the coefficients in $\mathbf{w}$ that satisfy a given optimization criterion. This is where the beamforming techniques come into play, as explained next.

3. Selected Beamforming Techniques

This section provides a brief description of a set of beamforming techniques that have been selected for their implementation in user terminals with a small number of antennas, and tested later in Section 4. Note that most of them rely on having prior knowledge about the DoA of the LOSS to be spatially filtered. Such knowledge can be obtained by resorting to maximum likelihood estimation [16], spectral estimation methods [17], or, in some applications such as in GNSS, this information can easily be obtained from the satellite ephemeris and the orientation of the user's terminal. In any case, determining the DoA of the LOSS remains out of the scope of the present paper, and it will be assumed to be a priori known henceforth.

3.1. Phased Array

The *phased array* beamformer (PHA), often referred to as *delay-and-sum* (DAS) in the literature, is the most straightforward form of beamforming. The underlying idea is to correct the diverse phase delays at which the signal will arrive at each antenna element. To this end, the weights $\mathbf{w}_{\mathrm{PHA}}$ become those that maximize the response of the array at the LOSS DoA, that is, those that maximize the power of the output beamformer $y(n)$. These weights are given by

$$\mathbf{w}_{\mathrm{PHA}} = \frac{1}{L}\mathbf{a}_0 \tag{5}$$

which are nothing but the steering vectors of the impinging LOSS signal. This beamformer can thus be understood as the spatial matched filter to the spatial signature of the received signal when the spatial noise is uncorrelated among antenna elements. However, its simplicity is counteracted by a poor performance when it comes to canceling the signals impinging on the array from particular directions.

3.2. Capon Beamformer

The *Capon* beamformer (CAP), also known as the *minimum variance distortionless response* (MVDR), is the most general beamforming approach. The beauty is that the array beam pattern is controlled by applying a series of constraints to the different directions to be either pointed at or nulled out. More particularly, the underlying idea of the CAP is to minimize the power at the beamforming output under the constraint that the LOSS be, conversely, unaltered. This is

$$\min_{\mathbf{w}} P_y = \min_{\mathbf{w}} \mathbf{w}^H \mathbf{R}_{\mathbf{X}} \mathbf{w} \text{ subject to } \mathbf{a}_0^H \mathbf{w} = 1 \tag{6}$$

where $\mathbf{R}_{\mathbf{X}} \doteq \mathrm{E}\left[\mathbf{X}\mathbf{X}^H\right] \in \mathbb{C}^{L \times L}$ is the auto-correlation matrix of the data in $\mathbf{X}$, and the term $\mathbf{a}_0^H \mathbf{w} = 1$ is the so-called distortionless response constraint. After some mathematical calculations with Lagrange multipliers, the weights $\mathbf{w}_{\mathrm{CAP}}$ become [5]

$$\mathbf{w}_{\mathrm{CAP}} = \frac{\mathbf{R}_{\mathbf{X}}^{-1} \mathbf{a}_0}{\mathbf{a}_0^H \mathbf{R}_{\mathbf{X}}^{-1} \mathbf{a}_0}. \tag{7}$$

3.3. Linear Minimum Mean Square Error Beamformer

The *linear minimum mean square error* (LMMSE) beamformer exploits the temporal diversity between the LOSS and the undesired reflections by comparing the received signal with some reference signal generated locally at the receiver. GNSS signals are a clear example of the latter, where such a reference signal is the local code replica used for despreading, that is, $\mathbf{s}(\tau, \varphi)$. The LMMSE beamformer aims at minimizing the MSE of the error between $\mathbf{s}(\tau, \varphi)$ and the array output $\mathbf{X}$. The solution to this problem is given by [6]

$$\mathbf{w}_{\mathrm{LMMSE}}(\tau_0, \varphi_0) = \mathbf{R}_{\mathbf{X}}^{-1} \mathbf{r}_{\mathbf{Xs}}(\tau_0, \varphi_0) \tag{8}$$

with $\mathbf{r}_{\mathbf{Xs}}(\tau, \varphi) \doteq \mathrm{E}[\mathbf{X}\mathbf{s}^*(\tau, \varphi)] \in \mathbb{C}^{L \times 1}$ being the cross-correlation between the array output and reference signals. The beauty of this approach is that it does not require knowledge on the DoA to be pointed at. On the downside, and as a result of the latter, severe LOSS degradation may be incurred.

3.4. LMMSE–Capon Hybrid Beamformer

The *hybrid beamformer* (HYB) circumvents the limitations of the LMMSE by combining the latter with the CAP exploiting the spatial reference of the LOSS. The beauty of the HYB is that it allows for an iterative implementation by making use of the recursive least squares (RLS) algorithm. It involves a computational load affordable for current microprocessors, thus making the technique suitable for handheld devices. The weights $\mathbf{w}_{\mathrm{HYB}}$ are computed as follows [7]:

$$\mathbf{w}_{\mathrm{HYB}}(\tau_0, \varphi_0) = \alpha_0^* \mathbf{w}_{\mathrm{LMMSE}}(\tau_0, \varphi_0) + \beta(\tau_0, \alpha_0) \mathbf{w}_{\mathrm{CAP}} \tag{9}$$

where $\beta(\tau_0, \alpha_0) \doteq 1 - \alpha_0^* \mathbf{a}_0^H \mathbf{w}_{\mathrm{LMMSE}}(\tau_0, \varphi_0)$.

3.5. Power-Based Capon Beamformer

In the presence of replicas that are correlated with the LOSS of interest, matrix $\mathbf{R}_{\mathbf{X}}$ is composed of the LOSS auto-correlation plus some cross-correlation terms between the LOSS and the replicas. As a consequence, the CAP beamformer $\mathbf{w}_{\mathrm{CAP}}$ can mix both contributions when minimizing the power at its output, thus eventually leading to an undesired cancellation of the LOSS. In that sense, the *power-based Capon* beamformer (PBC) circumvents this limitation by determining the cross-correlation terms in $\mathbf{R}_{\mathbf{X}}$ and eliminating their effect from the latter prior to applying beamforming onto $\mathbf{x}(n)$.

Briefly, we define $\boldsymbol{\kappa}(\tau_0, \varphi_0) \doteq \alpha_0 \mathbf{r}_{\mathbf{Xs}}(\tau_0, \varphi_0) - \alpha_0^2 \mathbf{a}_0$. The cross-correlation terms in $\mathbf{R}_{\mathbf{X}}$, denoted as $\boldsymbol{\Gamma}_{\mathbf{X}}(\tau_0, \varphi_0) \in \mathbb{C}^{L \times L}$, are as follows [8]:

$$\boldsymbol{\Gamma}_{\mathbf{X}}(\tau_0, \varphi_0) = \mathbf{a}_0 \boldsymbol{\kappa}(\tau_0, \varphi_0)^H + \boldsymbol{\kappa}(\tau_0, \varphi_0) \mathbf{a}_0^H. \tag{10}$$

Then, by defining a new auto-correlation matrix $\tilde{\mathbf{R}}_{\mathbf{X}}(\tau_0, \varphi_0) \doteq \mathbf{R}_{\mathbf{X}} - \mathbf{\Gamma}_{\mathbf{X}}(\tau_0, \varphi_0)$, the PBC beamformer is given by

$$\mathbf{w}_{\mathrm{PBC}}(\tau_0, \varphi_0) = \frac{\tilde{\mathbf{R}}_{\mathbf{X}}^{-1}(\tau_0, \varphi_0)\mathbf{a}_0}{\mathbf{a}_0^H \tilde{\mathbf{R}}_{\mathbf{X}}^{-1}(\tau_0, \varphi_0)\mathbf{a}_0} \quad (11)$$

which is the CAP in (6) after replacing $\mathbf{R}_{\mathbf{X}}$ with $\tilde{\mathbf{R}}_{\mathbf{X}}(\tau_0, \varphi_0)$.

4. Performance Analysis

This section provides the results of the performance assessment for the beamformers described in Section 3 in the context of a handheld device with either two- or four-antenna elements. Without loss of generality, GNSS signals are considered, where the despreading process leads to a correlation peak with time support of $\pm T_c$ around τ_0, with T_c termed *chip period*. In both urban and suburban scenarios considered herein, the received signal consists of the superposition of a LOSS and an undesired replica (i.e., $M = 1$), with carrier-to-noise ratios (C/N_0) of 44 dB-Hz and 41 dB-Hz, respectively. URA4 and ULA2 antenna distributions are considered, where the antennas are separated half a carrier wavelength apart from one another. The array is assumed to be perfectly calibrated for simplicity, and additive white Gaussian noise (AWGN)-independent antenna channels are adopted. The beamformers are implemented within a closed loop where the determination of τ_0 and φ_0, denoted as $\hat{\tau}_0$ and $\hat{\varphi}_0$, is carried out through a delay-locked loop (DLL) and a phase-locked loop (PLL), respectively, thus providing estimates in a timely manner and, consequently, allowing for some statistical averaging over time. For the correlation terms $\mathbf{R}_{\mathbf{X}}$ and $\mathbf{r}_{\mathbf{Xs}}(\tau_0, \varphi_0)$, we resort to the sample correlations $\hat{\mathbf{R}}_{\mathbf{X}} \doteq \frac{1}{N}\mathbf{X}\mathbf{X}^H$ and $\hat{\mathbf{r}}_{\mathbf{Xs}}(\hat{\tau}_0, \hat{\varphi}_0) \doteq \frac{1}{N}\mathbf{X}\mathbf{s}^*(\hat{\tau}_0, \hat{\varphi}_0)$, respectively, with $N = 49$ samples. Two different metrics are employed for the performance analysis. On the one hand, there is the cancellation of the replica, which measures the ratio between the amount of power of the replica canceled out by the beamformers and the LOSS power. On the other hand, there is the well-known *multipath error envelope* (MEE), which measures the errors induced by the computation of $\hat{\tau}_0$ compared to the true value τ_0. In GNSS, the MEE is directly linked to the error in the estimated receiver position.

4.1. Urban Scenario

Urban scenarios are characterized by the abounding presence of obstacles and tall buildings, thus having sky visibility reduced to a small region in the vertical direction. Hence, the signals are prone to be received from high elevation angles. In that sense, the LOSS and the replica are considered herein to impinge from 80° and 60° elevation, respectively. An azimuth angle of 30° is adopted for both.

Figure 1 illustrates the cancellation of the replica as a function of the delay spacing between the latter and the LOSS, normalized to T_c. Any spacing below one is considered as correlated. The top plot depicts the ULA2 case, and the bottom plot depicts the URA4 case. When it comes to conventional beamformers, the PHA presents very poor mitigation performance, which remains the same irrespective of the replica delay. The CAP is even observed to amplify the undesired replica when highly correlated with the LOSS, in an attempt to level the power of the replica to that of the LOSS. It is not until signals become uncorrelated that the CAP provides acceptable performance, thus making it unsuitable for coherent multipath and spoofing mitigation. The LMMSE seems to provide better performance compared to the PHA and CAP in terms of replica cancellation, but, in contrast, it is found to incur dramatic LOSS degradation, reaching power drops of up to 30 dB. These results apply to both ULA2 and URA4.

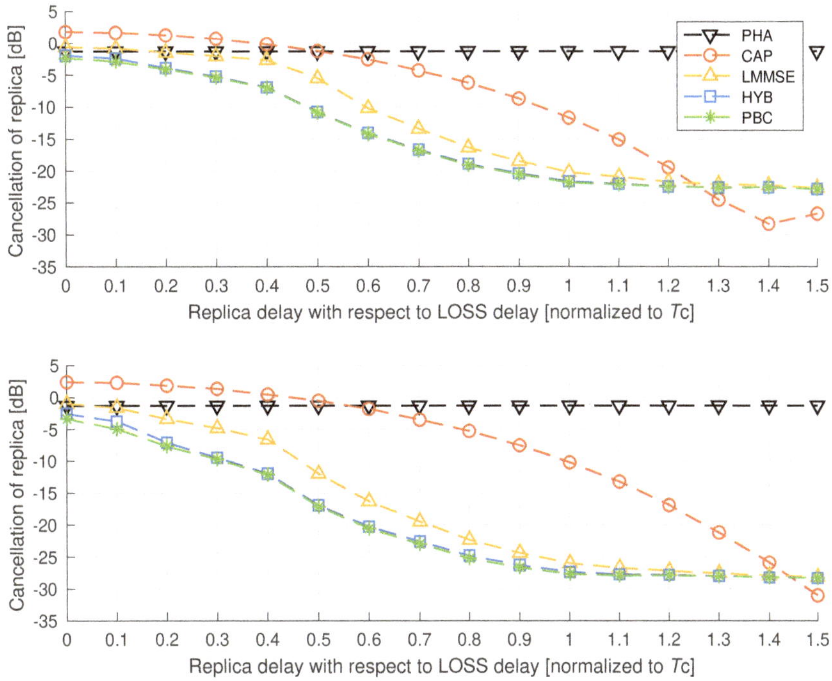

Figure 1. Cancellation of replica with respect to its delay in an urban scenario with ULA2 (**top**) and URA4 (**bottom**) array distributions.

Interestingly, the poor performances described above are overcome by the use of more sophisticated beamformers. This is the case for the HYB and PBC. They show the best replica cancellation performance even for highly correlated signals, and it improves rapidly when increasing the replica delay. The latter occurs particularly for the URA4 case, where the cancellation shows a steeper behavior than in ULA2. More specifically, a cancellation of more than 10 dB is achieved in URA4 for as few as 0.3 replica delay, whereas the same performance requires 0.5 replica delay in ULA2, a point where the cancellation in URA4 already exceeds 15 dB. In that sense, it is observed that the gain provided by the use of URA4 with respect to the smaller ULA2 lies in the range of 3–5 dB.

The above observations can also be inferred from the MEE results depicted in Figure 2, which ultimately unveil the impact of mitigation on the receiver performance. The CAP presents a detrimental behavior particularly when signals are highly correlated, with MEE values exceeding 60 m. The PHA provides a slight gain compared to when no beamforming (labeled as "No BF" in the figure) is performed, by decreasing the MEE from 25 m down to 20 m. However, acceptable MEE is not achieved until the signals become uncorrelated. In that sense, conventional beamformers are outperformed by the HYB and PBC, where the MEE barely exceeds 10 m and 5 m for ULA2 and URA4, respectively, and as few as 0.5 replica delay are required for the MEE to drop down below 5 m in both.

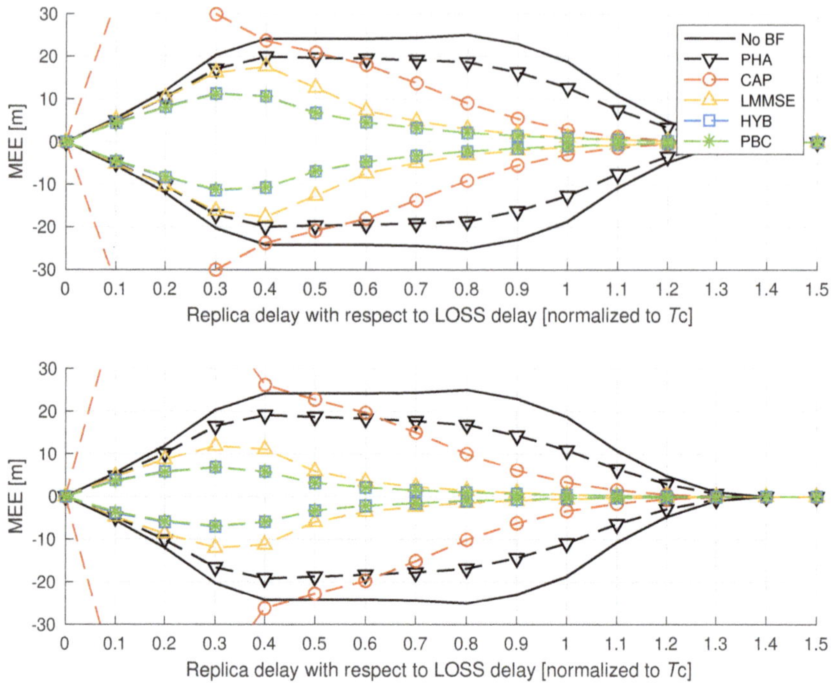

Figure 2. MEE in an urban scenario with ULA2 (**top**) and URA4 (**bottom**) array distributions.

4.2. Suburban Scenario

Suburban scenarios comprise those in the city outskirts and residential areas with small buildings, thus having a wider sky visibility than in urban environments. In this case, the LOSS and the replica are considered to impinge from 60° and 20° elevation, respectively. The azimuth angle is also kept here at 30° for both signals.

Figures 3 and 4 depict the replica cancellation and MEE results, respectively, for the suburban scenario. The very same conclusions to those in the urban scenario can be extracted here. Conventional techniques such as the PHA and CAP are outperformed by the HYB and PBC. In terms of replica cancellation, whereas the former two provide practically the same performance as in the urban case, the latter two present a steeper behavior and do provide a gain of $\sim$5 dB along all replica delays. This means that a cancellation of 10 dB is now achieved in ULA2 with as little as 0.3 delay, whereas at this point the cancellation already exceeds 15 dB in URA4. In terms of MEE, conventional beamformers exceed values of 10 m while requiring signals to become separated enough to improve these results. In contrast, the HYB and PBC present MEE no larger than 5 m in ULA2 with only 0.3 replica delay, a point at which the MEE is already below 5 m in URA4. Therefore, the MEE performance in the urban case is even improved in the suburban case by an amount of 3–5 m. These results conclude on the feasibility to perform multipath and spoofing mitigation in ULA2 and URA4 arrays by implementing advanced beamforming techniques such as the HYB and PBC.

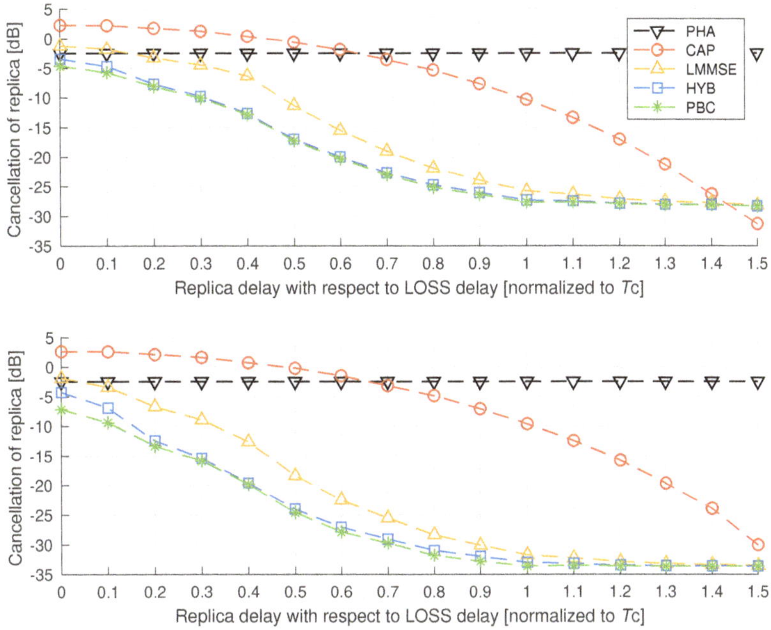

Figure 3. Cancellation of replica with respect to its delay in a suburban scenario with ULA2 (**top**) and URA4 (**bottom**) array distributions.

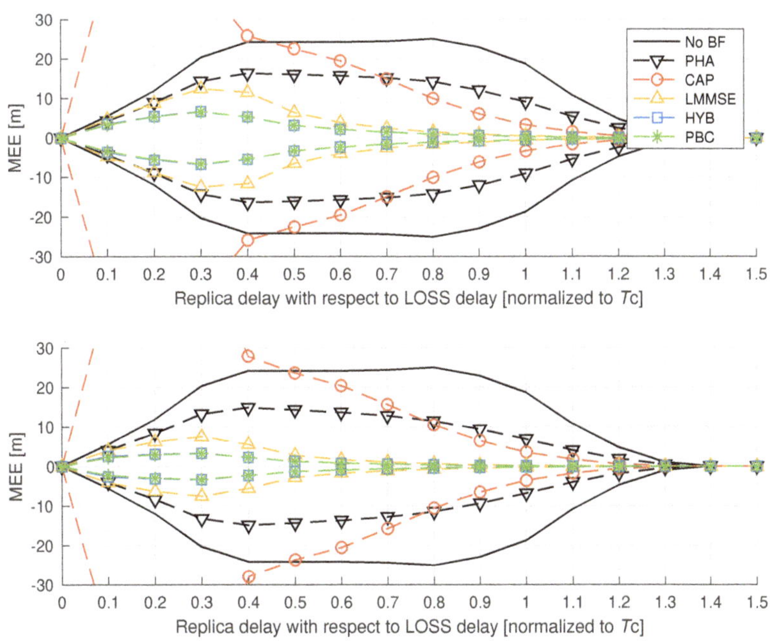

Figure 4. MEE in a suburban scenario with ULA2 (**top**) and URA4 (**bottom**) array distributions.

5. Conclusions

In this paper, we have evaluated the feasibility of performing robust multipath and spoofing mitigation with spatial filtering techniques in GNSS antenna arrays of a very small size. In particular, four-antenna rectangular and two-antenna linear arrays have been considered in a simulation campaign carried out for benchmarking different beamforming techniques in challenging environments, namely urban and suburban. The simulation results have shown the goodness of the LMMSE–Capon hybrid and the power-based Capon beamformers in such small arrays, particularly in the presence of coherent multipath and spoofing. Therefore, they have become potential candidates for signal mitigation purposes in decimeter-wave handheld devices, and their use over conventional beamformers has become mandatory to overcome the limitations of small arrays. Furthermore, although four-antenna arrays do provide better performance, comparable results have been obtained for the case of two-antenna arrays, thus confirming the suitability of the latter for the practical implementation of spatial filtering techniques in smaller devices such as smartphones.

Author Contributions: Conceptualization, S.L.-S. and J.A.L.-S.; methodology, L.P.-R. and S.L.-S.; software, L.P.-R., S.L.-S. and G.F.-C.; validation, L.P.-R., S.L.-S. and G.F.-C.; formal analysis, L.P.-R. and S.L.-S.; investigation, L.P.-R. and S.L.-S.; resources, G.S.-G. and J.A.L.-S.; data curation, L.P.-R. and S.L.-S.; writing—original draft preparation, L.P.-R. and S.L.-S.; writing—review and editing, all authors; visualization, L.P.-R. and S.L.-S.; supervision, S.L.-S. and J.A.L.-S.; project administration, J.A.L.-S.; funding acquisition, G.S.-G. and J.A.L.-S. All authors have read and agreed to the published version of the manuscript.

Funding: This work has been supported in part by the Spanish Agency of Research (AEI) under the Research and Development projects PID2020-118984GB-I00/AEI/10.13039/501100011033 and PDC2021-121362-I00/AEI/10.13039/501100011033.

Institutional Review Board Statement: Not applicable.

Informed Consent Statement: Not applicable.

Data Availability Statement: Data are contained within the article.

Conflicts of Interest: The authors declare no conflict of interest.

References

1. Sahmoudi, M.; Amin, M.G. Optimal Robust Beamforming for Interference and Multipath Mitigation in GNSS Arrays. In Proceedings of the IEEE International Conference on Acoustics, Speech and Signal Processing (ICASSP), Honolulu, HI, USA, 15–20 April 2007; Volume 3, pp. 693–696.
2. Psiaki, M.L.; O'Hanlon, B.W.; Powell, S.P.; Bhatti, J.A.; Wesson, K.D.; Schofield, T.E. GNSS spoofing detection using two-antenna differential carrier phase. In Proceedings of the 27th International Technical Meeting Satellite Division Inst. Navigation (ION GNSS+), Tampa, FL, USA, 8–12 September 2014; pp. 2776–2800.
3. Psiaki, M.L.; Humphreys, T.E. GNSS spoofing and detection. *Proc. IEEE* **2016**, *104*, 1258–1270. [CrossRef]
4. Swindlehurst, A.L.; Jeffs, B.D.; Seco-Granados, G. Applications of Array Signal Processing. In *Academic Press Library in Signal Processing. Array and Statistical Signal Processing*; Chapter 20; Academic Press: Cambridge, MA, USA, 2014; Volume 3, pp. 859–953.
5. Capon, J. High-resolution frequency-wavenumber spectrum analysis. *Proc. IEEE* **1969**, *57*, 1408–1418. [CrossRef]
6. Konovaltsev, A.; Antreich, F.; Hornbostel, A. Performance assessment of antenna array algorithms for multipath and interferers mitigation. In Proceedings of the Workshop on GNSS Signals & Signal Processing (GNSS SIGNALS), Noordwijk, The Netherlands, 24–25 April 2007.
7. Seco-Granados, G.; Fernández-Rubio, J.; Fernández-Prades, C. ML estimator and hybrid beamformer for multipath and interference mitigation in GNSS receivers. *IEEE Trans. Signal Process.* **2005**, *53*, 1194–1208. [CrossRef]
8. Mañosas-Caballú, M.; Swindlehurst, A.L.; Seco-Granados, G. Power-based Capon beamforming: Avoiding the cancellation effects of GNSS multipath. *Signal Process.* **2021**, *180*, 107891. [CrossRef]
9. Larsson, E.G.; Edfors, O.; Tufvesson, F.; Marzetta, T.L. Massive MIMO for next generation wireless systems. *IEEE Commun. Mag.* **2014**, *52*, 186–195. [CrossRef]
10. Rusek, F.; Persson, D.; Lau, B.K.; Larsson, E.G.; Marzetta, T.L.; Edfors, O.; Tufvesson, F. Scaling Up MIMO: Opportunities and Challenges with Very Large Arrays. *IEEE Signal Process. Mag.* **2013**, *30*, 40–60. [CrossRef]
11. Vook, F.W.; Ghosh, A.; Thomas, T.A. MIMO and beamforming solutions for 5G technology. In Proceedings of the IEEE MTT-S International Microwave Symposium (IMS2014), Tampa, FL, USA, 1–6 June 2014; pp. 1–4.
12. Chattha, H.T. 4-Port 2-Element MIMO Antenna for 5G Portable Applications. *IEEE Access* **2019**, *7*, 96516–96520. [CrossRef]
13. Parchin, N.O.; Al-Yasir, Y.I.A.; Ali, A.H.; Elfergani, I.; Noras, J.M.; Rodriguez, J.; Abd-Alhameed, R.A. Eight-Element Dual-Polarized MIMO Slot Antenna System for 5G Smartphone Applications. *IEEE Access* **2019**, *7*, 15612–15622. [CrossRef]
14. Vagle, N.; Broumandan, A.; Jafarnia-Jahromi, A.; Lachapelle, G. Performance analysis of GNSS multipath mitigation using antenna arrays. *J. Glob. Position. Syst.* **2016**, *14*, 4. [CrossRef]

15. Brenneman, M.; Morton, J.; Yang, C.; Van Graas, F. Mitigation of GPS multipath using polarization and spatial diversities. In Proceedings of the International Technical Meeting of the Satellite Division of The Institute of Navigation (ION GNSS 2007), Fort Worth, TX, USA, 25–28 September 2007; pp. 1221–1229.
16. Stoica, P.; Sharman, K.C. Maximum likelihood methods for direction-of-arrival estimation. *IEEE Trans. Acoust. Speech Signal Process.* **1990**, *38*, 1132–1143. [CrossRef]
17. Roy, R.; Kailath, T. ESPRIT-estimation of signal parameters via rotational invariance techniques. *IEEE Trans. Acoust. Speech Signal Process.* **1989**, *37*, 984–995. [CrossRef]

engineering proceedings

MDPI

Proceeding Paper

Performance Characterization of Hexagon | NovAtel's Robust Dual-Antenna Receiver (RoDAR) during the Norwegian Jamming Trial 2022 †

Ali Broumandan * and Sandy Kennedy

Hexagon | NovAtel, Calgary, AB T3K 2L5, Canada; sandy.kennedy@hexagon.com

* Correspondence: ali.broumandan@hexagon.com

† Presented at the European Navigation Conference 2023, Noordwijk, The Netherlands, 31 May–2 June 2023.

Abstract: NovAtel has recently leveraged its expertise in both receiver design and anti-jam technology to develop solutions for space- and weight-constrained applications in challenged GNSS environments. Robust Dual-Antenna Receiver (RoDAR), is based on a commercial dual-antenna receiver, originally designed for attitude determination, and employs special firmware to mitigate jammers and spoofers without an increase in size or power consumption. With RoDAR, the multi-frequency, multi-constellation dual-antenna receiver is capable of null-steering at two different frequency bands (e.g., L1 and L5). In September 2022, the Norwegian Public Roads Administration hosted JammerTest, a live, over-the-air broadcast jamming and spoofing test. This paper presents the jamming and spoofing detection and mitigation performance of RoDAR during this live broadcast test. The interference detection provides spectrum monitoring and jamming characterization on all GNSS bands. The mitigation is carried out by steering a null formed on-board the receiver towards a jamming/spoofing source at GPS L1 and L5 bands. The null steering performance is characterized as a function of signal and position availability compared to a non-protected NovAtel receiver. The effectiveness of the anti-jam and anti-spoofing technology is demonstrated using representative complex spoofing and jamming test cases during this event.

Keywords: detection; GNSS; high precision; jamming; null steering; mitigation; spoofing

Citation: Broumandan, A.; Kennedy, S. Performance Characterization of Hexagon | NovAtel's Robust Dual-Antenna Receiver (RoDAR) during the Norwegian Jamming Trial 2022. *Eng. Proc.* **2023**, *54*, 28. https://doi.org/10.3390/ENC2023-15470

Academic Editors: Tom Willems and Okko Bleeker

Published: 29 October 2023

1. Introduction

Global Navigation Satellite Systems are widely used in critical infrastructure and safety of life applications. With such widespread use, open signal descriptions, and a crowded RF spectrum, jamming and spoofing are well-known threats to GNSS. The most common type of intentional interference is jamming, which aims to prevent GNSS receivers from providing position and timing solutions. Interference can be suppressed in the time, frequency and/or spatial domains. Although methods for suppressing narrowband interference in the time/frequency domain have been widely studied [1,2], their performance degrades when presented with wideband interference. Spatial processing techniques can effectively mitigate both narrowband and wideband interference [3], for example, antenna array processing [4–10].

Spoofing is another form of intentional interference. Several spoofing detection methods have been proposed in the literature to distinguish counterfeit signals from authentic ones using both single-antenna and multi-antenna receivers [11]. In a single-antenna GNSS receiver, spoofing detection metrics are implemented in pre-despreading or post-despreading layers and are most effective when both spoofing and authentic signals are present. Pre-despreading and intermediate frequency signal monitoring metrics have been used to detect the presence of excessive amounts of power in GNSS bands [12]. Multi-antenna receivers can employ spatial processing techniques to defeat interfering signals regardless of their temporal or spectral characteristics, and are equally effective on narrow and wideband interferers including spoofed signals. Spatial processing is one of the most powerful countermeasures against various types of jamming and spoofing signals, but it typically comes with the burden of additional cost and power consumption.

NovAtel's GNSS Resilience and Integrity Technology (GRIT) is within the firmware of all OEM7 receivers for situational awareness and interference detection and mitigation tools. GRIT includes NovAtel's Interference Toolkit (ITK) and spoofing detection toolkit (SK) [12] to identify when a GNSS signal is under threat. For the dual-antenna variants of the OEM7 family, NovAtel's Robust Dual-Antenna Receiver (RoDAR) firmware option mitigates jammers and spoofers without an increase in size or power consumption. With RoDAR, the multi-frequency, multi-constellation dual-antenna receiver is capable of null-steering at two different frequency bands (e.g., L1 and L5). When not actively null-steering, RoDAR firmware also supports high-precision modes such as RTK, PPP and dual-antenna heading estimation.

2. GRIT

NovAtel's GRIT includes the toolkits discussed below.

Interference Toolkit (ITK): identifies when interference is present and characterizes the detected jammer in time and frequency domains. Mitigation options include implementing notch filters at detected frequencies, and High Dynamic Range (HDR) mode for wideband interference effect [13].

Spoofing Detection: further evaluates the incoming signals at different layers of GNSS receiver including IF sample, tracking and position level to characterize the radio frequency (RF) environment to identify when positioning, navigation and timing (PNT) is at risk from counterfeit signals [12].

RoDAR: offers an active anti-jamming tool that uses spatial processing to protect against all interference types, including the mitigation of wideband interference and spoofing signals. The operation of RoDAR is shown in Figure 1. The RF signals from two antennas are passed to the null-steering weight calculation unit after down conversion and digitization. Then, the second antenna signal undergoes phase rotation and gains compensation based on the calculated array weights and is removed from the first antenna signal. The resultant samples are jammer and spoofer free. The cleaned signals are then passed on to the tracking and position, velocity and time (PVT) solution module. RoDAR offers up to 30 dB of nominal protection compared to a non-protected receiver, and is classified as a commercial good for export control purposes.

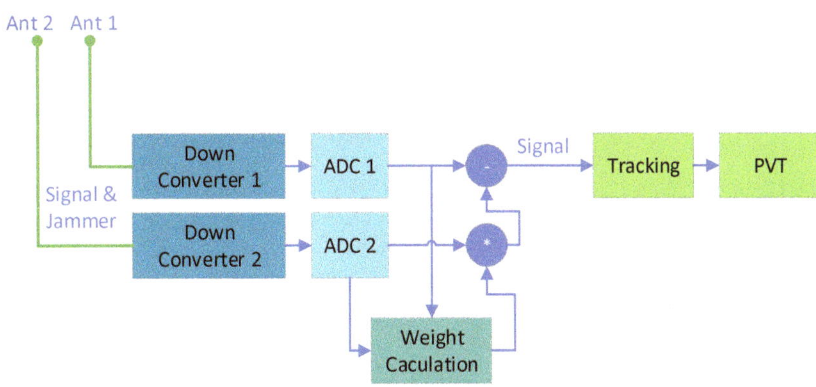

Figure 1. RoDAR operation.

OEM7's multi-constellation and multi-frequency support protects with diversity, and then if the situation turns very hostile, RoDAR provides active anti-jamming on two GNSS bands.

3. GRIT at Work: Norwegian Jamming and Spoofing Trial 2022

In September 2022, the Norwegian Public Roads Administration in conjunction with the Norwegian Communications Authority, Norwegian Metrology Service and the Norwegian Space Agency hosted JammerTest 2022, located at Bleik in Andøya. Live over-the-air broadcast jamming and spoofing tests were conducted over five days. This event provided a unique opportunity for international companies from varying sectors to step away from the safety of their simulators and controlled environments and test their ware's resiliency in real-world, hostile GNSS scenarios.

During JammerTest, dual-antenna and single-antenna OEM7 cards were evaluated, with different antennas representing different design priorities. A wide variety of jammers were tested, including single-band, dual-band, multi-band, high-power, directional, omni-directional, in-car, handheld, Norwegian Defense-owned, static and kinematic tests. RoDAR on OEM718D receiver cards, coupled to antennas from Hexagon | Antcom, were mounted to a test vehicle for the stationary jammer test series, along with PwrPak7 enclosed receiver and GNSS-850 geodetic grade antenna for reference [14].

Figure 2 shows NovAtel's test vehicle with the 1.2 G antennas from Antcom used as RoDAR arrays and the GNSS-850 antenna. The 1.2 G is a triple frequency GNSS antenna covering upper and lower bands including BDS B1, GPS L1 and GLO L1, and GPS L2, GPS L5/GalE5a and GalE5b bands. RoDAR on 718D protects the GPS L1CA and GPS L5 encoders and all other signals sharing those encoders. For instance, null steering at GPS L1CA band protects GalE1 and BDS B1C.

Figure 2. NovAtel's test vehicle with 4 RoDAR array and one GNSS-850 antenna on the roof.

The PwrPak7 receiver with GNSS-850 antenna was used for spectrum monitoring and situational awareness, while the RoDAR receivers were used for jamming and spoofing detection and mitigation. NovAtel's ITK (Interference Toolkit) and spoofing detection logs were collected to analyze data sets in real time and in post processing. ITK logs were used to detect jamming signals and characterize their parameters including power, bandwidth and their temporal and spectral behaviors. Spoofing detection logs were used to detect spoofing attacks and to distinguish spoofing events from jamming attacks.

The first three days of testing featured jamming attacks. Spoofing and jamming scenarios were propagated during the fourth day.

3.1. Jamming Test and Results

Different jamming types including single- and multi-band jamming scenarios were generated during this event. The single-band or multiple-band attacks with a mid-power jammer did not affect the position solution of the non-protected PwrPak7 or RoDAR. This is because OEM7 cards tracks all observable GNSS signals, providing resiliency against jamming and spoofing attacks via frequency diversity. Single-point positioning at the meter level of accuracy was of interest in these tests, rather than high-precision positioning at the centimeter level.

Next, the multi-frequency, high-power jamming attack is examined. Figure 3 shows the total in-band power as a function of time at L1, L2 and L5 during this jamming event, as measured by the OEM7 receiver itself [15]. The jammer power was gradually increased and then decreased in a ramping power test on all GNSS bands. As shown, in case of L1, the noise level is at −70 dBm when no jamming is present.

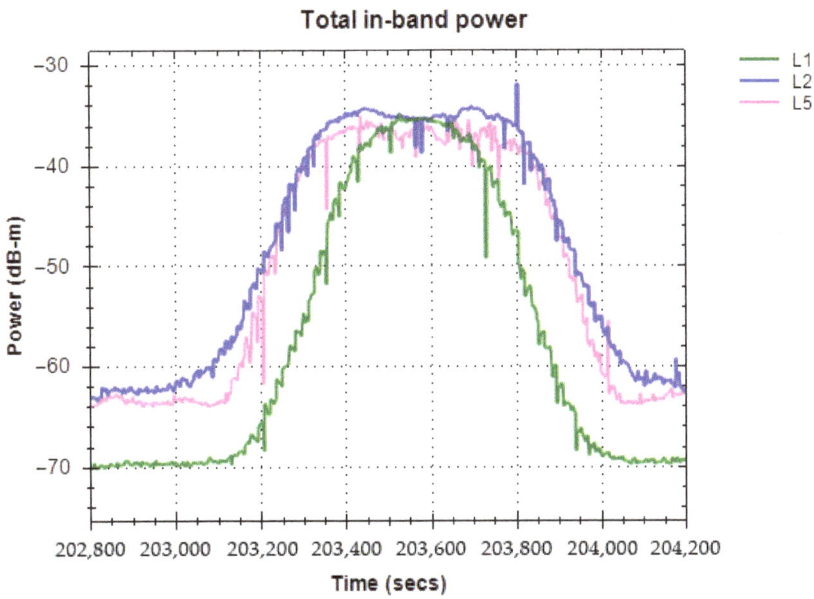

Figure 3. Input power measured by ITK during the ramp power jammer at L1, L2 and L5 bands.

Figure 4 shows mean CN0 values of GPS L1CA and L5 for the PwrPak7 and RoDAR units. GPS L1CA and L5 are protected by RoDAR with active null forming. The initial average CN0 of PwrPak7 is higher than RoDAR, because the GNSS-850 antenna paired with the PwrPak7 has a higher gain than the 1.2 G antenna used with the RoDAR unit. By introducing the jammer, the PwrPak7 CN0 values fell about 20 dB, and eventually, the receiver dropped the signals, while RoDAR maintained continuous tracking.

Figure 5 shows the reported position accuracy of RoDAR compared to the unprotected PwrPak7 under the high-power jammer. The single-point position standard deviation demonstrates the receiver's confidence in the solution. As shown, the PwrPak7 position solution degraded badly when the extra in-band power exceeded 15 dB. The RoDAR unit continued to provide a reliable position solution.

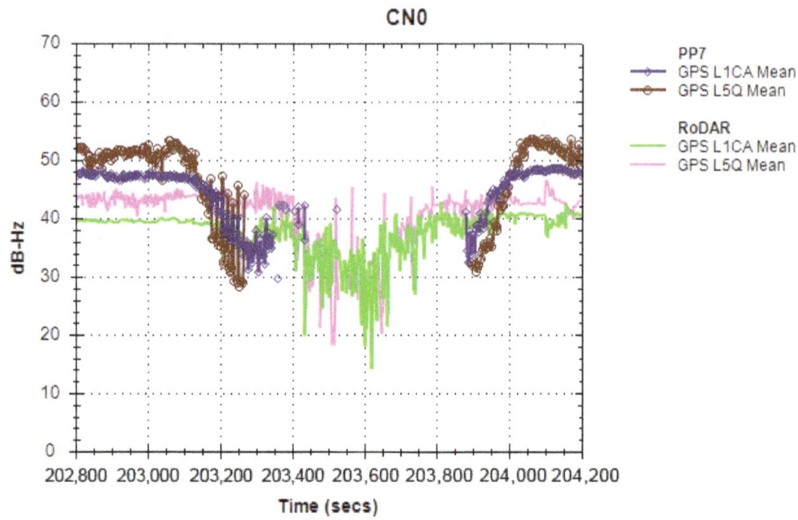

Figure 4. Average CN0 variation during the ramp power jamming for GPS L1 and L5.

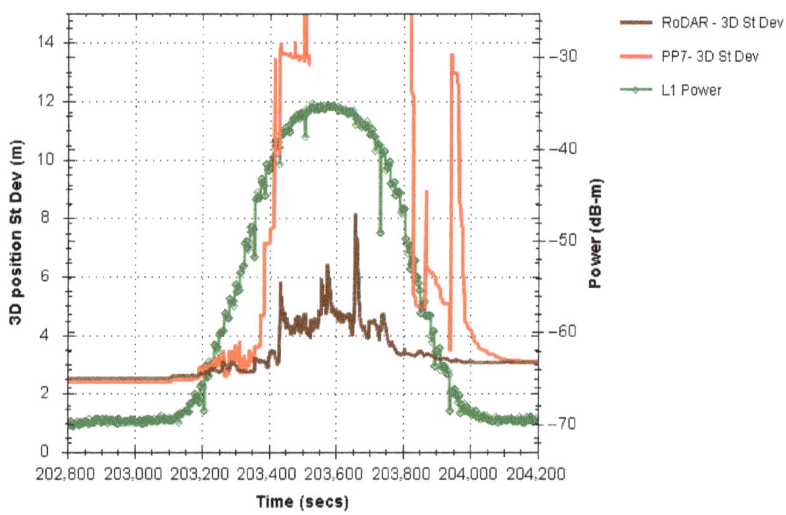

Figure 5. L1 Total in-band power, 3D position standard deviation for RoDAR and PwrPak7.

3.2. Spoofing Attack

The spoofing mitigation performance of RoDAR is evaluated and compared to the unprotected PwrPak7 receiver in this section. Each spoofing attack began with a period of high-power jamming at all GNSS bands followed by a spoofing transmission on GPS L1CA while jamming continued all other bands.

ITK logs enable visualization of jamming and spoofing attacks in the frequency and time domains. Figure 6 shows the Power Spectral Density (PSD) of all GNSS bands in clean, no-jamming conditions. This can be used to compare the jamming and spoofing power level and shape of the spectrum during the attacks.

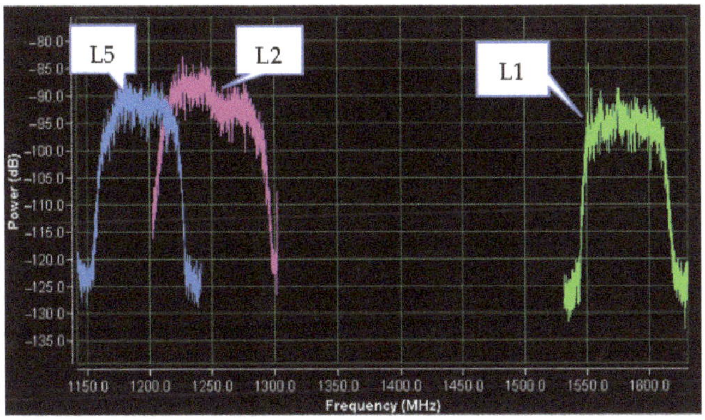

Figure 6. Spectrum of GNSS bands (L1, L2, L5) under clean, open-sky condition.

Figure 7 shows the PSD of L1 band (GPS L1, BDS B1, GLO L1) during the high-power jamming that began the attack. As shown, the input power is significantly increased compared to the plots in Figure 6 due to the jamming effect. A 20 MHz wideband jammer at the centre of GPS L1 (1575 MHz) is observable. BeiDou B1I at 1561 MHz and GLONASS L1 located at 1602–1615 MHz were also jammed.

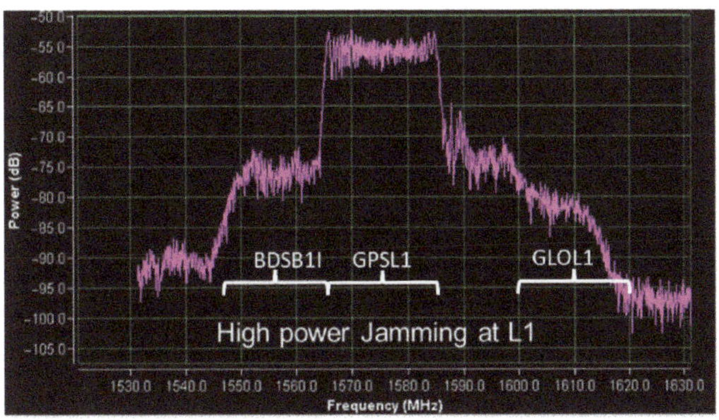

Figure 7. Spectrum of GNSS L1 band under high power jamming attack.

Figure 8 shows the PSD as the attack transitioned to spoofing. Comparing the PSD plots of Figures 7 and 8 reveals useful information. After the initial jamming period, the 20 MHz wideband jammer was moved to the GLO L1 band. The power of this jammer was also reduced. BDS B1I had a similar input power level compared to the benign condition (comparing Figures 6 and 8). This indicates this signal was not spoofed. GPS L1 input power was slightly increased. The spoofing detection toolkit detected a matched-power spoofing attack (where the spoofing power is marginally higher than the authentic signals) in this case.

Figure 9 shows the GPS L1 input power (blue) during the spoofing episodes. The scenario started with a clean, benign environment followed by knockout jamming (PSD was shown in Figure 7) at all GNSS bands and then a spoofing scenario (PSD was shown in Figure 8). The average CN0 values for GPS L1CA for RoDAR and PwrPak7 are also shown in Figure 9. The mean CN0 values in the benign stage were 43 and 48 dB-Hz for RoDAR and PwrPak7, respectively. The difference in CN0 values is due to higher grade antenna used for PwrPak7. During the jamming event, both receivers were overwhelmed and could not track anything.

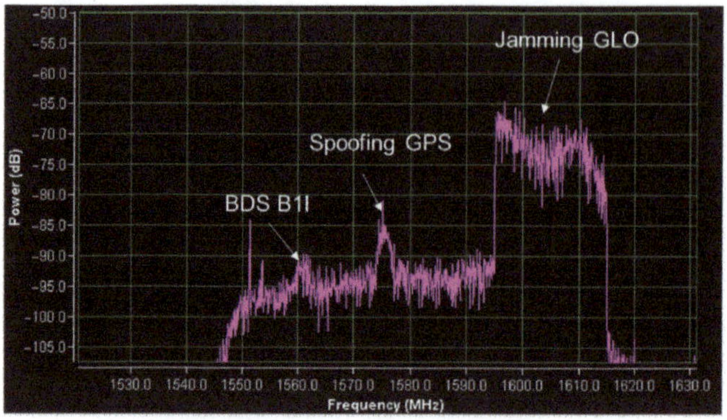

Figure 8. Spectrum of GNSS L1 band under jamming attack and spoofing attack.

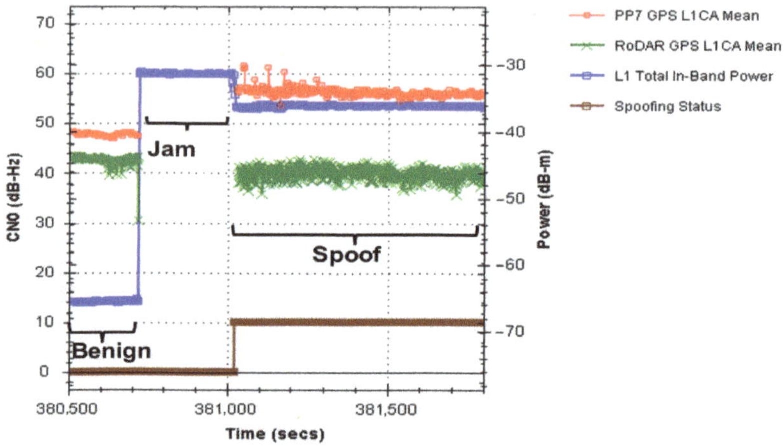

Figure 9. GPS L1 input power, mean CN0 for GPS L1CA for RoDAR and PwrPak7 units.

During the spoofing attack, the PwrPak7 mean CN0 jumped from 48 to 56 dB-Hz (red curve) due to tracking the strong spoofed signals, while RoDAR tracked authentic signals, which its lower CN0 reflects. The on-board spoofing detection status from the PwrPak7 is also shown in this figure.

Figure 10 shows all L1 band signals tracked by PwrPak7 during the spoofing attack. During the spoofing stage, PwrPak7 could only track GPS L1CA and BDS B1I. GLO L1 was jammed, GPS L1 was spoofed, and authentic BDS B1I slipped through because it was not jammed nor spoofed. The CN0 values of GPS L1CA were increased (spoofed), whereas the CN0 values of BDS B1I were reduced (no spoofed signal, higher noise floor from in-band jammer).

Figure 11 shows CN0 of L1 signals tracked during the spoofing stage by the RoDAR unit. RoDAR tracked authentic GPS L1CA, GalE1 and BDS B1I.

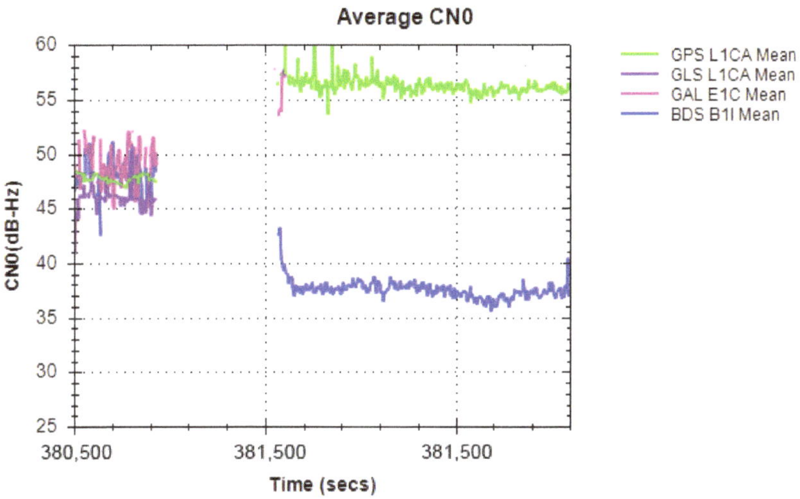

Figure 10. Average CN0 of signals at L1 band tracked by PwrPak7.

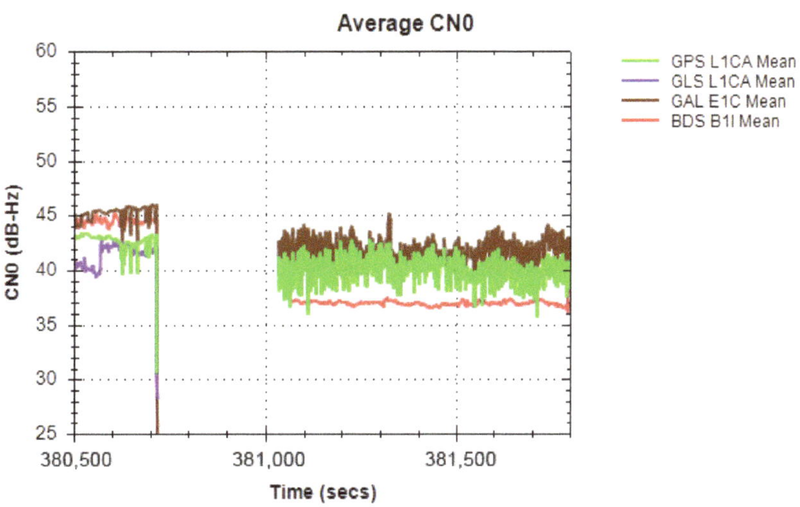

Figure 11. Average CN0 of signals at L1 band tracked by RoDAR.

4. Summary and Conclusions

During the Norwegian jamming and spoofing trials, NovAtel's GRIT features were used to reliably detect, monitor and mitigate interference and spoofing signals. These tests revealed very accurate spoofing detection results in real-world, over-the-air conditions that compared well with previous testing performed in controlled laboratory conditions. RoDAR provides dual-band spatial null-steering protection and multi-antenna, multi-constellation support. The experimental results in this trial demonstrated RoDAR's interference mitigation capabilities compared to the single-antenna PwrPak7 receiver under jamming and spoofing attacks. In the case of jamming, RoDAR was able to withstand 15–25 dB more jamming power than the single-antenna receiver in terms of position availability and accuracy. During the spoofing attack, the RoDAR unit successfully mitigated the spoofing signals, whereas the non-protected receiver tracked the spoofing signals. However, the non-protected receiver did correctly detect and report the spoofing. Neither RoDAR nor the PwrPak7 receivers provided a spoofed position during the spoofing trial.

For RoDAR, this resiliency was due to using null steering to suppress the spoofing attack. The PwrPak7's resiliency was due to signal diversity (multi-constellation support) and internal integrity checks.

Author Contributions: A.B. processed the data sets; A.B. and S.K. both contributed to the writing of the manuscript. All authors have read and agreed to the published version of the manuscript.

Funding: This research received no external funding.

Institutional Review Board Statement: Not applicable.

Informed Consent Statement: Not applicable.

Data Availability Statement: Data are contained within the article.

Conflicts of Interest: Authors Ali Broumandan and Sandy Kennedy were employed by Hexagon | NovAtel.

References

1. Pickholtz, R.L.; Schilling, D.L.; Milstein, L.B. Theory of spread-spectrum communications-a tutorial. *IEEE Trans. Commun.* **1982**, *30*, 855–884. [CrossRef]
2. Chang, C.L.; Juang, J.C. Performance analysis of narrowband interference mitigation and near-far resistance scheme for GNSS receivers. *Signal Process. Elseveir* **2010**, *90*, 2676–2685. [CrossRef]
3. Van Trees, H.L. *Optimum Array Processing, Detection, Estimation, and Modulation Theory, Part IV*; Wiley-Interscience: New York, NY, USA, 2002.
4. Fante, R.L.; Vaccaro, J.J. Cancellation of jammers and jammer multipath in a GPS receiver. *IEEE Aerosp. Electron. Syst. Mag.* **1998**, *13*, 25–28. [CrossRef]
5. Fante, R.L.; Vaccaro, J.J. Wideband cancellation of interference in a GPS receive array. *IEEE Trans. Aerosp. Electron. Syst.* **2000**, *36*, 549–564. [CrossRef]
6. Amin, M.G.; Sun, W. A novel interference suppression scheme for global navigation satellite systems using antenna array. *IEEE J. Sel. Areas Commun.* **2005**, *23*, 999–1012. [CrossRef]
7. Sun, W.; Amin, M.G. A self-coherence anti-jamming GPS receiver. *IEEE Trans. Signal Process.* **2005**, *53*, 3910–3915. [CrossRef]
8. Zoltowski, M.D.; Gecan, A.S. Advanced adaptive null steering concepts for GPS. In Proceedings of the Military Communications Conference, MILCOM 95, IEEE, San Diego, CA, USA, 5–8 November 1995; pp. 1214–1218.
9. Brown, A.; Gerein, N. Test results of a digital beamforming GPS receiver in a jamming environment. In Proceedings of the ION GPS 2001, Salt Lake City, UT, USA, 11–14 September 2001; pp. 894–903.
10. Seco-Granados, G.; Fernández-Rubio, J.A.; Fernández-Prades, C. ML estimator and hybrid beamformer for multipath and interference mitigation in GNSS receivers. *IEEE Trans. Signal Process.* **2005**, *53*, 1194–1208. [CrossRef]
11. Borio, D.; Gioia, C. A Dual-antenna Spoofing Detection System Using GNSS Commercial Receivers. In Proceedings of the 28th International Technical Meeting of the Satellite Division of The Institute of Navigation (ION GNSS+ 2015), Tampa, FL, USA, 14–18 September 2015.
12. Broumandan, A.; Kennedy, S.; Schleppe, J. Demonstration of a Multi-Layer Spoofing Detection Implemented in a High Precision GNSS Receiver. In Proceedings of the 2020 IEEE/ION Position, Location and Navigation Symposium (PLANS), Portland, OR, USA, 20–23 April 2020. [CrossRef]
13. Gao, F.; Kennedy, S. Demonstrated Interference Detection and Mitigation with a Multi-frequency High Precision Receiver. In Proceedings of the 29th International Technical Meeting of The Satellite Division of the Institute of Navigation (ION GNSS+ 2016), Portland, OR, USA, 12–16 September 2016; pp. 159–170.
14. Available online: https://novatel.com/products/receivers/enclosures (accessed on 27 November 2023).
15. Alves, P.; Himmelfarb, M. Interference Mapping Using Receiver Power. In Proceedings of the 31st International Technical Meeting of the Satellite Division of the Institute of Navigation (ION GNSS+ 2018), Miami, FL, USA, 24–28 September 2018.

engineering proceedings

MDPI

Proceeding Paper

E1-E6 SDR Platform Based on BladeRF for Testing Galileo-Assisted Commercial Authentication Service †

Rafael Terris-Gallego [1,*], Ignacio Fernandez-Hernandez [2], José A. López-Salcedo [1] and Gonzalo Seco-Granados [1]

1 Department of Telecommunications and Systems Engineering, School of Engineering, Universitat Autònoma de Barcelona (UAB), IEEC-CERES, 08193 Bellaterra, Barcelona, Spain; jose.salcedo@uab.cat (J.A.L.-S.); gonzalo.seco@uab.cat (G.S.-G.)
2 Directorate-General for Defence Industry and Space (DG DEFIS), European Commission, 1049 Brussels, Belgium; ignacio.fernandez-hernandez@ec.europa.eu
* Correspondence: rafael.terris@uab.cat
† Presented at the European Navigation Conference 2023, Noordwijk, The Netherlands, 31 May–2 June 2023.

Abstract: To enhance the robustness of the Global Navigation Satellite System (GNSS) against malicious attacks (e.g., spoofing), the European Galileo is working on new services, like Open Service Navigation Message Authentication (OSNMA), which provides authentication on the navigation bits, or Commercial Authentication Service (CAS), which aims to encrypt the spreading code chips. An assisted mode of the latter, named Assisted Commercial Authentication Service (ACAS), is currently under definition by the Galileo program. It uses the Timed Efficient Stream Loss-tolerant Authentication (TESLA) keys provided by OSNMA on the E1-B signal to re-encrypt some fragments of the encrypted E6-C signal, known as Re-Encrypted Code Sequences (RECSs), that are made available in the GNSS Service Centre (GSC). Once downloaded by a compatible receiver, they can be decrypted using the corresponding key and used to perform the correlation with the broadcasted E6-C signal. If that results in a correlation peak, the signal can be authenticated under certain circumstances. However, the probability of detecting this peak depends on the length of these fragments and their periodicity, since they are only provided for certain predefined instants. Indeed, if the receiver relies solely on E6-C signal and has no accurate time reference, this probability is severely degraded. This is why the nominal operating mode proposed for ACAS is to use the estimates provided by E1-B to reduce the uncertainty on the E6-C signal, so that the receiver can know precisely where these fragments are located. In the context of the PAULA project, we have developed a low-cost hardware platform based on bladeRF that allows acquiring both E1-B and E6-C samples synchronously. In this paper, we describe how to set up this platform and we characterise the alignment between the E1-B and E6-C estimates (code phase and Doppler frequency) using the real datasets obtained with such a platform, which is of key importance for the ACAS nominal mode. The results confirm the convenience of using the estimates from the E1-B signal for ACAS.

Keywords: GNSS; Galileo; ACAS; authentication; acquisition; snapshot; SDR; bladeRF; E1-B; E6-C

Citation: Terris-Gallego, R.; Fernandez-Hernandez, I.; López-Salcedo, J.A.; Seco-Granados, G. E1-E6 SDR Platform Based on BladeRF for Testing Galileo-Assisted Commercial Authentication Service. *Eng. Proc.* **2023**, *54*, 29. https://doi.org/10.3390/ENC2023-15428

Academic Editors: Tom Willems and Okko Bleeker

Published: 29 October 2023

1. Introduction

In recent years, Global Navigation Satellite System (GNSS) have become critical infrastructures for a wide range of applications, including transportation, agriculture, and telecommunications. However, these systems are vulnerable to malicious attacks like spoofing, which involve transmitting fake signals to deceive the receiver into computing incorrect positioning or timing information. To prevent these attacks, many techniques are being developed and implemented in GNSS receivers to verify the authenticity of the signals [1].

On the other hand, many proposals have emerged to integrate cryptographic protection into the signals themselves [2], allowing for an increase in the reliability and safety of GNSS-based applications. Galileo, the European GNSS, has been one of the most engaged entities in this area [3]. It was the first to develop a service to ensure the authenticity of the navigation message data, known as Open Service Navigation Message Authentication (OSNMA) [4]. Recently, this service has been made freely available to all users in the E1-B signal (operational capability).

Furthermore, Galileo is currently developing a new service named Commercial Authentication Service (CAS) [5], which is based on the encryption of the E6-C component at signal level to provide Spreading Code Authentication (SCA). This offers an enhanced level of security to the system, and when used together with OSNMA, is intended to provide a fully secure solution for GNSS users. These techniques are also being analysed in the forthcoming Global Positioning System (GPS) Chips-Message Robust Authentication (CHIMERA) [6].

To provide this type of authentication without modifying Galileo's current signal plan, and without the need to store any secret keys on the receiver, an assisted mode of CAS, known as Assisted Commercial Authentication Service (ACAS), is being defined. This is expected to be available by 2024, and involves selecting specific fragments of the E6-C signal (which is intended to be encrypted in the future) prior to its transmission. These fragments, known as Encrypted Code Sequences (ECSs), are then re-encrypted using the Timed Efficient Stream Loss-tolerant Authentication (TESLA) keys already provided by OSNMA on the E1-B signal, yet to be disclosed. The resulting Re-Encrypted Code Sequences (RECSs) are made publicly accessible through the GNSS Service Centre (GSC). Compatible receivers can download the necessary RECSs for their desired duration (e.g., several days or weeks) to operate autonomously.

Once the E6-C signal is broadcast, the receiver records a snapshot at the time where the ECS are expected. Later, when the corresponding key is disclosed in the E1-B signal, the receiver can decrypt the stored RECS and correlate it with the pre-recorded snapshot. A correlation peak indicates that the fragment is located where expected and, hence, the signal can be authenticated, provided that some conditions are met. For a more comprehensive explanation, interested readers can refer to [7], which also provides a description of the cryptographic operations involved in ACAS that fall outside the scope of this paper.

One critical factor influencing the performance of ACAS at the signal level is the length of the snapshot used for correlation with the decrypted RECS (i.e., the local replica). As discussed in [8], due to the nature of ACAS, relying solely on the E6-C signal can result in a significant degradation of the correlation peak detection probability, depending on the receiver's clock calibration. To mitigate this, the envisioned nominal operating mode for ACAS is to obtain the time reference from the E1-B signal and utilise it to reduce uncertainty in the E6-C signal, as elaborated in [9].

The proof-of-concept for this mode has been examined in [7]. To further validate the assumptions made in this proof-of-concept using real measurements, we have designed and implemented an evaluation platform using bladeRF Software Defined Radio (SDR) boards. This platform enables us to capture E1-B and E6-C samples synchronously. This allows the characterisation of the alignment between the E1-B and E6-C estimates, particularly concerning the code phase and Doppler frequency, which is crucial for the nominal operation of ACAS. The results obtained show that with just a few E6-C samples, the correlation can be successfully achieved. This confirms the feasibility of implementing ACAS authentication with minimal resources on a typical user receiver.

The rest of the paper is organised as follows. Section 2 presents the SDR evaluation platform used to acquire the E1 and E6 samples. Section 3 show the results obtained with the real data samples acquired using such a platform. Finally, Section 4 concludes the paper.

2. SDR Evaluation Platform Based on BladeRF

This section describes the evaluation platform and provides instructions on how to configure it to enable the synchronous acquisition of E1 and E6 samples.

2.1. Hardware Description

To shorten the development time and reduce costs, we implemented our evaluation platform using Commercial Off-The-Self (COTS) SDR devices available on the market. As both the E1-B and E6-C signals need to be processed in ACAS, we first review the main characteristics of these signals, which are detailed in [10] and summarised in Table 1.

Table 1. Signal characteristics.

Signal	Centre Frequency	Ref. Bandwidth	Code Frequency
E1-B	1575.42 MHz	24.552 MHz	1.023 MHz
E6-C	1278.75 MHz	40.920 MHz	5.115 MHz

In light of the range of existing technical solutions, we first determine a list of minimum requirements to narrow the search among the available SDR boards on the market:

- Frequency range including the L-band (1 to 2 GHz).
- Minimum sampling rate of 60 MSps (to provide enough accuracy for the envisaged tests).
- Minimum Analog to Digital Conversion (ADC) resolution of 8 bits (to provide enough dynamic ranges for the envisaged tests).
- Low-cost board (less than $500 per board).
- External clock reference input.

After a preliminary selection, which involved non-technical factors such as purchasing convenience, the solution retained for our platform was the bladeRF 2.0 micro board from Nuand, which is focussed on developing cost-effective SDR platforms. Its main specifications are summarised in Table 2, along with the other candidates analysed (LimeSDR [11] and Analog Devices [12]) that met all the requirements listed above. A comprehensive evaluation [13] of the available SDR boards can be found in [14], within the SDR Makerspace project [15]. For an up-to-date reference on GNSS SDR, including different receiver architectures and front ends, see [16,17]. Also, for comparison purposes, [18] presents a signal authentication device based on snapshot processing for ACAS.

Table 2. Candidate SDR boards specifications.

Board Model	Transceiver	Max. Sampling Rate	Max. RF Bandwidth	ADC Resolution	Device Connection	Price [1]
bladeRFmicro	AD9361	61 MSps [2]	56 MHz	12 bits	USB 3.0	$400
LimeSDR	LMS7002M	61 Msps	61 MHz	12 bits	USB 3.0	$300
ADRV9364	ADS9364	61 Msps	56 MHz	12 bits	USB 2.0 [3]	$800

[1] Approximative prices obtained in 2022, in USD. [2] This rate can be extended up to 122.88 MHz using the 8-bit mode support of the AD9361 transceiver integrated in the board, a feature released in February 2023 [19]. [3] PCIe and Ethernet interfaces with faster speeds are also available.

Going into further detail, the bladeRF 2.0 micro boards (Figure 1) are 2×2 Multiple-Input Multiple-Output (MIMO) SDR boards that offer a frequency range of 47 MHz to 6 GHz and a maximum sampling rate of 61.44 MHz using a resolution of 12 bits per sample. The core of the board is the latest generation Cyclone V Field Programmable Gate Array (FPGA) from Intel (formerly Altera). The FPGA size, measured in Logic Elements (LEs), depends on the bladeRF board model considered. The models used for our platform are the $\times$A4, with 49 K LEs, an the $\times$A5, 72 K LEs. The complete list of specifications can be found in [20].

Figure 1. bladeRF SDR board.

All the boards include an on-board PLL which allows for controlling its VCTCXO to a 10 MHz reference signal, but they can also use an external reference clock source through a dedicated surface-mounted U.FL connector. The board can be powered solely from a USB bus (USB 3.0 type), but an external power source can be supplied to ensure maximal linear performance of bias-tee peripherals if needed.

2.2. Platform Architecture

Even though the bladeRF micro 2.0 boards include 2 × 2 MIMO capabilities and therefore include two transmitters and two receiver modules, with both sharing the same oscillator. This prevents the use of the same board to acquire samples of both E1 and E6 bands simultaneously, as the former is located in the upper L-band and the latter in the lower L-band. Therefore, two boards need to record the E1 and E6 samples synchronously.

The two boards are connected to a (non-powered) Drotek multi-band antenna using a power splitter. One of the boards powers its use via its built-in bias-tee, while a DC-block is used in the other board. To prevent mismatches from using different clock sources, the boards are connected to a common external reference, specifically a Citrine Oven-Controlled Crystal Oscillator (OCXO) from Wenzel [21].

Finally, the J51 test points of each board are connected together with a jumper wire to allow the communication between boards, which permits performing the synchronisation tasks. A block diagram of the platform is shown in Figure 2, and a prototype of it is shown in Figure 3.

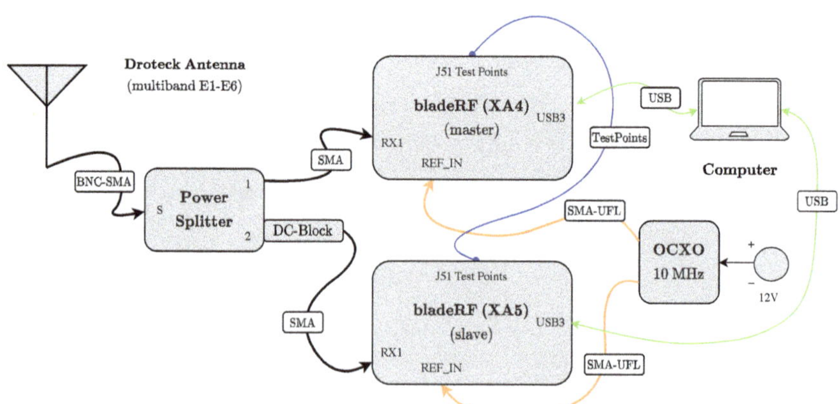

Figure 2. Block diagram of the SDR evaluation platform.

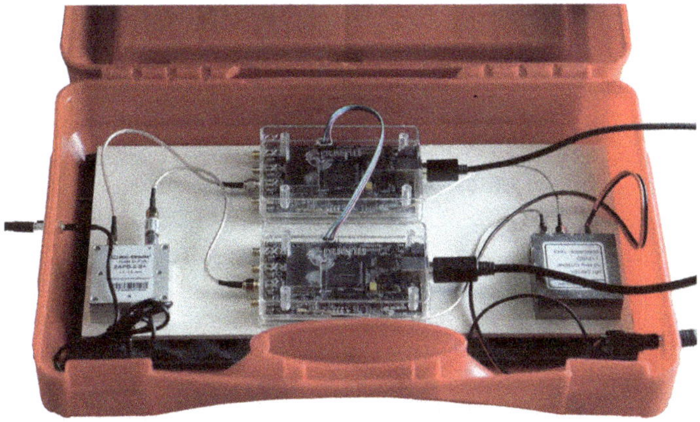

Figure 3. Prototype of the SDR evaluation platform.

2.3. Platform Configuration

The bladeRF boards of the platform were configured using the native libbladeRF library from Nuand, although higher-level tools like the open-source SoapySDR API [22], which includes a library for interfacing with other SDR devices, can also be used. Nuand provides a basic installation guide [23] for the libbladeRF library, which can be installed on various computer platforms. On the macOS platform chosen by the authors, the MacPorts utility [24] was used. Both Intel- and ARM-based platforms have been successfully tested.

The bladeRF library allows two modes of operation: interactive and non-interactive. The latter allows for batch-processing the configuration commands from the Command Line Interface (CLI), but the authors have found some inconsistencies during the operation with the current version, where some parameters keep getting misconfigured. Therefore, the more reliable interactive mode has been used, which is activated using the command `` `bladeRF-cli -interactive` `` from the host computer terminal.

The first step is to individually configure the boards. In our case, the xA5-model board, labeled as master, is assigned to acquire the E6-C samples, while the xA4-model board, labeled as slave, acquires the E1-B ones. After successfully connecting the boards to the host, which can be verified using the `` `probe` `` command, we can check the status of the FPGA. This is done using the `` `version` `` command, which shows the size of the FPGA and whether it is loaded. If not loaded, we can download the corresponding image from [25].

The results presented in this paper were obtained using version v0.14.0 of the software, which was released on 4 April 2021. At the time of writing, the latest version available is v0.15.0 (released on 13 February 2023), which mainly adds support for oversampling and, therefore, is not expected to affect the results presented here. To load the FPGA, the command `` `load-fpga [filename.rbf]` `` is used.

To avoid the need for manually loading the FPGA every time the system is powered on, Nuand provides an autoloading mechanism [26]. This mechanism can be host-software-based or firmware-based. In the former, the libbladerf library checks if the FPGA image file is available in a list of predefined folders in the host, and loads it if found. In the latter, which is host-independent but slightly slower, the FPGA bitstream needs to be written into the device's SPI flash. This is accomplished with the command `` `flash-firmware [filename.img]` ``. The latest FX3 firmware can also be downloaded from [27].

We can now configure the receiver parameters in the boards, either inputting the commands individually, or saving them in a script and running it with the command `` `run [filename.script]` ``. The contents of the scripts for both the master and slave boards are provided in Listing 1 and Listing 2, respectively.

Listing 1. blade-rf-conf-master.script.

```
set frequency rx 1278.75 M
set bandwidth rx 10 M
set samplerate rx 20 M
set agc rx off
set gain rx 45
set biastee on
set clock_ref enable
```

Listing 2. blade-rf-conf-slave.script.

```
set frequency rx 1575.42 M
set bandwidth rx 10 M
set samplerate rx 20 M
set agc rx off
set gain rx 46
set biastee off
set clock_ref enable
```

We can observe that the bias-tee is activated in the master board, as it is responsible for feeding the antenna. The Automatic Gain Control (AGC) is turned off, and the receiver gain is manually adjusted to a predetermined value that maximises the board ADC's dynamic range. Additionally, the gain is slightly increased in the slave board to compensate for the attached DC-block's losses. It is worth noting that the bladeRF micro 2.0 board converter has a native format of signed, complex 16-bit Q11, which implies that the values are within the range of $(-2048, 2048)$. Finally, the last command in the script allows the use of the Oven-Controlled Crystal Oscillator (OCXO) as the external clock reference.

The next step is to configure the boards so that they can acquire samples synchronously. To do so, we start defining the test points that will be used for triggering, first in the master board and then in the slave board. This is accomplished using the scripts provided in Listings 3 and 4. Finally, we need to fire the trigger back into the master board to start acquiring samples synchronously, which is done with the command `trigger j51-1 tx fire`.

Listing 3. blade_rf_rx_sync_master.script.

```
rx config file = master_e6.bin n = 200 M
trigger j51-1 tx master
trigger j51-1 rx slave
rx start
```

Listing 4. blade_rf_rx_sync_slave.script.

```
rx config file = slave_e1.bin n = 200 M
trigger j51-1 rx slave
rx start
rx wait
```

The `rx config` command allows for the specification of the filename where the samples will be recorded, as well as the total number of samples to be recorded. If not specified, the native binary format will be used. However, it is also possible to use the Comma-Separated Values (CSV) format using the option `format=csv` in the `rx config` command. In our case, since we have configured the boards with a sampling rate of 20 MHz, the 200 million samples (`n = 200 M`) to be acquired correspond to a 10 s snapshot.

Therefore, two files (one for E6, one for E1) will be generated, each approximately 800 MB in size, taking into account that each sample (I and Q) is coded in 2 bytes.

It is important to note that the synchronisation feature used in this evaluation platform is not documented in the official Nuand documentation for the bladeRF micro 2.0 boards. However, a similar feature is available for other boards from the same manufacturer [28], which has been used to infer the existence of this feature for the board used in our platform.

3. Results with Real Datasets

In this section, we use the real datasets obtained with the evaluation platform, which can be freely accessed (see the Data Availability section at the end of the paper), to analyse the alignment of E1 and E6 estimates. To accomplish this, the recorded snapshots are divided into smaller portions and processed individually using a custom-built MATLAB simulator to obtain these estimates. Then, a statistical comparison is performed by averaging roughly 2000 of these portions. A comparison of these estimates for a given portion is shown in Figure 4, where we can see that the width of the BOC (1,1) central peak used in E1-B is slightly wider but comparable to the BPSK(5) peak used in E6-C, as expected from the Autocorrelation Function (ACF). All the datasets used in this section were recorded under clear sky conditions, with a C/N_0 not lower than 45 dBHz, and that no ionospheric correction was applied.

It is worth noting that, in certain instances, one of the boards of the evaluation platform misses some samples in the recording process, so that the code phase difference is affected by this amount, which should be removed in the statics computation to obtain a fair comparison of the code phase estimates.

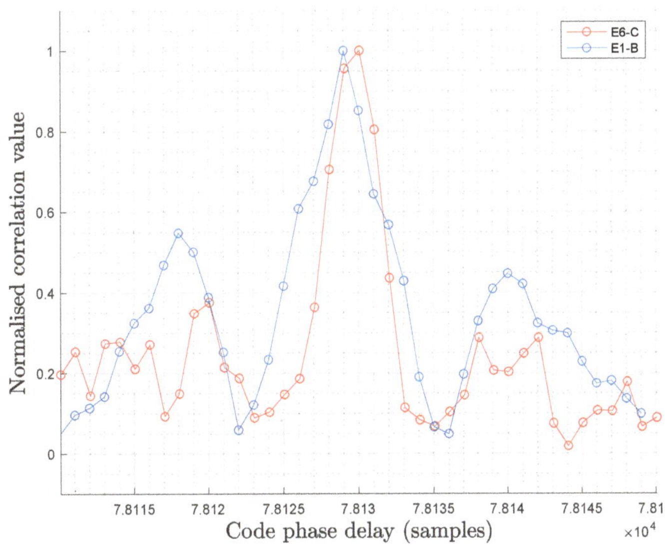

Figure 4. E1-E6 correlation peaks comparison.

3.1. E1-E6 Code Phase Delay Comparison

In Figure 5, we present a statistical comparison of the code phase delay difference between the E6-C and E1-B open signals. Four different real datasets are used, recorded approximately every 3 min, allowing for an analysis of the alignment with different power cycles by switching the platform on and off. Each snapshot lasts for 10 s, corresponding to 200 million samples at a sampling rate of 20 MHz. In this case, the snapshots are divided into portions of 4 ms, comprising 80,000 samples. Both the E6-C and E1-B estimates are obtained by processing a single spreading code, with durations of 1 ms and 4 ms, respectively. The obtained distribution exhibits, as expected, a non-centred Gaussian-like shape, the variance of which is related to the sample noise. A Gaussian curve (in red) is fitted to the experimental data (in blue) to interpolate the mean and variance for each snapshot.

The results demonstrate a high level of consistency with respect to the spread of the estimates, with a variance of around 2.5 m when only one E6-C spreading code is used for the acquisition. The observed bias primarily originates from the ionospheric effects and the hardware biases after power cycles resulting from the use of two distinct SDR boards in the platform.

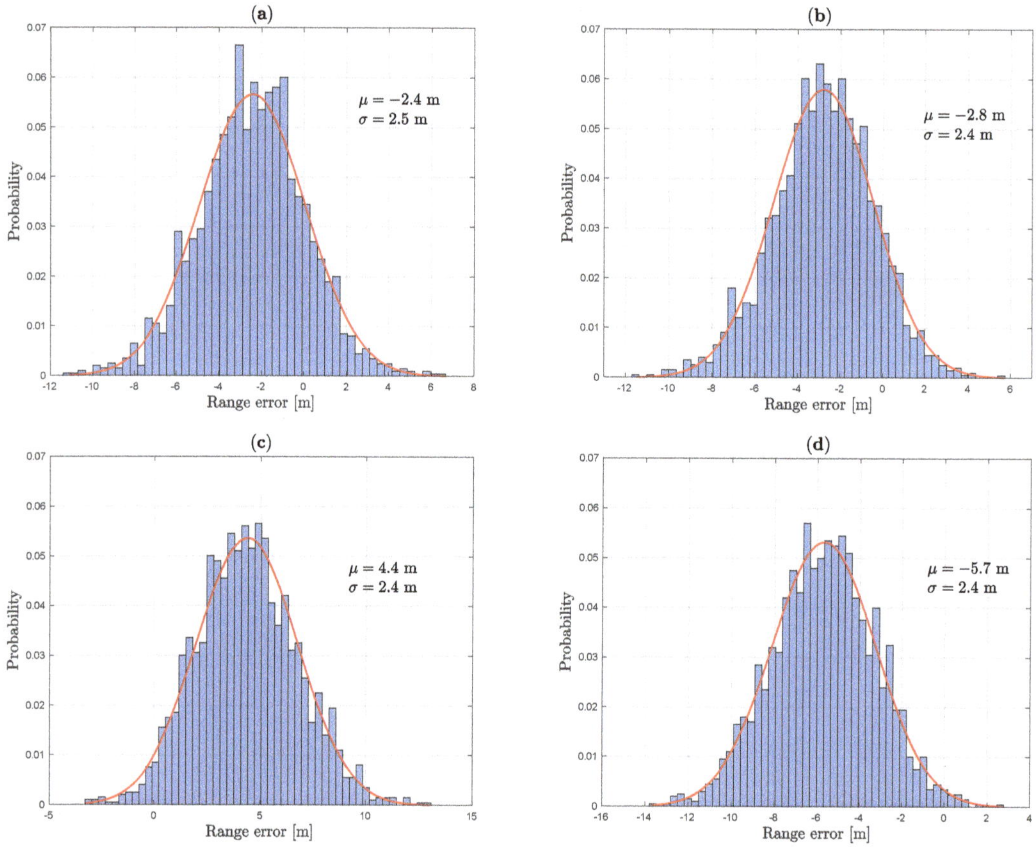

Figure 5. E1-B vs. E6-C code phase delay comparison ($F_s = 20$ MHz, 2000 portions averaged, different power cycles). (**a**) Dataset A. (**b**) Dataset B. (**c**) Dataset C. (**d**) Dataset D.

In Figure 6a, we compare the code phase delay between E1-B and E6-C using an additional dataset. To obtain more precise estimates, we use a sampling rate of 60 MHz, which is close to the maximum frequency supported by the bladeRF boards used in the platform for the nominal 12-bit mode. In this case, we use the same coherent integration time of 4 ms for both bands. Currently, this involves performing a secondary code acquisition on E6-C to coherently combine four 1 ms codes. However, in the nominal operating mode of ACAS, the receiver is synchronised with E1-B and, therefore, should know the location of the E6-C secondary code. Nevertheless, once the E6-C signal is encrypted and real ECSs are processed, the secondary code would not play any role, as the ECS provides the encrypted sequence after multiplying the keystream with the E6-C primary and secondary codes. As a result, the variance is reduced to less than 2 m.

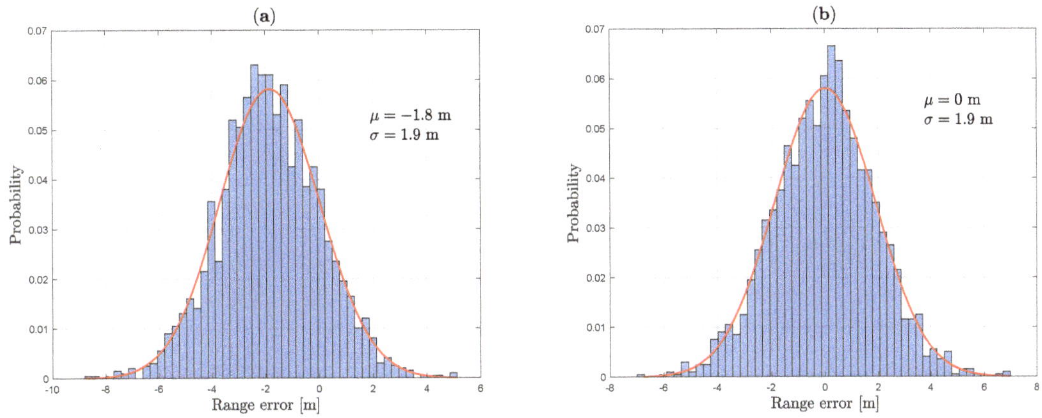

Figure 6. Code phase delay comparison, dataset E ($F_s = 60$ MHz, 2000 portions averaged, 4 ms of coherent integration time for both bands). (**a**) E1-B vs. E6-C. (**b**) E6-B vs. E6-C.

To diminish the hardware bias produced by using two distinct boards, we also provide a comparison of the code phase delay between the E6-B and E6-C using the latest dataset, but now processing a unique snapshot from the same board. As we can observe in Figure 6b, the obtained distribution exhibits a centred Gaussian-like shape in this case.

3.2. E1-E6 Doppler Frequency Comparison

In Figure 7, we compare the Doppler frequency estimates for both E6 and E1 bands using the latest dataset. For this comparison, we consider the relationship between both carrier frequency bands; hence, the Doppler frequency used for E6 is computed from the estimate obtained in E1 multiplied by the ratio f_{c_1}/f_{c_6}. We also use a coherent integration time of 4 ms for both bands. The obtained results indicate that the differences are sufficiently small to justify using a reduced frequency search space in E6 based on the E1 estimate.

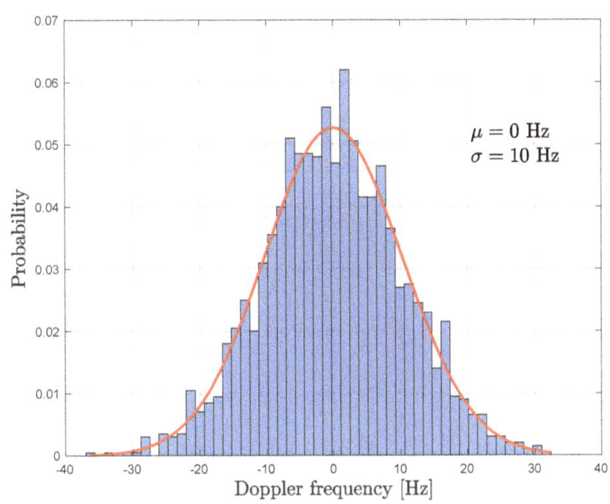

Figure 7. E1-B vs. E6-C Doppler frequency comparison, dataset E.

4. Conclusions and Further Work

ACAS aims to provide a mechanism to authenticate the Position, Velocity and Timing (PVT) using current Galileo signals. The OSNMA keys available in the E1-B signal are used to encrypt/decrypt some fragments of the E6-C signal, which are intended to be encrypted in the future. Unlike conventional approaches, in ACAS the receiver is required to record a snapshot at the time where these fragments are expected, as they are only broadcasted during certain instants. Therefore, the envisioned nominal operating mode is to first obtain a time reference from the open E1-B signal to reduce the uncertainty on the E6-C signal.

This is why analysing the alignment of the E1 and E6 estimates, which includes the code phase delay and Doppler frequency, is crucial for this mode. In this paper, we have developed a low-cost evaluation platform based on bladeRF SDR boards to examine this alignment using real datasets. The samples of E1 and E6 components were captured synchronously using an inferred, undocumented feature of the bladeRF boards. The results show the consistency between the E1 and E6 estimates, confirming the viability of ACAS using limited resources for typical GNSS receivers.

Further work will include the preliminary evaluation of ACAS using the existing E1-B and E6-C open signals using the evaluation platform used in this work [29].

Author Contributions: Software, validation, writing—original draft preparation, R.T.-G.; supervision, writing—review and editing, I.F.-H., J.A.L.-S. and G.S.-G. All authors have read and agreed to the published version of the manuscript.

Funding: This work was supported, in part, by the European Commission Defence Industry and Space Satellite Navigation (DEFIS) contract DEFIS/2020/OP/0002 and, in part, by the Spanish Agency of Research project PID2020-118984GB-I00 and by the Catalan ICREA Academia Programme. The views expressed herein can in no way be taken to reflect the official opinion of the European Commission.

Institutional Review Board Statement: Not applicable.

Informed Consent Statement: Not applicable.

Data Availability Statement: The datasets of real samples, recorded with the SDR evaluation platform used for this paper, can be downloaded from https://spcomnav.uab.es/resources/acas_datasets (accessed on 1 May 2023).

Acknowledgments: The authors would like to thank the PAULA project participants, J. Winkel, and C. O'Driscoll, for their discussions and contributions.

Conflicts of Interest: The authors declare no conflicts of interest.

References

1. Pozzobon, O.; Canzian, L.; Danieletto, M.; Chiara, A.D. Anti-spoofing and open GNSS signal authentication with signal authentication sequences. In Proceedings of the 5th ESA Workshop on Satellite Navigation Technologies and European Workshop on GNSS Signals and Signal Processing (NAVITEC), Noordwijk, The Netherlands, 8–10 December 2010; pp. 1–6. [CrossRef]
2. Scott, L. Anti-spoofing & authenticated signal architectures for civil navigation systems. In Proceedings of the 16th International Technical Meeting of the Satellite Division of The Institute of Navigation (ION GPS/GNSS), Portland, OR, USA, 9–12 September 2003; pp. 1543–1552.
3. Fernandez-Hernandez, I.; Vecchione, G.; Díaz-Pulido, F. Galileo authentication: A programme and policy perspective. In Proceedings of the 69th International Astronautical Congress, Bremen, Germany, 1–5 October 2018.
4. Fernandez-Hernandez, I.; Rijmen, V.; Seco-Granados, G.; Simon, J.; Rodríguez, I.; Calle, J.D. A navigation message authentication proposal for the Galileo open service. *Navig. J. Inst. Navig.* 2016, *63*, 85–102. [CrossRef]
5. European Commission. Call for Tenders (DEFIS/2020/OP/0002)—Test Platform on Galileo HAS/CAS/OSNMA—Tender Specifications, 2020. Available online: https://etendering.ted.europa.eu/cft/cft-display.html?cftId=6271 (accessed on 29 April 2023).
6. Anderson, J.M.; Carroll, C.K.L.; DeVilbiss, N.P.; Gillis, J.T.; Hinks, J.C.; O'Hanlon, B.W.; Rushanan, J.J.; Scott, L.; Yazdi, R.A. Chips-Message Robust Authentication (Chimera) for GPS Civilian Signals. In Proceedings of the 30th International Technical Meeting of the Satellite Division of the Institute 2388 of Navigation (ION GNSS+ 2017), Portland, OR, USA, 25–29 September 2017; Institute of Navigation: Portland, OR, USA, 2017.
7. Fernandez-Hernandez, I.; Winkel, J.; O'Driscoll, C.; Cancela, S.; Terris-Gallego, R.; Seco-Granados, G.; López-Salcedo, J.A.; Dalla Chiara, A.; Sarto, C.; Blonski, D.; et al. Semi-Assisted Signal Authentication for Galileo: Proof of Concept and Results. *IEEE Trans. Aerosp. Electron. Syst.* **2023**, *59*, 4393–4404. [CrossRef]

8. Terris-Gallego, R.; Fernandez-Hernandez, I.; López-Salcedo, J.A.; Seco-Granados, G. Guidelines for Galileo Assisted Commercial Authentication Service Implementation. In Proceedings of the International Conference on Localization and GNSS (ICL GNSS), Tampere, Finland, 7–9 June 2022.
9. Terris-Gallego, R.; López-Salcedo, J.A.; Seco-Granados, G.; Fernandez-Hernandez, I. Operating Modes and Performance Evaluation of Galileo Assisted Commercial Authentication Service. In Proceedings of the Institute of Navigation Conference (ION+ GNSS 2022), Denver, CO, USA, 19–23 September 2022.
10. European Commission. European GNSS (Galileo) Open Service—Signal-in-Space Interface Document v2.0, European Union 2021. Available online: https://www.gsc-europa.eu/sites/default/files/sites/all/files/Galileo_OS_SIS_ICD_v2.0.pdf (accessed on 29 April 2023).
11. LimeMicro Website—LimeSDR. Available online: https://limemicro.com/products/boards/limesdr/ (accessed on 29 April 2023).
12. Analog Devices Website—ADRV9364-Z7020. Available online: https://www.analog.com/en/design-center/evaluation-hardware-and-software/evaluation-boards-kits/ADRV9364-Z7020.html (accessed on 29 April 2023).
13. SDR Makerspace Wiki. Available online: https://gitlab.com/librespacefoundation/sdrmakerspace/sdreval/-/wikis/home (accessed on 29 April 2023).
14. Csete, A.; Christiansen, S. *Evaluation of SDR Boards and Toolchains—Final Report*; Technical Report; European Space Agency, Libre Space Foundation, Online, 2020. Available online: https://www.klofas.com/blog/2020/satnogs-station-and-minicircuits-lna-modifications/Evaluation_of_SDR_Boards-1.0.pdf (accessed on 29 April 2023).
15. SDR Makerspace Website. Available online: https://sdrmaker.space (accessed on 30 April 2023).
16. Borre, K.; Fernandez-Hernandez, I.; López-Salcedo, J.A.; Zahidul, M.; Bhuiyan, H. *GNSS Software Receivers*; Cambridge University Press: Cambridge, UK, 2022.
17. Akos, D.; Arribas, J.; Dovis, F.; Di Torino, P.; Fernandez-Hernandez, I.; Humphreys, T.; Salcedo, J.A.L.; Psiaki, M.L.; Tech, V.; Rügamer, A.; et al. GNSS Software Defined Radio: History, Current Developments, and Standardization Efforts. In Proceedings of the 35th International Technical Meeting of the Satellite Division of The Institute of Navigation (ION GNSS+), Online, 20–24 September 2021.
18. O'Driscoll, C.; Caparra, G. Nautilus: An embedded navigation authentication testbed. In Proceedings of the 34th International Technical Meeting of the Satellite Division of the Institute of Navigation, ION GNSS+, Online, 20–24 September 2021; pp. 3698–3710. [CrossRef]
19. Nuand Website—2023-02 Release. Available online: https://www.nuand.com/2023-02-release-122-88mhz-bandwidth (accessed on 1 May 2023).
20. Nuand Website—BladeRF 2.0 Micro. Available online: https://www.nuand.com/bladerf-2-0-micro/ (accessed on 29 April 2023).
21. Wenzel Citrine OCXO 501-24219. Available online: https://www.quanticwenzel.com/wp-content/parts/501-24219.pdf (accessed on 29 April 2023).
22. Soapy SDR Wiki. Available online: https://github.com/pothosware/SoapySDR/wiki (accessed on 30 April 2023).
23. Nuand Wiki—bladeRFInstallation. Available online: https://github.com/Nuand/bladeRF/wiki/ (accessed on 30 April 2023).
24. MacPorts Installation. Available online: https://www.macports.org/install.php (accessed on 1 May 2023).
25. Nuand Website—FPGA Images. Available online: https://www.nuand.com/fpga_images/ (accessed on 1 May 2023).
26. Nuand Website—FPGA Autoloading. Available online: https://github.com/Nuand/bladeRF/wiki/FPGA-Autoloading (accessed on 1 May 2023).
27. Nuand Website—FX3 Images. Available online: https://www.nuand.com/fx3_images/ (accessed on 1 May 2023).
28. Nuand Forum—Synchronized TRX. Available online: https://github.com/Nuand/bladeRF/tree/master/host/examples/bladeRF-cli/sync_trx (accessed on 30 April 2023).
29. Terris-Gallego, R.; López-Salcedo, J.A.; Seco-Granados, G.; Fernandez-Hernandez, I. Preliminary Evaluation of Galileo ACAS using Existing E1-E6 Open Signals and a Low-Cost SDR Platform. In Proceedings of the Institute of Navigation Conference (ION+ GNSS 2023), Long Beach, CA, USA, 25–27 January 2023.

engineering proceedings

MDPI

Proceeding Paper

S-TrackS: A Secure Snapshot-Based Solution for Positioning and Timing †

Aram Vroom, Tom van den Oever, Joaquín Gañez Fernandez, Nick van der Hijden, Alexandra Zevenbergen * and Bas van der Hoeven

CGI Nederland B.V., 3068 Rotterdam, The Netherlands
* Correspondence: alexandra.zevenbergen@cgi.com

† Presented at the European Navigation Conference 2023, Noordwijk, The Netherlands, 31 May–2 June 2023.

Abstract: With the large-scale usage of satellite navigation, spoofing and jamming are considerable threats to civilian society. Recent developments, such as Galileo's Open Service Navigation Message Authentication and GPS's Chimera, mitigate these risks. However, they provide authentication of the navigation message or ranging code, but not a true position in the case of interference. In critical applications, a protected navigation service is desired, such as Galileo's Public Regulated Service (PRS). PRS provides an access-controlled navigation service for authorized governmental users, with fully encrypted ranging codes and data channels, providing users with higher robustness against interference. The main challenge of implementing PRS on a large scale is the need to protect the cryptographic material that is required to access the PRS signals inside the receiver. For many applications, a stand-alone receiver solution is unnecessary. These applications could use a remote server for PRS. In this methodology, the end-user device has only a radio frequency front-end which sends short samples to a secure server. The (classified) signal processing is then carried out on this secure server, removing the need for the user device to protect cryptographic material. Besides decreasing the device's security requirements and power consumption, it also allows to utilize the advantages of PRS in applications that would otherwise not be able to use PRS. In this approach the PRS usage authorization would only be required for the server operations, and not for the end-user devices. It furthermore allows for using additional processing power for unaided PRS acquisition in case of interference. Within the Netherlands, a remote server solution is developed by CGI: S-TrackS, making PRS accessible. In this paper, the application of PRS and architecture for various use cases is presented. It is shown that PRS usage based on a remote server is feasible and can increase the robustness for governmental applications.

Keywords: GNSS; satellite navigation; Galileo; spoofing; jamming; snapshot positioning; encrypted signals; Public Regulated Service; authentication

Citation: Vroom, A.; van den Oever, T.; Gañez Fernandez, J.; van der Hijden, N.; Zevenbergen, A.; van der Hoeven, B. S-TrackS: A Secure Snapshot-Based Solution for Positioning and Timing. *Eng. Proc.* **2023**, *54*, 31. https://doi.org/10.3390/ENC2023-15457

Academic Editors: Tom Willems and Okko Bleeker

Published: 29 October 2023

1. Introduction

Since the development of GPS, GNSS receivers have been integrated in many aspects of modern day society. Satellite signals are used to determine position, velocity, and time (PVT) solutions, which are subsequently used to, for example, position ships, land aircraft, monitor remote assets, and synchronize system clocks. Society is therefore dependent on safe and robust GNSS signals.

Although GNSS receivers provide many benefits, the signals used for navigation are extremely weak, namely −128.5 dBm and −127.25 dBm for GPS L1 C/A and Galileo E1BC, respectively [1,2]. Consequently, a small jammer with a signal power of 10 dBm can disrupt PVT determination over a multi-km area. The disruption of a jammer can be noticed by the lack of proper PVT determination. Alternatively, a spoofer can overpower and falsify the satellite signals, convincing a receiver of a false position and/or time.

Galileo's Open Service Navigation Message Authentication [3] and GPS's Chimera mitigate some risks related to spoofing. They provide authentication of the navigation message or ranging code, but not a true position in the case of interference. In applications where continuity of service is critical, a protected navigation service is desired, such as Galileo's Public Regulated Service (PRS). PRS provides an access-controlled navigation service for authorized governmental users [4], with fully encrypted ranging codes and data channels. Hence, it is not possible for an attacker to spoof PRS, as the encrypted ranging codes and navigation data cannot be reproduced without the required cryptographic information. In addition, PRS is transmitted over a wider bandwidth compared to the public GPS and Galileo signals, increasing its resilience against jamming.

The challenge of utilizing PRS is that a stand-alone receiver will contain classified cryptographic information, which means that the receiver needs to adhere to strict security requirements. In order to mitigate this burden, a server-assisted PRS solution is proposed, where short RF recordings or snapshots are captured and processed on a secure server. Authorized use cases can therefore make use of the server-assisted PRS positioning service, without the need for hosting classified information in the user equipment. These PRS position–time solutions can then be used to authenticate the Open Service (OS) positions, and provide a true solution in case of spoofing or jamming. In this context, CGI Nederland B.V. developed S-TrackS [5], a snapshot-based authentication service based on the server-assisted PRS concept.

In this paper, the concept of snapshot processing techniques is introduced, together with the architecture of S-TrackS. Following, the test setup for a demonstration with S-TrackS is described. Finally, the results of a test campaign are shared, assessing the accuracy, the performance of the authentication service and the impact of metadata errors.

2. Snapshot Processing Techniques

In conventional GNSS receivers, both the radio frequency (RF) data collection and processing stages are carried out on the device. Conversely, novel techniques allow the separation of the processing stage from the user device. For instance, a Software-Defined Radio (SDR) can be used to collect the GNSS signals, which can be remotely post-processed to find a position–time solution.

The main disadvantage of the separated approach is that the file size of the RF recordings can be prohibitive for practical use since they can easily reach sizes of several gigabytes per recorded minute. Hence, techniques are developed to minimize the required recording time to find a position–time solution. If the ephemeris data are externally provided, it will not be necessary to decode the navigation message. By doing so, the total length of the recording can be reduced to well below 100 ms. These short recordings are called snapshots and contain a short snippet of the RF signals, which is used to find a single position–time solution.

Snapshot-based navigation offers the following main advantages:

1. The user device only needs to be enabled for the recording duration. If a use case requires one snapshot per minute, the duty cycle of the device can be reduced.
2. The full spectrum is available, enabling a wide variety of post-processing options, such as the analysis of interference.
3. The separation between user device and processing allows for the decryption of classified signals in a secure area, as utilized in the server-assisted PRS approach. Therefore, the security requirements of the user device are relaxed.
4. The server on which the application runs can have more processing power than a GNSS receiver. This enables more computationally intensive calculations and algorithms.

S-TrackS is CGI Nederland B.V.'s server-assisted PRS authentication service. For authorized use-cases, snapshots are processed in order to authenticate position–time solutions, detect spoofing and resolve the true position–time based on PRS. An overview of S-TrackS can be seen in Figure 1.

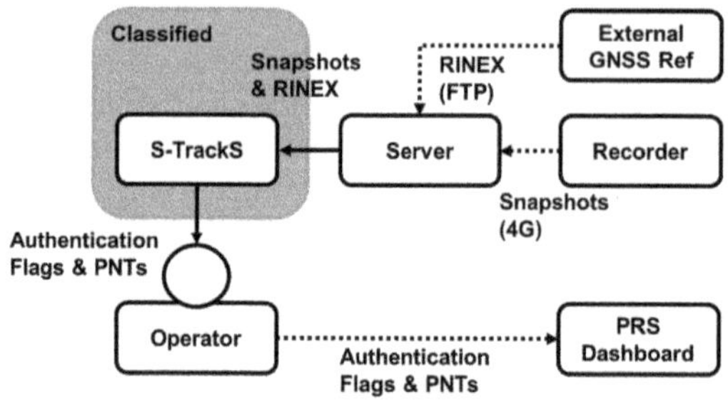

Figure 1. Overview of S-TrackS displaying the separated processing architecture [6].

An RF recorder can be used by a user in the field. For this purpose, prototype S-TrackS recorders are developed. S-TrackS is developed using open standards for RF recordings and is compatible with the ION GNSS-Metadata standard for Software-Defined Radio receivers [7]. Users of S-TrackS therefore have the possibility to utilize third party RF recording equipment and require no specific hardware from CGI Nederland B.V. The current S-TrackS recorder needs to be operated by a user, but commercially available RF front ends can be integrated in an existing track and trace chain. The collection of snapshots can then automatically be triggered based on events or timers.

The snapshots are automatically forwarded to a server through a REST API. The server then collects the snapshots and a RINEX navigation file from an external server. Within the classified environment, the snapshots are processed by S-TrackS, which includes the required infrastructure and classified information to process PRS.

Finally, the results can be inspected by an operator, and depending on the use case, shared with users through the PRS dashboard or interface of the user.

The architecture of S-TrackS is displayed in Figure 2.

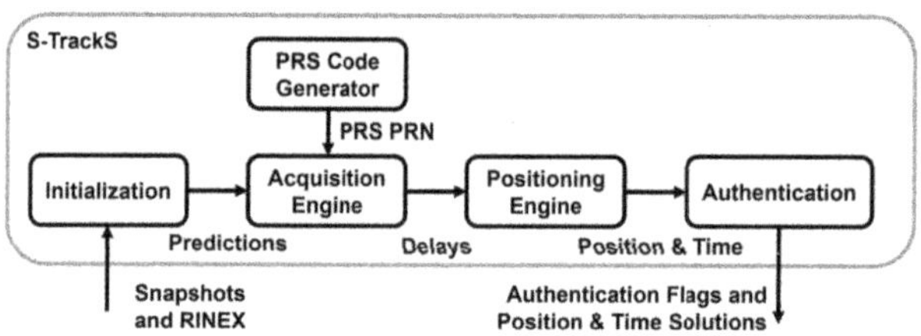

Figure 2. Architecture of S-TrackS.

The following building blocks can be identified:

- Initialization. During the initialization, the coarse position–time data from the metadata file are used together with a RINEX navigation file to predict the visible satellites and their Doppler frequencies. The metadata file also contains information on the recording configuration such as bit depth, sampling rate, bandwidth, and number of channels.
- Acquisition Engine. The acquisition engine searches for the transmitted signals. Here, the parallel code–phase search technique is used. As there is no tracking loop involved, it is important that the code delay and Doppler frequency found by the acquisition engine are refined. The acquisition engine supports the following open signals: Galileo E1BC, Galileo E6BC and GPS L1 C/A. It furthermore supports the encrypted Galileo E1A and E6A PRS signals.
- Positioning Engine. Pseudoranges are generated based on the code delays for each acquired signal in the acquisition engine. These are then used to find the position–time solutions for the various signals. Without tracking, the time of week (TOW) from the navigation message cannot be used to construct the pseudorange. Hence,

the number of integer PRN codes between the satellite and receiver is unknown. To solve this, coarse time navigation and millisecond integer ambiguity fixing have been implemented. The latter of these works by choosing a reference satellite and computing the integer number of PRN codes for the other satellites based on this [8].

- Authentication. The position–time authentication is performed based on both the acquired signals and the found position–time solutions. This authentication is performed on two levels. Firstly, it is carried out by comparing the delay for each acquired Galileo OS and PRS signal. Secondly, the obtained public position–time are compared to the PRS position–time and their differences are determined. Authentication is then based on whether these differences are within the configured thresholds.

An example of a successful authentication and spoofing detection based on a delay between the Galileo OS and PRS signals can be seen in Figure 3.

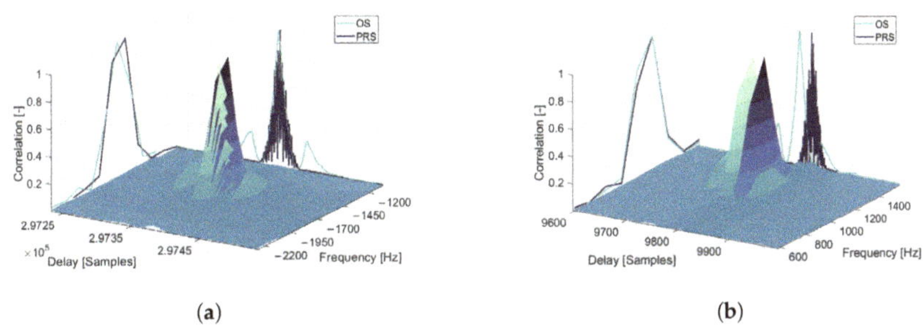

Figure 3. Authenticated (**a**) and spoofed (**b**) correlation peaks. (**a**) Authenticated correlation peak, with overlapping PRS and OS delays. (**b**) Spoofed correlation peak, with an offset between PRS and OS delays.

Once the processing has completed, S-TrackS returns a result string containing the authentication flag for the specific snapshot and a true position–time based on Galileo PRS. These results can be:

- Authenticated. This flag is returned if the Galileo PRS results matches the Galileo OS results and all position–time solutions match within set thresholds.
- Spoofed. This flag is returned if the PRS delays do not match the OS delays, or the position–time solutions do not match within set thresholds.
- Warning. This flag is returned if the PRS component is missing and no comparison can therefore be made.

3. Test Setup and Data Collection

In order to demonstrate S-TrackS as a position–time authentication service, a test campaign is carried out. The objective is to demonstrate S-TrackS' capability to authenticate the public signals and resolve PRS positions. Therefore, snapshots are statically collected under various spoofing conditions.

One hour of clean RF data are recorded using the Labsat3 Wideband with the Ardusimple ANT3B survey antenna, capturing both the OS and PRS signals in the E1 and E6 bands. The IQ data are captured using 2 bits of quantization and 58 MHz sampling rate.

The test data are collected in the Prinsenpark near the CGI Nederland B.V. Rotterdam office (51°55′56.7″ N, 4°32′57.9″ E). This is an open environment without any obstructions.

3.1. Nominal Scenario

First, a set of nominal snapshots is recorded. Here, the RF data collected with the Labsat is directly replayed and recorded with the S-TrackS recorder. The recorder is configured to capture the E1 and E6 bands once every 15 s. The recorder configuration can be seen in Table 1.

Table 1. Recording specifications.

Parameter	Value
Sample Rate	80 MHz
Format	IF
Recording Time	25 ms
Bands	2 (E1 and E6)
Quantization	2 bits
File Size	1000 kB

Using this setup, a total of 240 snapshots are collected, in order to evaluate the nominal performance of S-TrackS. The performance is computed by comparing the resulting position–time solutions with a precise point positioning (PPP) solution. The performance metrics consist of the Circular Error Probable (CEP) and 2D root mean square error (2DRMS). Furthermore, any potential bias is computed by comparing the mean of obtained positions with the reference PPP position.

3.2. Impact of Signal Spoofing

Secondly, sets of snapshots with spoofed signals are recorded. As in the nominal scenario, the aforementioned test data are replayed and recorded with the S-TrackS recorder. As it is illegal to spoof GPS or Galileo over the air, simulated signals are inserted over the cable. In order to simulate a realistic scenario, a HackRF One SDR [9] is used together with a signal combiner to simulate spoofing.

Two GPS and one Galileo spoofing scenarios are carried out. The datasets consists of 240 snapshots for each scenario. GPS L1 C/A is spoofed to a distance of 200 m and 200 km. These two cases evaluate the situation where Galileo OS signals can still be used to aid the acquisition of PRS signals. In the third case, Galileo OS signals are spoofed to a distance of 200 km and PRS needs to be acquired without any aiding data.

3.3. Impact of the Coarse Position–Time

Thirdly, a sensitivity analysis is carried out on the quality of the assistance data. As the coarse position–time in the metadata file are used within the processing, inaccurate assistance data can lead to errors. An analysis is therefore performed to investigate the impact of errors in the coarse position–time on the performance of S-TrackS.

For this analysis, one snapshot is taken from the nominal scenario. This snapshot is then reprocessed with varying position–time errors added to the assistance data. The obtained position–time solutions are then compared to the nominal solution. The position error is increased from 0 to 500 km in a Northward direction with a step size of 10 km. The time error is increased from 0 to 600 s, with a step size of 12 s.

4. Results

The results of this study are presented in this section. For each test, the data are processed using S-TrackS, and relevant statistics are derived. The results provide insight in the performance with respect to accuracy, spoofing detection, and sensitivity to metadata.

Due to the classification of PRS, no accuracy statistics of the actual signal are disclosed. As it is processed using a similar approach using code-based measurements from Galileo satellites, results comparable to those of E1BC and E6BC are found.

4.1. Nominal Results

The nominal results without spoofing are presented in this section. In total 240 positions are collected in the nominal dataset. The number and percentage of found position–time solutions per snapshot are shown in Table 2.

Table 2. Found position–time solutions within the nominal dataset.

Snapshots	GPS L1	E1BC	E1A	E6BC	E6A	Authenticated
Number	240	240	238	240	238	240
Percentage	100%	100%	99.17%	100%	99.17%	100%

In four snapshots, either an E1A or an E6A position is not found, but the other is still determined. All snapshots have therefore been authenticated. The accuracies for the L1 C/A and E1BC position solutions are shown in Figure 4a,b.

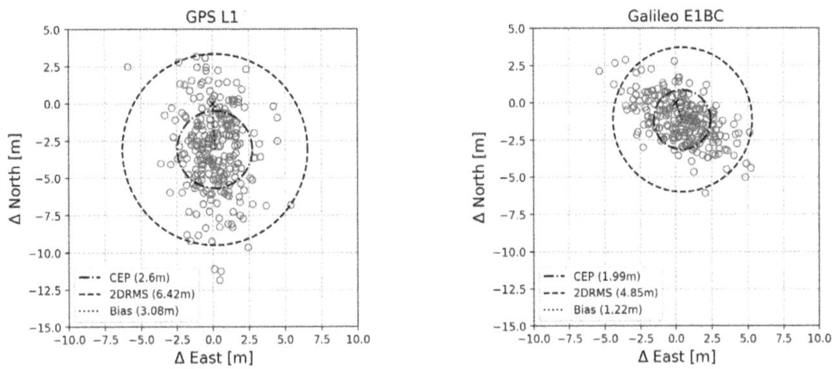

(**a**) GPS L1 accuracy. (**b**) Galileo E1BC accuracy.

Figure 4. Accuracies derived using S-TrackS for GPS L1 C/A (**a**) and Galileo E1BC (**b**).

The 2DRMS's for GPS L1 and E1BC are found to be 6.42 m and 4.85 m, respectively. In addition, a bias of 3.08 m is found for GPS and 1.22 m for Galileo. In order to improve the accuracy, the observations of both GPS and Galileo are combined in a single position–time solution as shown in Figure 5a. In addition, the associated cumulative density function (CDF) of the horizontal position errors (HPE) is shown in Figure 5b.

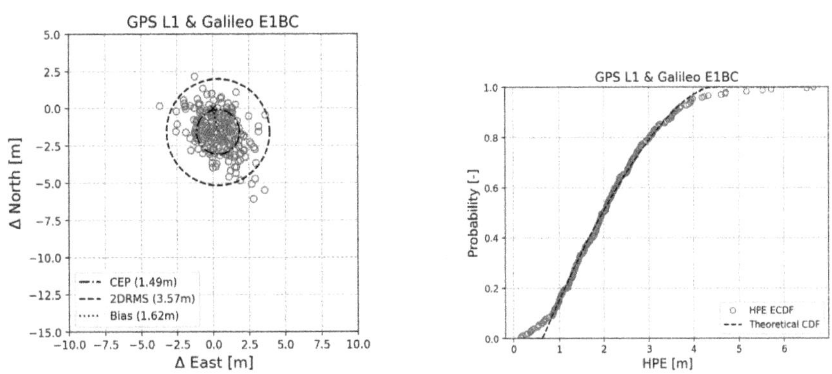

(**a**) (**b**)

Figure 5. Accuracies derived using S-TrackS for the combined Galileo E1BC and GPS L1 C/A position–time solutions (**a**) and the corresponding error distribution (**b**). (**a**) Combined GPS L1 and Galileo E1BC accuracy. (**b**) Distribution of errors for GPS L1 and Galileo E1BC.

When combining the GPS and Galileo observations in a single solution, it can be seen that the 2DRMS is reduced to 3.57 m. Furthermore, it can be seen that the bias is reduced from 3.08 m for GPS to 1.65 m for the combined GPS and Galileo solution.

It is found that the errors are not normally distributed and that the CDF therefore does not resemble a standard χ^2 distribution. This can also be seen from the difference between the mean position's bias in Figure 5a and the median shown in Figure 5b. This difference indicates a skewed distribution. In order to estimate the theoretical CDF more accurately, the generalized Pareto distribution (GPD) has therefore been chosen as in [10].

4.2. Impact of Signal Spoofing

The results of spoofed datasets are shown in Table 3, which includes whether or not the true PRS position–time were resolved, and which flag was returned by S-TrackS.

Table 3. Returned S-TrackS Results in various simulated scenarios.

Spoofing	PRS Position–Time Found (%)	Authenticated Flag (%)	Spoofed Flag (%)	Warning Flag (%)
Nominal	100%	100%	0%	0%
GPS (200 m)	98.75%	0%	98.75%	1.25%
GPS (200 km)	98.75%	0%	98.75%	1.25%
Galileo (200 km)	100%	0%	100%	0%

During both the first and second spoofing scenario, a true PRS position was found in 98.75% of the cases. The spoofing was therefore detected as well in 98.75% of the instances. In 1.25% of these instances, PRS signals were not found, and a warning flag was returned. In the third spoofing scenario, the Galileo E1BC signal was spoofed and could not be used to aid the acquisition of E1A PRS signals. Although this requires more processing time, the PRS positioning and time were resolved in 100% of the instances. This demonstrates that S-TrackS can successfully obtain PRS position–time solutions, independent on the state of public GPS and Galileo signals.

4.3. Impact of the Coarse Position–Time

As not all tracking devices are able to provide an accurate position–time in the metadata file, the sensitivity of S-TrackS to errors in these values is analyzed.

The robustness of the millisecond ambiguity fixing depends on the duration of the PRN codes. When a longer PRN code is used, a larger offset can be present in the estimated position–time before a different integer number of codes are present between the satellite and receiver. Hence, the longer codes of Galileo E1BC can aid in resolving the integer number of PRN codes for GPS L1 C/A when used together with Galileo E1BC measurements. By choosing one of the Galileo satellites as reference satellite in the millisecond ambiguity algorithm, the number of PRN codes can be estimated more accurately for GPS L1 C/A. The horizontal position–time differences compared to the case where no metadata errors are added can be found in Figure 6.

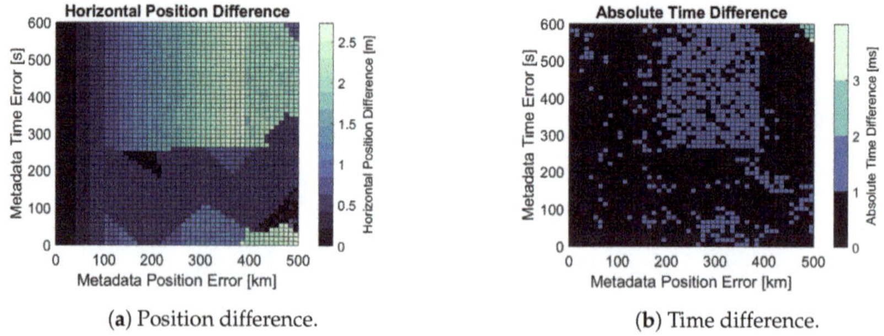

(**a**) Position difference. (**b**) Time difference.

Figure 6. Impact of errors in the coarse position–time on the position (**a**) and time (**b**) found for GPS L1 C/A and Galileo E1BC.

In Figure 6a, it can be seen that the horizontal position difference stays within 2.73 m of the nominal case. Furthermore, one can see in Figure 6b that the timing difference also stays within 3 ms. Based on these results, it can be concluded that errors up to 500 km and 10 min in the coarse position–time only have a marginal impact on the obtained position–time solution.

5. Discussion

It was found that the number of obtained E1A position–time solutions was higher than for E6A. As each Galileo satellite transmits both E1 and E6, this is an unexpected result of the study. Two possible explanations are considered. First of all, the signal sensitivity of the RF front end within the S-TrackS recorder might be lower for E6 than it is for E1. Secondly, more interference sources might be present within the E6 band. More research is required on the question of why the E6 sensitivity was found to be lower than E1.

There is a trade-off between the file size of a snapshot and the robustness of resolving a PRS position–time solution. Within this paper, a snapshot length of 25 ms was selected, as it provides an acceptable level of robustness while minimizing the file size of a snapshot. For dedicated use cases, a higher level of robustness might be required. In these cases a longer snapshot recording time can be used to increase the acquisition sensitivity.

This paper showed that the accuracy of the coarse position–time provided in the metadata file marginally influences the performance. As most IoT devices have access to a coarse position–time within a few kilometers and a few minutes, this data can be used to obtain reliable position–time solutions. Even without such assistance data, knowing the approximate time and the country that the device is in can be accurate enough. However, performing Galileo PRS acquisition on E1A or E6A without any Galileo OS signals on either frequency will be significantly faster if an accurate time is known.

With respect to the test setup, a RTK receiver could be used to increase the accuracy of the reference position. A bias was furthermore observed between this reference position and the positions found by S-TrackS. Preliminary tests have shown that the usage of precise products, such as precise ephemerides, in S-TrackS can decrease this bias.

6. Conclusions

Snapshot positioning in combination with PRS can provide a high level of security for users who do not require an instantaneous verification of their position–time. Snapshot positioning therefore offers a solution with eased requirements on user equipment compared to a stand-alone PRS receiver. This paper has demonstrated that currently available RF front ends can be utilized to deploy snapshot services. Furthermore, the snapshot based concept allows flexibility and can be configured to the users' needs.

During the nominal test, 100% of the snapshots were successfully authenticated, demonstrating the robustness of S-TrackS. In this test the accuracy of S-TrackS was evaluated. A 2DRMS of 3.57 m with a bias of 1.62 m was obtained using GPS L1 C/A and Galileo E1BC signals.

It has been demonstrated that S-TrackS can be used to authenticate position–time solutions under various spoofing conditions. During the GPS spoofing test, spoofing was detected and the true position was found with the aid of PRS in 98.75% of the snapshots. In the Galileo spoofing test, the true position was found in all the snapshots. This demonstrates that S-TrackS can successfully obtain PRS position–time solutions, independent on the state of the public GPS and Galileo signals.

Lastly, the influence of errors in the coarse position–time provided in the metadata file was analyzed. It was shown that metadata values within 500 km and 10 min of the truth only had a marginal impact on the obtained position–time solution. Hence, it is demonstrated that S-TrackS can reliably be used as an authentication service if the position–time are approximately known.

7. Patents

The developments as described in this paper have led to innovative solutions for the future of secure navigation technology. The solution described in this paper has been patented through the European Patent Office, published in Patent Number EP3923032 [11].

Author Contributions: Conceptualization, A.V. and T.v.d.O.; methodology, A.V., T.v.d.O. and J.G.F.; software, A.V., T.v.d.O., J.G.F. and N.v.d.H.; validation, T.v.d.O. and J.G.F.; formal analysis, A.V., T.v.d.O., J.G.F. and N.v.d.H.; investigation, A.V. and J.G.F.; resources, J.G.F.; data curation, T.v.d.O., J.G.F. and N.v.d.H.; writing—original draft preparation, A.V., T.v.d.O. and J.G.F.; writing—review and editing, N.v.d.H. and A.Z; visualization, A.V. and N.v.d.H.; supervision, A.Z. and B.v.d.H.; project administration, A.Z. and B.v.d.H.; funding acquisition, A.Z. and B.v.d.H.; All authors have read and agreed to the published version of the manuscript.

Funding: The research received no external funding.

Institutional Review Board Statement: Not applicable.

Informed Consent Statement: Not applicable.

Data Availability Statement: The data supporting this research is not available due to commercial restrictions.

Acknowledgments: The authors of this paper would like to thank the Ministry of Infrastructure and Water Management, the CPA and the Netherlands Space Office for enabling PRS developments in The Netherlands.

Conflicts of Interest: The authors of this paper were employed by CGI Nederland B.V.

References

1. Anthony, T. NAVSTAR GPS Space Segment/Navigation User Segment Interfaces, IS-GPS-200. 22 August 2022. Available online: https://www.gps.gov/technical/icwg/IS-GPS-200N.pdf (accessed on 2 May 2023).
2. European Union. European GNSS (Galileo) Open Service—Signal in Space Interface Control Document, OS SIS ICD. January 2021. Available online: https://www.gsc-europa.eu/sites/default/files/sites/all/files/Galileo_OS_SIS_ICD_v2.0.pdf (accessed on 2 May 2023).
3. European Union. Galileo Open Service Navigation Message Authentication Signal-in-Space Interface Control Document, OSNMA SIS ICD. January 2021. Available online: https://www.gsc-europa.eu/sites/default/files/sites/all/files/Galileo_OSNMA_SIS_ICD_v1.0.pdf (accessed on 4 May 2023).
4. European Union Agency for the Space Programme. PRS. June 2021. Available online: https://www.euspa.europa.eu/european-space/galileo/services/prs (accessed on 4 May 2023).
5. CGI. S-TrackS: Reliable Positioning Validation. Available online: https://www.cgi.com/en/media/brochure/s-tracks-reliable-positioning-validation (accessed on 8 May 2023).
6. European Union Agency for the Space Programme. Power-Efficient Positioning for the Internet of Things—White Paper. 2020; ISBN 978-92-9206-048-0. Available online: https://www.euspa.europa.eu/sites/default/files/uploads/gsa_internet_of_things_white_paper.pdf (accessed on 4 May 2023).
7. Institute of Navigation. GNSS Software Defined Receiver Metadata Standard. January 2020. Available online: https://sdr.ion.org/ (accessed on 3 May 2023).
8. van Diggelen, F.S.T. *A-GPS: Assisted GPS, GNSS, and SBAS*; Artech House: Boston, MA, USA, 2009.
9. Great Scott Gadgets. HackRF One. Available online: https://greatscottgadgets.com/hackrf/one/ (accessed on 8 May 2023).
10. Ahmad, K.; Sahmoudi, M.; Macabiau, C. Characterization of GNSS Receiver Position Errors for User Integrity Monitoring in Urban Environments. In Proceedings of the ENC-GNSS 2014, European Navigation Conference, Rotterdam, The Netherlands, 15–17 April 2014.
11. Vroom, A.; van den Berg, A.; van den Oever, T.D.; Ciuban, S. Method for Detecting Potential Tampering with Satellite Navigation Signals and/or for Determining a Position. European Patent EP 3923032, 10 June 2020.

engineering proceedings

MDPI

Proceeding Paper

Exploring the Contribution of PNT LEO Satellites to Precise Positioning Applications †

Jorge Durán, Damián Socías, Enrique Carbonell, Ana González, David Calle and Irma Rodríguez *

GMV, Madrid 28760, Spain; joduran@gmv.com (J.D.); damian.socias@gmv.com (D.S.); ecarbonell.ext@gmv.com (E.C.); agonzalez@gmv.com (A.G.); jdcalle@gmv.com (D.C.)

* Correspondence: irodriguez@gmv.com

† Presented at the European Navigation Conference 2023, Noordwijk, The Netherlands, 31 May–2 June 2023.

Abstract: Positioning services based on GNSS (Global Navigation Satellite Systems) have been using MEO satellites designed to guarantee Earth global coverage for decades. This satellite distribution is sometimes complemented with satellites in Inclined Geosynchronous (IGSO) and Geostationary (GEO) Orbits to improve satellite visibility in particular service areas. During recent years, with the advancements and reduction in costs in the deployment of LEO (Low Earth Orbit) constellations, the opportunity of using LEO satellites for PNT (Positioning, Navigation, and Timing) solutions is being studied. This concept opens the possibility to provide high accuracy positioning overcoming the typical drawbacks of RTK (Real-Time Kinematics) or PPP (Precise Point Positioning), such as the need for ground infrastructure or long convergence times. The high velocity dynamics of the low orbits can help to cancel the effect of the ionosphere in the signals to be processed at the user level. Therefore, the introduction of LEO satellites together with the classical MEO GNSS constellations could be a solution to reduce the dependency on dense station networks. The size of the LEO constellations and the design of their orbits are key factors to improve the PPP solution. Moreover, both the accuracy and the convergence time of the PPP solution depend also on the quality of the on-board equipment of the satellite, especially on the quality of the atomic clock in terms of stability and noise, and on the quality of the orbit and clock corrections sent to the PPP users. GMV has decades of experience in both GNSS and LEO precise orbit determination (POD) fields and in high-accuracy GNSS applications for different market domains. With this experience, several analyses have been carried out to assess the achievable performance when introducing the processing of LEO signals for high accuracy positioning solutions, contributing to the overall GNSS community. The objective of this paper is to describe the analysis run by GMV with the use of synthetic data simulating GNSS and LEO signals, showing results and the associated assessment of the achievable performance.

Keywords: LEO-PNT; high accuracy positioning; precise orbit determination (POD)

Citation: Durán, J.; Socías, D.; Carbonell, E.; González, A.; Calle, D.; Rodríguez, I. Exploring the Contribution of PNT LEO Satellites to Precise Positioning Applications. *Eng. Proc.* **2023**, *54*, 33. https://doi.org/10.3390/ENC2023-15426

Academic Editors: Tom Willems and Okko Bleeker

Published: 29 October 2023

1. Introduction

The emergence of LEO constellations is transforming an industry that was previously dominated by space agencies and public institutions. These large satellite constellations aim to push beyond the limitations of public GNSS systems, and there is a growing need to anticipate the entry of third-party players in this market, in addition to the potential developments for systems such as BeiDou or Galileo. The demand for LEO satellite applications is rapidly increasing, with communication constellations like Iridium, Starlink, Kuiper, OneWeb, and Telesat, as well as Earth observation systems like Spire, Planet, and BlackSky. Various companies like Xona Space, Trustpoint, and Satelles are already working on providing solutions for future user navigation technologies.

This promising development is paving the way for the challenging task of ensuring user interoperability among all LEO PNT players. As the industry continues to evolve, it is becoming clear that LEO constellations will continue to play a significant role in shaping the future of PNT applications.

The use of LEO constellations for PNT applications has been gaining momentum in recent years, as the demand for high-accuracy positioning and timing services continues to grow [1]. One of the main advantages of LEO constellations is their ability to provide high-power signals that can be received by GNSS-enabled devices with low power requirements. This makes them particularly suitable for applications such as asset tracking, IoT, wearables, and other similar applications. In addition, LEO constellations can provide enhanced satellite availability at polar regions, where the geometry of MEO and GEO constellations is not optimal for PNT solutions [. This can enable navigation and other applications in these areas, where traditional GNSS signals may not be available or may be unreliable.

Another advantage of LEO constellations is their potential to improve PPP convergence. Standard PPP algorithms typically require between 30 and 40 min to achieve a fully converged solution [2]. However, by using LEO satellites within the GNSS constellations, this convergence time can be reduced significantly. LEO satellites are closer to the Earth and have faster evolution dynamics, which means that they can potentially enable more accurate positioning solutions with almost immediate convergence. On one hand, the proximity of these orbits increases the signal strength received by the user. On the other hand, the fast change in geometry helps improve the estimation of those parameters not related with geometry and is crucial for faster convergence such as the ionospheric delay or the ambiguity. Furthermore, LEO constellations can be used to enhance the Galileo OS by providing additional measurements that can be used to enhance the ODTS (Orbit Determination and Time Synchronization) process and provide better estimations of ionosphere delays for mono-frequency users. This can improve the overall performance of the GNSS system and enable new applications that require high-accuracy positioning and timing services.

High accuracy services have become increasingly important, and different types of applications have strengths and weaknesses that need to be considered. Two main groups of applications can be analyzed based on their time to converge to sub-decimeter position accuracy, the required ground infrastructure, and the required bandwidth. RTK applications, based on differential positioning, can converge almost instantaneously but require a maximum separation of 25 km between a rover and a reference station [3], resulting in high infrastructure costs. Standard PPP algorithms rely solely on orbit and clock information and require between 30 and 40 min to achieve a fully converged solution. Techniques such as PPP-RTK require additional processing of local ionosphere corrections, which may lose some of the advantages of RTK. MEO, IGSO, and GEO satellites are far away from the Earth, requiring several minutes to achieve centimeter-level accuracy, primarily due to solving the carrier-phase ambiguity values. By using LEO satellites within the GNSS constellations, these issues can be addressed, potentially removing the precise atmospheric corrections' dependency for the RTK-PPP users. GMV has extensive experience in GNSS and LEO Precise Orbit Determination domains and high accuracy positioning user applications. This paper aims to bring these experiences together and validate the concept of the LEO PNT constellation's contribution to the GNSS community in the future [4,5].

This paper focuses on the determination of the optimal constellation definition that improves performance and convergence time for a PPP solution using the GSharp product [6] from GMV. The definition of the constellation also includes an analysis on the maximum degradation of the orbits and clocks of the LEO satellites that PPP can manage without degrading performances and the convergence time. The methodology employed to simulate the data, the different configurations of the LEO constellations under study, and the strategy to analyze their impact into the positioning solution are explained. In Section 3, the results obtained following the defined approach are presented.

2. Methodology

2.1. Data Simulation

For this paper, GNSS observations for MEO (GPS + GAL) and LEO satellites were synthetically generated for a given static position, simulating a GNSS ground station [7]. Signals generated for all the constellations are in frequency band L1 and L5 of GPS. GNSS measurements were generated considering various factors such as models for the ionosphere and troposphere, the effects of tides, and a Gauss–Markov noise model for situations with degraded accuracy, including multipath and white noise. The study did not account for any errors in the positions of the satellites or in the synchronization of their clocks. Therefore, pseudorange and phase observations are as follows:

$$R = \rho + c\,(dt_{rcv} - dt_{sat}) + Tr + I + Kp_{rec} - Kp_{sat} + \epsilon_R \quad (1)$$

$$\Phi = \rho + c\,(dt_{rcv} - dt_{sat}) + Tr - I + kL_{rec} - kL_{sat} + \lambda N + \epsilon_L \quad (2)$$

where

ρ is the geometric range between Antenna Phase Centers of the satellite and receiver.

dt_{rec} and dt_{sat} are the clock offsets between the receiver/satellite and the GNSS time used as the reference time. In this paper, clocks are assumed to be synchronized, and these terms are equal to zero.

Tr is the modelled tropospheric delay.

I is the modelled ionospheric delay.

Kp_{rec} and Kp_{sat} are the hardware delays of the receiver and satellite. This contribution to pseudorange measurement is assumed as zero for simplicity of the analysis performed in this paper.

λN is the integer ambiguity of the phase measurement.

kL_{rec} and kL_{sat} are the instrumental delays of the phase measurement and are assumed zero for the same reason as pseudorange instrumental delays.

ϵ_R and ϵ_L are pseudorange and phase measurements noise, respectively.

2.2. LEO Constellations Analyzed

The analysis of a set of five different constellation configurations has been conducted, with the aim of achieving optimal visibility and ensuring the best possible geometry for users. In order to attain these objectives, a combination of polar and inclined orbits has been tested. It has been observed that the increased visibility for the final user, considering polar orbits geometry, is lower when the user is located close to the equator, while inclined orbits exhibit poor geometries in the vicinity of the poles [8]. Thus, the mixing (M) of both polar (85° inclination) and inclined (55° inclination) orbits has been found to be effective in improving the final geometry of users at any point on the Earth's surface. By using a combination of polar and inclined orbits, it is possible to overcome the limitations of each type of orbit, thereby achieving a more robust and reliable solution for high-precision positioning and navigation applications.

For all the constellations studied, the orbital planes have an altitude of 1200 km. The near-polar orbits were simulated with 85° of inclination and the inclined orbits with 55° of inclination. The constellation tested are shown in Table 1.

Table 1. Definition of constellations tested.

Name	Number of SVs	Sats in Polar Orbit	Sats in Inclined Orbit	Number Planes
M-200	200	100	100	20
55-200	200	0	200	20
85-200	200	200	0	20
M-400	400	200	200	40
55-400	400	0	400	40

2.3. PPP Performance and Convergence Analysis

The LEO constellations will be added to 24 GPS and 24 Galileo satellites. In order to verify the initial convergence improvement, the analysis will involve 20 Precise Point Positioning (PPP) executions using GMV GSharp HA solution and its online PPP service [9], each lasting 20 min. The convergence will be deemed successful when the horizontal positioning error achieves a stable accuracy of 0.1 m. To ensure the accuracy of the solutions, they will be compared with the PPPs obtained using only GPS and Galileo constellations, which will be used as reference points. The analysis has been performed in a simulated GNSS station located in Madrid, Spain.

This experiment aims to test the feasibility and efficiency of using multiple constellations to improve the PPP convergence. The addition of GPS and Galileo satellites to the existing constellations is expected to provide more precise positioning and navigation solutions. By comparing the solutions obtained using the new constellations with those using the reference constellations, the accuracy of the former can be evaluated. If the results show a significant improvement in accuracy and convergence time, it will demonstrate the potential of using multiple constellations for PPP and could pave the way for future developments in this field.

2.4. Orbits and Clocks Degradation Analysis

After selecting the optimal constellation configuration for the current case study, a thorough analysis of the 55° inclination and 400 satellite constellation's degradation was conducted. The objective of this analysis was to evaluate the performance of the chosen constellation under various degraded orbits/clocks products.

To achieve this, the 55° inclination and 400 satellites constellation clocks were subjected to several degradation factors of a Gauss–Markov noise model to simulate degraded corrections. The objective of this analysis is to determine the maximum error in corrections that can be managed by the PPP algorithm without significant degradation of PPP performances in terms of error and convergence time. The different levels of noise added to the clock correction are shown in Table 2.

Table 2. Noise level used for the performance of degraded corrections scenarios.

Test	Noise Level
a	2.5 cm
b	5 cm
c	10 cm
d	15 cm
e	25 cm

Once again, the analysis involved running simulations of 20 PPP executions, each lasting 20 min, to assess the 55-400 constellation's convergence performance under degraded conditions. Convergence was considered to have been achieved when the horizontal positioning error reached a stable accuracy level of 0.1 m. The results of the simulations were then compared with those obtained from PPP executions without any degradation in orbits and clocks.

3. Results

3.1. Optimal Constellation Definition

Figures 1–3 depict the horizontal error values obtained during the 20 convergence runs for each of the constellations under analysis, as well as for the reference scenario, which employed only GPS and GAL for PPP. To compare these results and select the most favorable configuration for the present case study, Table 3 presents the corresponding numerical outcomes, such as the RMS error and convergence time. The wave pattern shown in all the figures is due to the correlation between the different consecutive runs.

Every run is only delayed 1 min with respect to the previous run, so the starting point is changing, but the measurements between different runs are correlated.

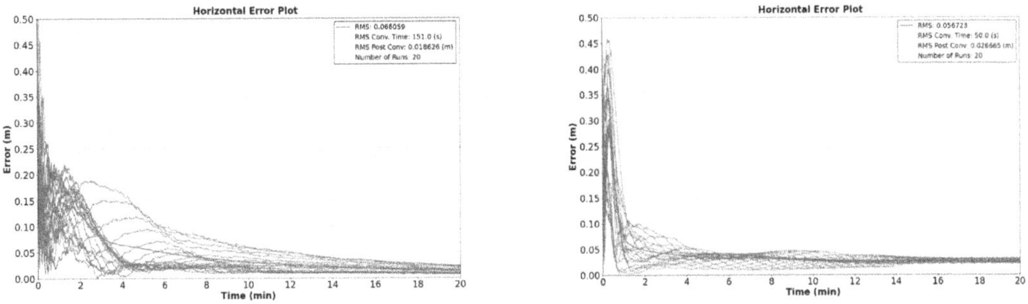

(a) (b)

Figure 1. Overlap of PPP convergences. (**a**) GPS + GAL. (**b**) GPS + GAL + M200.

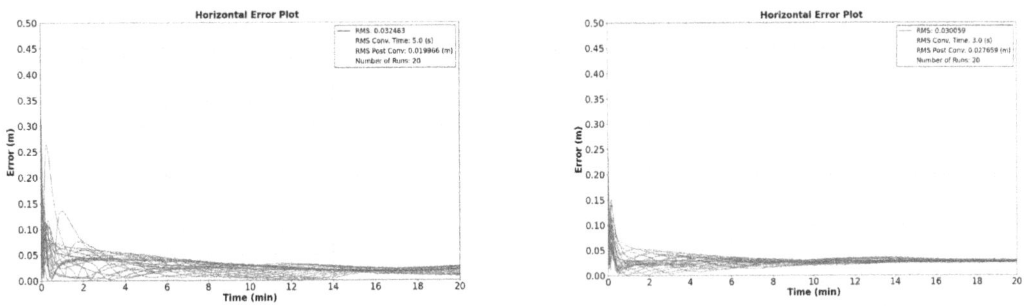

(a) (b)

Figure 2. Overlap of PPP convergences. (**a**) GPS + GAL + 55-200. (**b**) GPS + GAL + 85-200.

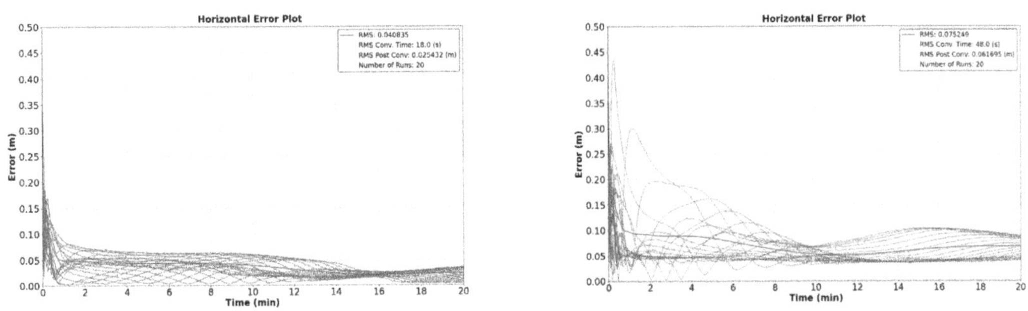

(a) (b)

Figure 3. Overlap of PPP convergences. (**a**) GPS + GAL + M400. (**b**) GPS + GAL + 55-400.

Table 3. RMS horizontal error and RMS of convergence time of different constellation configurations.

	RMS Error (m)	Convergence Time (s)
GPS + GAL	0.066	151
M200	0.057	50
55-200	0.041	18
85-200	0.075	48
M400	0.032	5
55-400	0.030	3

A significant bias can be seen in the results due to the synthetic simulated data. The mixed 85° and 55° degrees of inclination constellations show a greater bias, giving the highest RMS of all the solutions (including only MEO). However, the convergence time is reduced compared to the only MEO solution.

3.2. Orbits and Clocks Degradation Analysis

The horizontal errors for each of the degradation levels analyzed, as well as the reference scenario using 55-400 constellation + GPS + GAL for PPP, are presented in Figures 4–6. Table 4 contains numerical data that compare the RMS error and convergence time for all the scenarios, which help in determining the largest error in orbits and clock products that PPP can manage for the current case study without degrading performances.

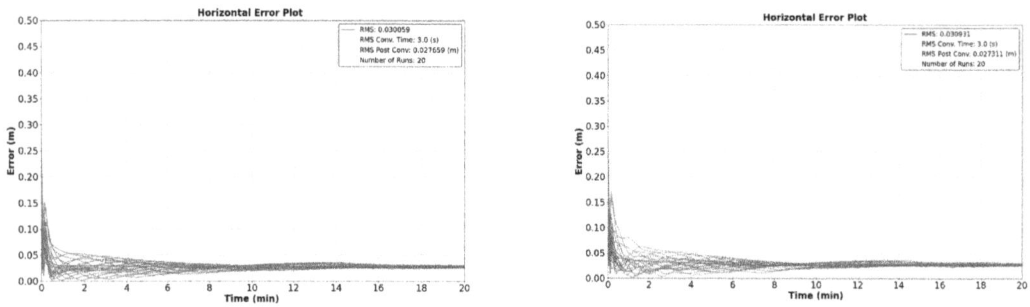

(**a**) (**b**)

Figure 4. Overlap of PPP convergences. (**a**) No Error. (**b**) 2.5 cm error.

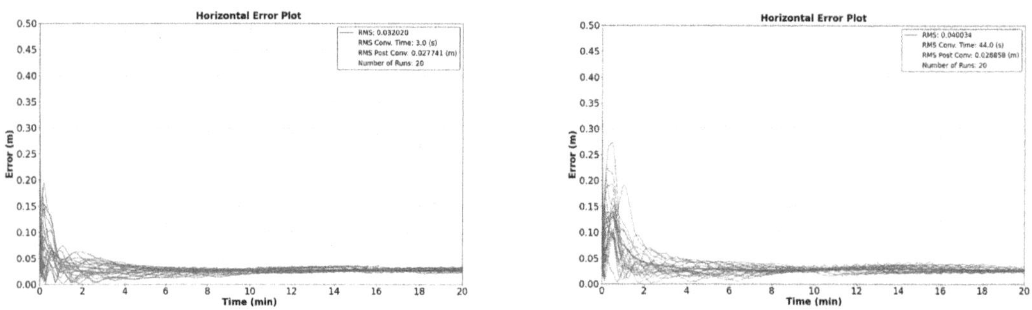

(**a**) (**b**)

Figure 5. Overlap of PPP convergences. (**a**) 5 cm error. (**b**) 10 cm error.

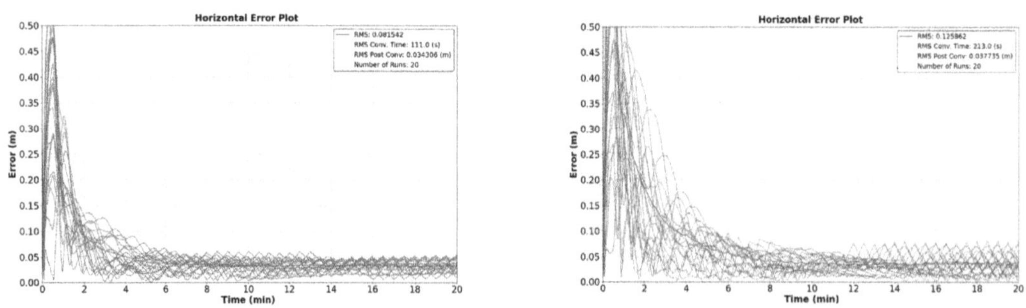

(**a**) (**b**)

Figure 6. Overlap of PPP convergences. (**a**) 15 cm error. (**b**) 25 cm error.

Table 4. RMS horizontal error and RMS of convergence time of different levels of correction degradation.

	RMS Error (m)	Convergence Time (s)
No error	0.030	3
2.5 cm	0.031	3
5 cm	0.032	3
10 cm	0.040	44
15 cm	0.082	111
25 cm	0.126	213

4. Conclusions

After reviewing several constellation configurations, it has been observed that the LEO + MEO satellite configuration provides the optimal results in terms of cold convergence. It has been observed that the utilization of mixed polar and inclined orbits in such a configuration yields superior performance. For mid-latitudinal regions such as Spain, the optimal constellation configuration consists of approximately 400 satellites, with a mixture of polar and inclined orbits.

This configuration not only ensures high accuracy positioning solutions but also reduces the dependency on dense station networks. The use of LEO satellites, with their high velocity dynamics and advanced onboard equipment, can effectively mitigate the effects of ionospheric interference in signals processed at the user level. This, in turn, enhances the accuracy and stability of the Positioning, Navigation, and Timing (PNT) solutions offered to users.

On the other hand, the impact of the clock noise stability has been assessed by simulating different levels of noise; validating the a priori assumption of the relative importance that good quality atomic clocks might have in the deployment of these kinds of constellations.

There are several opportunities for future work that can build upon the findings of this study. Firstly, it would be valuable to replicate this study for different latitudes to evaluate how well the LEO and MEO constellation configuration performs in different regions of the world. Secondly, it would be interesting to investigate the potential for precise orbit determination of LEO satellites using only ground-based observations, without relying on measurements from MEO satellites in an embedded receiver onboard. This would be a challenging task due to the low altitude of LEO satellites, which can result in limited visibility and make clock determination more difficult with sparse reference station networks. Therefore, future work could focus on developing new techniques and methodologies to overcome these challenges and improve the accuracy of clock determination in LEO satellite GNSS constellations.

Author Contributions: Conceptualization, E.C.; methodology, J.D.; software, D.S.; formal analysis, D.S.; resources, A.G.; writing—original draft preparation, D.S.; writing—review and editing, J.D.; supervision, D.C. and I.R.; project administration, D.C. All authors have read and agreed to the published version of the manuscript.

Funding: This research received no external funding.

Institutional Review Board Statement: Not applicable.

Informed Consent Statement: Not applicable.

Data Availability Statement: Data are unavailable due to privacy restrictions.

Conflicts of Interest: The authors declare no conflict of interest.

References

1. Crapart, R.; Ahmad, R.A.; Lannelongue, S.; Rouanet-Labe, A.; Blaudin-De-The, Y.; Seco-Granados, G.; Fabra, F. PNT user-segment optimization integrating LEO components. In Proceedings of the ESA Workshop on Satellite Navigation User Equipment Technologies (NAVITEC), Online, 5–7 April 2022.
2. Precise Point Positioning. Available online: https://gssc.esa.int/navipedia/index.php?title=Precise_Point_Positioning (accessed on 15 January 2023).
3. Real-Time Kinematic (RTK). Available online: https://www.u-blox.com/en/technologies/rtk-real-time-kinematic (accessed on 15 January 2023).
4. Su, M.; Su, X.; Zhao, Q. BeiDou Augmented Navigation from Low Earth Satellites. *Sensors* **2019**, *19*, 198. [CrossRef] [PubMed]
5. Kulu, E. Satellite Constellations—2021 Industry Survey and Trends. In Proceedings of the [SSC21-XII-10] 35th Annual Small Satellite Conference, Logan, UT, USA, 7–12 August 2021.
6. GSharp Webpage. Available online: https://www.gmv.com/en/products/space/gmv-gsharp (accessed on 20 March 2023).
7. Focussuite Webpage. Available online: https://www.gmv.com/es-es/productos/espacio/focussuiter (accessed on 15 January 2023).
8. González, A.; Tobías, M.; Rodriguez, I.; Navarro, P.; Sobrero, F.; Carbonell, E.; Calle, D.; Fernández, J. LEO Satellites for PNT, the Next Step for Precise Positioning Applications. In Proceedings of the 35th International Technical Meeting of the Satellite Division of The Institute of Navigation (ION GNSS+ 2022), Denver, CO, USA, 19–23 September 2022; pp. 2573–2581.
9. magicPPP Webpage. Available online: https://www.gmv.com/en/Products/magicPPP/ (accessed on 15 January 2023).

engineering proceedings

Proceeding Paper

Analysis of Multipath Code-Range Errors in Future LEO-PNT Systems †

Sibren De Bast [1,2,*], Jean-Marie Sleewaegen [1] and Wim De Wilde [1]

1 Septentrio NV, 3001 Leuven, Belgium
2 Networked Systems, Department of Electrical Engineering (ESAT), KU Leuven, 3001 Leuven, Belgium
* Correspondence: sibren.debast@septentrio.com
† Presented at the European Navigation Conference 2023, Noordwijk, The Netherlands, 31 May–2 June 2023.

Abstract: In recent years, low-Earth-orbit (LEO) constellations have been proposed for Positioning, Navigation and Timing (PNT) applications. Moreover, a couple of test satellites have already been launched and many more are scheduled in the near future. LEO constellations are characterised by their rapid change in geometry in comparison to the current medium-Earth-orbit (MEO) Global Navigation Satellite Systems (GNSSs). In this study, we analyse the impact of this high geometry change rate on the code-range error induced by multipaths. We develop a simulation environment with a static receiver and a nearby large building. We track the multipath signal using classical delay- and phase-locked loops (DLL and PLL). Multiple scenarios are simulated and analysed, comparing different orbit heights, MEO and LEO, and carrier frequencies (L-, S- and C-band). The LEO scenarios show up to 96% less code-range error for fast-changing multipath components. We show that this phenomenon is linked to the large phase delay rate between the direct signal and the multipath components, which is up to 75 times higher for LEO satellites when compared to MEO satellites. The phase delay rate reaches values higher than the DLL bandwidth. As a result, the DLL filters out the errors induced by fast-changing reflected signals, partially eliminating the multipath-induced code-range errors. The presented effect is coupled to the wavelength of the used carrier frequency. Our simulations show a reduction in multipath-induced code-range error for S- and C-band LEO-PNT signals in comparison to L-band signals.

Keywords: GNSS; multipath; LEO; PNT

Citation: De Bast, S.; Sleewaegen, J.-M.; De Wilde, W. Analysis of Multipath Code-Range Errors in Future LEO-PNT Systems. *Eng. Proc.* **2023**, *54*, 34. https://doi.org/10.3390/ENC2023-15453

Academic Editors: Tom Willems and Okko Bleeker

Published: 29 October 2023

1. Introduction

Global Navigation Satellite Systems (GNSSs) are a cornerstone of modern life. They provide location and timing services that play a crucial role in the localisation of several-billion GNSS-enabled mobile devices. Moreover, they are used to regulate and synchronise critical infrastructure such as telecommunication networks and power grid infrastructure. Current GNSS systems are MEO constellations, consisting of tens of satellites. They can provide meter-level accuracy in stand-alone mode or centimetre-level when using advanced algorithms and augmentation signals.

Interest in LEO-PNT systems has skyrocketed in recent years. LEO-based systems promise a couple of benefits, for example, higher resilience to jamming due to a higher received power and a faster convergence time for Precise Point Positioning (PPP) [1].

In addition, the global space economy has changed recently, lowering the cost of launching satellites to low-Earth orbit. This has given rise to the deployment of several telecommunication LEO mega-constellations. Moreover, several LEO constellations have been proposed specifically for PNT services. Several commercial companies have launched initial test satellites, for example, the PULSAR system from Xona Space Systems [2], the GeeSAT satellites from GeeSpace and the CentiSpace satellites from Future Navigation [3]. Governmental institutes are also interested, as the European Space Agency

(ESA) is planning to launch several LEO-PNT test satellites in the near future as part of the FutureNAV program.

The recent literature on LEO-PNT often mentions a decrease in multipath errors when using LEO in comparison to MEO satellites. Lawrence et al. [4] argue that when using LEO satellites, the multipath will whiten. In other words, the rapidly changing geometry of the LEO satellites makes the phase offset of the reflected signals resemble white noise and will be treated as that by the receiver. As a result, the induced multipath error will decrease.

Iannucci et al. [5] also mention that the rapidly changing geometry of LEO satellites will "average out" the multipath error. The paper gives a rough estimate of a 100 times larger geometry change in comparison to MEO counterparts. However, the exact value of this factor depends on the orbit height of the LEO satellite as this dictates the geometry change rate. In addition, they argue that the shorter wavelength of the used signals in LEO satellites will further increase the multipath decorrelation effect. The shorter wavelength increases the sensitivity of the relative phases of the multipath components to path-length differences. However, the authors focus on using broadband LEO satellites, which often operate at much higher frequencies than traditional GNSS satellites. The real impact of this effect will be determined by the used frequency of the LEO-PNT system.

Until now, the literature only provides back-of-the-envelope calculations. This work tries to overcome this by performing an analysis of the experienced multipath errors. A simulated LEO-PNT multipath environment is presented, using realistic signal values and material properties to study the multipath code-range errors in more depth.

This paper is organised as follows: Section 2 describes the system model. In Section 3, the simulation scenario is introduced. Afterwards, the results are discussed in Section 4. Section 5 concludes this study.

2. System Model

In this paper, we study a PNT signal affected by multipath propagation. The signal at the receiver is composed of $K+1$ signal components and noise. The first component is the direct signal. This signal can freely propagate from the transmitting satellite to the receiver without encountering objects. The other K components are Non-Line-of-Sight (NLoS) components. These signals arrive delayed at the receiver through reflections. The resulting received signal is presented as follows:

$$s_{rx}(t) = \left(\sum_{k=0}^{K} A_k s(t - \Delta t_k) e^{j2\pi f_c (t - \Delta t_k) + \phi_k} \right) + n(t) \tag{1}$$

A_k is the amplitude of the signal component, $s(t)$ is the time-variant baseband signal, Δt_k is the time delay in comparison to the direct component, f_c is the carrier frequency and $n(t)$ is the noise. ϕ_k is the phase offset.

The amplitude A_k of each component relies on three factors: the Received Isotropic Power (RIP), the receiver antenna gain, and the polarisation change and attenuation induced by each reflection.

2.1. Received Isotropic Power (RIP)

The RIP dictates how much power is received when using an isotropic receiving antenna. The RIP is defined by the transmit power of the satellite, its antenna gain pattern and the distance between the satellite and the receiver. The RIP depends on the boresight angle of the receiver as seen from the satellite's antenna; hence, on the receiving side, the RIP is dependent on the elevation angle θ_{el} of the satellite.

2.2. Receiver Antenna Gain Pattern

The gain pattern of the receiving antenna will boost or decrease the signal amplitude, depending on the direction of the incoming wave and its polarisation. A good GNSS antenna focuses on signals coming from a direction with a positive elevation angle while rejecting signals coming from the ground. In this way, ground reflections are partially mitigated. In addition, the antenna gain is dependent on the polarisation of the incoming wave. A receiving GNSS antenna has different gain characteristics for Right-Hand Circularly Polarised (RHCP) waves and Left-Hand Circularly Polarised (LHCP) waves. The signals transmitted by GNSS satellites are RHCP waves; therefore, GNSS antennas are optimised to boost RHCP waves and mitigate LHCP waves as these are the result of reflected signals.

2.3. Reflections and Signal Polarisation

The electric field of circularly polarised waves is constantly rotating in the direction perpendicular to the propagation direction of the wave. The rotation direction decides if it is right-handed- or left-handed-polarised.

The normalised electrical field of an RHCP wave is presented as a Jones vector as follows: [6]

$$E = \frac{1}{\sqrt{2}}\begin{bmatrix} 1 \\ -j \end{bmatrix} \tag{2}$$

The amplitude and polarisation of the wave change due to reflections, depending on the material properties of the reflection interface and the incidence angle of the incoming wave. A part of the incoming power is reflected back, while the rest is transmitted through the material. Reflections can be described by means of reflection coefficients, which represent how much of the power is reflected. A set of two reflection coefficients must be calculated, r_p and r_s. r_p corresponds to waves with an electric field in the plane of incidence and r_s to waves with an electric field normal to the plane of incidence. They can be calculated using the Fresnel equations. We note the m-th reflection in Jones vector notation as follows:

$$R_m = \begin{bmatrix} -r_{p,m} & 0 \\ 0 & r_{s,m} \end{bmatrix} \tag{3}$$

with

$$r_{p,m} = \frac{n_{1,m}\sqrt{1-\left(\frac{n_{1,m}}{n_{2,m}}\sin\theta_{i,m}\right)^2} - n_{2,m}\cos\theta_{i,m}}{n_{1,m}\sqrt{1-\left(\frac{n_{1,m}}{n_{2,m}}\sin\theta_{i,m}\right)^2} + n_{2,m}\cos\theta_{i,m}}, \tag{4}$$

$$r_{s,m} = \frac{n_{1,m}\cos\theta_{i,m} - n_{2,m}\sqrt{1-\left(\frac{n_{1,m}}{n_{2,m}}\sin\theta_{i,m}\right)^2}}{n_{1,m}\cos\theta_{i,m} + n_{2,m}\sqrt{1-\left(\frac{n_{1,m}}{n_{2,m}}\sin\theta_{i,m}\right)^2}}. \tag{5}$$

$\theta_{i,m}$ is the incidence angle of reflection m; this is the angle between the direction of the incoming wave and the normal on the reflective surface. $n_{1,m}$ and $n_{2,m}$ are the refractive indices of the two materials. $n_{1,m}$ relates to the medium in which the incoming wave is propagating, while $n_{2,m}$ relates to the material on which the wave reflects.

The resulting reflection coefficient R after M reflections is calculated as follows:

$$R = \prod_{m=1}^{M} R_m \tag{6}$$

The reflected electric field is calculated by multiplying E and R of Equations (2) and (6). To calculate the impact of receiving reflected waves using a circularly polarised antenna, the reflection coefficient can be decomposed into an RHCP and an LHCP component [7]:

$$R_R = \frac{r_s + r_p}{2},\ R_L = \frac{r_s - r_p}{2} \tag{7}$$

Each component has to be multiplied by the respective antenna gain pattern of the receiving antenna. In general, the amplitude of a multipath signal component can be written as follows:

$$A_k = \sqrt{\text{RIP}(\theta_{el})}\sqrt{\left(\frac{R_R}{\sqrt{2}}G_{rx,R}(\theta_{rx})\right)^2 + \left(\frac{R_L}{\sqrt{2}}G_{rx,L}(\theta_{rx})\right)^2} \quad (8)$$

The RIP is a function of the elevation angle of the satellite θ_{el}. R is the reflection coefficient of the electric field of the signal, while the subscript $\{R, L\}$ denotes the polarisation of the component, RHCP and LHCP, respectively. The receiver antenna gain G_{rx} is dependent on the polarisation and on the angle of incidence of the signal θ_{rx}.

2.4. Signal Tracking

To estimate the code-ranging error, the signal has to be acquired and tracked. As the acquisition does not influence the multipath error, we assume a perfect acquisition at the start of tracking the signal. The signal tracking is performed using a classical delay-locked loop and phase-locked loop (DLL and PLL). Each of the loops has been tuned for MEO GNSS systems and is used without modification to track the LEO signals. Unless otherwise specified in Section 4, all results presented in this paper are generated using the following bandwidth (BW) values for the locked loops: BW_{DLL} = 0.25 Hz; BW_{PLL} = 15 Hz.

2.5. Limitations of the Model

Several important error sources are not included in the used model. For example, atmospheric effects, ephemeris errors and satellite clock offset errors are not included in this study. In addition, the environment presented in Section 3 and the propagation of the waves are kept simple. As such, we do not consider internal structures, conductive media, diffraction or transmittance. In this way, we are keeping to analytic insight into the main contributions. As a result, the presented results in Section 4 do not represent the expected ranging errors for LEO-PNT satellites; they should only be compared with other results in this work to understand the multipath behaviour of LEO-PNT systems.

3. Simulation Scenario

This section delineates the details of the specific simulated scenario used to analyse multipath errors in LEO-PNT systems. First, we introduce the geometry and material properties of the scenario. Next, the geometric properties of the different considered reflections are presented. Afterwards, the satellite orbits and signal properties are discussed.

3.1. Multipath Environment

The scenario considered in this study is illustrated in Figure 1. A static receiver is placed at a height h_{rx} at a distance d from a tall building with height h_b. The wall from the building on which the rays will reflect is made of concrete, with a refractive index n_c. We do not consider internal reflections as the attenuation of L-, S- and C-band signals in concrete is high [8], rendering the internal reflections insignificant. The ground is considered dry, with refractive index n_g. The air has a refractive index of n_a. The values used in the simulation [9] can be found in Table 1.

Table 1. The used values of the parameters in the presented scenario.

Parameters	d	h_{rx}	h_b	n_a	n_c	n_g
Value	20 m	2 m	50 m	1	3	4

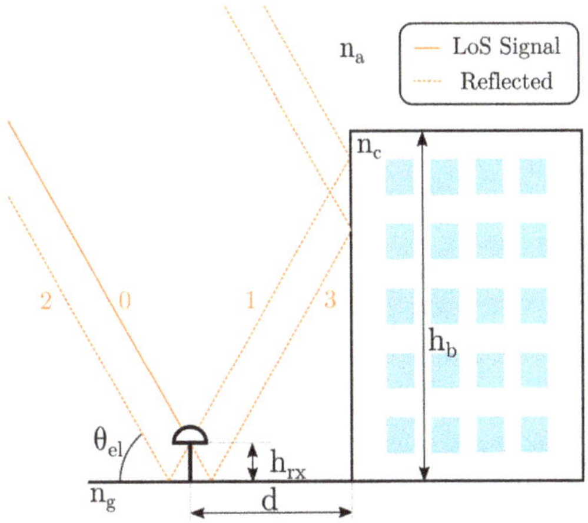

Figure 1. The scenario studied in this paper. A static receiver is placed above the ground next to a high building made of concrete. The satellites rise on the left of the scenario and move overhead. Together with the direct signal, we consider three reflected multipath components.

3.2. Signal Components

In this study, we employ a four-ray model. We consider the direct signal, a ground reflection, a wall reflection and a double reflection using the wall and ground.

To calculate the influence of each of the components, several parameters have to be known. The first parameter is the multipath delay Δt of the reflected signals. The multipath delay is related to the extra distance Δd the reflected signal travels before reaching the receiver. To calculate the delay, the distance must be divided by the propagation speed of the waves, i.e., the speed of light. The next parameter is the elevation cut-off angle θ_c, above which there is no reflection. This cut-off is only present for reflections on the building. In addition, to calculate the reflection coefficients, the incidence angle θ_i has to be determined. And to end, the angle of the incoming wave at the receiver θ_{rx} is needed to calculate the antenna gain. Table 2 lists for each signal component the discussed geometric parameters.

Table 2. Geometric characteristics of the different signal components.

Signal	Direct	Ground	Wall	Wall-Ground
k	0	1	2	3
Δd	0	$2h_{rx}\sin\theta_{el}$	$2d\cos\theta_{el}$	$2h_{rx}\sin\theta_{el} + 2d\cos\theta_{el}$
θ_c	/	/	$\arctan\left(\frac{h_b-h_{rx}}{d}\right)$	$\arctan\left(\frac{h_b+h_{rx}}{d}\right)$
θ_i	/	$90-\theta_{el}$	θ_{el}	θ_{el}, $90-\theta_{el}$
θ_{rx}	θ_{el}	$-\theta_{el}$	$180-\theta_{el}$	$180+\theta_{el}$

To calculate the amplitude of the signal, the RIP of a receiver on Earth needs to be known. Braasch et al. [10] report the RIP of typical MEO GNSSs. There is no public RIP model for LEO-PNT satellites, so we assume the same RIP for both the MEO and LEO systems. In addition, the antenna gain pattern is needed to compute the signal amplitude. The antenna pattern for RHCP and LHCP signals presented by Yang et al. is used [11]. The resulting C/N_0 value for the LoS signal ranges from 40 dB-Hz at the horizon to 57 dB-Hz when the satellites are directly overhead.

The receiving antenna is mainly sensitive to RHCP waves, so the RHCP component of the reflections will be the main influence on the localisation performance of the GNSS receiver. In the presented scenario, at low elevation, the ground reflection is dominant (Figure 2). At higher elevations, the wall reflection becomes the dominant reflection. The incoming wave from the wall reflection arrives with a positive elevation angle. Therefore, the antenna will not be able to mitigate the RHCP component of this reflected signal.

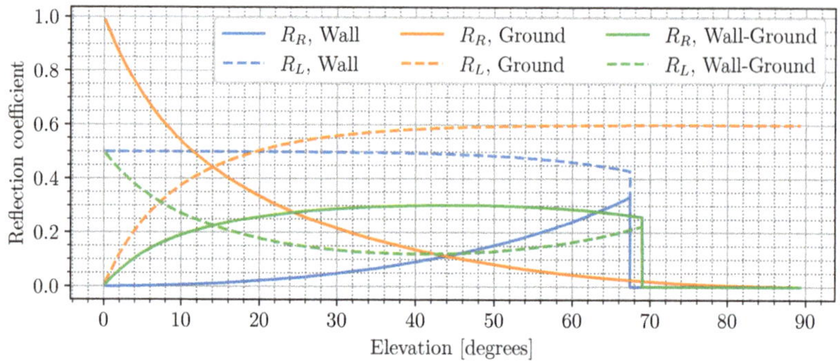

Figure 2. The reflection coefficient R for the different reflected signals and polarisations. Note the cut-off angle of 67 degrees for the wall reflection and 69 degrees for the wall–ground reflection.

3.3. Satellite Orbits

As the focus of this study is on the influence of the height of the satellite orbits, the orbits are simply modelled as perfect circles around the centre of the Earth. We consider four different orbit heights. Three of them are LEO orbits, with a height of 500 km, 1000 km and 1500 km. The fourth orbit is an MEO orbit at a height of 23,222 km, the same height as the Galileo constellation. During the simulation, only a half-pass is simulated, where the satellite rises from the horizon to an elevation of 90 degrees.

3.4. PNT Signals

During the simulation, only the output value of the Early (E), Prompt (P) and Late (L) correlators is computed. The E-L spacing is 1 chip, with P located in the middle of E and L. The correlation peak is based on a GPS L5-like signal with a chipping rate of 10.23 MHz. The receiver has a sampling frequency of 20.46 MHz and a bandwidth of 51.15 MHz.

The values of the E, P and L outputs depend on several factors. The first factor is the code offset between the sampling point and the signal. The second factor is the phase offset between the signal and the estimated phase at the PLL. The third factor is the frequency offset between the estimated frequency and the actual signal. The last factor is the amplitude of the signal component. We do not consider navigation bits.

The E, P and L values are calculated for all different multipath components. Next, they are summed together to create the E, P and L values of the composite signal. Afterwards, noise is added. The correlator outputs are used to calculate the discriminator outputs, which in turn are used by the non-coherent DLL and PLL to update their states.

We consider three carrier frequencies. The first frequency is the GPS L1 frequency. This frequency is the oldest frequency in use by current MEO GNSS systems. We expect that upcoming LEO-PNT systems will transmit a signal in the neighbourhood of the L1 frequency. In addition, new frequency bands are proposed for the upcoming LEO-PNT systems, both in the S- and C-band. Therefore, we also analyse S- and C-band signals at 2491 MHz and 5020 MHz, respectively.

4. Results

This section discusses the results of the presented simulations. To analyse the impact of the multipath error when using LEO satellites, several experiments are designed. We will discuss for each experiment the simulation settings and the corresponding results.

4.1. Influence of the Orbit Height

During the first experiment, the influence of the orbit height of the satellite is analysed. The rate at which the elevation angle varies depends on the orbit height. One pass of an MEO GNSS satellite is in the order of hours, while for an LEO satellite it is in the order of minutes. This large difference in the orbit period translates into a large difference in the change in the elevation angle. The hypothesis is that the multipath error decreases due to the faster change in the phase difference between the LoS signal and the reflected signals.

For this experiment, the orbit height is varied, $h = \{500, 1000, 1500, 23{,}222\}$ km, while the L-band is used as carrier frequency. The code-range error is shown in Figure 3. The first observation is that the code-range error induced by multipath effects is strongly decreased for LEO satellites in comparison to their MEO counterparts. For low elevations, we observe multipath errors that are equally large for both LEO and MEO satellites. However, when the satellites' elevation rises, a clear difference between the different orbit heights becomes visible. A strong reflection of the signal on the nearby building induces a large error in the MEO scenario. Due to the large incidence angle, a significant RHCP component is still present in the signal after the reflection. At an elevation of 66 degrees, an RMSE value of up to 2.40 m is recorded, while for the LEO satellites, the RMSE is much smaller, from 0.096 m to 0.303 m (Table 3). This is a decrease of up to 96%. Over the full simulation, the RMSE of the code range is reduced by up to 87% by employing LEO satellites.

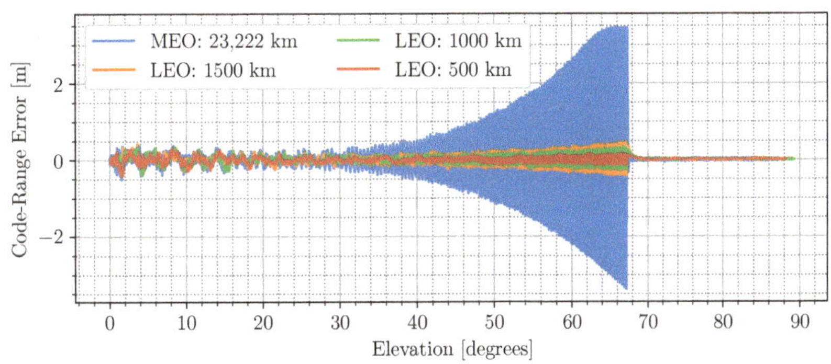

Figure 3. Code-range error versus orbit height, L-band. The strong reflection on the building as the elevation angle of the satellite increases only strongly impacts the MEO system. For LEO systems, the multipath errors are suppressed when the elevation angle increases. Note that above 67 degrees elevation, the wall reflection disappears.

Table 3. Comparing the influence of the orbit height.

Orbit Height	23,222 km	1500 km	1000 km	500 km
Overall RMSE (m)	0.756	0.120	0.111	0.098
Reduction in comparison to MEO	0%	84.1%	85.4%	87.0%
RMSE at $\theta_{el} = 66°$ (m)	2.397	0.303	0.199	0.096
Reduction in comparison to MEO	0%	87.4%	91.7%	96.0%
Max phase delay rate (Hz)	0.042	0.992	1.540	3.155
Increase factor in comparison to MEO	1	23.62	36.67	75.12

This behaviour can be explained when we calculate the phase delay rate. The phase delay is the difference in phase between the LoS component and an indirect multipath component. In this case, we will only study the wall reflection as this has the largest influence in this scenario. The phase delay rate is the derivative of the phase delay over time. It illustrates how fast the phase is rotating the correlation peak of the indirect signal in the complex plane. Figure 4 shows the phase delay rate for the wall reflection as a function of the elevation angle. For LEO systems, the phase delay rate reaches frequencies higher than the bandwidth (BW) of the DLL, which was set to 0.25 Hz for this study. Our simulations show that the maximum phase delay rate is increased up to 75 times by employing LEO satellites (Table 3). As a result, the DLL starts treating the multipath components as noise and filters out their influence. This behaviour is only present above a certain elevation angle, which depends on the orbit height of the satellite.

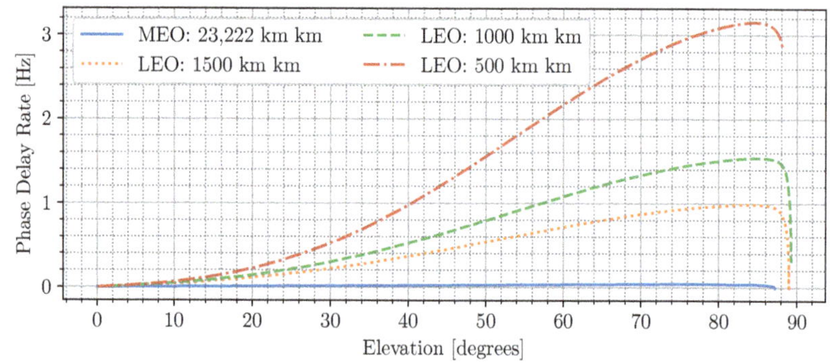

Figure 4. Phase delay rate of the wall reflection for the different orbit heights. Due to the faster geometry change of lower orbits, the phase delay rate becomes larger. This effect is increased for high elevations as LEO satellites have higher perceived mobility at high elevations for ground users.

4.2. Influence of the DLL Bandwidth

In the second experiment, we investigate the influence of the DLL BW on the code-range error. Three different values of the DLL BW are evaluated: $DLL_{BW} = \{1, 0.25, 0.05\}$ Hz. The satellite orbits at a height of 1000 km using an L-band signal. The code-range error is shown in Figure 5. At low elevations, the amplitude of the error is only slightly decreased by decreasing the DLL BW; however, at higher elevations, the error amplitude is strongly reduced as the filtering effect has become larger.

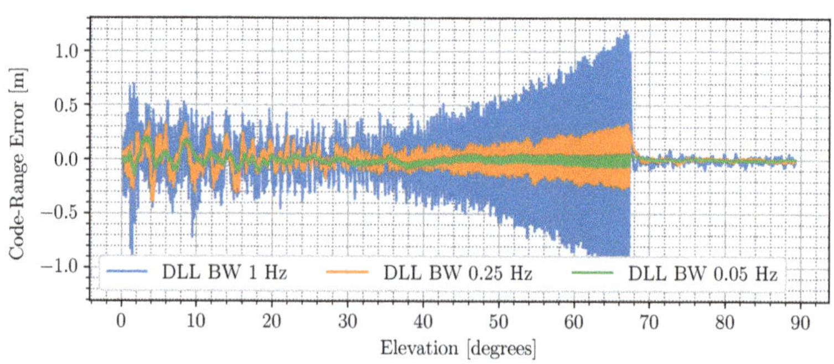

Figure 5. Code-range error versus DLL bandwidth. The considered satellite orbits are at a height of 1000 km. When lowering the DLL bandwidth, the code range error decreases and vice versa.

4.3. Influence of Carrier Frequency

As the reported effect is coupled to the phase delay, the wavelength of the carrier has an influence as well. To illustrate this effect, the scenario was simulated using an orbit height of 1000 km while using three different carrier frequencies: L1 (1575.42 MHz), S-band (2491 MHz) and C-band (5020 MHz). For this simulation, we assume that the material properties are equal for the considered frequencies. At elevations below 30 degrees, no significant difference is noticeable; above this elevation, the S- and C-band signals have a decreased ranging error (Figure 6). In comparison to the L-band, the ranging error decreases by 10.8% for the S-band and 24.3% for the C-band (Table 4).

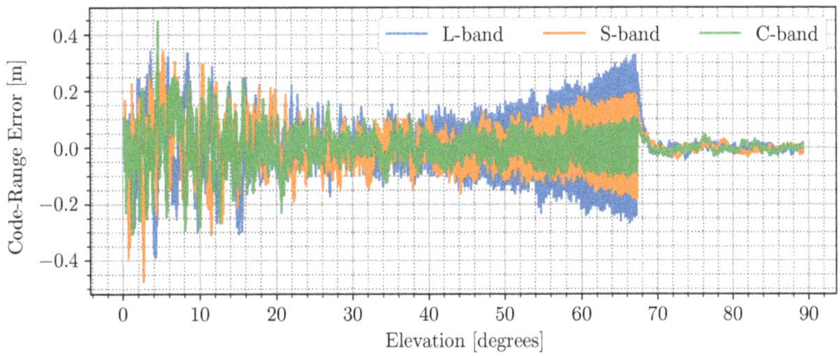

Figure 6. As the reported effect is coupled to the wavelength of the carrier frequency, the multipath error decreases with increasing frequency.

Table 4. Root mean squared error of the code-range error as a function of the frequency band.

Frequency	L-Band	S-Band	C-Band
Overall RMSE (m)	0.111	0.099	0.084
Reduction in comparison to L-band	0%	10.8%	24.3%
RMSE at $\theta_{el} = 66°$ (m)	0.199	0.123	0.064
Reduction in comparison to L-band	0%	38.2%	67.8%
Reduction in wavelength λ	0%	36.8%	68.6%

5. Conclusions

Multipath-induced code-range errors are strongly reduced in LEO-PNT systems for fast-changing multipath components. Our simulations show an overall reduction in the code-range error of 87% when comparing an MEO satellite to an LEO satellite at a height of 500 km. We attribute this reduction to the phase delay rate, which rises above the DLL BW for fast-changing multipath components. The phase delay rate is up to 75 times larger for LEO than for MEO signals. As a result, the DLL BW has a large influence on the experienced multipath code error in LEO-PNT systems. In addition, including higher carrier frequencies in LEO-PNT can further reduce multipath code-range errors. In comparison to L-band signals, our simulations show a decrease of 10.8% for S-band signals and 24.3% for C-band signals.

Author Contributions: Conceptualization, S.D.B., J.-M.S. and W.D.W.; methodology, S.D.B. and W.D.W.; software, S.D.B.; validation, J.-M.S. and W.D.W.; formal analysis, S.D.B.; investigation, S.D.B.; resources, J.-M.S.; data curation, S.D.B.; writing—original draft preparation, S.D.B.; writing—review and editing, S.D.B.; visualization, S.D.B.; supervision, W.D.W.; project administration, W.D.W.; funding acquisition, S.D.B. and W.D.W. All authors have read and agreed to the published version of the manuscript.

Funding: This research was partially funded by VLAIO, grant number: HBC.2022.0169.

Institutional Review Board Statement: Not applicable.

Informed Consent Statement: Not applicable.

Data Availability Statement: The data presented in this study are available on request from the corresponding author. The data are not publicly available due to strict IP management.

Conflicts of Interest: Septentrio is a GNSS receiver manufacturer. As such research into novel GNSS opportunities and GNSS receiver technology is of high interest to the company.

References

1. Li, X.; Ma, F.; Li, X.; Lv, H.; Bian, L.; Jiang, Z.; Zhang, X. LEO constellation-augmented multi-GNSS for rapid PPP convergence. *J. Geod.* **2019**, *93*, 749–764. [CrossRef]
2. Reid, T.; Banville, S.; Chan, B.; Gunning, K.; Manning, B.; Marathe, T.; Neish, A.; Perkins, A.; Sibois, A. PULSAR: A New Generation of Commercial Satellite Navigation. In Proceedings of the ION GNSS+ 2022, Denver, CO, USA, 19–23 September 2022.
3. Yuan, J.; Zhou, S.; Hu, X.; Yang, L.; Cao, J.; Li, K.; Liao, M. Impact of Attitude Model, Phase Wind-Up and Phase Center Variation on Precise Orbit and Clock Offset Determination of GRACE-FO and CentiSpace-1. *Remote Sens.* **2021**, *13*, 2636. [CrossRef]
4. Lawrence, D.; Cobb, H.S.; Gutt, G.; O'Connor, M.; Reid, T.G.; Walter, T. Navigation from LEO, GPS World, July 2017. Available online: https://web.stanford.edu/group/scpnt/gpslab/pubs/papers/Lawrence_GPSWorld_July2017.pdf (accessed on 27 November 2023).
5. Iannucci, P.A.; Humphreys, T.E. Economical Fused LEO GNSS. In Proceedings of the 2020 IEEE/ION Position, Location and Navigation Symposium (PLANS), Portland, OR, USA, 20–23 April 2020; pp. 426–443. [CrossRef]
6. Peatross, J.; Ware, M. Physics of Light and Optics. 2015. Available online: https://optics.byu.edu/ (accessed on 8 March 2023).
7. Smyrnaios, M.; Schön, S.; Nicolás, M.L. Multipath Propagation, Characterization and Modeling in GNSS. In *Geodetic Sciences*; Jin, S., Ed.; IntechOpen: Rijeka, Croatia, 2013; Chapter 2. [CrossRef]
8. Stone, W. *Electromagnetic Signal Attenuation in Construction Materials*; NIST: Gaithersburg, MD, USA, 1997. [CrossRef]
9. Hannah, B.M. Modelling and Simulation of GPS Multipath Propagation. Ph.D. Thesis, Queensland University of Technology, Brisbane, QLD, Australia, 2001.
10. Braasch, M.; Dempster, A. Tutorial: GPS receiver architectures, front-end and baseband signal processing. *IEEE Aerosp. Electron. Syst. Mag.* **2019**, *34*, 20–37. [CrossRef]
11. Yang, N.; Freestone, J. High-performance GNSS antennas with phase-reversal quadrature feeding network and parasitic circular array. In Proceedings of the 29th International Technical Meeting of the Satellite Division of The Institute of Navigation (ION GNSS+ 2016), Portland, OR, USA, 12–16 September 2016; pp. 364–372.

engineering proceedings

MDPI

Proceeding Paper

Performance Analysis of the Pilot and Data Component of a CSS Signal for LEO-PNT †

Daniel Egea-Roca *, José A. López-Salcedo and Gonzalo Seco-Granados

Department of Telecommunication and Systems Engineering, Universitat Autònoma de Barcelona (UAB), 08193 Bellaterra, Spain; jose.salcedo@uab.es (J.A.L.-S.)
* Correspondence: daniel.egea@uab.es
† Presented at the European Navigation Conference 2023, Noordwijk, The Netherlands, 31 May–2 June 2023.

Abstract: From the analysis of chirp spread spectrum (CSS) signals in radar and communication applications, we can distinguish the cases of estimating only data or time-delay propagation or Doppler frequency when all the other parameters are known or set to zero. The case of estimating all three parameters at the same time has not been treated in the literature of radar or communications. We consider in this paper the use of a CSS signal for a LEO-PNT signal, which perfectly fits with the case of estimating the three parameters mentioned above. Specifically, the main goal of this paper is to explain the design of a CSS signal for PNT. This will include the design of the ranging component and the data component. The performance evaluation of both components will also be analyzed and linked. It is important to note that the focus of this paper will be the data component, and we link it with the pilot component through sensitivity performance. The results given in this paper show the capabilities of CSS for PNT. Previous works have shown accuracies on the level of meters with achievable data rates shown in this paper in the order of tens of kbps.

Keywords: chirp spread spectrum; LEO-PNT; low complexity

Citation: Egea-Roca, D.; López-Salcedo, J.A.; Seco-Granados, G. Performance Analysis of the Pilot and Data Component of a CSS Signal for LEO-PNT. *Eng. Proc.* **2023**, *54*, 35. https://doi.org/10.3390/ENC2023-15425

Academic Editors: Tom Willems and Okko Bleeker

Published: 29 October 2023

1. Introduction

The extended use of battery-powered devices, such as smartphones, tablets, portable computers, and wearable and IoT devices, is changing the way our technology is designed. In the last decade, there has been a trend of devices with low-power consumption and low-complexity processing. To do so, we must consider simple signal structures leading to simple receiver processing architectures. In addition, there is a trend toward targeting everywhere connectivity by relying on the satellite segment. A proof of that is the extended list of low Earth orbit (LEO) satellite constellations for internet services or IoT connectivity in the NewSpace landscape [1,2]. The opportunities of LEO constellations deal with the main challenges of IoT communication and localization systems [3,4]. One clear example joining both communication and localization is a positioning, navigation, and timing (PNT) system [5], and there is a current interest in LEO-PNT. Indeed, the first PNT results from the Starlink constellation were published in [6], and it opened a new era of alternative PNT solutions based on signals of opportunity (SoO) from LEO constellations or by the development of an alternative LEO PNT system [7].

During the past decades, the use of global navigation satellite systems (GNSSs) has become the primary and sometimes only way of providing a global PNT solution for many outdoor applications [8]. Unfortunately, GNSS is a very power-hungry technology that is vulnerable to several threats [9]. Thus, it is important to provide alternative PNT solutions to back up GNSS in case of failure or denial of service [9,10]. In this sense, the use of LEO constellations has been considered in the literature to provide a global PNT solution, but more importantly, because it will bring some benefits with respect to MEO constellations [6,7]. In this paper, we focus on the signal design for a LEO-PNT system. Different options can be adopted for the LEO-PNT signal, but if we aim to reduce complexity with respect to current GNSS, we should change the current GNSS signal structure [4,11]. Otherwise, the signal acquisition would be prohibitive for low-complexity devices. Recall that the high dynamic in LEO constellations will produce a large search space for the acquisition process of a direct-sequence spread spectrum (DSSS) signal (i.e., current GNSS signal structure) [4]. One alternative with lower acquisition complexity can be based on a chirp spread spectrum (CSS) signal.

The CSS signal has extensively been used in radar [12] since a long time ago, and it is reaching the realm of low-power wide-area networks (LPWANs) by the deployment of LoRa networks [13]. From the analysis of CSS signals in radar and communication applications, we can distinguish the cases of estimating only data or time-delay propagation or Doppler frequency when all the other parameters are known or set to zero. The case of estimating all three parameters at the same time has not been treated in the literature of radar or communications. The first time a CSS signal was considered for a PNT signal was given in [4]. Indeed, the complexity of the CSS signal was analyzed in [4], and it was compared with a DSSS signal. Improvements in terms of complexity from 1 and up to 2 orders of magnitude can be obtained using CSS with respect to DSSS. Furthermore, ref. [11] provides details on the key performance indices (KPIs) for PNT of the CSS signal. We consider in this paper the use of a CSS signal for a LEO-PNT signal, which perfectly fits with the case of estimating the three parameters mentioned above.

The goal of this paper is to explain the design of a CSS signal for PNT. This will include the design of the ranging component and the data component. The performance evaluation of both components will also be analyzed and compared. Therefore, the contribution of this paper is to complete the design of a CSS-based PNT signal by analyzing both the ranging and data component. Note that the ranging component was already introduced and evaluated in [4] and [11], respectively. In this paper, we include the design and evaluation of the data component, and we relate it with the design of the ranging component. To do so, the rest of this paper is organized as follows: Section 2 explains the signal model for the CSS-based PNT signal by showing the general expression of a CSS signal, the challenges we find in PNT, and how to solve them by including a pilot and data component. Then, Section 3 goes into detail about the signal processing and design of the CSS-based PNT signal, and Section 4 provides the performance analysis of the proposed CSS signal. Finally, Section 5 concludes the paper.

2. Signal Model

In this section, we introduce the signal model for the proposed CSS-based PNT signal in this paper. For that, let T_c be the chirp period or duration and B be the signal bandwidth. Then, a general expression for a CSS signal is given by the following waveform:

$$s(t) = e^{j2\pi \int_0^t [f_0+\mu u]_B du}, \tag{1}$$

with f_0 as the initial frequency, $\mu \doteq T_c/B$ as the chirp rate, and $[\cdot]_B$ as the modulus-B operator. For a PNT signal, we have to consider the presence of data information (e.g., represented by f_0), time delay (τ), and frequency Doppler (f_D). The challenges of a CSS-based PNT signal comes from the traditional de-chirp process used in radar and communications [4]. Traditionally, the frequency of the de-chirped signal has been used to estimate either time delay, frequency Doppler, or data information. In a PNT scenario, all the parameters of interest (i.e., $\{f_0, f_D, \tau\}$) are embedded on the de-chirped frequency [4]. This scenario leads us to the transmission of two separate components: the pilot and data component. The pilot component does not carry data information, and it is used to synchronize the signal. Once the signal is synchronized, the data component is used for the reception of the useful information needed for PNT. Second, in the act of synchronizing the signal, we will extract the ranging information (i.e., time delay) needed to localize the user.

Finally, it is worth noting that the design of a GNSS signal must include a multi-satellite access (MSA) scheme. These concepts are introduced in the following.

2.1. Ranging Capabilities

The most important task of a PNT signal is to provide the means for ranging. The CSS signal considered in this paper is a complex signal so that it needs both I&Q components for the transmission and reception of the signal. Therefore, the transmission of a CSS signal can be used for either ranging (i.e., $\{\tau, f_D\}$ estimation) or data transmission. This is in contrast with traditional GNSS [14], in which the transmitted signal is usually real and then a pilot/data scheme is transmitted through the I&Q components. Then, in our case, we should consider some way to make the ranging (pilot) component coexist with the data component, e.g., waveform multiplexing [15] or time multiplexing [4,11]. Note that for the pilot component, we have absence of data, so that we would be able to obtain the time-delay and Doppler frequency estimation using a BOK-chirp signal as defined in [4,11]. Therefore, in the following, we will assume that a BOK-chirp signal is transmitted from the satellites for the purpose of ranging/synchronization. The signal processing and design of the pilot component was studied in [4,11]. Here, in Section 3.1, we give the main ideas needed to understand the design of the data component.

2.2. Data Transmission

Once the signal is synchronized, we can assume $\tau = f_D = 0$, so that data reception can be achieved through a simple de-chirp process when transmitting the general CSS waveform in (1). The details for the data component will be given in Section 3.2. Here, let us give a summary of the advantages and drawbacks of different CSS modulations from a PNT signal design's point of view:

- BOK-CSS: A digital modulation scheme sometimes considered in practice for chirp signals is the so-called BOK for communications [16]. Note that although considering up- and down-chirp signals, the BOK chirp scheme considered here is different from the one used in the pilot component [4]. In this case, the transmitted signal is written as

$$s_{\mathrm{CBOK}}(t) = b(t)e^{j2\pi \int_0^t [\mu t]_B \mathrm{d}u} + (1 - b(t))e^{-j2\pi \int_0^t [\mu t]_B \mathrm{d}u}, \quad (2)$$

with $b(t) = \{0, 1\}$ as the data bit. Note that only one bit per T_c can be sent, which limits the data-rate capability.

- PSK-CSS: A phase shift keying (PSK)-CSS modulation is given by [17]:

$$s_{\mathrm{PSK}}(t) = e^{j(2\pi \int_0^t [\mu u]_B \mathrm{d}u + \theta(t))}, \quad (3)$$

where $\theta(t)$ stands for the phase corresponding to the transmitted symbol at time t. This modulation could be useful to increase the data-rate capabilities with respect to the BOK-CSS option when considering a higher-order modulation than BPSK.

- FSK-CSS: As long as the initial frequency of (1) is not used for other purposes (e.g., satellite identification), it can be used for data delivery as performed in LoRa [13]. In that way, each symbol in an M-ary constellation is assigned to a given initial frequency, s_k. Then, the transmitted signal is written as

$$s_{\mathrm{FSK}}(t) = e^{j2\pi \int_0^t f_k(u; s_k) \mathrm{d}u}, \quad (4)$$

for $(k-1)T_c \leq t \leq kT_c$, and the instantaneous frequency given by

$$f_k(t; s_k) = [s_k + \mu t]_B. \quad (5)$$

This option can be useful to increase the data rate with respect to BOK-CSS, and it can provide a better alternative to PSK-CSS in terms of HW impact on current GNSS receivers. Note, once synchronized, that the demodulation of frequency shift keying (FSK)-CSS can be based on a simple de-chirp process.

2.3. Multi-Satellite Access (MSA)

Multiple satellite access with CSS can be based on two main parameters given in (1): the slope, μ, and the initial frequency, f_0. Both parameters can be used to identify the different users. Unfortunately, there are some consequences on the selection of the MSA scheme that have to be considered when dealing with PNT. On the one hand, the use of multiple slopes may be limited when the number of satellites increases due to interference between satellites. On the other hand, when considering different frequencies for different satellites, we are intrinsically discarding an FSK-CSS modulation for the data transmission. Different schemes for MSA in PNT were proposed and analyzed in [4]. The conclusion is that the so-called multi-dual-slope (MDS) scheme provides the shortest chirp period possible among the analyzed MSA schemes in [4]. As a consequence, the MDS scheme allows us to reduce the processing complexity and to target high data-rate configurations for the data component. For the above reasons, in this paper we use the MDS scheme for the pilot component of the proposed CSS-based PNT signal. The philosophy of the MDS scheme is to use the two following chirp rates: $\mu_i^{(1)} = i\frac{2\mu}{N_{\text{sat}}}$ and $\mu_i^{(2)} = 2\mu - \mu_i^{(1)}$. The first one is used for $t \leq T_c/2$, and the other one for $T_c/2 < t \leq T_c$.

3. Signal Processing and Design

In this section, we show the receiver architecture and processing as well as the signal design of the proposed CSS-based PNT signal. Specifically, Section 3.1 deals with the pilot component and the design of the chirp period, whereas Section 3.2 deals with the data component and data transmission capabilities. Before going into detail, it is important to describe the receiving process of a CSS signal as given in (1), that is, the de-chirp process. The de-chirp process is illustrated in Figure 1, and it consists of mixing the received signal with the corresponding local replica and looking for the frequency of the maximum value of the resulting fast Fourier transform (FFT). This process is useful to receive a CSS signal and to estimate the data symbol when the signal is synchronized. Furthermore, this process is the cornerstone block for the design of both the pilot component used for ranging and synchronization [4,11].

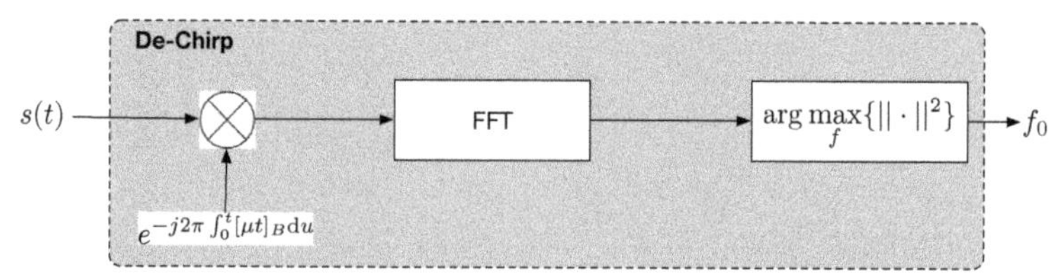

Figure 1. De-chirp process to receive a CSS signal: data demodulation block for FSK-CSS when synchronized.

3.1. Pilot Component

For the design process of the pilot component of a PNT system, we refer the reader to the study logic followed in [11]. This study looks for the shortest chirp period possible in order to minimize complexity. Unfortunately, we must fix a chirp period large enough to satisfy different requirements in terms of different KPIs. In particular, the KPIs that drive the design of the pilot component are the sensitivity and the accuracy. Since the target design is the complexity, we should look for the minimum chirp period (T_c) that provides the target sensitivity. Then, with this fixed chirp period, the ranging accuracy can be evaluated. Recall that the pilot component uses a BOK-chirp signal, so the acquisition and estimation blocks will deal with 4 de-chirp processes as the one in Figure 1 [11].

In this section, we show the results of the sensitivity analysis useful to design the chirp period and further evaluate the performance of the data component. For the analysis of the accuracy performance, we refer to [11]. For the sensitivity, we define a sensitivity performance given by the probabilities of false alarm (PFA) and detection (PD) of the acquisition module, e.g., PD $\geq$ 0.9 when PFA = 10^{-5}. Then, the main target of a sensitivity analysis is to find the minimum chirp period that provides the target sensitivity performance for different values of carrier-to-noise ratio (CN0), constellation size (N_{sat}), and visible satellites (N_{vis}). For the results shown in this paper, we will consider $\{N_{\text{sat}}, N_{\text{vis}}\} = \{50, 5\}$ satellites. An important analysis for the design of the CSS signal is the trend of T_c as a function of CN0. For instance, Figure 2 shows this analysis in the range of CN0 $= [35, 50]$ dB-Hz and different bandwidths in the range of 100 kHz to 10 MHz. We take these results as a reference for the analysis of the data component. More details are given in [11].

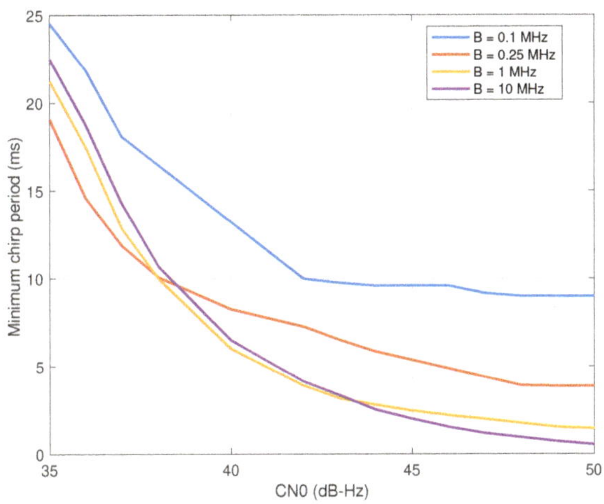

Figure 2. Minimum chirp period needed to obtain target sensitivity performance (PD > 0.9 @ PFA = 10^{-5}) for different bandwidths as a function of CN0 and $\{N_{\text{sat}}, N_{\text{vis}}\} = \{50, 5\}$ satellites.

3.2. Data Component

Let us focus now on the signal processing and design of the data component of a PNT system. First of all, it is important to note that the primary goal of a PNT signal is the ranging capabilities needed to obtain the user's PNT solution. Therefore, the signal design parameters should be optimized, targeting the sensitivity and accuracy performance required for the (pilot) ranging component. Data dissemination capabilities are dependent on the system. For instance, for standalone GNSS receivers [18], a very small amount of data is needed to solve PNT (e.g., 50 bps is enough for GPS L1/CA), but a larger amount of data is needed for an SBAS receiver [14] (e.g., 250 bps for SBAS L1). Furthermore, different requirements are needed in terms of data latency, data rate, or quality depending on the application [14]. When targeting low-complexity GNSS, we should follow the same signal design followed in [11]; that is, in this section, we consider a chirp period design given by the sensitivity analysis (see Figure 2). Then, we define the relevant KPIs for the data component, namely, the bit error rate (BER) and achievable bit rate, R_b. We do so for two different CSS modulations: the BOK- and FSK-CSS signals.

3.2.1. Bit Rate

To quantify the amount of data we can transmit over a CSS signal, we need to evaluate the number of bits transmitted per chirp period. This depends on the selected CSS modulation, e.g., BOK-CSS transmits one bit per (symbol) chirp period. Meanwhile, with a FSK-CSS signal, we work with a constellation of $M = 2^{\text{SF}}$ symbols, with SF as the spreading factor. The spreading factor denotes the number of bits per chirp period, and in a LoRa-like manner, it is given by [19]

$$\mathrm{SF} \doteq \log_2(BT_c), \tag{6}$$

where T_c is the chirp period, denoting the symbol duration and selected to achieve sensitivity performance. Then, the BOK- and FSK-CSS bit rates are respectively given by

$$R_b^{(b)} \doteq \frac{1}{T_c} \quad \text{and} \quad R_b^{(f)} \doteq \frac{\mathrm{SF}}{T_c}. \tag{7}$$

3.2.2. Bit Error Rate (BER)

For the evaluation of the BER of the considered CSS modulations, we use available formulas in the literature. On the one hand, the BER of BOK-CSS can be evaluated as follows [20]:

$$\mathrm{BER}_{\mathrm{BOK}} = \frac{1}{2}\mathrm{erfc}\left(\sqrt{0.25 \cdot \mathrm{MSI} \cdot \mathrm{CN0} \cdot T_c}\right), \tag{8}$$

with $\mathrm{erfc}(\cdot)$ as the complementary error function and MSI as the multisatellite interference accounting for the interference between satellites, as given in [11]. On the other hand, the BER for an FSK-CSS signal is approximated as [21]

$$\mathrm{BER}_{\mathrm{FSK}} \approx \frac{1}{4}\mathrm{erfc}\left(\sqrt{\mathrm{MSI} \cdot \mathrm{CN0} \cdot T_c} - \gamma(B, T_c)\right), \tag{9}$$

with $\gamma(B, T_c) = \sqrt{0.693 \cdot \log_2(BT_c) + 0.577}$.

4. Performance Analysis

In this section, we provide the numerical analysis to evaluate the performance of the considered CSS signal in this paper. Recall that the performance evaluation of the pilot component was already studied in [11]. In this section, we focus on the performance of the data component and its relationship with the pilot component. To do so, we first provide in Section 4.1 the analysis of the achievable data rate for the two considered modulations (i.e., BOK- and FSK-CSS). Second, Section 4.2 studies the BER for the two considered modulations. For these two analyses, we consider different bandwidth and CN0 values. Furthermore, to link the analysis of the data component with the pilot component, we consider the sensitivity analysis given in the left plot of Figure 2 as a reference for the design of the chirp period and the corresponding MSI as computed in [11].

4.1. Bit Rate Analysis

The analysis of the bit rate for the two considered modulations is shown in the left plot of Figure 3. For both modulations, as expected, we obtain an improvement of the achievable bit rate with the increment of CN0. We see that the bit rate moves to a saturation level at a given CN0. For instance, in the left plot, we see a saturation level of 1 kbps at 42 dB-Hz for the $B = 0.1$ MHz curve (i.e., blue curve). Both the saturation level and the CN0 value at which it is reached are increased with the bandwidth value. The second observation we can extract from the left plot of Figure 3 is that we obtain less and less gain in bit rate for larger bandwidths. For instance, from $B = 0.1$ MHz to $B = 0.25$ MHz, we obtain an improvement in bit rate that is proportional to the bandwidth increment (i.e., $\times 2.5$). However, the improvement with $B = 10$ MHz is just $\times 20$ bit rate (instead of $\times 100$) with respect to $B = 0.1$ MHz. Finally, another important point extracted from the results of bit rate is the difference between achievable values for FSK- and BOK-CSS. The former provides an improvement of 1 order of magnitude for all bandwidth and CN0 values.

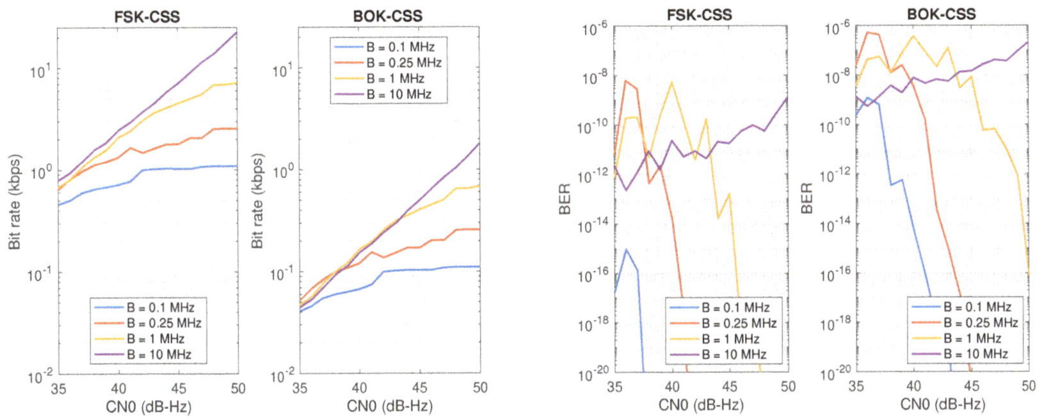

Figure 3. Performance analysis of the data component when linked to the pilot component with the sensitivity performance: (**Left**) bit rate analysis and (**right**) BER analysis.

4.2. BER Analysis

The evaluation of the BER for the two considered modulations is shown in the right plot of Figure 3. In both cases, the design of the chirp period is given by the sensitivity analysis as given in Figure 2. First, for both modulations, for a given CN0 value, we obtain an improvement of the BER when increasing the chirp period, T_c. Note from Figure 2 that T_c increases as the bandwidth value is reduced. Second, analogously, the narrower the bandwidth value, the better BER we obtain at a given CN0 value. Finally, from the results of the right plot of Figure 3 for a given bandwidth and CN0 value, we see an improvement of FSK-CSS with respect to BOK-CSS in terms of BER. The improvement varies with the bandwidth value, but we obtain values ranging from 2 to 7 orders of magnitude. The results provided in this section for the data component are very interesting for a practical implementation of a PNT signal. This is so for both FSK- and BOK-CSS signals. The reason is that they provide a good demodulation performance in comparison with current GNSS signals. Actually, the BER values obtained in this section could be modified in order to reach higher values by using smaller T_c values and then providing higher bit rates than in the current configuration. Unfortunately, by doing so, we would incur a modification of the sensitivity performance to a poorer level. Finally, we have not considered in this section the application of channel coding techniques that may allow better demodulation performance, and thus a better achievable bit rate for a given CN0 value.

5. Discussion

This work has investigated the use of CSS for PNT. This allowed us to show the main challenges and KPIs useful for the design of a CSS-based PNT signal to be operated in LEO constellations and targeting low complexity. Since a PNT signal is equivalent to a joint ranging and communication signal, during the elaboration of the PNT signal design, we are evaluating the ranging and communication capabilities of the signal. Note that today GNSS technology is power hungry in the sense that it needs a high computational load, so the design of a CSS-based PNT signal might provide the solution to this problem. Furthermore, we see a trend of LEO-PNT systems. Unfortunately, the power consumption problem of GNSS is aggravated when moving from MEO to LEO. The reason is that the acquisition search space is increased in LEO due to the increased Doppler frequency with respect to MEO. In this paper, we have shown the different challenges and how to solve them for the design of a low-complexity LEO-PNT signal based on CSS. The first challenge comes when dealing with the estimation of the triple needed for PNT (time delay, Doppler, data) from the CSS signal. To do so, a pilot/data scheme is proposed based on a BOK-

CSS and FSK structure for the pilot (i.e., time delay and Doppler estimation) and data (i.e., data demodulation) components, respectively. The second challenge to be dealt with is the design of an MSA scheme useful for PNT. Different schemes can be considered based on two main parameters of the CSS signal: the chirp rate and initial frequency. The performance analysis of the data component is given when linked with the design of the pilot component. The link is given by the design of the chirp period according to the sensitivity performance. The achievable data rate and BER of the considered modulations have been analyzed. Based on the results analyzed in this paper, we see how the proposed CSS signal is a good candidate for a PNT signal. First, it provides valuable results to be operated in a PNT system. For instance, as shown in previous works, it provides accuracies on the level of meters. Furthermore, we show in this paper achievable data rates in the order of tens of kbps. These are metrics of interest for a PNT system. On the other hand, we see that the CSS signal provides a better trade-off between ranging accuracy and achievable data rate than traditional DSSS signals used in GNSS. Note that current GNSSs provide accuracies in the order of meters or tens of meters with just a few tens of bps.

Author Contributions: Conceptualization, methodology, and writing—original draft preparation, D.E.-R.; validation, G.S.-G. and J.A.L.-S.; supervision, G.S.-G. All authors have read and agreed to the published version of the manuscript.

Funding: This research was partly supported by the Spanish Ministry of Science and Innovation under Grant PID2020-118984GB-I00 and by the TRP programme in the framework of the INNUENDO project, ESA Contract No. AO/1-9585/19/NL/CRS.

Institutional Review Board Statement: Not applicable.

Informed Consent Statement: Not applicable.

Data Availability Statement: Data are contained within the article.

Conflicts of Interest: The authors declare no conflict of interest.

References

1. Lawrence, D.; Cobb, H.S.; Gutt, G.; O'Connor, M.; Reid, T.G.R.; Walter, T.; Whelan, D. Navigation from LEO: Current capability and future promise. *GPS World Mag.* **2017**, *28*, 42–48.
2. Reid, T.G.; Neish, A.M.; Walter, T.; Enge, P.K. Broadband LEO Constellations for Navigation. *NAVIGATION J. Inst. Navig.* **2018**, *65*, 205–220. [CrossRef]
3. Qu, Z.; Zhang, G.; Cao, H.; Xie, J. LEO Satellite Constellation for Internet of Things. *IEEE Access* **2017**, *5*, 18391–18401. [CrossRef]
4. Egea-Roca, D.; López-Salcedo, J.A.; Seco-Granados, G.; Falletti, E. Comparison of Several Signal Designs Based on Chirp Spread Spectrum (CSS) Modulation for a LEO PNT System. In Proceedings of the 34th International Technical Meeting of The Satellite Division of the Institute of Navigation (ION GNSS+ 2021), St. Louis, MO, USA, 20–24 September 2021; pp. 2804–2818. [CrossRef]
5. Egea-Roca, D.; Arizabaleta-Diez, M.; Pany, T.; Antreich, F.; Lopez-Salcedo, J.A.; Paonni, M.; Seco-Granados, G. GNSS User Technology: State-of-the-Art and Future Trends. *IEEE Access* **2022**, *10*, 39939–39968. [CrossRef]
6. Neinavaie, M.; Khalife, J.; Kassas, Z.M. Exploiting Starlink Signals for Navigation: First Results. In Proceedings of the 34th International Technical Meeting of The Satellite Division of The Institute of Navigation (ION GNSS+ 2021), St. Louis, MO, USA, 20–24 September 2021; pp. 2766–2773. [CrossRef]
7. Reid, T.G.; Chan, B.; Goel, A.; Gunning, K.; Manning, B.; Martin, J.; Neish, A.; Perkins, A.; Tarantino, P. Satellite Navigation for the Age of Autonomy. In Proceedings of the IEEE/ION Position, Location and Navigation Symposium (PLANS), Portland, OR, USA, 20–23 April 2020; Institute of Electrical and Electronics Engineers Inc.: Piscataway Township, NJ, USA, 2020; pp. 342–352. [CrossRef]
8. Seco-Granados, G.; López-Salcedo, J.A.; Jiménez-Baños, D.; López-Risueño, G. Challenges in indoor global navigation satellite systems. *IEEE Signal Process. Mag.* **2012**, *29*, 108–131. [CrossRef]
9. Hansen, A.; Mackey, S.; Wassaf, H.; van Dyke, K. Complementary PNT Technology Demonstration. In Proceedings of the 34th International Technical Meeting of The Satellite Division of The Institute of Navigation (ION GNSS+ 2021), St. Louis, MO, USA, 20–24 September 2021; pp. 3127–3141. [CrossRef]
10. Del Peral-Rosado, J.A.; Raulefs, R.; López-Salcedo, J.A.; Seco-Granados, G. Survey of Cellular Mobile Radio Localization Methods: From 1G to 5G. *IEEE Commun. Surv. Tutor.* **2018**, *20*, 1124–1148. [CrossRef]
11. Egea-Roca, D.; Lopez-Salcedo, J.; Seco-Granados, G.; Falletti, E. Performance Analysis of a Multi-Slope Chirp Spread Spectrum Signal for PNT in a LEO Constellation. In Proceedings of the 10th Workshop on Satellite Navigation Technology (NAVITEC), Virtual, 5–7 April 2022; Institute of Electrical and Electronics Engineers Inc.: Piscataway Township, NJ, USA, 2022. [CrossRef]

12. Barrick, D.E. *FM/CW Radar Signals and Digital Processing*; Technical Report; Defense Advanced Research Projects Agency: Boulder, CO, USA, 1973.
13. Bor, M.; Vidler, J.E.; Roedig, U. LoRa for the Internet of Things. In Proceedings of the International Conference on Embedded Wireless Systems and Networks (EWSN), Graz, Austria, 15–17 February 2016.
14. Teunissen, P.J.G.; Montenbruck, O. *Handbook of Global Navigation Satellite Systems*; Springer: Berlin/Heidelberg, Germany, 2017; Volume 10.
15. Yao, Z.; Lu, M. Signal Multiplexing Techniques for GNSS: The Principle, Progress, and Challenges within a Uniform Framework. *IEEE Signal Process. Mag.* **2017**, *34*, 16–26. [CrossRef]
16. Winkler, M.R. *Chirp Signals for Communications*; Convention Record Paper; WESCON: Wichita, KS, USA, 1962; Volume 14.
17. Hengstler, S.; Kasilingam, D.P.; Costa, A.H. A novel chirp modulation spread spectrum technique for multiple access. In Proceedings of the IEEE International Symposium on Spread Spectrum Techniques and Applications, Prague, Czech Republic, 2–5 September 2002; Institute of Electrical and Electronics Engineers Inc.: Piscataway Township, NJ, USA, 2002; pp. 73–77. [CrossRef]
18. Spilker, J.J., Jr.; Axelrad, P.; Parkinson, B.W.; Enge, P. *Global Positioning System: Theory and Applications*; American Institute of Aeronautics and Astronautics: Reston, VA, USA, 1996; Volume 1.
19. Goursaud, C.; Gorce, J.M. Dedicated networks for IoT: PHY/MAC state of the art and challenges. *EAI Endorsed Trans. Internet Things* **2015**, *1*, 150597. [CrossRef]
20. Springer, A.; Gugler, W.; Huemer, M.; Reindl, L.; Ruppel, C.C.; Weigel, R. Spread spectrum communications using chirp signals. In Proceedings of the IEEE/AFCEA—EUROCOMM 2000: Information Systems for Enhanced Public Safety and Security, Munich, Germany, 19 May 2000; Institute of Electrical and Electronics Engineers Inc.: Piscataway Township, NJ, USA, 2000; pp. 166–170. [CrossRef]
21. Elshabrawy, T.; Robert, J. Closed-Form Approximation of LoRa Modulation BER Performance. *IEEE Commun. Lett.* **2018**, *22*, 1778–1781. [CrossRef]

engineering proceedings

Proceeding Paper

Analysis of Baseband Algorithms for LEO PNT †

Fran Fabra [1,*], José A. López-Salcedo [1,2] and Gonzalo Seco-Granados [1,2]

1 Institut d'Estudis Espacial de Catalunya (IEEC), 08034 Barcelona, Spain; jose.salcedo@uab.cat (J.A.L.-S.); gonzalo.seco@uab.cat (G.S.-G.)

2 Department of Telecommunication and Systems Engineering, Universitat Autònoma de Barcelona (UAB), 08193 Bellaterra, Spain

* Correspondence: fabra@ieec.cat

† Presented at the European Navigation Conference 2023, Noordwijk, The Netherlands, 31 May–2 June 2023.

Abstract: This paper provides an analysis of the impact of transmitter dynamics on baseband algorithms for position, navigation and timing applications using a low Earth orbit constellation. In particular, the acquisition and tracking of signals with high dynamics have larger Doppler and Doppler drift values that might impair the use of standard methods. Our analysis combines a theoretical assessment of acquisition and tracking thresholds with a performance evaluation by means of a simulation. Three frequency bands are tested (UHF, S and Ka) and an open-source software receiver has been adapted for such a purpose. The results obtained show the potential feasibility of LEO-PNT at baseband level under minimal considerations.

Keywords: LEO-PNT; GNSS; acquisition; tracking

Citation: Fabra, F.; López-Salcedo, J.A.; Seco-Granados, G. Analysis of Baseband Algorithms for LEO PNT. *Eng. Proc.* **2023**, *54*, 36. https://doi.org/10.3390/ENC2023-15458

Academic Editors: Tom Willems and Okko Bleeker

Published: 29 October 2023

1. Introduction

The high demand on Position Navigation and Timing (PNT) solutions, plus the advent of several Low Earth Orbit (LEO) constellations, has brought attention to the use of these signals for PNT applications [1–3]. Compared with traditional Global Navigation Satellite Systems (GNSS), LEO signals benefit from higher power transmission, a greater number of satellites, faster multipath conditions, and a lower cost of deployment, which enables the more frequent replacement of constellation generation.

From a baseband point of view, LEO-based PNT is characterized by high dynamics signals due to the larger Doppler shift induced by the relatively high velocity of the transmitter [4]. This means an increase in computation time during acquisition due to the larger search space. Moreover, in severe Doppler conditions, code-frequency offset (also referred to as code Doppler) can also be significant and needs to be compensated [5]. Considering that the visibility time window of a LEO satellite is also smaller, external aiding might be a mandatory condition for acquiring signals in certain frequency bands. Regarding tracking, the increase of Doppler-rate under high dynamics might cause phase, frequency and code deviations that cannot be followed using traditional tracking configurations in standard GNSS receivers (there is a larger probability of tracking loss due to dynamic stress error).

The analysis presented in this paper is split in two parts. First, a theoretical assessment is carried out to evaluate the acquisition and tracking thresholds for a set of signals considered (ranging from UHF to Ka bands) under the corresponding Doppler and Doppler rates from a LEO constellation. In the second part, the previous results are compared against a simulation analysis based on Monte Carlo realizations of the different signals under the selected scenarios.

2. Theoretical Assessment

In order to evaluate the potential frequency diversity of LEO-PNT systems, three bands have been preliminary selected in the present study: UHF, S and Ka. For each case, spread-spectrum GNSS-like signals have been considered, and no navigation message has been included.

In a standard LEO use case scenario, the dynamics will be mainly driven by the satellite motion. Table 1 shows the dynamics considered in the following analysis derived from the assumed LEO constellation (maximum values and comparison with GPS [6] as a reference).

Table 1. User–satellite range dynamics for LEO scenario and GPS.

Range Derivative	LEO	GPS
$\dot{\rho}$	7000 m/s	800 m/s
$\ddot{\rho}$	90 m/s^2	0.14 m/s^2
$\dddot{\rho}$	1 m/s^3	0.000122 m/s^3

2.1. Acquisition Thresholds

We will evaluate here the impact of LEO dynamics on signal acquisition and compare the results obtained against those from GPS L1, which will serve as a reference. The first parameter that we can check is the number of Doppler bins needed to cover the whole dynamic range. For a given code length, such a parameter gives the number of cells that need to be scanned during a cold start, thus providing some information about its feasibility as a real-time application for a given receiver. Table 2 provides the number of Doppler bins (computed as one half of the inverse of the coherent integration time) for two values of coherent integration time (T_{co}): the code period of the corresponding component for each signal and 10 ms. The next step is to check the impact of the Doppler change during the coherent integration. For this purpose, Table 2 also provides the amount of Doppler drift as a percentage of the Doppler bin. The results obtained for 10 ms of coherent integration reveal the problems that would be faced when trying to acquire signals at high frequency bands, where the coherent accumulation would not efficiently increase the signal-to-noise ratio without external assistance to correct the Doppler drift.

Table 2. Acquisition parameters for the different signals under analysis.

Frequency Band	GPS L1	UHF	S	Ka
Doppler bins ($T_{co} = T_{code}$)	17	30	117	115
Doppler drift over bin [%]	0.0001	0.015	0.037	0.003
Doppler bins (T_{co} = 10 ms)	169	382 [1]	2328	28627
Doppler drift over bin [%]	0.01	2.55 [1]	14.96	184.03

[1] Actual T_{co} for UHF is 10.4 ms to be an integer multiple of the corresponding T_{code}.

Finally, we follow the procedure summarized in [7] to compute the probability of detection during the acquisition process as follows:

$$P_d = 1 - F_{(\chi^2(2K,\beta))}(T_h^{acq}), \tag{1}$$

where $F_{(\chi^2(2K,\beta))}$ is the cumulative distribution function of a non-central χ^2 distribution with $2K$ degrees of freedom (with K being the number of incoherent accumulations) and a non-centrality parameter β, which is approximated, because the delay and Doppler uncertainties are not accounted for:

$$\beta = \frac{C}{N_0} K T_{co}. \tag{2}$$

The threshold value T_h^{acq} is computed from a desired probability of false alarm P_{fa} as follows:

$$T_h^{\text{acq}} = F^{-1}_{(\chi^2(2K))}(1 - P_{\text{fa}}). \quad (3)$$

By combining the previous expressions and fixing desired values for P_{d} and P_{fa}, we can scan over C/N_0 to obtain the acquisition threshold. Table 3 shows the results obtained when considering the whole search space for each type of signal according to the code length and Doppler range.

Table 3. Acquisition threshold to achieve $P_{\text{d}} = 0.95$ with a $P_{\text{fa}} = 0.001$ [dB-Hz].

Frequency Band	GPS L1	UHF	S	Ka
$T_{co} = T_{code}$	47.6	48.9	51.2	62.1
T_{co} = 10 ms	37.9	38.2 [1]	38.7	38.9

[1] Actual T_{co} for UHF is 10.4 ms to be an integer multiple of the corresponding T_{code}.

2.2. Tracking Thresholds

We will evaluate here the tracking performance of the different signals under analysis. A set of representative configurations of the delay, frequency and phase lock loops will be analysed by following the procedures from [8].

In general, the rule-of-thumb for setting the tracking threshold is to apply the following:

$$\sigma_{\text{track}} + e_d/3 \leq T_h^{\text{track}}, \quad (4)$$

where the particular values of tracking jitter σ_{track}, dynamic stress error e_d and threshold limit T_h^{track} for each type of loop are given in Table 4. Details on the jitter computation for delay (DLL), frequency (FLL) and phase lock loops (PLL) will be described later. The impact of these dynamics is mainly determined by the derivatives of the line-of-sight range ρ and the natural frequency of the loop filter of n-order ω_0, which is proportional to the loop noise bandwidth B_n. We take $B_n = 0.53\omega_0$ for a second-order filter and $B_n = 0.7845\omega_0$ for a third-order filter. Finally, the threshold limits for DLL depend on the early-to-late distance D of the discriminator. In the case of FLL, this value is a function of the integration time T, while for PLL it is fixed to 30° because we consider a data-less four-quadrant arctangent discriminator.

Table 4. Terms for tracking threshold computation depending on type of loop.

Term	DLL	FLL	PLL
σ_{track}	σ_{DLL}	σ_{FLL}	σ_{PLL}
e_d	$\rho_e = \frac{d^n\rho/dt^n}{\omega_0^n}$	$f_e = \frac{d^{n+1}\rho/dt^{n+1}}{\lambda\omega_0^n}$	$\theta_e = \frac{360}{\lambda}\frac{d^n\rho/dt^n}{\omega_0^n}$
T_h^{track}	$D/6$	$1/12T$	30°

The general expression for thermal noise code tracking jitter σ_{DLL} is given by the following:

$$\sigma_{\text{DLL}} = c\sqrt{\frac{B_n\int_{-B_{fe}/2}^{B_{fe}/2} S_s(f)\sin^2(\pi f D T_c)df}{(2\pi)^2 C/N_0\left(\int_{-B_{fe}/2}^{B_{fe}/2} f S_s(f)\sin(\pi f D T_c)df\right)^2}\left(1+\frac{\int_{-B_{fe}/2}^{B_{fe}/2} S_s(f)\cos^2(\pi f D T_c)df}{T C/N_0\left(\int_{-B_{fe}/2}^{B_{fe}/2} f S_s(f)\cos(\pi f D T_c)df\right)^2}\right)}, \quad (5)$$

where c is the speed of light, B_{fe} is the front-end bandwidth, $S_s(f)$ is the power spectral density of the signal, normalized to unit area over infinite bandwidth, T is the pre-detection integration time and $T_c = 1/R_c$ is the chip period (the inverse of the chipping rate).

Then, the tracking thresholds can be computed by determining the C/N_0 value required to obtain an equality in the aforementioned rule-of-thumb for the stress error level given by the LEO dynamics. Table 5 provides the results for DLL, where we can see the impact of dynamics in the improvement obtained when using third-order loop architectures in the LEO cases. The integration time is set to the corresponding T_{code} (the wide-band component for the LEO cases).

Table 5. Tracking threshold for DLL [dB-Hz].

Frequency Band	GPS L1	UHF	S	Ka
$B_n = 2$ Hz, 2nd order	22.7	20.7	25.9	-
$B_n = 2$ Hz, 3rd order	22.7	18.6	20.0	25.0
$B_n = 18$ Hz, 2nd order	27.9	24.5	25.6	30.1
$B_n = 18$ Hz, 3rd order	27.9	24.5	25.6	29.7

The frequency tracking jitter due to thermal noise is given by the following:

$$\sigma_{\mathrm{FLL}} = \frac{1}{2\pi T}\sqrt{\frac{4FB_n}{C/N_0}\left(1+\frac{1}{TC/N_0}\right)}, \tag{6}$$

where the term F is set to 1 and changes to 2 when the result is near the threshold. The results obtained for the different signals and configurations are shown in Table 6. Again, we can see the improvement obtained when increasing the order of the filter in LEO cases, especially for low values of B_n.

Table 6. Tracking threshold for FLL [dB-Hz].

Frequency Band	GPS L1	UHF	S	Ka
$B_n = 2$ Hz, 1st order	24.4	23.5	41.5	-
$B_n = 2$ Hz, 2nd order	24.4	21.5	22.7	27.2
$B_n = 18$ Hz, 1st order	30.2	28.6	29.7	33.1
$B_n = 18$ Hz, 2nd order	30.2	28.3	29.0	32.4

Finally, the phase tracking jitter due to thermal noise is given by the following:

$$\sigma_{\mathrm{PLL}} = \frac{360}{2\pi}\sqrt{\frac{B_n}{C/N_0}\left(1+\frac{1}{2TC/N_0}\right)}. \tag{7}$$

It is important to remark that we are neglecting two other relevant components of the total phase jitter: the Allan deviation oscillator noise and the vibration-induced noise. The reason for this is that we prefer to focus on the impact of dynamics in the present study. However, as described in [9], these terms are directly proportional to the carrier frequency, so their impact will be especially relevant for high frequency bands and thus should be carefully analyzed in the implementation of LEO PNT test platforms.

Table 7 provides the results obtained for the different signals and configurations. We can see how the dynamics in LEO disable the possibility of reaching acceptable values in most of the scenarios and that second-order loops are not a valid option.

Table 7. Tracking threshold for PLL [dB-Hz].

Frequency Band	GPS L1	UHF	S	Ka
$B_n = 2$ Hz, 2nd order	24.0	-	-	-
$B_n = 2$ Hz, 3rd order	21.4	16.8	-	-
$B_n = 18$ Hz, 2nd order	27.2	24.2	-	-
$B_n = 18$ Hz, 3rd order	27.1	20.8	21.8	26.2

3. Simulation Analysis

Each of the LEO signals described have been simulated assuming a constellation at an altitude of 600 km. With the aim of covering most of the Doppler and Doppler drift ranges at each frequency band, seven satellites have been simultaneously generated at different elevations to build 4 s of raw samples. This datasets are then injected into a modified version of FGI-GSRx [10,11], an open-source GNSS receiver. In addition to the adaptation of the particular characteristics of our LEO signals (modulation and code properties), the modification includes code Doppler compensation in terms of acquisition and a new tracking loop engine that implements a second-order DLL with a third-order PLL aided by a second-order FLL.

3.1. Results for Acquisition

The acquisition analysis consists in performing 1000 realizations of data, with added white Gaussian noise (adapted to a given C/N_0 value), to obtain the corresponding metrics employed to determine the signal presence during the acquisition process. The value of the fifth percentile (5%), which would correspond to a probability of detection of 0.95, is compared to the acquisition threshold for a probability of false alarm of 0.001 (the computation of this term takes into account the total number of code-Doppler cells required during a full acquisition process). Figure 1 provides the results obtained for the same LEO cases shown previously in Table 3. We can see that there is a generally good agreement with the theoretical assessment, including the case at the Ka band with longer integration. Therefore, the lack of Doppler drift aiding does not have a strong impact in these scenarios.

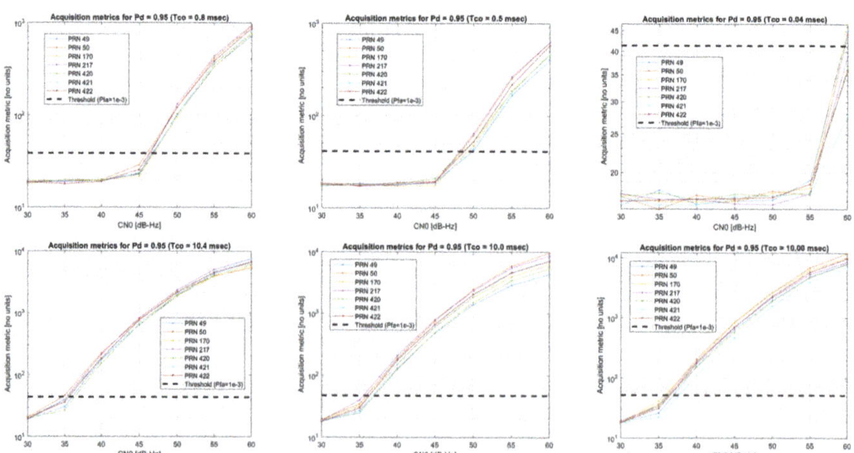

Figure 1. Acquisition results from simulation analysis. A dashed line indicates the corresponding threshold limit for a probability of false alarm of 0.001. Columns from left to right: results from UHF, S and Ka bands. **Upper row:** $T_{co} = T_{code}$. **Bottom row:** T_{co} = 10 ms (10.4 ms for UHF).

3.2. Results for Tracking

For tracking analysis, the procedure consists in performing the acquisition of noise-free data and starting tracking with a wide loop bandwidth (40 Hz) in order to safely reach a steady state in a short time interval and to then include additive white Gaussian noise and to set a given loop bandwidth. Then, the resultant code, phase and Doppler evolution are compared against the corresponding models to estimate the error values as the standard deviation of such differences over the given interval (from seconds 3 to 4) and averaged after 10 different iterations. Figures 2–4 provide the code, phase and Doppler output error results for the loop bandwidth values employed during the theoretical assessment. In order to limit the number of different combinations, the same loop bandwidth is applied to the three different loops of the implemented architecture. The integration time is set to the code period of the corresponding wide band component. Regarding the obtained results, the code-tracking thresholds that can be derived from the graphs are well aligned with the theoretical values for third-order loops given in Table 5 (recall that our second-order DLL is aided by a third-order PLL/FLL loop). On the other hand, the frequency results seem to benefit from the combination with the phase discriminator in the loop architecture. However, such a combination also has an impact on the phase, which provides more robust results for $B_n = 2$ Hz (tracking is achieved), while worsening the expectation when using a larger loop bandwidth value (the thresholds seem to be bounded by the performance of the FLL). It is also relevant to point out the results in frequency and phase from PRN 421, which are probably due to the fact that this was the satellite with the highest Doppler drift.

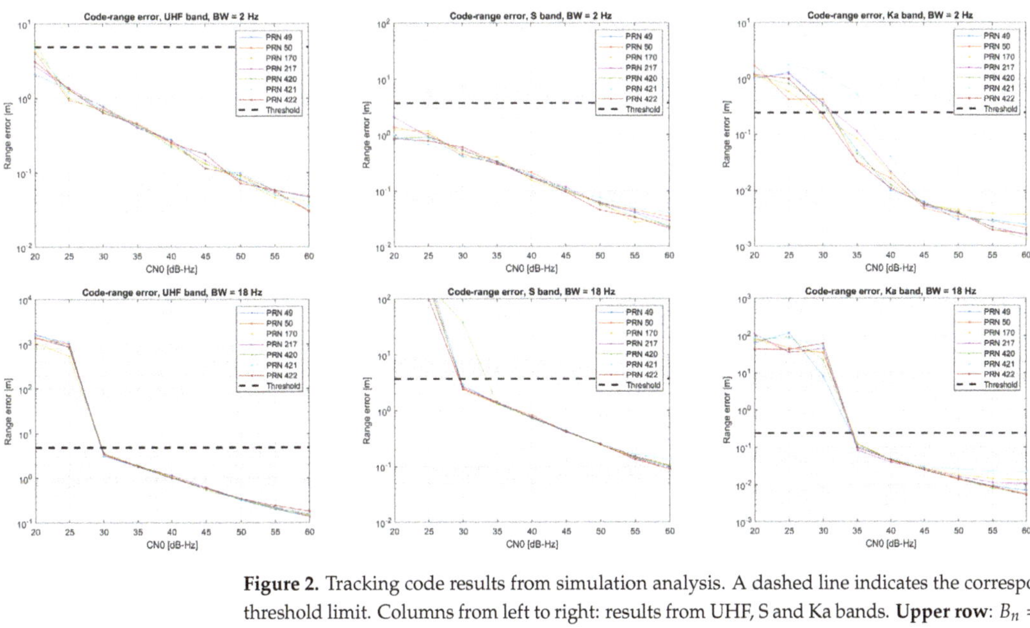

Figure 2. Tracking code results from simulation analysis. A dashed line indicates the corresponding threshold limit. Columns from left to right: results from UHF, S and Ka bands. **Upper row**: $B_n = 2$ Hz. **Bottom row**: $B_n = 10$ Hz.

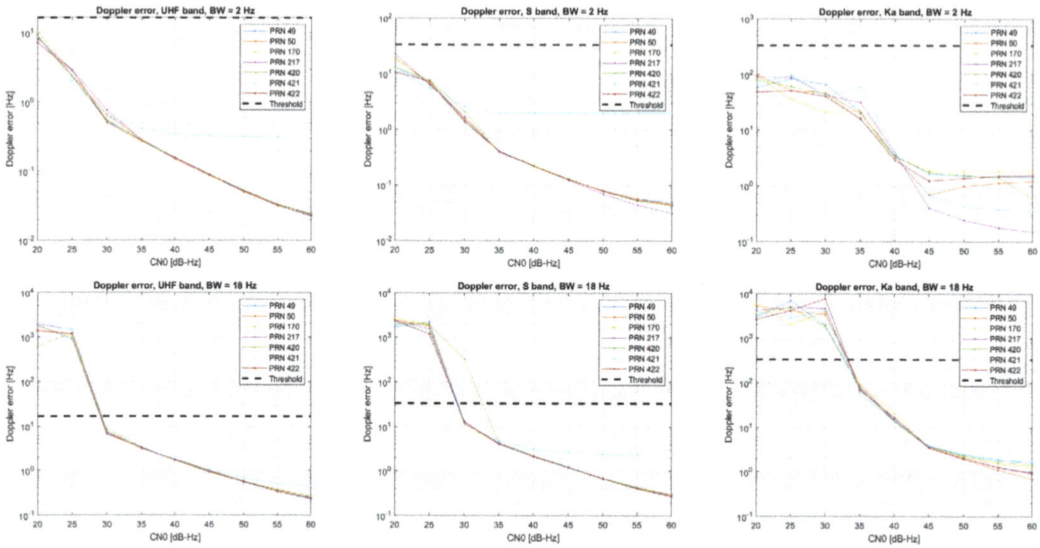

Figure 3. Tracking frequency results from simulation analysis. A dashed line indicates the corresponding threshold limit. Columns from left to right: results from UHF, S and Ka bands. **Upper row**: $B_n = 2$ Hz. **Bottom row**: $B_n = 10$ Hz.

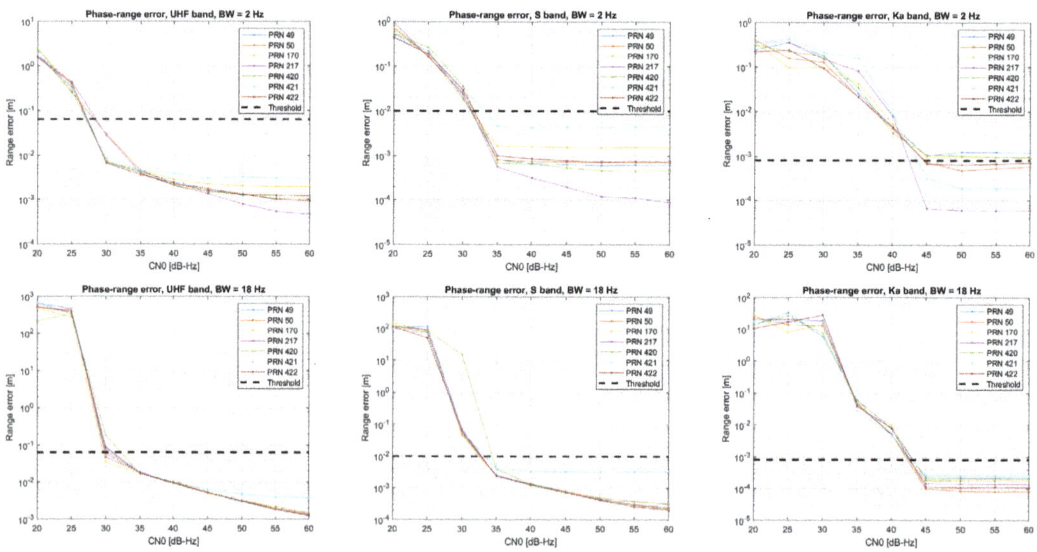

Figure 4. Tracking phase results from simulation analysis. A dashed line indicates the corresponding threshold limit. Columns from left to right: results from UHF, S and Ka bands. **Upper row**: $B_n = 2$ Hz. **Bottom row**: $B_n = 10$ Hz.

4. Discussion

The analysis presented in this paper shows the potential feasibility of LEO-PNT at baseband level. Compared with the standard GNSS, LEO-PNT will typically have higher demanding acquisition thresholds which can be compensated for with higher received power levels. The large Doppler ranges imply a higher number of Doppler cells, thus significantly increasing the total number of search cells. Although this might not be a limitation in terms of acquisition thresholds, it can have a strong impact on real-time implementations. However, this effect could be compensated for by means of Doppler aiding. The most relevant aspect in our software-based GNSS receiver is the necessity of including code Doppler correction to enable longer integration times.

Regarding tracking, the theoretical assessment shows that only high-order loops can be used to properly track the LEO dynamics (even considering a static receiver). Despite this, the tracking loop architecture implemented has shown the capability of successfully tracking the selected signals in the different bands under analysis. However, it is worth mentioning that neither clock- or vibration-induced errors have been included, which have an important impact on high frequency bands.

Author Contributions: Conceptualization, F.F., J.A.L.-S. and G.S.-G.; methodology, F.F.; software, F.F.; formal analysis, F.F.; writing—original draft preparation, F.F.; writing—review and editing, J.A.L.-S. and G.S.-G.; project administration, G.S.-G.; funding acquisition, J.A.L.-S. and G.S.-G. All authors have read and agreed to the published version of the manuscript.

Funding: This research was carried out under a programme funded by the European Space Agency (ESA): project ESA AO/1-10372/20/NL/CRS "LEO Positioning Navigation and Timing from User Equipment". The views expressed herein can in no way be taken to reflect the official opinion of the European Space Agency.

Institutional Review Board Statement: Not applicable.

Informed Consent Statement: Not applicable.

Data Availability Statement: Data sharing is not applicable due to privacy restrictions.

Acknowledgments: The authors would like to acknowledge the contributions of Lea Castel, Rami Ali-Ahmad, Romain Crapart and Simon Naws from Thales Alenia Space France in providing the raw data samples from the different LEO signals.

Conflicts of Interest: The authors declare no conflict of interest.

Abbreviations

The following abbreviations are used in this manuscript:

LEO	Low Earth Orbit
PNT	Positioning, Navigation and Timing
UHF	Ultra-High Frequency
GNSS	Global Navigation Satellite System
GPS	Global Positioning System
DLL	Delay Lock Loop
FLL	Frequency Lock Loop
PLL	Phase Lock Loop

References

1. Benzerrouk, H.; Nguyen, Q.; Fang, X.; Amrhar, A.; Rasaee, H.; Landry, R. LEO satellites Based Doppler Positioning Using Distributed nonlinear Estimation. *IFAC-PapersOnLine* **2019**, *52*, 496–501. [CrossRef]
2. Cheng, L.; Dai, Y.; Guo, W.; Zheng, J. Structure and Performance Analysis of Signal Acquisition and Doppler Tracking in LEO Augmented GNSS Receiver. *Sensors* **2021**, *21*, 525. [CrossRef] [PubMed]
3. Kassas, Z.; Morales, J.; Khalife, J. New-age satellite-based navigation–STAN: Simultaneous tracking and navigation with LEO satellite signals. *Inside GNSS Mag.* **2019**, *14*, 56–65.
4. Ali, I.; Al-Dhahir, N.; Hershey, J.E. Doppler Characterization for LEO Satellites. *IEEE Trans. Commun.* **1998**, *46*, 309–313. [CrossRef]
5. Cheng, U.; Hurd, W.J.; Statman, J.I. Spread-Spectrum Code Acquisition in the Presence of Doppler Shift and Data Modulation. *IEEE Trans. Commun.* **1990**, *38*, 241–250. [CrossRef]
6. Misra, P.; Enge, P. *Global Positioning System, Signals, Measurements, and Performance*, 2nd ed.; Ganga-Jamuna Press: Lincoln, MA, USA, 2006.
7. Foucras, M.; Julien, O.; Macabiau, C.; Ekambi, B.; Bacard, F. Assessing the Performance of GNSS Signal Acquisition: New Signals and GPS L1 C/A Code. Inside GNSS, Inside GNSS Media LLC. 2014; pp. 68–79. Available online: https://insidegnss.com/wp-content/uploads/2018/01/IGM_julaug14-WP.pdf (accessed on 10 January 2023).
8. Kaplan, E.D.; Hegarty, C.J. *Understanding GPS: Principles and Applications*, 2nd ed.; Artech House: Boston, MA, USA, 2006.

9. Irsigler, M.; Eissfeller, B. PLL tracking performance in the presence of oscillator phase noise. *GPS Solut.* **2002**, *5*, 45–57. [CrossRef]
10. FGI-GSRx Software Receiver. Available online: https://github.com/nlsfi/FGI-GSRx (accessed on 15 May 2023).
11. Borre, K.; Fernández-Hernández, I.; Lopez-Salcedo, J.A.; Bhuiyan, M.Z.H. *GNSS Software Receivers*; Cambridge University Press: Cambridge, UK, 2022, *in press*.

engineering proceedings

MDPI

Proceeding Paper

Lunar Navigation System ODTS Signal in Space Error Analysis †

Cosimo Stallo [1,*], Carmine Di Lauro [1], Mattia Carosi [1], Laura De Leo [1], Martina Cappa [1], Daniele Musacchio [1], Daniele Cretoni [1], Henno Boomkamp [2], Pietro Giordano [3], Richard Swinden [3] and Javier Traveset [3]

1 Thales Alenia Space Italy (TAS-I), 00131 Rome, Italy; carmine.dilauro@thalesaleniaspace.com (C.D.L.); mattia.carosi@thalesaleniaspace.com (M.C.); laura.deleo@thalesaleniaspace.com (L.D.L.); martina.cappa@thalesaleniaspace.com (M.C.); daniele.musacchio@email.com (D.M.); daniele.cretoni@thalesaleniaspace.com (D.C.)
2 Telespazio Germany (TPZ), 64293 Darmstadt, Germany; henno.boomkamp@telespazio.de
3 European Space Agency (ESA), 2200 Noordwijk, The Netherlands; pietro.giordano@esa.int (P.G.); richard.swinden@esa.int (R.S.); javier.ventura-traveset@esa.int (J.T.)
* Correspondence: cosimo.stallo@thalesaleniaspace.com

† Presented at the European Navigation Conference 2023, Noordwijk, The Netherlands, 31 May–2 June 2023.

Abstract: In recent years, the Moon has gained renewed interest in terms of human exploration for scientific purposes. In this context, Thales Alenia Space is leading a consortium to define the main concepts for a Lunar Radio Navigation System (LRNS) in terms of Orbit Determination and Time Synchronization (ODTS) as part of an ESA Technology Development Element (TDE) programme. This work focuses on the latest performance results achieved through a dedicated simulator in terms of Signal In Space Error (SISE).

Keywords: Signal In Space Error (SISE); Orbit Determination and Timing Synchronisation (ODTS); Lunar Radio Navigation System (LRNS)

Citation: Stallo, C.; Di Lauro, C.; Carosi, M.; De Leo, L.; Cappa, M.; Musacchio, D.; Cretoni, D.; Boomkamp, H.; Giordano, P.; Swinden, R.; et al. Lunar Navigation System ODTS Signal in Space Error Analysis. *Eng. Proc.* **2023**, *54*, 37. https://doi.org/10.3390/ENC2023-15468

Academic Editors: Tom Willems and Okko Bleeker

Published: 29 October 2023

1. Introduction

In recent years, lunar missions have been attracting renewed interest, in the clear perspective of achieving the required technologies and capabilities necessary to expand the human presence in the solar system [1,2]. Unlike the first race to the Moon in the 1960s, this return is aimed to be effective and permanent, so that more than a hundred missions are planned in the upcoming years, opening up new opportunities for both space agencies and private companies. Despite the high number of planned missions, no global lunar communication and navigation infrastructure is yet available, so every mission is currently planning its own custom solution to communicate and navigate using Earth-ranging or dedicated in situ relay capabilities. For this reason, the European Space Agency's (ESA) Moonlight initiative [3,4] aims to provide navigation and communication capabilities and services to establish a reliable, sustainable and scalable network, compatible with the LunaNet Framework [5].

This contribution will present an ODTS concept to implement the satellite synchronization and generation of the clock and ephemeris data required for the navigation message, as well as the performances achievable. To accomplish this, an ODTS tool capable of simulating the generation of the broadcasted navigation message, which contains information about the satellites' ephemeris and clock, has been developed. This study assumes a constellation of four satellites in Elliptical Lunar Frozen Orbits (ELFO) in order to achieve the highly accurate and reliable positioning of a user on the Moon South Pole (such as in [6,7]). The proposed baseline for LRNS orbit determination consists of Earth-based tracking techniques through Telemetry, Tracking, and Command (TT&C) links [8,9]. On the other hand, time synchronization will rely on the observation of on-board time provided by space-qualified atomic clocks with respect to a Master Clock on Earth through the means of Two-Way Time Transfer (TWTT) from exploiting the same TT&C link.

For what concerns the Time Synchronization (TS) analysis, a set of Monte Carlo (MC) simulations has been run to estimate the LRNS satellite onboard clock offset using two-way time transfer observables in a Kalman filter. On the other side, the orbit determination (OD) contribution has been evaluated in two ways: a first approach makes an assumption on the outputs of the OD process, while the second one performs a Monte Carlo analysis. The final results are then compared with the requirement to drive conclusions.

The work performed in this study has been financed under the ESA TDE program and applies to a generic lunar radio navigation system, thus it is not necessary to represent what the Moonlight programme will implement. In addition, it does not apply the Moonlight service performance requirements defined by ESA as part of the Moonlight programme invitation to tender.

The paper is structured as follows: Section 2 provides a description of the system architecture of this satellite system; Section 3 provides a description of the SISE error and details the simulations performed to assess its 95% value as a function of the Age Of Data (AoD). Finally, Section 4 reports the main conclusions.

2. LRNS ODTS System Architecture

On the basis of the previous results achieved in [8,9], the final ODTS system baseline foresees the following elements for Space Segment and Ground Segment.

2.1. Space Segment

For the OD process, LRNS satellites will carry onboard an X-band multi-functional Code Division Multiple Access (CDMA) transponder that will be used for TT&C, Ranging and Time Synchronization, one High Gain Antenna (HGA) for normal mode and two Low Gain Antennas (LGA) for contingency operations, and a passive Satellite Laser Ranging (SLR) retroreflector for independent validation.

Concerning time keeping, three mini-RAFS (Rubidium Atomic Frequency Standard clocks) with a combination unit will be available onboard. These clocks will generate a free-running time scale initialized to the LRNS Reference Time (LRT), while asynchronous two-way time transfer is exploited on the same TT&C link used for OD.

The key advantage of this concept is that both OD and TS make use of the same TT&C Deep Space Transponder (DST) and related antennas and HPA.

2.2. Ground Segment

The LRNS ground segment includes three dedicated TT&C stations (a 90 cm diameter dish for normal mode in Multiple Spacecraft Per Antenna (MSPA) approach [9], and at least a 5 m diameter dish for contingency mode per site) at a longitudinal separation of around 120° to obtain global coverage. The stations must be endowed with an internal delay calibration system in the X band to calibrate the excess delay due to the ground station ranging measurement equipment. Moreover, SLR tracking will be provided by the International Laser Ranging Service on a best-effort basis.

The ground TS architecture is based on a network of ground stations synchronized via the Stand-Alone-All-in-View GNSS TS technique [10,11], including a master clock station in charge of generating the LRNS Reference Time.

3. SISE Analysis

An ODTS simulator has been developed to reproduce the generation and transmission of the broadcasted navigation message, containing information about the satellites' ephemeris and clock. Within the LRNS project context, the SISE is defined in line with the LunaNet Interoperability Specification [5] as the instantaneous difference between the position, velocity, and time of an LRNS satellite as broadcasted by the LRNS navigation message, and the true satellite position, velocity, and time, respectively, expressed in the lunar reference frame and the lunar time system standard. The SISE can be expressed for convenience in two parts:

1. Signal-In-Space Error for positioning ($SISE_{pos}$), according to Equation (1), where x, y, z, and t are the true position and time, while the corresponding tilde represent the values broadcasted in the navigation message.
2. Signal-In-Space Error for velocity ($SISE_{vel}$) according to Equation (2), where $\dot{x}$, $\dot{y}$, $\dot{z}$, and $c\dot{t}$ represent the velocity and clock drift, while the corresponding tilde are the values broadcasted in the navigation message.

Within this project, an initial SISE target value of 25 m at 95% (this activity has been performed prior to the publication of the ESA Moonlight ITT and the related ESA Service Requirement Document (ESRD), so the SISE target value here considered is not the final one in the Moonlight ESRD) has been proposed by ESA and will be used as reference for the performance assessment.

$$SISE_{pos} = \sqrt{\left(x-\tilde{x}\right)^2 + \left(y-\tilde{y}\right)^2 + \left(z-\tilde{z}\right)^2 + \left(ct-\widetilde{ct}\right)^2}, \quad (1)$$

$$SISE_{vel} = \sqrt{\left(\dot{x}-\tilde{\dot{x}}\right)^2 + \left(\dot{y}-\tilde{\dot{y}}\right)^2 + \left(\dot{z}-\tilde{\dot{z}}\right)^2 + \left(c\dot{t}-\widetilde{c\dot{t}}\right)^2}, \quad (2)$$

The following sections will present the approach proposed for the time synchronization and the orbit determination simulations.

3.1. TS Error Budget for SISE Derivation

A TS simulator is implemented around the TS baseline system architecture described in [8–10] and summarized in previous paragraphs, to simulate the TS concept designed and compute the accuracies achieved. To reach its purpose, it reproduces the real satellite on-board clock offset and the ones resulting from the TWTT observables collection. These are then exploited in the implemented Kalman Filter (KF) to derive the three clock corrections (time offset—a_{f0}[s], frequency offset—a_{f1}[s/s], and frequency drift—a_{f2}[s/s^2]) which, broadcasted to a user in the navigation message, allow it to predict the satellite on-board clock offset. Based on them, the simulator computes the SISE Clock Error, the difference between the real and estimated clock offset. The TS simulator process is summarized in the following steps:

1. Simulation of the real satellite on-board clock offset w.r.t. Lunar Reference Time (LRT).

The on-board clock behaviour is reproduced by the time-dependent Equation (3). The model includes: a deterministic timing error expressed by a quadratic polynomial equation, a random noise error generated according to [11], and a thermal sensitivity variation result of a sinusoidal temperature variation profile.

$$x_0(t) = a_{f0} + a_{f1}(t-t_0) + \frac{1}{2}\cdot a_{f2}(t-t_0)^2 + \varepsilon(t-t_0) + Th_v(t-t_0), \quad (3)$$

The clock offset coefficients, the clock noise parameters and the thermal sensitivity intervening, respectively, in the three frequency deviations are selected properly to reproduce miniRAFS technology behaviour. Moreover, the clock offset is reported with respect to the Master Clock timescale which defines the LRT for the constellation. Typical a_{f0}, a_{f1}, and a_{f2} parameters characteristic of an Active Hydrogen Maser (AHM) are considered to generate the reference.

2. Simulation of the Two-Way Time Transfer noise based on error budget.

To reproduce the residual error due to the TWTT, each contribution of noise on top the satellite on-board clock offset measurements is simulated as a Gaussian distribution with mean and standard deviation defined by a TWTT error budget. Table 1 summarizes the errors assessment conducted within this project [8–12].

Table 1. Two-Way Time Transfer residual error budget. The delay due to relativistic effects, crossing of troposphere, and earth rotation are considered as negligible.

Type of Residual Error	Description	Bias or Mean [ns]	Standard Deviation [ns]
Epoch-mismatch	Misalignment between the epochs timestamped on the basis of the local time reference. It is derived from the difference between ground and space clock error assuming conservative values of frequency offset: Ground Clock of 1×10^{-11} and Space Clock of 1×10^{-9}.	0.012	-
Ionospheric	Delay due to crossing of ionosphere. Computed through the equation below with f_{gs} = 7.22 GHz, f_{sg} = 8.45 GHz and an assumed worst case TEC = 1×10^{18} electrons/m^2. $\varepsilon_{iono} = 40.3 \times \frac{TEC}{c} \times \left(\frac{1}{f_{gs}^2} - \frac{1}{f_{sg}^2}\right)$	0.7	-
LRNS-Orbit	OD error in LRNS satellite position estimation, depending on OD error statistics. The proportionality is derived from the mean and standard deviation of the ephemeris error in the radial direction resulting from OD tool simulations.	$\propto$0.2	$\propto$4
Calibration	On-board group delay of hardware components. It is the accuracy reached considering the state-of-the-art TWSTFT protocol ($1.4\sqrt{2}$ ns = 2 ns) [13,14].	2	-
TWTT noise	Background white noise due to time transfer. Typical values of noise affecting the Time Transfers through a signal in space with the Two-Way technique [15].	-	0.75
Ground stations synchronization	Residual delay due to ground-to-ground TS to a Master Clock Station by GNSS timing RX [16]. Not included in measurements w.r.t. Master Clock Station. The assessed values are based on TASI background experience on GNSS timing RX available on the TASI timing laboratory (RF path timing chain) and by the on-going Galileo program.	7	0.5

The sum of error contributions are then exploited to derive three different overall TWTT random noise errors, one per each Ground Station.

3. Simulation of TWTT observables.

The TWTT observables result from the sum of the TWTT noises to the real clock offset. To be in line with the mission selected Time Transfer (TT) sessions, satellite visibility, and ground-to-ground calibration aspects, the 1s sampled components of the simulated measurements are further processed as follows:

- The observables of the 5 min time transfer session are averaged to provide one sample each 45 min (according to the baseline TT session, [12]);
- They are filtered through the visibility matrix that considers the time interval in which the selected satellite is in view of the selected ground station.
- They are pre-processed in a calibration window to remove the residual ground calibration bias from the observables collected w.r.t. Ground Clock Station.

4. Kalman Filter estimation of satellite on-board clock offset.

Due to its proved effectiveness in predicting and compensating clock offsets [17], a Kalman Filter is implemented in the simulator to obtain from noisy TWTT observables, which are known to have residual errors, the searched states (a_{f0}, a_{f1}, and a_{f2}).

After an initialization phase and fine tuning, a Dynamic State Model equation describing the dynamical evolution of the physical system in time is implemented, representing the prediction equation of the filter. At each step, the predicted states are updated through a weighted correction extrapolated from the comparison with the measurements (the collected TWTT observables). Since the TWTT baseline considers the collection of three observables, one per each Ground Station, a Kalman filter for multiple sensor system is implemented: the Parallel Kalman Filter [18,19]. It updates the states once all measurements are processed. This computation gives, in output, the best reachable estimation of the satellite clock offset states (a_{f0}, a_{f1}, and a_{f2}) in the whole time window of simulation. These states are the clock parameters sent in the Navigation Message to the User Segment.

5. SISE Clock Error computing.

The SISE Clock Error is computed as the difference between the real and the estimated clock offset at a specific prediction time. The estimated value is derived by the clock offsets estimate, obtained as a product of the Kalman Filter, with the procedure described below. The computation starts at the end of the Kalman filter initialization process window and discards the first solution data comprised in the imposed Kalman filter warm-up period, since they are affected by high instabilities.

A configurable time window of 12 h is fixed to establish the amount of clock offset data for the least-squares estimation of the a_{f0}, a_{f1}, and a_{f2} parameters. The clock coefficients are then exploited to propagate in future batch of data the predicted clock offset. The propagation is realized at different prediction time, from 1 s to 12 h with a time span of 15 min. The predicted value at the end of the coverage time is finally compared with the real clock offset for the computation of the SISE clock prediction error.

The fitting and prediction time window is then translated to the next sample along the time window and the steps are repeated. SISE computation logic is summarized in Figure 1.

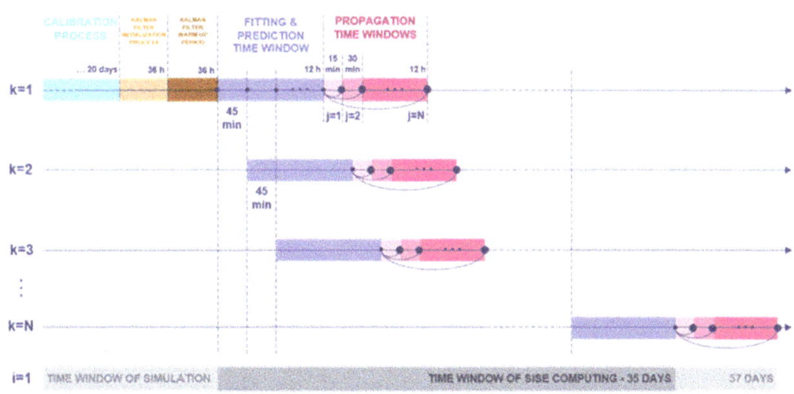

Figure 1. SISE Clock Error computation logic.

SISE Clock Error Results

The SISE Clock Error with a different Prediction Time of propagation (from 1 s to 12 h with a time span of 15 min) is obtained in Figure 1. A Monte Carlo based on 150 simulations applied on 1120 samples (one sample each 45 min in a time window of 35 days) for 49 different prediction times (for a total of 8.23×10^6 realizations) is applied.

The SISE Clock Error (shown in Figure 2) at 0 AoD is mainly due to the TWTT errors affecting the measurements. This noise distribution has a no zero mean value that has the major impact.

The SISE Clock Drift Error is computed with the same steps of the SISE Clock Error computing applied to the clock offset drift, a_{f1}. The Clock Drift Error, propagated at a specific prediction time, is compared to the real Clock Drift calculated as the first derivative of the simulated real on-board clock offset. The result is obtained in Figure 3.

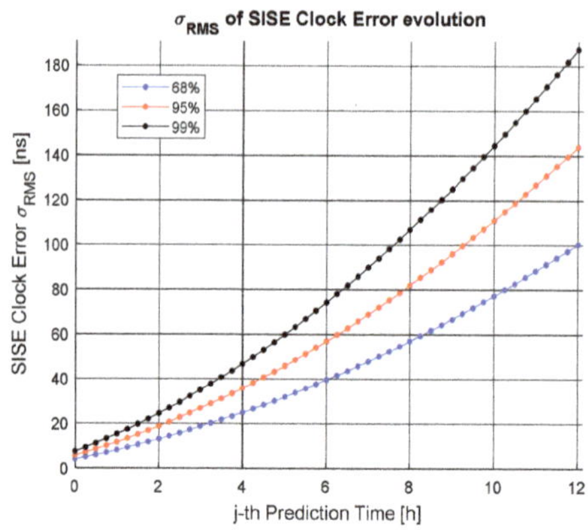

Figure 2. SISE Clock Error evolution from TS software v4.0 Monte Carlo simulation.

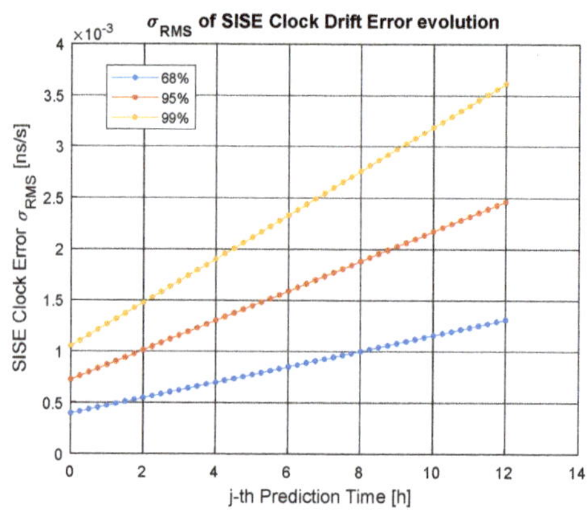

Figure 3. SISE Clock Drift Error evolution from TS software v4.0 Monte Carlo simulation.

Its value indicates how accurately we can predict the SISE Clock Drift at that specific propagation time.

These results are then merged with the OD error budget for the overall SISE derivation, as shown in Section 3.3.

3.2. OD Error Budget for SISE Derivation

In the framework of the LRNS project, the OD contribution to the SISE is estimated, acknowledging that it will vary under the influence of a large number of contributing factors such as the unpredictable variability in solar radiation pressure, deviations due to the difference between the real spacecraft attitude and the modelled one, unmodelled biases, and orbit estimation accuracy and mismodelling errors. Moreover, given that the navigation message broadcasted to the user is based on an orbit prediction, it is based on the numerical integration of the satellites' equations of motion, which suffer from input errors in the initial position X_0, initial velocity V_0, and acceleration model.

Two analysis methods will be shown to assess the SISE error, as defined above:

1. The first approach makes assumptions on the errors on X_0, V_0, and acceleration and computes the associated range of SISE errors from these, i.e., it makes assumptions directly on the outputs of the ODTS process.
2. The second approach makes an assumption on the input conditions to the ODTS process and runs a series of Monte Carlo simulations to vary the range of the SISE values.

3.2.1. SISE Derived from Assumptions on the Outputs of the ODTS Process

The orbit prediction is a fully deterministic process that follows the Newtonian equation of motion (Equation (4)):

$$X(t) = X_0 + V_0 t + \iint A(t) dt^2, \tag{4}$$

Therefore, the error in the predicted position is the linear sum of the errors in each of the three input errors presented above. Values for the three errors, based on early ODTS simulator results and data arcs with a length from 4 up to 6 days, have been provided as follows:

- Error in the initial position X_0 of 2 m
- Error in the initial velocity V_0 of 0.0002 m/s
- Error in the acceleration A(t) of 5×10^{-9} m/s^2

Monte Carlo simulations allow for re-assessing the assumptions, especially for X_0 and V_0.

As an example, Figure 4a shows how the initial position factor is varied over a range to obtain a bundle of curves with the "typical" value curve in the middle of each plot. Taking the lowest assumption for the three errors, a "best case" curve can be constructed, as well as a "worst case curve" can be obtained with the worst values for each errors.

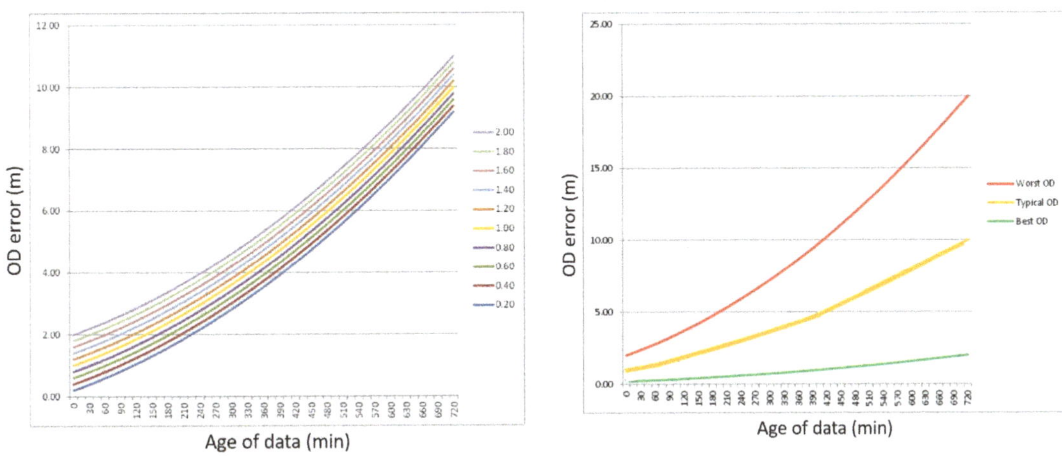

(a) (b)

Figure 4. (**a**) Orbit prediction error shown as function of different position errors X_0 and (**b**) SISE error as a function of the age of data for the OD process.

Finally, the total SISE error is obtained in Figure 4b. Derived from this analysis and assuming a normal distribution of errors varying between the worst case and best case, the expected SISE errors (3D position and clock combined) are reported in Table 2, from which we can derive that the SISE requirement of 25 m seems feasible with this approach.

Table 2. 95% and 99% SISE error levels as a function of the Age of Data (AoD).

	AoD 6 h	AoD 12 h
95% error level	7.6 m	18.4 m
99% error level	9.1 m	22.6 m

3.2.2. SISE Derived from Monte Carlo Analysis

As an alternative to the previous analysis, it is possible to vary some input assumptions to the OD simulator, simulate a truth orbit and tracking data, and run a series of OD solutions that cover the same period. The used assumptions are summarized in Table 3.

Table 3. Assumptions* made and impact on number of MC cases and actual number of simulations performed.

	Assumption	Units	Low	Typical	High	MC Factor	Accum.
Truth orbits	Solar activity	SFU	-	-	350	1	1
	Arc length	days	-	4.0	-	1	1
Data simulation	Lunar reference frame error	masec	0	-	1	2	2
	Range noise	m	0.05	-	0.5	2	4
	Range-rate noise	mm/s	0.00	-	0.10	2	8
	Range bias	m	-	1.00	-	1	8
	Data gaps (nr/max duration)	-/hrs	0/0	5/3	-	2	16
ODTS configuration	Arc length (overlap)	days	-	4.0/0.5	-	1	16
	Prediction interval	days	-	0.5	-	1	16
	SRP validity (Bernese 9p model)	days	1	daily	-	2	32
	One or both data types	-	-	Range only	both	2	64
	Observation sample interval	minutes	-	5	-	1	64
	Number of solution arcs	-	-	10	-	10	640
Message fit	Typical fit error RMS	m	-	0.2	-	1	640

* Most OD solutions without Doppler were ignored leading to a total number of simulations executed of 528.

In all cases, truth orbits were propagated with the best known dynamic models, using IERS standards and/or best practices (IGS/IERS). Furthermore, the errors in time synchronization do not have any impact on the orbit determination results, although they do affect the SISE. The overall length of a simulation in the analysis is given by the product of the number of solution arcs (10) by the length of each arc, which is assumed to be equal to 3.5 days. This means that a single simulation lasts 35 days and is made up of arcs of 3.5 days length. In order to evaluate the SISE, the last 12 h of each arc is assumed to be the predicted orbit part from which, in reality, the navigation message is generated.

The comparisons between the truth orbit and orbit available to the user were made in the three directions (radial, along-track, and cross track) separately, and 3D norm for position and velocity. With respect to the previous case, the orbit model is updated including a set of empirical harmonic accelerations at a frequency of one cycle per orbital revolution, to better simulate the dynamic model's imperfections. This practice is very common for spacecrafts that have a less than perfect dynamic model, such as uncalibrated surface force properties, and is therefore used for LRNS satellites. The result is that, if two orbit solutions for the same period are compared, the dominant error signal is typically an oscillation with a period of one orbit, because almost all dynamic models have some form of periodicity that is in phase with the orbit (e.g., the gravity field, the solar radiation pressure, and other surface forces).

In Figure 5, it is possible to see the SISE evaluated for position and velocity, respectively, with the 95% and 99% error levels, that follow the error envelope imposed by some of the least accurate cases. The figures show the point clouds resulting from the MC simulations described in Table 3 as a function of the age of data, that is, the time elapsed within the prediction interval upon which the orbit information is based.

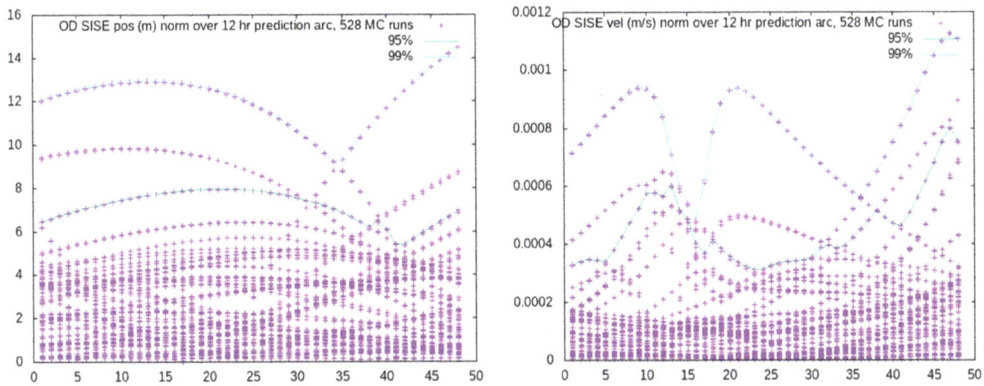

(**a**) OD $SISE_{pos}$ (m) (**b**) OD $SISE_{vel}$ (m/s)

Figure 5. OD SISE plots: (**a**) OD SISE for position (m) as a function of age-of-data (hrs) and (**b**) OD SISE for velocity (m/s) as a function of age-of-data (hrs).

Table 4 summarizes the expected SISE error levels for 95% and 99%, for position and velocity with OD and TS errors combined.

Table 4. SISE summary 95% and 99% error levels. Combination of OD&TS: 1 ns = 0.3 m.

		OD		TS		OD & TS Combined	
AOD	Error Level	Position (m)	Velocity (mm/s)	Bias (ns)	Drift (s/s)	Pos + Bias (m)	Vel + Drift (mm/s)
6 h	95%	7.9	0.33	57	2.20×10^{-3}	18.8	0.74
	99%	11.9	0.85	74	3.00×10^{-3}	25.2	1.24
12 h	95%	7.0	0.61	140	5.30×10^{-3}	42.6	1.70
	99%	14.5	1.10	182	7.40×10^{-3}	56.5	2.48

3.3. SISE Results Obtained by Combination of OD and TS Errors

The OD SISE results do not depend on the TS SISE values, but the clock estimation depends to some extent on the orbit accuracy. In fact, the TS algorithm receives, as inputs, the value of the OD error at the initial epoch of each simulation arc and the visibility matrix computed for each satellite and station pair. Nonetheless, the results for OD and TS are combined by the means of a root-sum-square approach, because a combination of the simulation results in a more direct way was not practical.

The SISE 95% and 99% are also shown in Figure 6, from which a 95% error level of about 6 m is obtained at the initial time, assessing that the initial assumption of a 2 m error was too optimistic. Because the OD errors now contain the once-per-orbit signals, the shape of these figures is no longer as simple as those in Figure 4, which are based on the assumption that the acceleration error is constant during the prediction interval. However, the most significant conclusion that can be drawn from the comparison of the two analysis methods is that they result in reasonably similar error levels, in the order of 20 m after 6 h, depending on the assumptions and input conditions.

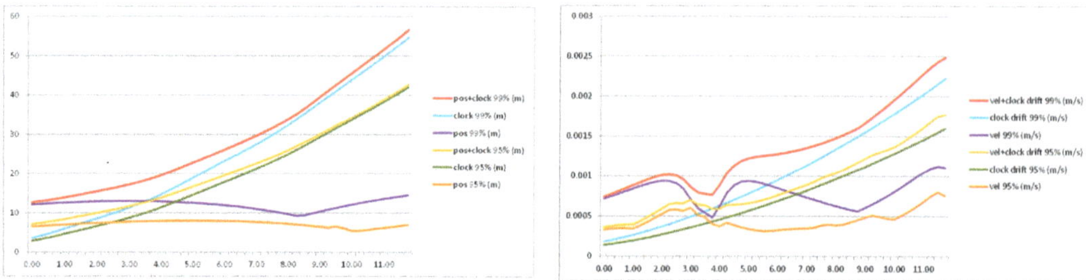

(**a**) (**b**)

Figure 6. SISE 95% and 99% error levels: (**a**) error levels for position (m) as a function of age-of-data (hours) and (**b**) error levels for velocity and clock drift (m/s) as a function of age-of-data (hours).

4. Conclusions

This study aimed at demonstrating the feasibility of the SISE requirements in terms of achievable ODTS performances, showing that good performances for target value of 25m at 95% are achieved in the first 6 h of Age of Data. In the frame of this study, an approach based on the outputs of the OD process and Monte Carlo style simulations showed the consistency of the employed simulator, since they achieved similar error levels. Concerning the TS error, the Monte Carlo approach allowed for obtaining a sufficiently large number of simulations to be used to drive a statistic and obtain a reliable error contribution. Although several factors still have to be considered at this stage of detailed design, the proposed ODTS baseline is considered as the best candidate for potential future lunar radio navigation systems.

Author Contributions: Conceptualization, C.S., C.D.L., M.C. (Mattia Carosi), L.D.L., D.M., P.G., R.S. and J.T.; methodology, C.S., C.D.L., M.C. (Mattia Carosi), L.D.L., D.M.; software, M.C. (Martina Cappa), D.C. and H.B.; validation, L.D.L.; formal analysis, C.S., D.C., M.C. (Martina Cappa) and H.B. ; investigation, C.S., C.D.L., M.C. (Mattia Carosi) and D.M.; resources, C.S., P.G., R.S. and J.T.; data curation, C.S., M.C. (Martina Cappa) and H.B.; writing—original draft preparation, L.D.L. and M.C. (Martina Cappa); writing—review and editing, L.D.L.; visualization, L.D.L.; supervision, C.S.; project administration, C.S. All authors have read and agreed to the published version of the manuscript.

Funding: This research was funded by ESA TDE study LRNS contract No. 4000135400/21/NL/CRS.

Institutional Review Board Statement: Not applicable.

Informed Consent Statement: Not applicable.

Data Availability Statement: Further data used in this study cannot be shared due to policy agreement between TAS-I and the other members of the consortium that have carried out this study.

Conflicts of Interest: The authors declare no conflict of interest.

References

1. International Space Exploration Coordination Group. The Global Exploration Roadmap. *January 2018*. Available online: https://www.globalspaceexploration.org/wordpress/wp-content/isecg/GER_2018_small_mobile.pdf (accessed on 28 October 2023).
2. Kaplan, S. *Eyes on the Prize. The Strategic Implications of Cislunar Space*; Center for Strategic & International Studies: Washington, DC, USA, 2020.
3. Grenier, A.; Giordano, P.; Bucci, L.; Cropp, A.; Zoccarato, P.; Swinden, R.; Ventura-Traveset, J. Positioning and Velocity Performance Levels for a Lunar Lander using a Dedicated Lunar Communication and Navigation System. *NAVIGATION J. Inst. Navig.* **2022**, *69*, navi.513. [CrossRef]
4. Schönfeldt, M.; Grenier, A.; Delépaut, A.; Swinden, R.; Giordano, P.; Ventura-Traveset, J. Across the Lunar Landscape: Towards a Dedicated Lunar PNT System. *Inside GNSS*. 2020. Available online: https://insidegnss.com/across-the-lunar-landscape-towards-a-dedicated-lunar-pnt-system/ (accessed on 27 November 2023).
5. LunaNet Interoperability Specifications. Available online: https://esc.gsfc.nasa.gov/static-files/LunaNet%2520Interoperability%2520Specification.pdf (accessed on 28 October 2023).
6. Ely, T.A. Stable constellations of frozen elliptical inclined lunar orbits. *J. Astronaut. Sci.* **2005**, *53*, 301–316. [CrossRef]

7. Ely, T.A.; Lieb, E. Constellations of elliptical inclined lunar orbits providing polar and global coverage. *J. Astronaut. Sci.* **2006**, *54*, 53–67. [CrossRef]
8. Stallo, C.; Zini, E.E.; Cretoni, D.; Tomasicchio, G.; Di Lauro, C.; Musacchio, D.; Boomkamp, H.; Ventura-Traveset, J.; Giordano, P.; Carosi, M.; et al. Candidate System concepts for a Lunar Satellite Navigation System. In Proceedings of the NAVITEC 2022, Online Event, 5–7 April 2022.
9. Stallo, C.; Di Lauro, C.; Carosi, M.; Zini, E.; Musacchio, D.; Cretoni, D.; De Leo, L.; Cappa, M.; Laurenti, M.; Boomkamp, H.; et al. Orbit Determination and Time Transfer in Selenodetic reference frames for a Lunar Radio Navigation System. In Proceedings of the 8th GNSS Colloquium, Sofia, Bulgaria, 14 September 2022.
10. Lombardi, M. Fundamentals of Time and Frequency. In *The Mechatronics Handbook 1.0-8493-0066-5*; CRC Press: Boca Raton, FL, USA, 2002; pp. 341–358.
11. Riley, W.J. *Handbook of Frequency Stability Analysis*; Special Publication 1065; NIST: Boulder, CO, USA, 2008.
12. Stallo, C.; Di Lauro, C.; Carosi, M.; Zini, E.; Musacchio; De Leo, L.; Cretoni, D.; Laurenti, M.; Boomkamp, H. Performance Analysis of Lunar Radio Navigation ODTS System. In Proceedings of the 2023 International Technical Meeting of the Institute of Navigation, Long Beach, CA, USA, 24–26 January 2023; pp. 190–203. [CrossRef]
13. Zhang, V.; Jiang, Z.; Fujieda, M.; Piester, D.; Esteban, H.; Matsakis, D.; Naumov, A. CCTF Working Group on Two-Way Satellite Time and Frequency Transfer. In: CCTF/17-23 (2017). Available online: https://www.bipm.org/documents/20126/28431178/working-document-ID-10203/527b697a-9394-f690-6cb5-c0992e201de7/ (accessed on 27 November 2023).
14. Piester, D.; Achkar, J.; Becker, J.; Blanzano, B.; Jaldehag, K.; de Jong, G.; Koudelka, O.; Lorini, L.; Ressler, H.; Rost, M.; et al. Whibberley: Calibration of Six European TWSTFT Earth Stations Using a Portable Station. In Proceedings of the 20th European Frequency and Time Forum, Braunschweig, Germany, 27–30 March 2006; pp. 460–467.
15. Michito, I. Precise Time and Frequency Transfer. *J. Natl. Inst. Inf. Commun. Technol.* **2003**, *50*, 105–112.
16. Bhamidipati, S.; Mina, T.; Gao, G. A Case Study Analysis for Designing a Lunar Navigation Satellite System with Time-Transfer from Earth-GPS. In Proceedings of the 34th International Technical Meeting of ION GNSS+, Online, 20–24 September 2021; pp. 950–965.
17. Shin, M.Y.; Park, C.; Lee, S.J. Atomic clock error modeling for GNSS software platform. In Proceedings of the IEEE/ION Position, Location and Navigation Symposium, Monterey, CA, USA, 5–8 May 2008; pp. 71–76. [CrossRef]
18. Galleani, L.; Tavella, P. Time and the Kalman Filter. *IEEE Control. Syst. Mag.* **2010**, *30*, 44–65.
19. Willner, D.; Chang, C.; Dunn, K. Kalman filter algorithms for a multi-sensor system. In Proceedings of the 1976 IEEE Conference on Decision and Control including the 15th Symposium on Adaptive Processes, Clearwater, FL, USA, 1–3 December 1976; pp. 570–574. [CrossRef]

engineering proceedings

Proceeding Paper

Cooperative Swarm Geometry Optimization for Assured Navigation with Range Radios in GNSS-Denied Environments †

Mats Martens * and Maarten Uijt de Haag

Institute of Aeronautics and Astronautics, Technical University of Berlin, 10587 Berlin, Germany
* Correspondence: martens@tu-berlin.de

† Presented at the European Navigation Conference 2023, Noordwijk, The Netherlands, 31 May–2 June 2023.

Abstract: This paper addresses cooperative navigation using range radios to enable absolute positioning of low-flying UAS (LF) while operating in GNSS-denied environments. High-flying UAS (HF) are positioned above the denied area and broadcast position reports. These reports, in combination with range measurements from the LF to the HF, enable absolute positioning of the LF. (1) Methods: As the navigation performance is directly influenced by the geometry of both LF and HF's relative positions, HF positions shall be optimized such that the Dilution of Precision (DOP) becomes minimal. The authors derive optimal azimuth angle combinations, which guarantee a minimal Horizontal Dilution of Precision (HDOP), and show the error characteristic for sub-optimal configurations, which enables the formulation of multi-vehicle-constrained optimization problems for specific combinations of numbers of HF and LF. (2) Results: An optimization problem is derived and solved for two HF aiding three LF as an example application for the derived rules. (3) Conclusions: The resulting geometry has yielded promising HDOP values, improving navigation performance in GNSS-denied environments.

Keywords: range radios; DOP optimization; cooperative navigation

Citation: Martens, M.; de Haag, M.U. Cooperative Swarm Geometry Optimization for Assured Navigation with Range Radios in GNSS-Denied Environments. *Eng. Proc.* **2023**, *54*, 38. https://doi.org/10.3390/ENC2023-15418

Academic Editors: Tom Willems and Okko Bleeker

Published: 29 October 2023

1. Introduction

Unmanned Aircraft Systems (UAS) have gained much attention in many fields including infrastructure monitoring and support for emergency services. One of the most important operating environments are urban areas, where high population density and the accumulation of infrastructure requires more efficient solutions. Here, UAS—especially small UAS (sUAS)—are equipped with sensors to observe the environment and allow operators to make decisions faster. To increase efficiency and effectivity of the UAS operation, swarms of UAS can split work, e.g., to reduce the weight of individual drones by spreading the sensors over the swarm instead of equipping one drone with all required sensors.

To reduce the ground risk, which is imposed by the operation of swarms of UAS, collision avoidance is required to keep the UAS clear of obstacles and other aircrafts. Typically, to avoid collisions with the environment, tactical collision avoidance is implemented, which has widely been elaborated in the field of robotics. However, to geo-reference sensor information with the UAS' position, UAS must be capable of estimating their absolute position reliably. In open field conditions, UAS typically rely on GNSS (Global Navigation Satellite System) as the UAS is likely to receive more than the minimum number of four Line-of-Sight (LOS) signals to the orbiting satellites to calculate an independent position estimate. In urban areas and below the elevation of the structures, an important problem is the degradation of GNSS due to shadowing, non-line-of-sight reception, and multipath [1]. Here, non-GNSS alternatives can be used to improve the reliability of the navigation solution, such as Visual Inertial Odometry (VIO) or Simultaneous Localization and Mapping (SLAM) using LiDAR scanners. In this paper, the authors build on their already proposed sensor framework, enabling UAS swarm member navigation with range radios in the absence of GNSS signals as shown in Figure 1 [2].

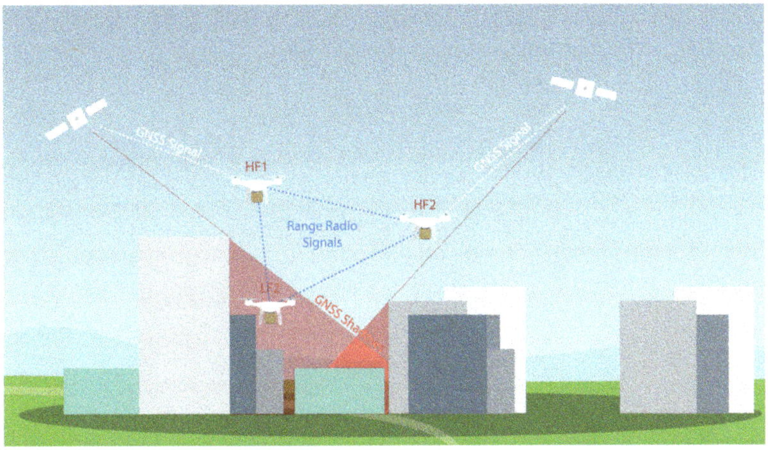

Figure 1. Swarm geometry of High Flying UAS (HF) with good GNSS reception and Low Flying UAS (LF) in a GNSS shadowed region, all connected via range radios.

The proposed scenario includes high-flying UAS (HF) equipped with GNSS receivers in open-sky conditions and low-flying UAS (LF) that perform a swarm mission close to obstacles, e.g., infrastructure monitoring; all UAS are equipped with range radios. Furthermore, the LF follow a set of swarm behavioral rules, e.g., Reynolds Rules or free operation in limited regions, while HF can move freely.

The goal of all swarm member behaviors is to simultaneously meet the mission- and navigation-performance requirements, where the latter is dependent on the geometry between the LF and HF. This paper formalizes the navigation performance of the swarm as a function of its geometry and mission parameters and finds the optimal swarm geometry while retaining Line-of-Sight (LOS) conditions between its members.

2. Methodology

2.1. Range Radio Measurements

Different techniques exist to measure the range between two transceivers. Here, we assume our lightweight swarm members to be equipped with Ultra-Wide-Band transceivers. Existing work has shown that this technology enables reliable and accurate ranging. For example, in [3], the standard deviation of Ultra-Wide-Band (UWB)-based symmetric double-sided two-way ranging (SDS-TWR) measurements was as low as 6 cm at ranges up to 5 m. This can also be seen in [4], where a ground vehicle is equipped with UWB beacons to aid the localization of a flying UAS, thus enabling relative navigation at distances of over 30 m. The standard deviation of the UWB range was found to be around 10 cm, which is at least one decade better than the uncorrected pseudorange measurement from GNSS. In the following sections and mathematical formulations, UWB range measurements are assumed to be i.i.d., specifically characterized by range noise v_ρ that follows a bias-free normal distribution, i.e., $v_\rho \sim \mathcal{N}(0, \sigma_r^2)$.

2.2. Problem Formulation

To assess the navigation performance, the measurement equations are presented. The range $\rho_{i,j}$ between the unknown position of the i-th LF p_i and the known position of the j-th HF q_j is $h(p_i, q_j) = ||q_j - p_i||_2$. For the i-th LF, the measurement vector z_i is formed as a function of M HF, where $j = \{1, \ldots, M\}$:

$$z_i = \begin{pmatrix} \rho_{i,1} \\ \cdots \\ \rho_{i,M} \end{pmatrix} = \begin{pmatrix} h(p_i, q_0) \\ \cdots \\ h(p_i, q_M) \end{pmatrix} \quad (1)$$

To solve this equation, it is typically linearized around the current estimate. One can then solve for the position using the Least-Squares method. The linearized measurement equation is a function of the measurement matrix H, whose rows are the normalized vectors e_{ij} from the HF to one LF:

$$z_i = H_i p_i = [e_{i1} e_{i2} \dots e_{iN}]^T p_i. \tag{2}$$

In order to assess the navigation performance, one is interested in the covariance matrix of the position estimate to evaluate the accuracy:

$$\mathrm{Cov}(\delta x) = H^{+}\mathrm{Cov}(\delta z)H^{+T} \tag{3}$$

Here, H^{+} is the pseudoinverse $(H^T H)^{-1} H^T$ as H is generally not invertible. As the range measurements are assumed to be independent, the covariance matrix of the range measurements $\mathrm{Cov}(\delta z)$ is assumed to be diagonal, i.e., $I\sigma_r^2$. For the presented measurement matrix H, it follows directly:

$$\begin{aligned}
\mathrm{Cov}(\delta x) &= H^{+}\mathrm{Cov}(\delta z)H^{+T} && (4)\\
&= (H^T H)^{-1} H^T \mathrm{Cov}(\delta z)((H^T H)^{-1} H^T)^T && (5)\\
&= (H^T H)^{-1} H^T I\sigma_r^2 H (H^T H)^{-1} && (6)\\
&= (H^T H)^{-1} H^T H (H^T H)^{-1} \sigma_r^2 && (7)\\
&= (H^T H)^{-1} \sigma_r^2 && (8)\\
&= G \quad \sigma_r^2 && (9)
\end{aligned}$$

The covariance of the position estimate, and hence the navigation accuracy, is dependent on the range measurement accuracy σ_r in δz and the geometry of the ranging partners to the LF, which is captured via G. As G is dependent on H, which is formed using the geometry of the HF to the LF as shown in Equation (2), G is also referred to as the geometry matrix.

As a quality metric, the Dilution of Precision (DOP) is calculated to characterize the effect of errors in the range domain on the position estimate as a function of the geometry. For example, the Horizontal Dilution of Precision is calculated as

$$\mathrm{HDOP} = \sqrt{G_{1,1} + G_{2,2}}. \tag{10}$$

As initially assumed, all UWB measurements $\rho_{i,j}$ are i.i.d and stationary; thus, the navigation accuracy can only be maximized by optimizing the geometry of the HF positions relative to the LF positions. The optimization of geometry is a well-known research field within the domain of optimal satellite selection for GNSS positioning. For example, in [5], the authors found the relation between minimizing GDOP and maximizing the determinant of $H^T H$, which is related to the volume of the resulting polytope. However, in [6], the authors conclude that with an increasing number of ranging sources, the correlation between GDOP and volume decreases significantly. An additional constraint should be added as satellites closer to the regular polyhedron should be favored.

Unlike the GNSS satellite selection problem, here the satellite position, i.e., the HF position, is not given, but the optimization variable is. For UAS applications, most related work concentrates on the optimal positioning of UWB ground beacons. For example, in [7], denied environments are explored with fixed ground stations whose configuration is chosen manually. To investigate the optimal positioning of ranging sources, this paper relies on a couple of assumptions. First, the vertical position can be estimated, in addition to a GNSS- and range-radio-based estimation, with LiDAR scanners, laser altimeters, and baro-altimeters. Thus, for further optimization, the optimization goal is restricted to the horizontal 2D plane; therefore, the problem reduces to a minimization of HDOP.

Furthermore, the HF should be as low as possible to increase the observability of the LF position within the horizontal plane. Given this assumption, UAS positions are also restricted to the 2D plane, i.e., $r \in \mathbb{R}^{2x1}$. As LF operate close to obstacles, e.g., facades of buildings, the HF need to be positioned within an azimuth range of $\sim 180°$, which further reduces the region of feasible HF locations.

2.3. Single Low-Flier Optimization

Before covering the problem of optimizing the geometry of multiple LF, the optimal positioning of HF for a single LF is derived. As shown in Equation (2), the H-matrix consists of unit-length vectors from the HF to the LF, i.e., polar coordinates with an azimuth θ and a range of 1.

As H is real valued, it can be decomposed into $H = U\Sigma V^T$ using the Singular Value Decomposition, where Σ is a diagonal matrix containing the singular values and V^T contains the related right-singular vectors. As $H^T H$ can be decomposed into $V\Sigma^2 V^T$, the geometry matrix can be rewritten as

$$\begin{aligned} G &= (H^T H)^{-1} && (11)\\ &= (V\Sigma^2 V^T)^{-1} && (12)\\ &= V^{-T}\Sigma^{-2}V^{-1} && (13)\\ &= V\Sigma^{-2}V^T. && (14) \end{aligned}$$

Here, it is used that V and U are unitary matrices, i.e., $V^T = V^{-1}$. The geometry matrix can therefore be interpreted as a Gram matrix, whose diagonal entries G_{ii} are formed of $<v_i, 1/s_i^2 v_i>$. The squared HDOP can therefore be rewritten as

$$\begin{aligned} HDOP^2 &= \sum_{i=1}^{2} \frac{1}{s_i^2} v_i^T v_i && (15)\\ &= \frac{1}{s_1^2} + \frac{1}{s_2^2}. && (16) \end{aligned}$$

For M to be linearly independent unit-length vectors in H, the rank of $H^T H$ is assumed to be M. As the squared sum of singular values is equal to the the trace of $H^T H$, the squared singular values sum up to M: $s_1^2 + s_2^2 = M$. To minimize HDOP, HDOP^2 shall be minimized, which results in the following minimization problem:

$$\min_{s_1, s_2} \quad \text{HDOP}^2 = \frac{1}{s_1^2} + \frac{1}{s_2^2} = \frac{M}{s_1^2(M - s_1^2)} \tag{17}$$

The function can be minimized by maximizing $s_1^2(M - s_1^2)$, which yields $s_1 = \sqrt{M/2}$, and most importantly, $s_1 = s_2$. This means the HDOP minimization problem can be thought of as the problem to distribute the HF directions, such that the covariance ellipse of $H^T H$ becomes a circle. Then, the major axis equals the minor axis and the largest singular and the smallest singular value are identical, which minimizes the HDOP.

For two HF, the problem of HF positioning is trivial as the HF should be positioned such that they form an orthogonal basis. The angle between two HF and an LF should therefore be 90 degrees to maximize the HDOP for that LF.

For the number of HF being larger than two and even, pairs of HF can be symmetrically and evenly distributed around the $\theta_{opt} = \{-45°, 45°\}$ directions, which also directly follows from the symmetry conditions. An overview is given in Figure 2.

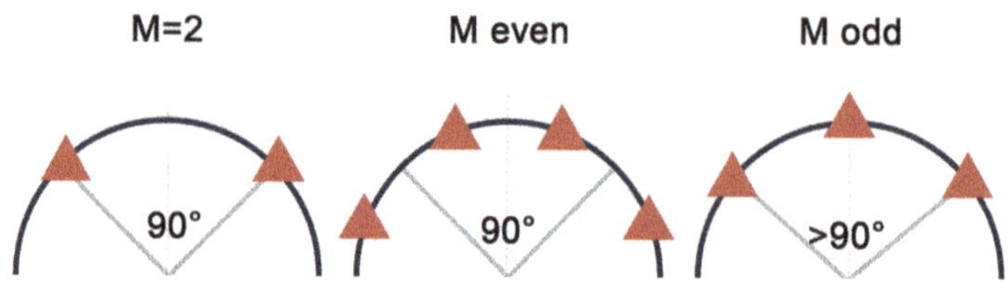

Figure 2. In 2D, optimal Horizontal Dilution of Precision (HDOP) is achieved depending on the number of HF, M; for M being even, the HF can be located at or symmetrically around an orthogonal basis; for M being odd, the optimal azimuth is a function of M.

For an odd number of HF, one HF is placed at the symmetry line between the two circle's quarters. It can be seen that for $\theta \in [0°, 180°]$ one right-singular value will point towards this direction. Assuming this direction to be the y-axis and the orthogonal axis to be the x-axis, the problem can be formalized as

$$E[(X-\mu_X)^2] = \mathrm{Var}(X) \overset{!}{=} \mathrm{Var}(Y) = E[(Y-\mu_Y)^2] \tag{18}$$

As all vectors are centered, the mean value μ is zero. Transforming the equation to polar coordinates and using that, the radius equals to one, and the equation becomes

$$\sum_{i=1}^{M} \sin(\theta_i)^2 = \frac{M}{2} \tag{19}$$

Applying the knowledge that one HF is located on the symmetry axis, i.e., $x = \theta = 0$, Equation (20) shows the optimal azimuth angle θ (angle between the mirror axis and the HF) for M being odd:

$$\theta_{opt} = \sin^{-1}\left(\sqrt{\frac{M}{2(M-1)}}\right) \tag{20}$$

For 3 HF, i.e., $M = 3$, the optimal azimuth angle between the symmetrically chosen HF and the one on the symmetry line is 60°, for $M = 5$ θ_{opt} is 52.2°.

The LF operate close to obstacles. Especially when flying around corners, the feasible area for the HF positioning decreases as it is essential to keep the Line-of-Sight condition between the HF and all LF. Being at orthogonal sides of a building, the overlapping area, and thus, the feasible area for HF positions, is half the size as before. To asses the effect of obstacles reducing the possible azimuth range, a Monte Carlo simulation was conducted for $M = 2$ HF. The feasible θ range was restricted from 10° to 110°. For each scenario, one million uniformly distributed independent position combinations were generated for the two HF. The combination with the best HDOP was saved and is shown in Figure 3 as a function of the maximum allowable azimuth. It can be seen that for smaller azimuth ranges, the problem tends to become more singular, resulting in an increase in HDOP. For ranges greater than 90°, two vectors can always be found, which are an orthonormal basis for the LF. For two HF, this is also the optimal condition, which means the HDOP is $\sqrt{2}$.

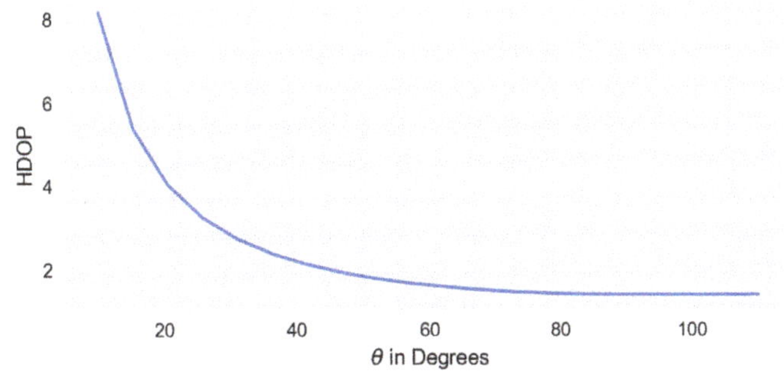

Figure 3. Influence of the maximum allowed azimuth range on the HDOP for two HF; towards zero, the problem tends to become singular, while for the range being greater than 90°, two vectors can always be found, which are an orthonormal basis for the LF.

2.4. Multi-User Optimization

To optimize the geometry for multiple LF, an objective function must be derived that can be minimized using the optimization algorithm. An important property is the impact of a suboptimal geometry on the HDOP. For $M = 2$ HF, HDOP is shown in Figure 4. Additionally, a threshold is set at HDOP = 10. During operation, the HF are affected from GNSS-based position estimation uncertainty and they may show unmodeled dynamics. For a position uncertainty of $\sigma_{r,GNSS} = 2.5$ m and a safety factor of $s = 3$, the feasible azimuth range is further reduced on both sides by

$$\Delta_\theta = \tan^{-1}(\frac{\sigma_{r,GNSS} \cdot s}{R_{min}}) \tag{21}$$

Here, R_{min} is the minimum distance from the HF to an LF, which comes from the assumption that HF should have an elevation angle as low as possible to increase the observability of the LF position within the horizontal plane. Furthermore, it can be seen that within the safety buffer interval, the HDOP can be approximated using a quadratic function.

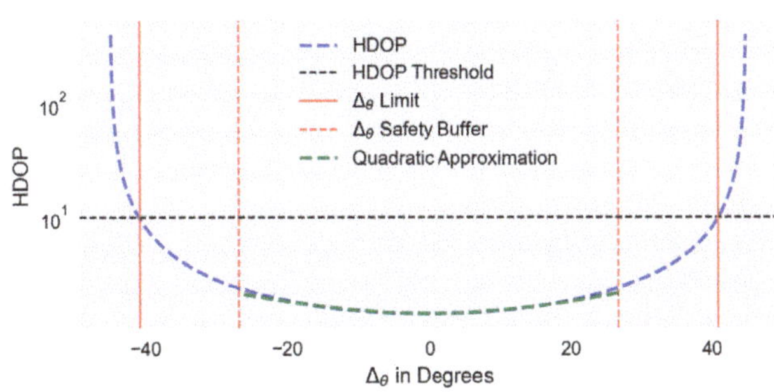

Figure 4. The effect of a suboptimal geometry for M = 2 HF on HDOP; threshold set at HDOP = 10, safety buffer with $R_{min} = 20.0$, $\sigma_{r,GNSS} = 2.5$ m, and $s = 3$.

To visualize the effect of the safety buffer on the feasible area for HF positioning, the limits are shown in polar coordinates in Figure 5. It can be seen that rather than a direction, the feasible area is now a valid range in which the HDOP is likely to be below the value of $HDOP = 10$.

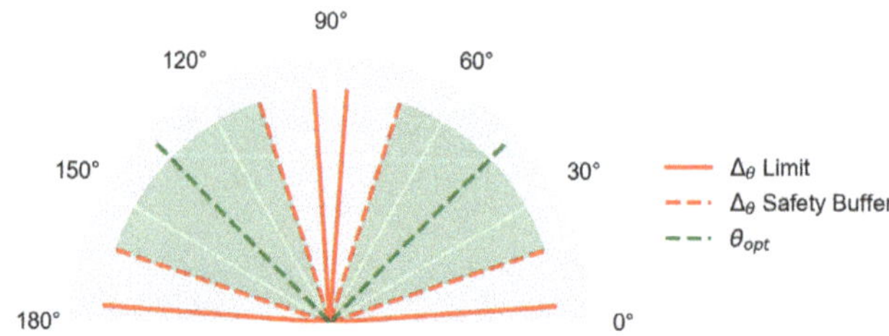

Figure 5. Valid azimuth ranges for $M = 2$ HF; Δ_θ limit found for $HDOP = 10$; safety buffer includes position and maneuvering uncertainty.

3. Results

To illustrate the effectiveness of the feasible azimuth areas, an optimization example is given for $M = 2$ HF and $N = 3$ LF. For efficient computation, the problem is formulated using Cartesian coordinates. Instead of calculating the non-linear value of Δ_θ, the normal distance to the θ_{opt} vector is calculated, which underestimates the HDOP when close to the LF and overestimates it when further away. Note that the feasible azimuth areas are defined relative to each other. Here, both θ_{opt} lines are separated by 90° from each other. By rotating the symmetry axis slightly in one direction, one could improve the HDOP as the θ_{opt} vectors might form a better joint geometry with those of other LF. Therefore, the optimization variables do not only consist of the HF positions, q_i, but also of the change in the symmetry axis–angle from 90°, ϕ_j. The orthogonal distance to the θ_{opt} vector is then given as

$$d_{n,ij} = (q_i - p_j)^T (I + [\phi_i]_x) w. \tag{22}$$

Here, $w(\theta_{opt})$ is the normal vector of the θ_{opt} vector and $[\phi_i]_x$ the skew-symmetric matrix of ϕ. The objective function is then formed as a sum of squared errors (SSE), which is supported via the quadratic approximation of the HDOP error as a function of Δ_θ, as shown in Figure 4. Together with the minimum and maximum range constraints and the restriction of the feasible area to the valid azimuth areas, the optimization problem is defined as follows:

$$\begin{aligned} \min_{q_i,\phi_i} \quad & \sum_{i=1}^{M}\sum_{j=1}^{N}(q_i - p_j)^T (I + [\phi_i]_x) w(\theta_{opt}) \\ \text{s.t.} \quad & (q_i - p_j)^T (I + [\phi_i]_x) w(\theta_{opt} + \Delta_{\theta,max}) < 0 \\ & (q_i - p_j)^T (I + [\phi_i]_x) w(\theta_{opt} - \Delta_{\theta,max}) > 0 \\ & (q_i - p_j)^T (q_i - p_j) > R_{min}^2 \\ & (q_i - p_j)^T (q_i - p_j) < R_{max}^2 \\ & -10^\circ \leq \phi_i \leq 10^\circ \end{aligned} \tag{23}$$

The optimization result is shown for one scenario in Figure 6. All LF operate close to an obstacle while two HF must be positioned such that the HDOP is optimized for all LF. The left two UAS tend to tilt their mirror axis towards the further distant LF on the right, while that UAS tilts the symmetry axis to the left such that the θ_{opt} vectors are nearly intersecting. Exactly at that point is when the optimization finds the optimal HF positions. For two HF, the minimum HDOP is $\sqrt{2} \sim 1.414$, while the found solution yields 0.039%, 0.054%, and 0.044% higher HDOP values for each of the LF.

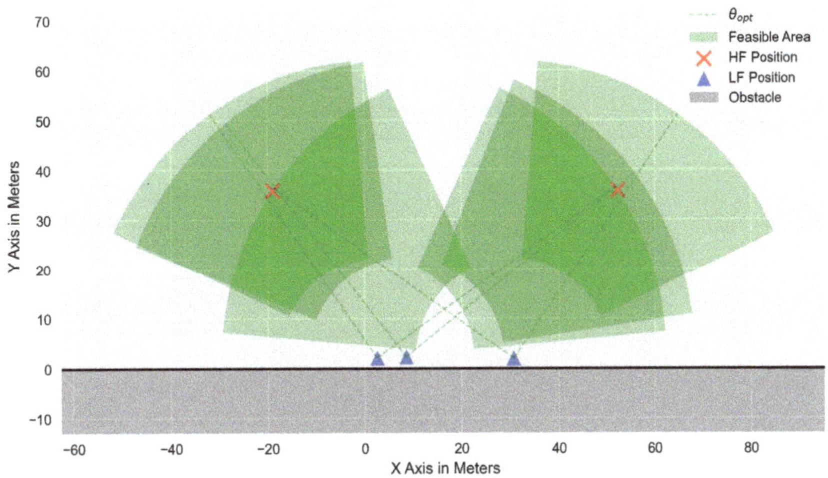

Figure 6. Solution to the quadratic minimization problem of minimizing HDOP for all LF close to an obstacle; solution is constrained within the feasible azimuth areas; HDOP error $\sim 0.05\%$.

A less favorable scenario is shown in Figure 7. Here, the most outer LF are close to losing connectivity as they operate at R_{max}. Additionally, the middle LF is located at a further distance to the wall. This geometric deficiency of the LF configuration results in a worse HDOP, which is also reflected by the larger distance of the θ_{opt} vectors. Here, they are further from intersecting than in the first scenario. Compared to the minimum achievable HDOP, here they are 3.68%, 5.36%, and 3.72% higher.

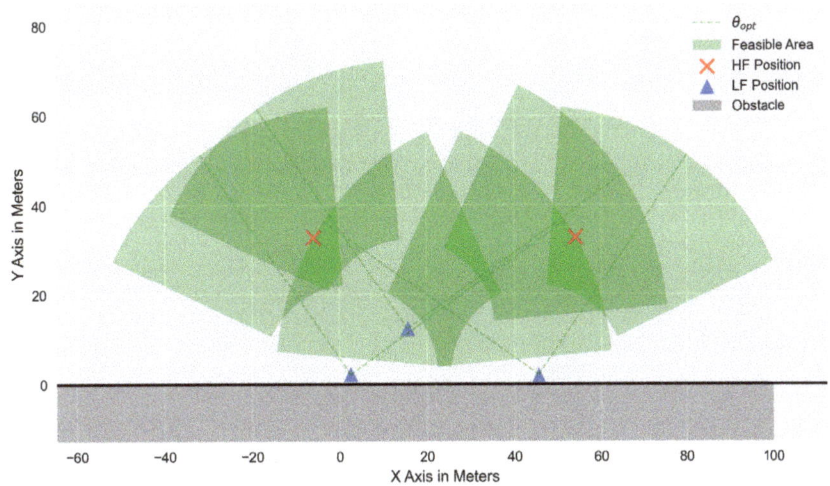

Figure 7. Solution to a less favorable scenario; sub-optimal geometry visible via intersection of all θ_{opt} vectors of one direction; HDOP error $\sim 5.4\%$.

For other configurations, a similar optimization could be applied. The presence of additional obstacles or non-convex obstacles can be incorporated by limiting the feasible θ-range even further.

4. Conclusions

This paper provides insights into the optimal positioning of High-Flying UAS (HF) to aid the absolute horizontal navigation of Low-Flying UAS (LF) operating in GNSS-denied environments. The problem is formalized and is transformed into polar coordinates so it can be interpreted as an azimuth selection problem. For different numbers of HF, optimal azimuth angles are presented to minimize the Horizontal Dilution of Precision (HDOP) of an LF. As an example, an optimization problem is derived and solved for two HF aiding three LF. The resulting geometry has yielded promising HDOP values, guaranteeing an improved navigation performance in GNSS-denied environments. Future work will include online optimization for the multi-LF case as well as an extension to three-dimensional DOP. In conclusion, HF positioning constraints have been derived in a closed form which optimize the HDOP for one LF. For multiple LF, an algorithm is presented which improves the horizontal navigation performance of low-flying UAS in GNSS-denied environments.

Author Contributions: M.M. contributed to this work mainly; supervision, M.U.d.H. All authors have read and agreed to the published version of the manuscript.

Funding: This research received no external funding.

Institutional Review Board Statement: Not applicable.

Informed Consent Statement: Not applicable.

Data Availability Statement: Data are contained within the article.

Conflicts of Interest: The authors declare no conflict of interest.

References

1. Braasch, M.S. Multipath. In *Springer Handbook of Global Navigation Satellite Systems*; Teunissen, P.J., Montenbruck, O., Eds.; Springer International Publishing: Cham, Switzerland, 2017; pp. 443–468. [CrossRef]
2. de Haag, M.U.; Martens, M.; Huschbeck, S. Urban Swarm Navigation Exploiting sUAS Collaboration and Cognition. In Proceedings of the 2022 International Conference on Unmanned Aircraft Systems (ICUAS), Dubrovnik, Croatia, 21–24 June 2022; pp. 319–328. [CrossRef]
3. Tiemann, J.; Schweikowski, F.; Wietfeld, C. Design of an UWB indoor-positioning system for UAV navigation in GNSS-denied environments. In Proceedings of the 2015 International Conference on Indoor Positioning and Indoor Navigation (IPIN), Banff, AB, Canada, 13–16 October 2015; pp. 1–7. [CrossRef]
4. Yu, X.; Li, Q.; Queralta, J.P.; Heikkonen, J.; Westerlund, T. Cooperative UWB-Based Localization for Outdoors Positioning and Navigation of UAVs aided by Ground Robots. In Proceedings of the 2021 IEEE International Conference on Autonomous Systems (ICAS), Montreal, QC, Canada, 11–13 August 2021; pp. 1–5. [CrossRef]
5. Kihara, M.; Okada, T. A Satellite Selection Method and Accuracy for the Global Positioning System. *Navigation* **1984**, *31*, 8–20. [CrossRef]
6. Blanco-Delgado, N.; Nunes, F.D.; Seco-Granados, G. On the relation between GDOP and the volume described by the user-to-satellite unit vectors for GNSS positioning. *GPS Solut.* **2017**, *21*, 1139–1147. [CrossRef]
7. Yang, B.; Yang, E.; Yu, L.; Loeliger, A. High-Precision UWB-Based Localisation for UAV in Extremely Confined Environments. *IEEE Sens. J.* **2022**, *22*, 1020–1029. [CrossRef]

engineering proceedings

Proceeding Paper

Quantum-Based Relative Inertial Navigation with Velocity-Aided Alignment and Initialization †

Pieter Simke de Vries [1,*], Jim Rojer [2] and Floris E. Kalff [3]

[1] Intelligent Autonomous Systems (IAS), Netherlands Organization for Applied Scientific Research (TNO), Oude Waalsdorperweg 63, 2597 AK The Hague, The Netherlands
[2] Heat Transfer and Fluid Dynamics (HTFD), Netherlands Organization for Applied Scientific Research (TNO), Oude Waalsdorperweg 63, 2597 AK The Hague, The Netherlands
[3] Electronic Defence (ED), Netherlands Organization for Applied Scientific Research (TNO), Oude Waalsdorperweg 63, 2597 AK The Hague, The Netherlands
* Correspondence: pieter.devries@tno.nl
† Presented at the European Navigation Conference 2023, Noordwijk, The Netherlands, 31 May–2 June 2023.

Abstract: Quantum sensors are expected to offer significant advantages in magnetic- and gravity-aided navigation. However, these techniques depend on the local environmental conditions and require an alternative solution to ensure high position accuracy on shorter timescales. This paper explores the potential of a hypothetical quantum Inertial Measurement Unit (IMU) with much better performance than classical IMUs when used for dead reckoning position estimation preceded by an alignment period to achieve a relative inertial navigation solution. We show that significantly reduced noise levels (velocity random walk) for the accelerometer and/or gyroscope cannot be automatically exploited. In practice, axis misalignment and initial state errors in orientation limit performance. Incorrect orientation causes errors in compensating for the gravity vector, having a dominant effect. This research proposes to introduce an alignment period with a reliable velocity sensor, prior to starting the mission relying solely on inertial navigation. The orientation errors can be estimated with numerical optimization, in which we match the dead reckoning estimate to the reference velocity signal. The fact that quantum IMUs have a much more accurate measurement means that these orientation errors could be compensated much more accurately. When mitigating these dominant error sources, there is a significant benefit to using a quantum IMU for inertial navigation. The initial position error growth is significantly smaller than existing systems. Some estimates are given for the necessary quality of the IMUs to benefit from these protocols and offer a similar position solution quality to high-end or military-grade IMUs. In the future, military operations that do not rely on Global Navigation Satellite System (GNSS) for their Positioning, Navigation and Timing (PNT) information may be performed with unprecedented position accuracy when using a quantum IMU complemented with velocity-aided alignment period, thereby increasing the effectiveness and dependability of the operation.

Keywords: inertial navigation; quantum sensors; sensor fusion

Citation: de Vries, P.S.; Rojer, J.; Kalff, F.E. Quantum-Based Relative Inertial Navigation with Velocity-Aided Alignment and Initialization. *Eng. Proc.* **2023**, *54*, 39. https://doi.org/10.3390/ENC2023-15437

Academic Editors: Tom Willems and Okko Bleeker

Published: 29 October 2023

1. Introduction

The main aim of this research is to find 'calibration' methods and facilitate sensor (suite) design to optimize short-term (<4 h) GNSS-denied operational capability in military scenarios by improving the position drift performance of TNO's Navigation Solution Estimator (NSE): the Timing, Orientation and Positioning Service (TOPS) Fusion Engine. This is in contrast to other proposed contemporary GNSS-denied navigation methods such as the magnetic-aided inertial navigation system (MAINS) and the gravity-aided inertial navigation system (GAINS), which guarantee inertial navigation system (INS) accuracy over longer time-spans, but may not be suited for all operational scenarios since accurate magnetic or gravity anomaly maps of the operational areas are not available (making high-resolution maps requires low-altitude surveying with advanced (military or otherwise technologically sensitive) equipment, which is often not possible in contested or non-friendly territory).

The potential of an INS with quantum-based IMU is shown as a means of GNSS-denied navigation for short (<4 h) mission durations. First, a significantly reduced sensor measurement noise was investigated, and major error sources identified. In a second phase, the effects of these major error sources was mitigated by the use of a dynamic alignment and initialization (DAI) procedure. This procedure uses a velocity sensor such as a Doppler velocity log (DVL) for alignment in a suitable environment before relying solely on pure inertial navigation. With this DAI procedure, the position drift can be significantly (10–100×) reduced, and military operational capability in hostile environments can be extended.

1.1. Inertial Navigation Basics

The unavailability, being denied or even spoofed, of the GNSS poses a significant threat to military operational capability. In the 1940s, a steady development in inertial navigation using (electronic) sensors began. Since then, mathematical algorithms, in computer systems, have been used to translate the measurements of the inertial instrumentation to orientation (pitch, roll and yaw), as well as position and velocities in the three axes. Today, readily available strong computers with excellent size, weight and power (SwaP) characteristics allow for high-quality positioning and orientation solutions.

1.2. Advanced Inertial Navigation

More advanced and expensive INSs offer better physical sensing elements. Their principles, being fiber-optic gyroscope (FOG)-based or ring laser gyroscope (RLG)-based, offer superior sensing performance with respect to less expensive microelectricalmechanical system (MEMS) IMUs. Typically, also more advanced processing algorithms are used in these INSs, as well as better processing hardware. This results in a better navigation performance of these INSs.

1.3. Major Error Sources

Inertial navigation is sensitive to disturbances such as sensor bias (and stability/flicker noise) as well as measurement noise (thermo-mechanical white noise, or angle random walk) [1]. While more expensive sensors typically perform better on these characteristics, most of their advantages would be negligible if major error sources are not considered:

1. The misalignment between IMU measurement axes; since these sensors are not manufactured perfectly, the axes have an unknown residual misalignment.
2. The initial orientation of the vehicle with attached INS, when starting the inertial navigation.

State-of-the-art sensor characteristics would result in a position drift in the order of 1100 m per hour [2]. As will be discussed in Section 2.1, the significance of the effect of these error sources resulted in the development of the DAI routine, introduced in Section 1.5 and further explained in Section 2.4.

1.4. Quantum Potential

Over the past decade, quantum sensing principles have been explored, mostly in a laboratory environment. These offer great potential for inertial navigation due to their characteristics [3]. Sensors utilizing a nitrogen-vacancy center (NV center) offer very low measurement noise while cold atom interferometers (CAIs) offer a bias-free, although somewhat noisy and irregular, measurement frequency-wise [4].

1.5. Mitigation of the Effects of the Error Sources

The above sections consider the limitations of inertial navigation subject to the major error sources. This chapter will concern the mitigation of the effects of these error sources.

A DAI procedure is proposed, in which a velocity sensor will be used to estimate the initial orientation error. The velocities of the vehicle are estimated in its INS. The initial velocity (typically zero), initial orientation as well as measured accelerations and angular velocities are used to achieve this. In the DAI routine, the initial orientation angle is optimized such that the weighted mean square error of the velocity measured by the DVL and estimated by the INS are minimized. After the routine, inertial navigation (even without the availability of the velocity sensor) can be used to obtain far greater accuracy, thereby scaling with the improvement in sensor characteristics such as measurement noise. A quantum-based sensor, having a measurement noise 10–100 times lower than the current state-of-the-art, may be able to navigate with a drift performance that improves by a factor 10–100, reducing the drift to 110–11 m/h, respectively. More advanced contemporary INSs, such as those possessing high-end IMUs based on a RLG, may also benefit from a DAI. However, these INSs also use other, proprietary methods to improve dead reckoning position estimation accuracy over longer periods of time. It is beyond the scope of current and near-future research to reverse-engineer these methods. Lower-end INSs are limited by their sensor performance, as discussed further on in Section 3.2.

2. Materials and Methods

This section details the methods used in this research. The materials used are solely software programs written in Python and C++, contained in TNO's PNT software suite: TOPS. Especially within the the TOPS research tool, the simulation enhanced sensor designer (SENSED) is extensively used and improved through the course of this research.

2.1. Phase 1: Identification of Major Error Sources

The major error sources of using dead reckoning inertial navigation are identified through simulating three different scenarios:

1. Using a FOG IMU and two potential hypothetical quantum IMUs during a 10 s flight, evaluated using the Euclidian position error at the final time.
2. Using a FOG IMU and two potential quantum IMUs during a 10 s flight, evaluated by the Euclidian position error at the final time. IMU axis non-orthogonalities (in pitch, yaw and roll), generated from a normal distribution with zero mean and 0.1 degree standard deviation, are considered.
3. Using a FOG IMU and two potential quantum IMUs during a 10 s flight. The evaluation criterion is the Euclidian position error at the final time. Multiple combinations of initial orientation errors (in pitch, yaw and roll) are considered:
    (a) A range of initial orientation errors in only yaw;
    (b) A range of combination of initial orientation errors in both pitch and roll;
    (c) A range of combination of initial orientation errors in pitch, roll and yaw.

An evaluation period of 10 s is chosen. Since the effects in this elementary analysis are linear, they can easily be extrapolated. Longer periods do not provide extra insights.

2.2. State Estimation with TOPS Fusion Engine

At the center of TNO's PNT software suite is the fusion engine. It allows multiple types of interfaces with other (online) system modules and middleware, such as Robot Operating System (ROS) [5] or other common communication protocols. It incorporates a multitude of vehicle and sensor models. At the core, an unscented Kalman filter (UKF) [6] is used to use vehicle models and sensor measurements (either synthetic or real world) to perform sensor fusion and state estimation.

The TOPS Fusion Engine has been used in various projects at TNO and has recently been applied in a successful field test using a BlueROV unmanned underwater vehicle [7]. Showing minimal deviation from the reference real-time kinematic (RTK) positioning Global Positioning System (GPS), it allows for autonomous underwater navigation [8]. Previously existing systems were an order of magnitude larger in terms of SwaP than this vehicle, required a far larger vessel and significantly more crew to deploy and are around twenty times more expensive in capital expense. This all results from the TOPS Fusion Engine's capability to leverage the sensor qualities to a far greater extent than existing PNT systems do.

2.3. SENSED Software Tool and Workflow

Within this research, another part of the TOPS software suite, SENSED, was used. A significant amount of time was spent to make the code base understandable and re-useable for future research. That is, the code was explicitly designed such that it can be connected to different simulation platforms, fusion engines and can utilize different sensor models. This allows the code to be used for new use-cases with different vehicles and different sensor suites.

The workflow in SENSED was designed as a robust and general method. The method is repeatable and easily adjusted to incorporate other sensors, other waypoints or even another vehicle. By configuring a plant model (=vehicle) and selecting which sensors are used, a workflow is easily configurable.

In essence, SENSED is a procedural fusion engine configuration manipulator. A researcher can formulate a 'research plan' in the form of a simple Python script. In this script, they can define what sensor(s) (combination/characteristics) to evaluate or optimize. Its genesis was driven by the realization that the utility of improving sensor characteristics in (more) expensive sensors (such as IMUs, but also magnetometers, barometers, among others) in a sensor fusion framework, such as a Kalman filter, can only be evaluated by simulating the sensors (or using measurement data) and then passing it through the Kalman filter-based fusion engine. The pragmatic goal was to find the least expensive sensors that would still achieve the position drift performance required by the customer. At the core of SENSED is a sensor simulator which generates sensor time-histories based on sensor specification sheets. Using a multitude of realizations (=parameters such generated from a stochastic distribution described by the sensor specifications), the performance of a set of sensors and their characteristics in a certain operational scenario may be evaluated and optimized using the TOPS Fusion Engine.

2.4. Phase 2: Alignment and Initialization

Most current INS systems utilize a stand-alone alignment and initialization period in which they require a period of ≈5 min under standstill conditions. During this phase, the gravitation vector measured by the accelerometers is utilized to establish a local vertical and establish roll and pitch angles. In the same period, the north-finding algorithm tries to detect the Earth's rotation in the gyroscope's measurement to establish a heading estimate. In practice, the vehicle cannot be fully at a standstill and the high number of degrees of freedom with sensor-biases and angle errors cause inaccuracies in the order of 0.005° [9].

This research proposes to perform a dynamic, velocity (measurement)-aided alignment and initialization phase to specifically minimize the orientation errors of the initial state, the unknown residual within-sensor non-orthogonality and manufacturing/installation misalignments. The advantage of DAI is that the vehicle does not have to be at a standstill and no errors will be made by violating such assumptions. Furthermore, the movement of the vehicle allows the effects of the different error sources more distinguishable. Therefore, the orientation errors can be estimated with higher accuracy and precision.

The DAI procedure proposed consists of three phases as seen in Figure 1. These are:

Perform DAI track (1. in Figure 1): In this phase, a figure-8-like track is covered in which all axes are excited to make the orientation errors around each axis observable. The DAI track should be executed such that the velocity measurement can be taken reliably.

Orientation estimation (2. in Figure 1): In this phase, the orientation errors are estimated by minimizing the difference between the dead reckoning velocity estimate and the measured velocity on the alignment track data through optimization.

Continue mission with dead reckoning (3. in Figure 1): In this phase, the dead reckoning is continued considering the orientation errors estimated in the previous phases.

This means that the dead reckoning should perform significantly better, and navigation solely based on the inertial measurements should be possible.

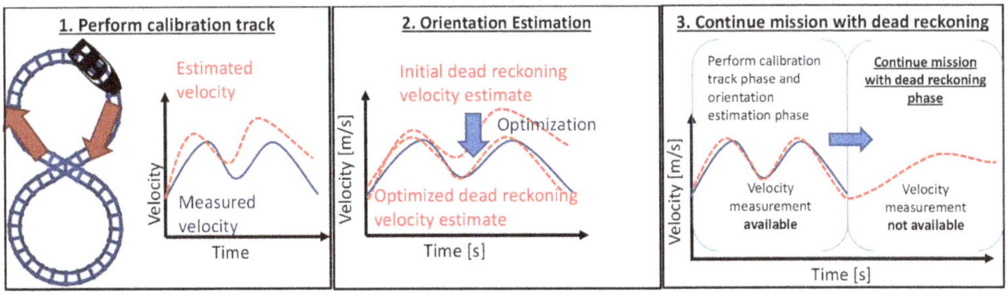

Figure 1. Dynamic alignment and initialization procedure. From left to right: 1. record measured velocity by DVL and estimated velocity by INS; 2. use optimization to find lumped non-orthogonality errors; 3. with optimized error estimate, correct for these and perform (short, <4 h) mission with adequate positioning quality.

For this research, the maritime use case was selected with a frigate or submarine in mind. This use-case was selected as these vehicles are often fitted with high-end INSs [10] and typically also have a DVL sensor fitted that can reliably measure their speed when the local seafloor is of adequate quality. However, the use of DVL in mission conditions can be limited, as its accuracy can be affected by gradients in the sea floor, and its acoustic signature may be undesirable.

It is envisioned that phase 1 and 2 would be executed in a friendly harbor or other benign environment, after which the mission can venture into more challenging or even hostile environments.

2.5. Optimization

This research aims to make a DAI procedure utilizing a quantum IMU and a velocity measurement to minimize orientation errors. This means that an optimization approach is needed that can estimate orientation errors given the DAI time-history track data. However, the question at hand is more complicated, as it also requires insight in in what sensor specifications are needed to successfully perform a DAI procedure, or how long this procedure needs to be.

The approach utilizes an inner and outer loop optimization strategy, cf. Figure 2.

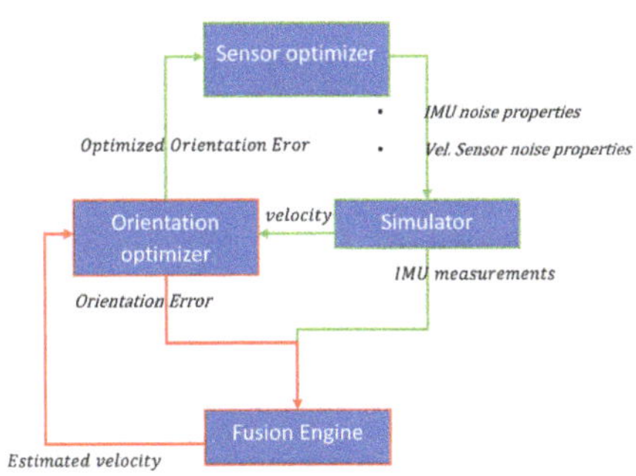

Figure 2. Optimization setup for SENSED program.

Inner Loop: Outlined in red is the inner loop optimization, the inner loop optimizes the orientation error, ϵ_θ, to minimize the dead reckoning (DR) velocity with regard to the DVL velocity measurement, as described in Equation (1). All velocities are estimated in a local reference frame. Since the length of the evaluation error due to Schuler oscillations is minimal, their influence was briefly assessed and deemed negligible. Future research will include longer simulations and therefore revisit the influence of Schuler oscillations.

$$J = (v_{DVL} - v_{DR}(\epsilon_\theta)) \quad (1)$$

The inner loop functions on a single DAI track dataset. The fusion engine block handles the dead reckoning processing. A two-stage optimization strategy was selected. In the first stage, a Bayesian optimization method is applied. Followed by a second stage, in which a gradient descent optimization method is used. The Bayesian method was selected to avoid local minima, as the non-linear nature of orientation errors can lead to non-convex landscapes. Bayesian optimization approximates the objective function by a Gaussian process model, which can approximate any smooth function. However, many point evaluations were found to be needed to find the exact minimum. Therefore, the number of iterations of the Bayesian method was limited. Subsequently, its best solution was used as a seed for a gradient descent method, which can converge very quick to the nearby minimum. The Bayesian optimization iterations were selected to be 50, with initial test runs showing subsequent solutions all being in close to the same minimum.

Outer Loop: The outer loop is outlined in green and has two main functionalities: first, it can optimize the simulation parameters for better orientation optimization. This can both be use-case specific settings like the calibration period, as well as sensor parameters like noise power. Secondly, it can also use multiple noise realizations to avoid overfitting of the optimizer and to obtain insight into the performance spread that can be achieved in realistic conditions.

The philosophy was to implement the outer loop also as a Bayesian optimizer that would optimize the parameters of the IMU and calibration procedure. This would allow the entire calibration concept to be optimized. However, due to budget limitations, the outer loop was only implemented as a simple grid search for pre-selected parameters. This means that it is not possible to optimize the combinations of sensor properties that have interdependencies with the dead reckoning performance.

2.6. Simulation

This study aimed to analyze what parameters would induce the best orientation estimation within the inner loop. Table 1 shows the considered parameter variations for the outer loop. With the current developments in quantum measuring techniques, the accuracy (noise level) of both the accelerometer and the gyroscope are considered. Furthermore, the calibration time was expected to be of influence. The DVL accuracy could also be of influence but is expected to be less critical as there will be no cumulative errors as in the dead reckoning velocity estimate.

Table 1. Sensor reference noise values and tested ranges.

Parameter	Reference Value	Tested Range	Note
Accelerometer Noise	$\sigma = 10^{-5}\ [\frac{m}{s^2}]$	$[10^{-6}, 10^{-4}]$	Honeywell HG9900
Gyroscope Noise	$\sigma = 10^{-6}\ [\frac{rad}{s}]$	$[10^{-8}, 10^{-6}]$	Honeywell HG9900
DVL Noise	$\sigma = 10^{-3}\ [\frac{m}{s}]$	-	Nortek 500

Each of the three parameters is evaluated separately, meaning that when the accelerometer noise is varied, the other parameters are kept at their reference values, which are based on the current high-end sensor specifications. The reference value for calibration time was made similar to typical calibration times in inertial navigation (inertial navigation requires calibration for heading estimation). The mean value on 25 noise realizations will be used to judge the performance on each tested optimization parameter value.

3. Results

3.1. Phase 1 Results

With increasing pitch initialization error, the position error increases far more strongly than with increasing yaw initialization error, as can be seen in Figure 3. This is as expected: the pitch (and roll) angles produce an error in the perceived acceleration while the yaw only influences the (horizontal) direction.

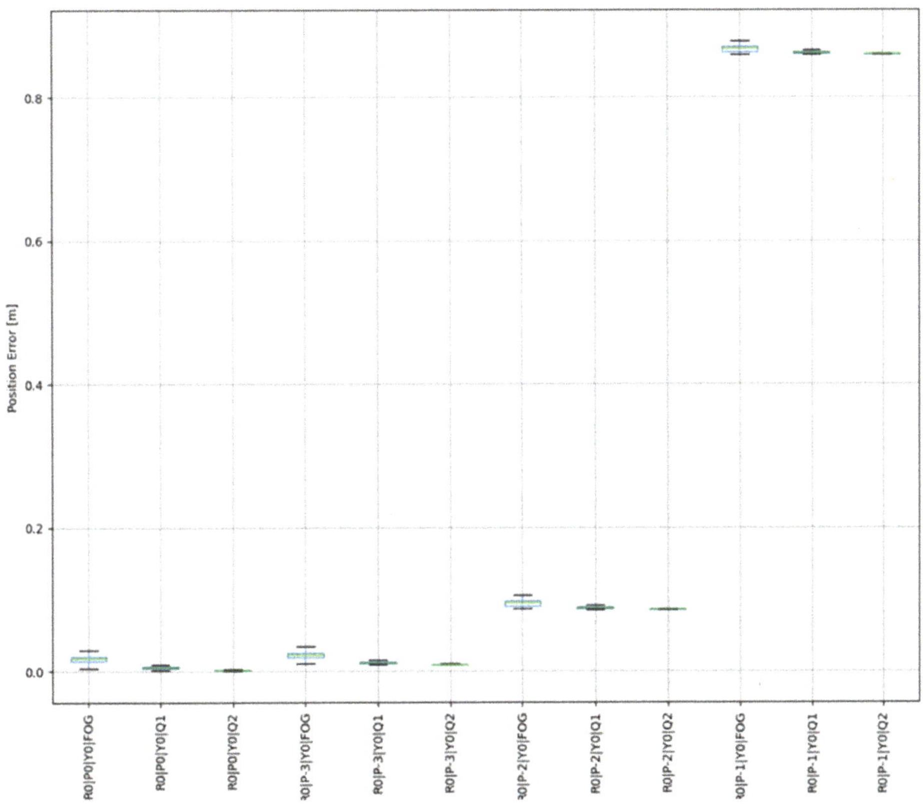

Figure 3. Phase 1 results: Increasing pitch initialization error for all IMUs, a large influence in dead reckoning performance can be seen, with the sensor noise characteristic improvements only amounting to a very small improvement in dead reckoning performance. Note the significant differences in position error between P0, P3 and P2 and P2.

From Figure 3, it can be seen that for a 10 s dead reckoning period, small initialization errors in pitch of 0.01° are already dominant over sensor noise specifications.

In real-life situations, it is possible, though difficult, to establish an initial estimate within 0.01° accuracy [11].

3.2. Phase 2 Results

The results are limited to the analysis of the gyroscope influence, since the accelerometer influence was negligible due to the one-dimensional scenario.

3.3. Gyroscope

Figure 4 shows the mean angle estimate for various gyroscope noise levels. In this case, there is clearly an improved result going for lower noise levels. The result now clearly converges to the simulated angle error of 0.045° and reaches an accuracy of about

$5 \times 10^{-5}[°]$ at $\sigma^2 = 2.5 * 10^{-15}$ [rad/s]^2. This accuracy would yield an 11 m position drift in the first hour, which is improvement by a factor of 100 compared to the current state-of-the-art dead reckoning performance. The limiting factors beyond this point were not investigated due to time and budget constraints.

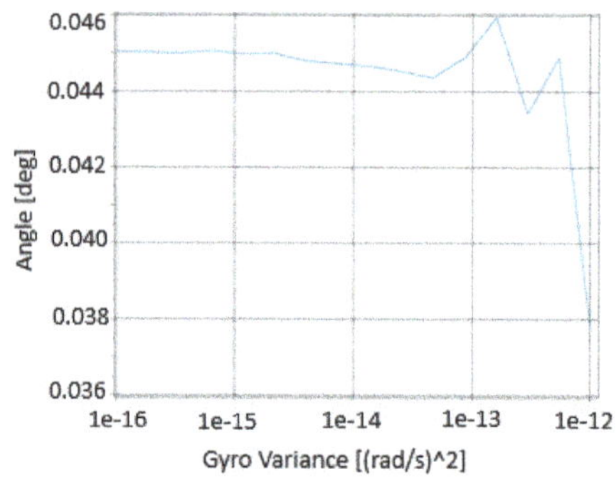

Figure 4. Varying gyroscope variance plotted against estimated pitch initialization error angle.

The noise of the gyroscope is of more importance than that of the accelerometer as the integration errors of the gyroscope lead to incorrect compensation of gravity, which is the current dominating error in dead reckoning systems.

Figure 5 shows the mean cost function (Equation (1)) and values for the different gyroscope noise levels. For values below $\sigma^2 = 10^{-13}$ [rad/s]^2, the cost approaches the simulated noise level of the DVL. This is the minimum value expected out of the cost function, as lower values would indicate overfitting.

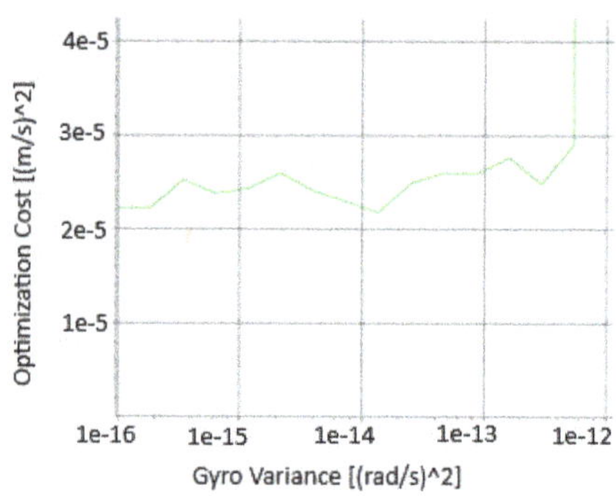

Figure 5. Varying gyroscope variance plotted against cost.

4. Discussion

In this study, only dead reckoning performance is considered. Dead reckoning is the main method in inertial navigation and is utilized by most high-end systems like the iXblue Phins [10], Boreas D90 [12], HG9900 [13] and many more.

In the first phase of this research, the major error sources were identified, and if these major error sources are not mitigated, no advantage can be observed by using better sensors. The above commercial off-the-shelf (COTS) INSs do not employ further methods to improve on this performance. The inertial navigation performance claimed by these products is in line with the simulated results presented in this study. Other systems claim better performance, incorporating other methods to improve error growth over longer periods of time, which will be investigated in future research.

In the second phase, the analysis was reduced to one dimension. The inclusion of the DAI procedure meant that the major error sources can be mitigated by the inclusion of a velocity sensor such as a DVL. The current state-of-the-art IMU could achieve an orientation accuracy of $5 * 10^{-3} [^\circ]$. This yields a 1100 m drift in the first hour. It was found that improvement in accelerometer precision does not yield better orientation estimation. However, an improvement in in gyroscope precision does yield better orientation estimation. Reducing the gyroscope measurement noise level down to $\sigma = 5 * 10^{-8}$ [rad/s] would yield orientation estimation accuracy of $5 * 10^{-5} [^\circ]$, resulting in a 11 m drift in the first hour.

No sensitivity towards the calibration time was found; however, these results seem unreliable as the performance for the 100 s period did not match with the other runs. This is assumed to be a coding bug that we could not solve within budget constraints; therefore, the calibration period results will not be further discussed.

The fact that the accelerometer precision is of little influence may be due to the fact that only one dimension is considered.

5. Conclusions

This paper investigated the potential of using quantum-based IMUs for the purpose of inertial navigation. This is particularly relevant for military use-cases, where it is likely that GNSS-based positioning methods are not available. In the first part, IMUs with quantum-based accelerometers and with significantly reduced measurement noise were considered. These sensors were used for the purpose of dead reckoning positioning. From a perfect starting point, with no error sources other than measurement noise, unknown residual misalignment was added in the accelerometer's measurement axes. An initial orientation error was also considered. It was found that both these error sources can have a detrimental effect on the strongly improved performance (with a measurement noise a factor 10–100. lower than a baseline FOG IMU). Improvements in other sensor characteristics cannot be exploited if the effects of these major error sources, present in all IMUs, are not mitigated.

In the second part, it was investigated how the effects of these major error sources could be mitigated. The problem was reduced to one dimension, and the equivalent to pitch/roll error was considered. A velocity sensor was added to the INS, previously only based on an IMU, and a DAI procedure was proposed. In this procedure, the vehicle covers a track, and the resulting dead reckoning velocity estimate is optimized through varying the estimated lumped axes' orientation angle error. After the most likely initial orientation angle is found, the vehicle may continue navigation without the presence of a velocity measurement, relying solely on inertial navigation. With high-quality (quantum) IMUs, the accuracy of a state-of-the-art DVLs may be achieved during this post-DAI inertial navigation period. This would reduce the need for the availability of GNSS while providing similar positioning accuracy and resulting operational capability, ultimately extending military operational capability in hostile environments.

6. Future Work

Future research will include more simulations, as well as more realistic sensor characteristics, including, but not limited to:

1. Colored noise;
2. Bias and bias stability;
3. Analysis in terms of Allan variance;
4. Scale factor and stability.

The above hold for conventional as well as quantum sensors. Modeling of cold atom interferometers (CAIs) and nitrogen cacancy-centers (NV-centers) and their characteristics could also be performed:

1. NV-centers [4]:
    (a) Low noise;
    (b) Not bias free;
2. CAIs [3]:
    (a) Slow frequency/batch-wise measurements;
    (b) Bias-free;

(c) 'High' noise.

Furthermore, the analysis will be extended to depict in detail the error buildup, in terms of Allan variance [1,14]. The new analysis will consider the problem in three dimensions instead of the current one-dimensional approach.

Author Contributions: Data Curation: P.S.d.V.; Formal Analysis: J.R.; Investigation: P.S.d.V. and J.R.; Methodology: P.S.d.V. and J.R.; Supervision: F.E.K.; Validation: P.S.d.V. and J.R.; Visualization: P.S.d.V. and J.R.; Writing—Original Draft: P.S.d.V. and J.R.; Writing—Review and Editing: P.S.d.V., J.R. and F.E.K. All authors have read and agreed to the published version of the manuscript.

Funding: This research was funded through TNO's internal funding program for Risk-bearing Explorative Research RVO (Dutch: Risicodragend Verkennend Onderzoek).

Institutional Review Board Statement: Not applicable.

Informed Consent Statement: Not applicable.

Data Availability Statement: Data is withheld due to the nature of funding (defense research).

Conflicts of Interest: The authors declare no conflict of interest. The funders had no role in the collection, analyses, or interpretation of data.

References

1. Woodman, O.J. *An Introduction to Inertial Navigation*; University of Cambridge Computer Laboratory: Cambridge, UK, 2007.
2. El-Sheimy, N.; Youssef, A. Inertial sensors technologies for navigation applications: State of the art and future trends. *Satell. Navig.* **2020**, *1*, 2. [CrossRef]
3. Travagnin, M. *Cold Atom Interferometry for Inertial Navigation Sensors*; JRC Technical Reports; European Comission: Brussels, Belgium, 2020.
4. Ajoy, A.; Cappellaro, P. Stable three-axis nuclear-spin gyroscope in diamond. *Phys. Rev.* **2012**, *86*, 062104. [CrossRef]
5. Robotic Operating System. Available online: https://www.ros.org/ (accessed on 4 May 2023).
6. van der Merwe, R. *Sigma-Point Kalman Filters for Probabilistic Inference in Dynamic State-Space Models*; Oregon Health & Science University: Portland, OR, USA, 2004.
7. BlueROV2, BlueRobotics. Available online: https://bluerobotics.com/store/rov/bluerov2/ (accessed on 12 May 2023).
8. Subsea Autonomous Navigation Capabilities for Small-Sized Underwater Vehicles. Available online: https://www.hydro-international.com/content/article/subsea-autonomous-navigation-capabilities-for-small-sized-underwater-vehicles (accessed on 12 May 2023).
9. Chatfield, A.B. *Fundamentals of High Accuracy Inertial Navigation*; AIAA: Reston, VA, USA, 1997.
10. Inertial Navigation Solutions for Navies. Available online: https://www.ixblue.com/defense/naval-navigation/inertial-navigation-solutions-for-navies/ (accessed on 27 September 2022).
11. Marins Series. Available online: https://www.ixblue.com/sites/default/files/2021-05/marins-series_2020.pdf (accessed on 4 May 2021).
12. Boreas D90. Available online: https://www.advancednavigation.com/wp-content/uploads/2021/08/Boreas-D90-Datasheet.pdf (accessed on 27 September 2022).
13. HG9900 Inertial Measurement Unit. Available online: https://aerospace.honeywell.com/content/dam/aerobt/en/documents/learn/products/sensors/brochures/N61-1638-000-000-hg9900inertialmeasurementunit-bro.pdf (accessed on 27 September 2022).
14. IEEE Standards Board. *IEEE Standard Specification Format Guide and Test Procedure for Single-Axis Interferometric Fiber Optic Gyros*; IEEE: Piscataway, NJ, USA, 1998.

engineering proceedings

MDPI

Proceeding Paper

Local Differential GNSS Augmentation for Integration into Urban Air Mobility †

Daniel Gerbeth *, Maria Caamano and Chen Zhu

German Aerospace Center (DLR), 51147 Cologne, Germany; maria.caamanoalbuerne@dlr.de (M.C.); chen.zhu@dlr.de (C.Z.)
* Correspondence: daniel.gerbeth@dlr.de
† Presented at the European Navigation Conference 2023, Noordwijk, The Netherlands, 31 May–2 June 2023.

Abstract: This paper describes a concept for local GNSS (Global Navigation Satellite System) augmentation derived from the established Ground Based Augmentation System (GBAS) in civil aviation. The aim of this concept is to provide reliable and accurate GNSS measurements for integration into a redundant, safe and reliable integrated navigation architecture tailored to serve Urban Air Mobility (UAM). The proposed concept for local GNSS augmentation addresses the specific challenges of UAM, in particular for take-off and landing operations, and ensures safe separation between UAM vehicles en route between vertiports. By using less expensive hardware compared to the traditional GBAS, the concept aims to make the integration into future urban airspace easier and more cost-effective, both in terms of ground infrastructure demands and on-board navigation hardware. In addition to the high-level system concept and considerations, we present an initial nominal performance assessment of local augmentation using lower-cost airborne and ground hardware. This assessment is based on actual UAV (Unmanned Aerial Vehicle) flight trials conducted in different urban scenarios, as well as long-term rooftop measurements.

Keywords: urban air mobility; GNSS; navigation; local augmentation; GBAS

Citation: Gerbeth, D.; Caamano, M.; Zhu, C. Local Differential GNSS Augmentation for Integration into Urban Air Mobility. *Eng. Proc.* **2023**, *54*, 40. https://doi.org/10.3390/ENC2023-15442

Academic Editors: Tom Willems and Okko Bleeker

Published: 29 October 2023

1. Introduction

Urban Air Mobility (UAM) has emerged as a promising solution to the growing demand for efficient and sustainable transport in densely populated urban areas. Especially using electric vertical take-off and landing (eVTOL) vehicles, UAM offers the potential to reduce traffic congestion and commute times while minimising the environmental impact associated with traditional ground transport. Despite the many benefits, the implementation of UAM presents several challenges that must be overcome to ensure safe and effective operations. These challenges include ensuring safe, reliable and accurate navigation in complex urban environments, seamless integration with existing air traffic management systems and the development of necessary infrastructure, such as vertiports [1].

In recent years, significant developments have been made to address these challenges. Advances in GNSS (Global Navigation Satellite System) augmentation, sensor fusion and other navigation technologies have contributed to enhancing the accuracy and reliability of UAM navigation systems. In the ongoing DLR (German Aerospace Center) projects, safe and reliable navigation, also considering cyber–physical threats, is one of the researched aspects [2].

Even though it is not considered a stand-alone solution for the challenge of reliable navigation in UAM, especially for the critical phases of flight, such as during take-off and landing within densely populated areas, the GNSS is still one of the cornerstones of future solutions. While not providing the required accuracy for safety-critical operations in urban centres on its own, the idea of using local differential augmentation inspired by the Ground Based Augmentation System (GBAS) used in classical aviation for precision approach guidance has been around for at least a decade [3]. Over the years, a variety of studies researched the implications of the new environment, hardware, operational aspects and constraints, and developed accordingly, adapting its means to ensure system integrity [4,5].

In this research, we build upon this foundation and present an approach to local augmentation aimed at integration into urban air mobility. We discuss the integration of ground infrastructure and ground-based monitoring systems to improve the accuracy and reliability of the GNSS in densely populated regions. This work also explores the airborne integration of the developed solution into a multi-sensor approach, exploiting sensor redundancy and enhancing navigational accuracy and integrity. We conclude with the results of initial flight tests using a simplified version of the Urban GBAS (U-GBAS) hardware which has shown promising results, providing nominal performance in urban environments and successfully integrating noise and multipath propagation models.

2. System Concept

2.1. Overall Concept

The core idea of the suggested urban augmentation system (U-GBAS) is a network of local reference receivers located mainly at vertiports within an urban area. It is logical for the main driver of such a system to ensure safe and reliable operations, especially for passenger flights (air taxis), placing references at the locations where the strictest requirements are placed upon the operation (take-off/landing).

Equipping each of the vertiports within an urban area with a U-GBAS reference would create a network spanning over an urban or metropolitan area, typically up to the size of tens of kilometres. The core service volume of the system would be defined by a polygon spanned by the reference locations, and thus by the vertiports. Within this service volume, many integrity aspects would be covered/supported by the ground network. Only when leaving this volume, additional onboard monitoring would be required and either availability or provided accuracy would be reduced. Figure 1 shows an exemplary scenario how such a network could look like and evolve over time for the city of Hamburg, Germany. These scenario simulations were performed in the work of Niklaß et al. [6].

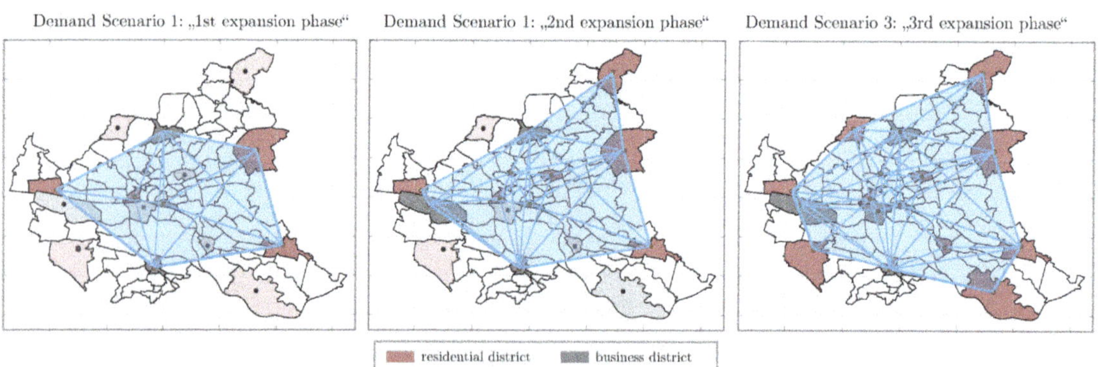

Figure 1. Assumed initial vertiport locations (black dots) for city of Hamburg based on a demand simulation. Figure adapted from [6]. The core service volume covered by the system is shaded in blue for each of the scenarios.

At the same time, it can be assumed that most passenger traffic that would occur between vertiports in the same metropolitan area would automatically happen within the core service volume. Outside, traffic can be assumed to be less dense, potentially in higher altitudes and, therefore, less demanding in terms of accuracy and separation between vehicles. Thus, operational demand would match very well with the system's performance.

2.2. Ground Infrastructure

In terms of ground infrastructure, the proposed system differs significantly from what we know from classical GBAS [7]. GBAS, as a stand-alone system that has to cover all potential threats and faults with a single, self-sufficient installation at an airport, relies on particular, very expensive antennas installed in open-sky environments as well as on receivers.

Both from a siting and a financial perspective, such hardware, in the range of several hundred thousand Euros, is unfavourable for the deployment of larger numbers within a U-GBAS network. Thus, we envision the use of smaller and cheaper hardware, especially in terms of antennas, while still covering all relevant civil frequency bands and multiple constellations to make full use of geometry diversity, especially in areas where parts of the sky are obstructed.

While this comes, of course, at the cost of nominal performance, we envision the primary mode of operation to be an integral part of an onboard multi-sensor navigation and integrity system (see also Figure 2 in Section 2.5). While technically being able to provide stand-alone integrity, the U-GBAS would typically augment the GNSS measurements used in a sensor fusion system. Depending on the instantaneous demands, the user could decide what systems to use to meet the current requirements in terms of accuracy while maintaining a continuous operation.

2.3. Integrity Monitoring

For brevity in the scope of this paper, in this section, we focus on integrity monitoring aspects that are particularly different from the classical GBAS in the envisioned urban system. The different conditions under which the proposed U-GBAS must operate, compared to the classical GBAS, pose new challenges but also provide freedom for new solutions. On one side, certain monitoring cannot be implemented with the foreseen lower-cost hardware in less optimal siting conditions. On the other hand, the aspect of a regional network of connected receivers covering an area of several kilometres opens up new possibilities to compensate for the typically lower-quality observations provided by the individual reference receivers.

2.3.1. Ionosphere

Monitoring for ionospheric threats in currently operational classical GBASs is limited to the use of L1 signals. Thus, either extensive computations to exclude potentially dangerous geometries [8] or precise carrier-phase-based monitoring within the ground station are required to mitigate ionospheric errors [9]. Both approaches are hardly applicable to the envisioned urban augmentation.

Recently we proposed a new airborne monitoring approach for the traditional GBAS that leverages the use of a second frequency to assess the ionospheric threat at the user side [10], deriving ionospheric protection levels based on the used satellite geometry. This approach is currently under consideration for future dual-frequency GBAS in classical aviation [11,12] but also for use within augmentation systems in the context of urban air mobility [13].

When applied to urban air mobility, and especially in a scenario with a network of ground reference stations at vertiports, additional benefits can be leveraged. With a variety of baselines between stations, ionospheric disturbances and gradients observable within the service volume can be identified by monitoring in the ground stations and potentially harmful satellites can be excluded from use.

2.3.2. Ephemeris

In a similar manner to ionospheric monitoring, mitigating the threat of ephemeris errors becomes significantly easier when measurements from multiple receivers spread over an area can be utilized. Thus, a proposed network of receivers spread over tens of kilometres in an urban area allows for the monitoring of according threats within the serviced area based on differential code measurements. An according methodology has recently been presented, also deriving an associated user-specific ephemeris protection level [13].

2.3.3. Interference

With a network of receivers within the service area, the option to monitor for intentional and unintentional interference with the GNSS signals can also be leveraged. Especially in dense urban areas, it is much more likely to experience interference events for a variety of reasons, be it hardware malfunctions, intentional disturbance of operations or side-effects of the use of personal privacy devices. With receivers close to vertiports, and thus within the city at least to some extent, monitoring can be implemented to warn UAM users and locate an interference source within a few kilometres.

While the use of array receivers would allow us to go even one step further and identify the origin of an interference to potentially counteract [14]. Even with conventional single-element antennas, whenever an interference is detected on the ground a warning can be issued to vehicles in the surrounding airspace or planning to land at the affected vertiport. Thereby, air traffic management can react to such disturbances earlier and re-route traffic to avoid critical situations.

2.3.4. Other Threats

Aside from the previously described threat-monitoring approaches that require a totally new approach compared to classical GBAS when transferring the system into the urban environment, further signal-in-space threats require monitoring. Among those are, for example, cycle slips during tracking, which are especially likely to happen on the airborne side when operating under severe multipath conditions. Additionally, code-carrier divergence, excessive acceleration and ground station multipath have to be monitored. Given the space limitations, we will not go into detail here but a good overview can be found in the literature, for example by Pullen et al. [3].

2.4. Data Link

In the classical GBAS, a one-directional VHF-Uplink with very limited capacity (max ~1.7 kB per second) is used to uplink local corrections and integrity parameters to airborne users. In the context of urban air mobility, providing a higher data-rate, as well as sufficient security, is a crucial aspect. To partially compensate for the degraded performance of received signals, as well as potentially blocked signals due to urban obstacles, the uplink should provide corrections for all major GNSS constellations, maximizing the diversity of usable satellite geometries. Thus, the uplink needs to support up to 50 satellites provided with corrections for at least two frequencies. Additionally, to further improve performance in urban scenarios, the provision of corrections for multiple code-carrier smoothing time constants is foreseen. Especially for low-elevation satellites, shorter-filter time constants, and therefore shorter waiting periods before a satellite can be used, are favorable. For high-elevation satellites on the other hand, long time constants minimize residual noise and multipath. With currently three foreseen time constants between 20 s and 200 s to be provided with corrections, the total data rate increases by a factor of two to three.

The currently planned solution within our developments is a C-Band data link [15]. Broadcast messages are utilized to provide all necessary information to operating UAVs in fixed and rotating fields. To ensure system integrity on all levels, these broadcast messages are authenticated using the Timed Efficient Stream Loss-tolerant Authentication (TESLA) protocol. The suitability of this method, especially with regard to latency and throughput, has already been demonstrated in real-time during flight tests [16,17]. An additional downlink channel is considered to allow further integrity monitoring.

2.5. Airborne Integration

As previously outlined, the core concept revolves around integrating a multi-sensor onboard system, which may include an Inertial Navigation System (INS), barometric sensors, cameras (especially for take-off and landing) and other sensors tailored to the vehicle's specific needs and operational requirements, such as Lidar for night operations. The primary mode of operation would be a tight integration with an INS, providing corrected ranges together with integrity information. This aims towards a sensor fusion approach, an integrated integrity concept [18] and a risk-sharing methodology across different subsystems. Additionally, the design would also allow for loose coupling, which facilitates the provision of a U-GBAS position with integrity information to the sensor fusion.

Furthermore, when the instantaneous accuracy is sufficient, a standalone U-GBAS position can provide vertical and horizontal protection levels comparable to the (anticipated) positioning services within the classical GBAS [19]. A high-level overview of the interaction between ground and airborne subsystems, as well as the onboard integration, is given in Figure 2.

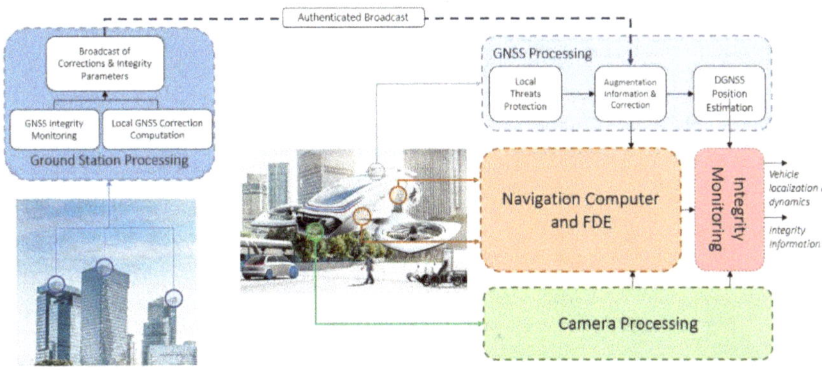

Figure 2. High-level system overview of the integration of the U-GBAS on ground within a multi-sensory onboard unit providing positioning as well as integrity information to the flight control system as well as air traffic management.

3. Experimental Results and Discussion

In this section, we give a brief overview of flight experiments conducted for initial, post-processing validations of the described concept. From multiple tests and experiments carried out in the course of two DLR internal projects, we focus on two scenarios. In 2019, flights were conducted in the city of Hamburg, in northern Germany. A detailed description of the flights can be found in [20]. Herein, we focus on the aspects of U-GBAS navigation. The trajectories flown by our drone during the tests are depicted in Figure 3a. All flights were below the street level of the bridge, some of them partially under the bridge as well as very close to the western bridge pier. During the campaign, three mobile GNSS reference stations were distributed across the city of Hamburg, two of them on rooftops and a third close to the take-off/landing area next to the bridge. Figure 3b shows one of the receivers installed on rooftops, about 4–5 km away from the area where the flights were conducted. In this campaign, no real-time processing took place, but all the data were collected and used afterwards in post-processing.

In the second scenario, flights were conducted on the DLR campus in Oberpfaffenhofen, near Munich. The flight paths for these flights are shown in Figure 4a. At different heights, several flights were carried out along the façade of an office building, which represents a particularly challenging scenario in terms of signal blockage and multipath errors. In this case, a single portable local reference receiver of a similar type was placed nearby (~100 m) in a grass field, thus representing a best-case scenario in terms of local corrections.

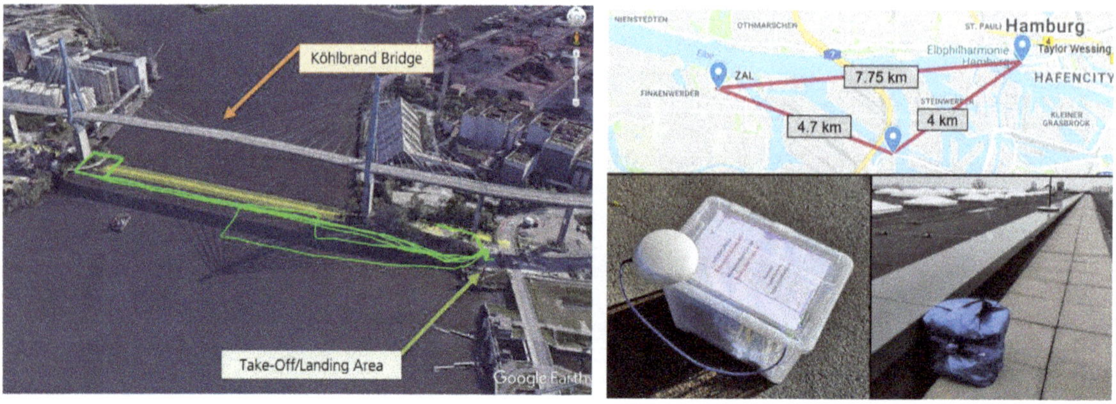

(**a**) (**b**)

Figure 3. Flight trials in the city of Hamburg: (**a**) trajectory of the four flights conducted next to and below the bridge; (**b**) one of the three portable reference stations distributed across the city as shown in the map.

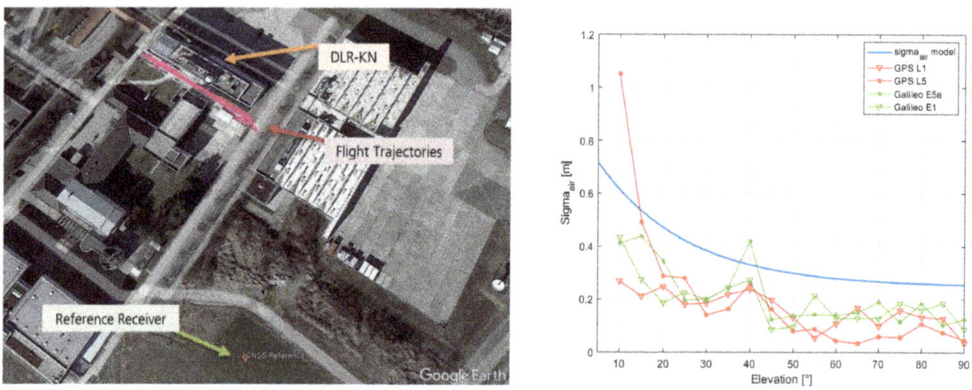

(**a**) (**b**)

Figure 4. (**a**) Flight trajectories conducted in different heights (10–40 m) between two office buildings at the DLR campus Oberpfaffenhofen; (**b**) derived preliminary model for residual airborne noise and multipath (σ_{air}) for 100 s code-carrier smoothing derived from around 4 h of GNSS measurements from the drone (blue) as well as measured residual errors over elevation for GPS and Galileo.

Based on the flights (conducted with a DJI Spreading Wings S900 frame, DJI Ltd., Shenzhen, China as well as a Tallysman TW3972 antenna, Tallysman Wireless Inc., Ottawa, Canada) described here, we derived initial airborne noise and multipath models, as illustrated in Figure 4b. The measured residual noise and multipath errors considering 100 s code-carrier smoothing are shown in red and green for GPS and Galileo in the plot. Due to the limited number of samples, the differences between the constellations and the frequencies can be seen to be rather insignificant. Especially the expected lower multipath on L5/E5a, due to the different signal modulation/higher bandwidth, is only visible over parts of the satellite elevation. Thus, for initial evaluations, a uniform, in most cases conservative model was therefore fitted to the data.

Even though our final goal is an integrated solution in which augmented GNSS is on piece to enable a robust, accurate and reliable means of navigation during all phases of flight in UAM, we also want to present preliminary results of the stand-alone U-GBAS for the scenario conducted in the urban canyon between office buildings. Figure 5a shows the number of tracked and used satellites during the different flights. Typically, 15–25 satellites were tracked, of which 8 to 17 could be used during the flights. In Figure 5b, the achieved position errors (based on RTK-post-processing), as well as nominal (H0) protection levels in the vertical domain, are presented.

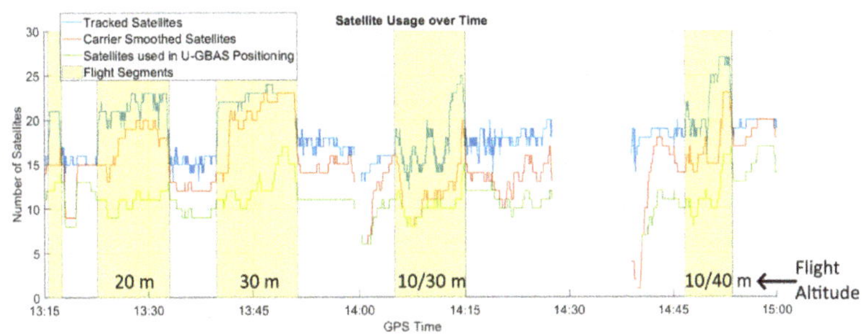

(**a**)

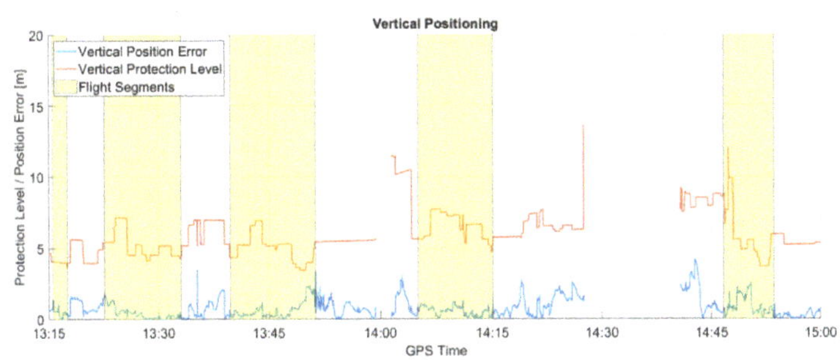

(**b**)

Figure 5. Results of flights between office buildings. (**a**) Satellite tracking and usage statistics during flights and on the ground; (**b**) achieved nominal (H0) protection level using a close-by reference station. Protection levels are derived for an L1/E1 100 s smoothing processing mode using GPS, Galileo and the nominal airborne noise model shown in Figure 4b.

During flights, even in deep urban scenarios with rather limited satellite visibilities, errors typically stayed below 2.5 m, and for most of the higher flights even below 1 m. As expected, the largest errors were seen when the drone was static on the ground. Nominal protection levels typically ranged from 4 to 8 m, bounding the errors during flight.

Finally, for the flights along the bridge in Hamburg, we want to give a first idea regarding the suitability of the corrected ranges for integration with, e.g., an INS within a navigation filter. In Figure 6, we show the root mean squared error over approximately 90 min of flights and separated for individual tracked satellites during that time period. As expected, the errors for stand-alone GNSS measurements reach up to several meters, largely depending especially on the satellite elevation. After applying local corrections, those errors typically drop to less than half a meter. Worth noting is that this also applies to shorter code-carrier smoothing times, where only the residual noise is larger. While for the provision of a stand-alone U-GBAS position smoothing is necessary, potentially shrinking the number of usable satellites due to obstructions and cycle slips, in case of an integrated solution with subsequent filtering in an integrated GNSS/INS filter, this allows for potentially more usable satellite and ultimately even better performance under such challenging conditions.

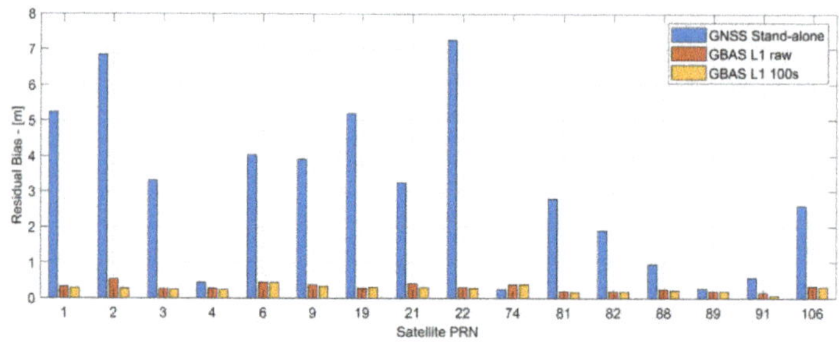

Figure 6. RMS error (bias) on ranges after application of corrections for the flight conducted in Hamburg.

4. Conclusions and Outlook

Based on the concept of classical Ground Based Augmentation Systems and incorporating recent advancements, we developed and presented a U-GBAS concept for local GNSS augmentation tailored to Urban Air Mobility. The cornerstone of this concept lies in its integration with other navigation methods, yielding a multisensory solution that includes integrity information, thereby enhancing reliability and precision. In preliminary flight trials, we assessed the nominal performance of this concept using representative hardware and obtained promising initial results. Notably, these trials were conducted in realistic urban scenarios involving flights near and between office buildings and near a major bridge in the city of Hamburg.

This research lays the foundation for further developments and enhancements of UAM navigation systems with applications beyond the urban domain. Initial real-time demonstrations are underway using an internally developed TESLA-authenticated transmission of local corrections and integrity parameters to the drones. Theoretical developments towards the ultimate goal of integration with INS, camera, barometer and possibly other sensors in a comprehensive integrity concept are underway.

Author Contributions: Conceptualization, campaign execution and evaluation and writing/original draft preparation, D.G.; writing/review, M.C.; writing/review, project management and campaign execution, C.Z. All authors have read and agreed to the published version of the manuscript.

Funding: This research received no external funding.

Institutional Review Board Statement: Not applicable.

Informed Consent Statement: Not applicable.

Data Availability Statement: The data presented in this study are available on request from the corresponding author.

Acknowledgments: The authors would like to thank Dennis Becker, Lukas Schalk, Omar Garcia Crespillo, Wenhan Hao and Young-Hee Lee as additional participants in the flight tests conducted to collect relevant measurements for this work.

Conflicts of Interest: The authors declare no conflict of interest.

References

1. Schuchardt, B.I.; Becker, D.; Becker, R.G.; End, A.; Gerz, T.; Meller, F.; Metz, I.C.; Niklaß, M.; Pak, H.; Schier-Morgenthal, S.; et al. Urban Air Mobility Research at the DLR German Aerospace Center—Getting the HorizonUAM Project Started. In Proceedings of the AIAA Aviation 2021 Forum, Virtual Event, 2–6 August 2021.
2. Torens, C.; Volkert, A.; Becker, D.; Gerbeth, D.; Schalk, L.; Garcia Crespillo, O.; Zhu, C.; Stelkens-Kobsch, T.; Gehrke, T.; Metz, I.C.; et al. HorizonUAM: Safety and security considerations for urban air mobility. In Proceedings of the AIAA Aviation 2021 Forum, Virtual Event, 2–6 August 2021.
3. Pullen, S.; Enge, P.; Lee, J. Local-Area Differential GNSS Architectures Optimized to Support Unmanned Aerial Vehicles (UAVs). In Proceedings of the International Technical Meeting of the Institute of Navigation (ION ITM 2013), San Diego, CA, USA, 28–30 January 2013.

4. Kim, M.; Kim, K.; Lee, J.; Pullen, S. High Integrity GNSS Navigation and Safe Separation Distance to Support Local-Area UAV Networks. In Proceedings of the 27th International Technical Meeting of the Satellite Division of The Institute of Navigation (ION GNSS+ 2014), Tampa, FL, USA, 8–12 September 2014.
5. Kim, D.; Lee, J.; Kim, M.; Lee, J.; Pullen, S. High-Integrity and Low-Cost Local-Area Differential GNSS Prototype for UAV Applications. In Proceedings of the 30th International Technical Meeting of the Satellite Division of The Institute of Navigation (ION GNSS+ 2017), Portland, OR, USA, 25–29 September 2017; pp. 2031–2054.
6. Niklaß, M.; Dzikus, N.; Swaid, M.; Berling, J.; Lührs, B.; Lau, A.; Terekhov, I.; Gollnick, V. A Collaborative Approach for an Integrated Modeling of Urban Air Transportation Systems. *Aerospace* **2020**, *7*, 50. [CrossRef]
7. *RTCA DO-245A*; Minimum Aviation System Performance Standards for the Local Area Augmentation System (LAAS). RTCA: Washington, DC, USA, 2004.
8. Lee, J.; Seo, Y.S.; Park, Y.S.; Pullen, S.; Enge, P. Ionospheric threat mitigation by geometry screening in ground-based augmentation systems. *J. Aircr.* **2011**, *48*, 1422–1433. [CrossRef]
9. Khanafseh, S.; Pullen, S.; Warburton, J. Carrier phase ionospheric gradient ground monitor for GBAS with experimental validation. *Navigation* **2012**, *59*, 51–60. [CrossRef]
10. Gerbeth, D.; Caamano, M.; Circiu, M.-S.; Felux, M. Airborne Ionospheric Gradient Monitoring for Dual-Frequency GBAS. In Proceedings of the 2022 International Technical Meeting of The Institute of Navigation, Long Beach, CA, USA, 25–27 January 2022.
11. Murphy, T.; Harris, M.; McGraw, G.; Wichgers, J.; Lavik, L.; Topland, M.; Tuffaha, M.; Saito, S. Alternative Architecture for Dual Frequency Multi-Constellation GBAS. In Proceedings of the 34th International Technical Meeting of the Satellite Division of The Institute of Navigation (ION GNSS+ 2021), St. Louis, MO, USA, 20–24 September 2021.
12. Caamano, M.; Gerbeth, D.; Sato, H.; Circiu, M.S.; Felux, M. Performance Evaluation of the Ionospheric Threat Mitigation Strategies in Dual-Frequency Multi-Constellation GBAS. In Proceedings of the 2023 IEEE/ION Position, Location and Navigation Symposium (PLANS), Monterey, CA, USA, 24–28 April 2023.
13. Lee, J.; Nam, G.; Min, D.; Sun, A.; Pullen, S. Network-Based Augmentation System (NBAS) Architectures Optimized to Support Urban Air Mobility (UAM). In Proceedings of the 2023 International Technical Meeting of The Institute of Navigation, Long Beach, CA, USA, 25–27 January 2023; pp. 441–453.
14. Hehenberger, S.P.; Elmarissi, W.; Caizzone, S. Design and Installed Performance Analysis of a Miniaturized All-GNSS Bands Antenna Array for Robust Navigation on UAV Platforms. *Sensors* **2022**, *22*, 9645. [CrossRef] [PubMed]
15. Mielke, D.M. Frame structure of the C-band digital aeronautical communications system. In Proceedings of the Integrated Communications, Navigation, Surveillance Conference (ICNS), Herndon, VA, USA, 10–12 April 2018.
16. Mäurer, N.; Gräupl, T.; Bellido-Manganell, M.A.; Mielke, D.M.; Filip-Dhaubhadel, A.; Heirich, O.; Gerbeth, D.; Felux, M.; Schalk, L.M.; Becker, D.; et al. Flight Trial Demonstration of Secure GBAS via the L-band Digital Aeronautical Communications System (LDACS). *IEEE Aerosp. Electron. Syst. Mag.* **2021**, *36*, 8–17. [CrossRef]
17. Mäurer, N.; Caamano, M.; Gerbeth, D.; Gräupl, T.; Schmitt, C. A secure broadcast service for LDACS with an application to secure GBAS. In Proceedings of the IEEE/AIAA 40th Digital Avionics Systems Conference (DASC), San Antonio, TX, USA, 3–7 October 2021.
18. Crespillo, O.G.; Grosch, A.; Adjroloh, P.H.; Zhu, C.; Capua, R.; Frittella, F.; Kutik, O. Multisensor Localization Architecture for High-Accuracy and High-Integrity Land-based Applications. In Proceedings of the 35th International Technical Meeting of the Satellite Division of The Institute of Navigation (ION GNSS+ 2022), Denver, CO, USA, 19–23 September 2022; pp. 1873–1889.
19. *RTCA DO-253D*; Minimum Operational Performance Standards for GPS Local Area Augmentation System Airborne. RTCA: Washington, DC, USA, 2019.
20. Kern, S.; Geister, D.; Korn, B. City-ATM—Demonstration of Traffic Management in Urban Airspace in case of bridge inspection. In Proceedings of the 2019 IEEE/AIAA38th Digital Avionics Systems Conference (DASC), San Diego, CA, USA, 8–12 September 2019.

engineering proceedings

MDPI

Proceeding Paper

IMU and GNSS Postprocessing for High-Resolution Strapdown Airborne Gravimetry †

Vadim Vyazmin * and Andrey Golovan

Navigation and Control Laboratory, Lomonosov Moscow State University, Leninskie Gory 1-52, 119991 Moscow, Russia; aagolovan@yandex.ru
* Correspondence: vadim.vyazmin@math.msu.ru

† Presented at the European Navigation Conference 2023, Noordwijk, The Netherlands, 31 May–2 June 2023.

Abstract: Strapdown airborne gravimeters based on a navigation-grade inertial measurement unit (IMU) are highly sensitive to perturbations during aircraft flight, especially in the case of flights in draped mode (at a constant altitude above the terrain) or drone-based flights. This implies the crucial importance of postprocessing, including determination of the IMU and GNSS navigation solutions, IMU/GNSS integration, and gravity estimation. In the paper, we briefly outline the key aspects of the developed postprocessing methodology. By processing raw data from two surveys (one is based on a small aircraft and the other on a drone), we investigate the best achievable spatial resolution of strapdown airborne gravimetry. We show that high-accuracy gravity estimates (at sub-mGal level) at a half-wavelength spatial resolution of 1 km can be obtained in the considered surveys.

Keywords: airborne gravimetry; strapdown IMU; GNSS; postprocessing; Kalman filter

Citation: Vyazmin, V.; Golovan, A. IMU and GNSS Postprocessing for High-Resolution Strapdown Airborne Gravimetry. *Eng. Proc.* **2023**, *54*, 41. https://doi.org/10.3390/ENC2023-15455

Academic Editors: Tom Willems and Okko Bleeker

Published: 29 October 2023

1. Introduction

Airborne gravimetry is a well-established technique for measuring Earth's gravity in remote areas, which is able to provide gravity data at a spatial resolution reaching 2–5 km [1]. Commonly, stabilized-platform inertial gravimeters are used in airborne gravimetry [2], but strapdown instruments are currently used as well [3]. The latter are based on a navigation-grade inertial measurement unit (IMU) and are attractive for aerogravimetry due to less weight and compact size. This allows one to install them onboard a small slow-moving aircraft or unmanned aerial vehicle (UAV), which leads to a higher spatial resolution of gravity data.

However, since strapdown airborne gravimeter measurements are highly sensitive to disturbances during flight, it is of crucial importance to develop adequate methodology and algorithms for raw data postprocessing.

Over the past three years, Lomonosov MSU has developed postprocessing algorithms for strapdown airborne gravimetry [4]. The objective of this study is to determine the best spatial resolution of aerogravity data with the developed algorithms in surveys based on a small aircraft and UAV.

2. Postprocessing Methodology

We formulate the problem of strapdown airborne gravimeter data postprocessing as determining gravity disturbance δg on the aircraft flight path from raw data collected by the gravimeter's IMU inertial sensors (accelerometer and gyroscope readings) and global navigation satellite system (GNSS) receivers.

The developed postprocessing methodology includes several stages, which are briefly outlined below:

1. GNSS raw data postprocessing;
2. IMU raw data quality control;

3. IMU initial and final alignments and pre- and post-flight calibration of IMU's accelerometers;
4. IMU/GNSS integration;
5. Gravity disturbance estimation on the flight path.

In the first step, high-accuracy velocity and position of the onboard GNSS receiver antenna are determined from the multifrequency pseudorange, Doppler pseudorange rate and carrier-phase measurements recorded by the onboard and reference (ground-based) receivers [5]. The carrier-phase differential mode is used. This allows us to reach accuracy of GNSS velocity solution at the level of several cm/s. The developed GNSS postprocessing algorithms (and software) were widely used for GT-2A airborne gravimeter data processing [1].

In step 3, the IMU attitude is determined at aircraft standstill before and after the flight. The developed alignment algorithm admits angular motion of the gravimeter during aircraft standstill (caused by wind at the airfield or other external disturbances). During this step, the biases of IMU accelerometers are also determined using a calibration procedure [4].

In step 4, the integration of IMU and GNSS data is performed using the horizontal channels only. The integration algorithm is the Kalman filter and smoothing (open-loop estimation) [6].

In step 5, the gravity disturbance δg and residual instrumental errors of the gravimeter are estimated jointly using the following equation:

$$\Delta \dot{v}_{Up} = -\delta g + k_E f'_N - k_N f'_E + \mathbf{n}^T \mathbf{q}_f, \quad (1)$$

where Δv_{Up} is the IMU vertical velocity error, k_E and k_N are the residual attitude errors (misalignment of the vertical axis of the computed geodetic frame), f'_E and f'_N are the "horizontal" accelerometer measurements, $\mathbf{q}_f$ is the vector of accelerometer noises and $\mathbf{n}$ is the unit vector of the vertical accelerometer sensitivity axis expressed in the navigation (geodetic) frame. Other residual systematic errors of the gravimeter (time-synchronization error of IMU and GNSS data, lever-arm effect) [7] are not shown in (1) for brevity. The estimates of δg, k_E and k_N are obtained simultaneously through Kalman filtering and smoothing.

3. Results

We present the numerical results from two airborne gravimetry campaigns carried out in 2022 by Aerogeophysica JSC (Russia), who owns state-of-the-art strapdown airborne gravimeters (manufactured by iMAR and a domestic company). The raw data were kindly provided to Lomonosov MSU.

3.1. Fixed-Wing Aircraft Flights

The survey was based on a small fixed-wing aircraft (Antonov An-3T) flown in draped mode over a mountainous area. The flight height variations reached 600 m at survey lines, and the average flight height above the WGS-84 ellipsoid was about 1300 m. The average aircraft speed at survey lines was 43 m/s.

The flights were characterized by extreme vertical accelerations up to 2.5 g (Figure 1), which appears to be one of the first cases in airborne gravimetry (see also results in [4]). In Figure 1, the gravimeter's IMU roll and pitch angles (estimated in stage 4 of the methodology) are also shown, indicating exceptionally strong dynamic conditions during the flight. As can be seen in Figure 1, the pitch angle varied from −30 up to 30 deg at several lines. This implies large specific force values in the vertical accelerometer measurements visible in Figure 1.

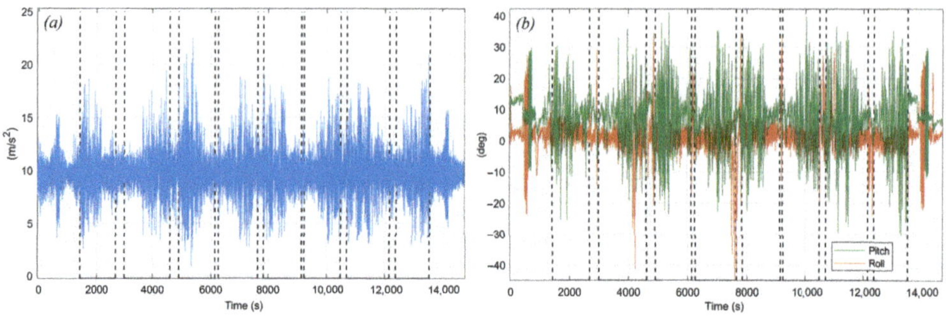

Figure 1. (**a**) Gravimeter's IMU vertical accelerometer readings (fixed-wing aircraft flight), m/s^2. (**b**) The estimates of IMU pitch and roll angles, deg. The dotted lines mark the time intervals of survey lines.

After the IMU accelerometer calibration (stage 3), the autonomous (stand-alone) IMU navigation solutions were computed to analyze the quality of IMU calibration. The autonomous IMU position errors (w.r.t. the GNSS position solution computed at stage 1) are shown in Figure 2. The Schuler oscillations [8] can be seen in both position errors (in the east and north directions). The position errors do not exceed 1 km during the 6.5 h flight. The IMU relative velocity errors are less than 0.5 m/s. These results show excellent quality of calibration as well as high accuracy (navigation grade) of the IMU.

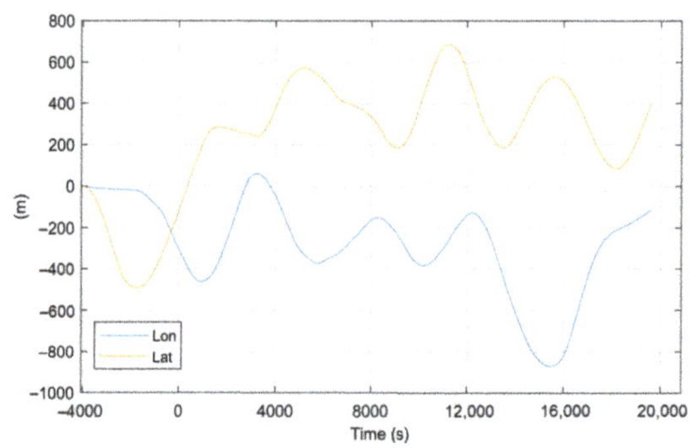

Figure 2. The IMU position errors in the autonomous mode (w.r.t. GNSS position solution), m.

The gravity disturbance was estimated (stage 5) using the gravimetric filter based on (1). Different cutoff frequencies of the filter were tested in order to obtain high-accuracy gravity data (within 1 mGal accuracy) at the best possible spatial resolution (Table 1). For this, the gravity disturbance estimates at repeat lines flown over the control track were analyzed. From the statistics presented in Table 1, it follows that the gravimetric filter with the cutoff frequency of 1/80 Hz provides the best result (1 mGal accuracy at a half-wavelength spatial resolution of 1.7 km). The obtained gravity disturbance estimates are shown in Figure 3. Also shown is the elevation profile, which illustrates the correlation of the obtained gravity estimates with topography.

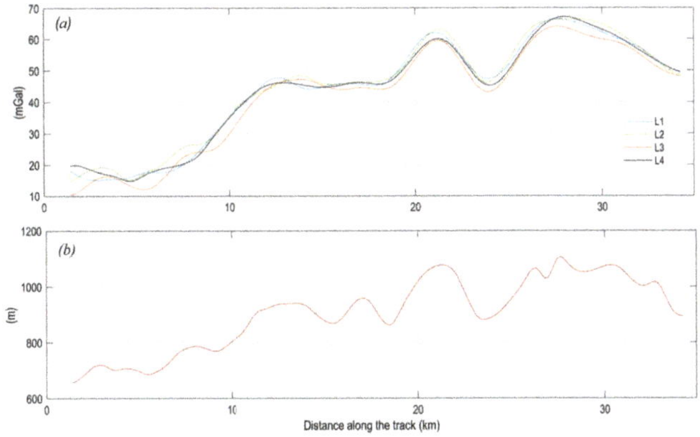

Figure 3. (**a**) Gravity disturbance estimates at repeat lines from the fixed-wing aircraft flights, mGal. The filter cutoff frequency is 1/80 Hz. (**b**) Elevation profile, m.

The gravity disturbance estimates at survey lines obtained with the same cutoff frequency of the gravimetric filter (1/80 Hz) can be seen in Figure 4. The circles in the left side of the plot are due to the maneuvering of the aircraft when trying to follow more closely to the terrain. This was necessary for other onboard measurements (electromagnetometry, gamma-ray spectrometry) collected simultaneously with gravimetry.

Table 1. Statistics for gravity estimation accuracy from four repeat lines (fixed-wing aircraft flights).

Filter Cutoff Frequency (Hz)	Spatial Resolution (km)	RMS (mGal)
1/70	1.5	1.14
1/80	1.7	0.97
1/100	2.2	0.83
1/120	2.6	0.81

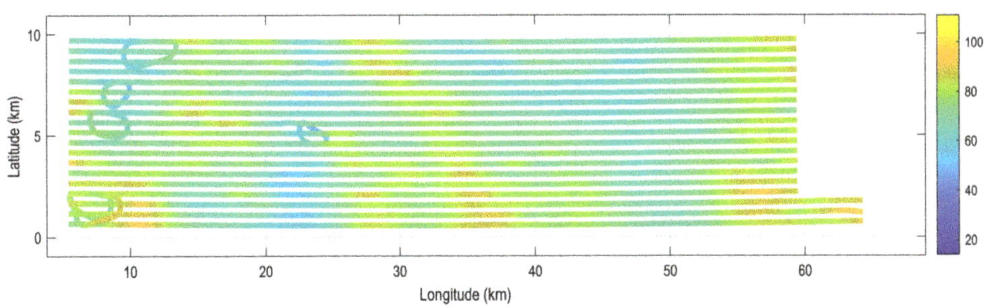

Figure 4. Gravity disturbance estimates at survey lines from the fixed-wing aircraft flights, mGal. The filter cutoff frequency is 1/80 Hz.

3.2. UAV Flights

The campaign was based on using a helicopter-type UAV (the photo is presented in [4]). Several flights were flown at a constant height of 340 m above the WGS-84 ellipsoid at an average speed of 25 m/s. The flights were flown in strong dynamic conditions due to wind (changes in roll and pitch angles reached 5–7 deg) and vibrations produced by the carrier (oscillations in the vertical velocity with an amplitude of 1 m/s).

The postprocessing strategy was the same as in the case of fixed-wing aircraft flights. In Figure 5, the gravity disturbance estimates at repeat lines are shown. The filter cutoff frequency is 1/100 Hz, which is equivalent to the half-wavelength spatial resolution of

1.25 km. The lower cutoff frequency was chosen to suppress the effect of vibrations in the gravimeter data. The estimation accuracy (from repeat line analysis) is 0.59 mGal (RMS). Figure 6 shows the gravity estimates at survey lines from two flights flown with this UAV (the line length is about 20 km).

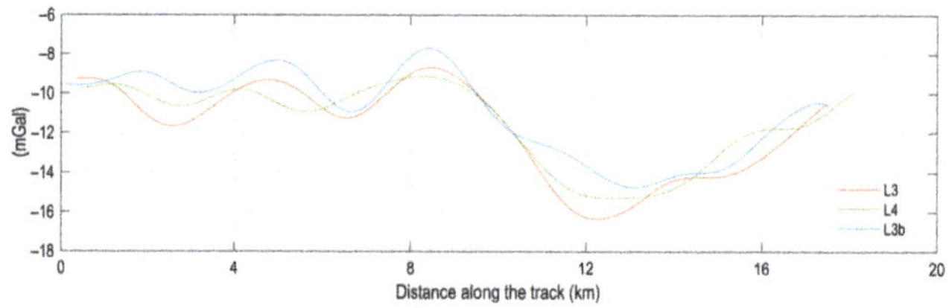

Figure 5. Gravity disturbance estimates at repeat lines from the UAV test, mGal. The filter cutoff frequency is 1/100 Hz.

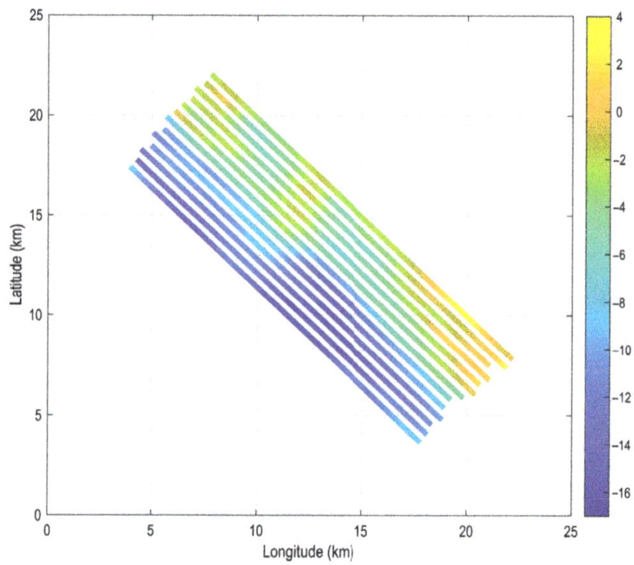

Figure 6. Gravity disturbance estimates at survey lines from the UAV test, mGal. The filter cutoff frequency is 1/100 Hz.

4. Discussion

From Table 1, it follows that the spatial resolution of gravity estimation in the fixed-wing aircraft (An-3T) flights is limited by the gravimeter filter cutoff frequency of 1/80 Hz. Any higher cutoff frequency value will lead to larger errors in the gravity estimates and, therefore, to RMS values larger than 1 mGal (from repeat line analysis). The estimation errors are mainly caused by (but not limited to) high-frequency noise in the GNSS solutions [1]. Another error source may be an underestimation of the instrumental errors of the IMU inertial sensors, especially in the case of draped flights. Recall that the vertical accelerations reached 2.5 g in the considered survey flights. To investigate this in more detail, flights in benign dynamic conditions (flights at a constant height above the ellipsoid) should also be carried out and analyzed. A laboratory calibration of the gravimeter will also help analyze the IMU instrumental errors.

As for the UAV test, a lower filter cutoff frequency (1/100 Hz) was chosen to suppress strong vibrations of the carrier. The achieved gravity estimation accuracy (0.6 mGal) is excellent. However, it should be noted that the magnitude of gravity disturbance was not large in this test (the measured gravity values were in the range from −16 mGal up to −8 mGal). Nevertheless, these results seem promising, as this test was one of the first surveys based on a UAV.

Thus, the achieved half-wavelength spatial resolution of gravity estimates in the two surveys is 1.7 km (with 1.0 mGal accuracy) in the case of An-3T flights in the draped mode and 1.3 km (with 0.6 mGal accuracy) in the case of the UAV test.

5. Conclusions

The objective of this study is to determine the best achievable spatial resolution of strapdown airborne gravimetry using the developed postprocessing algorithms and slow-moving carriers (a small aircraft and UAV).

We show that high-accuracy IMU/GNSS solutions are of crucial importance for obtaining high-quality gravity data, especially in the case of draped flights (i.e., following the terrain). Using the proposed postprocessing methodology, high-accuracy gravimetry results (at sub-mGal level) were obtained with the half-wavelength spatial resolution down to 1.3 km in a drone-based survey. This result is promising and can be regarded as an improvement in the achievable spatial resolution in airborne gravimetry (typically 2–5 km). This also confirms a good potential of strapdown airborne gravimetry for providing high-resolution gravity data.

Author Contributions: Conceptualization, V.V. and A.G.; methodology, V.V. and A.G.; software, V.V. and A.G.; formal analysis, V.V. and A.G.; investigation, V.V.; data curation, V.V.; writing—original draft, V.V.; writing—review and editing, V.V. and A.G.; visualization, V.V.; supervision, A.G.; project administration, V.V. and A.G.; funding acquisition, A.G. All authors have read and agreed to the published version of the manuscript.

Funding: This research received no external funding.

Institutional Review Board Statement: Not applicable.

Informed Consent Statement: Not applicable.

Data Availability Statement: The raw data are not publicly available due to privacy restrictions.

Acknowledgments: We thank the reviewers for the valuable comments.

Conflicts of Interest: The authors declare no conflict of interest.

References

1. Krasnov, A.A.; Sokolov, A.V.; Bolotin, Y.V.; Golovan, A.A.; Parusnikov, N.A.; Motorin, A.V.; Nosov, A.; Stepanov, O.A.; Yurist, S.; Vyazmin, V. Data processing methods for onboard gravity anomaly measurements. In *Methods and Technologies for Measuring the Earth's Gravity Field Parameters*; Peshekhonov, V.G., Stepanov, O.A., Eds.; Springer: Berlin/Heidelberg, Germany, 2022; pp. 63–150. [CrossRef]
2. Studinger, M.; Bell, R.; Frearson, N. Comparison of AIRGrav and GT-1A airborne gravimeters for research applications. *Geophysics* **2008**, *73*, 151–161. [CrossRef]
3. Jensen, T.E.; Olesen, A.V.; Forsberg, R.; Olsson, P.-A.; Josefsson, Ö. New results from strapdown airborne gravimetry using temperature stabilisation. *Remote Sens.* **2019**, *11*, 2682. [CrossRef]
4. Golovan, A.A.; Vyazmin, V.S. Methodology of airborne gravimetry surveying and strapdown gravimeter data processing. *Gyroscopy Navig.* **2023**, *14*, 36–47. [CrossRef]
5. Golovan, A.A.; Vavilova, N.B. Satellite navigation. Raw data processing for geophysical applications. *J. Math. Sci.* **2007**, *146*, 5920–5930. [CrossRef]
6. Kailath, T.; Sayed, A.H.; Hassibi, B. *Linear Estimation*; Prentice Hall: Upper Saddle River, NJ, USA, 2000; 854p.
7. Vavilova, N.B.; Golovan, A.A.; Kozlov, A.V.; Papusha, I.A.; Zorina, O.A.; Izmailov, E.A.; Kukhtevich, S.E.; Fomichev, A.V. INS/GNSS integration with compensated data synchronization errors and displacement of GNSS antenna. Experience of practical realization. *Gyroscopy Navig.* **2021**, *12*, 236–246. [CrossRef]
8. Farrell, J.A. *Aided Navigation Systems: GPS and High Rate Sensors*; McGraw-Hill: New York, NY, USA, 2008; 552p.

engineering proceedings

MDPI

Proceeding Paper

Experimentation of Monocular Visual-Aided Inertial Navigation on Fixed-Wing Unmanned Aerial Vehicle †

Baheerathan Sivalingam * and Ove Kent Hagen

Norwegian Defence Research Establishment (FFI), 2027 Kjeller, Norway; ove-kent.hagen@ffi.no
* Correspondence: baheerathan.sivalingam@ffi.no

† Presented at the European Navigation Conference 2023, Noordwijk, The Netherlands, 31 May–2 June 2023.

Abstract: This paper presents an experimental study utilizing the vision-aided inertial navigation system (VaINS) on a fixed-wing unmanned aerial vehicle (UAV). VaINS operates as a data-logging box, collecting sensor data, processing them, and transmitting the navigation solution to an operator terminal in real time. The flight data contain buildings, vegetation, a river, a lake, and mountains. The camera's pitch angle was tested at both a grazing angle of 22° and a nadir angle of 90°. The aim of this experiment was twofold: to explore the application of VaINS on a fixed-wing UAV and to assess the performance of the image-aided navigation system. The image-aided inertial navigation system (INS) solution was compared with a reference solution during GNSS-denied time intervals. The camera's pitch angle at 90° exhibited a slight advantage over the grazing angle at 22°. Moreover, in the linear flight segments, the image-aided navigation system demonstrated significantly better performance compared to the free inertial solution during the GNSS-denied periods.

Keywords: navigation; tracking; image feature points; INS; fixed-wing UAV

Citation: Sivalingam, B.; Hagen, O.K. Experimentation of Monocular Visual-Aided Inertial Navigation on Fixed-Wing Unmanned Aerial Vehicle. *Eng. Proc.* **2023**, *54*, 42. https://doi.org/10.3390/ENC2023-15476

Academic Editors: Tom Willems and Okko Bleeker

Published: 29 October 2023

1. Introduction

Unmanned aerial systems (UAS) predominantly utilize the global navigation satellite system (GNSS) in conjunction with an INS for navigation. If the GNSS signal is denied, it results in an unacceptable position error, even over a brief time span. One potential solution to the problem is to use additional aiding sensors, e.g., LiDAR, radar, radio, or visual-based navigation [1]. In the VaINS, we use image-aided navigation, which can be implemented in various forms. A comprehensive literature review of the image-assisted methods for UAV navigation is presented in [2]. The techniques are categorized based on using a map as prior information, building the map while navigating as in visual simultaneous localization and mapping (SLAM) techniques, or operating without maps, such as with feature tracking and optical flow techniques.

The Norwegian Defense Research Establishment (FFI) has been investigating an image-aided inertial navigation system to cope with GNSS failure that is operating without maps. The system [3] is based on tightly integrating inertial sensor data with pixel position data of image feature points over an image sequence in an error-state extended Kalman filter (EKF).

The system's earlier versions underwent testing using image data obtained from a manned helicopter [4], an octocopter [3,5], and a hex-copter. The collected data were processed and analyzed offline. Later, a real-time navigation demonstration box called a visual-aided inertial navigation system (VaINS) was developed [6]. This VaINS box, as shown in Figure 1a, collects sensor data, processes them, and transmits the navigation solution to an operator terminal in real time.

An extensive experiment was conducted at Elvenes airstrip in the Salangen municipality area, Norway, using VaINS. It was used as a data-logging box on a fixed-wing UAV, Scout (Figure 1b) from NORCE [7]. NORCE was responsible for mounting the VaINS box and operating the aircraft.

(a)

(b)

Figure 1. (**a**) A real-time navigation box named VaINS. (**b**) The Scout UAV.

There is related work on large-scale tests with image-aided navigation on UAVs. In [8], an autonomous route-following system of visual teach and repeat is coupled with a downward-facing camera on a fixed-wing UAV as a backup navigation when GNSS is unavailable. The system was evaluated using the data from the fixed-wing UAV flying a 1200 m trajectory at an altitude of 80 m. An optical-aided navigation method employing a camera and a LIDAR or radar is used in [9] when the GNSS is unavailable. It was tested with data sets from crewed helicopters and fixed-wing UAVs and achieved an accumulation error of 1–6% of the distance traveled, depending on the flight scenario.

This work herein builds on the technical report of the extensive experiment [10]. The purpose of this experiment was twofold: first, to explore the application of the VaINS box on a fixed-wing UAV, and second, to assess the performance of the image-aided INS. This document describes the developed image-aided inertial navigation system in brief, experimental setup, data collection, and results and summarizes the experiences and the performance.

2. Image-Aided Inertial Navigation System

The concept of the image-aided inertial navigation system is illustrated in Figure 2. An image feature point is the projection of a landmark, which is a static point in the scene, onto the image plane via a camera model. At each time step, the INS computes an estimate of the UAV's state, and this estimate is used together with the estimated landmark positions by the Kalman filter to predict the positions of the image feature points.

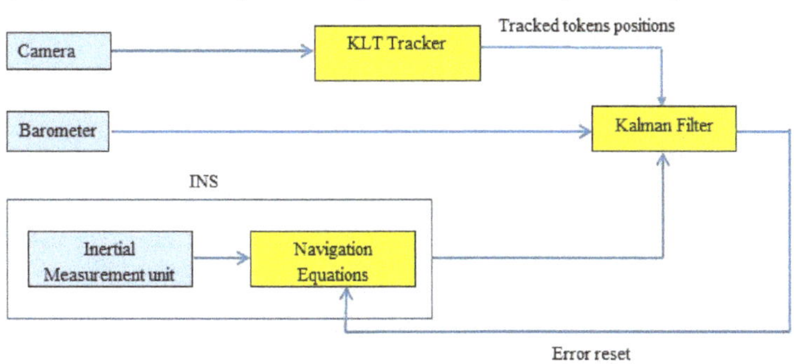

Figure 2. Image-aided inertial navigation system.

Image feature points are tracked by using the KLT algorithm [11]. The difference between the predicted and tracked positions are used as measurements by the Kalman filter, which estimates the INS errors. This process is repeated for each image time stamp. Details of the system are described in [3,5].

When using a monocular camera system, as in our case, the range to a tracked landmark is initially unknown and has to be guessed or externally estimated somehow. The tight integration with the EKF, therefore, uses a variant of the inverse depth parameterization [3,12], which is robust to large initial range errors. With the combination of accelerated motion and feature tracking, these ranges become observable through the EKF [13], which results in a lower position error drift compared to a free inertial INS. There still is, however, a sensitivity to the level of error of our initial estimate for the range.

In [14], they suggest using a near range to the camera, setting the initial range to twice the near range with a 95% confidence level from the near range to infinity. Another strategy uses some initial feature measurements to pre-estimate the range before actually using them in the estimator [15]. In the tests herein, we estimated an effective near-range from coarse estimates of the UAV's height above ground and the camera view angle, assuming nothing was tracked in between.

3. Experimental Setup

VaINS consists of three sensors and two processing units. The sensors are the MTi-100 MEMS Inertial Measurement Unit (IMU), a u-blox GPS, and an embedded Basler da1600 monochrome camera. The processing units are a Jetson TX2 running the KLT-tracker and a TE0720 with a Xilinx Zynq 7020 running the INS on its CPU and interfacing and time stamping the sensors, including the camera through an FPGA. The technical documentation of VaINS is described in [6].

This experiment was a collaborative experiment involving both FFI and NORCE. NORCE was in charge of mounting the VaINS box on the aircraft (see Table 1 for aircraft specification) and planning the flight operations, while FFI took responsibility for devising the test plan, conducting data logging, and performing data analysis.

Table 1. Aircraft specification.

Parameter	Value
Cruise speed	15–20 m/s
Wingspan	2.62 m
Maximum endurance	3 h
Standard empty weight	4.3 kg
Batteries—Li-Ion	2.6 kg
MTOW	10 kg
Maximum useful load	3.1 kg

For practical implementation, the camera was securely mounted outside the VaINS box at the front of the payload bay, using a fixture that regulated various pitch angles. The VaINS box found its position at the back of the bay. After the aircraft was launched from a launching device, it was monitored from a mobile ground control station. More details are given in [10].

4. Data Collection

Data collection took place in the vicinity of an airstrip located in Salangen municipality, Norway. The flight trajectory remained consistent within a predefined area, viewing buildings, vegetation, a river, a lake, and mountains. The camera's pitch angle was set at a grazing angle of 22° and 90°.

The flight details are tabulated in Table 2. The initial segment of the flight altitude is approximately 80 m, while the altitude of the primary portion of the flight is around 180 m. The camera captures monochrome images at an 8-bit depth, with a frame rate of 10 Hz and a resolution of 1600 × 1200 pixels. Figure 3 shows an example of an image for each pitch angle of the camera.

Table 2. Flight details.

Flight #	Camera's Pitch	Time after Launch (s)	Total Time (s)
1	22°	903	1653
2	90°	742	992

(a) (b)

Figure 3. Example image from each pitch angle. (**a**) 22° (**b**) 90°.

Flight time, after launch, for the flights was around 16–28 min. However, data collection stopped around 12–15 min after launch due to a system setup error.

5. Results

An INS post-processing tool developed by FFI, NavLab4 [16], is used to process and analyze the data from VaINS offline and gives an indication of the expected real-time performance since the INS algorithms in VaINS are based on the implementation in NavLab4. The aircraft's reference position and orientation are determined using the optimal smoother in NavLab4, which utilizes IMU and GPS data for the whole flight. To test the image-aided INS, image feature point data (12 points at 10 Hz) is combined with the IMU data and GPS height during the simulated GPS-denied intervals. Alternatively, barometer height measurements from NORCE can replace GPS height.

The performance of the image-aided INS solution is assessed by comparing it separately with the reference solution during linear flight (straight-line) and banking maneuvers. The root-mean-square (RMS) error between the image-aided INS and the reference solution at the end of each GPS-denied interval serves as the evaluation metric. For the straight-line portions, the comparison considers both the percent of distance traveled (%DT) and the average speed error (ASE), while for the banking portions, only the ASE is taken into account.

From the initial analysis in NavLab4, when including all the sensors, we identified and coarsely estimated a significant misalignment of the camera with respect to the IMU [10]. The camera and the box were remounted prior to each flight and without guaranteeing the same mounting alignment. Thus, we estimated the misalignment for each flight.

5.1. Flight 1

The camera's pitch angle of this flight is 22 degrees, and the flight time after launching is 903 s. This flight has four linear flights and four banking maneuvers.

Figure 4 presents the trajectory of the first two linear flights and the four banking maneuvers. The image-aided navigation when GPS data are disabled is presented with a green line. The red line presents the reference trajectory. The rest of the maneuvers are similar to the presented ones.

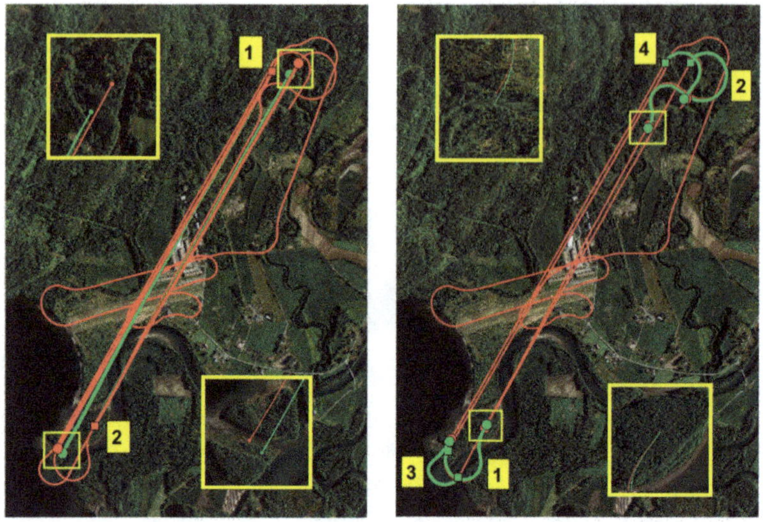

(**a**) (**b**)

Figure 4. The trajectories of the first flight with image-aided navigation when GPS data are disabled (green line) and the reference trajectory (red line). (**a**) Linear flights (about 2400 m) (**b**) and banking maneuvers.

Tables 3 and 4 tabulate the duration, distance traveled, and drift compared to the reference data in the straight-line parts and banking parts, respectively. When comparing the drift (ASE), the drift in linear flights is slightly lower than the drift in banking maneuvers.

Table 3. Duration of image-aided navigation (flight 1) when GPS is disabled, distance traveled within the duration, and the drift in RMS error, in ASE, and in %DT.

Duration (s)	Distance (m)	RMS (m)	ASE (m/s)	%DT
136	2439	37.6	0.28	1.54
130	2338	82.4	0.63	3.52
122	2231	50.3	0.41	2.25
133	2411	56.7	0.43	2.35

Table 4. Duration of image-aided navigation (flight 1) and the drift in RMS error and ASE in banking maneuvers.

Duration	RMS (m)	ASE (m/s)
33	18.4	0.56
53	14.1	0.27
26	18.9	0.73
24	15.0	0.63

5.2. Flight 2

Flight 2 has the camera's pitch angle at 90° nadir, and flight time after launching is 742 s. This flight has three linear flights and two banking maneuvers. The first two trajectories of the linear flights and the banking maneuvers are shown in Figure 5.

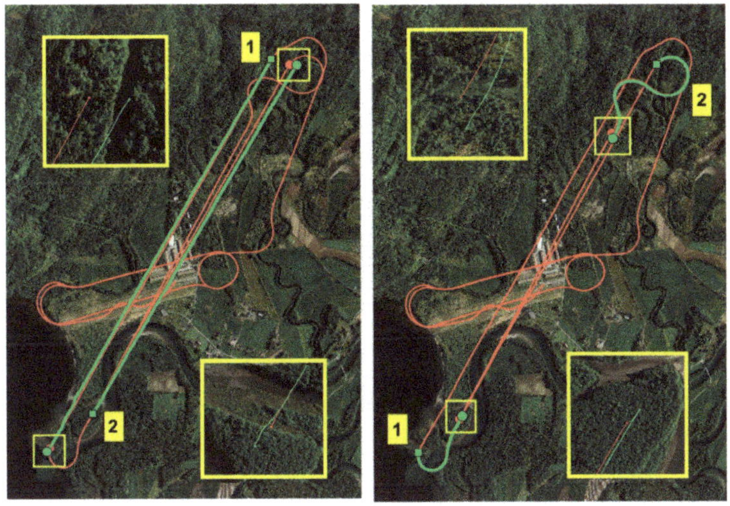

(**a**) (**b**)

Figure 5. The trajectories of the second flight with image-aided navigation (green line) and the reference trajectory (red line). (**a**) Linear flights (about 2400 m) (**b**) and banking maneuvers.

The green line represents the image-aided navigation when the GPS is disabled, and the red lines represent the reference trajectories. Table 5 tabulates the duration, distance traveled, and the drift of the linear flights, and Table 6 tabulates the drift data for the banking maneuvers. In the linear parts, the ASE is slightly lower than the banking parts.

Table 5. Duration of image-aided navigation (flight 2), distance traveled within the duration, and the drift in RMS, in ASE, and in %DT.

Duration (s)	Distance (m)	RMS (m)	ASE (m/s)	%DT
158	2576	20.0	0.13	0.78
123	2310	36.9	0.30	1.60
67	1090	20.5	0.31	1.88

Table 6. Duration of image-aided navigation (flight 2) and the drift in RMS error and in ASE.

Duration	RMS (m)	ASE (m/s)
30	16.3	0.54
57	24.4	0.43

5.3. Free Inertial

Within the simulated GPS-denied time intervals, the performance of free inertial is analyzed using the RMS error and ASE for straight-line laps and for the banking laps separately (see Tables 7 and 8).

In the linear parts, the free inertial ASE ranges from 0.20 to 8.31 m/s, while image-aided navigation ASE ranges from 0.13 to 0.73 m/s. In the banking parts, however, the results are more similar: from 0.37 to 5.32 m/s for free inertial and from 0.27 to 0.73 m/s for image-aided navigation.

Table 7. The RMS error and ASE at the end of the GPS-denied intervals for straight-line laps under free inertial.

Flight	Lap	Duration (s)	Distance (m)	RMS (m)	ASE (m/s)
1	1	136	2439	340	2.50
	2	130	2338	855	6.58
	3	122	2231	627	5.14
	4	133	2411	1105	8.31
2	1	158	2576	960	6.08
	2	123	2310	26	0.20
	3	67	1090	59	0.88

Table 8. The RMS error and ASE at the end of the GPS-denied intervals for banking laps under free inertial.

Flight	Lap	Duration (s)	RMS (m)	ASE (m/s)
1	1	33	68.9	2.12
	2	53	39.2	0.74
	3	26	9.5	0.37
	4	24	10.3	0.43
2	1	30	159.7	5.32
	2	57	39.5	0.69

6. Discussion

The results obtained from flights 1 and 2, as discussed here, are indicative of the other flights in the experiment. For further details, refer to [10]. It was observed that the image-aided INS exhibited slightly larger drift during banking maneuvers compared to linear flights. This phenomenon can be attributed to the image feature points being selected from distant regions in the scene during banking maneuvers due to the combination of low pitch angle and high roll angle. Consequently, these feature points rapidly disappeared from the camera's field of view, leading to increased drift in the image-aided navigation.

When comparing the two pitch angles (90° nadir and 22° grazing), it was observed that the nadir configuration performed slightly better than the grazing configuration, as depicted in Figure 6. The advantage of the nadir configuration lies in the utilization of the entire image scene, with similar ranges to all landmarks, and the ease of obtaining a good initial estimate of altitude. However, a drawback is that the image feature points are not tracked for an extended period.

On the other hand, the grazing configuration has certain downsides, such as a portion of the image above the horizon that is never used, resulting in significant variation in the range to landmarks and making it more challenging to establish a good initial estimate. To address this, an approach is proposed to dynamically estimate the initial range from the first few frames of each image-feature-point track before initializing it in the EKF, as described in [13]. The advantage of the grazing configuration is the potential for tracking landmarks over a longer duration. Nonetheless, during banking maneuvers, almost all landmarks are lost.

Comparing image-aided navigation with free inertial, image-aided navigation gives a great overall improvement in linear flights. Over greater time intervals, the free inertial error drift becomes non-linear and increases faster, as seen in the long straight-line parts. In the banking turns, there are two effects that make the free inertial error lower. The time period is shorter, so the non-linear parts of the error drift are less, and errors may cancel due to the turn. In these parts, the image-aided navigation seems to primarily introduce increased robustness (see Figure 6).

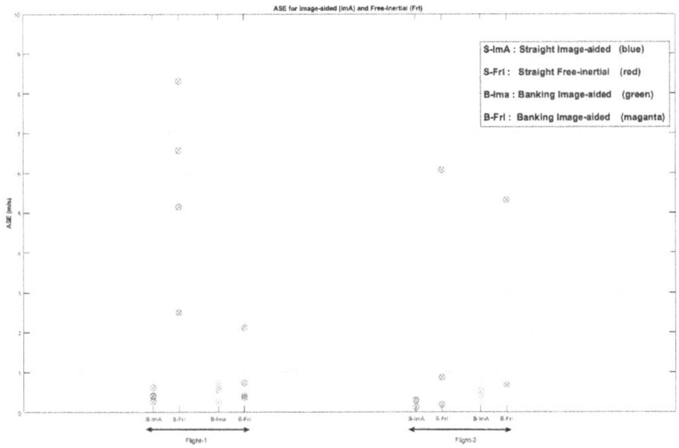

Figure 6. Comparison of ASE between image-aided INS and free-inertial for flight 1 and flight 2 under straight-line and banking laps.

The navigation solution is sensitive to the alignment between the IMU and the camera. Thus, the current installation, with an unknown misalignment for each flight, is not ideal. An improvement is to fix the IMU and the camera on a bracket placed outside the VaINS box. By doing so, the alignment can be estimated once, streamlining the setup process.

7. Conclusions

This document outlines an experiment involving the VaINS box. The VaINS box provides a navigation solution in real time. It integrates inertial sensor data with position data of image feature points over an image sequence employing an error-state extended Kalman filter. This experiment's aim was to explore the application of the VaINS box on a fixed-wing UAV and to assess the image-aided navigation solution.

The data were collected from flights with camera pitch angles of 22° and 90° and viewing buildings, vegetation, a river, a lake, and mountains. The aircraft's reference navigation solution is calculated using the optimal smoother, utilizing IMU and GPS data. For the image-aided navigation solution, image-feature-point data are combined with IMU data and GPS height during GPS-denied intervals. The flights with the camera pitched at the nadir demonstrated an overall better performance compared to the flights with a grazing angle. The drift during banking maneuvers was slightly larger than during straight-line maneuvers. During banking maneuvers, the image-aided navigation performed slightly better than the free inertial navigation within the same GPS-denied time intervals. Remarkably, it performed significantly better during straight-line maneuvers.

In this experiment, the IMU and the camera were loosely mounted together, which led to additional uncertainty in misalignment. This could be solved by fixing the camera and IMU on a rigid fixture out of VaINS.

Author Contributions: B.S. and O.K.H. together contributed in all aspects of the article. All authors have read and agreed to the published version of the manuscript.

Funding: This research received no external funding.

Institutional Review Board Statement: Not applicable.

Informed Consent Statement: Not applicable.

Data Availability Statement: Data are available when requested.

Conflicts of Interest: The authors declare no conflict of interest.

References

1. Gyagenda, N.; Hatilima, J.V.; Roth, H.; Zhmud, V. A review of GNSS-independent UAV navigation techniques. *Robot. Auton. Syst.* **2022**, *152*, 104069. [CrossRef]
2. Lu, Y.; Xue, Z.; Xia, G.; Zhang, L. A survey on vision-based UAV navigation. *Geo-Spat. Inf. Sci.* **2018**, *21*, 21–32. [CrossRef]
3. Baheerathan, S.; Hagen, O.K. Image-aided inertial navigation for an Octocopter. In Proceedings of the SPIE Defense + Security 2018, Orlando, FL, USA, 3 May 2018.
4. Baheerathan, S.; Hagen, O.K. Image-aided inertial navigation system based on image tokens. In Proceedings of the SET-168 NATO Symposium on Navigation Sensors and Systems in GNSS Denied Environments, Imzir, Trucky, 8–9 October 2012.
5. Baheerathan, S.; Hagen, O.K. Experiment of Image-aided inertial navigation with multiple cameras in Octocopter. In Proceedings of the SPIE Defense + Security 2020, Anaheim, CA, USA, 26–30 April 2020.
6. Baheerathan, S.; Hagen, O.K. *VaINS Technical Documentation*; FFI Report 22/01877; Norwegian Defense Research Establishment: Kjeller, Norway, 2022.
7. NORCE Norwegian Research Center. Available online: https://www.norceresearch.no (accessed on 15 August 2023).
8. Warren, M.; Paton, M.; MacTavish, K.; Schoellig, A.P.; Barfoot, T.D. *Towards Visual Teach and Repeat for GPS-Denied Flight of a Fixed-Wing UAV, Field and Service Robotics*; Springer: Berlin/Heidelberg, Germany, 2018; pp. 481–498.
9. Andert, F.; Ammann, N.; Krause, S.; Lorenz, S.; Bratanov, D.; Mejias, L. Optical aided aircraft navigation using decoupled visual SLAM with range sensor augmentation. *J. Intell. Robot. Syst.* **2017**, *88*, 547–565. [CrossRef]
10. Baheerathan, S.; Hagen, O.K. *Experimentation of Vision-Aided Inertial Navigation System on a Small Fixed-Wing UAV*; FFI Report 23/00148; Norwegian Defense Research Establishment: Kjeller, Norway, 2023.
11. Thomasi, C.; Kanade, T. *Detection and Tracking of Point Features*; Tech. Rep. CMU-CS-91-132; Carnegie Mellon University: Pittsburg, PA, USA, 1991.
12. George, M.; Sukkarieh, S. Inertial navigation aided by monocular camera observations of unknown features. In Proceedings of the 2007 IEEE International Conference on Robotics and Automation, Rome, Italy, 10–14 April 2007.
13. Martinelli, A. Vision and IMU data fusion: Closed form solutions for attitude, speed, absolute scale and bias determination. *IEEE Trans. Robot.* **2012**, *28*, 44–60. [CrossRef]
14. Civera, J.; Davison, A.J.; Montiel, J.M.M. Inverse depth parametrization for monuclar SLAM. *IEEE Trans. Robot.* **2008**, *24*, 932–945. [CrossRef]
15. Munguia, R.; Antoni, G. Delayed inverse depth monocular SLAM. *IFAC Proc.* **2008**, *41*, 2365–2370. [CrossRef]
16. Gade, K. NAVLAB: A generic simulation and post-processing tool for navigation. *Eur. J. Navig.* **2004**, *2*, 51–59. [CrossRef]

engineering proceedings

MDPI

Proceeding Paper

Galileo Receiver Performance Analysis with New I/NAV Improvements Live Data †

Rui D. Nunes [1,*], Mahamoudou Ouedraogo [2], Susanne Schlötzer [2], Nityaporn Sirikan [3] and Stefan Wallner [3]

1 Deimos Engenharia, 1070-061 Lisbon, Portugal
2 Airbus Defense and Space GmbH, 81663 Munich, Germany; mahamoudou.ouedraogo@airbus.com (M.O.); susanne.schloetzer@airbus.com (S.S.)
3 European Space Agency, 2200 AG Noordwijk, The Netherlands; nityaporn.sirikan@esa.int (N.S.); stefan.wallner@esa.int (S.W.)
* Correspondence: rui.nunes@deimos.com.pt
† Presented at the European Navigation Conference 2023, Noordwijk, The Netherlands, 31 May–2 June 2023.

Abstract: This work presents the results showing the improvement of the performance of a Galileo receiver processing the new I/NAV words 16–20 and SSP in comparison to the performance using the previous legacy I/NAV configuration. The results were obtained by replaying GNSS signals recorded in Urban and Open Sky mobile environments when at least four satellites broadcasting the new words were visible. The results show an improvement of Time to First Fix values using RedCED (word 16) and FEC2 (words 17–20) in cold start conditions and a reduction in time needed to converge using SSP when coarse time (within 3 s error) is available.

Keywords: Galileo; E1 band; I/NAV improvements; secondary synchronization pattern; Reed–Solomon; reduced CED; Time to First Fix

1. Introduction

In order to compute its time and position, a Galileo receiver needs to decode the Clock and Ephemeris Data (CED) information of the satellites present in the message obtained via the received signals from space. The speed of the retrieval of CED from four different satellites will have a direct impact on the Time to First Fix (TTFF). The legacy I/NAV message in Open Service E1B and E5B [1] contains the CED information on words 1–4, and all need to be decoded in the same Issue of Data (IoD) to provide the necessary information of the satellite. These words are distributed in the subframes according to Table 1, meaning that, in perfect conditions (with no data loss), the minimum time needed to retrieve CED is 14 s. With four satellite channels tracked in parallel, the TTFF computation is straightforward, although, in environments with less visibility and/or degraded reception, the computation of the CED of four different satellites will take significantly longer.

The inclusion of Reduced CED (hereafter RedCED) word 16 and Outer Forward Error Correction (FEC2) based on Reed–Solomon (hereafter RS) words 17–20 [2,3] was proposed in order to improve the conditions of CED information retrieval. The RedCED word 16 contains the information of legacy words 1–4 compressed in a single word with reduced bits (and, therefore, reduced accuracy), while RS words 17–20 complement words 1–4 with information protected via the FEC2 RS algorithm that will complement words 1–4, as shown in Table 1. By computing the new words, the data decoding and TTFF value are greatly improved in comparison to the legacy configuration.

The introduction of the Secondary Synchronization Pattern (hereafter SSP) [2] is intended to improve the time needed for time convergence when a coarse estimate is available (within +/− 3 s), as it will resolve the ambiguity when the receiver retrieves SSP bits pattern every 2 s among three possible patterns. The SSP information is present on the raw navigation data bits with no need to decode further data.

Citation: Nunes, R.D.; Ouedraogo, M.; Schlötzer, S.; Sirikan, N.; Wallner, S. Galileo Receiver Performance Analysis with New I/NAV Improvements Live Data. *Eng. Proc.* **2023**, *54*, 43. https://doi.org/10.3390/ENC2023-15422

Academic Editors: Tom Willems and Okko Bleeker

Published: 29 October 2023

Table 1. I/NAV message subframe layout (Blue = Legacy CED, Green = RS CED, Orange = Reduced CED).

T_0 (s)	Legacy E1B I/NAV	Improved E1B I/NAV
1	W02	W02
3	W04	W04
5	W06	W06
7	W07/W09	W07/W09
9	W08/W10	W08/W10
11	W00	W17/W19
13	W00	W18/W20
15	W00	W16
17	W00	W00
19	W00	W00
21	W01	W01
23	W03	W03
25	W05	W05
27	W00	W00
29	W00	W16

In this paper, the improvements of RedCED, RS, and SSP are shown by comparing the performance of the configuration using the new words with the configuration using the legacy I/NAV message with live signal data. For further reading on Reduced CED and FEC2 RS CED, refer to [4,5].

2. Materials and Methods

The first step is the collection of GNSS signals via a Radio Frequency recorder in a mobile testbed vehicle in both Urban and Open Sky environments (Figure 1). The time windows were chosen in order to have a minimum of 4 visible satellites with the new I/NAV Improvement features. The data collection resulted in a total of 2 h of Urban environment data collection and 4 h of Open Sky environment with 4 to 6 satellites visible in the sky broadcasting the new I/NAV improvements. It is important to note that since the I/NAV update in the constellation is still being deployed, the performance characterization is based on live data recorded from a reduced subset of satellites. The recorded GNSS signals were replayed in the ESA Navigation Laboratory connected to a Septentrio Test User Receiver (TUR-N) able to process the new I/NAV Improvement features. All tests were performed in Single Frequency (SF) mode using E1B signal. In order to properly compute the improvement of the new features with the same satellite constellation geometry and visibility, all tests were conducted using only the satellites broadcasting the new I/NAV features.

Figure 1. Vehicle used for mobile data collection.

For RedCED and RS performance tests, the receiver is reset via an automatic setup in cold start in order to evaluate individual satellite data demodulation time (excluding acquisition and tracking time) through a statistically significant number of points. Figure 2 presents both Data Demodulation and TTFF time, but the results in Section 3 focus only on Data Demodulation improvement using the new I/NAV features, as the overall TTFF is also impacted by the acquisition and tracking time.

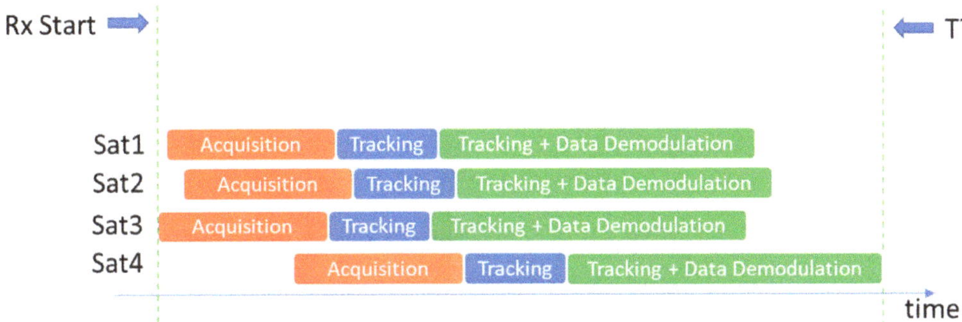

Figure 2. Data Demodulation and TTFF visual scheme.

For SSP, the receiver is reset via an automatic setup in warm start where a coarse time estimate is provided to the receiver after each reset (for the legacy time retrieval, the provided coarse time estimate has no effect). As these upgraded satellites broadcast both legacy and new words, it is possible for the receiver to choose which words to process. The RS, RedCED, and SSP tests' reset starting times are chosen randomly in order to evenly distribute the entry points of the I/NAV subframe.

3. Results

This section covers the results of the time needed for satellite data demodulation using the new words RS (words 17–20) and RedCED (word 16) in the cold start and the time needed for time convergence using SSP in the warm start.

3.1. Reed–Solomon and Reduced CED Results

The results in this section present the statistical value of Data Demodulation by computing the needed time for each satellite to retrieve the necessary CED words. The values of Time to Data (TTD) were grouped per scenario case (Urban or Open Sky) for the first four satellites retrieved per reset point for all reset points within each scenario. For each satellite, the legacy configuration computes CED 1–4, while the RS configuration computes any four different words from CED 1–4 + CED 17–20, and the RedCED + RS configuration computes either RedCED W16 or the same words as RS configuration. In order to fully benefit from RS CED, the receiver is not waiting for W05, which does contain Signal-In-Space (SIS) health status information, Broadcast Group Delay (BGD), Galileo System Time (GST), and Ionospheric correction data:

- Confirmation of healthy SIS status can be derived from the presence of W16. In case the SIS is flagged as unhealthy or marginal, W16 is not broadcast via the Galileo satellite and is replaced with W05.
- BGD information is generally stable over a long time and can be reused from information retrieved earlier and stored in the receiver.
- Galileo System Time can be recovered from SSP, assuming the receiver has a residual time uncertainty of +/− 3 s or can be retrieved from W05 decoded from any other satellite.
- Ionospheric correction data can be retrieved from W05 decoded from any other satellite or can be omitted at the cost of higher ranging error.

Figure 3a shows the results in the Open Sky environment. Looking at the 95 percentile of the data demodulation in Figure 3a, we see an improvement of 10 s (33.3%) using RS and 16 s (53.3%) using RS + RedCED. The Open Sky results are deterministic since, in ideal conditions, the retrieval of the CED words is a function of the I/NAV sub-frame layout. Therefore, the obtained results are useful to confirm the obtained theoretical values.

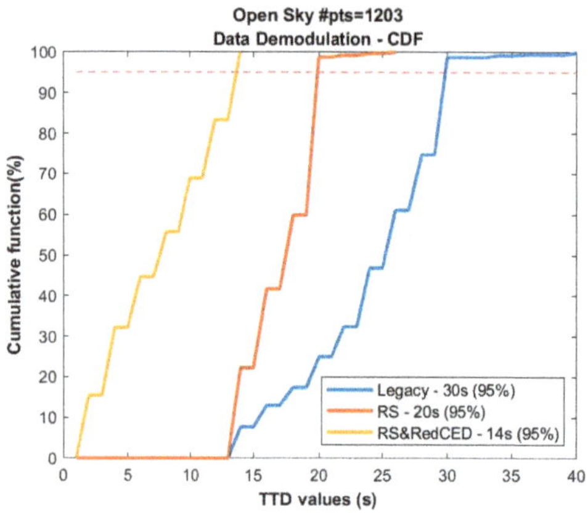

(**a**)

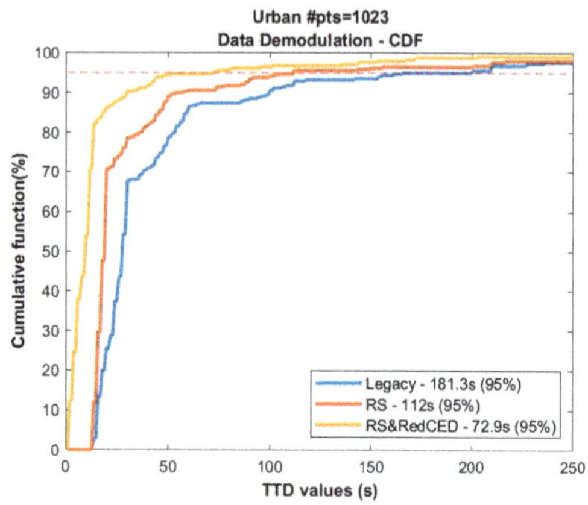

(**b**)

Figure 3. Comparison of legacy I/NAV to RS (I/NAV 17–20) and RedCED (I/NAV 16) Data Demodulation: (**a**) Open Sky CDF of Data Demodulation and (**b**) Urban CDF of Data Demodulation.

Figure 3b shows the results in an Urban environment. Compared to the Open Sky results, the increase in the data demodulation time in the Urban case is caused by the loss of words due to the loss of satellite visibility. Looking at the 95 percentile of the data demodulation in Figure 3b, we see an improvement of 69.3 s (38.2%) using RS and 108.4 (59.8%) using RS + RedCED.

As a complement to the displayed results, we have included the maximum and average data demodulation values for both Open Sky and Urban scenarios. It is important to note that the results for the Urban scenario are highly dependent on the specific dataset used and should be considered illustrative examples:

- Open sky:
    - Legacy: max = 40 s, average = 24.5 s.
    - RS: max = 26 s, average = 17.5 s.
    - RS + RCED: max = 14 s, average = 8 s.
- Urban
    - Legacy: max = 448 s, average = 48.1 s.
    - RS: max = 404 s, average = 35.5 s.
    - RS + RCED: max = 400 s, average = 18.9 s.

3.2. Secondary Synchronization Pattern Results

Figure 4a,b show the results in the Open Sky environment. Looking at the 95 percentile value of the time needed for time convergence in Figure 4a, we see an improvement of 5 s (35.7%) using SSP, while Figure 4b shows the histogram where the distribution of the values is represented.

Figure 4c,d show the results in an Urban environment. Looking at the 95 percentile value of the time needed to time convergence in Figure 4c, we see an improvement of 4.2 s (23.1%) using SSP, while Figure 4d shows the histogram where the distribution of the values is represented.

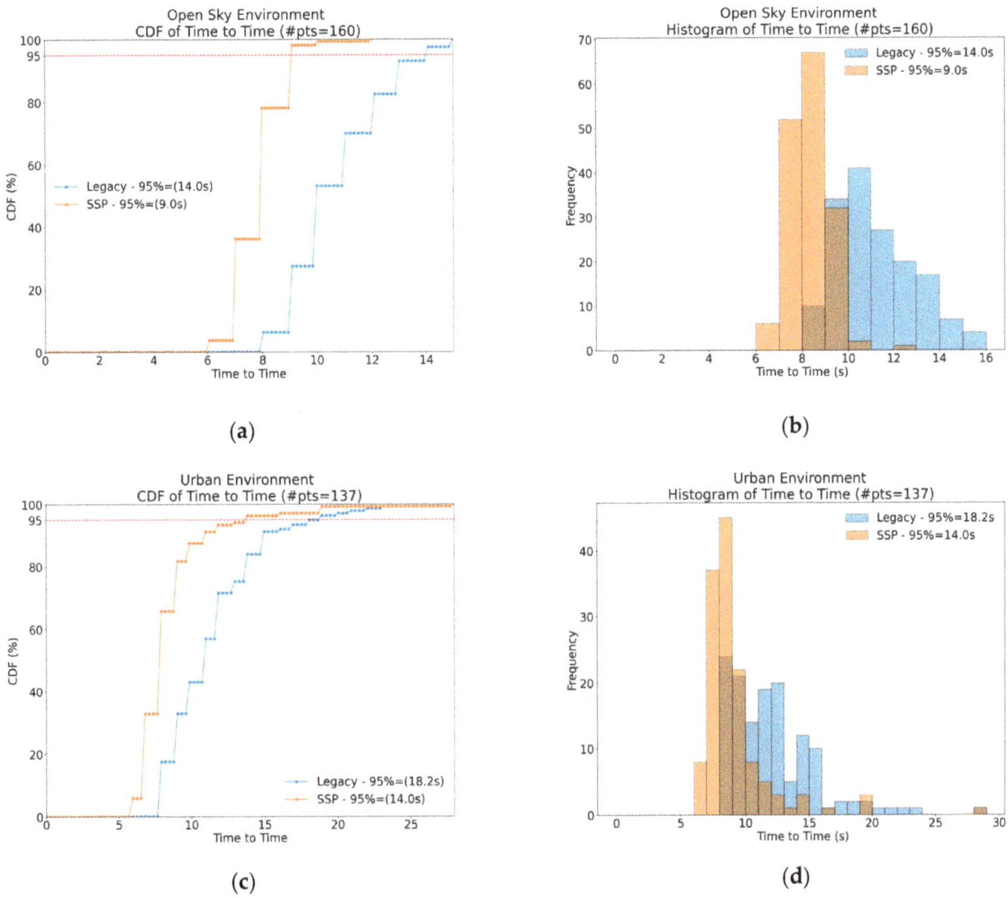

Figure 4. Comparison of legacy I/NAV to SSP time-to-time convergence: (**a**) Open Sky CDF, (**b**) Open Sky histogram, (**c**) Urban CDF, (**d**) Urban histogram.

4. Conclusions

Table 2 presents the summary of the results of RS and RedCED features in receiver cold start conditions. Overall, there is an improvement in data demodulation time from 33–38% using RS and 53–60% using RS + RedCED in Open Sky and Urban conditions, which results in an improvement in the TTFF value. The improvements are particularly important in Urban environments where the visibility is reduced and the signal is lost, resulting in receiver position reset and re-fix. Note that, as described previously, the number of satellites used was filtered in order to compare coherently the performance of the new features with the legacy features using the same broadcasting satellites. Therefore, particularly in urban conditions, the visibility was reduced to 4–6 simultaneous satellites, which has a big impact on CED computations, resulting in an increase in the absolute value of TTFF. Nonetheless, this exercise is very useful for retrieving the relative improvement when using the new features.

Comparing the obtained results with previously simulated data in [3], the values are in the expected range, confirming the simulated data with real data broadcast via the upgraded satellites.

Table 3 presents the summary of the results of time convergence using SSP when time is provided within 3 s error. The results vary from 23 to 36% in improvement in both Open Sky and Urban environments. The new SSP feature will result in faster re-synchronization with the Galileo System Time (GST) for the connected (or assisted) users.

Table 2. Results of 95 percentile Data Demodulation (DD) using RS and RedCED.

Test Case	DD: Legacy (s)	DD: RS (s)	DD: RS + RedCED (s)
Open Sky	30.0	20.0 (−33.3%)	14.0 (−53.3%)
Urban	181.3	112.0 (−38.2%)	72.9 (−59.8%)

Table 3. Time convergence results using SSP with coarse time provided.

Test Case	Legacy (s)	SSP (s)
Open Sky	14.0	9.0 (−35.7%)
Urban	18.2	14.0 (−23.1%)

Additionally, the presented results utilize a single-frequency signal, GAL E1B, which currently broadcasts the new features. However, please note that E5B is not currently broadcasting the new I/NAV features. If you use dual-frequency with E5B, it will enhance the performance of the legacy I/NAV system. This, in turn, will result in a reduced relative improvement when using RS and RCED.

Author Contributions: Conceptualization, R.D.N.; methodology, R.D.N.; software, R.D.N.; validation, R.D.N. and S.W.; formal analysis, R.D.N.; investigation, R.D.N.; resources, R.D.N.; data curation, R.D.N.; writing—original draft preparation, R.D.N.; writing—review and editing, S.W., N.S., S.S. and M.O.; visualization, R.D.N.; supervision, S.W., M.O. and N.S.; project administration, S.W., M.O. and N.S. All authors have read and agreed to the published version of the manuscript.

Funding: This research received no external funding.

Institutional Review Board Statement: Not applicable.

Informed Consent Statement: Not applicable.

Data Availability Statement: The data presented in this study are available on request from the corresponding author. The data cannot be made publicly accessible as they are subject to internal restrictions imposed by a dedicated test campaign.

Acknowledgments: The material and laboratory activities were supported by the ESA Navigation Laboratory. This work was also possible due to the support of the Galileo Receiver Team in Estec: Xurxo Otero, Luciano Musumeci, Matteo Fè, and César Vázquez.

Conflicts of Interest: The authors declare no conflicts of interest.

References

1. European GNSS (Galileo) Open Service Signal-In-Space Interface Control Document (OS SIS ICD), Issue 2.0, January 2021. Available online: https://www.gsc-europa.eu/sites/default/files/sites/all/files/Galileo_OS_SIS_ICD_v2.0.pdf (accessed on 28 November 2023).
2. Paonni, M.; Anghileri, M.; Burger, T.; Ries, L.; Schlötzer, S.; Schotsch, B.E.; Ouedraogo, M.; Damy, S.; Chatre, E.; Jeannot, M.; et al. Improving the Performance of Galileo E1-OS by Optimizing the I/NAV Navigation Message. In Proceedings of the 32nd International Technical Meeting of the Satellite Division of The Institute of Navigation (ION GNSS+ 2019), Miami, FL, USA, 16–20 September 2019; pp. 1134–1146. [CrossRef]
3. Peres, T.; Jimenez, D.; Burger, T.; Schotsch, B.E.; Anghileri, M.; Ouedraogo, M. A Novel Approach for Real-World Performance Assessment of Galileo Navigation Message Evolution. In Proceedings of the 31st International Technical Meeting of the Satellite Division of The Institute of Navigation (ION GNSS+ 2018), Miami, FL, USA, 24–28 September 2018; pp. 4079–4093. [CrossRef]

4. Schotsch, B.E.; Anghileri, M.; Ouedraogo, M.; Burger, T. Joint Time-to-CED Reduction and Improvement of CED Robustness in the Galileo I/NAV Message. In Proceedings of the 30th International Technical Meeting of the Satellite Division of The Institute of Navigation (ION GNSS+ 2017), Portland, OR, USA, 25–29 September 2017; pp. 1544–1558. [CrossRef]
5. Schlötzer, S.; Söllner, M.; Burger, T.; Ouedraogo, M.; Anghileri, M.; Bey, T. Reduced Clock and Ephemeris Data making use of equinoctial parameterization. In Proceedings of the Navitec 2018, ESA/ESTEC, Noordwijk, The Netherlands, 5–7 December 2018.

engineering proceedings

MDPI

Proceeding Paper

Reliability of Smartphone Positioning in Harsh Environment †

Giovanni Cappello [1], Ciro Gioia [2], Antonio Angrisano [3], Silvio Del Pizzo [4], Gabriele Portelli [4] and Salvatore Gaglione [1,4,*]

1 International Ph.D. Programme UNESCO Chair "Environment, Resources and Sustainable Development", Department of Science and Technology, "Parthenope" University of Naples, 80143 Naples, Italy; giovanni.cappello001@studenti.uniparthenope.it
2 Independent Researcher, 21020 Brebbia, Italy; cirogioia@tin.it
3 Department of Engineering, Messina University, 98166 Messina, Italy; angrisanoa@unime.it
4 Department of Science and Technology, "Parthenope" University of Naples, 80143 Naples, Italy; silvio.delpizzo@uniparthenope.it (S.D.P.); gabriele.portelli001@studenti.uniparthenope.it (G.P.)
* Correspondence: salvatore.gaglione@uniparthenope.it
† Presented at the European Navigation Conference 2023, Noordwijk, The Netherlands, 31 May–2 June 2023.

Abstract: Since 2016, it has been possible to record and collect GNSS (Global Navigation Satellite System) raw data on Android devices. This has been a game changer in low-cost device applications. Researchers in the GNSS field have tried to answer new questions regarding Android device positioning performance, the quality of their measurements, positioning techniques that could be applied, and methods to increase the accuracy and the reliability of the PNT (Position Navigation Timing) solution. Several research groups have demonstrated accurate positioning using smartphones while also showing the potential limitations of such devices. In particular, problems related to the GNSS antenna performance have been reported; indeed, in urban scenarios, where users typically operate, the presence of multiple outliers could make the navigation solution inaccurate, if not unfeasible. Hence, techniques useful for verifying the reliability of the navigation solutions have become fundamental. The reliability of the PNT solution provided by smartphones is an open research question. In this study, traditional RAIM (Receiver Autonomous Integrity Monitoring) algorithms were adapted to the case of smartphones. Navigation solution algorithms, including FDE (Fault Detection and Exclusion), were tested using a long data collection made by a smartphone located in a harsh environment in static mode. The performance of the proposed approaches was assessed in terms of horizontal and vertical errors, solution reliability, and residual distribution.

Keywords: RAIM; integrity; smartphone; subset; Forward–Backward; disturbed scenario

Citation: Cappello, G.; Gioia, C.; Angrisano, A.; Del Pizzo, S.; Portelli, G.; Gaglione, S. Reliability of Smartphone Positioning in Harsh Environment. *Eng. Proc.* **2023**, *54*, 44. https://doi.org/10.3390/ENC2023-15429

Academic Editors: Tom Willems and Okko Bleeker

Published: 29 October 2023

1. Introduction

The accurate and reliable positioning of low-cost devices is a current challenge in the Global Navigation Satellite System (GNSS) research field due to the low quality of the collected measurements. In particular, smartphones, the principal device of interest of this study, are used in everyday life for personal navigation, which is usually conducted in urban areas, wherein the presence of buildings can limit signal availability and introduce gross errors in observations, leading to unreliable position calculations. However, in 2016, thanks to the release of Android 7 (Nougat), users became able to collect raw GNSS data [1,2], making it possible for researchers to direct their efforts to the enhancements of smartphone positioning. Several research groups have studied and assessed the quality of observations and the positioning performance of different smartphone devices [3–5]. The low-cost receivers and antennas embedded in smartphones constitute the principal limitation to the positional accuracy of such devices, especially in signal-degraded environments where the presence of recurrent multipath phenomena strongly impacts the navigation solution. In such environments, the identification and exclusion of outliers becomes fundamental. A possible solution to this issue is the application of Receiver

Autonomous Integrity Monitoring (RAIM) algorithms. The benefits of these algorithms have been demonstrated in [6,7], and they can play key roles in detecting and excluding blunder-affected measurements. RAIM is a term that includes a family of user algorithms that, by comparing the consistency between measurements, calculate the integrity of the GNSS solution [8]. Thanks to a residual analysis, RAIM algorithms are capable of determining the most probable satellite whose measurement is blunder-affected [8] with a minimum of five visible satellites and to exclude it with a minimum of six visible satellites [9] in case of a single constellation.

RAIM was born out of aeronautics applications [9,10], but it has also been used in other fields, such as personal navigation [11,12]. The objective of this study was to apply classical RAIM algorithms to smartphone GNSS observables in order to detect and exclude faulty measurements and guarantee the reliability of the navigation solution. Specifically, two Fault Detection and Exclusion (FDE) algorithms were tested: Forward–Backward and Subset.

The integrity algorithm has been tested using real data in a typical signal-degraded scenario. The device adopted for this study was a Xiaomi Mi8 equipped with a Broadcom BCM47755 dual-frequency chip and able to acquire L1/E1 and L5/E5 GNSS signal frequencies [13]. For the proposed study, the introduced device was located in an obstructed environment using an ad hoc setup.

The performances of Single Point Positioning (SPP) with and without the application of RAIM are assessed in terms of horizontal and vertical error, availability (without RAIM), and reliable availability (with RAIM) percentage and residual distribution.

The remainder of this paper is structured as follows: Section 2 details the fundamentals of the SPP technique and RAIM algorithms and provides a brief overview on the main differences between the two adopted algorithms. Section 3 describes the conducted experiment. Section 4 discusses the results, and Section 5 concludes the paper.

2. Positioning and Integrity

SPP is a basic technique largely employed in the satellite navigation field. It enables one to estimate the position of a receiver exploiting pseudorange measurements, the classical equation of which is reported in (1):

$$\rho = d + c\delta t_r + \epsilon \quad (1)$$

where ρ is the pseudorange, d is the satellite-receiver geometrical distance, $c\delta t_r$ is the receiver clock offset (δt_r) (expressed in meters by multiplying it with the speed of light c), ϵ is a term that contains modeled (atmospheric, relativistic, satellite clock errors) and unmodeled errors (multipath, receiver hardware, noise).

Equations such as (1) are linearized around a set of n unknown parameters (whose number varies according to the number of the involved GNSSs), and through acquiring m measurements, the linearized equations can be expressed in a matrix notation as:

$$\underline{z} = H\underline{\Delta x} + \underline{\varepsilon} \quad (2)$$

where $\underline{z}$ represents the vector containing the m differences between observed and corrected pseudoranges (referred to as measurement vector), H is the $m \times n$ design matrix, $\underline{\Delta x}$ is the unknown vector containing the corrections needed to update the position estimated at the previous step, $\underline{\varepsilon}$ is the vector containing the unmodeled errors.

$\underline{\Delta x}$ is estimated using Weighted Least Squares (WLS) [9,14], the solution of which is reported in (3):

$$\underline{\Delta x} = \left(H^T W H\right)^{-1} H^T W \underline{z} \quad (3)$$

where W is the weighting matrix containing diagonal elements that are the inverse of the variance of each measure and whose determination is dependent on the adopted weighting strategy, which, in this work, is based on User Range Accuracy (URA), a parameter provided in the navigation message, and the satellite elevation.

The principal limitation of Single Point Positioning pertains to the reachable accuracy, which, in nominal conditions, is on the metric order [9]. In disturbed environments, where multipath phenomena are frequent events and the satellite visibility is limited, such accuracy can be worsened because the position could be computed considering blunder-affected measurements. In order to guarantee the integrity of the solution, the use of RAIM algorithms is necessary in critical environments in order to detect and eventually exclude erroneous measurements that lead to an inaccurate position. So, after the solution estimation, residuals are analyzed using the RAIM algorithm in order to perform "fault detection" to verify the presence of anomalous measurements and, with sufficient redundancy, "fault exclusion". Classical RAIM algorithms perform, as a first step, a Global Test (GT) that is based on the use of a decision variable D depending on the residual r and the weighting matrix W:

$$\begin{aligned} \underline{r} &= \underline{z} - H\underline{\Delta x} \\ D &= \underline{r}^T W \underline{r} \end{aligned} \tag{4}$$

D is compared with a threshold T, whose values depend on the required performance of RAIM, on the redundancy, and on the assumed behavior of D. If $D > T$, a probable outlier presence is assumed. So, in function of the adopted RAIM algorithms, a Local Test (LT) wherein the standardized residuals $\underline{w}$ are analyzed can be carried out:

$$w_i = \left| \frac{r_i}{\sqrt{CV_{ii}}} \right| \tag{5}$$

where CV_{ii} refers to the respective diagonal elements of the covariance matrix of the residuals. w_i is compared to a local threshold, and the largest one overpassing the threshold is assumed to be a blunder and excluded after a separability check. In the separability check, a coefficient based on the variance of the standardized residuals is analyzed and compared to a specific value (0.9 for this study) in order to avoid erroneous exclusions [15–17].

For this study, independently of the RAIM algorithm, a geometry check was performed before and after the application of RAIM using a threshold of 30 for PDOP, 25 for HDOP, and 20 for VDOP in order to declare the solution unreliable if the DOP (Dilution of Precision) conditions are not satisfied and to not perform the FDE process in bad geometrical conditions. Regarding integrity, the geometry can be further analyzed using the ARP method (the Approximate Radial-error Protected method) [14,18]. Furthermore, if the redundancy is not sufficient to perform fault detection, the solution is declared to be impossible to check and RAIM cannot be applied.

In Sections 2.1 and 2.2, the main differences between the RAIM Subset and RAIM Forward–Backward (FB) algorithms are assessed.

2.1. RAIM Subset

The RAIM Subset algorithm only carries out the Global Test. So, if a set of measurements is declared inconsistent, a series of subsets, in all the possible combinations, are re-checked. The subset passing the GT is declared consistent and, through it, the position is re-computed and declared reliable if the geometry check is passed. If any subset passes the GT, the solution is declared unreliable. The principal drawback of this algorithm is its computational heaviness. In fact, a large number of combinations could require a longer time of computation, especially in cases involving a high number of measurements [15–17].

2.2. RAIM Forward–Backward

RAIM FB performs both GT and LT, and it is based on two steps. The first one, referred to as "Forward", carries out the GT, and if the set is declared inconsistent, the LT is performed to identify and, after the separability check, remove erroneous measurements. The Forward step is repeated recursively until no erroneous measurements remain (or until the chosen maximum number of exclusions). If, after those steps, the solution is declared reliable and more than one blunder has been excluded, the second step, referred to as "Backward", is carried out in order to reinclude any measurements excluded in an erroneous way [15–17].

3. Test Setup

For the conducted experiment, about 14 hours of data collected from a Xiaomi Mi8, manufactured by Xiaomi Corporation [19] smartphone was used. The smartphone was equipped with a Broadcom BCM47755 receiver manufactured by Broadcom Inc [20], able to acquire double-frequency (L1 and L5) measurements from GPS, Glonass, Galileo, BeiDou. With SPP being a single-frequency and code-based positioning technique, only pseudorange observations on the first main frequency are used: L1 for GPS and E1 for Galileo. The Android application used to collect raw GNSS data was "rinex ON". The environment surrounding the device was obstructed, as shown in Figure 1, where, in the upper box, the PDOP (Position Dilution of Precision) values are not optimal, especially for the GPS only configuration from 04:41 to 07:01 of UTC time, where the value is within 2 and 6. PDOP is enhanced in the multi-constellation GPS/Galileo configuration, but its evolution in time clearly indicates an obstructed environment. In the lower box, the satellite visibility is reported. A satellite is considered visible only if its SNR is larger than 20 dB-Hz and its elevation is larger than 15 degrees. For GPS, the number of visible satellites varies between 5 and 11; in addition, rapid variation during the entire test and a fall in the last part was observed. For Galileo, the number of tracked satellites is more limited than GPS; indeed, the number of visible satellites was between 1 and 6.

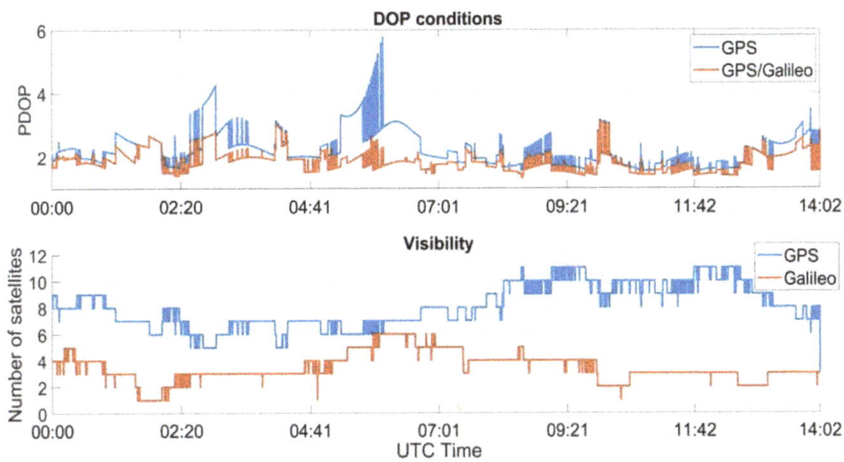

Figure 1. PDOP evolution (**upper box**) and satellite visibility (**lower box**) for the single and double constellation cases.

In Figure 2, the average C/N0 computed for each epoch is shown. From the figure, it can be noted that the parameter has a fast variability, confirming the presence of an obstructed context.

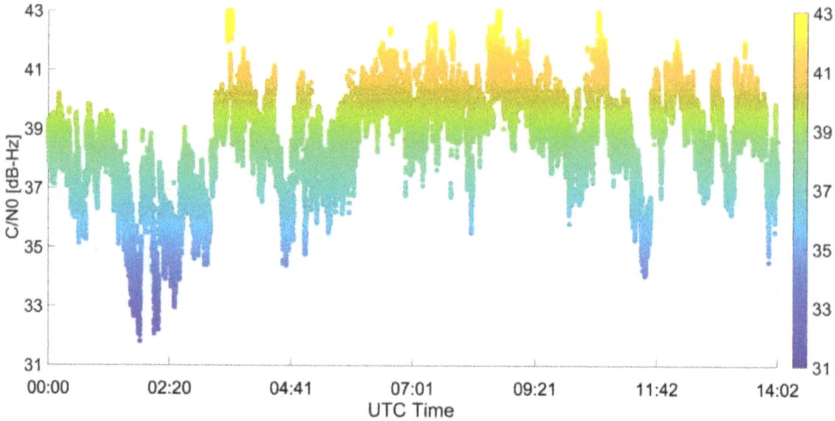

Figure 2. Average C/N0.

4. Results and Discussion

In this section, the results are discussed. The adopted configurations are referred to as "GPS No RAIM" and "GPS/Galileo No RAIM" for cases where RAIM was not applied, "GPS + SS" and "GPS/GAL + SS" for cases where the Subset (SS) algorithm was applied, and "GPS + FB" and "GPS/GAL + FB" for cases where the Forward–Backward (FB) algorithm was applied. Figure 3 shows a horizontal scatter plot for all the considered configurations. From the figure, it can be noted that the configurations with RAIM (yellow markers) have a significant reduction of horizontal errors. In addition, when the integrity algorithms are not able to identify and reject the erroneous measurements, they provide useful information to the user; no trust should be placed in the estimated solution. The unreliable solutions are marked with red markers; for the SS cases, a larger number of unreliable solutions are visible, leading to a decrease in reliability. The 95th percentiles are reported in both figures in black for the "No RAIM" case and in magenta for the "RAIM" case, while their values are reported in Table 1. For all configurations with RAIM, the 95th percentile (computed considering only reliable solutions) is smaller than the no RAIM case. This clearly shows the benefits of the application of the FDE algorithms. For the horizontal channel, the 95th percentile for the SS case is slightly lower than the FB case.

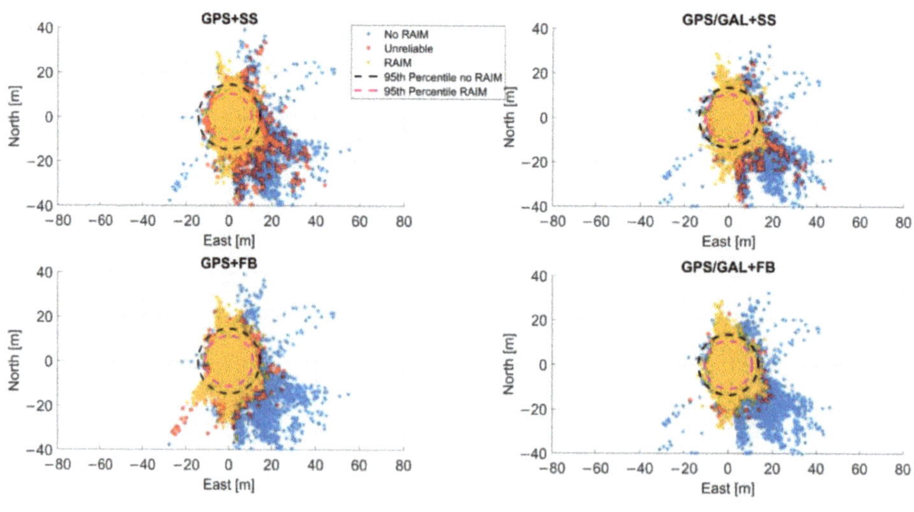

Figure 3. Horizontal scatter plot for all the considered configurations.

Table 1. The 95th Percentile of the horizontal and vertical errors for all the configurations.

Configuration	95th Percentile [m] Horizontal Error	95th Percentile [m] Vertical Error
GPS No RAIM	14.55	22.95
GPS + SS	10.43	17.16
GPS + FB	11.26	18.03
GPS/GAL No RAIM	13.53	19.52
GPS/GAL + SS	10.62	16.07
GPS/GAL + FB	10.74	15.62

In Figure 4, the time evolution of the vertical error for the four configurations is shown. Also, in the vertical channel, the benefits of the application of the FDE techniques are evident. In all the cases, the yellow lines are lower than the blue lines, demonstrating a reduction in the vertical error when RAIM is applied.

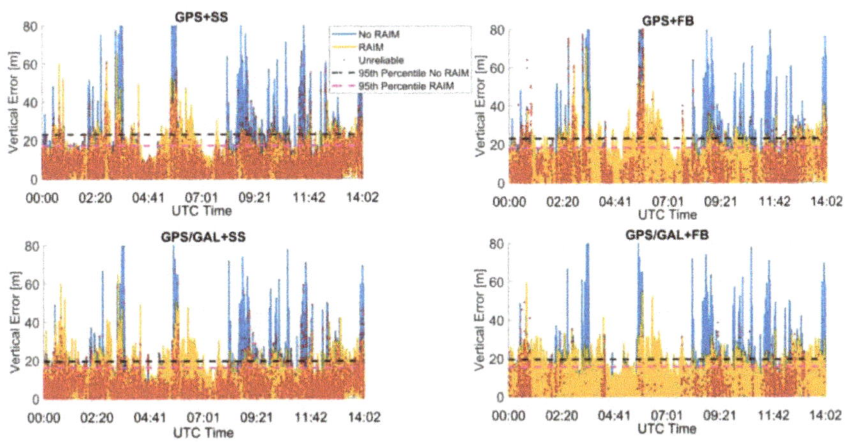

Figure 4. Vertical error as a function of time for all the considered configurations.

The reliable availability (or reliability) of the adopted configurations is shown is Figure 5, where the blue bars indicate the percentage of reliable solutions, and the yellow bars indicate the percentage of unreliable solutions. The "unreliable" solutions pertain to the following cases:

- Insufficient redundancy;
- Failure of the geometry check executed before and after the application of RAIM;
- Exceeding the horizontal and vertical alarm limits;
- The failure of the RAIM statistical tests (which, in this study, was the main cause of the solutions unreliability).

As can be seen from Figure 5, both in single and dual constellations, the FB algorithm provided better reliability (89.58% for GPS and 96.28% for GPS/Galileo). The RAIM Subset cases reached a reliable availability of 70.43% for GPS and 82.28% for GPS/Galileo, probably because of the drawback that characterizes this algorithm. The Subset scheme, in the presence of an high number of measurements, is not able to find a unique subset passing the GT.

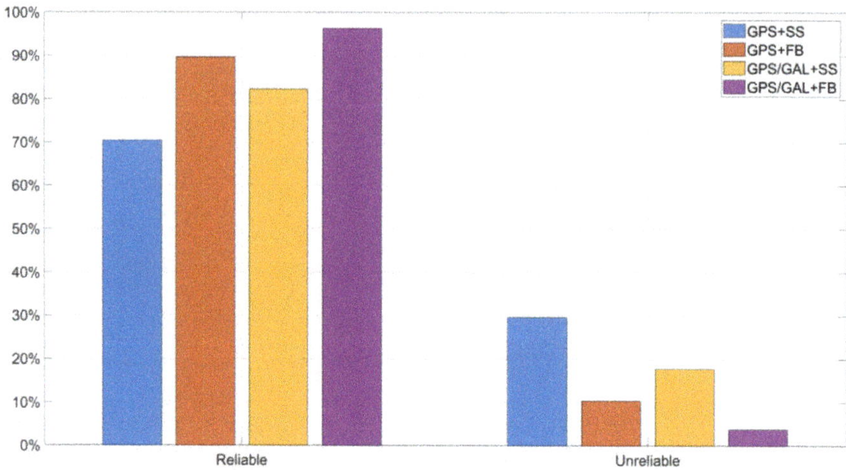

Figure 5. Reliable availability for all the considered configurations.

In order to better evaluate the performance of the used integrity algorithms, Figure 6 reports the maximum residual distribution with and without RAIM (on the left) and Cumulative Distribution Function (CDF) (right boxes) for the four considered configurations. It is important to specify that, for such graphics, the absolute value of maximum residuals vector $\underline{r}$, calculated as in (4), is considered. The residual distribution is closer to zero and with a reduced variation in the Forward–Backward case both in single and dual constellations, as also confirmed by CDF. Minor but present improvements can be noticed also for the Subset case, whose CDF is closer to the no RAIM case, showing its minor effectiveness in this test.

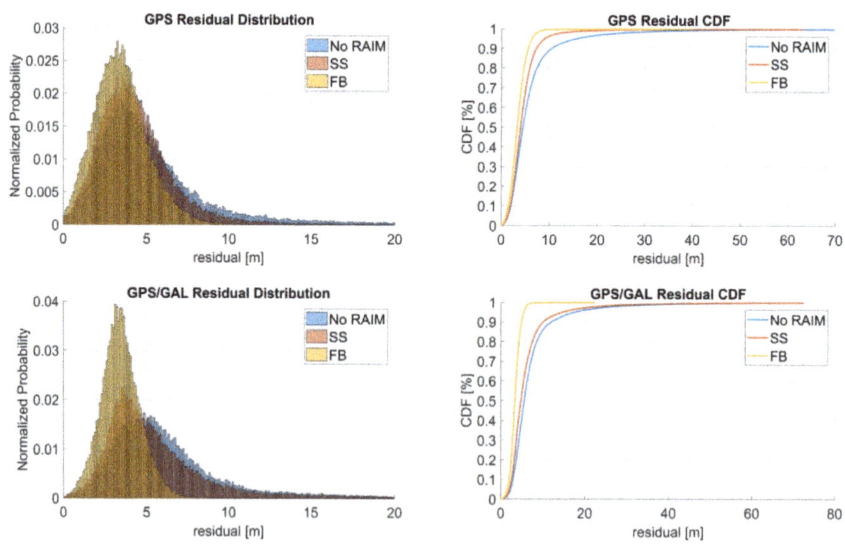

Figure 6. Residual distribution (**on the left**) and CDF (**on the right**) for all the adopted configurations.

5. Conclusions

This study involved analyzing the performance of two traditional RAIM algorithms (i.e., Subset and Forward–Backward) on the positioning of a smartphone in an obstructed environment in static mode. About 14 hours of data were processed, with and without the application of RAIM, for a total of six configurations, including GPS and Galileo measurements. From the test, it emerged that RAIM benefits are evident in difficult scenarios thanks to the exclusion of blunder-affected measurements. For the SS algorithm, a high number of unreliable solutions were observed. In the performed test, better results were provided by the FB algorithm due to its higher reliability percentage and lower horizontal and vertical errors.

Possible extensions of this study could involve simultaneous comparisons of different devices and the adoption of additional integrity algorithms.

Author Contributions: Conceptualization, A.A., G.C., S.D.P., S.G., C.G. and G.P.; methodology, A.A., G.C., S.D.P., S.G., C.G. and G.P.; software, A.A., G.C., S.D.P., S.G., C.G. and G.P.; validation, A.A., G.C., S.D.P., S.G., C.G. and G.P.; formal analysis, A.A., G.C., S.D.P., S.G., C.G. and G.P.; investigation, A.A., G.C., S.D.P., S.G., C.G. and G.P.; resources, A.A., G.C., S.D.P., S.G., C.G. and G.P.; data curation, A.A., G.C., S.D.P., S.G., C.G. and G.P.; writing—original draft preparation, A.A., G.C., S.D.P., S.G., C.G. and G.P.; writing—review and editing, A.A., G.C., S.D.P., S.G., C.G. and G.P.; visualization, A.A., G.C., S.D.P., S.G., C.G. and G.P.; supervision, A.A., G.C., S.D.P., S.G., C.G. and G.P. All authors have read and agreed to the published version of the manuscript.

Funding: This research received no external funding.

Institutional Review Board Statement: Not applicable.

Informed Consent Statement: Not applicable.

Data Availability Statement: The data presented in this study are available on request from the corresponding author.

Conflicts of Interest: The authors declare no conflict of interest.

References

1. European Union Agency for the Space Programme (EUSPA). GNSS Raw Measurements Delivering Greater Accuracy. Available online: https://www.euspa.europa.eu/newsroom/news/gnss-raw-measurements-delivering-greater-accuracy (accessed on 3 May 2023).
2. GPS World. Google Opens up GNSS Pseudoranges. Available online: https://www.gpsworld.com/google-opens-up-gnss-pseudoranges/ (accessed on 3 May 2023).
3. Zhang, X.; Tao, X.; Zhu, F.; Shi, X.; Wang, F. Quality assessment of GNSS observations from an Android N smartphone and positioning performance analysis using time-differenced filtering approach. *GPS Solut.* **2018**, *22*, 70. [CrossRef]
4. Lambert, W.; Anja, H. GNSS code and carrier phase observations of a Huawei P30 smartphone: Quality assessment and centimeter-accurate positioning. *GPS Solut.* **2020**, *24*, 64.
5. Robustelli, U.; Paziewski, J.; Pugliano, G. Observation Quality Assessment and Performance of GNSS Standalone Positioning with Code Pseudoranges of Dual-Frequency Android Smartphones. *Sensors* **2021**, *21*, 2125. [CrossRef] [PubMed]
6. Angrisano, A.; Gaglione, S. Smartphone GNSS performance in an urban scenario with RAIM application. *Sensors* **2022**, *22*, 786. [CrossRef] [PubMed]
7. Angrisano, A.; Gaglione, S.; Gioia, C. RAIM algorithms for aided GNSS in urban scenario. In Proceedings of the 2012 Ubiquitous Positioning, Indoor Navigation, and Location Based Service (UPINLBS), Helsinki, Finland, 3–4 October 2012; pp. 1–9.
8. ESA Navipedia. RAIM. Available online: https://gssc.esa.int/navipedia/index.php/RAIM#RAIM_Concept (accessed on 3 May 2023).
9. Kaplan, E.D.; Hegarty, C.J. *Understanding GPS. Principles and Applications*, 2nd ed.; Artech House: Norwood, MA, USA, 2006.
10. Teunisses, P.J.; Montenbruck, O. *Springer Handbook of Global Navigation Satellite Systems*; Springer International Publishing: Cham, Switzerland, 2017.
11. Zishen, L.; Liang, W.; Ningbo, W.; Ran, L.; Ang, L. Real-time GNSS precise point positioning with smartphones for vehicle navigation. *Satell. Navig.* **2022**, *3*, 19.
12. Amarildo, H.; Harris, P.; Thanassis, M.; Vassilis, G. Testing of a Combined Hatch Filter / RAIM Algorithm for SPP Smartphone Kinematic Positioning in GNSS Harsh Environments. In Proceedings of the 2nd Symposium of IAG Commission 4 "Positioning and Applications", Potsdam, Germany, 5–8 September 2022.
13. EUSPA—European Union Agency for the Space Programme. «World's First Dual-Frequency GNSS Smartphone Hits the Market» 2018. Available online: https://www.euspa.europa.eu/newsroom/news/world-s-first-dual-frequency-gnss-smartphone-hits-market (accessed on 15 May 2023).
14. Parkinson, B.W.; Spilker, J.J.J. *Global Positioning System: Theory and Applications*; American Institute of Aeronautics and Astronautics, Inc.: Reston, VA, USA, 1996.

15. Kuusniemi, H. User-Level Reliability and Quality Monitoring in Satellite-Based Personal Navigation. Ph.D. Thesis, Tampere University of Technology, Tampere, Finland, 2005.
16. Kuusniemi, H.; Wieser, A.; Lachapelle, G.; Takala, J. User-level reliability monitoring in urban personal satellite navigation. *IEEE Trans. Aerosp. Electron. Syst.* **2007**, *43*, 1305–1318. [CrossRef]
17. Angrisano, A.; Gioia, C.; Gaglione, S.; Del Core, G. GNSS Reliability Testing in Signal-Degraded Scenario. *Int. J. Navig. Obs.* **2013**, *2013*, 870365. [CrossRef]
18. Chin, G.Y.; Kraemer, J.H.; Brown, R.G. GPS RAIM: Screening Out Bad Geometries Under Worst-Case Bias Conditions. *Navigation* **1992**, *39*, 407–427. [CrossRef]
19. Xiaomi. About Us. Available online: https://www.mi.com/global/about/ (accessed on 29 November 2023).
20. Broadcom. Company History. Available online: https://www.broadcom.com/company/about-us/company-history (accessed on 29 November 2023).

engineering proceedings

MDPI

Proceeding Paper

A Comparative Experimental Performance Assessment of RTK+OSNMA-Based Positioning for Road Vehicle Applications †

Roman Lesjak [1,*], Susanne Schweitzer [1,‡], José M. Vallet García [2], Karl Diengsleder-Lambauer [3] and Selim Solmaz [3]

[1] DIGITAL—Institute for Digital Technologies, JOANNEUM RESEARCH Forschungsgesellschaft mbH, A-8010 Graz, Austria; susanne.schweitzer@joanneum.at
[2] Finnish Geospatial Research Institute, National Land Survey of Finland, FI-02150 Espoo, Finland; jose.vallet@nls.fi
[3] Virtual Vehicle Research GmbH, A-8010 Graz, Austria; karl.lambauer@v2c2.at (K.D.-L.); selim.solmaz@v2c2.at (S.S.)
* Correspondence: roman.lesjak@joanneum.at; Tel.: +43-316-876-5318
† Presented at the European Navigation Conference 2023, Noordwijk, The Netherlands, 31 May–2 June 2023.
‡ Presented the paper at the conference.

Abstract: To realize the societal need for greener, safer, and smarter mobility, ambitious technical challenges need to be addressed. With this aim, the H2020-EUSPA project ESRIUM investigates various aspects of highly accurate, reliable, and assured EGNSS localization information for road vehicles with a particular focus on automated vehicles. To analyze the achievable accuracy, reliability, and availability of multi-frequency and multi-GNSS mass-market receivers, we have conducted test drives under different GNSS reception conditions. In the tests, special focus was placed on using the Galileo Open Service Navigation Message Authentication (OSNMA) service, offering an additional feature for assured PVT (position, velocity, and time) information with respect to spoofing. We analyzed the performance of three Septentrio Mosaic-X5 receivers operated with different OSNMA settings. It could be shown that strict use of OSNMA provides very good positioning accuracy as long as sufficient suitable satellites are available. However, the overall performance suffers from a reduced satellite number and is therefore limited. The performance of a receiver using authenticated Galileo with GPS signals (final status of Galileo OSNMA) is very good for a mass-market receiver: 92.55% of the solutions had a 2D position error below 20 cm during 8.5 h of driving through different environments.

Keywords: EGNSS localization; navigation; Galileo Open Service Navigation Message Authentication (OSNMA); automated driving

Citation: Lesjak, R.; Schweitzer, S.; Vallet García, J.M.; Diengsleder-Lambauer, K.; Solmaz, S. A Comparative Experimental Performance Assessment of RTK+OSNMA-Based Positioning for Road Vehicle Applications. *Eng. Proc.* **2023**, *54*, 45. https://doi.org/10.3390/ENC2023-15434

Academic Editors: Tom Willems and Okko Bleeker

Published: 20 December 2023

1. Introduction

Signals from the Global Navigation Satellite Systems (GNSS) are used in many modern devices and various applications, such as navigation, surveying and mapping, satellite communication, emergency response, location-based services, and precise time services, among many others. The robustness and resilience of the GNSS signals are very important for those applications, especially in safety-critical applications. There are particular concerns regarding the spoofing of signals, which means that a GNSS receiver can be deceived by fake GNSS signals in an attempt to modify the Position, Velocity, and Time (PVT) solution of the target receiver.

One method for preventing spoofing is the authentication of signals, which enables ensuring that a message is identical to the one transmitted at its origin and that it was generated by a trusted source [1]. The Galileo Open Service Navigation Message Authentication (OSNMA) is the first service providing this authenticity check in the civilian sector.

It has currently been in the Public Observation (PO) test phase for slightly longer than a year. It uses a navigation message authentication scheme based on the Timed Efficient Stream Loss-tolerant Authentication (TESLA) protocol [2] and enables authentication at the receiver end in a manner that virtually eliminates the need for a chain of trust with dependence on third-party services. This offers the opportunity for Galileo to become the reference for many diverse new applications requiring authenticated positions.

Galileo OSNMA signals may also play a vital role in the field of road vehicle navigation. In this respect, the ESRIUM project, funded in the context of Horizon 2020, aims at creating road wear maps with accurate information about the position and shape of road damage, and to send prompt and real-time notifications to drivers and autonomous vehicles with instructions to avoid the damaged areas and route recommendations (e.g., change of lanes and driving offsets with respect to the lane center) to even out the road wear [3]. ESRIUM relies in this regard on Galileo services for increasing the positioning accuracy of vehicles receiving these notifications, and for authenticating the position estimates using Galileo OSNMA. The overall goal is to increase the safety and robustness of the localization solution and to reduce road maintenance costs as well as improve road safety and comfort at the same time [4].

Within the scope of ESRIUM, the Finnish Geospatial Research Institute (FGI) developed an open-source implementation of Galileo's OSNMA, called *FGI-OSNMA* [5], which distinguishes itself, e.g., by modularity and real-time usability. Other implementations also exist already, for example *Osnmalib* [6] and *galileo-osnma* [7]. The Septentrio company included support for OSNMA in some of their products, such as the multi-constellation GNSS receiver module Mosaic-X5 [8]. Theoretical work regarding OSNMA is provided by several authors, e.g., [2,9–13]. Practical performance assessments are reported by, e.g., [14–25]. From an experimental performance assessment of Galileo OSNMA within the scope of ESRIUM, it was observed in a four day-long static observation that the receiver could produce authenticated signals 99.74% of the time under open sky conditions [26]. The authentication status could not be attained in cases with a low number of satellites transmitting OSNMA data to the receiver. Therefore, it was concluded that the authentication performance will significantly decrease in environments with degraded satellite visibility, such as urban environments. Hence, with regard to OSNMA, it is beneficial to also use data from satellites with low elevation, as this increases the chance to cross-authenticate visible satellites.

In the current paper, we analyze the experimental performance of RTK+OSNMA-based positioning for road vehicle applications by comparing the PVT solutions of three different Mosaic-X5 receivers that were operated using three different OSNMA configurations (for details see Section 2). We conducted about 8.5 h of test drives under different driving environments, including open sky, motorways, rural roads, hilly and forested environments, and suburban and urban regions. We analyze the overall performance aggregated over all the environments regarding the proportion of Real-Time Kinematics (RTK) solutions and the horizontal position error. A detailed description of the experimental settings is provided in Section 2 and the results are shown in Section 3.

2. Experimental Settings

A sensor car shown in Figure 1 provided by Virtual Vehicle Research GmbH (Graz, Austria) was equipped by the study consortium members with three Septentrio Mosaic-X5 GNSS receivers in different configurations and an iMAR iNAT-FSLG-01 inertial measurement system (IMU) with gyro compassing capability (high-end fiber-optic gyroscope) and an integrated GNSS receiver (Novatel OEM729-2.01). All GNSS receivers were connected to the same antenna using an active one-by-four antenna splitter. The used antenna model was a NavXperience 3G+C multi-band GNSS antenna. The architecture of this setup is shown in Figure 2. The test vehicle was equipped with the DataSpeed drive-by-wire kit (https://www.dataspeedinc.com/, accessed on 15 May 2023), which enables access to most CAN data as well as the control actuators for ADAS/AD system-development purposes.

The vehicle was also equipped with additional sensors, which could be modified depending on the measurement or the use-case requirements.

Figure 1. Test vehicle from Virtual Vehicle Research GmbH and the measurement setup visuals.

The GNSS receivers were connected to the so-called car PC (industrial PC installed in the vehicle) for power supply and data monitoring. The car PC was powered over an inverter belonging to the vehicle's DataSpeed drive-by-wire kit. The connection between the PC and the X5 receivers was achieved via USB and ethernet (cf. Figure 2). A control laptop was also connected via ethernet.

All three Mosaic-X5 receivers received the GNSS signals via the same antenna and operated in the RTK mode using the incoming RTCM messages via 4G mobile internet. To guarantee that all receivers are receiving identical RTCM data with the same latency, Septentrio Data Link software was used to act as an RTCM splitter, acting as an NTRIP client to receive correction data in the form of virtual reference station data and forward the data to all receivers via a TCP stream. GNSS receivers operating in theh RTK mode can either output RTK fixed solutions, where the carrier phase ambiguities could be solved (estimated as an integer value), which provides the best positioning accuracy, or RTK float solutions, where the carrier phase ambiguities could be estimated as float values only, resulting typically in less accurate position information. When the receiver could not compute RTK positions, it fell back to so-called Single Point Positioning (SPP), where only data from the satellites were used in the PVT computation. The resulting positions were saved together with relevant information such as the type of solution attained, the quality of the solution, the satellites that were used for the position estimation, and so forth. The data were stored using the SBF protocol with a sampling rate of 10 Hz.

The Mosaic-X5 receivers are one of the first mass-market receivers natively supporting Galileo OSNMA. For the performance analysis, one receiver was configured to use all GPS L1/L2 + Galileo E1/E5a satellite signals (called *OSNMA off*) for the PVT solution. The second receiver was configured in the OSNMA loose configuration. According to Septentrio, this means using all GPS signals and Galileo signals from satellites with authenticated navigation messages or unknown status. The third receiver was configured to use no GPS signals and only authenticated Galileo signals (called *OSNMA strict*). The reason for not using either Glonass or Beidou is that in the final stage, the Galileo OSNMA service will support Galileo and GPS, and hence, the OSNMA loose configuration should give an estimate of the final performance of Galileo OSNMA. The Galileo navigation data were recorded including I/NAV messages containing the OSNMA-related data bits.

The measurement data from the FOG-based EGNSS/IMU iMAR iNAT FSLG-01 system were post-processed to serve as a highly accurate reference solution for data analysis. Therefore, the IMU data rate was recorded at 500 Hz and the GNSS data at 1 Hz.

The sensor car with the installed equipment described above was then used in several test drives. In total, during four test drives, we recorded about 8.5 h of measurement data using the receiver settings as described above. For the study at hand, we analyzed the data in post-processing to investigate the accuracy of the PVT solution under the different OSNMA settings. The results are shown in Section 3.

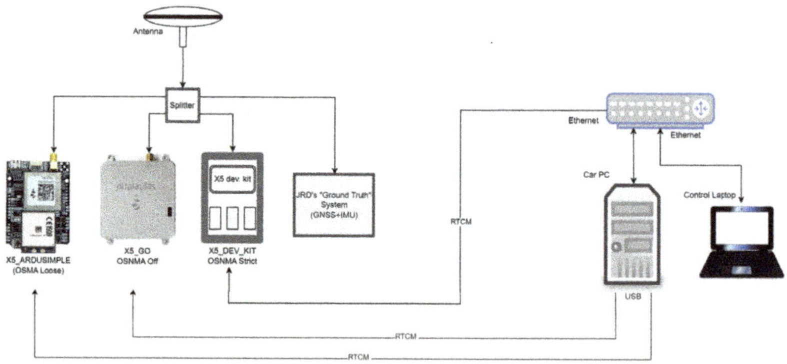

Figure 2. Scheme of the setup for OSNMA performance measurements.

3. Results and Discussion

In the analysis, the performance of the three solutions in terms of the 2D position error and fix type was compared, also in relation to the number of satellites. The current limitations of receivers using signals from only OSNMA-authenticated satellites in terms of availability and accuracy are discussed in relation to potential use cases involving highly automated driving.

Figure 3 shows the proportion of position types encompassing all driving conditions, ranging from open sky in motorways and rural roads, through hilly and forested environments, to suburban and urban areas. It can be seen that the percentage of fixed solutions (RTK fix) was 89.92% when all GPS L1/L2 and Galileo E1/E5a satellite signals were used (OSNMA off). The percentage was slightly lower in the OSNMA loose configuration (89.59%).

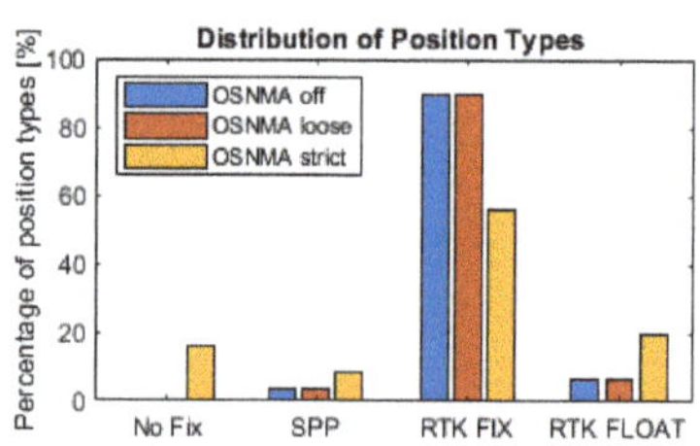

Figure 3. Proportions of position types.

Regarding the receiver with OSNMA strict settings (use of authenticated Galileo signals only, no GPS signals), the percentage of (now authenticated) RTK-fixed solutions was dramatically reduced to only 55.77%. This is obvious since the number of available satellites is significantly reduced, resulting in a significant reduction of the percentage of authenticated fixes (RTK fix under the OSNMA strict settings). Moreover, and as analyzed in [26], the use of cross-authentication in OSNMA can lead to having visible but not authenticated Galileo satellites, which therefore will not take part in the PVT computation. Although to a lesser extent, this can also contribute to the overall satellite visibility reduction when OSNMA is in strict mode.

In 16.17% of the strict cases even no PVT solution at all (no-fix) could be achieved, which was most probably due to the reduced number of visible satellites. The probability of an SPP solution was 8.65% for the OSNMA strict case. For the other OSNMA configurations, the probability for an SPP solution was only about half as small (3.72% for OSNMA off, 3.53 for OSNMA loose). These results clearly underline that it is very important to expand the network of authenticated satellites so that the OSNMA service can be used reliably.

When considering Figure 4, one can derive that the percentage of RTK fixes increases with the number of satellites. Simultaneously, the percentage of RTK float solutions decreases. The percentage of authenticated fixes (OSNMA strict) is best when 9 to 10 authenticated satellites are available. There were no cases with more then 10 authenticated satellites. More than 10 satellites could only be used by applying the OSNMA off and loose configurations. When further increasing the number of satellites, the percentage of RTK fix solutions improved slightly and the percentage of SPP solutions decreased for both configurations. This can even be seen when increasing the number of satellites from 13–14 to 15–18. The increase in this case accounts for 1.52% in the OSNMA off configuration, and for 2.24% in the loose configuration. Most RTK-fix solutions could be achieved in the OSNMA loose configuration, with a probability of 97.06%. This value displays the performance that can presumably be achieved with the final Galileo OSNMA service. This is a very promising result in view of autonomous driving vehicles.

In general, the performance of the receivers with the OSNMA off and loose configurations was very similar, regardless of the number of visible satellites.

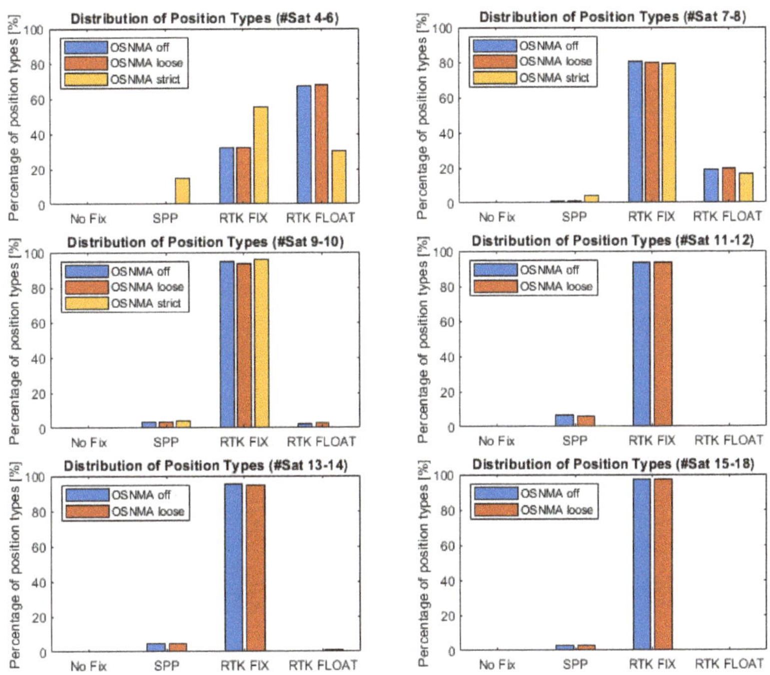

Figure 4. Proportion of position types depending on the number of satellites. (**Top left**): 4 to 6 satellites; (**top right**): 7 to 8 satellites; (**middle left**): 9 to 10 satellites; (**middle right**): 11 to 12 satellites; (**bottom left**): 13 to 14 satellites; (**bottom right**): >=15 satellites.

The localization accuracy was investigated by looking at the distribution of the two-dimensional (horizontal) localization error for the different OSNMA configurations. The left side of Figure 5 shows the Cumulative Distribution Function (CDF) for an open sky case. An autonomous vehicle was driving there, and all three receivers could achieve RTK-fix for the whole time. As can be seen, the 2D localization accuracy was always better than 5 cm. Totals 72.62% of the OSNMA strict solutions, 86.75% of the OSNMA loose solutions, and 86.86% of the OSNMA off solutions had an accuracy of better than 2 cm. These results show that under open sky conditions a vehicle can safely be guided within a traffic-lane, and this is also in agreement with [27–29].

The right hand side of Figure 5 shows CDFs for data aggregated over different driving environments. This is a result from drives where we passed motorways, hilly and forested rural roads, a mountain, and suburban and urban areas. The 2D errors are obviously significantly larger due to the more challenging environments. Hence, an RTK-fixed solution was not always possible, but all solution types occurred (no-fix, SPP, RTK float and fix). The worst performance was achieved in the OSNMA strict mode, which is mainly due to the limited satellite visibility and availability of authenticated Galileo satellites. Only 73.2% of the data had an error of less than 20 cm. For the OSNMA off case, 92.47% of the PVT solutions were better than 20 cm, and the OSNMA loose case again performed slightly better with 92.55% of the solutions being better than 20 cm. This is a very good performance for a mass-market receiver.

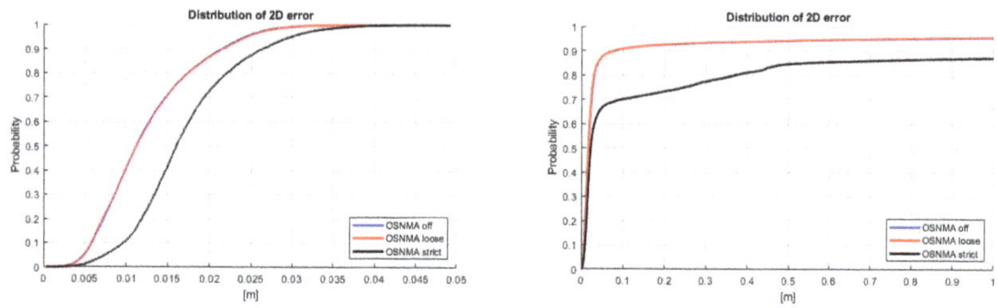

Figure 5. CDF functions of the 2D localization error. (**Left**): open sky (special training area); (**right**): different driving environments (open sky in motorways and rural roads, hilly and forested, urban and suburban).

Looking more into detail, one can observe that the localization accuracy typically increases with the number of satellites used during the PVT calculation. The general behavior of the 2D errors is shown in Figures 6 and 7. When the satellite number was between 4 to 6 (top left), the solutions for all OSNMA configurations performed moderately well. The OSNMA strict solution performed slightly better in the relevant accuracy range (below 50 cm). This most probably resulted from the fact that the satellite visibility is better in situations where 4–6 Galileo satellites are visible instead of when 4–6 GPS+Galileo satellites are visible (comparing solutions with a different number of visible satellites does not mean comparing the performance at the same locations!). In addition, Galileo signals are typically better than GPS ones, especially in multipath conditions. For the OSNMA strict configuration, 65.37% of the 2D errors were smaller than 20 cm, compared to 61.16% for OSNMA loose and 58.19% for OSNMA off. The authors would expect that the advantage of the OSNMA strict solution over the others should be maintained, but this was not the case for the group of 7 to 8 visible satellites. The reason for it is unclear. In summary, the performance significantly improved when having 7–8 satellites. For the use of 9 to 10 satellites, the OSNMA strict performance was extraordinarily good (98.00% of the errors were below 20 cm). This is obvious because having 9 or 10 satellites of one constellation means that the receiver was in a very open area compared to having 9 or 10 satellites of two constellations.

Further increasing the satellite number to more than 10 satellites only slightly improved the positioning accuracy. In principle, one can see nearly no difference between the OSNMA loose and off configurations. This indicates that nearly all Galileo satellites could be authenticated (or the status was unknown). The combined GPS+Galileo solutions with 15 or more satellites have a comparable performance to that of the OSNMA strict solution with 9 or 10 satellites, which is obvious since most probably, the receiver was in a very open region.

The performance of the OSNMA loose configuration gives an indication of what most probably can be achieved with the Galileo OSNMA service when it is fully operational. This is a very promising result in view of highly automated vehicles.

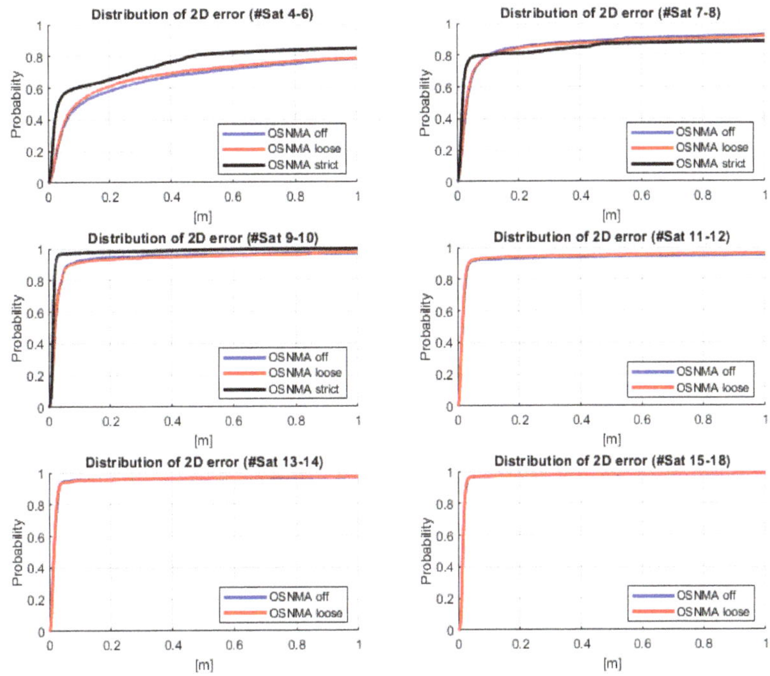

Figure 6. CDF functions of the 2D localization error depending on the number of satellites. (**Top left**): 4 to 6 satellites; (**top right**): 7 to 8 satellites; (**middle left**): 9 to 10 satellites; (**middle right**): 11 to 12 satellites; (**bottom left**): 13 to 14 satellites; (**bottom right**): >=15 satellites.

# of used satellites	strict	loose	off
4–6	65.4%	61.1%	58.2%
7–8	81.3%	84.4%	85.1%
9–10	98.0%	93.5%	94.5%
11–12	N/A	94.3%	93.6%
13–14	N/A	95.8%	95.9%
15–18	N/A	97.7%	97.7%

Figure 7. Probabilities of 20 cm 2D position error depending on the number of satellites.

Due to some issues with the wireless connection in other tests, we were also able to investigate the position accuracy during periods without correction data. For that matter, the receiver was configured to use base station data with ages up to 50 s. In Figure 8, on the left we see the 2D position error against the mean correction's age showing only RTK-fix solutions, while on the right side, one can see the 2D position errors for RTK-float solutions. In both figures, within the first seconds, one can see the errors that are typically encountered in urban and suburban situations. Afterwards, one can see the influence of older correction data on the RTK solution (mostly for highway and rural road conditions). As long as an integer fix is possible, one can be quite sure that the 2D position error stays below 20 cm up to 50 s, which was not expected. For float solutions, it can stay as good as 50 cm if the initial float solution is stable, and otherwise the errors can be as large as 22 m (maximum observed 2D error of float solution).

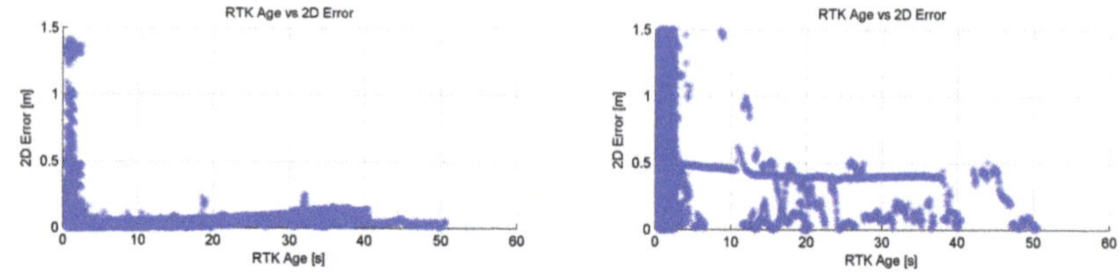

Figure 8. Influence of RTK correction data on the achievable 2D position accuracy. (**Left**): 2D localization errors of RTK-fix solutions; (**right**): 2D localization errors of RTK-float solutions.

4. Conclusions

Summarizing, within the H2020-EUSPA project ESRIUM, test drives were performed to investigate the achievable localization accuracy in different Galileo OSNMA configurations for road vehicles with a particular focus on automated vehicles. Therefore, three mass-market GNSS receivers of the type Septentrio Mosaic-x5 were installed in a test vehicle and configured differently with respect to Galileo OSNMA. The results show that the performance of the receiver is quite good for a mass-market product. Currently, OSNMA can only authenticate Galileo satellites. Using only authenticated Galileo satellites (OSNMA strict mode), 73.2% of the solutions had a 2D position error below 20 cm during 8.5 h. For the OSNMA loose configuration (using in addition GPS satellites), by contrast, 92.55% of the solutions had a 2D localization error below 20 cm. The probability of achieving an RTK-fix solution in the OSNMA loose configuration was 97.06%. The OSNMA loose configuration is probably the performance that can be achieved in common environments with the Galileo OSNMA service when it is fully operational. In total, the results coincide with the theory that the number of satellites plays an important role regarding the accuracy of the PVT solutions that can be achieved. A medium number of satellites, around 10, using GPS and Galileo, showed already quite good performance in the results, while more are obviously preferable. The investigations showed that even longer RTK correction data outages did not necessarily lead to critical situations in terms of 2D positioning accuracy, since as long as the RTK fix can be kept, errors below 20 cm can be achieved. Therefore, it makes sense to increase the standard setting of the maximum RTK age in the receiver.

Author Contributions: Conceptualization, all authors; methodology, all authors; software, all authors; validation, R.L. and S.S. (Susanne Schweitzer); formal analysis, R.L. and S.S. (Susanne Schweitzer); investigation, all authors; resources, all authors; data curation, all authors; writing—original draft preparation, S.S. (Susanne Schweitzer) and R.L.; writing—review and editing, J.M.V.G., K.D.-L., Selim Solmaz (S.S.); visualization, R.L. and S.S. (Susanne Schweitzer). All authors have read and agreed to the published version of the manuscript.

Funding: This work was conducted within the scope of the ESRIUM Project, which has received funding from the European Union Agency for the Space Program under the European Union's Horizon 2020 research and innovation program and under grant agreement No 101004181. Details of the project can be found under esrium.eu. The content of this paper reflects only the authors' view. Neither the European Commission nor the EUSPA are responsible for any use that may be made of the information it contains. Virtual Vehicle Research GmbH has received funding within COMET Competence Centers for Excellent Technologies from the Austrian Federal Ministry for Climate Action, the Austrian Federal Ministry for Labour and Economy, the Province of Styria (Dept. 12) and the Styrian Business Promotion Agency (SFG). The Austrian Research Promotion Agency (FFG) has been authorized for the program management.

Institutional Review Board Statement: Not applicable.

Informed Consent Statement: Not applicable.

Data Availability Statement: Data sharing is not applicable.

Acknowledgments: This project was supported by the Austrian RTK service provider EPOSA (www.eposa.at (accessed on 15 May 2023)) providing their service free of charge. We would also like to thank EUSPA.

Conflicts of Interest: The authors declare no conflict of interest.

References

1. Petovello, M.; O'Driscoll, C. What is Navigation Message Authentication? *Inside GNSS Magazine*, January/February 2018, pp. 26–31. Available online: https://insidegnss-com.exactdn.com/wp-content/uploads/2018/04/janfeb18-SOLUTIONS.pdf (accesssed on 15 May 2023).
2. Fernández-Hernández, I.; Rijmen, V.; Seco-Granados, G.; Simon, J.; Rodríguez, I.; David Calle, J. A Navigation Message Authentication Proposal for the Galileo Open Service. *J. Inst. Navig.* **2016**, *63*, 85–102. [CrossRef]
3. ESRIUM Project Webpage: EGNSS-Enabled Smart Road Infrastructure Usage and Maintenance for Increased Energy Efficiency and Safety on European Road Networks. Available online: https://esrium.eu/ (accessed on 15 May 2023).
4. Rudigier, M.; Nestlinger, G.; Tong, K.; Solmaz, S. Development and Verification of Infrastructure-Assisted Automated Driving Functions. *Electronics* **2021**, *10*, 2161. [CrossRef]
5. Hammarberg, T.; Vallet García, J.M.; Alanko, J.N.; Bhuiyan, M.Z.H. FGI-OSNMA: An Open Source Implementation of Galileo's Open Service Navigation Message Authentication. In Proceedings of the 36th International Technical Meeting of the Satellite Division of The Institute of Navigation (ION GNSS+ 2023), Denver, CO, USA, 11–15 September 2023, pp. 3774–3785. [CrossRef]
6. Galan, A.; Fernandez-Hernandez, I.; Cucchi, L.; Seco-Granados, G. OSNMAlib: An Open Python Library for Galileo OSNMA. In Proceedings of the 2022 10th Workshop on Satellite Navigation Technology (NAVITEC), Noordwijk, The Netherlands, 5–7 April 2022; pp. 1–12. [CrossRef]
7. Galileo-Osnma. Available online: https://github.com/daniestevez/galileo-osnma (accessed on 15 May 2023).
8. Septentrio. Release Notes and Installation Guide, Mosaic-X5 Firmware Package v4.12.1. 2023. Available online: https://www.septentrio.com/system/files/support/Mosaic-x5_firmware_v4.12.1_release_notes.pdf (accessed on 15 May 2023).
9. Fernández-Hernández, I.; Ashur, T.; Rijmen, V.; Sarto, C.; Cancela, S.; Calle, D. Toward an Operational Navigation Message Authentication Service: Proposal and Justification of Additional OSNMA Protocol Features. In Proceedings of the 2019 European Navigation Conference (ENC), Warsaw, Poland, 9–12 April 2019; pp. 1–6. [CrossRef]
10. Margaria, D.; Marucco, G.; Nicola, M. A first-of-a-kind spoofing detection demonstrator exploiting future Galileo E1 OS authentication. In Proceedings of the 2016 IEEE/ION Position, Location and Navigation Symposium (PLANS), Savannah, GA, USA, 11–14 April 2016; pp. 442–450. [CrossRef]
11. Motella, B.; Nicola, M.; Damy, S. Enhanced GNSS Authentication Based on the Joint CHIMERA/OSNMA Scheme. *IEEE Access* **2021**, *9*, 121570–121582. [CrossRef]
12. Motallebighomi, M.; Sathaye, H.; Singh, M.; Ranganathan, A. Cryptography Is Not Enough: Relay Attacks on Authenticated GNSS Signals. *arXiv* **2022**, arXiv:2204.11641. https://doi.org/10.48550/arXiv.2204.11641.
13. O'Driscoll, C.; Fernández-Hernández, I. Mapping Bit to Symbol Unpredictability in Convolutionally Encoded Messages with Checksums, with Application to Galileo OSNMA. In Proceedings of the 33rd International Technical Meeting of the Satellite Division of The Institute of Navigation (ION GNSS+ 2020), Online, 22–25 September 2020; pp. 3751–3765. [CrossRef]
14. Nicola, M.; Motella, B.; Pini, M.; Falletti, E. Galileo OSNMA Public Observation Phase: Signal Testing and Validation. *IEEE Access* **2022**, *10*, 27960–27969. [CrossRef]
15. Gamba, M.T.; Nicola, M.; Motella, B. Computational Load Analysis of a Galileo OSNMA-Ready Receiver for ARM-Based Embedded Platforms. *Sensors* **2021**, *21*, 467. [CrossRef] [PubMed]
16. Motella, B.; Gamba, M.T.; Nicola, M. A Real-time OSNMA-ready Software Receiver. In Proceedings of the 2020 International Technical Meeting of The Institute of Navigation, San Diego, CA, USA, 21–24 January 2020; pp. 979–991.
17. Gamba, M.T.; Nicola, M.; Motella, B. Galileo OSNMA: An implementation for ARM-based embedded platforms. In Proceedings of the 2020 International Conference on Localization and GNSS (ICL-GNSS), Tampere, Finland, 2–4 June 2020; pp. 1–6. [CrossRef]
18. Cucchi, L.; Damy, S.; Paonni, M.; Nicola, M.; Gamba, M.T.; Motella, B.; Fernandez-Hernandez, I. Assessing Galileo OSNMA Under Different User Environments by Means of a Multi-Purpose Test Bench, Including a Software-defined GNSS Receiver. In Proceedings of the 34th International Technical Meeting of the Satellite Division of The Institute of Navigation (ION GNSS+ 2021), St. Louis, MO, USA, 20–24 September 2021; pp. 3653–3667. [CrossRef]
19. Sarto, C.; Pozzobon, O.; Fantinato, S.; Montagner, S.; Fernández-Hernández, I.; Simon, J.; Calle, J.D.; Díaz, S.C.; Walker, P.; Burkey, D.; et al. Implementation and Testing of OSNMA for Galileo. In Proceedings of the 30th International Technical Meeting of the Satellite Division of The Institute of Navigation (ION GNSS+ 2017), Portland, Oregon, USA, 25–29 September 2017; pp. 1508–1519. [CrossRef]
20. Cucchi, L.; Damy, S.; Paonni, M.; Nicola, M.; Motella, B. Receiver Testing for the Galileo E1 OSNMA and I/NAV Improvements. In Proceedings of the 35th International Technical Meeting of the Satellite Division of The Institute of Navigation (ION GNSS+ 2022), Denver, Colorado, USA, 19–23 September 2022; pp. 808–819. [CrossRef]

21. Damy, S.; Cucchi, L.; Paonni, M. Impact of OSNMA Configurations, Operations and User's Strategies on Receiver Performances. In Proceedings of the 35th International Technical Meeting of the Satellite Division of The Institute of Navigation (ION GNSS+ 2022), Denver, Colorado, USA, 19–23 September 2022; pp. 820–827. [CrossRef]
22. Cancela, S.; Calle, J.D.; Fernández-Hernández, I. "CPU Consumption Analysis of TESLA-based Navigation Message Authentication. In Proceedings of the 2019 European Navigation Conference (ENC), Warsaw, Poland, 9–12 April 2019; pp. 1–6. [CrossRef]
23. Seco-Granados, G.; Gómez-Casco, D.; López-Salcedo, J.A.; Fernández-Hernández, I. Detection of replay attacks to GNSS based on partial correlations and authentication data unpredictability. *GPS Solut.* **2021**, *25*, 33. [CrossRef]
24. Cancela, S.; Navarro, J.; Calle, D.; Reithmaier, T.; Chiara, A.D.; Broi, G.D.; Fernández-Hernández, I.; Seco-Granados, G.; Simón, J. Field Testing of GNSS User Protection Techniques. In Proceedings of the 32nd International Technical Meeting of the Satellite Division of The Institute of Navigation (ION GNSS+ 2019), Miami, FL, USA, 16–20 September 2019; pp. 1824–1840. [CrossRef]
25. Štern, A.; Kos, A. Positioning Performance Assessment of Geodetic, Automotive, and Smartphone GNSS Receivers in Standardized Road Scenarios. *IEEE Access* **2018**, *6*, 41410–41428. [CrossRef]
26. Hammarberg, T.; Vallet García, J.M.; Alanko, J.N.; Bhuiyan, M.Z.H. An Experimental Performance Assessment of Galileo OSNMA. In Proceedings of the International Conference on Localisation and GNSS (ICL-GNSS 2023), Castellon, Spain, 6–8 June 2023.
27. More Precise Positioning Needed for Vehicle Autonomy. Available online: https://insidegnss.com/more-precise-positioning-needed-for-vehicle-autonomy/ (accessed on 15 May 2023).
28. Dai, M.; Li, H.; Liang, J.; Zhang, C.; Pan, X.; Tian, Y.; Cao, J.; Wang, Y. Lane Level Positioning Method for Unmanned Driving Based on Inertial System and Vector Map Information Fusion Applicable to GNSS Denied Environments. *Drones* **2023**, *7*, 239. [CrossRef]
29. Du, J.; Masters, J.; Barth, M. Lane-level positioning for in-vehicle navigation and automated vehicle location (AVL) systems. In Proceedings of the 7th International IEEE Conference on Intelligent Transportation Systems (IEEE Cat. No.04TH8749), Washington, WA, USA, 3–6 October 2004; pp. 35–40. [CrossRef]

engineering proceedings

MDPI

Proceeding Paper

Advanced Receiver Autonomous Integrity Monitoring (ARAIM) for Unmanned Aerial Vehicles †

Merle Snijders [1,*], Heiko Engwerda [1], Javier Fidalgo [2,*], Enrique Domínguez [2], Ginés Moreno [2], Fulgencio Buendía [2], Joao Pedro Duque [2], Jorge Martínez [2], Ilaria Martini [3], Matteo Sgammini [3] and Juan Pablo Boyero [4]

[1] Department of Aerospace Systems, Royal Netherlands Aerospace Centre (NLR), 1059 CM Amsterdam, The Netherlands
[2] NUSP Business Unit, GMV, 28760 Tres Cantos, Madrid, Spain
[3] Joint Research Centre (JRC), European Commission, 21027 Ispra, Italy
[4] Defence Industry and Space (DEFIS), European Commission, Rue de la Loi 130/Wetstraat 130, 1049 Brussel, Belgium; juan-pablo.boyero@ec.europa.eu
* Correspondence: merle.snijders@nlr.nl (M.S.); jfidalgo@gmv.com (J.F.)
† Presented at the European Navigation Conference 2023, Noordwijk, The Netherlands, 31 May–2 June 2023.

Abstract: Advanced Receiver Autonomous Integrity Monitoring (ARAIM) is an evolution of the currently used aviation-focused Global Navigation Satellite System integrity service, Receiver Autonomous Integrity Monitoring (RAIM). Where RAIM supports only lateral navigation, with its adaptations including multiple frequencies and constellations, and with the use of Integrity Support Messages (ISMs), ARAIM also supports vertical guidance. Although these techniques were designed to serve the aviation community, ARAIM could be used in a wide range of applications, especially safety-critical applications. With further evolutions, ARAIM could also be extended to cover more demanding applications in various sectors. This work reports the outcomes of the study of the applicability of ARAIM for the Unmanned Aerial Vehicle (UAV) sector.

Keywords: integrity; ARAIM; GNSS; UAV; HAS; sensor fusion

Citation: Snijders, M.; Engwerda, H.; Fidalgo, J.; Domínguez, E.; Moreno, G.; Buendía, F.; Duque, J.P.; Martínez, J.; Martini, I.; Sgammini, M.; et al. Advanced Receiver Autonomous Integrity Monitoring (ARAIM) for Unmanned Aerial Vehicles. *Eng. Proc.* **2023**, *54*, 46. https://doi.org/10.3390/ENC2023-15435

Academic Editors: Tom Willems and Okko Bleeker

Published: 29 October 2023

1. Introduction

The UAV sector represents a rapidly growing market; according to different reports, its value will triple between 2020 and 2030 [1]. The navigation system within the majority of these vehicles relies on GNSS as a primary means of positioning. Many operations are expected to be conducted in densely populated urban environments, which means that the requirements of navigation accuracy and especially integrity are stringent [2].

ARAIM is a promising integrity technique which could meet the requirements of the UAV sector. As shown in the ARAIMTOO project [3], the ARAIM concept for aviation can already meet the requirements of some non-aviation applications, in particular those operated in open-sky conditions. For the most demanding applications, operated in non-open sky conditions or with requirements more stringent than those in aviation, the evolution of the ARAIM concept is needed.

For the UAV sector, the following high-level ARAIM evolutions have been investigated within the ARAIMTOO project: the combination of dual constellation (GPS + Galileo) multi-frequency GNSS signals and Precise Point Positioning (PPP) techniques with an integrity algorithm at user level as ARAIM with hybridization with IMU, which could allow it to cope with harsh environments typical of urban areas and the stringent accuracy and integrity requirements of UAVs in urban environments.

To prove the concept of the ARAIM evolutions, an experiment was conducted based on the adapted Matlab Algorithm Availability Simulation Tool (MAAST), developed by Stanford [4]. In this adaptation, the ARAIM algorithms implemented are aligned with the

ARAIM ADD version 4.0, as implemented in the tool, and the proposed ARAIM evolutions have been prototyped.

Secondly, a MATLAB post-processing tool was developed to analyse collected flight data. With the use of this data, the fault detection and exclusion (FDE) and protection levels (PLs) were analyzed for real-life user scenarios.

2. Architecture of the Algorithm Evolution

The proposed evolution of the ARAIM algorithm for UAV implementations consists of the augmentation and extension of the GNSS signal by means of the recently rolled-out High Accuracy Service (HAS) as well as hybridization with an Inertial Measurement Unit (IMU). This combination shall result in high-accuracy PVT (position, velocity and time) solutions in combination with tight integrity budgets associated with the demanding urban user environment.

An overview of the implementation of this architecture is shown in Figure 1. The measurements and information are retrieved as displayed on the left side. The core of the ARAIM evolution is a Kalman filter followed sequentially by the ARAIM (snapshot) algorithm. The filter is fed with IMU measurements: accelerations and angular rates applicable to the inertial frame. As this frame has a certain offset in both its position and attitude from the GNSS antenna phase center, a set of corrections has to be provided to the filter as well. The second set of measurements is obtained from the GNSS receiver, in the figure displayed as the Multi-Frequency Multi-Constellation Galileo and GPS data. This data is augmented by the PPP corrections through the HAS. In order to apply this model, the ionospheric free combinations have to be determined based on the multi-frequency signals. Both the carrier phase and code phase measurements are implemented, with which the filter can model the state variables. The additional error states, in combination with the IMU error model and performance specifications, are integrated in the Kalman filter. Within the filter, the variances and error states are monitored and evaluated in order to provide a first FDE attempt, referred to as the FDE_{KALMAN}. Based on the increased navigation performance of the hybridization scheme, faulty satellites can be detected and excluded. The set of remaining satellites is then communicated to the ARAIM algorithm, as the 'all-in-view' solution. Other data that need to be transmitted from the filter to the ARAIM algorithm are the carrier phase measurements corrected for atmospheric disturbances and the solved ambiguities. As the ARAIM error model requires variances to calculate the PL, the KF variances are also shared, including the ambiguities and atmospheric variances. Furthermore, based on the external (sensor) input or from a monitoring scheme based on the state prediction residuals, the variance of the error due to the user environment can be estimated. Based on the ISM data fed to the ARAIM algorithm, the subsets of to-be-monitored solutions can be obtained. The same strategy to determine the PL is followed as in the legacy ARAIM algorithm. FDE_{ARAIM} can still be performed, although the effects are negligible, as the PPP KF already implements an FDE_{ARAIM} algorithm based on the hybridized solution.

In parallel to the ARAIM algorithm, the navigation solution can be propagated relying on the higher-frequency IMU data. Based on the IMU performance specifications, the PL can be inflated over time to account for drift. Careful consideration is required here, as long periods without GNSS cause an exponential drift. For shorter intervals, the behaviour can be expected to be close to linear.

Once the GNSS and a new ARAIM-calculated PL become available, the propagated PL and new PL can be compared and a new PL can be selected. This approach can be helpful in bridging the GNSS data gaps and cases where PPP convergence is lost.

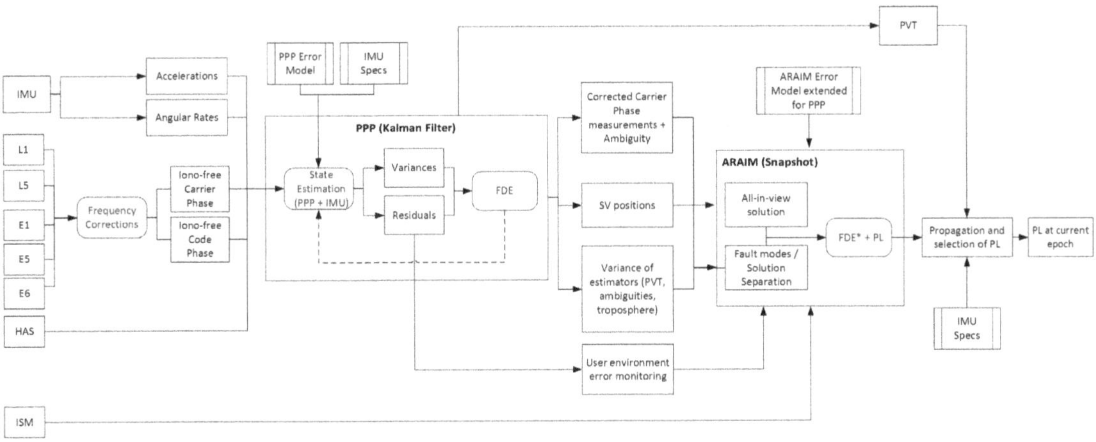

Figure 1. ARAIM evolution architecture for UAV applications. Note that the FDE* in the snapshot ARAIM block is different from the legacy FDE implementation.

3. Test Cases

By means of a series of test cases, the performance and behaviour of the ARAIM evolution can be tested. In this section, two test cases will be evaluated; first, a simulated set where spatial and temporal variation are covered, and a second set, where actual flight data is evaluated with feared events added.

3.1. Test Case 1

In this test case, service volume simulations using the modified MAAST code are performed. With such simulations, the global protection levels and availability performance are assessed for the ARAIM evolution for UAVs. This evolution implements both PPP through the HAS as well as IMU through a hybridized solution. In order to simulate the effects of PPP, an improved set of parameters has been supplied to the simulator:

- P_{sat}: the prior probability of a satellite fault, reduced to account for the monitoring efforts in supplying the PPP correction data.
- P_{const}: the prior probability of a constellation-wide fault, reduced to account for the monitoring efforts in supplying the PPP correction data.
- B_{nom}: the effects of the nominal bias are compensated with the PPP input data (precise data).
- URA: the PPP corrections will provide a better ranging accuracy. Note that a reduced overbounding distribution results in an increased probability of an error being considered a fault, which is assumed to be covered by the prior probability of a satellite fault.

The HAS service does not provide integrity monitoring commitment. Therefore, two different P_{sat} and P_{const} values, the original values and improved values, are simulated. In this way, the potential improvement due to HAS can be shown. Furthermore, the error budget is augmented with the following parameters:

- σ_{IMU}: the user error budget must be extended with the errors due to the propagation with IMU measurements.
- σnonlos: in an urban environment, this sigma bounds the errors due to non-line-of-sight signals not being detected by the FDE and entering the solution.
- $P_{FDE\text{-}MD}$: the probability of missed detection by the FDE algorithm is introduced to represent the occurrences where a non-line-of-sight signal is used by the receiver, resulting in an increased user variance, as explained in the previous bullet point.
- $P_{cycleslip}$: the probability of a non-detected cycle slip is added to the prior fault probability of the satellite.

3.2. Test Case 2

For this test case, the FDE capability of the ARAIM evolution for the UAV sector is central. In order to test its ability to exclude erroneous measurements, a set of input data should be used that includes such events. For this purpose, UAV flight data (IMU + dual antenna GNSS) is used with a trajectory following waypoints and manual flights, without any masking or interference, as the input data. A set of feared events is simulated by adding observation errors to this input data. These events mimic the conditions of the urban environment, and provide a robustness check against feared events such as ionospheric scintillation. These events are introduced in five specific intervals, as shown in Table 1.

Table 1. Table with inserted feared events.

Feared Event	Value/Description	Comments and Justification (If Needed)	Time Window
Slowly accumulating fault	Single satellite fault 10 cm per second increment	A slowly increasing pseudorange and phase measurement is modified in the observation file.	12:17:35–12:18:00
Multipath	Pseudorange: 1–5 m, phase measurement: negative of multipath pseudorange effect converted to cycles, 10–16.5 dB signal reduction	Multipath is applied to both pseudorange and phase measurement, for all applied frequencies and constellations. The exact value is drawn from a random uniform distribution. Reference values taken from [5].	12:22:00–12:22:30
Blanking	Multiple satellites unavailable	Based on the service volume simulations conducted, the difference between open-sky and urban environments can result in a loss of 6 visible satellites when utilizing two constellations. These satellites are chosen from the lowest elevation PRNs visible.	12:27:00–12:28:00
Non-line-of-sight	5–10 m pseudorange, 10–16.5 dB signal reduction	Values are chosen to represent non-line-of-sight signals arriving at the receiver while the direct path is obstructed. The exact value is drawn from a random uniform distribution. Reference values are taken from [6].	12:31:00–12:31:30
Multipath and blanking combined	1–5 m pseudorange, 10–16.5 dB signal reduction.Multiple satellites unavailable	This scenario is a combination of the multipath and blanking (4 satellites, based on elevation) described above.	12:33:00–12:33:30

In order to verify the ability of the FDE algorithms to mitigate feared events in this scenario, ideally, a reference truth trajectory is available that is not correlated with the GNSS errors. As such a dataset was not available, the approach is taken to introduce simulated feared events to a dataset where faults are introduced during the campaign. The open sky dataset is best suited for this approach.

4. Experimentation and Simulation Results

4.1. Test Case 1

4.1.1. TC-1A: Open Sky with PPP and IMU

By reducing the URA to 20 cm and satellite nominal bias (B_{nom}) to 10 cm, as a result of the correction from the HAS, combined with the PL propagation from the IMU and Kalman filter, the following simulation results are obtained (see Figure 2). The availability map below is the result for the open-sky scenario, where a VAL of 30 m and HAL of 20 m was applied with the original P_{sat} and P_{const} values (From stakeholder interviews, the requirements have been determined for operational scenarios such as U-Space, geofencing and search and rescue. In open sky, these operations require an accuracy of 10 m (1 m in urban environment), 95% availability (99.5% in urban), a continuity of 10^{-5}/hour, integrity risk of 10^{-7}, time-to-alert of 6 s and alert limits of 20 m horizontally and 30 m vertically). As can be seen, most of the globe exceeds 99.9% availability. In fact, the 95% availability is reached for the entire volume simulated. Without the need for integrity commitment by HAS, the open sky requirements for integrity can therefore be reached.

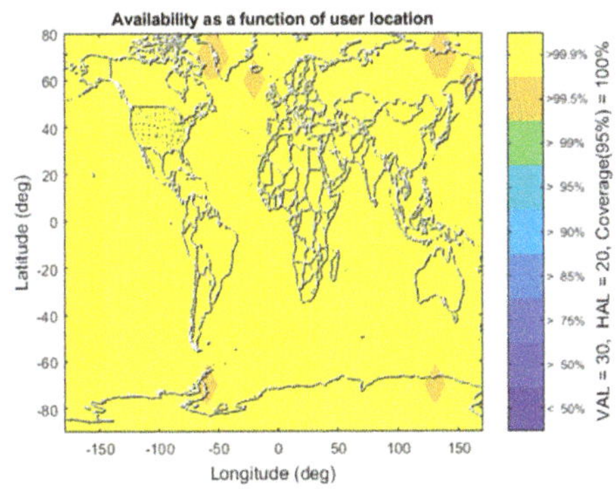

Figure 2. Open sky with PPP and IMU availability map indicating 95% values.

4.1.2. TC-1B: Open Sky with IMU and without PPP

In case of an HAS outage (URA = 2 m (GPS) 6 m (GAL), bnom = 0.75 m), or when other reasons not to apply the PPP corrections exist, the performance of the ARAIM-based integrity determination will decrease. This is a result of the deteriorating accuracy, whereas the number of visible satellites does not differ. It can be seen in Figure 3, that the AL requirements cannot be met and the availability is close to 0%. The best performance is achieved around the equator, but it is still significantly below the required 95% availability threshold.

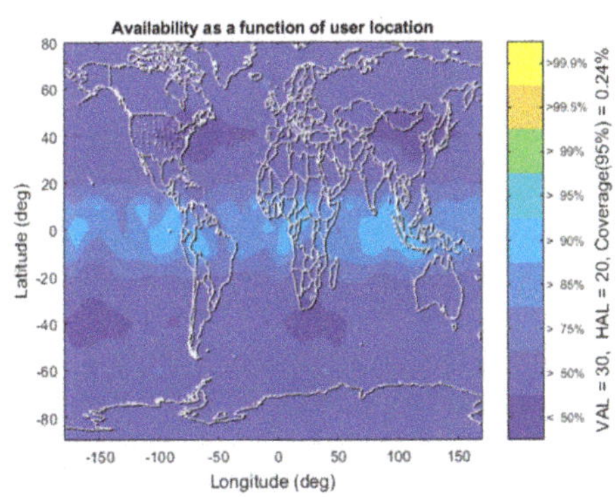

Figure 3. Open sky with IMU and without PPP availability map indicating 95% values.

4.1.3. TC-1C: Urban with PPP and IMU

In this test, an urban scenario with buildings blocking the horizon has been applied to determine the multipath, masking and non-line-of-sight signals. This scenario, based on the Schuman Roundabout in Brussels, has been applied to all the different grid locations.

The result of this simulation can be found in Figure 4. Here, a latitude effect becomes visible due to the orientation of the simulated building masks, which has a different obstruction in the northern direction as it has from the southern. The number of satellites visible is therefore lower in the European service area; this is completely dependent on the building geometry and orientation.

As a result of the poor conditions, the protection levels have increased with respect to the urban scenario and the availability has degraded.

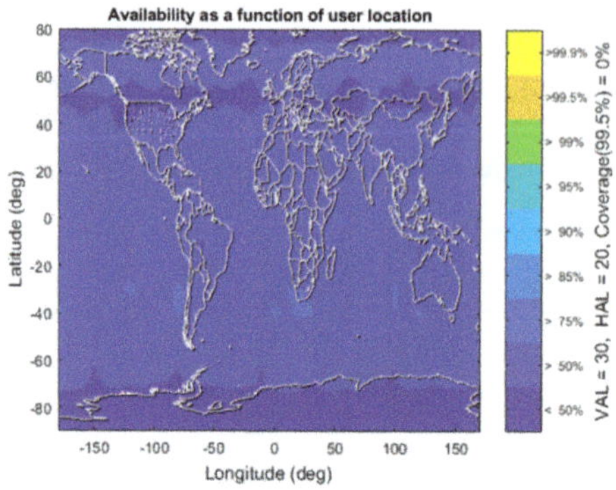

Figure 4. Urban with PPP and IMU availability map indicating 95% values.

4.1.4. TC-1D: Suburban with PPP and IMU

In the previous tests, UAV operations in open-sky and (deep) urban conditions have been evaluated. For this scenario, a set of parameters is modified with respect to the open-sky and urban conditions. The same elevation mask is used as for the urban scenario (Schuman Roundabout), but now with only a third of the signals being blocked. This approach is chosen to remove the bias from selecting a specific azimuth range for which signals are blocked and the impact this has on the geometry and GDOP. Furthermore, an elevation mask of 5 degrees is applied to remove the influence of low elevation satellite signal distortions. The multipath model is chosen in between that of open-sky and urban environments, as obtained from previous work [7].

The results of the simulation with these settings can be found in Figure 5, where the availability is displayed with coverage metrics.

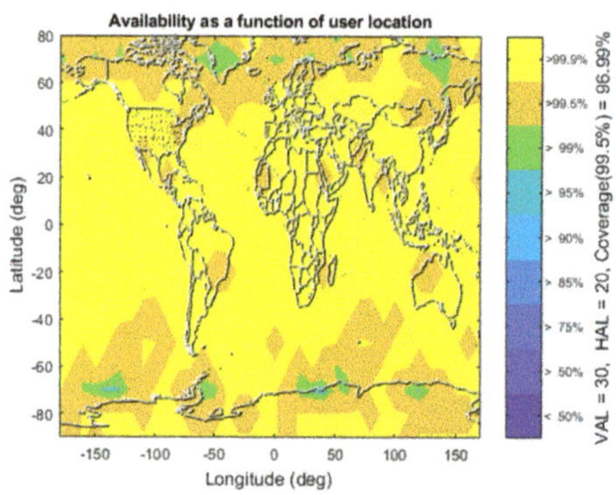

Figure 5. Suburban with PPP and IMU availability map indicating 95% values.

Without the need for integrity commitment, i.e., reduced satellite and constellation prior fault probabilities, as well as further accuracy improvements, the availability requirement is achieved for most of the service volume. Only at some high (70 degrees) latitudes, further improvements are required. At these locations, the suburban assumption is questionable, however, and open-sky performance is more likely. Therefore, the displayed results were deemed satisfactory and no further parameter assessment was performed.

Some remarks have to be stated regarding these results, however. First of all, as the number of available satellites with a direct line of sight is critical to the displayed results, the signal mask and building or environment model has a big influence on the results. Depending on the user location, these parameters may change resulting in significant variations in the availability. However, the results displayed here are a strong indicator that the proposed ARAIM evolution allows for use of UAVs in suburban environments.

4.2. Test Case 2

The influence of the feared events is best tested with the protection levels provided by the ARAIM algorithm under urban settings. The results are shown in Figure 6, the horizontal error and protection levels are displayed on the left, while the right hand side figure outlines the results in the vertical direction. In grey, the five distinct feared event scenarios are overlaid, matching the protection level increments. The dashed line is an indicator of the alert limit; the protection levels should remain below this threshold for the availability of the algorithm. The error with respect to the 'truth track', the combined forward and backward solution of the open sky (i.e., without feared events), is displayed in blue. The red dots display the results of the protection level outputted by the snapshot ARAIM algorithm modified with the settings just described. After propagation with the Kalman filter, the results in yellow are obtained.

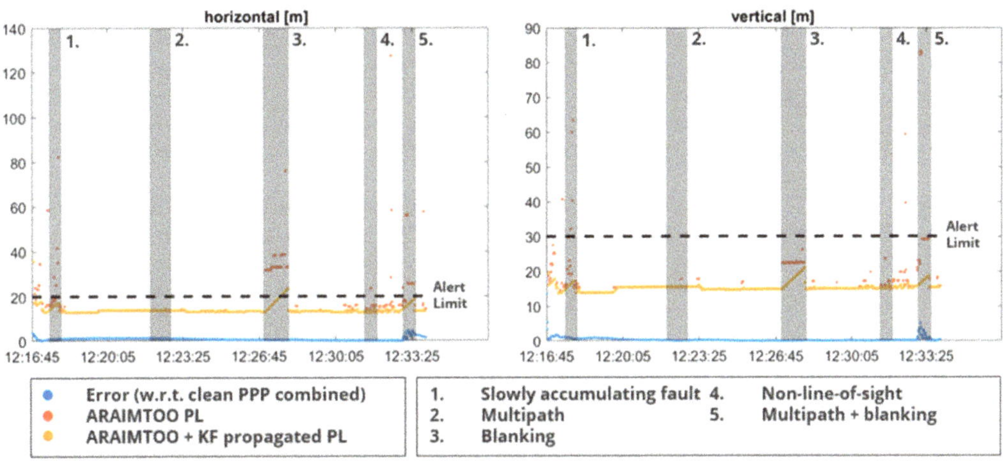

Figure 6. Protection level and positioning error for the urban UAV scenario.

Starting with the most significant impact visible, the protection level is severely impacted by the number of satellites visible due to blanking (indicated by the grey Sections 3 and 5). Even though the snapshot algorithm causes the horizontal protection level to exceed the alert limit immediately, the fusion with the IMU data allows for an extended duration of the availability. Such behavior allows for instance availability while moving through an underpass or through deep urban canyons for short durations. After roughly 45 s (during feared event 3), the alert limit is exceeded, which is a duration depending on the specifications of the IMU and implementation of the Kalman filter. Furthermore, it is interesting to note that especially during feared event 3, the actual error with respect to the open-sky 'truth track' is very small. This indicates there is more room for algorithm improvement in both the calculation of the protection level, for instance with tighter bounding, as well as the propagation algorithm.

Another interesting observation is made with the help of the RTKLib output, which is used for post-processing the data with PPP [8]. During the slowly accumulating fault period, the built-in checks already provide FDE functionality, which means not all introduced fault are observed in these results.

5. Conclusions

The proposed solution provides the following advantage, over stand-alone GNSS (using legacy ARAIM)

- A higher accuracy through HAS/PPP processing in combination with hybridization through filtering techniques, including IMU data.

- A reduced vulnerability to local disturbances such as multipaths in the urban environment. This results in a higher availability and continuity of service.
- Tighter integrity budgets and protection level bounding through reduced p_{sat}, p_{const} and b_{nom}, as well as added p_{nonlos} and $p_{cycleslip}$. This is further enabled through the comparison of the protection level calculated by ARAIM and the propagated state.
- More robust to GNSS outages due to hybridization with IMU.
- An improved FDE through residual monitoring within the Kalman filter.
- An improved time-to-alert due to filtering techniques leveraging the IMU information.

5.1. Conclusions from Service Volume Simulations

The combination of the GNSS dual frequency dual constellation measurements, with PPP processing based on the High Accuracy Service and hybridized with IMU measurements, results in ARAIM's eligibility for the open-sky UAV user sector. The significant reduction in, amongst others, the URA and the improvement in the available satellites with respect to the legacy ARAIM implementation for aviation, results in reduced protection levels over the globe. Reducing the URA significantly should be analyzed further for its consequences, as this could affect the P_{sat} and P_{const} values. It might be more feasible to reduce the URE instead, but this is beyond the scope of the presented work. Without PPP, the accuracy required for open-sky applications cannot be met, and other means of accuracy improvement have to be applied.

For UAV operations in the urban environment, the baseline settings considered do not result in acceptable availability. When considering the lower fault probabilities of both the satellite and constellation, the results are on the border of providing adequate availability. In the European service area, the requirements can be met with such settings. Although HAS does not commit on integrity, there is room for a discussion on implementing such values, especially considering the extra fault detection capabilities that the Kalman filter offers.

For global acceptance, also outside the European service area, some further improvements have to be considered. The improvement of the tropospheric variance due to PPP processing, as well as the reduced URA to 10 cm do not provide enough improvement. It should be noted that, since most of the area does not meet the availability requirements, the urban scenario may not be appropriate. For instance, better performance over the ocean could be expected, thereby closing in on the global availability requirement.

5.2. Conclusions on Feared Events Performance

The added benefit of hybridizing the ARAIM algorithm with PPP input and IMU data is clear. A significant improvement in the accuracy and integrity bounding is obtained.

6. Discussion and Recommendation

The proposed ARAIM evolution exceeds the 95% availability required for open-sky flights. This would allow, for example, for geo-referencing, drone operation under U-space, and search and rescue applications in open-sky settings. When considering UAV applications in the suburban environment, no further developments are required to meet the availability requirements. It should be noted, as the line between suburban and urban is blurred, thorough further studies with real flight data are required, evaluating the settings and assumptions applied in this chapter. It should be noted that the assumptions made in this work are very conservative. Some of the assumptions could be relaxed in a future analysis.

An even better performance could be achieved by further developing the following points:

- The use of more than two constellations will increase the satellite availability and therefore decrease the protection levels.
- A longer duration of the availability during lower satellite visibility can be achieved with further improved Kalman filter implementations.
- A longer duration of the availability during lower satellite visibility can be achieved with higher-grade IMUs.

- Tighter ARAIM budgets are possible, considering the propagation mitigates outages and rapid variations in the calculated PL.
- The further enhancement of PPP can result in lower protection levels which are more in line with the actual positioning error after filtering.

And, most importantly, further analysis of flight data is required to test under actual urban conditions. Statistically significant conclusions require multiple flights to be recorded, such as those displayed here, which requires a proper amount of testing.

Author Contributions: M.S. (Merle Snijders) and H.E. were responsible for the design and testing of the ARAIM evolution for unmanned aerial vehicles. E.D., G.M., F.B., J.M. and J.P.D. were responsible for the other sectors (rail and maritime) in the project and J.F. led the project. The EC (I.M., M.S. (Matteo Sgiammini) and J.P.B.) provided guidance, funding and reviewing. All authors have read and agreed to the published version of the manuscript.

Funding: The work is performed in the frame of the ARAIMTOO (ARAIM for Applications Beyond the Aviation Sector) project, funded by the EC (European Commission) as part of the Horizon 2020 R&D Programme and aimed at studying the use of ARAIM in sectors beyond aviation. The project's consortium is led by GMV, with VVA and NLR as partners.

Institutional Review Board Statement: Not applicable.

Informed Consent Statement: Not applicable.

Data Availability Statement: Data sharing is not applicable to this article.

Conflicts of Interest: The authors declare that this study received funding from the European Commission. The funder had the following involvement with the study; guidance and reviewing.

References

1. Precedence Research. *Unmanned Aerial Vehicle (UAV) Market (By Class: Tactical UAVs, Small UAVs, and Strategic UAVs; By Technology: Fully-Autonomous, Semi-Autonomous, and Remotely Operated; By System: UAV Payloads, UAV Airframe, UAV Avionics, UAV Software, and UAV Propulsion; By Application: Commercial, Military, and Recreational)-Global Industry Analysis, Size, Share, Growth, Trends Analysis, Regional Outlook and Forecasts, 2021–2030. Report Code 1269*; Precedence Research Pvt. Ltd.: Ottawa, ON, Canada, 2020.
2. Isik, O.K.; Hong, J.; Petrunin, I.; Tsourdos, A. Integrity analysis for GPS-based navigation of UAVs in urban environment. *Robotics* **2020**, *9*, 66. [CrossRef]
3. Fidalgo, J.; Domínguez, E.; Moreno, G.; Buendía, F.; Duque, J.P.; Martínez, J.; Snijders, M.; Engwerda, H.; Burdash, O.; Luciano, A.; et al. ARAIM for Applications Beyond Aviation. In Proceedings of the 35th International Technical Meeting of the Satellite Division of The Institute of Navigation, ION GNSS+ 2022, Denver, CO, USA, 19–23 September 2022; pp. 725–751.
4. Jan, S.-S.; Chan, W.; Walter, T.; Enge, P. Matlab simulation toolset for SBAS availability analysis. In Proceedings of the 14th International Technical Meeting of the Satellite Division of The Institute of Navigation (ION GPS 2001), Salt Lake City, UT, USA, 11–14 September 2001; pp. 2366–2375.
5. WG-C Advanced RAIM Technical Subgroup Reference Airborne Algorithm Description Document. 2017. Available online: https://web.stanford.edu/group/scpnt/gpslab/website_files/maast/ARAIM_TSG_Reference_ADD_v3.0.pdf (accessed on 19 December 2021).
6. Kaplan, E.D.; Hegarty, C. (Eds.) *Understanding GPS/GNSS: Principles and Applications*; Artech House: New York, NY, USA, 2017.
7. Martini, I.; Susi, M.; Paonni, M.; Sgammini, M.; Fernandez-Hernandez, I. Satellite Anomaly Detection with PPP Corrections: A Case Study with Galileo's High Accuracy Service. In Proceedings of the 2022 International Technical Meeting of The Institute of Navigation, Long Beach, CA, USA, 25–27 January 2022; pp. 1246–1262.
8. Takasu, T.; Yasuda, A. Development of the low-cost RTK-GPS receiver with an open source program package RTKLIB. In Proceedings of the International Symposium on GPS/GNSS, Seogwipo-si, Republic of Korea, 4–6 November 2009; Volume 1, pp. 1–6.

engineering proceedings

MDPI

Proceeding Paper

Precise Positioning for Mass-Market: Optimal Data Dissemination DAB+ Demonstrator †

Delphine Isambert [1,2,*], Paul Chambon [2], Raphaël Baucry [2], Christopher Perschke [3], Jannes Wübbena [3], Xavier Leblan [4], Giuseppe Rotondo [4], Fabien Galtier [4], Samar Kaddouri [5] and Florin-Catalin Grec [6]

1 SAMOVAR, Télécom SudParis, Institut Polytechnique de Paris, 91120 Palaiseau, France
2 Exagone Teria, 94400 Vitry-sur-Seine, France; paul.chambon@reseau-teria.com (P.C.); raphael@reseau-teria.com (R.B.)
3 Geo++, 30827 Garbsen, Germany; christopher.perschke@geopp.de (C.P.); jannes.wuebbena@geopp.de (J.W.)
4 Guide-Gnss Laboratory, 31400 Toulouse, France; xavier.leblan@guide-gnss.com (X.L.); giuseppe.rotondo@guide-gnss.com (G.R.)
5 ATDI, 75008 Paris, France; s.kaddouri@atdi-group.com
6 European Space Agency (ESA), 2200 Noordwijk, The Netherlands; florin-catalin.grec@esa.int
* Correspondence: delphine.isambert@hotmail.fr
† Presented at the European Navigation Conference 2023, Noordwijk, The Netherlands, 31 May–2 June 2023.

Abstract: The main issue in the development of precise positioning is the way in which to send GNSS corrections. To handle this case, a terrestrial broadcast mechanism based on DAB+ (digital audio broadcasting) signals was chosen to overcome the scalability challenges, as it would allow a service provider to transfer assistance data based on a "one-to-many" scheme. DAB is easily widely deployed, with large coverage predictions carried using ATDI's HTZ simulator. The DAB+ reception demonstrator uses SSRZ, a compressed and compact state space representation corrections format developed by Geo++. SSRZ corrections were generated using data from Teria's CORS network and have been broadcasted over the air with DAB+ signals. To validate the SSRZ corrections, the "Record&Replay" technique was used. It consists of digitizing GNSS signals during a normal field test. Once data were collected and validated, replays were executed in the laboratory. Different scenarios were set up to validate the accuracy of the position with SSRZ corrections sent via DAB+ and the convergence to the corrected position after ambiguity loss. The results show that the accuracy was acceptable when the user remained close to the transmitting DAB+ antenna (<10 km). The lower performance might result from broadcast limitations in urban environments.

Keywords: SSR; DAB+; NRTK; PPP-RTK

Citation: Isambert, D.; Chambon, P.; Baucry, R.; Perschke, C.; Wübbena, J.; Leblan, X.; Rotondo, G.; Galtier, F.; Kaddouri, S.; Grec, F.-C. Precise Positioning for Mass-Market: Optimal Data Dissemination DAB+ Demonstrator. *Eng. Proc.* **2023**, *54*, 47. https://doi.org/10.3390/ENC2023-15446

Academic Editors: Tom Willems and Okko Bleeker

Published: 29 October 2023

1. Introduction

Precise positioning can be achieved using one of two techniques: real-time kinematic (RTK) and precise point positioning (PPP). These two solutions were mainly developed for professional applications; the costs of dual-frequency GNSS receivers did not allow their use for mass-market applications. Currently, thanks to the commercialization of low-cost, multi-frequency GNSS chipsets, high-accuracy techniques have become ever more attractive for several applications, like the automotive industry, agriculture, and drones [1].

The main issue in the current condition is represented by the way in which GNSS corrections are sent to the user. The GNSS corrections used to compute the position are typically sent to the user via the NTRIP protocol [2] (networked transport of RTCM via Internet protocol). NTRIP is an open, non-proprietary protocol designed about 15 years ago to distribute the GNSS streaming data to a stationary or mobile receiver over the Internet. NTRIP dissemination is a proven technique, with worldwide adoption, that is based on existing internet standards. Its success is partly due to the HTTP and TCP protocols. This communication protocol, however, has a significant drawback because it involves exactly two endpoints (each characterized by a globally unique IP address), and HTTP is a standard unicast application. Thus, the current approach adopted in the GNSS community has limited scalability and, therefore, poses some limitations for mass-market applications. Currently, more and more use cases are emerging, most of them demanding accurate positioning information such as drones, Internet of Things (IoT), and autonomous driving. In this scenario, traditional GNSS data dissemination solutions may face challenges, especially when tens of thousands of devices are simultaneously requesting a response [3]. To handle this case, service operators can set up a terrestrial broadcast mechanism based on cellular signals or DAB+ signals. It can overcome the scalability challenges, as it would allow a service provider to transfer GNSS assistance data based on a "one-to-many" scheme. In this project, a DAB+ demonstrator is developed to receive GNSS corrections. Firstly, we present this demonstrator and the way to receive GNSS corrections. Then, the outcomes of different experimentations are described. In the last section, the results are discussed and analyzed for future work on the subject.

2. Materials and Methods

The aim is to directly broadcast the GNSS corrections from a DAB+ transmitter to a large area. During the experimentation phase, SSR corrections generated with data from Teria's own CORS (continuously operating reference station) network were broadcasted wirelessly using a DAB+ signal. DAB+ was selected because this technology has proven to be widely deployed, and because coverage predictions using ATDI's HTZ simulator have shown that the DAB+ signals have a much larger reach in rural and urban areas compared to LTE. Moreover, DAB+ has another advantage: it is a numerical signal with a data channel, and it is easier to broadcast data on DAB+ signals. The detailed configuration of the DAB+ reception is shown in Figure 1.

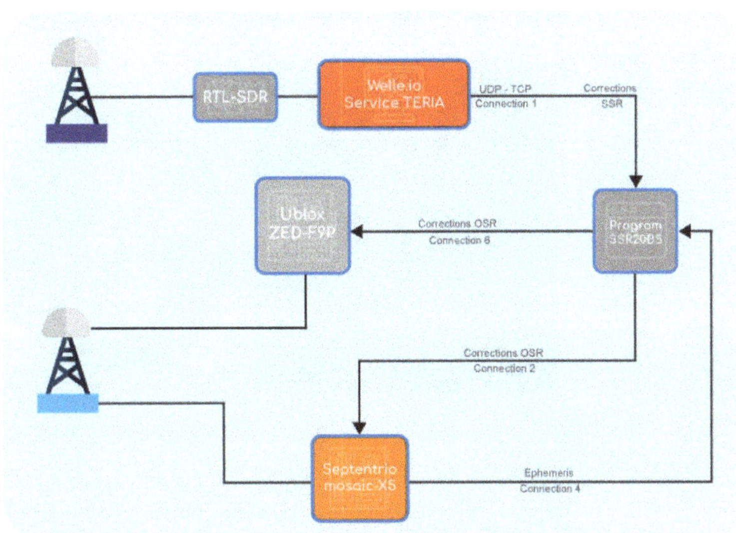

Figure 1. Scenario for record and replay using correction through DAB.

The DAB+ transmitter uses SSRZ, a compressed and compact state space representation corrections format developed by Geo++ [4]. The main advantage of using this type of corrections is that there are no problems with two-way server–user communication, as in NRTK. Moreover, using SSRZ corrections aims to solve the computation of the position, even if it is with latency.

In this configuration, there are three different receivers: RTL-SDR (for the DAB+ reception), Ublox, and Septentrio mosaic (GNSS receivers). Two ways to compute the position were used: with NRTK and with SSRZ corrections. The first step was to receive the data service "teria" in DAB+ with the USB stick RTL-SDR. The data service "teria" is broadcast on channel 9B (204,352 MHz). This data service contains SSRZ corrections. The next step is to convert the SSRZ corrections into OSR (observation space representation) corrections. Indeed, to compute the position with the receiver Ublox or Septentrio, OSR corrections are needed. But, for this conversion, the ephemeris from the Septentrio mosaic-X5 is necessary. For this reason, a backup link was set up between the mosaic-X5 and the SSR2OBS program. Then, the corrections were sent to Ublox and Septentrio receivers to compute the position.

To validate the proposed corrections solution, the "Record & Replay" technique was used [5]. It consisted of digitizing GNSS signals during a normal field test. During the test, a high-grade inertial navigation system was embedded to calculate, in post-processing with post-processing Kinematic (PPK) and inertial hybridization, a high-accuracy trajectory. Once the data were collected and validated, replays were executed in the laboratory. GUIDE compared the performances obtained using classical NRTK corrections and the ones obtained with the new SSRZ corrections. For this phase, corrections were replayed and broadcast using the NTRIP protocol. Once validation proved that the SSRZ corrections were able to provide similar accuracy and availability of high-precision PVT solution as the RTK, the real-life test took place.

3. Tests and Results of Two Scenarios

Different scenarios were set up to validate not only the accuracy of the position with GNSS corrections sent via DAB, but also the convergence to the corrected position after ambiguity loss. In this case, SSRZ corrections were broadcasted and converted on user equipment to be compatible with the RTK receiver. One scenario (Figure 2) consisted of moving in an urban environment around the transmitting antenna to observe the evolution of the position accuracy with respect to the distance from the DAB+ emitter.

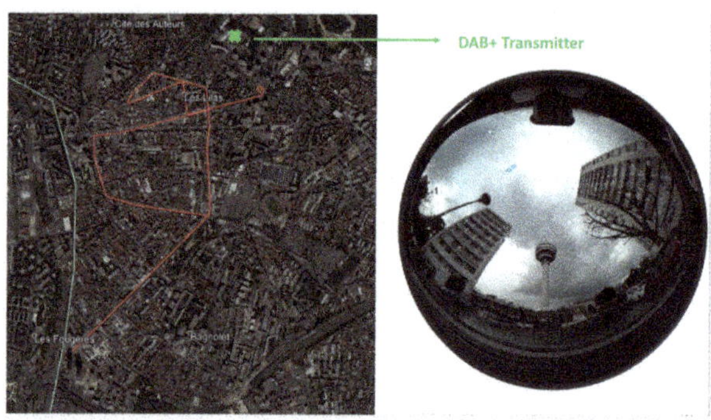

Figure 2. Scenario A for record and replay using correction through DAB.

The session lasted 6 km around the transmitting antenna, which was located in Romainville (France). This test session lasted 30 min. The results (Figure 3) show that, in some sections, the accuracy was acceptable when the user stayed close to the transmitting DAB+ antenna (<10 km). Lower performances might result from broadcast limitations in urban environments.

Horizontal errors were between 3 cm and 50 cm, and vertical errors were between 5 cm and 80 cm. These errors were due to the quality of the DAB+ signal. Indeed, in this scenario, corrections on DAB+ signals are often lost in urban environments. This loss is mostly due to the non-optimization of the DAB+ prototype antenna for the good reception of signals. Furthermore, the use of a single transmitter for DAB+ limited the signal coverage area and caused gaps. This can explain the different errors in the position. However, the accuracy of the position was around a few centimeters, with a convergence time of less than 5 min.

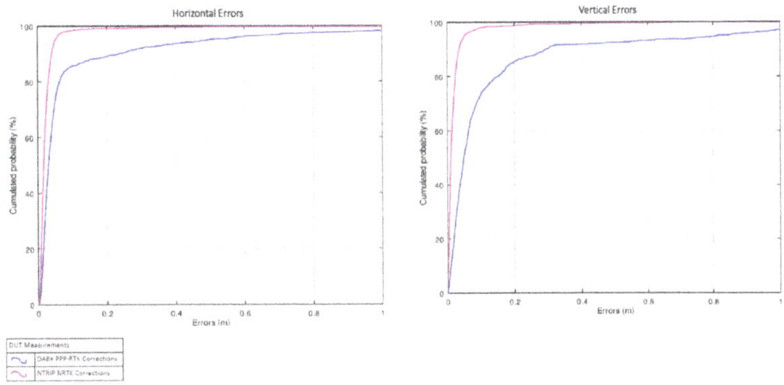

Figure 3. Horizontal and vertical errors in scenario A for DAB+ and NRTK solutions.

In the second scenario, B, one moved away from the antenna gradually over 28 km for 100 min (Figure 4).

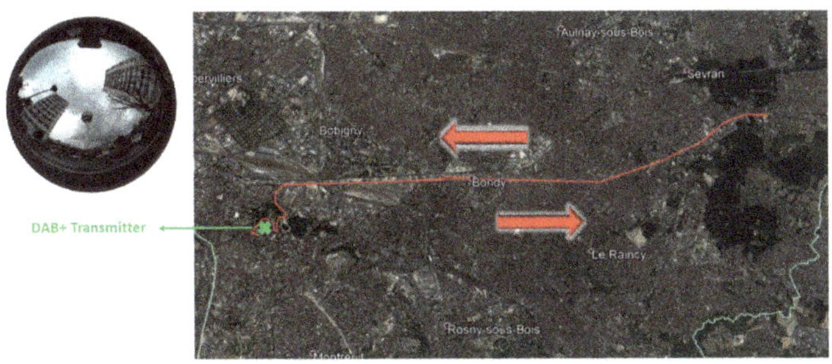

Figure 4. Scenario B for record and replay using correction through DAB+ compared with NRTK solutions. In red the trajectory of the car.

In scenario B, GNSS corrections on the DAB+ signal were lost at a distance of approx. 10 km (Figure 5). The horizontal and vertical accuracy with SSRZ corrections, broadcast through DAB+ signals, was on the centimeter level whenever phase measurement ambiguities could be solved. The accuracy was about the same as for the NRTK service. The results obtained are in line with those obtained in the preliminary study, which was conducted by relaying GNSS signals and sending both NRTK and PPP RTK corrections to an NTRIP server. DAB+ transmitter networks are an interesting way to broadcast PPP RTK corrections for all devices without LTE connection, such as IoT.

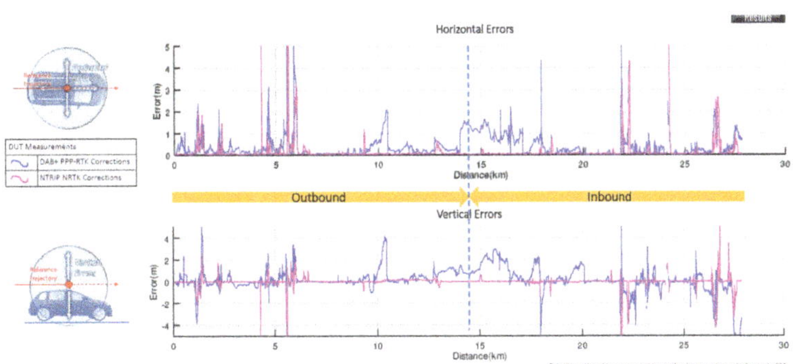

Figure 5. Scenario for record and replay using correction through DAB.

4. Discussion and Conclusions

The reception of DAB+ signals, in all scenarios, was strongly impacted by different factors. Firstly, the hardware and software utilized had purely a demonstrator scope. They were not optimized to be used in dynamic conditions and/or far from the transmitter. Moreover, a unique transmitter was used for the broadcast of corrections. It limited the coverage of DAB+ signals and the quality of the data once the distance was greater than 10 km. However, the outcomes of both scenarios are quite encouraging in terms of convergence time and accuracy. Indeed, the accuracy was centimetric, and when the ambiguity was lost, the convergence was fast: just a few minutes. This experimentation reveals the possibility to send GNSS corrections differently with the same performance in NRTK solution. The main advantage is the ease of broadcasting GNSS data in DAB+ over longer ranges. In addition, the DAB+ broadcast of GNSS corrections would allow several devices to calculate their positions directly without a bi-directional link between the transmitter and the receiver. In the future, it could be possible to implement this GNSS broadcast system with several antennas to improve the availability of the DAB+ signal and to improve the position calculation. In addition, other channels, such as LTE via 3GPP protocol, could be used to broadcast GNSS corrections.

Author Contributions: Conceptualization, P.C. and X.L.; methodology, P.C.; software, D.I. and C.P.; validation, F.-C.G. and D.I.; formal analysis, S.K., G.R., R.B. and F.G.; investigation, D.I.; resources, P.C.; data curation, P.C.; writing—original draft preparation, D.I., S.K., G.R. and C.P.; writing—review and editing, F.-C.G., G.R. and P.C.; visualization, D.I.; supervision, P.C.; project administration, P.C., X.L. and J.W.; funding acquisition, P.C. All authors have read and agreed to the published version of the manuscript.

Funding: This research is funded by the NAVISP project N° 4000129782/20/N/LW from ESA.

Institutional Review Board Statement: Not applicable.

Informed Consent Statement: Not applicable.

Data Availability Statement: The data presented in this study are available upon request to EXAGONE TERIA, by email at: paul.chambon@reseau-teria.com.

Conflicts of Interest: This research is funded by the NAVISP project N° 4000129782/20/N/LW from ESA. The authors declare that the research was conducted in the absence of any commercial or financial relationships that could be construed as a potential conflict of interest.

References

1. de Salas, J.; Torroja, M. Carrier phase positioning experiences in consumer GNSS devices. In Proceedings of the 2016 International Conference on Localization and GNSS (ICL-GNSS), Barcelona, Spain, 28–30 June 2016; pp. 1–6. [CrossRef]
2. Weber, G.; Dettmering, D.; Gebhard, H. Networked Transport of RTCM via Internet Protocol (NTRIP). In *A Window on the Future of Geodesy. International Association of Geodesy Symposia*; Springer: Berlin/Heidelberg, Germany, 2005. [CrossRef]
3. Darugna, F.; Wübbena, J.; Ito, A.; Wübbena, T.; Wübbena, G.; Schmitz, M. RTK and PPP-RTK Using Smartphones: From Short-Baseline to Long-Baseline Applications. In Proceedings of the 32nd International Technical Meeting of the Satellite Division of The Institute of Navigation (ION GNSS+ 2019), Miami, FL, USA, 16–20 September 2019; pp. 3932–3945.
4. Wabbena, G.; Schmitz, M.; Bagge, A. PPP-RTK: Precise Point Positioning Using State-Space Representation in RTK Networks. In Proceedings of the 18th International Technical Meeting of the Satellite Division of The Institute of Navigation (ION GNSS 2005), Long Beach, CA, USA, 13–16 September 2005; pp. 2584–2594.
5. Leblan, X. GUIDE GNSS METROLOGIE GNSS Validation de la Représentativité et Reproductibilité de la Technique de Rejeu des Signaux GNSS, Dénommée «Record & Replay». 2020. Available online: https://guide-gnss.com/wp-content/uploads/2020/01/Guide-gnss-record-replay-representativite-reproductivite.pdf (accessed on 10 May 2023).

engineering proceedings

MDPI

Proceeding Paper

European Geostationary Navigation Overlay Service Performance Prediction †

Pieter Bastiaan Ober [1,*], Jiri Doubek [2], Tamás Prajczer [3], János Nemes [3], Csongor Fekete [3], Mickael Dall'Orso [4] and Guillaume Buscarlet [4]

1 Integricom, Tjalkenwerf 30, 2317 DD Leiden, The Netherlands
2 Iguassu Software Systems, Evropská 120, Dejvice,160 00 Praha, Czech Republic
3 GeoX Kft, Huszti út 32, 1033 Budapest, Hungary
4 ESA/EGNOS Project Office, Center Spatial de Toulose, 31401 Toulouse, France
* Correspondence: integricom.enc2023@gmail.com

† Presented at the European Navigation Conference 2023, Noordwijk, The Netherlands, 31 May–2 June 2023.

Abstract: The paper describes the development of a machine learning-based system to predict the performance of EGNOS. The system is shown to perform considerably better than existing simple geometry-based 'macro' models developed for the same purpose in the past.

Keywords: EGNOS; GPS; machine learning; performance prediction; integrity

Citation: Ober, P.B.; Doubek, J.; Prajczer, T.; Nemes, J.; Fekete, C.; Dall'Orso, M.; Buscarlet, G. European Geostationary Navigation Overlay Service Performance Prediction. *Eng. Proc.* **2023**, *54*, 48. https://doi.org/10.3390/ENC2023-15456

Academic Editors: Tom Willems and Okko Bleeker

Published: 29 October 2023

1. Introduction

The European Geostationary Navigation Overlay Service (EGNOS) is a European satellite-based augmentation system (SBAS) that improves the performance of GPS by providing range corrections and protection levels as upper bounds of the position errors. It has been deployed to provide safety of life navigation services to aviation, maritime, and land-based users over most of Europe.

In aviation, pilots have, for two decades, been required to confirm that the (predicted) GPS protection levels will be sufficient to support their operations. For EGNOS, performance prediction could be equally valuable for aviation and other user communities, both from a safety and an economic standpoint. Prediction capabilities can, for example, be integrated in flight-planning services but can also serve as part of an EGNOS simulation service, for example to investigate the performance impact of changes in the EGNOS infrastructure or in the core GNSS constellations.

A decade ago, the European Organisation for the Safety of Air Navigation (EUROCONTROL) developed an EGNOS performance prediction tool ('EURONOTAM') to provide a similar prediction for EGNOS-based operations, based on a simplified so-called 'meta model' with a limited accuracy [1]. The advent of machine-learning techniques allows EGNOS performance prediction to be predicted more accurately by moving away from the simplified macro-model (which is based largely on satellite and EGNOS reference station (RIMS) network geometries only) and incorporating other factors that are known to impact performance. This has, for example, been performed in [2], which uses a similar approach to ours, but with significant differences, for example, in terms of the selected feature sets and the algorithms used.

Funded by the European Space Agency under a NAVISP contract, we have implemented an EGNOS performance prediction system that employs state-of-the-art machine-learning techniques to forecast EGNOS performance using a wide variety of parameters, including weather and space-weather. As we will show, our prediction system yields significantly more accurate predictions than the currently employed macro models, which is not unexpected as the approach in [2] already yielded similar outcomes. Note that at the time of writing, the project was still ongoing and the reported results reflect the minimum achievable performance.

2. EGNOS Performance Prediction

2.1. The EGNOS System

EGNOS improves the performance of GPS L1 signal accuracy by providing range corrections and associated quality metrics to achieve the required level of reliability for Safety of Life applications. To do so, it provides users with protection levels as upper bounds of their position errors. The current version of EGNOS is called EGNOS v2 and has been declared 'Safety of Life' capable since 2011. Since then, several EGNOS v2 releases have been deployed to improve the EGNOS service. The next version of EGNOS called EGNOS v3 is under development and will provide Dual-Frequency Multi Constellation (DFMC) services by augmenting GPS L1/L5 and Galileo E1/E5a signals to enable more demanding service levels beyond what EGNOS v2 can provide. The work presented in this paper addresses EGNOS v2 and not EGNOS v3.

EGNOS v2 uses L1C/A and L2P(Y) GPS measurements, collected by accurately located reference stations (RIMS) deployed across Europe and remotely. All GPS measurements are transferred to a central computing centre hosting the Central Processing Facility (CPF), which runs the navigation algorithms to calculate differential corrections and integrity messages are calculated. Two types of corrections are provided: corrections for satellite orbit and clock errors, and corrections for ionospheric delays. Consequently, two types of integrity indicators are provided:

- The User Differential Range Error Index (UDREI). An integer in the range [0–15], which characterizes the residual range errors due to orbit and clock inaccuracies after applying the clock and orbit corrections. Below index 13, a lower UDREI value means that the GPS signal is more accurate, while a higher UDRE value means that there is a greater chance of error in the GPS signal. When the satellite is not observed by sufficient RIMS, a UDREI value of 14 is broadcast to indicate that the satellite is Not Monitored, while a UDREI value of 15 indicates that the satellite is not to be used by the users (Do Not Use status as per MOPS) because a problem has been detected on this satellite (for example, the presence of a failure);
- The Grid Ionospheric Vertical Error Index (GIVEI), an integer in the range [0–15], which characterizes the residual error of the ionospheric delays after application of the ionospheric corrections. As for the UDREI, The lower the GIVEI value, the more accurate the ionosphere corrections and vertical position measurement from the GNSS signal.

These corrections and integrity indicators are broadcast to users using geostationary satellites that serve as an augmentation to the original GPS signals [3]. Over the period considered in this paper, between 34 and 38 Reference and Integrity Monitoring Stations (RIMS) were deployed to monitor the GPS constellation. This number varies because of new RIMS deployments or RIMS decommissioning over the last decade.

2.2. EGNOS Outputs

Users of EGNOS use the additional information to improve their positioning accuracy by applying the corrections and excluding Not Monitored or Do Not Use satellite measurements. In addition, they can use the GIVEI and UDREI to compute horizontal and vertical protection levels (HPLs and VPLs) that can be seen as a measure of the integrity of their position [4]. Protection levels can be seen as upper bounds to the position error and are typically expressed as a distance, such as a radius, within which the actual position of the user is expected to be located with a high level of probability. The size of the protection levels thus characterizes the overall EGNOS performance.

The protection levels are updated regularly based on the GIVE and UDRE values received from the EGNOS system. Users can determine in real-time whether their operational needs are met based on the protection levels, which should be sufficiently low to allow for specific safe operations.

The prediction of the protection levels in the future (in the next one or two days) thus serve the purpose of allowing for users to anticipate whether EGNOS will provide sufficient performance. This prediction can be particularly interesting in the case of planed operations (satellite decommissioning, satellite maneuvers, RIMS reparation, etc.) that may impact the performance of EGNOS over a specific area.

To predict protection levels, it is required to predict what values of the UDREI and GIVEI will be broadcast by EGNOS. Based on these predicted values, as well as the predicted satellite positions, predicted protection levels can be obtained as a function of time and user location. The GPS satellite positions can simply be computed from the GPS almanacs published by the U.S. Coast Guard. With the most recent almanac, the satellite positions can be determined quite accurately and predicted for the next few days. Hence, the remainder of this paper only focuses on the prediction of the UDREI and GIVEI values, which can benefit from machine learning (ML) using a variety of inputs.

3. EGNOS Performance Prediction

Predicted EGNOS performance relies on predicting the UDREI and GIVEI values as computed by the EGNOS ground segment's processing facilities. These are complicated data processors of which only some characteristics are public [5,6], while many implementation details are proprietary. Prediction models thus need to mimic the internal workings of the ground segment's processing, as well as consider the expected status of the GPS space segment and the expected signal reception conditions to the extent possible. An overview of a prediction system's architecture is shown in Figure 1.

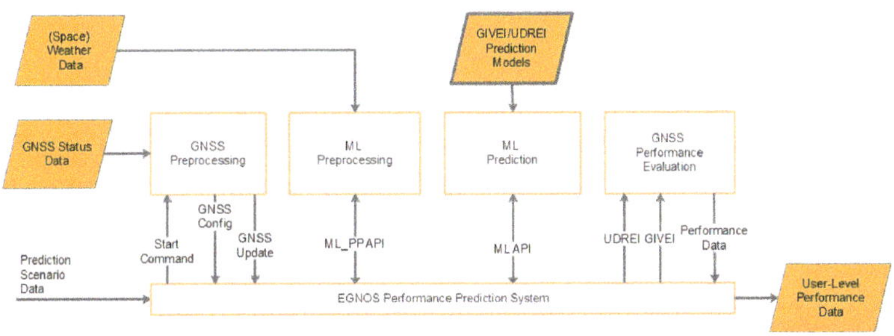

Figure 1. The operational EGNOS Performance Prediction system uses trained GIVEI and UDREI prediction models in combination with preprocessed GNSS and (space)weather data to produce predicted UDREI and GIVEI timeseries. These are then translated into the user-level performance indicators such as protection levels and EGNOS availability for the user's operations.

The Use of Macro Models

Several macro models are in use to predict EGNOS performance. We used a proprietary macro model provided to us by the ESA to compare our results against. The European Satellite Services Provider (ESSP) also employs a (proprietary) macro model [2], which possibly builds on the model from [1]. Finally, our partner Iguassu employs a macro-model in their SBAS simulator [7]. These models mainly rely on satellite and ground segment geometry, and a simple relation (e.g., in the form of tables) with the UDREI and GIVEI to the set of RIMS stations tracking the satellites and the set of ionospheric piercing points (IPPs) around the ionospheric grid point (IGP) to which the GIVEI relates. Due to their simplicity, these models are easy to implement and fast to run. Their simplicity is expected to limit their performance, and we would expect that machine learning can improve prediction performance considerably.

4. Machine Learning Approach

We followed an agile CRISP-DM method to find the best models with the best hyper parameters, by starting with simple models and then gradually increasing the number of features and the model's complexity. The key components of CRISP DM-based agile holistic machine learning approach include the following steps:

1. The *Domain and Data Understanding* step involves gaining understanding of the EGNOS performance prediction problem. See also Section 5.1 below;
2. The *Data Collection and Preprocessing* step gathers relevant data for the problem domain and preparing it for use in a machine learning model by cleaning the data and transforming them into a suitable format, see also Section 5.1;
3. *Feature Engineering* is then performed by selecting relevant input features that are expected to help the machine learning model make accurate predictions. See also Section 5.2;
4. *Model Selection and Training* involves selecting an appropriate machine learning algorithm and optimizing its parameters to achieve the best performance on the training data. More information is found in Section 5.3;
5. *Evaluation*: once a model has been trained, it needs to be evaluated on a separate testing set to determine its performance. More information is found in Section 5.4;
6. *Deployment*: the trained model is finally deployed in a production environment and monitored to ensure that it continues to perform well over time.

5. Machine Learning-Based Performance Prediction for EGNOS

Supervised machine learning is exploited to build three separate models: one to predict the UDREI, and two to predict the GIVEI models under nominal and active ionospheric conditions, respectively, as in the latter case the ionospheric corrections are known to be less accurate and higher GIVEI values can be expected. The use of two different GIVEI models allows for assessment of the expected performance of EGNOS under different ionospheric conditions. It is left up to the users to use one model or the other, depending on their assessment of the likelihood and potential impact of ionospheric storms. In the remainder of this paper, only the nominal model is considered. The nominal model is also expected to provide valid results in case of strong ionosphere since the machine learning modules have been trained during active ionosphere periods, and features related to the ionosphere activity are known by the machine learning modules.

5.1. Machine Learning Labels

To train our models, we have used data from the EGNOS Data Access Service (EDAS) offers free access to historical EGNOS data from 2013 up to the current day, including the ephemerides and range measurements collected by the RIMS, as well as the EGNOS messages that were broadcast. The latter contain the UDREI and GIVEI timeseries (labels) which the (supervised) machine learning can use to train models targeted at reproducing these labels.

Investigations of the data showed that UDREI values of 11 and 12 were hardly ever broadcast. Hence, even though we had many years of historical broadcast data available, there were insufficient cases to train the machine learning effectively. We therefore decided to define a single class to represent UDREI values of 11, 12 and 14.

5.2. Selecting Machine Learning Features

The first step in any machine learning-based model development is the careful selection of the input data features that will drive these models. Ideally, the data should contain all information that is available to the EGNOS system when it is in operation. However, there are some limitations, as the features need to be available for the period for which to predict. As a result, unpredictable quantities or events (such as the raw RIMS observations of the unexpected occurrences of faults in satellites or RIMS) cannot be incorporated.

Important factors in determining the UDREI and GIVEI that can be predicted are the quality of the range measurements, as well as geometric aspects which impact the observability of satellite orbit and clock errors or ionospheric delays.

For UDREI prediction, the following features have been selected:

- A list of locations of the RIMS tracking the satellite;
- Satellite position;

- Geometry metrics, such as DOP, number of RIMS in view;
- Elevation-dependent multipath variances;
- Weather indicators (from OpenWeather);
- Local space-weather indicators, including vertical TEC values and its gradients over time and space (IONEX files from CDDIS and CODE);
- Satellite and satellite clock generation.

For GIVEI, the following features have been selected:

- IGP location;
- A list of IPP locations of IPP near the IGP;
- Elevation-dependent multipath variances;
- Satellite positions.
- Satellite block and satellite clock generation;
- Global space-weather indicators (from the NOAA Space Weather Prediction Center, Space Weather Canada, GFZ German Research Centre for Geosciences, Royal Observatory of Belgium);
- Local space-weather indicators (as for UDREI).

To train our models, we obtained most feature data, as well as UDREI and GIVEI labels from EDAS. For actual predictions, the machine learning model can use forecasted equivalents, based on GPS almanacs and weather and space-weather predictions.

5.3. Algorithm Selection and Training

Figure 2 shows the training process, for which we used over five years of data. The features (preprocessed historical GNSS and (space)weather data) were used in conjunction with the labels (historical UDREI and GIVEI broadcast values) to train different prediction algorithms, of which the best performing one was eventually selected for the product to be developed in the frame of this project.

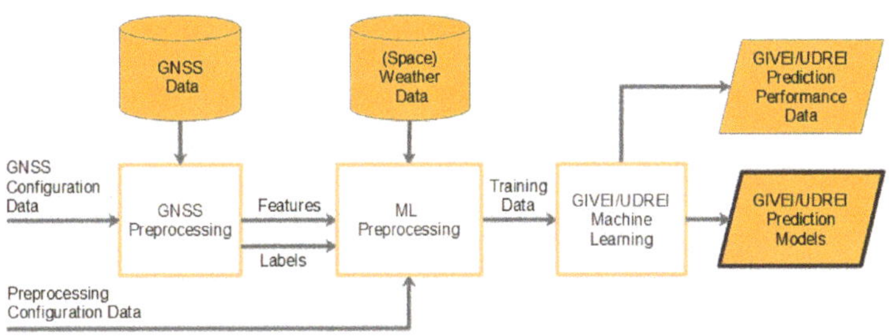

Figure 2. The training process. GNSS and (space)weather data are gathered and pre-processed before being fed to the training module. The training produces both GIVEI and UDREI prediction models that will become part of the operational prediction system and related performance data.

Machine learning algorithms are quickly evolving and there is an obvious 'best' algorithm for predicting the UDREI and GIVEI values based on our selected input features. Both UDREI and GIVEI values are categorical labels representing a given (discrete) class, the prediction of which can be based on regression, time-series, and classification algorithms. Because the 'unmonitored' UDREI and GIVEI values are assigned based on special rules, regression algorithms were deemed inappropriate. Similarly, the use of time-series was excluded due to the impact of sudden changes in geometry when satellites go in- and out of view of the RIMS stations. We therefore decided to focus only on multiclass classification algorithms. The multiclass classification problem type narrowed down the potential algorithms that are applied as part of best practices.

There are two main approaches for classification: decision tree-based methods and (deep) neural network-based algorithms. We tested both types of algorithms, as well as mixed architectures running deep-learning models, and ended up trying the following algorithms: decision tree, gradient boosted decision trees, XGBoost, LightGBM, CatBoost, CNN and CNN + MLP. For the deep-learning models, which were tried for GIVEI only, we have tested multiple values for number of layers, number of units per layer, pool types, padding, momentum, learning rates, optimizers, and architectures.

For both GIVEI and UDREI, the decision tree algorithms were fast to train, but their accuracy figures are not impressive. LightGBM provided the best performance overall and an acceptable training time and therefore was the algorithm we ended up using for both UDREI and GIVEI prediction. Catboost, XGBoost, and deep-learning methods also provided acceptable performances and training times. The LightGBM models were eventually trained using over 5 years of training data, which were subsampled to provide 50,000 examples of each UDREI and GIVEI outcome.

5.4. Performance Evaluation

The performance of our LightGBM models was evaluated using k-folding cross-validation using 30% of the training data. LightGBM's performance can be compared to that of the macro models provided by ESA by computing confusion matrices that show to what extent actual (observed) GIVEI and UDREI values match the predicted values for the validation data. Ideally, only the diagonals of the matrices would be populated, see Figures 3 and 4. The figures clearly show that our model outperforms the macro model significantly and the outcomes are clustered significantly more around the diagonal. The percentual improvements on the correct estimations are also provided in Tables 1 and 2.

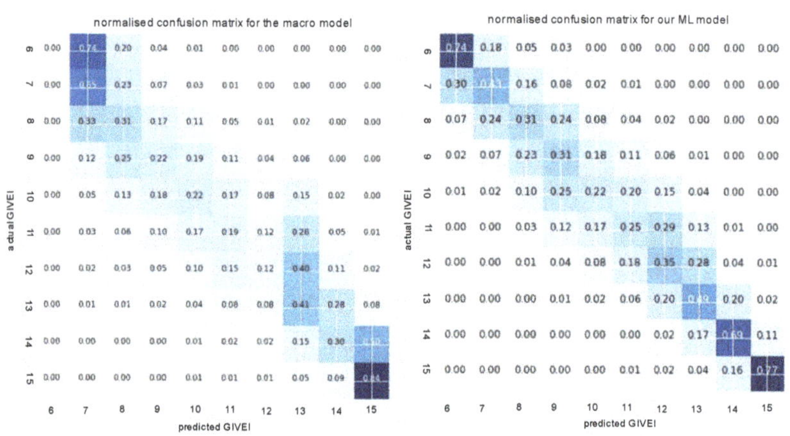

Figure 3. Confusion matrices for GIVEI prediction of the macro model and our ML model. Darker colors correspond to higher percentages of outcomes falling into the tile.

Table 1. GIVEI prediction: percentages of correct outcomes for each class.

GIVEI	Macro Model	ML
6	0	82
7	72	44
8	30	34
9	18	21
10	18	27
11	15	24
12	10	37
13	35	46
14	28	69
15	81	72

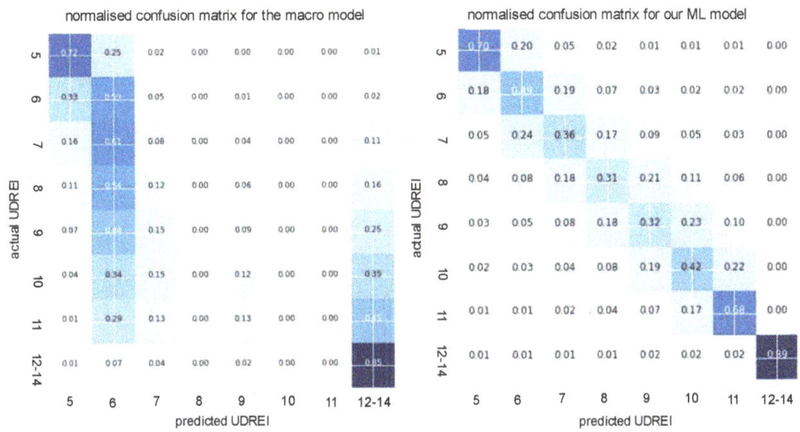

Figure 4. Confusion matrices for UDREI prediction of the macro model and our ML model. Darker colors correspond to higher percentages of outcomes falling into the tile.

Table 2. UDREI prediction: percentages of correct outcomes for each class.

UDREI	Macro Model	ML
5	72	70
6	59	49
7	8	36
8	0	31
9	9	32
10	0	42
11	0	68
12–14	85	89

6. User-Level Results

An assessment was conducted for three example scenarios to evaluate the accuracy of the machine learning application for different ground- and space-segment configurations. The metric used is the protection level, as calculated from the UDREI and GIVEI, for users located on a regular grid across Europe:

- Scenario 1 ("nominal case") runs from the 17th to the 19th of September 2016, with a nominal ionospheric environment and no RIMS or GPS satellite outages;
- Scenario 2 ("GPS outages") equals Scenario 1 but with GPS satellite outages (loss of G18 on 17/9, loss of G18 and G20 on 18/9, and loss of G18, G20 and G21 on 19/9);
- Scenario 3 ("RIMS outages") equals Scenario 1 but with RIMS outages (loss of Golbasi on 17/9, Golbasi and Sofia on 18/9 and Golbasi, Sofia and Athens on 19/9).

To investigate cases with assumed GPS and RIMS outages, the GIVEI and UDREI values actually broadcasted from EDAS are invalid, and reference GIVEI and UDREI values were therefore produced by the European Space Agency using their EGNOS v2 Performance Engineering Platform (SPEED), which uses the same algorithms as the actual EGNOS ground segment.

To compare how close the machine learning predictions are to the real EGNOS v2 performances, we first compare predicted and actual the EGNOS LPV200 availability service areas. This is the area where the protection levels are below the Alert Limits (40m horizontally and 35m vertically). This criterion is quite sensitive to UDREI and GIVEI changes. The resulting availability map comparisons are shown in Figure 5. The availability prediction at the centre of the service area is excellent, but towards the borders of the coverage the performance degrades somewhat and the availability is both over- and underestimated in some areas. When the satellite constellation is degraded, the availability at the edges becomes underestimated in more cases and some degradation near the core of the coverage area can be seen. A reduced number of RIMS makes the predictions only slightly worse at the edges of the coverage, with some additional over- and underestimated regions of limited size.

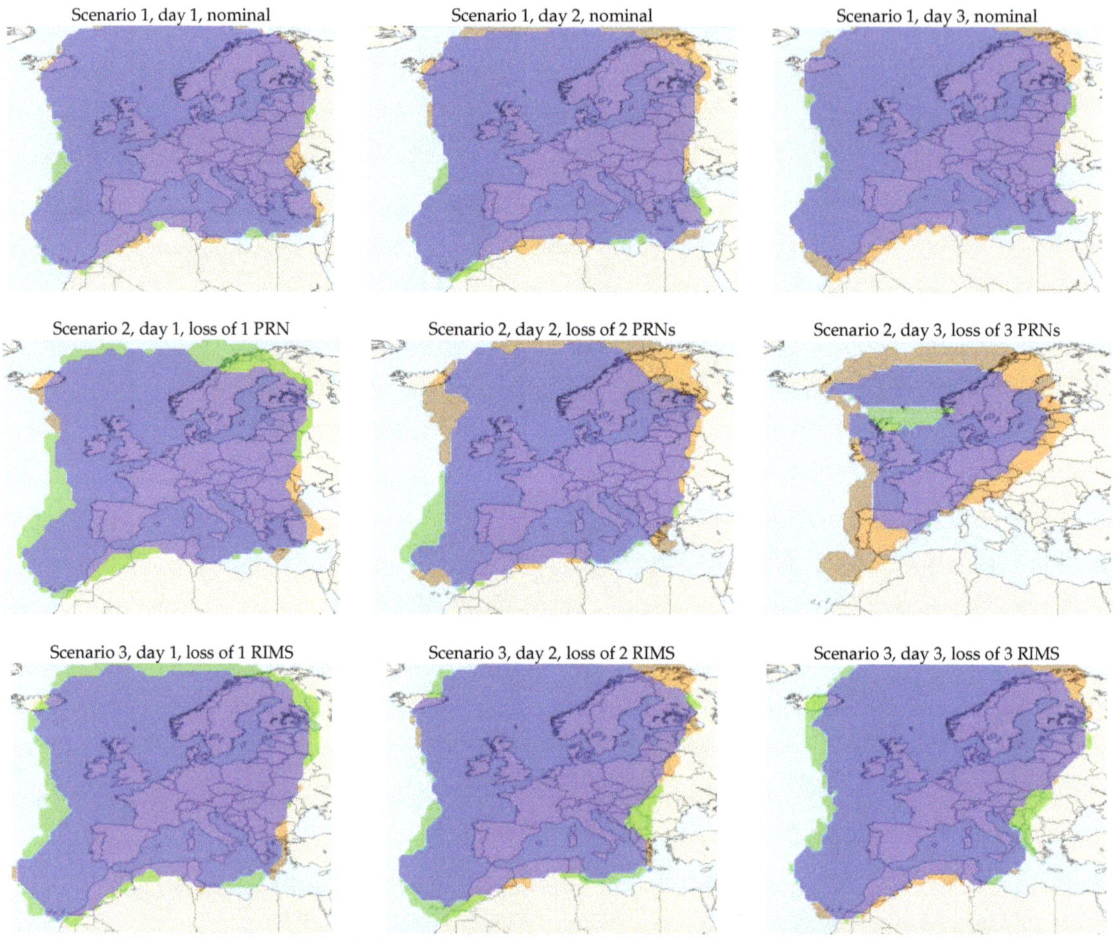

Blue: ML Availability > 0.99 and Actual EGNOS v2 Availability > 0.99 (predicted matches actual availability)
Green: ML Availability > 0.99 and Actual EGNOS v2 Availability < 0.99 (predicted exceeds actual availability)
Orange: ML Availability < 0.99 and Actual EGNOS v2 Availability > 0.99 (predicted falls behind actual availability)

Figure 5. Machine learning vs. EGNOS v2 (LPV200 99% availability differences). In the green areas, EGNOS performance is overestimated, while in the orange areas it is underestimated.

As a second comparison, some time series of vertical protection levels for users near the east and north borders of the service area (Bulgaria and the north of Finland), as well as the centre of this area (South of France) have been produced, see Figure 6. The protection levels from the core of the service area can be predicted well, as was also found in [2]. This is likely because low values of UDREI and GIVEI are common and these are predicted with high accuracy as evident from Figures 3 and 4. Towards the edges of the coverage area, our algorithm tends to be somewhat conservative and predict larger-than-actual protection levels. The conservatism increases somewhat in the presence of satellites and RIMS outages. In some cases, the actual protection level is unavailable, while the prediction still produces valid protection levels. This has been investigated by ESA and corresponds to rapid

GIVEI variations with short periods of 'Not Monitored' status, which is not reproduced by the machine learning module due to the relatively large sampling time used in these simulations. A reduction in the sampling time could improve the prediction of such events.

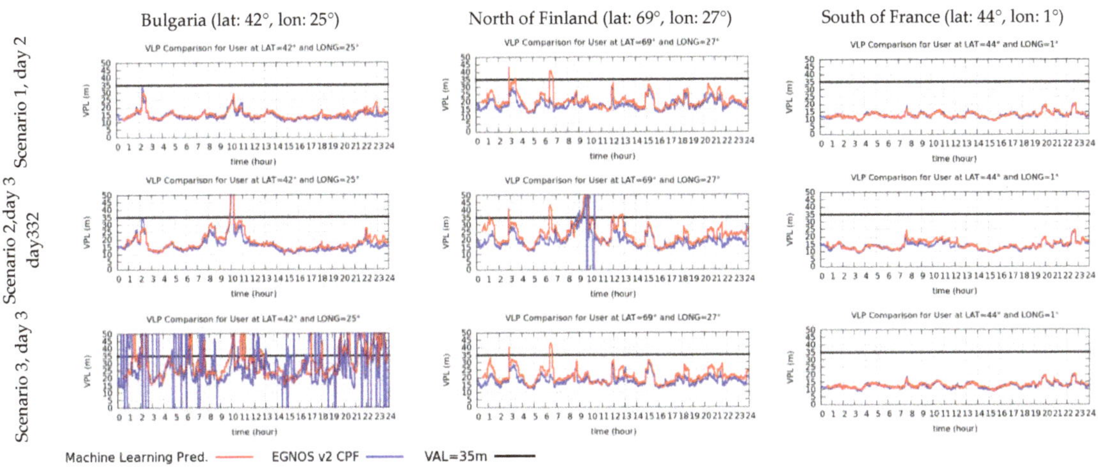

Figure 6. Daily VPL time series for three different locations: ML versus EGNOS v2.

7. Conclusions

We have presented a machine learning-based solution to predict the performance of the current version of the EGNOS system called EGNOS v2. Our solution provides a considerably better performance than the macro model that we compared it against, in terms of the accuracy of the predicted UDREI and GIVEI. To what extent this translates into operational gains and reduced risks for actual users of EGNOS will be investigated in the next (and final) phase of the project, during which the user-level performances of the macro model and our machine learning approach will be assessed more thoroughly.

Author Contributions: Conceptualization and methodology, P.B.O., J.D., T.P., J.N.; C.F., M.D. and G.B.; software, J.N. and C.F.; validation, P.B.O. and M.D.; writing—original draft preparation, review and editing, P.B.O.; visualization, P.B.O. and M.D.; project administration, P.B.O.; funding acquisition, P.B.O. All authors have read and agreed to the published version of the manuscript.

Funding: The reported work was carried out under ESA NAVISP Element 1, which is devoted to the development of innovative PNT, systems, technologies, algorithms, and techniques.

Institutional Review Board Statement: Not applicable.

Informed Consent Statement: Not applicable.

Data Availability Statement: Data sharing is not applicable to this article.

Acknowledgments: Thanks to our anonymous reviewer for the constructive suggestions.

Conflicts of Interest: The authors declare no conflicts of interest. The research received funding from the European Space Agency, who have assisted in providing the visualization of the results.

References

1. Vázquez, J.; Sánchez, M.A.; Cegarra, J.; Tejera, P.D.; Martínez, P.G. The EGNOS NOTAM Proposals Service: Towards Full ICAO Compliance. In Proceedings of the ION GNSS+ 2013, Nashville, TN, USA, 16–20 September 2013; pp. 301–315.
2. Pérez, R.R.; Moreno, V.C.; García, J.M.; Rojo, E.R.; Sánchez, M.Á. Enhanced EGNOS NOTAM Predictions Through Machine Learning Techniques. In Proceedings of the ION GNSS+ 2021, St. Louis, MO, USA, 20–24 September 2021; pp. 364–384.
3. ESSP Website. Available online: https://egnos-user-support.essp-sas.eu/services/safety-of-life-service/forecast-performance/ (accessed on 11 September 2023).

4. *RTCA DO-229E*; Minimum Operational Performance Standards (MOPS) for Global Positioning System/Satellite-Based Augmentation System Airborne Equipment. RTCA, Inc.: Washington DC, USA, 2020.
5. Westbrook, J.; Rerolle, A.; Blomenhofer, H.; McAnany, I.; Cosmen, J.; Werner, W. An Introduction to the EGNOS Central Processing Facility (CPF). In Proceedings of the IAIN World Congress and the 56th Annual Meeting of the Institute of Navigation (2000), San Diego, CA, USA, 26–28 June 2000; pp. 6–13.
6. Sardon, E.; de la Fuente, J.A.; Zarraoa, N.; Nieto, J.; Cosmen, J. UDRE computation: A key issue for space-based augmentation system performance. In Proceedings of the 11th International Technical Meeting of the Satellite Division of The Institute of Navigation (ION GPS 1998), Nashville, TN, USA, 15–18 September 1998; pp. 2091–2100.
7. SBAS Simulator. Available online: http://iguassu.cz/wp-content/uploads/2015/10/SBAS-Simulator2-brochure.issue0rev1.pdf (accessed on 11 September 2023).

engineering proceedings

MDPI

Proceeding Paper

Resilience in Maritime Transport for the Next Decade †

Alina Colling [1] and Stephan Procee [2,3,*]

1 ABB Marine and Ports, 49460 Hamina, Finland; alina.colling@no.abb.com
2 Academie Maritiem Instituut Willem Barentsz, 8881 EG Terschelling, The Netherlands
3 NHL Stenden University of Applied Sciences, 8900 CB Leeuwarden, The Netherlands
* Correspondence: stephan.procee@nhlstenden.com

† Presented at the European Navigation Conference 2023, Noordwijk, The Netherlands, 31 May–2 June 2023.

Abstract: This vision paper covers the increased complexity of navigational environments and presents an innovative navigation system that can support the human in this task. Shipping in general also requires a mean to support navigators reduce the risk of accidents due to smaller passing distance or a shorter reaction time. An innovative approach called Velocity Obstacles might help the operator to avoid intruding into an object's protected zone by visualizing the problem space in relation to the maneuvering potential of the Own Ship.

Keywords: velocity obstacles; protected zone; transparency; collision avoidance systems

Citation: Colling, A.; Procee, S. Resilience in Maritime Transport for the Next Decade. *Eng. Proc.* **2023**, *54*, 49. https://doi.org/10.3390/ENC2023-15416

Academic Editors: Tom Willems and Okko Bleeker

Published: 14 December 2023

1. Introduction

Conventional vessel navigation thus far has always made use of the various tools at hand to determine the best navigational solution for an individual vessel. Limited consideration is taken of the entire situation as this is a very complex task to perform. With more computational power available and navigational environments becoming more multifaceted with not only an increasing number of vessels, but also additional obstacles such as wind farms becoming more prominent on navigational routes, a need for improved situational awareness is rising. To optimize the navigational performance of an individual vessel and reduce the navigation risk, a shift of the navigation perspective from an own vessel-centric perspective to a situation-centric perspective, taking into consideration the constraints of all participants, has to take place.

This article advocates for this shift by explaining the problem of stress creation during conventional sailing by guiding the reader through a real scenario of Automatic Identification System (AIS) recordings. It then explains the concept of the Protective Zone around vessels and their use within Velocity Obstacle (VO) diagrams, as a stepwise solution to gain a better visualizing and understanding the consequences of navigational changes in complex navigational situations. To do so, this article provides a step-by-step time-lapse of a real navigational recording from two different vessel perspectives and explains how VO could have helped reduce the navigational risk. It is also addressed how such visualization tools enable transparent monitoring, supporting the integration of automated or autonomous navigational systems.

2. Velocity Obstacles a Novel Approach to Avoiding Danger for Collision

2.1. Analysis of Conduct

The conduct of vessels is regulated by the IMO's COLREG [1]. Due to the absence of external traffic control, i.e., outside Vessel Traffic Service (VTS) areas in a nation's territory, this conduct can be regarded as self-regulated. Based on formal education, training and licensing, the professional mariner behaves responsible and prudent, and applies the rules. Hence, it can be expected that danger for collision is mitigated at all times.

However, individual ships have their individual constraints and each Officer On Watch (OOW) has their individual interpretation of a safe passing distance. The dimension of a Protected Zone (PZ) or a safety limit, like the Distance Closest Point of Approach (dCPA), is neither exchanged nor mutually agreed. This can result in a stressful situation for the stand-on vessel that has to wait for measures from the give-way vessel with a smaller safety limit.

Although some general idea about safe passing distance exists can be analyzed that in reality, this distance shows quite some variation. From observed AIS data during four months in 2021, a selection of 113 encounters in the area east of the F-3 buoy were analyzed. The F-3 buoy functions as a central point in the Belgium–UK bound traffic and the North Sea–English Channel traffic. All of these encounters consisted of two vessels with a dCPA less than three miles, and one of the two, the Northbound vessel, showing an action. (see Figure 1).

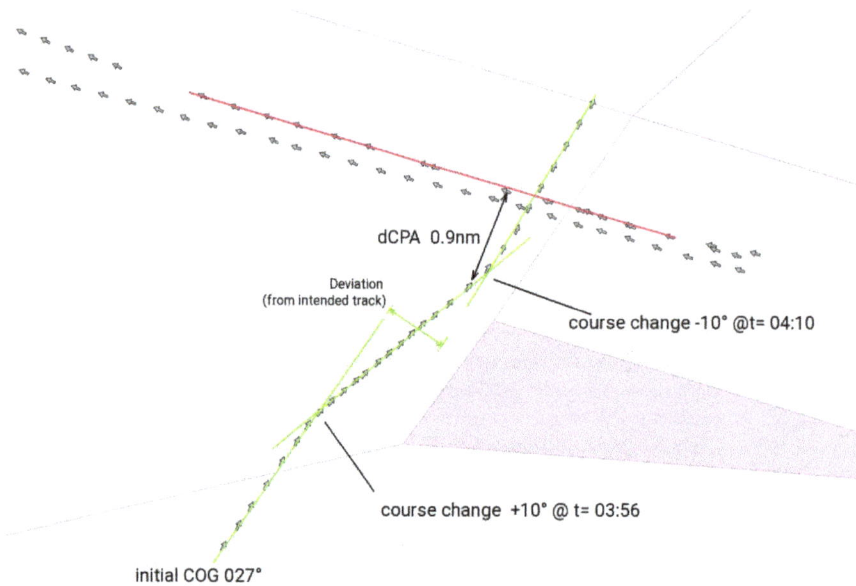

Figure 1. Typical Selected Encounter.

It was found that practically all reacting vessels applied only a course change to starboard. This change varied from 5° to 60° (m = 15°). The resulting safe passing distance (dCPA) varied from a third nautical mile (about 660 m.) to two miles. The amount of course change and the safe passing distance appeared not to be correlated. From this analysis, it might be concluded that neither a common safe passing distance nor a common course change seems to exist. This makes predicting the other vessel's behavior difficult. The assumption that the reaction time, i.e., the time between the moment of course change and the moment of CPA, relates to the amount of course change (e.g., a late reaction requiring a drastic maneuver) could not be proven. This corroborates the hypothesis that a common method to conduct in accordance with the rules does not exist.

2.2. Dimension of PZ

Many studies have been spent on the required protected zone, also referred to as the ship domain. Apart from these theoretical approaches, there actually exists an area around the Own Ship (OS) that inhibits an emotion of uneasiness by the OOW. From research carried out by cadets at sea, it was found that the often intuitive ship domain has a dimension of about one and a half mile in front of OS, half a mile astern of it and one mile abeam. This research was carried out on merchant vessels underway with lengths between 150 and 200 m.

Analyzing traffic based on AIS broadcasts can also determine the area that a ship apparently wants to keep free from other traffic. Several studies, Fujii and Tanaka [2], Goodwin [3], Pietrzykowski and Magaj [4], Procee, Borst, van Paassen, and Mulder [5] a.o., show all slightly different dimensions than the intuitive area based on the cadets' research. They all, however, provide proof of the very existence of an area around the OS that the OOW wants to keep free from dangers.

It might be striking to observe that, in the marine world, collision avoidance as well as the ship's domain is almost uniquely approached from the OS's perspective, meaning, the self-imposed obligation to keep traffic outside OS's own domain. Aviation, on the other hand, uses a uniformly defined horizontal Protected Zone (PZ) of five nautical miles radius around the airplane in which another should not intrude. This different perspective is interesting for marine navigation and might promise a solution for the stressful situation where the stand-on vessel is forced to wait for action of the give-way vessel. In the case that the latter vessel uses a much smaller domain than the stand-on vessel, it might enter, unintentionally, the other's domain and cause great stress for the waiting OOW, potentially resulting in unexpected action or illogical behavior. An example of this can be found in [6].

Ships are, different from airplanes, not highly standardized. Applying a uniformly defined protected zone for shipping might not be feasible due to a lack of acceptance, nor be effective due to the variety of ships and their individual operations. However, if the required PZ of a ship would be shared with the environment directly around the OS, e.g., by a dedicated field in the AIS broadcast, than a novel approach for resolving danger for collision becomes available. This approach visualizes the required PZ of the target with which one is engaged with and predicts whether one is going to intrude into that PZ. In the case that intrusion is imminent, then action is required, if not, then course and speed can be maintained by all vessels involved. The tempting simplicity of this novel approach is that when intrusion can be avoided by e.g., a minor course change long before any engagement can be defined, then the size of the required maneuvers will be minimal, which impacts efficiency, and traffic shows a less chaotic, i.e., more predictable, pattern. Although in this paper it is called a novel approach, a taught strategy at nautical colleges, e.g., in the Netherlands, also refers to taking a preemptive measure, i.e., a minor course change long before the situation starts, to be 'collision avoidance', in order to avoid that situation at all. So, this principle is known and has been effectively used for a long time, while the planning by visualizing the PZ of the other vessel is not. The latter is also known as Velocity Obstacles (VO). In the next paragraph, the method of visualizing is explained based on some realistic examples.

As pointed out in this paragraph, the dissemination of the dimension of one's PZ to vessels that are directly involved in the situation, e.g., within a radius of three to six miles, would provide a number of opportunities. First, it avoids the situation that a vessel with a relatively small PZ unintentionally enters the PZ of a vessel that has a much larger PZ. Because the PZ's dimension of one's own ship is usually not formalized, i.e., intuitively present only, the awareness of a PZ of the target is even less likely. By visualizing the target's PZ and its associated conflict zone it is expected that stressful situations will be reduced because ships can avoid, unintentionally, intruding into the target's PZ. It is expected that this reduced stress level leads to better decision-making. An example of a stressful situation leading to a collision can be found in [6]. Secondly, by visualizing the solution space, the time it takes to decide and reach a resolving action is expected to diminish, hence providing opportunity to increase the quality of decision-making. Based on Wickens, Hollands, Banbury, and Parasuraman [7], their information processing model of decision-making it can be inferred that a repetition of the Diagnosis- > Confirmation- > Selective Attention- > Cue Filtering- > Diagnosis cycle, enhances the quality of the assessment and its subsequent decision-making process. Kahneman's [8] 'slow thinking' refers to a similar process in decision-making.

3. Planning with Velocity Obstacles

Solution Space vs. Problem Space

In their initial paper, Degré and Lefèvre [9] pointed out that predicting the OS's position relative to the target's domain, even based on the simplest form of a circular zone, will provide useful information for OS's OOW. Their visualization differentiates between the 'room to manoeuvre', also known as the solution space and the 'danger zone', which is commonly referred to as the 'problem space'. Visualizing the solution space provides multiple combinations of course and speed that will guarantee a passing distance greater than the critical distance, i.e., the circumference of the target's domain. In particular, when two or more targets are to be dealt with, the prediction of the combined solution space helps to choose an evasive course and speed that solves all problems at the same time. Also, Huang, van Gelder, and Wen [10], and Westrenen & Ellerbroek [11] provide evidence that the method of VO, i.e., visualizing the problem space and solution space, enables the OOW to resolve complicated traffic situations with multiple targets involved.

An example of traffic conduct that is observed during the four months of AIS data collection is shown in Figure 2.

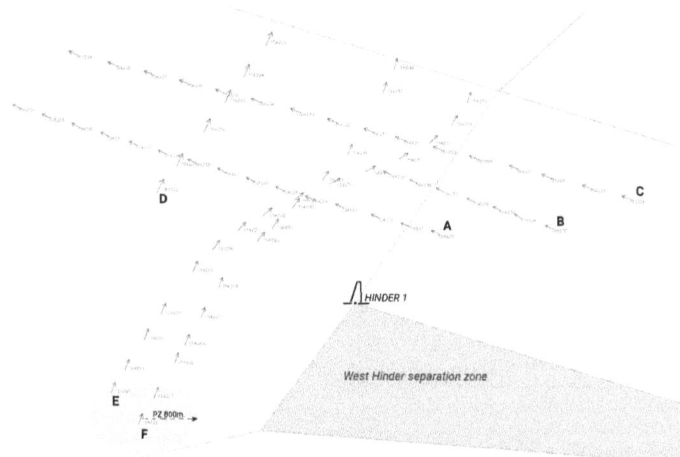

Figure 2. Example Maneuver showing Westbound vessels A, B and C crossing the northbound traffic lane. Vessels D, E and F are Northbound. The shaded circular protected zone (PZ) of 800 m. radius is shown around vessel F.

In the given example, there are six vessels sailing near the exit of the Traffic Separation Scheme (TSS) West Hinder into the Dover Strait TSS. Every AIS broadcast is considered to be a vessel underway, its position is reflected in the position of the arrow (in blue). The direction of the arrow corresponds with the course over ground and the speed over ground is expressed by the length of the arrow. The time interval is limited to one minute, the relative time stamp is shown as number in red with one minute time resolution.

Three vessels, i.e., A, B and C, are leaving the West Hinder TSS on a westerly course, and three vessels, D, E and F, are heading for the North Sea on a northerly course. According to collision regulations, the three ships on the northerly course are to give way when risk for collision is deemed to exist. Apparently, the north bound vessel sailing in the western part of the area, vessel D, did not assess the situation as 'risk for collision'. The other north bound vessel, vessel F, assesses the situation as risky and decides to change course to starboard at about time stamp 154228. The third northbound vessel, vessel E initially time stamped 154220, assessed the situation as such that a drastic course change, i.e., 35°, to starboard was executed, which resulted in a bow crossing of the second northbound vessel. Ultimately both vessels, E and F, passed the three westbound vessels astern. This situation developed in the early morning around half past three GMT. Weather analysis showed there was a declining ridge of high pressure from Scandinavia to the Iberian peninsula. As the wind gradient was weak at the F-3 location, there is no reason to expect much wind. Some disturbances in the southern North Sea and NW Europe hint at vertical movement of the atmosphere, hence, a small chance for low visibility at the surface. This means that it is likely that the ships in this example sailed in sight of each other and the normal rules, i.e., not Rule 19, were applicable.

It is assumed that each of the ships in the given example has a circular protected zone with a tentatively chosen radius of four cables, i.e., ±800 m. This is illustrated as the gray-shaded oval in Figure 2 for vessel F only once. The explanation for the apparent oval shape is the scale difference between the major and minor axis at this latitude in the Mercator projection. Buoys and fairway markers, usually referred to as Aids To Navigation (ATON), can be dedicated to have a protected zone as well. The dimension of this dedicated PZ depends on the navigator because no formal rule applies. In the given example, the Hinder-1 buoy is to be avoided by applying a PZ with a tentatively chosen distance of 185 m., i.e., a tenth of a mile.

A third feature that determines the navigable space is Routing Measures, also known as Traffic Separation Schemes. In this case, the Separation Zone, the light magenta shaded area in Figure 2 is a forbidden area, the northbound vessels are not allowed to enter the exit of the West Hinder TSS, i.e., the traffic lane where the west bound vessels are coming from. This fairway restriction is not shown in the following example.

The developing traffic situation is illustrated in Figures 3 and 4. Each of these Figures shows the same situation albeit from two different perspectives. Figure 3 refers to vessel E which is represented by the pink protected zone. Figure 4 refers to vessel's C perspective. Vessel C is represented by the green oval.

The top half of each row shows the developing situation in the chart, the lower half of the row shows the orthogonal projection of the conflict zones of each target from either, vessel's E perspective (Figure 3), or vessel's C perspective (Figure 4).

At t = 22, vessel E is northbound underway with a speed of 14 knots over ground. It has one northbound vessel, i.e., vessel F with blue shaded PZ, on its starboard quarter. Also, there is a northbound vessel on E's port bow, i.e., vessel D with a yellow shaded PZ, and there is one westbound crossing vessel, i.e., A with the purple PZ. The Hinder-1 buoy is on the northwestern corner of the West Hinder Separation zone (See Figure 2).

From the perspective of vessel E, the Conflict Zones (CZ) of vessel F, vessel A and the Hinder buoy (Hby) are shown (See Figure 3 t = 22 lower half). The CZs are displayed relative to the OS's bow, also known as the Head Up orientation, the red arrow in this diagram indicates the OS's speed and course. From this diagram it becomes clear that vessel E has no conflict with vessel F as long as E keeps its speed. It also is clear that E does not have a conflict with A as long as E does not speed up. There is neither a conflict with the Hinder buoy at this course. Hence there is no need to take action at the present moment (t = 22), and if vessel E should want to change course, it finds potential combinations of course and speed to do so, i.e., without intruding in someone's PZ, in the solution space. The vessels B, C, D are not shown in the diagram for the reason that their CZs are outside the area of interest which in this case is limited to 3 miles.

At t = 24 the CZ of vessel C shows up in the diagram, and from the position of the OS's arrowhead, that is located inside C's CZ it becomes clear that intrusion into C's PZ will happen at some time. At t = 25, vessel E has started to change course to starboard, at t = 27 the effect of this course change becomes noticeable as the OS's arrow head shifts gradually to the border of the CZ. At t = 29, it becomes obvious that vessel E will not intrude into C's PZ because of its course change. The course change was found in the solution space, meaning that no conflict arose as a result of the action to resolve the first conflict.

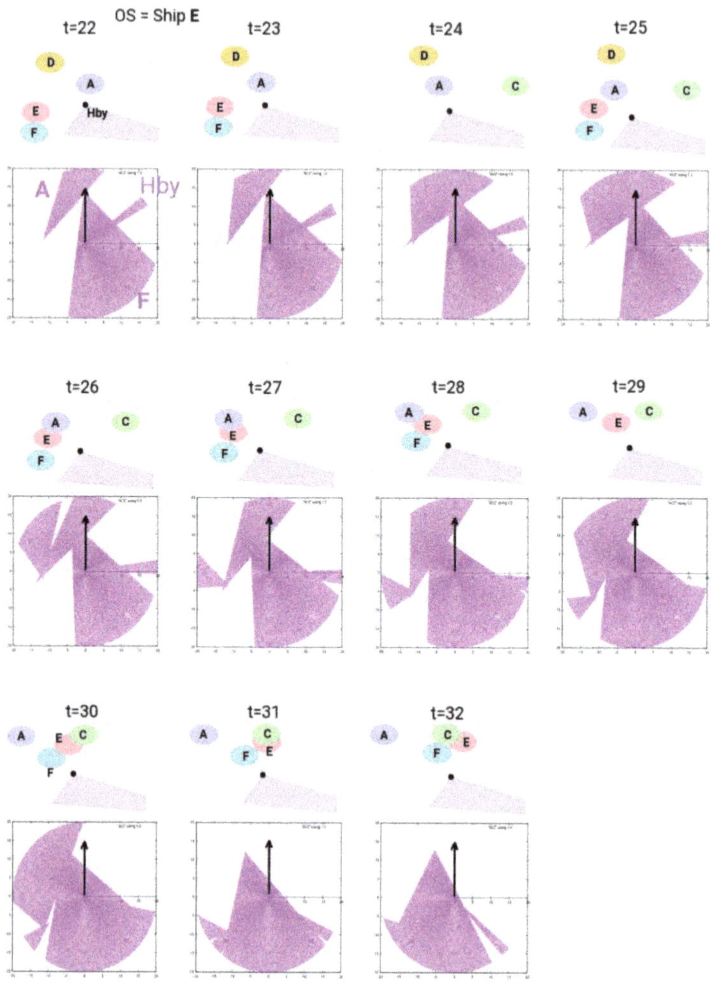

Figure 3. Velocity Obstacles from Vessel E's perspective (vessels B C D are not shown).

The identical situation is presented from the perspective of vessel C in Figure 4. At t = 22, the CZ is shown for vessel D (to starboard of OS) and for vessel E. Also, the CZ of the Hinder buoy (Hby) is shown, to port, originating from OS. From the head of OS's vector (red) it appears that intrusion in the PZ of vessel E is not expected. As the situation develops over time, the CZ of vessel D shifts and increases in dimension. At t = 24, the head of OS's vector is inside the CZ of vessel E, indicating that intrusion in its PZ is expected (in time). At t = 26, a part of the CZ of vessel F appears separate from the overlapping part with E's CZ. It also appears that the head of OS's vector shifts gradually toward the margin of E's CZ indicating that E's course change to starboard (at t = 25) has an effect. The apparent continuation of the effect is illustrated in the subsequent fragments until t = 29, where it can be seen that OS is neither expected to intrude into the PZ of vessel E nor into the PZ of vessel F.

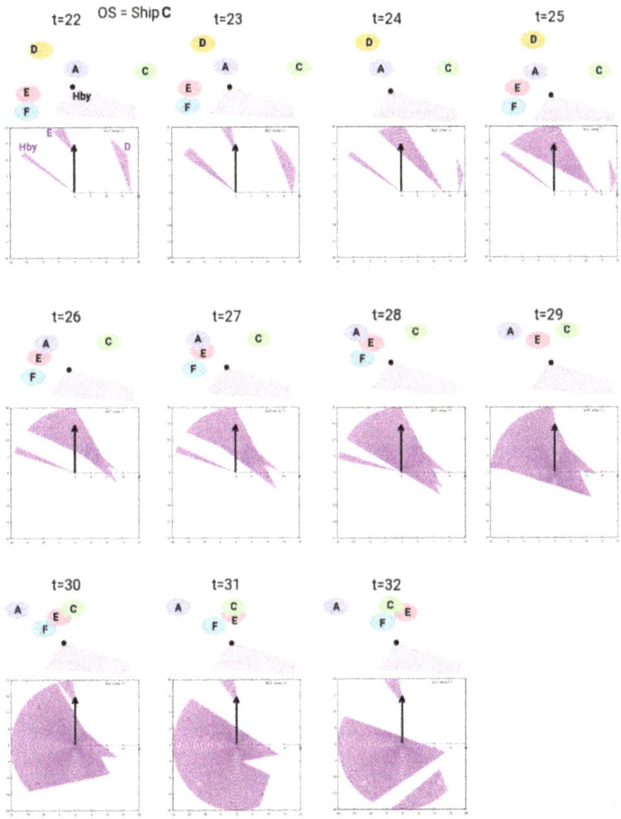

Figure 4. Velocity Obstacles from Vessel C's perspective.

If OS (vessel C) assesses the situation as danger for collision, e.g., at t = 24, it might be tempted to take action. Objectively, OS might be tempted to find any combination of course and speed that will result in a position of OS's vector head outside the presented CZs, i.e., inside the solution space. Changing course, however, might have the greatest effect in a limited time due to the maneuvering potential of vessels. As shown in the fragments, a course change to port is also part of the solution space. However, this would result in a heading towards the target (s) that ship C wants to avoid, hence the preferred change to starboard if such a maneuver is desired by the OOW of ship C. From the viewpoint of ColRegs there would only be reason for vessel C to maneuver when danger for collision is deemed to exist and the give way vessel, i.e., ship E here, is deemed to be unable to avoid the collision on its own.

4. Using the VO in Automated System Settings

The use of tools such as the VO play a key role in conveying trust to automated and future autonomous systems. They enable sufficient interpretation and transparency in the decision-making process of the system. It is vital for the users to judge whether they are willing to trust the system and not simply intervene at every instance where the system does not perform the same way a human operator would. The visual simplification of the situation with the VO is needed to allow effective monitoring, and also enables a faster situational awareness overview when coming from a temporary mind-off navigation task. In an emergency situation, the effective human-to-machine takeover capability forms the foundation to demonstrate safe operation a pre-requisite to regulatory compliance of highly automated or autonomous systems.

The accurate visualization of the collision scenarios surrounding the vessel is reliant on the sensor data available. The merger of different sensors information isn't new, as superimposing radar and AIS on the ECDIS is a common practice on most vessels, however, the processing of this data for the decision-making process of systems is the new aspect in navigation and control systems. The uncertainty of sensors measurements in, for instance, GPS positioning, is commonly accounted for through Kalman filters in control systems. Yet, for situational awareness systems the strengths and weaknesses of different target identification methods AIS, X-/S-/W-band radars, LIDAR or cameras need to be carefully considered for different operating conditions. Not only are the HW components of varying cost levels thereby not making each applicable for all applications, yet even if they are all available, they each have varying update rates and operational envelope/performance which affects their data reliability. For instance, in time-critical situations in busy waters surrounded by vessels, not all vessels have AIS. Also, the 25 rotations per minute for standard marine radar and the target following algorithm translates to an ARPA target in 1–2 min. This is too slow to perform reliable risk assessments based on only this sensor data for the moving targets that surround the vessel. Even though one could interpolate the estimated position based on the data history, there is no certainty that this data of this interpolation is accurate. In these situations, the continuous object detection using machine vision can support. In turn, the machine vision system has limited operational envelopes based on the visibility conditions and is not as accurate when for long-distance target estimations. Thus, based on the type and the movement of the targets the sensor fusion needs to be able to decide which data information is the most trustworthy.

5. Conclusions

This article has provided a concrete example of stress creation in a commonly encountered situation and explained the importance of Protective Zones for not only manual operations, but also as a commonly used principle in autonomous collision avoidance. While the topic of PZ and VO are not new in research and are used in the aerospace industry, they have not yet taken off in the maritime context. The authors believe that the increased technology available, combined with the greater navigational complexity for today's navigational needs shines a new light on the concepts making their benefits more visible and tangible to support the navigational crew in their tasks.

To illustrate this use, a collision avoidance scenario was analyzed from two different vessel's perspective using the aid of VO diagrams to emphasize how such visual support can reduce stress and collision risks and, in the future, become a vital monitoring tool for partially unmanned bridges. To enable such concepts to be realized the authors suggest a technical adaptation to the AIS messaging or any future kind of communication technologies to include Protective Zone dimensions of individual vessels.

It is acknowledged that this proposed solution is not a silver bullet. There are numerous valid weaknesses of the AIS including reliability that are not to be denied. Additionally, as was addressed by other authors (Zhang, Kujala & Wang [12], Rawson & Brito [13]), PZ have their limitations, being influenced by the geographical factors surrounding the vessels as well as the operating mode the vessels are operating in. More limitations will reveal themselves as the application rolls out and becomes common use in the daily navigation, some of which will be addressable with machine learning technologies individualizing the parameters for different navigational participants. Yet, the step of the AIS message adaptation can be the first step in creating a mindset change, and enables advisory systems such as VOs to become more reliable than being based only on the assumption of the technology developers.

Author Contributions: Conceptualization, A.C. and S.P.; methodology, S.P.; software, S.P.; validation, A.C. and S.P; formal analysis, S.P.; investigation, S.P.; resources, S.P.; data curation, S.P.; writing—original draft preparation, A.C. and S.P.; writing—review and editing, A.C.; visualization, S.P.; supervision, A.C. and S.P.; project administration, S.P.; All authors have read and agreed to the published version of the manuscript.

Funding: This research received no external funding.

Institutional Review Board Statement: Not applicable.

Informed Consent Statement: Not applicable.

Data Availability Statement: Data sharing is not applicable to this article.

Conflicts of Interest: Author Alina Colling was employed by the company ABB. The remaining author declares that the research was conducted in the absence of any commercial or financial relationships that could be construed as a potential conflict of interest.

References

1. International Maritime Organization IMO (I.M.O.). *The International Regulations For Preventing Collisions At Sea (COLREGS)*; International Maritime Organization IMO (I.M.O.): London, UK, 1972.
2. Fujii, Y.; Tanaka, K. Traffic Capacity. *J. Navig.* **1971**, *24*, 543–552. [CrossRef]
3. Goodwin, E.M. A Statistical Study of Ship Domains. *J. Navig.* **1975**, *28*, 328–344. [CrossRef]
4. Pietrzykowski, Z.; Magaj, J. Ship Domain as a Safety Criterion in a Precautionary Area of Traffic Separation Scheme. *Int. J. Mar. Navig. Saf. Sea Transp.* **2017**, *11*, 93–98. [CrossRef]
5. Procee, S.; Borst, C.; van Paassen, R.; Mulder, M. Using Augmented Reality to Improve Collision: Avoidance and Resolution. In Proceedings of the 17th International Conference on Computer and IT Applications in the Maritime Industries, Pavone, Italy, 14–16 May 2018; pp. 237–249.
6. Transport Malta. Marine Safety Investigation 201512/005. *December 2016.* Available online: https://maritimesafetyinnovationlab.org/incident-reports/ (accessed on 14 February 2023).
7. Wickens, C.; Hollands, J.; Banbury, S.; Parasuraman, R. *Engineering Psychology and Human Performance*; Routledge: New York, NY, USA, 2016.
8. Kahneman, D. *Thinking Fast and Slow*; Penguin Books: London, UK, 2012.
9. Degré, T.; Lefèvre, X. A Collision Avoidance System. *J. Navig.* **1981**, *34*, 294–302. [CrossRef]
10. Huang, Y.; van Gelder, P.; Wen, Y. Velocity obstacle algorithms for collision prevention at sea. *Ocean Eng.* **2018**, *151*, 308–321. [CrossRef]
11. Westrenen, F.; Ellerbroek, J. The Effect of Traffic Complexity on the Development of Near Misses on the North Sea. *IEEE Trans. Syst. Man Cybern.* **2017**, *47*, 432–440. [CrossRef]
12. Zhang, W.; Goerlandt, F.; Kujala, P.; Wang, Y. An advanced method for detecting possible near miss ship collisions from AIS data. *Ocean Eng.* **2016**, *124*, 141–156. [CrossRef]
13. Rawson, A.; Brito, M. Developing contextually aware ship domains using machine learning. *J. Navig.* **2021**, *74*, 515–532. [CrossRef]

engineering proceedings

MDPI

Proceeding Paper

Smart Port Shuttle: Sensor-Based Navigation for Inland Waterway Transportation †

Katrin Dietmayer [1,*,‡], Jochen Schwenninger [2,‡], Himanshu Gupta [1,‡], Muhammad Saad [1,‡], Melanie Lipka [3,‡], Matthias Overbeck [1,‡] and Wolfgang Felber [1,‡]

1 Satellite Based Positioning Systems Department, Fraunhofer Institute for Integrated Circuits IIS, 90411 Nuremberg, Germany; himanshu.gupta@iis.fraunhofer.de (H.G.); muhammad.saad@iis.fraunhofer.de (M.S.); matthias.overbeck@iis.fraunhofer.de (M.O.); wolfgang.felber@iis.fraunhofer.de (W.F.)

2 Bertrandt Technologie GmbH, 93053 Regensburg, Germany; jochen.schwenninger@bertrandt.com

3 Indie Semiconductors/ Symeo GmbH, 85579 Neubiberg, Germany; melanie.lipka@indiesemi.com

* Correspondence: katrin.dietmayer@iis.fraunhofer.de

† Presented at the European Navigation Conference 2023, Noordwijk, The Netherlands, 31 May–2 June 2023.

‡ These authors contributed equally to this work.

Abstract: Germany has a wide-meshed waterway network. Although inland navigation is one of the most efficient transportation methods, its share of transport is steadily decreasing. The research project Smart-Port-Shuttle Hildesheim aims to strengthen the waterway system both from the aspect of inland navigation and from the aspect of ports. For this purpose, a holistic approach is chosen, which includes technological innovations on the one hand and integration into a logistical concept on the other. This paper focuses on the first point: a technological approach for a (partially) automated inland vessel. It will present the hardware setup and an evaluation of the sensors at a selection of critical points, such as in a lock, at a bridge passage, or during the passing of an obstacle. Different methods, such as Galileo E5a measurements for positioning or Simultaneous Localization and Mapping (SLAM) for constructing and updating the map of the branch canal, are analyzed. The data structure of the Robot Operating System (ROS) is presented, as the raw data are available to the audience.

Keywords: global navigation satellite system (GNSS); light detection and ranging (LiDAR); radar; robot operating system (ROS); inland waterways

Citation: Dietmayer, K.; Schwenninger, J.; Gupta, H.; Saad, M.; Lipka, M.; Overbeck, M.; Felber, W. Smart Port Shuttle: Sensor-Based Navigation for Inland Waterway Transportation. *Eng. Proc.* **2023**, *54*, 50. https://doi.org/10.3390/ENC2023-15438

Academic Editors: Tom Willems and Okko Bleeker

Published: 29 October 2023

1. Introduction

Germany's inland waterways are an extensive network of rivers and canals that provide a crucial transportation infrastructure. They play a vital role in linking Germany's industrial centers with its ports and other transportation hubs. Although inland navigation is one of the most efficient transportation methods, its share of transport is steadily decreasing [1]. In recent years, there has been a growing trend toward the use of partially automated transportation. This is seen as a promising solution to the challenges faced by the shipping industry, such as a shortage of skilled workers, increasing fuel costs, and the need to reduce emissions.

With their reduced complexity—no flowing water, only navigable in one lane, easily separable by the locks—demarcated branch canals in particular offer ideal conditions for the gradual automation of ship navigation. The challenges in planned (partially) autonomous shipping with regard to positioning are similar in the context of sensors for autonomous driving. The research project Smart-Port-Shuttle Hildesheim (SPS) [2] aims for a holistic approach that includes technological innovations on the one hand and the integration of a logistical concept on the other. This paper focuses on the first point: a technological approach to a (partially) automated inland vessel. The aim is to adapt existing sensors from the automotive industry to a vessel and to show technological ways that allow for partially autonomous shipping on German canals. For this purpose, the ship's own control systems have to be analyzed, which include other timing aspects, for example. Therefore, a sensor setup was integrated on a test ship, see Figure 1, in order to run various test scenarios and to record and analyze the data.

Figure 1. The test ship of the SPS project consisting of a head barge, one barge, and a pusher.

Currently, the German inland navigation industry is undergoing a transformation, driven by factors such as digitization, automation, and sustainability. One of the main trends in the industry is the development of partially automated and autonomous inland vessels, which can help to increase efficiency, safety, and environmental sustainability. The Central Commission for the Navigation of the Rhine (CCNR) currently lists 40 pilot and research projects in the field of automation in inland navigation [3]. They cover a variety of aspects of the automation task, like positioning, navigation, motion control, control center development, establishing digital test fields, or the mobility chain.

The feasibility study "Autonomes Fahren in der Binnenschifffahrt" [4] showed the primarily associated opportunities for automated and (partially) autonomous driving in bin shipping. The researchers conclude that it can help to overcome the shortage of skilled workers, reduce costs, increase the level of safety, better link transport carriers to form intermodal and integrated transport chains, and include harbours more in the automatization while developing new processes. A test field is proposed to showcase and test automated and (partially) autonomous driving for ships and to use substantial synergy effects from other research projects. Regarding the sensor setup, they stated that the following sensors are suitable: echo sounders, radar, Global Navigation Satellite System (GNSS) receivers, cameras, Light Detection and Ranging (LiDAR), compasses, inertial sensors, and ultrasonic sensors. This is similar to the SPS setup, which utilizes radar, LiDAR, cameras, and GNSS receivers.

The SciPPPer project [5] aims to develop an assistance system for automated lock entry based on Precise Point Positioning (PPP) and a VHF Data Exchange System (VDES) for inland waterways. Their sensor setup includes the GNSS and near-field sensors (LiDAR), as well as reference systems on the waterway. To compensate for the shadowing effects and guarantee highly accurate positioning, the project team developed a new PPP protocol with a VDES. The SciPPPer technologies are now being further developed and the team is currently building a digital test field for inland vessels on the Spree–Oder waterway [6]. The project encountered some of the same difficulties that SPS was facing, but had a different approach due to their use of reference stations.

The SPS project did not plan to autonomously steer a vessel, but rather to define and showcase a sensor setup which is practical (since ships can be upgraded with it) and enables vessels to navigate in inland waterways and detect obstacles in the water as well as height restrictions, e.g., due to very low bridges. The recorded sensor data add a great value, as they can be reused and different approaches can be tested further.

This paper presents the hardware setup and the evaluation of the sensors at a lock in the Hildesheim canal. Different methods for positioning with GNSS or a SLAM for constructing and updating a map of the branch canal are analyzed. A discussion of the results and their implications and the outlook for the future conclude the paper.

2. Hardware Description and Design

This section provides an overview of the hardware specification and description, as well as the system architecture.

2.1. Hardware Specification and Description

In order to calculate a global position and time, the GNSS is the appropriate solution. Most vessels are already equipped with GNSS receivers to broadcast their position over the Automatic Identification System (AIS). In the scope of this work, a GNSS Receiver with Open Software Interface (GOOSE)© (software version: 1.26.0-71) [7] is used, which was developed by Fraunhofer IIS. This platform records raw GNSS measurements and calculates the corresponding Position, Velocity, and Time (PVT) solution. This receiver allow users to test different configurations according to the requirements.

During the measurement tests, two different data streams were recorded simultaneously. One contains the position and timing solution recorded via an ROS node and the PVT data in an Open GNSS Receiver Protocol (OGRP) format developed by Fraunhofer IIS. This approach allows us to visualize and analyze the data in post-processing. A multi-system and multi-frequency configuration of GPS L1/L5 and Galileo E1/E5a was used in testing. In order to determine the heading of the vessel, two GOOSE© [7] receivers were used, one placed on the head barge and one on the pusher. Each of them was connected to a Tallysman VSS6037L VeroStar™ [8] GNSS antenna.

In order to provide distances to potential obstacles during the ride as well as for docking and coupling maneuvers, LiDAR was selected as the sensor technology. These sensors allow not only the provision of 2D information like monocular cameras, but output a complete 3D representation of the sensor's surroundings. The chosen Velodyne Puck VLP-16 [9] provides a 360° field of view and has a maximum range of approximately 100 m. Since a complete coverage of at least the entire bow of the vessel is required, one sensor was placed at either side of the bow. Using slightly tilted mounting angles for the two sensors, the detection accuracy in the area inside both fields of view was enhanced. After fusion of the individual point clouds, this representation could be used to either generate/enhance a map of the vessel's surrounding and thus improve the vessel's navigation or to detect obstacles on the planned route and thus enable an emergency response.

A radar system was deployed on the front and back of the vessel to detect obstacles and map the surroundings, such as bridges and walls. An AVR Quad DigiMMIC radar from Symeo GmbH [10] was mounted at the front of the vessel. This is a 77 GHz multiple-input-multiple-output (MIMO) evaluation platform based on four cascaded monolithic microwave-integrated circuits (MMICs) and has a total of 12 transmitter and 16 receiver channels. The propagation characteristics of electromagnetic waves at around 77 GHz come in especially handy in difficult visibility conditions like fog and drizzle. Depending on the operation mode, maximum measurement ranges of 250 m or more in a field of view of $\pm 40°$ are possible. It is capable of deriving the distance from the round-trip time-of-flight to the target, the velocity from the Doppler shift, and the angle in the x-plane by evaluating the incident wave at the antenna array of the module. Furthermore, a Halo20 radar from SIMRAD [11] was deployed at the back of the vessel. It is a compact dome radar with pulse compression technology that detects collision hazards and other targets at close range and up to 24 nautical miles away.

In addition to the sensors from above, a monocular camera was installed at the front of the vessel, enabling visual inspection of the situation during the measurement campaign.

2.2. Hardware Architecture

Figure 2 shows the sensor setup on the test vessel. At the pusher, one Car-PC is connected to the HALO20 radar, to one GOOSE© receiver, and one of the Wireless Fidelity

(WiFi) bridges. The GNSS antenna is centrally mounted on the rooftop of the pusher. The setup on the head barge consists of another Linux computer (Car-PC) for data processing, using a second WiFi bridge to connect the two PCs and thus all sensors to one network. The Symeo radar, a second GNSS antenna, and the camera were placed in the front-center of the head barge, whereas two LiDARs were mounted on the left and right of the head barge to enable a 360° view. All sensors on the head barge were aligned in the direction of motion. The Car-PC of the head barge as well as the GOOSE receiver were mounted in a waterproof box. The energy supply of the setup was ensured with a gasoline generator, since the head barge used for the measurement campaign does not provide electricity by itself.

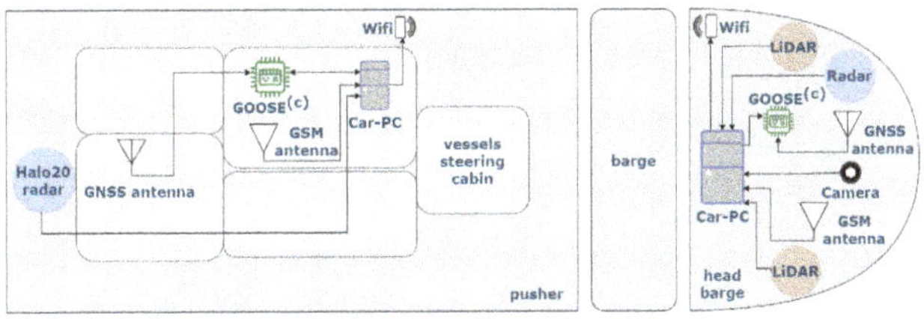

Figure 2. The hardware architecture on the test vessel. Sizes are not to scale.

2.3. System Architecture

Individual ROS nodes were developed for each sensor to convert raw sensor readings into ROS-compliant message types. The ROS framework [12] is a combination of a messaging middleware, predefined messages for various information related to sensor data and actor commands, and an expanding set of software modules (nodes) capable of generating, processing, and visualizing messages geared towards real-time robotic applications. As this open-source framework eliminates the need for implementing all required components from scratch, the development speed is greatly enhanced and project resources can be used for tuning the performance or enhancing critical components with alternative approaches.

One benefit of using an ROS is the ability to record all messages being passed in the system with temporal consistency in archives that can be used to replay the actual situation and test or debug each component, at least in an open-loop scenario where the scenario is not influenced by the processing in an ROS. One key concept of an ROS is the use of multiple coordinate systems called frames that allow for easy integration of different sensor measurements if the transformations between coordinate frames are known. Since the scenarios in the measurement campaign exclude dynamic vessel configurations, the transformations between sensors are assumed to be static and dynamic transformations are used to locate the vessel in its environment. The setup uses the latest version of an ROS, the Noetic Ninjemys [13] ROS; the main components are shown in Figure 3.

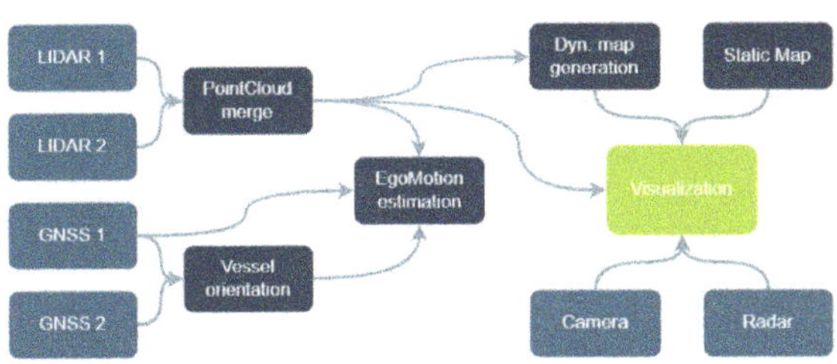

Figure 3. ROS components and data flow.

In the initial phase of semi-automatic operation, visualization of sensor data can assist the captain for an improved situational awareness by providing information about obstacles on a map. Towards the intermediate stages, sensors can not only visualize but also use the real-time data to track the current vessel state and compare it to its ideal trajectory to assist in steering the vessel. With full automation, the system not only collects data but also interprets and acts upon them without human intervention. This includes a full scan of the surroundings, identifying obstacles to avoid collisions, steering the vessel to follow an ideal trajectory on the canal, interpreting signs, connecting with locks and ports, and more.

3. Measurement Campaign

Based on the hardware architecture mentioned in Section 2.2, the setup was prepared and installed on the vessel as shown in Figure 4. The sub-figures clearly show how the various sensors were mounted on the vessel.

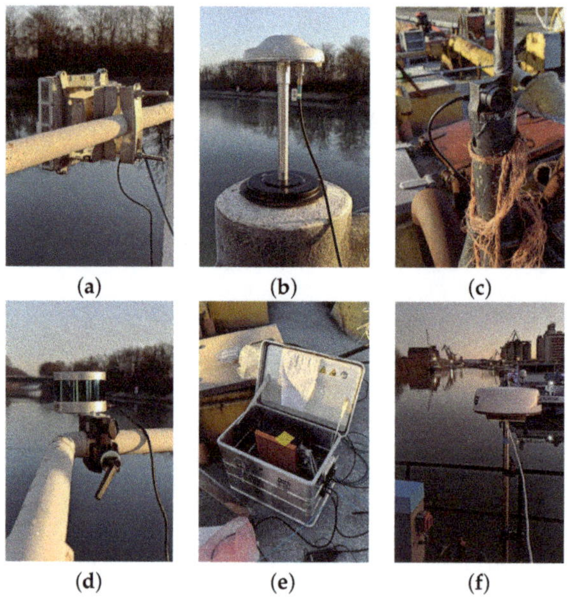

Figure 4. The sensor setup on the head barge, (**a**) Symeo radar, (**b**) GNSS antenna, (**c**) camera, (**d**) LiDAR sensors, (**e**) control box with the PC and GOOSE© receiver, and (**f**) Halo20 radar.

For the measurement campaign, different test scenarios were defined to cover the most challenging aspects for a cruise on the canal. The following enumeration summarizes the experiments and their goals.

1. Lock: Detect the position where the vessel has to be secured.
   The lock, especially when its not flooded, is a challenging environment for GNSS due to the high multipath and low satellite visibility. Local sensors need to precisely detect where the vessel has to stop to be secured for the lock process.
2. Bridge: Detection of bridge heights.
   Bridges on the canal have different heights. Due to the water level, the ship, and the cargo of the ship, it is necessary to detect the height to adjust the cabin on the main ship. This needs to be carried out as soon as possible to stop the vessel if it is to high to pass under the bridge.
3. Canal: Detect the shoreline and possible obstacles.
   The vessel ideally cruises in the middle of the channel for efficiency. Therefore, the detection of the shoreline with sensors is required in order to keep the vessel in the center of the channel. Additionally, any obstacles, like barrels or other vessels, need to be detected.

In this paper, just a few examples are presented in order to show the possibilities of visualization and fusion of sensor data. However, the recordings of all experiments and the complete Hildesheim canal, from the harbour to the lock, are available and analyzed.

4. Results

A functioning hardware setup and data recording was one of the goals of the SPS project. Additionally, the recorded data were analyzed and different post processing methods were applied to showcase some approaches for future positioning in inland waterway canals. First, the GNSS data were processed and different frequencies were used to calculate the position. Second, all sensor data were used in a SLAM approach to fuse them for an overview of the vessel's position and surroundings.

4.1. GNSS Positioning

Figure 5a shows the horizontal position solutions of the GOOSE© receiver using GPS L1 C/A and Galileo E1 combined measurements, and Galileo E5a. E5a results have fewer outliers compared to L1/E1 because of its signal characteristics [14]; it is more robust to multipath signals and interference. Here, the vessel moves into the lock from the south (bottom of the figure). The vessel enters the lock with the water level raised. Due to the high satellite visibility and little to no multipath effects, the position solution has no major outliers. After the vessel enters the lock, it is tied up to the left side of the lock. The lock process begins, and the water levels slowly drops. Due to the higher outer walls of the lock, interference and multipath effects occur, leading to outliers in the position solution. This scenario also has lower satellite availability since the walls limit the direct path of the satellites to the GNSS antenna. This effect continues to be observed when the vessel starts to move out of the lock.

A comparison of the skyplots in Figure 5b,c explains the low visibility of the satellites. The upper skyplot shows the satellites tracked for the whole cruise in the lock, whereas the lower one only shows the satellites tracked when the vessel is at the bottom of the lock. Low-elevation satellites in the upper plot can be seen as missing from the lower plot due to blockage from the walls. Nevertheless, for the whole duration, the GNSS position solution is available.

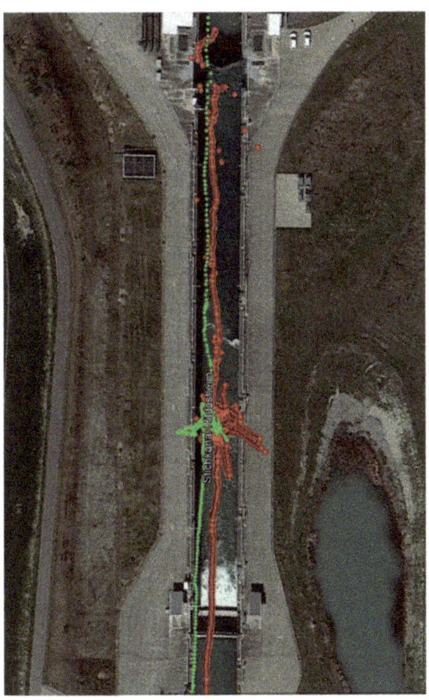

(a)

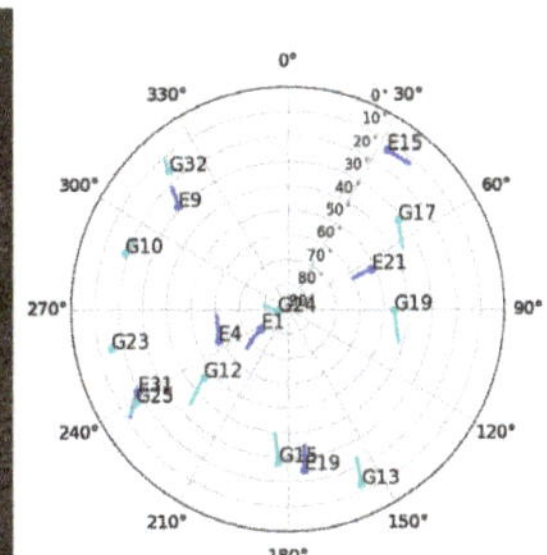

(b)

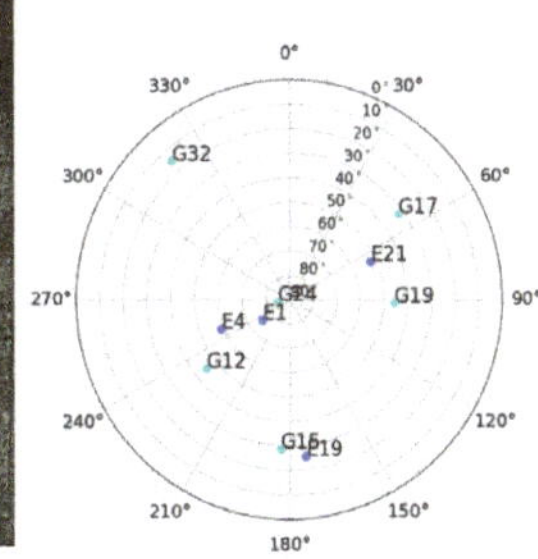

(c)

Figure 5. (**a**) L1/E1-based PVT solution (red) vs. Galileo E5a-based PVT solution (green), (**b**) skyplot during the whole cruise in the lock, (**c**) skyplot for the entire duration when the vessel is at the bottom and moves out of lock, Gx for GPS satellites (green) and Ex for Galileo satellites (blue).

4.2. SLAM

Figure 6 shows the sub-meter resolution of the LiDAR point cloud when entering the lock, where small structures in the wall can be identified and the concrete shoreline is clearly visible. The grid for point cloud visualization uses 1×1 m cells; the sensor distance resolution is well below 10 cm.

Figure 6. Screenshot of LiDAR and camera data during lock entry.

Figure 7 displays a dynamic 3D map generated during cruising via OctoMap [15] on top of static map information. The gray rectangle in the center represents the vessel; the colors are mapped to the height information. Such a map can be refined by multiple passages and provides a high-resolution image of the canal and its surroundings, including vegetation and infrastructure. By using this map, control algorithms are able to plan well beyond the actual sensor range.

Figure 7. Visualization of SLAM-generated 3D map and static map background.

In the case of previously unknown obstacles in the water, the vessel can only adapt its course as soon as the sensors detect the objects. Depending on the vessel's agility, additional long-range sensors might be needed to ensure successful collision avoidance. The sensor set used during the test campaign was reliably able to detect buoys used to mark narrow passages in the canal up to a distance of 75 m, see Figure 8.

The vessel's speed and braking distance are highly dependent on the length and weight of the vessel and on the canal (width, depth, flow). In still waters—like the Hildesheim canal—the standstill will be against the water over a distance, measured against the land, of no more than 350 m for vessel convoys longer than 110 m and wider than 11.45 m, and 305 m for shorter and narrower convoys. Hence, a mixed sensor setup with, e.g., long range LiDAR, is needed to detect obstacles a higher distances.

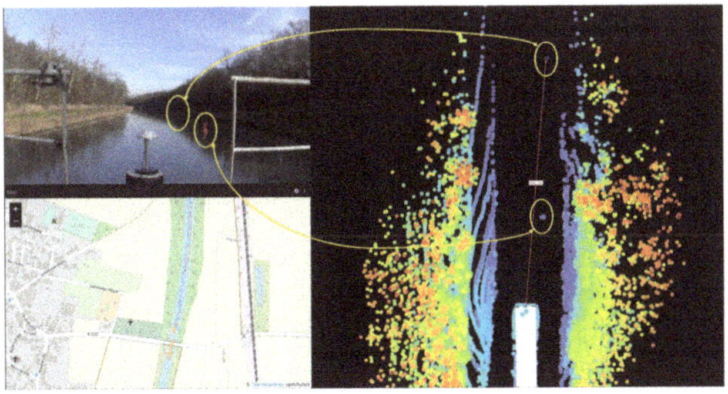

Figure 8. Three-dimensional pointclouds showing buoys in the canal.

5. Discussion

In order to fulfill the requirements for automated inland vessels, it is important to precisely and reliably locate the position of a vessel in a channel. Furthermore, the timely detection of surroundings and obstacles plays a vital role. For this project, sensors were used that have also been used in the development of systems for autonomous driving of cars. The used hardware setup is highly flexible and can be integrated and tested in any kind of vessel. The final hardware setup presented in this paper is the result of several iterations of tests in the canal and for each test cruise, a different vessel was provided. Each test cruise was used to optimize the setup and methodology. As far the project team knows, the sensors connected to one ROS have not been not mentioned by others. This setup enables synchronized recording on one centralized node for future data usage.

The test cruise recordings were used to evaluate positioning solutions with complex Galileo E5a signals. A dynamic 3D map was generated by fusing all sensor data. This generated map can be used in addition to a static map, in which the vessel is located purely via the GNSS. The exact requirements for using the dynamic map for later control purposes remain unclear, since, e.g., accuracy constraints heavily depend on the approach to controlling the vessel. For tasks like automated docking, a full map might not be needed, and instead, the free space or objects detected by LiDAR sensors can be used directly, since the required surroundings of the vessel can be observed by the sensors. This is also true for automated assistance in situations where dynamic/floating objects are present on the waterways and need to be avoided. Further analysis and visualization were performed for all defined experiments. An examination of how the data can be used for autonomous control of the vessel was outside the scope of this paper.

To keep inland shipping attractive and competitive, automation is an important step. Similar challenges to autonomous driving on roads also apply to inland navigation. Sensors may fail, the required infrastructure may not be available, or interfering signals may interrupt the reception of the GNSS or other signals. During the test cruises, a poor to non-existent network reception and interference in the E5 frequency band were the main problems encountered. However, it must be possible to reliably cover all these problems in the future. The sensor setup used in this project is modular and can be enhanced to have fall-back solutions, to fuse inertial sensors, and to use possible existing methods, like correction data from a reference station or Digital Audio Broadcasting (DAB)+, including 5G if available, e.g., at the port. The team is aiming to test the method to showcase the visualization in real time.

Author Contributions: Conceptualization: M.S. and J.S.; Methodology: M.S.; Software: J.S.; Validation: M.S.; Formal analysis: J.S. and H.G.; Investigation: J.S. and H.G.; Writing—original draft preparation: K.D.; Writing—review and editing: K.D., J.S., H.G. and M.L.; Visualization: J.S. and H.G.; Project administration: K.D. and M.O.; Funding acquisition: W.F. and K.D. All authors have read and agreed to the published version of the manuscript.

Funding: This project was funded by the Federal Ministry of Transport and Digital Infrastructure under the funding code 45DTWV009E. The responsibility for the content of this publication lies with the author.

Institutional Review Board Statement: Not applicable.

Informed Consent Statement: Not applicable.

Data Availability Statement: The data presented in this study are available on request from the corresponding author. The data are not publicly available due to the data size and further explanations may be needed.

Acknowledgments: We would like to acknowledge all project partners of Smart-Port-Shuttle Hildesheim for their contributions, such as the provision of the test ship and help with the planning of the measurement campaigns and realization. We would also like to thank the crew of the test ship for their interest, patience, and time. We also want to express our gratitude to Fabio Garzia and Muhammad Hamza Khalil for their help in preparing the sensor setup.

Conflicts of Interest: The authors declare that this study received public funding from Federal Ministry of Transport and Digital Infrastructure. The funder was not involved in the study design, collection, analysis, interpretation of data, the writing of this article or the decision to submit it for publication.

References

1. Granzow, A. Binnenschifffahrt im Jahr 2050: Automatisierung und Digitalisierung können vieles ändern. Available online: https://www.dvz.de/rubriken/land/binnenschifffahrt/detail/news/binnenschifffahrt-im-jahr-2050-automatisierung-und-digitalisierung-koennen-vieles-aendern.html (accessed on 14 May 2023).
2. Smart Port Shuttle (SPS). Available online: https://www.iis.fraunhofer.de/en/ff/lv/lok/proj/sps.html (accessed on 14 May 2023).
3. Automation: Listing of Pilot and Research Projects in the Field of Automated Navigation. Available online: https://automation.ccr-zkr.org/1000-en.html (accessed on 18 April 2023).
4. Industrie- und Handelskammer im Ruhrgebiet; DST Entwicklungszentrum für Schiffstechnik und Transportsysteme e.V. Autonomes Fahren in der Binnenschifffahrt, Machbarkeitsstudie für ein Testfeld im Ruhrgebiet. Available online: https://www.dst-org.de/autonomes-fahren-in-der-binnenschifffahrt/ (accessed on 10 May 2023).
5. SciPPPer. Available online: http://www.scippper.de/ (accessed on 10 May 2023).
6. Friedrich List. Mit Scipper sicher in die Schleuse: Elektronische Einparkhilfe für Binnenschiffe. Available online: https://innovationorigins.com/de/mit-scipper-sicher-in-die-schleuse-elektronische-einparkhilfe-fuer-binnenschiffe/ (accessed on 10 May 2023).
7. Seybold, D. GOOSE (GNSS Receiver with Open Software Interface), TeleOrbit GmbH. 2020. Available online: https://teleorbit.eu/en/satnav/goose/ (accessed on 10 May 2023).
8. VSS6037L VeroStar™. Available online: https://www.tallysman.com/product/vss6037l-verostar-surface-mount-full-gnss-precision-antenna-l-band/ (accessed on 10 May 2023).
9. Velodyne Puck VLP-16 Datasheet. Available online: https://velodynelidar.com/wp-content/uploads/2019/12/63-9229_Rev-K_Puck-_Datasheet_Web.pdf (accessed on 27 April 2023).
10. AVR-QDM-110 Product Datasheet. Available online: https://www.symeo.com/cms/upload/pdf/de/Datenblaetter/DOC.DBL.000444.LATEST.Symeo_Datenblatt_AVR-QDM-110.pdf (accessed on 12 May 2023).
11. Halo20 Radar. Available online: https://www.simrad-yachting.com/de-de/simrad/type/radar/halo20simradradar-9207f6f6/ (accessed on 13 May 2023).
12. ROS Noetic Ninjemys. Available online: http://wiki.ros.org/noetic (accessed on 10 May 2023).
13. ROS. Available online: https://www.ros.org/ (accessed on 10 May 2023).
14. Galileo Signal Plan. Available online: https://gssc.esa.int/navipedia/index.php/Galileo_Signal_Plan (accessed on 15 May 2023).
15. OctoMap. Available online: https://octomap.github.io/ (accessed on 14 May 2023).

engineering proceedings

MDPI

Proceeding Paper

Required Navigation Performances for Drone Flight Operations †

Pablo Haro *, Carmen Aguilera and Giovanni Lucchi

Market Downstream and Innovation Department, European Union Agency for the Space Programme (EUSPA), 170 00 Prague, Czech Republic; carmen.aguilera@euspa.europa.eu (C.A.); giovanni.lucchi@euspa.europa.eu (G.L.)
* Correspondence: pablo.haro@euspa.europa.eu
† Presented at the European Navigation Conference 2023, Noordwijk, The Netherlands, 31 May–2 June 2023.

Abstract: Drone flight operations require ensuring the containment of the drone along the desired flight path in the domain of the Total System Error (TSE). The paper proposes a number of Required Navigation Performance (RNP) specifications tailored to drones by assessing the contributing error sources: the definition of the desired flight path, flyability, autopilot envelope, positioning errors, and operational environment (bottom-up approach). Ongoing work with drone operators is collecting their navigation requirements (top-down approach) in order to determine to which extent RNP navigation specifications meet their requirements along the flight profile.

Keywords: RNP; TSE; NSE; FTE; PDE; SAIL; GNSS; EGNOS; Galileo

Citation: Haro, P.; Aguilera, C.; Lucchi, G. Required Navigation Performances for Drone Flight Operations. *Eng. Proc.* **2023**, *54*, 51. https://doi.org/10.3390/ENC2023-15459

Academic Editors: Tom Willems and Okko Bleeker

Published: 29 October 2023

1. Introduction

This study describes ongoing research by the European Union Agency for the Space Programme (EUSPA) on navigation specifications' concepts for drone flight operations. The general principles are based on Performance-Based Navigation (PBN), as described by ICAO Doc 9613 [1] and the EUROCAE ED-75E [2]. A variety of drone use cases, including multi-copters and fixed wing drones, are being assessed from which typical mission requirements are derived. Those requirements are technology-agnostic and commensurate with the risk of the operation (based either on mission or safety arguments). The desired flight path must be entirely contained within the drone's flight geography, both in the horizontal and vertical planes. When flying that path, there will be errors in the horizontal and vertical components. In both cases, the TSE includes three error sources, i.e., the Navigation System Error (NSE), Flight Technical Error (FTE), and Path Definition Error (PDE). Therefore, a careful allocation of the TSE to these error sources is necessary to keep it under control, considering the environmental aspects from the mission such as spatial constraints, RF interference, and meteo conditions (e.g., wing and gust). The paper outlines a number of RNP specifications for drones, including an On-Board Performance Monitoring and Alerting (OBPMA) mechanism. The proposed concept is motivated as a response to the requirements stated by EASA in the Special Condition (SC) for Light UAS-Medium Risk [3], applicable to SAIL III and IV of the specific category of operation, in particular, Light-UAS.2510 Equipment, Systems, and Installation, Light-UAS.2511 Containment, and Light-UAS.2529 UAS Navigation Function. The study focuses on the bottom-up approach, i.e., the aggregation of error sources to derive the resulting TSE (95%) or RNP values, which were discussed at the User Consultation Platform held in October 2022 in the frame of the EU Space Week with positive feedback. Ongoing work with drone operators is collecting their navigation requirements (top-down approach) that will allow for validating a set of RNP navigation specifications covering the whole drone flight profile. Following the introduction, the concepts of RNP for drones and TSE containment are presented in Sections 2 and 3, respectively. Section 4 describes the TSE error sources and Section 5 presents the RNP calculations. The GNSS performance requirements (focusing on accuracy and integrity) are described in Section 6. Finally, Section 7 presents the conclusions and way forward.

2. RNP Concept for Drones

Within PBN, the proposed concept relies on area navigation, including the requirement for OBPMA. Therefore, it is built on RNP navigation specifications tailored to drones. The aim is to ensure the containment of the drone within a narrow 3D corridor along the desired flight trajectory, i.e., in the domain of the TSE.

2.1. Total System Error

The TSE includes three error components: NSE, FTE, and PDE.

2.1.1. Navigation System Error

The NSE depends on a number of factors, such as the GNSS (it is assumed here that GNSS is the only source for position determination (position fixing), but other sources are possible as well) constellations and frequencies used by the GNSS receiver, the absence of RFI, and the number and geometry of satellites in view from the drone's GNSS antenna (affected by terrain or obstacles that may mask satellites signals or generate multipath), as well as the attitude of the drone during manoeuvres. The drone operator may consult the user manual (typically, open sky conditions are assumed) for the specific drone and GNSS sensor to determine the a priori expected horizontal and vertical NSE in each case.

2.1.2. Flight Technical Error

Since the flight dynamics of multi-copters and fixed wing drones are different, the ability of the drone's autopilot to fly the defined path, i.e., the FTE, will also differ. Experience shows that the FTE is generally the dominant error source, in particular, for fixed-wing drones that cannot hover on waypoints or perform instantaneous heading changes. Additionally, it is essential to define a trajectory that is actually flyable by the drone, and this requires operation within the operational envelope (ground speed, wind, and gust conditions) and a careful definition of the turns at the waypoints, as well as the minimum stabilisation distance in the straight segments. The behaviour of autopilots and therefore the FTE in drone flight operations needs to be characterised. A relevant reference for this exercise is the FAA AC 20-138D "Airworthiness Approval of Positioning and Navigation Systems" [4], where several sections might be tailored for new drone-specific RNP navigation specifications, in particular, Table 9 'RNP FTE Performance' in Chapter 17 'Installation Considerations-RNP'. In the absence of validated values on the characterisation of the FTE, it may be assumed that (in manned aviation) the autopilot will be able to ensure that:

$$\text{FTE}(95\%) \leq \frac{1}{2}\text{RNP value} \tag{1}$$

Such a condition does not necessarily hold for drone flight operations, since FTE is typically the dominant error source. It is also assumed that the desired flight path is properly designed (it is flyable), coded, and processed by the autopilot, considering the type of drone, its dynamics, and the environmental conditions. When the FTE deviates (exceeds) the expected behaviour, a specific alerting mechanism on the FTE may be convenient.

2.1.3. Path Definition Error

The PDE, i.e., defined vs. desired flight path, depends on the quality of the database of the terrain, as well as the vertical reference to define the altitude/height along the trajectory. In manned aviation, the PDE is typically disregarded with respect to the other error sources, but this is not necessarily the case for UAS and has to be confirmed.

2.2. Longitudinal Navigation

Longitudinal (along track) performance implies navigation against a position along the track (e.g., 4D control). However, UAS missions requiring 4D control are not considered is this study, so there is no FTE in the longitudinal dimension. Therefore, along-track accuracy only includes NSE and PDE (not necessarily negligible). The along-track accuracy affects position reporting, i.e., the awareness of reaching a waypoint (WP) or a certain lead distance before a WP when flying a turn.

2.3. Lateral Navigation

Figure 1 shows the lateral TSE and its components: NSE and FTE, where it has been assumed that the PDE is negligible (in such a case, the defined path is the desired path).

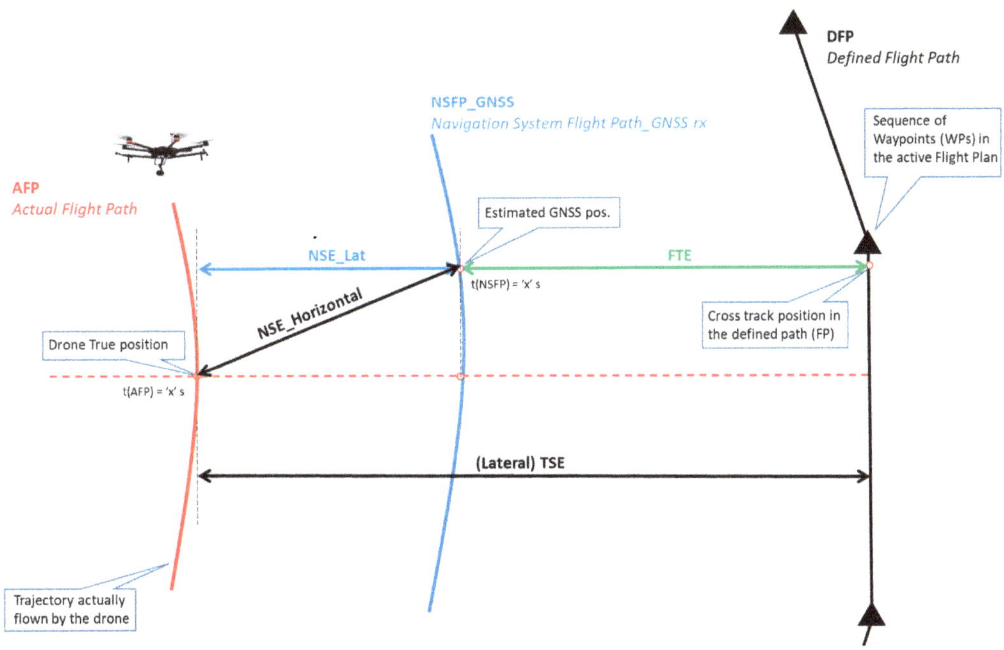

Figure 1. Lateral TSE.

To be noted that the sequence of WPs from the active flight plan is not the same as the reference trajectory offered to the guidance and control functions (from which FTE results), particularly during the turns, even though it is governed by those WPs.

3. TSE Containment Concept

To ensure the containment of the drone in a 3D corridor along the desired flight path, the TSE must lie within suitable limits. To this end, the TSE must meet two requirements derived from manned aviation, [1] ICAO Doc 9613 (II-A-2-6) and [2] EUROCAE ED-75E (1.7.3 Lateral Containment Concept):

- TSE accuracy requirement: the TSE remains equal to or less than the required RNP value (accuracy value) for 95 per cent of the flight time; i.e.,

$$\text{TSE} \leq \text{RNP value for 95\% of flight time} \tag{2}$$

- TSE integrity requirement: the probability that the TSE exceeds the specified TSE limit (equal to two times the RNP value) without annunciation is less than the applicable Integrity Risk (IR).

$$P(\text{TSE} > 2 \times \text{RNP value, without alert}) < \text{IR per flight hour} \tag{3}$$

For drones, the first requirement is taken as it is, but the second one needs to be adjusted to drone flight operations in the specific category, i.e., the IR is not necessarily

1×10^{-5}, since other risk levels might be acceptable, in particular for low- or medium-risk operations. Note that the concept of 95% containment volume as well as the 'deviation threshold' are also referred to in the EASA AMC and GM to Regulation (EU) 2021/664 [5].

4. TSE Error Sources

4.1. Horizontal TSE

The horizontal subspace has lateral (cross-track) and longitudinal (along-track) components, sub-indexed as "Lat" and "Long".

4.1.1. Horizontal NSE

HNSE is radial, i.e., it has both lateral and longitudinal components. It can be modelled as a bivariate normal distribution.

$$\mathrm{HNSE} = \left(\mathrm{HNSE_{Lat}}, \mathrm{HNSE_{Long}}\right) \sim \mathrm{N}(0, \Sigma_{\mathrm{HNSE}}) \tag{4}$$

where the covariance matrix is given by:

$$\Sigma = \begin{bmatrix} \mathrm{Var(X)} & \mathrm{Cov(Y,X)} \\ \mathrm{Cov(X,Y)} & \mathrm{Var(Y)} \end{bmatrix} \quad \mathrm{X} = \mathrm{HNSE_{Lat}}, \mathrm{Y} = \mathrm{HNSE_{Long}} \tag{5}$$

HNSE is assumed to have a circular distribution, i.e., the covariance matrix is diagonal with variances $\sigma_x^2 = \sigma_y^2$. If GNSS is the PVT source, this hypothesis does only hold if certain conditions are imposed on the mission trajectory to ensure a good visibility of satellites above the local horizon. The R95 for the HNSE is the radial quantile within which navigation accuracy lies for 95% of the flight time. Its value can be found in the corresponding GNSS documentation. Three different GNSS sensors onboard the drone are considered and the corresponding values of the horizontal NSE are shown in Table 1:

(a) GPS-EGNOS: data are taken from EGNOS SoL SDD [6] (Table 6-2); HNSE (R95) = 3 m.
(b) GPS-Galileo: A GNSS uBlox receiver is used as representative for the multi-constellation mode, where accuracy values are found as CEP (circular error probable, CEP value is the 50% confidence interval) in NEO-M9N-00B data sheet (Ublox NEO-M9N-00B Standard precision GNSS module, Table 1, page 5, CEP position accuracy (GPS, Galileo) = 2 m). The conversion factor from CEP to R95 is 2.08. In this case, HNSE (CEP) = 2 m, or HNSE (R95) = 4.16 m.
(c) GPS: data from GPS SPS Performance Standard [7] (Table 3.8-3), HNSE (R95) = 8 m.

Table 1. $\sigma_{HNSE\ Lat}$ for three GNSS receiver configurations.

	HNSE (R95) [m]	$\sigma_{HNSE\ Lat}$ **[m]**
GPS-EGNOS	3	1.23
GPS-Galileo	4.16	1.70
GPS	8	3.27

The standard deviation of the lateral component of the HNSE is given by:

$$\sigma_{\mathrm{HNSE\ Lat}} = 0.4085\ \mathrm{HNSE(R95)} \tag{6}$$

It is stressed that those figures require open sky conditions. Moreover, in the case of GPS-EGNOS, it is necessary to ensure that the EGNOS signal broadcast by geostationary satellites is received at the drone's GNSS antenna, which results in certain limitations such as a maximum bank angle depending on the latitude.

4.1.2. Lateral FTE

The HFTE is lateral by definition and can be modelled as 1D normal distribution.

$$\mathrm{HFTE} = \mathrm{HFTE_{LAT}} = \mathrm{Lat\ FTE} \sim \mathrm{N}\left(0, \sigma^2_{\mathrm{Lat\ FTE}}\right) \quad (7)$$

The instantaneous FTE is known in real time, as it is the cross-track distance between the defined flight path (DFP) and the estimated position. Nonetheless, typical values of the Lat FTE can be indicated based on experience for fixed and rotary wing drones. Values of 25 m and 10 m are the 95% confidence values chosen, respectively, as shown in Table 2.

$$\sigma_{\mathrm{Lat\ FTE}} = \frac{\mathrm{Lat\ FTE}(95\%)}{1.96} \quad (8)$$

Table 2. Lat FTE(95%) and $\sigma_{\mathrm{Lat\ FTE}}$ for fixed-wing and rotary-wing drones.

	Lat FTE (95%) [m]	$\sigma_{Lat\ FTE}$ **[m]**
Fixed wing	25	12.76
Rotary wing	10	5.10

4.1.3. Horizontal TSE-Lateral

The NSE and FTE are assumed to be independent, zero-mean, and Gaussian distributions. The TSE is also Gaussian with standard deviation equal to the root sum square (RSS) of the standard deviations of the error components.

$$\mathrm{Lat\ TSE} = \mathrm{Lat\ FTE} + \mathrm{HNSE_{Lat}} \sim \mathrm{N}\left(0, \sigma^2_{\mathrm{Lat\ FTE}} + \sigma^2_{\mathrm{HNSE\ Lat}}\right) \quad (9)$$

4.2. *Vertical TSE*

VNSE and VFTE are also assumed to be independent and Gaussian.

$$\mathrm{VTSE} = \mathrm{VNSE} + \mathrm{VFTE} \sim \mathrm{N}\left(0, \sigma^2_{\mathrm{VNSE}} + \sigma^2_{\mathrm{VFTE}}\right) \quad (10)$$

As in the horizontal subspace, the values of VTSE (95%) can be derived from its error components and considering that:

$$\sigma_{\mathrm{VNSE}} = \frac{\mathrm{VNSE}(95\%)}{1.96} \text{ and } \sigma_{\mathrm{VFTE}} = \frac{\mathrm{VFTE}(95\%)}{1.96} \quad (11)$$

4.2.1. Vertical NSE

The VNSE (95%) values for the three GNSS systems considered are shown in Table 3. The sources are the same as in the previous section, except for GPS-Galileo, as it is not provided in the manual. Thus, it is estimated following the rule of thumb that the vertical accuracy is roughly 1.5 times the horizontal one.

Table 3. σ_{VNSE} for three GNSS receiver configurations.

	VNSE (95%) [m]	σ_{VNSE} **[m]**
GPS-EGNOS	4	2.04
GPS-Galileo	6.24	3.18
GPS	13	6.63

4.2.2. Vertical FTE

The VFTE (95%) values shown in Table 4 are estimated from flight trials:

Table 4. VFTE (95%) and σ_{VFTE} for fixed wing and rotary wing drones.

	VFTE (95%) [m]	σ_{VFTE} [m]
Fixed wing	10	5.10
Rotary wing	5	2.55

5. RNP Lateral/Vertical Calculation

Once assumptions on how to model the errors are made, it is possible to derive the accuracy or RNP values ("RNP Lateral" and "RNP vertical") and the containment regions that characterize the 3D corridor where the drone is confined throughout its trajectory.

5.1. RNP Lateral

$$\text{RNP lateral} = \text{Lat TSE}(95\%) = 1.96\ \sigma_{\text{HTSE Lat}} = 1.96\sqrt{\sigma^2_{\text{HNSE Lat}} + \sigma^2_{\text{Lat FTE}}} \quad (12)$$

The driving component in the lateral HTSE is the ability to follow the defined trajectory, i.e., the FTE is typically greater than the NSE. Values for the RNP lateral are presented below in Table 5 for both fixed and rotary-wing drones, as well as for the three GNSS sensors studied.

Table 5. RNP Lateral values.

Drone	Sensor	$\sigma_{HNSE\ Lat}$ [m]	$\sigma_{Lat\ FTE}$ [m]	RNP Lateral [m]
Fixed wing	GPS-EGNOS	1.23	12.76	25.13~25
	GPS-Galileo	1.70		25.23~25
	GPS	3.27		25.82~26
Rotary wing	GPS-EGNOS	1.23	5.10	10.28~10
	GPS-Galileo	1.70		10.54~11
	GPS	3.27		11.87~12

The results show that the "RNP Lateral" values differ significantly depending on the type of drone, whereas the influence of the GNSS sensor is noticeably smaller than the FTE, based on the assumptions taken.

5.2. RNP Vertical

$$\text{RNP vertical} = \text{VTSE}(95\%) = 1.96\ \sigma_{\text{VTSE}} = 1.96\sqrt{\sigma^2_{\text{VNSE}} + \sigma^2_{\text{VFTE}}} \quad (13)$$

Values for the RNP vertical are presented in Table 6.

Table 6. RNP Vertical values.

Drone	Sensor	σ_{VNSE} [m]	σ_{VFTE} [m]	RNP Vertical [m]
Fixed wing	GPS-EGNOS	2.04	5.10	10.77~11
	GPS-Galileo	3.18		11.78~12
	GPS	6.63		16.39~16
Rotary wing	GPS-EGNOS	2.04	2.55	6.40~6
	GPS-Galileo	3.18		7.99~8
	GPS	6.63		13.92~14

6. GNSS Performance Requirements

6.1. GNSS Accuracy

The GNSS position error is the difference between the estimated position and the actual position, as defined in ICAO Annex 10 [8] (Attachment D). HNSE (R95) values are shown in Table 1 and VNSE (95%) values in Table 3 for the three GNSS sensor configurations considered.

6.2. GNSS Integrity

Drone operations in the specific category may require an integrity (OBPMA) mechanism. The drone's navigation system should raise an alert in the case of exceeding the containment limits in either the lateral or vertical dimensions, and the probability of doing so without alerting should be lower than the applicable IR. To this purpose, an integrity mechanism is defined including Integrity Risk, Alert Limits, and Time to Alert. Protection Levels (xPL), computed in real time by the onboard GNSS sensor, are upper confidence bounds on position errors (xNSE), which are unknown in real time (HPL bounds the HNSE and VPL the VNSE). A graphical depiction of the integrity mechanism is shown in Figure 2.

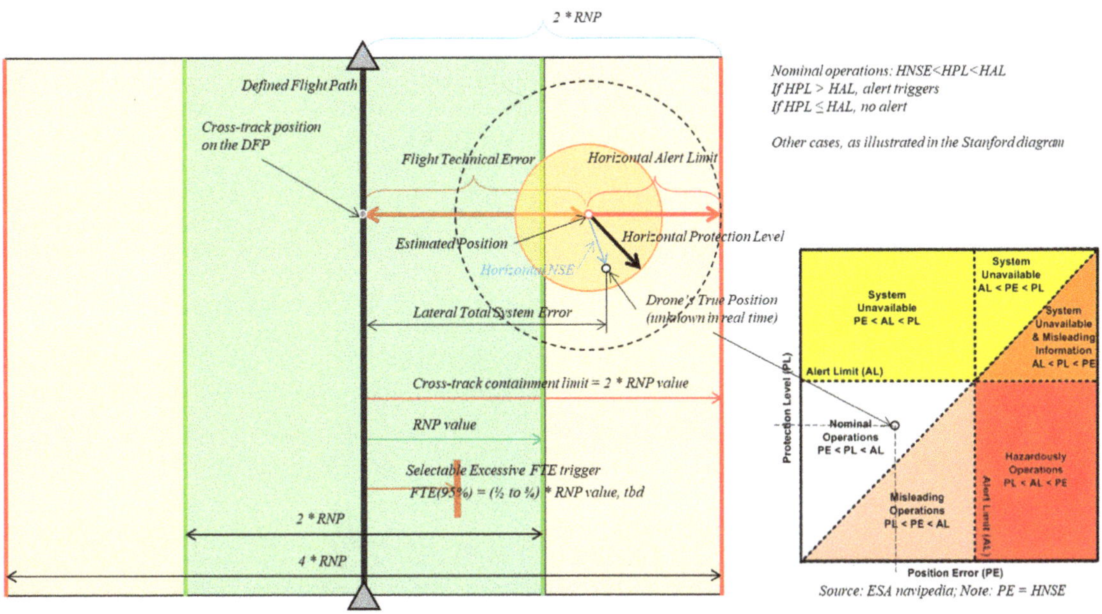

Figure 2. GNSS integrity mechanism (Lateral)-RNP value (green corridor) and Containment region (green and orange corridors, with semi width equal to twice the RNP value, i.e., $2 * \text{RNP}$ in this Figure, resulting in a total width of four times the RNP value, i.e., $4 * \text{RNP}$ in the Figure), nominal operation.

When considering the application of the integrity mechanism, it is important to distinguish between VLOS and BVLOS operations. Such a mechanism is especially relevant for BVLOS, as in VLOS operations, the remote pilots can also work as an external integrity mechanism themselves, by taking control over the drone if they detect visually that it deviates from the expected trajectory or is about to leave the flight geography.

The HPL is assumed to be isotropic with a circular distribution. This assumption requires ensuring enough numbers of GNSS satellites are in view with a good geometry. Otherwise, the horizontal region where the true position lies around the estimated position becomes an ellipse instead of a circle. Moreover, this condition has an impact on the definition of the desired flight path (minimum height and ground track). Figure 2 above shows the integrity mechanism in the lateral dimension. The green corridor represents the RNP value or 'accuracy region' within which the drone lies at least for 95% of the flight time. For the remaining 5%, it may be located outside the green corridor. The red lines delimit the containment region with a width of four times the RNP value. The integrity mechanism works as follows in the lateral dimension:

$$\text{P (LatTSE} > 2 \times \text{RNP}_{\text{Lat}}, \text{without alert)} = \text{P(HNSE} > \text{HAL} = 2 \times \text{RNP}_{\text{Lat}} - \text{Lat FTE, without alert)} < \text{IR per flight hour} \quad (14)$$

In the vertical dimension, an equivalent mechanism is implemented with the difference that the VNSE is a 1D distribution.

6.2.1. Integrity Risk

Special Condition (SC) Light UAS defines objective requirements for UAS operated in the specific category and is limited to specific assurance and integrity level (SAIL). The maximum allowable rate of loss of control of the operation per flight hour (FH) is linked with the SAIL and achieved by the means of Operational Safety Objectives (OSOs).

$$\mathrm{P_{loc}} < \frac{10^{-\mathrm{SAIL}}}{\mathrm{FH}} \tag{15}$$

According to SORA, "The Loss of control of operation corresponds to situations: where the outcome of the situation highly relies on providence; or which could not be handled by a contingency procedure; or when there is grave and imminent danger of fatalities". A loss of control of the operation means that the drone may be flying out of the operational volume, potentially leading to harm to third parties in the air or on the ground. A loss of control condition represents an emergency situation that is potentially caused by the occurrence of a contingency event. A number of contingency events can be identified as directly related to the occurrence of the out-of-control condition: C2 link loss, GNSS loss of performances, or loss of control in flight, among others. Within the GNSS loss of performance, a dangerous situation may occur, as determined by a Functional Hazard Assessment, in the case of non-integrity of the PVT solution, i.e., not valid for the intended operation without annunciation. Such a situation is referred to as Integrity Risk (IR) and it is assumed that its allowable occurrence is at most one order of magnitude lower than the P_{loc}.

$$\mathrm{IR(GNSS\ PVT)} = \frac{1}{10}\mathrm{P_{loc}} < \frac{10^{-(\mathrm{SAIL}+1)}}{\mathrm{FH}} \tag{16}$$

Table 7 presents the integrity risks per flight hour for SAIL III and IV, the equivalent number of flight hours and flight days or years per failure, giving a more understandable figure to assess the acceptability of such IRs for drone flight operations. This has to be assessed considering the typical duration of a flight operation and the number of flight operations performed by a drone during its entire operational life. For instance, for a drone package delivery mission, a flight duration of 10 min may be assumed. In this case, for SAIL III: IR < 10^{-4} per flight hour, resulting in >10,000 flight hours per failure or >60,000 missions per failure. The integrity of a GNSS solution requires (a) strict application of the SiS ICDs and the implementation of an augmentation scheme (e.g., RAIM, SBAS, or GBAS) to mitigate the SiS risk, (b) development constraints (DAL) to protect against the SW or HW design/development errors, and (c) integration/architecture constraints (equipment redundancy and cross-checks between equipment) to protect at least against HW failures. Receivers have to be duly qualified to obtain the pertinent approval or certification.

Table 7. IR per flight hour for SAIL III and IV.

SAIL	IR (GNSS PVT)	Hours per Failure	Days or Years per Failure
III	<10^{-4} per FH	>10,000	>417 days = 1.14 years
IV	<10^{-5} per FH	>100,000	>4167 days = 11.4 years

6.2.2. Alert Limits

Alert limits define the error tolerance not to be exceeded without issuing an alert, i.e., the maximum position error allowed for an operation, [8] 3.7 and [9] (2.2.2).

Horizontal Alert Limit

Figure 2 shows that the Horizontal Alert Limit (HAL) is given by:

$$HAL(t) = 2 \times RNP_{Lat} - Lat\ FTE\ (t) \quad (17)$$

HAL is the maximum radius of the circle with the centre at the estimated position and radius HPL, until this circle is tangential to the cross-track containment limit. If HPL > HAL, then an alert must be triggered before TTA, since the drone's true position might be outside the containment region. Figure 2 shows a case with HNSE < HPL < HAL and the system is available in nominal operation. If HNSE < HPL, but HPL > HAL, the system is unavailable (yellow zone). Finally, if HNSE > HPL, whereas HPL < HAL, the system is still available, but the HPL is not bounding the HNSE, then if HNSE > HAL, it is an HMI event (red zone) that has not been detected by the integrity mechanism. This is the dangerous case that shall occur with a very low probability, less than the IR per flight hour.

If the Lat FTE is not characterised, i.e., Lat FTE(95%) is unknown, the equation above for HAL still holds and is applicable in real time. In such a case, the HAL is a dynamic alert limit, HAL(t). The instantaneous value of the Lat FTE(t) is known in real time since it is the cross-track distance between the estimated position (output of the GNSS sensor) and the active segment of the defined flight path. In case the Lat FTE is characterised, the Lat FTE(95%) is a known value (e.g., those in Table 8) and a fixed value of the HAL may be derived for a particular drone operation or phase of flight, as shown in Table 8.

$$HAL = 2 \times RNP_{Lat} - Lat\ FTE\ (95\%) \quad (18)$$

Table 8. HAL for fixed and rotary wing drones and different GNSS sensors.

Drone	Sensor	RNP Lat [m]	Lat FTE (95%) [m]	HAL [m]
Fixed wing	GPS-EGNOS	25	25	25
	GPS-Galileo	25		25
	GPS	26		27
Rotary wing	GPS-EGNOS	10	10	10
	GPS-Galileo	11		12
	GPS	12		14

Vertical Alert Limit

Similarly to HAL, the Vertical Alert Limit (VAL) is given by:

$$VAL\ (t) = 2 \times RNP_{Ver} - VFTE\ (t) \quad (19)$$

where the integrity mechanism for the vertical dimension is based on 1D distributions. If the VFTE is not characterised, i.e., VFTE (95%) is unknown, the equation above for VAL still holds and is applicable in real time. In case the VFTE is characterised, the VFTE (95%) is a known value (Table 9) and a fixed value of the VAL may be derived for a particular drone operation or phase of flight as shown in Table 9.

$$VAL = 2 \times RNP_{Ver} - VFTE\ (95\%) \quad (20)$$

Table 9. VAL for fixed and rotary wing drones and different GNSS sensors.

Drone	Sensor	RNP Ver [m]	VFTE (95%) [m]	VAL [m]
Fixed wing	GPS-EGNOS	11	10	12
	GPS-Galileo	12		14
	GPS	16		22
Rotary wing	GPS-EGNOS	6	5	7
	GPS-Galileo	8		11
	GPS	14		23

6.2.3. Time-to-Alert

The Time-to-alert (TTA) is the maximum allowable time elapsed from the onset of the navigation system being out of tolerance (the drone could be located at the boundary of the containment region in the lateral or vertical dimensions, xPL > xAL) until the equipment enunciates the alert (to the onboard autopilot or to the remote human pilot) [8] (3.7).

In the case of drones, TTAs have been derived in both horizontal and vertical dimensions. The resulting values are typically in the order of few seconds, ranging from 1 to 2 or 3 s depending on the case.

7. Conclusions and Way Forward

RNP navigation specifications tailored to drones have been studied following a bottom-up approach, i.e., based on the aggregation of error components within the TSE. The methodology and results were positively discussed at the User Consultation Platform 2022. The 2023 Report on Aviation and Drones User Needs and Requirements [10] includes GNSS requirements for drone en-route missions as shown in Table 10, including also figures for continuity of service and availability.

Table 10. GNSS performance requirements for drone en-route.

RNP Lat/Ver [m]	HNSE (95%) [m]	VNSE (95%) [m]	IR (GNSS PVT)	TTA [s]	HAL/VAL [m]
Fixed-wing RNP 26/16 Rotary-wing RNP 12/14	3–8	4–13	SAIL III/IV: $10^{-4}/10^{-5}$ per FH	1–3	25–27/12–22 10–14/7–23

Ongoing work with drone operators is collecting their mission requirements (top-down approach) that will allow for consolidating and validating RNP navigation specifications, based mainly on GNSS PVT, covering the whole drone flight profile.

Author Contributions: Main author P.H.; validation, P.H., C.A. and G.L. All authors have read and agreed to the published version of the manuscript.

Funding: This research received no external funding.

Institutional Review Board Statement: Not applicable.

Informed Consent Statement: Not applicable.

Data Availability Statement: Data sharing is not applicable to this article.

Conflicts of Interest: The authors declare no conflict of interest.

References

1. ICAO. *Performance-Based Navigation (PBN) Manual*, 4th ed.; ICAO Doc 9613; ICAO: Montreal, QC, Canada, 2013.
2. *ED-75E*; MASPS: RNP for RNAV. EUROCAE: Saint-Denis, France, 2022.
3. EASA. *Special Condition (SC) for Light UAS—Medium Risk*; Issue 01; EASA: Cologne, Germany, 2020.
4. *AC 20-138D*; Airworthiness Approval of Positioning and Navigation Systems. FAA: Washington, DC, USA, 2016.
5. EASA. *AMC and GM to Regulation (EU) 2021/664 on a Regulatory Framework for the U-Space*; Issue 1; EASA: Cologne, Germany, 2022.

6. EGNOS. *Safety of Life (SoL) Service Definition Document*; Issue 3.4; EUSPA: Prague, Czech Republic, 2021.
7. GPS. *SPS Performance Standard*, 5th ed.; Department of Defense: Washington, DC, USA, 2020.
8. ICAO. *Annex 10 Aeronautical Telecommunications, Volume I—Radio Navigation Aids*, 7th ed.; ICAO: Montreal, QC, Canada, 2018.
9. ICAO. *Doc 9849, Global Navigation Satellite System (GNSS) Manual*, 3rd ed.; ICAO: Montreal, QC, Canada, 2017.
10. EUSPA. Report on Aviation and Drones User Needs and Requirements. Available online: https://www.euspa.europa.eu/sites/default/files/report_on_aviation_and_drones_user_needs_and_requirements.pdf (accessed on 1 May 2023).

engineering proceedings

MDPI

Proceeding Paper

Development of a Global Navigation Satellite System Receiver for Specific Category Unmanned Aerial Systems Operations †

Sergi Dueñas Pedrosa [1,*], Samuele Fantinato [1], Laura García-Junceda del Rio [2], Esther Lopez [3], David Abia [3], Jacopo Marangoni [4] and Sara Molinari [4]

1 Qascom SRL, 36061 Bassano del Grappa, Italy; samuele.fantinato@qascom.it
2 UAVNavigation Grupo OESIA, 28703 San Sebastián de los Reyes, Spain; lgarciajunceda@uavnavigation.com
3 Acorde Technologies, 39011 Santander, Spain; ester.lopez@acorde.com (E.L.); david.abia@acorde.com (D.A.)
4 EuroUSC Italia, 00185 Roma, Italy; jacopo.marangoni@eurousc-italia.it (J.M.); sara.molinari@eurousc-italia.it (S.M.)
* Correspondence: sergi.duenaspedrosa@qascom.it

† Presented at the European Navigation Conference 2023, Noordwijk, The Netherlands, 31 May–2 June 2023.

Abstract: The DEGREE project is focused on the development of a cutting-edge dual-frequency GNSS receiver intended to achieve optimum performance and take advantage of the EGNSS (European Global Navigation Satellite System), which allows the leveraging of several differentiators in order to safely integrate UASs into non-segregated airspace and into the U-Space. In order to meet the requirements for commercial operations purposes, the objective of the DEGREE project is to develop a receiver that could meet the requirements for the high level of robustness required for SAIL IV and beyond, thus unlocking all possible risk levels for operations in the specific category.

Keywords: UAS; GNSS; specific category; OSNMA; HAS; SDR; SORA

Citation: Dueñas Pedrosa, S.; Fantinato, S.; del Rio, L.G.-J.; Lopez, E.; Abia, D.; Marangoni, J.; Molinari, S. Development of a Global Navigation Satellite System Receiver for Specific Category Unmanned Aerial Systems Operations. *Eng. Proc.* **2023**, *54*, 52. https://doi.org/10.3390/ENC2023-15467

Academic Editors: Tom Willems and Okko Bleeker

Published: 29 October 2023

1. Introduction

As a result of the increasing demand for commercial applications for UASs (Unmanned Aerial Systems) and in order to harmonize the U-Space, technology development and EU investments on GNSS (Global Navigation Satellite System) receivers are now focusing on the requirements related to the specific category of UAS operations in terms of the security and safety of the navigation subsystems [1].

To this purpose, the current European regulatory framework for specific category operations suggests assessing navigation precision and safety features through a SORA (Specific Operations Risk Assessment) methodology developed by JARUS (Joint Authorities for Rulemaking on Unmanned Systems) [2]. This procedure quantifies the overall risk level (in the air and on the ground) that a certain operation holds for third parties through a parameter defined as SAIL (Specific Assurance and Integrity Level), which has an impact on the robustness at which safety requirements, namely OSOs (Operational Safety Objectives), need to be fulfilled to demonstrate that risk has been adequately mitigated and made acceptable. In this context, the definition of requirements for the development of the GNSS receiver lies in the identification of the applicable OSOs.

Considering the aforementioned situation, the DEGREE (DronEborne Galileo ReCEivEr) project is focused on the development of an innovative GPS/Galileo dual-frequency GNSS receiver intended to take advantage of the EGNSS (European Global Navigation Satellite System), which allows the leverage of several differentiators in order to safely integrate UASs into non-segregated airspace and into the U-Space.

The technology developed during the DEGREE project guarantees a TRL (Technology Readiness Level) of seven within European Union standards introduced for the Horizon

2020 project [3]. Therefore, a completed prototype is provided in addition to the corresponding integration and testing on a UAV (Unmanned Aerial Vehicle) in a relevant operational scenario within the specific category.

2. User and Industry Needs for UAV PNT Technology

It has been evidenced that the emergence of VLL (Very Low Level) operations is one of the main trends affecting the UAS community, and efforts shall be made towards the integration of such systems with other airspace users and as part of the U-Space concept. Particularly, primary areas of work focus on BVLOS (Beyond Line of Sight) and urban environment operations [1], which are expected to emerge in the near future and bring strict requirements in terms of robust navigation, especially for autonomous operations and safe flight termination systems.

The introduction of unmanned technology that coexists with other users and manned aircraft creates a tight dependency on GNSSs as a source of PNT (Positioning, Navigation, and Timing) information. In this regard, when we analysed the market demands for a mid-grade, specific category GNSS receiver, it was found that accuracy and integrity are the most demanded features. As a summary, Table 1 shows the main user and industry needs for a specific category GNSS receiver.

Table 1. User and industry needs for specific category UAS.

User/Industry Need	Rationale
Improved accuracy	As suggested in [2], although there are no current general regulations, almost 50% of drone users expect an achieved horizontal accuracy below 10 cm and 38% of them a vertical accuracy below 10 cm. Regarding maximum tolerable horizontal positioning error, 40% of drone users agreed that it should not exceed 1 m, with 35% holding a similar conviction about the vertical error level. Indeed, while these values are highly desired by the community according to [2], each mission shall perform an operation analysis to determine the needed accuracy and the tolerable position errors as in RNP to enable commercial aviation operations.
Improved availability and continuity	A first discussion for performance requirements in an urban environment which came up during EUSPA 2020 UCP [1] suggests that for urban canyon operations availability should be granted within 99% to 99.9%. On the other hand, continuity, defined as the capability of the system to perform its function without unscheduled interruptions, should be threatened less than once per year [4].
Improved Integrity	According to recent EUSPA's surveys [3], the majority of UAS operators require a warning (usually a visual and/or audible alarm message on the Ground Control Station) in case of a failure or GNSS coverage loss. According to [3], this warning/alarm shall be raised within one to a few seconds from the GNSS outage event. GNSS integrity is important because it ensures that GNSS positioning information is reliable, accurate, and safe for use in critical applications.
Improved robustness and authentication capabilities	Specific category UASs are meant to develop critical operations both in safety and economic terms. While a disruption of the service could cause huge economic loses for the business, cybersecurity RF threats could even target the safety and integrity of people and the equipment itself. Being robust against spoofing and jamming attacks could even become a requirement imposed by insurance companies and authorities when developing an activity in which a specific category UAV is involved.

In recent years, the UASs market has experienced exponential growth with the appearance of new autonomous medium-sized systems between 10 Kg and 50 Kg [3]. These systems allow for lifting heavier payloads with greater flight autonomies, thus increasing the number of functionalities and applications in different sectors. By using UASs, an outperforming of current operation performances in terms of time saving, safety considerations, cost efficiency, reliability, and repeatability is expected.

Some of the operational scenarios in which DEGREE will be deployed are:

- Precision agriculture;
- Infrastructure monitoring;

- Public safety;
- Delivery and transport;
- Aerial photography;
- Surveying and mapping.

3. Guidelines for Required Target Navigation Performance Level

One of the main goals of the DEGREE project is to standardize the implementation of specific category GNSS receivers. The guidelines for the implementation of EGNSS differentiators in UASs' receivers and navigation solutions are drafted in order to achieve the defined performance levels as potential inputs to future standard to operations of Medium/High Risk (SAIL III and IV) in the specific category, based on the SORA (v.2.0) methodology [5].

Operations in urban environments will likely require UASs to follow predefined routes in order to keep them separated from obstacles and away from high-risk areas. This requirement emerges as the outcome of a risk assessment and is defined in terms of RNP. For typical urban applications, RNP 0.01 (approximately 18 m) and/or RNP 0.005 (approximately 9 m) could be needed for the en-route flight phase and thus may become standard for small UAS navigation requirements.

The baseline approach for the guidelines definition for the required target navigation performance level is focused on the assessment of the level of RNP and the technologies and tools necessary to achieve it. The objective requirements, independent of the specific technology, which must be met to develop receivers with adequate performance, are indicated in the guidelines. These guidelines will then be shared in the meetings of the standardization bodies to ensure that they become the heart of the standards being developed.

The preliminary activity that is envisioned in order to complete the guidelines consists of an analysis of the general context in which the guidelines shall be developed, focusing on the applicable and ongoing regulations, relevant stakeholders, and the relationships among them. A key step is the identification of the gaps that need to be discussed to make explicit all assumptions that need to be made in order to provide such guidelines. Then, the SORA methodology is analysed in detail, focusing on the role of the Operational Safety Objectives and, in particular, on the OSO#13 "External services supporting UAS operations are adequate for the operation", with the aim to fulfil the criteria of High Robustness (i.e., for both Integrity and Assurance) for the GNSS external service for High Risk Operations. However, the key step consists of drafting usable and practical specific guidelines for the UAS operators to comply with the robustness criteria of the OSO#13 for GNSSs based on the previous analysis.

Finally, the last step consists of providing the UAS operator with recommendations about mechanisms to detect potential GNSS service performance degradation, as well as potential actions to mitigate that degradation.

4. DEGREE Receiver Overview

The DEGREE receiver is based on an SDR platform, providing substantial amounts of flexibility to quickly adapt and upgrade the receiver functionalities to current and future developments and upcoming services, just by applying software upgrades and keeping the hardware unchanged.

The receiver is composed of two stacked boards, called the RF board and digital board (see Figure 1). The RF board includes the RF front end of the receiver, ADCs (Analog to Digital Converters), saw filters, and RF splitters (Figure 2a). The receiver is prepared to receive GNSS signals from two different antennas for heading purposes. While the primary RF chain is L1/E1/L5/E5a/E6, the second one is L1/E1 only, for true heading purposes. A third RF chain tuned at 700 MHz is 5G ready for signal-of-opportunity positioning and timing.

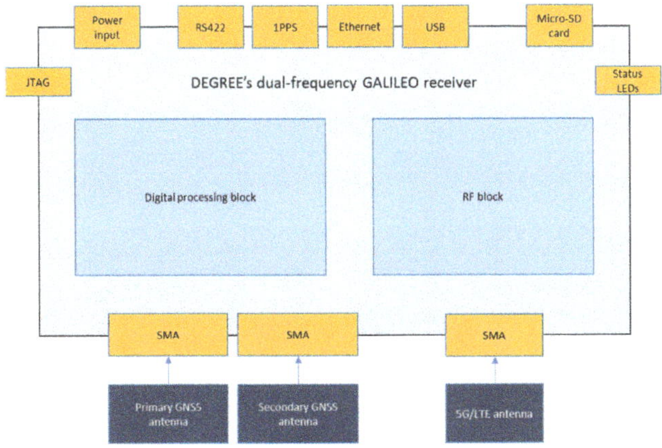

Figure 1. DEGREE receiver high-level block diagram.

(**a**) (**b**)

Figure 2. DEGREE receiver prototype. (**a**) RF board. Inside the red rectangle, the IMU and the COTS GNSS receiver. (**b**) Digital board prototype integrating QN400's Zynq 7000 SOC.

The RF board integrates a COTS GNSS receiver. This secondary receiver provides redundancy and integrity. While redundancy is achieved by means of a second set of GNSS measurements and PVT information, integrity is achieved by comparing the PVT estimations of the primary and secondary GNSS receiver. It is worth noting that with it having two separate receivers with different clocks, care must be taken when comparing data products from both receivers.

The last component of the RF board is the IMU. This IMU offers the capability to be PPS synchronized with external sources. In DEGREE, the primary or secondary GNSS receiver provides a PPS signal, so the IMU is synchronized with the receivers. The loosely coupled sensor fusion algorithm runs inside the IMU, providing fused measurements based on the EKF (Extended Kalman Filter) algorithm, using PVT estimations from the main or secondary receiver and inertial measurements from the IMU itself.

On the other hand, the digital board includes the QN400's Zynq 7000 SoC (CPU + FPGA), as well as interfaces to interconnect with the RF board, user interfaces such as ethernet, RS-422, or JTAG, and status LEDs (Figure 2b).

The major features of the DEGREE receiver are shown in Table 2.

Table 2. DEGREE Rx features.

Feature	Development Status
Multi-constellation (GPS, Galileo, Beidou, Glonass) and dual-frequency (L1/E1+L5/E5a or L1/E1+E6) GNSS receiver.	Completed
Signal of opportunity PNT ready, based on 700 MHz 5G signals	Under consolidation
EGNOS DFMC L1/L5 support	SBAS L1 completed SBAS L5 guidelines under consolidation
High-accuracy navigation based on Galileo E6 HAS (High Accuracy Service)	Under development
Robust FPGA-based anti-jamming algorithms	Completed
Robust FPGA/CPU-based anti-spoofing algorithms	Completed
Support for loosely coupled GNSS+INS, based on EKF	Completed
Assured Navigation Engine, implementing fine-tuned acquisition and tracking loops	Completed
Support for Galileo E1 OSNMA	Completed
Support for true heading using dual antenna architecture	Under development
State-of-the-art RAIM algorithm based on random sample consensus	Under consolidation

Although the DEGREE receiver is a brand-new development in the frame of the EUSPA (EU Agency for the Space Programme)'s DEGREE project, it receives all of its heritage from Qascom's state-of-the-art QN400 GNSS SDR-based receiver [6] The DEGREE receiver has been consolidated as a strategic product inside Qascom's business roadmap.

5. Test Campaign and Scenarios

Two types of tests are defined in order to validate the DEGREE receiver.

The first batch of tests are a mixture between simulated and real scenarios. Simulated scenarios are executed using a Spirent GSS7000 GNSS simulator (Spirent, Crawley, UK), foreseeing both static and kinematic scenarios. In these scenarios, it is envisaged to simulate spoofing attacks at a signal and data level to validate the anti-spoofing capabilities of the receiver. Jamming attacks are also foreseen.

For kinematic tests, the receiver is installed in a car. The scenario is a mixture of urban and rural environments, with narrow streets and 90 degree turns, trees, light poles, and other cars parked on roadsides, but also wider, open-sky roads with taller trees and more dense vegetation, bridges, and industrial buildings. With this scenario, it is intended to evaluate receiver tracking robustness when working on severe multipath scenarios. Moreover, this scenario allows us to evaluate the performance of the sensor fusion solution (GNSS+INS) against the GNSS-only solution and to characterize system continuity and availability in challenging environments.

The second batch of tests consists of real flight tests in real drones.

The receiver will be integrated with the flight control and navigation solution of the drone. The navigation control solution consists of an on-board segment, consisting of the autopilot, the sensor suite, the actuators, as well as the on-board payloads and a ground segment that, in turn, is made up of a ground control station, a joystick, and a mission planning, monitoring, and management SW.

In order to achieve the best possible results, the tests will be conducted in both a fixed-wing and rotary-wing platform. Both platforms are contained within the specific category, in accordance with EASA (European Union Aviation Safety Agency) legislation. These platforms are capable of mounting all the necessary devices for the flight tests and can reproduce a more stable flight with higher inertial movements.

Table 3 summarizes the characteristics of the two test platforms in which the DEGREE receiver will be installed.

Table 3. Flight test platform specifications.

Type of Platform	Fixed Wing	Rotary Wing
Maximum receiver size (mm)	160	160
Maximum weight (g)	Empty weight = 7 kg MTOW = 9 kg The 2 kg of difference shall be shared between fuel and payloads.	Empty weight = 10 kg MTOW = 14 kg The 4 kg of difference shall be shared between fuel and payloads.
Maximum receiver volume (mm^3)	500,000	500,000
Platform dimensions	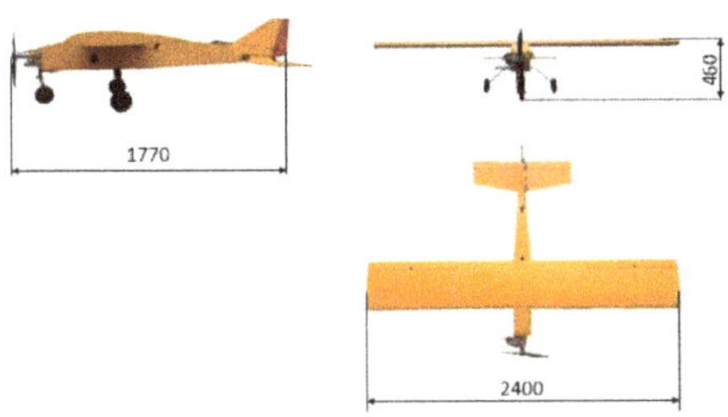	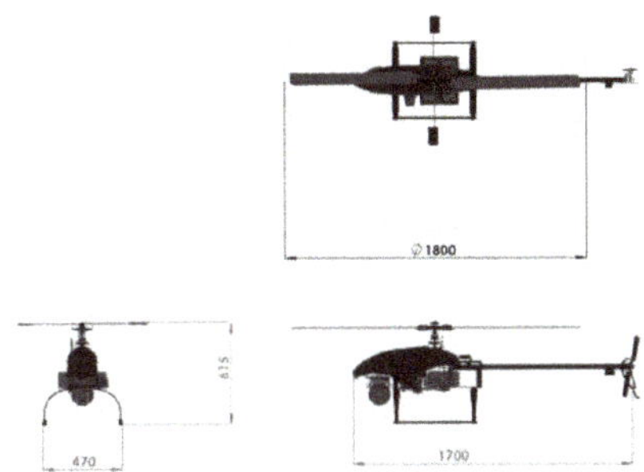

6. Galileo Services on UAVs

6.1. E1 OSNMA

By default, the DEGREE receiver is configured to execute authentication operations using OSNMA protocol. This allows the receiver to authenticate Galileo E1 I/NAV messages. Four different authentication modes are defined for the DEGREE receiver.

In the first mode, called "OFF" mode, the navigation engine does not discard any ranging source that fails the authentication. In other words, the DEGREE receiver uses all ranging sources available, providing the best position accuracy possible. Indeed, there is a major risk that spoofed satellites will penetrate the navigation engine, faking the PNT (Positioning Navigation Timing).

The second mode is called "LOOSE MODE". In this mode, the navigation engine discards all those sources that explicitly fail the authentication process through OSNMA. In the first version of the DEGREE receiver, only Galileo E1 satellites can be authenticated using OSNMA; thus, only Galileo E1 satellites can fail the authentication. For the rest of the sources, i.e., GPS L1/L5 and Galileo E5a/E6, the authentication status is unknown and, hence, is valid to be used by the navigation engine since these sources do not explicitly fail OSNMA authentication. Future versions of Qascom's OSNMA implementation will add support for GPS and Galileo L5 authentication.

The third mode is called "STRICT MODE". This is the safest mode since the DEGREE receiver only uses authenticated sources. This means that in this version of the DEGREE receiver, only Galileo E1-authenticated sources are used in the navigation engine to compute the PVT (Position Velocity Time). While this is the most robust mode in terms of safety against navigation data spoofing, it is the mode that offers less position accuracy since only Galileo E1 satellites are used to compute the PVT. Figure 3 shows the position performance when working in LOOSE MODE (orange plot) and STRICT MODE (blue plot).

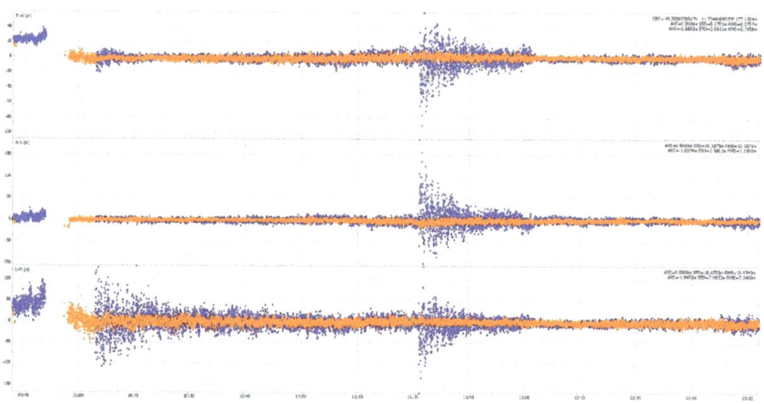

Figure 3. Position accuracy comparison between DEGREE OSNMA LOOSE MODE (blue plot) and STRICT MODE (orange plot).

While these modes can be selected by the user through receiver configuration, a fourth mode called "AUTOMATIC MODE" can also be selected to improve the flexibility of the system. If no spoofing is detected while working in STRICT MODE, the receiver changes to LOOSE MODE to increase positioning accuracy. Instead, if the receiver is working in LOOSE MODE and an authentication failure occurs, the receiver switches to STRICT MODE, assuming there might be other ranging sources spoofed.

6.2. E6 HAS

One of key parameters evaluated when deploying the DEGREE receiver in an operation is the accuracy required by the operation itself. Certain operations may require sub-meter accuracy, and while RTK or standard PPP can achieve centimetre accuracy, they rely on either base stations and radio links with the rover station or external products such as precise orbits and clocks produced by third-party entities.

The solution to this requirement, while operating in real time and in standalone mode, is the usage of Galileo-provided PPP correction, transmitted in HA messages in an E6 signal. To that purpose, the DEGREE receiver reports if HAS corrections are received and applied to the satellite orbits, clocks, and biases. Hence, a dedicated flag is issued to report the previous conditions. Note that the flag is only issued if the Galileo E6 signal is set to be used in the DEGREE receiver, otherwise the flag is not issued. Moreover, the flag is issued for every satellite in view since PPP corrections are available for GPS L1/L5/L2 and Galileo E1/E5a/E5b/E6 frequencies.

7. Towards a Standardized Product

The DEGREE project is targeting TRL 7 for the DEGREE receiver [7]. Although it is not foreseen to achieve enough maturity for a commercial product, the DEGREE project works in this direction. By the end of the project, a specific category SAIL IV GNSS receiver will be designed, developed, and extensively tested through test campaigns, including real flight tests, as explained in Section 5.

While the manufacturing of the DEGREE receiver and the integration of latest Galileo services such as E1 OSNMA and E6 HAS is of utmost importance for the project, the DEGREE consortium also tackles the standardization of UAS receivers for specific category operations. To do that, the consortium is actively collaborating with EUROCAE to consolidate a new standard yet to be released by 2024. With this effort, the DEGREE consortium expects to help the industry and operators to embrace a common standard to simplify the development and deployment of GNSS receivers for specific category UAS operations.

Indeed, while it is difficult for the consortium to adapt the design of the DEGREE receiver to a standard that is yet being defined, we are providing all the knowhow, knowledge, and lessons acquired during the project to EUROCAE working groups, bearing in mind a possible DEGREE project continuation to refine the TRL 7 DEGREE receiver towards a TRL 9 commercial product that implements the standard we are helping to establish.

Author Contributions: Conceptualization, S.D.P., S.F., L.G.-J.d.R., E.L., D.A., J.M. and S.M.; writing—review and editing, S.D.P., S.F., L.G.-J.d.R., E.L., D.A., J.M. and S.M.. All authors have read and agreed to the published version of the manuscript.

Funding: The DEGREE project is funded by European Union/EUSPA (under the grant "Development of a Droneborne Double Frequency Galileo receiver" GSA/GRANT/07/2019).

Institutional Review Board Statement: Not applicable.

Informed Consent Statement: Not applicable.

Data Availability Statement: Data is commercial in confidence.

Conflicts of Interest: The authors declare no conflict of interest. The view expressed herein can in no way be taken to reflect the official opinion of the EU Agency for the Space Programme.

References

1. User Consultation Platform 2020—Minutes of Meeting of the Aviation Panel (GSA-MKD-AV-MOM-A09839). Available online: https://www.euspa.europa.eu/euspace-applications/euspace-users/user-consultation-platform-2020 (accessed on 20 January 2022).
2. Report on Aviation User Needs and Requirements (EUSPA-MKD-AV-UREQ-250287). Available online: https://www.gsc-europa.eu/sites/default/files/sites/all/files/Report_on_User_Needs_and_Requirements_Aviation.pdf (accessed on 20 January 2022).
3. EUSPA Market Report. 2022. Available online: https://www.euspa.europa.eu/sites/default/files/uploads/euspa_market_report_2022.pdf (accessed on 20 January 2022).
4. Updates Following the User Consultation Platform 2020 (EUSPA-MKD-AV-UREQ-250287). Available online: https://www.euspa.europa.eu/sites/default/files/uploads/annex7_aviation.pdf (accessed on 20 January 2022).
5. JARUS Guidelines on SORA. Available online: http://jarus-rpas.org/content/jar-doc-06-sora-package (accessed on 20 January 2022).
6. Qascom News. Available online: https://www.qascom.it/news_POR.php (accessed on 20 January 2022).
7. EU TRL Levels. Available online: https://ec.europa.eu/research/participants/data/ref/h2020/wp/2014_2015/annexes/h2020-wp1415-annex-g-trl_en.pdf (accessed on 20 January 2022).

engineering proceedings

MDPI

Proceeding Paper

Automated Route Planning from LiDAR Point Clouds for Agricultural Applications †

Fabian Theurl *, Christoph Schmied, Eva Reitbauer and Manfred Wieser

Institute of Geodesy, Graz University of Technology, 8010 Graz, Austria; schmied@tugraz.at (C.S.); eva.reitbauer@tugraz.at (E.R.); manfred.wieser@tugraz.at (M.W.)

* Correspondence: fabian.theurl@tugraz.at

† Presented at the European Navigation Conference 2023, Noordwijk, The Netherlands, 31 May–2 June 2023.

Abstract: This paper develops an algorithm to compute optimal routes for an autonomous compost turner. In commercial composting, the material to be composted is piled up in large heaps called windrows and turned regularly by compost turners. The environment at the composting site is constantly changing, as the locations of the windrows change with each turning procedure. Therefore, we propose a novel method that automatically computes routes on a composting plant from LiDAR data. The LiDAR is mounted on the compost turner together with a dual-antenna GNSS receiver, an IMU, and rotary encoders. An extended Kalman filter is used to obtain the vehicle's pose. Through direct georeferencing, a global point cloud is obtained. The routing algorithm crops, segments, and filters the point cloud until the points along the ridge of each windrow remain. These points are used to compute the optimal routes along each windrow. Furthermore, a user can select the windrows which need to be turned and the algorithm then computes the most efficient path for the compost turner, which also includes the passages between the windrows. The method was tested within a simulation environment using a 3D model of the composting site. The results show that the algorithm detects the windrows and computes the routes with sufficient accuracy for autonomous compost turning.

Keywords: mobile mapping; routing; point cloud processing; LiDAR; simulation; digital twin; 3D model; autonomous agriculture; multi-sensor fusion; Kalman filtering

Citation: Theurl, F.; Schmied, C.; Reitbauer, E.; Wieser, M. Automated Route Planning from LiDAR Point Clouds for Agricultural Applications. *Eng. Proc.* **2023**, *54*, 54. https://doi.org/10.3390/ENC2023-15448

Academic Editors: Tom Willems and Okko Bleeker

Published: 29 October 2023

1. Introduction

In 2020, the European Commission introduced a circular economy action plan [1] as part of its European Green Deal. Within the Waste Framework Directive [2], it specified the target that by the year 2035, a minimum of 65% of municipal waste should be reused or recycled. As bio-waste amounts to 34% of all municipal waste [3], recycling bio-waste plays a key role in meeting the EU's recycling targets.

The most common recycling techniques for bio-waste are composting and anaerobic digestion. According to [4], over 90% of the bio-waste that is separately collected in Europe is composted. Therefore, composting can be considered the most important bio-waste recycling technique in Europe.

In composting, the organic material is turned into compost, a biochemically stable product containing microorganisms that can improve soil properties, such as organic matter content, water retention capacity, and pH buffer capacity [3].

The most common method of commercial composting is windrow composting, where the organic material is stacked into long heaps that are turned mechanically by compost turners to promote aerobic microbial activity, moisture release, heat removal, and aeration. However, the job of turning windrows is monotonous and the operator of a compost turner is exposed to high temperatures, water vapor, and gases. Therefore, it is desirable to develop autonomous compost turners.

In recent years, a research team at Graz University of Technology has focused on the development of a navigation module for autonomous compost turners. A concept for an automatic steering system was first presented in [5]. In [6], tests are conducted to select a suitable set of navigation sensors. Two different multi-sensor fusion architectures for fusing GNSS, IMU, wheel odometry, and stereo camera are presented and evaluated in [7]. Ref. [8] presents an odometry model specifically tailored to tracked compost turners that allows for bridging GNSS outages. In [9], a real-time positioning module to estimate the position, velocity, and attitude of a compost turner is developed and tested in an automatic steering system at a composting plant.

In [10], a concept for autonomous compost plant management is proposed. An autonomous plant management not only requires an automatic steering system that steers the compost turner through a single windrow, but requires more complex maneuvers such as turning and driving from one windrow to the next. To achieve these tasks in an automated manner, a complex routing algorithm is required. However, no study has yet focused on routing algorithms for autonomous compost turners. This is exactly where this paper comes in: it develops an algorithm to compute optimal routes from a mobile mapping system mounted on the compost turner.

A major challenge for developing routing algorithms for a composting plant is the constantly changing environment, as the locations of the windrows change with each turning procedure. Therefore, we propose a novel method that automatically computes routes on a composting plant from LiDAR data. In this paper, the concept is presented and tested within a simulated composting plant environment.

Figure 1 shows an overview of the simulated composting plant where several windrows can be seen. The aim of the routing algorithm is to first detect each separate windrow (cyan) in an initialization phase. Afterward, an operator can specify which windrows shall be turned before the compost turner computes an optimal route (orange) through them.

Figure 1. Composting plant in the simulation environment: The routing algorithm first computes the optimal path through each windrow (cyan) and then the final path along which the autonomous compost turner is steered (orange).

Path planning in environments that are observed with a LiDAR scanner is a common task in mobile robotics. For planning global paths, commonly used algorithms are the A* algorithm, which was first published in [11] and then constantly improved (e.g.: [12–14]), and the Dijkstra algorithm [15]. For high-level planning that considers the robot configuration, a local path planner is used. Examples of such local planners are the Dynamic Window Approach algorithm [16] and the Timed Elastic Band (TEB) algorithm [17]. All planners, global and local, use cost maps to define the traversability of different areas within the map.

This paper uses the Dijkstra algorithm as a global and the TEB algorithm as a local planner. However, the main difference to conventional routing applications is that windrows, which would be classified as obstacles by conventional algorithms, are in our case traversable and part of the desired route. This means that each cluster of points within the point cloud that represents such a windrow must not be classified as an obstacle but must be processed to determine the optimal route through it. To the authors' knowledge, no similar system has yet been developed.

2. Materials and Methods

This chapter describes the steps of the routing algorithm and its validation in the simulation environment. Section 2.1 gives an overview of the developed software and the used software packages. Section 2.2 outlines how the global point cloud of the composting plant is obtained from LiDAR data. Section 2.3 describes how the global point cloud is segmented into individual windrow point cloud clusters. Note that point cloud clusters will be referred to as *clusters* throughout this paper. Section 2.4 explains how the individual routes through each windrow cluster are calculated and Section 2.5 outlines how the final route for the compost turner is computed using input from an operator. Section 2.6 describes the simulation environment that is used to validate the developed algorithm.

2.1. Software Development

The software for the routing algorithm is developed in C++. For the point cloud processing operations, the openly available Point Cloud Library [18] is used. The Robot Operating System (ROS) [19] provides the appropriate framework for the development of the robotic application. The ROS navigation stack offers the 2D navigation capabilities needed for route planning and following.

2.2. Obtaining the Global Point Cloud from the Windrow Area

In the first step, an up-to-date global point cloud of the composting plant must be generated. This cloud is acquired in an initialization phase where the compost turner is steered manually around the windrows at the composting plant. For this task, a simulated Velodyne Ultra Puck LiDAR scanner was used. Given the scanner's limited vertical field of view of 40°, it is mounted on top of the compost turner with a forward tilt of 15° to better cover the area in front of the machine.

The full procedure to incrementally build the global point cloud in the initialization phase is shown in Figure 2. Each incoming LiDAR point cloud must first be georeferenced to be in the same coordinate frame as the global point cloud. This is achieved through direct georeferencing using the position and attitude results (*Robot Pose*) from a previously developed Error State Kalman Filter (ESKF). The filter fuses the odometry data of the compost turner with data from an Inertial Measurement Unit (IMU) and a Real-Time Kinematic (RTK) dual antenna Global Navigation and Satellite System (GNSS) receiver to calculate the state vector of the robot in real-time. For a detailed description of the positioning module, the reader is referred to [9].

To keep the size of the global point cloud as small as possible in order to reduce computational load, several further operations are performed:

- **Windrow Area Filter:** The windrows are always located within the same area on the composting plant. Therefore, the extent of this area is defined using a confining polygon (*Windrow Area Polygon*) created from known 2D coordinates. All points outside this polygon are simply filtered out during the initialization.
- **Local Box Filter:** Points are only added to the global point cloud when they are within a certain vertical range from the ground and within a certain horizontal distance from the LiDAR. The vertical range ensures that ground points are removed. The horizontal distance ensures that points that are further away from the LiDAR—and, therefore, have a lower accuracy—are not included.
- **Downsampling:** The constant addition of the current LiDAR point cloud to the global cloud leads to areas that have a very high point density. In a downsampling step, the total amount of points can be reduced, especially in these dense areas, by making sure that only one point is kept within a specified *Downsampling Radius*.

After the initialization phase, the *Global Windrow Point Cloud* contains all points that belong to windrows and provides the basis for extracting the individual windrow clusters.

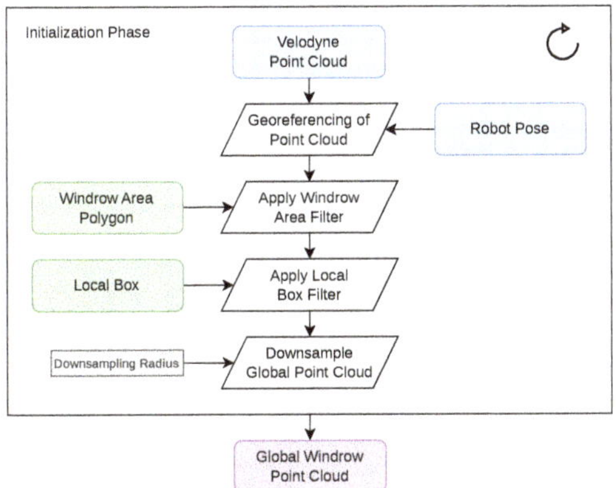

Figure 2. Incremental procedure to build the global windrow cloud during an initialization phase.

2.3. Extraction of the Windrow Clusters

From the *Global Windrow Point Cloud*, the individual windrow clusters can now be extracted. An overview of the clustering process is given in Figure 3. In the first step, an outlier removal using Random Sample Consensus (RANSAC) is performed, since outliers can negatively influence the clustering result. Therefore, RANSAC with a planar model is used to estimate the best-fitting plane through all points. Points further than a specified *Distance Threshold* away from this plane, i.e., the outlier points, are removed.

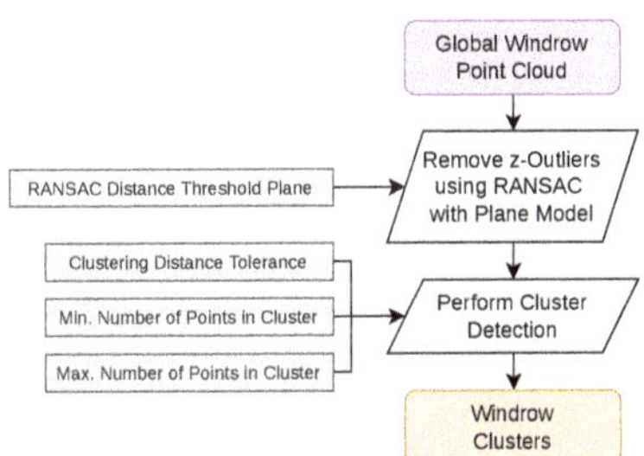

Figure 3. Filtering and clustering of the global point cloud.

After the outlier removal, the clustering is done using Euclidean Clustering provided by the Point Cloud Library (PCL). The *Clustering Distance Tolerance* parameter specifies the maximum distance between points to be considered part of the same cluster. Additionally, the cluster size limits (*minimum and maximum number of points* within a single cluster) can be set. These parameters depend on the density of the point cloud. If the *Downsampling Radius* (see Section 2.2) is large, a larger *Clustering Distance Tolerance* and lower cluster size limits are necessary, and vice versa. The resulting clusters each represent one windrow.

2.4. Finding the Optimal Route through each Windrow Cluster

In the next step, each *Windrow Cluster* is analyzed and processed to find the optimal route for the compost turner through the windrow. The optimal route follows the ridge of each windrow and thus requires the extraction of the ridge points from the *Windrow Cluster*. A simple line is estimated, as it guarantees a relatively straight windrow after the compost turner finishes moving through the windrow. In Figure 4, the full process of finding this optimal line through each windrow is outlined. The individual steps are explained in the following.

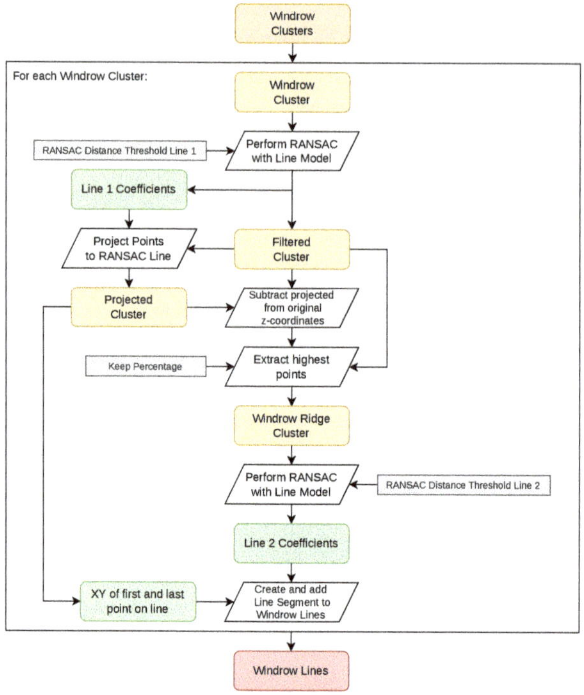

Figure 4. Overview of the processing steps needed to obtain the optimal route through a Windrow Cluster.

First, a RANSAC line-fitting algorithm is used to obtain the *Line Coefficients* for the best-fitting line through the cluster points and to further eliminate outliers (*Filtered Cluster*). Projecting the *Windrow Cluster* to the RANSAC line (*Projected Cluster*) assigns every point i a corresponding z-coordinate on the line (z^i_{line}). Then, z^i_{line} is subtracted from the original z-coordinate, $z^i_{original}$, of the point:

$$z^i_{rel} = z^i_{original} - z^i_{line}. \tag{1}$$

This is performed for all n points. The resulting set $Z_{rel} = \{z^i_{rel}\}^n_{i=1} = \{z^1_{rel}, z^2_{rel}, \ldots, z^n_{rel}\}$, which represents the z-coordinates of the points with respect to the line, is distributed around zero. From this set, the points with the highest z-coordinates relative to the line can be extracted. The *Keep Percentage* parameter specifies the percentage of points extracted. These points then represent the ridge of the windrow (*Windrow Ridge Cluster*).

Finally, a second RANSAC line-fitting algorithm can be applied to obtain the coefficients for the optimal line through the *Windrow Ridge Cluster*. From the *Projected Cluster*, the first and last points on the line are extracted. These points are used together with the *Line Coefficients* from the second RANSAC to obtain the start and endpoints of a line segment—the optimal route through the windrow.

2.5. Route Planning Algorithm for the Compost Turner

After determining the optimal route through each windrow (*Windrow Lines*), the routing algorithm for the global path at the composting plant can be developed. The aim is to find a route through user-selected windrows, which takes maneuvers on the composting plant into account. Such maneuvers include moving from the end of one windrow to the start of another windrow or moving to a specified target with a desired final orientation (e.g., to the charging station).

A user can select which windrows should be turned by the autonomous compost turner. The full route consists of a list of waypoints. One windrow is stored as two separate waypoints, i.e., the start and endpoints of the *Windrow Line*. Depending on which point is closer to the compost turner, the start and endpoints of the estimated route through the windrow can be switched around.

When starting the algorithm, the first stored waypoint is passed to the route planner. The route is planned with the ROS navigation stack, specifically with the *move_base* package which incorporates a global and a local planner. For the global path planning, the *global_planner* package, which contains the Dijkstra algorithm, was used. Local planning was integrated using the *teb_local_planner* plugin [20], which is based on the timed elastic band (TEB) method. The planner then computes the optimal route to the waypoint and outputs linear and angular velocity commands to the control module of the robot. This is repeated until the last waypoint is successfully reached.

2.6. Simulation Environment

The developed routing algorithm is tested and evaluated within Gazebo. Gazebo is a simulation environment that allows for ROS integration. Several plugins exist that enable the simulation of navigation sensors such as GNSS receivers, IMUs, or LiDARs. Within Gazebo, the compost turner is simulated and can be steered manually using a controller. Outputs from a differential drive controller are used and translated to steering commands for the robot.

The environment of the composting plant is simulated from a mesh file. This 3D model was created using data from UAV photogrammetry. The environment contains an area of approximately 50 by 70 m. In total, there are 16 windrows, each with a width of approximately 3 m and a height of 1.5 m. The windrows are 40 m long, except for the last two, which are only 15 m long. To properly test the windrow detection algorithm that estimates the optimal routes through each windrow, several tests were conducted within the simulation environment (see Section 3).

3. Results

This chapter presents the results of the validation of the routing algorithm in the simulation environment. First, the processing steps for detecting a single windrow are visualized, and then the results for multiple initialization rounds are presented and compared.

3.1. Detection Results for a Single Windrow

Figure 5 shows the result of the windrow detection algorithm for a single windrow. The simulated windrow (A) is observed in the initialization phase as a point cloud (B). Then the ridge points of the windrow are extracted (C). Finally, a line is estimated through the ridge points (D), from which the start and endpoints of the windrow can be derived as a result.

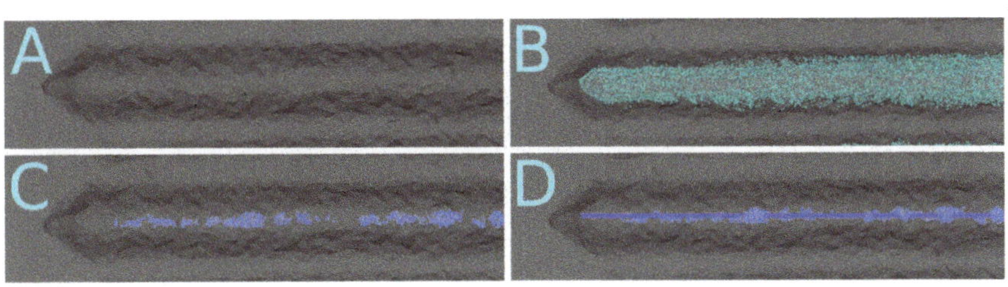

Figure 5. Visualization of the processing steps for detecting a single windrow. Simulated windrow (**A**); Windrow after initialization (**B**); Extracted ridge points (**C**); Estimated line through ridge points (**D**).

3.2. Detection Results for Multiple Initialization Rounds

To validate the windrow detection, three sets of ten individual initialization runs were executed in the simulation environment. The three sets differ regarding the input settings (see Table 1). The *Downsampling Radius* was varied from 0.10 m (Set A) to 0.30 m (Set C), which also has an effect on the other parameters for the cluster detection (see Section 2.3).

Table 1. Windrow detection settings for the different test sets.

Setting	Set A	Set B	Set C
Downsampling Radius	0.10 m	0.20 m	0.30 m
Clustering Distance Tolerance	0.15 m	0.30 m	0.45 m
Min. Number of Points in Cluster	1000	500	300
Max. Number of Points in Cluster	100,000	50,000	30,000
Local Crop Box: Horizontal Distance from LiDAR Vertical Range above Ground		 25 m 0.9–1.8 m	
RANSAC Distance Thresholds: Plane Line 1 Line 2		 0.5 m 0.7 m 0.5 m	
Ridge Points—Keep Percentage		40%	

Each initialization run was performed manually by steering the compost turner around the windrows of the simulated composting plant, meaning the routes are always similar but never exactly the same. Each time, all 16 windrows were successfully detected. To validate the results numerically, the estimated start and endpoints as well as the orientation of each windrow were compared. Since each windrow is detected ten times, the standard deviation of the distance from the mean start and endpoint is given in Table 2. Keep in mind that these results show the quality of the detected *Windrow Lines* compared to one another and not to an actual reference position of the windrow.

Table 2. Windrow detection results for the three different test sets.

ID	Set A Standard Deviations Points (2D) [m]	Set A Standard Deviations Orientation [°]	Set B Standard Deviations Points (2D) [m]	Set B Standard Deviations Orientation [°]	Set C Standard Deviations Points (2D) [m]	Set C Standard Deviations Orientation [°]
1	0.10	0.43	0.07	0.41	0.16	0.41
2	0.05	0.16	0.15	0.35	0.11	0.34
3	0.08	0.29	0.09	0.25	0.16	0.39
4	0.06	0.15	0.06	0.18	0.12	0.35
5	0.07	0.36	0.10	0.28	0.11	0.13
6	0.06	0.28	0.12	0.28	0.13	0.50
7	0.09	0.18	0.08	0.26	0.08	0.14
8	0.10	0.08	0.13	0.23	0.11	0.29
9	0.07	0.24	0.08	0.12	0.13	0.36
10	0.08	0.20	0.11	0.31	0.14	0.43
11	0.06	0.11	0.04	0.14	0.10	0.12
12	0.05	0.16	0.08	0.43	0.10	0.32
13	0.07	0.15	0.10	0.27	0.09	0.11
14	0.04	0.07	0.08	0.22	0.16	0.26
15	0.09	0.38	0.11	0.62	0.10	1.03
16	0.10	1.13	0.09	0.90	0.11	1.45
Average	**0.07**	**0.27**	**0.09**	**0.33**	**0.12**	**0.41**

The results show that a higher *Downsampling Radius* increases the standard deviations. On average, the standard deviations for Set A of the estimated start and endpoints was 0.07 m and the average standard deviation of the orientation was 0.27°. They increase to 0.12 m and 0.41° when looking at the results of Set C. However, the results for Set C are still sufficiently accurate to navigate the compost turner to the start of a windrow. Additionally, increasing the *Downsampling Radius* (Set C) reduces the size of the point cloud by a factor of 8. This minimizes the computational load and can reduce the hardware requirements for the mobile processing unit of the compost turner.

The last two windrows (IDs 15 and 16) are shorter than the other windrows. Since the orientation is computed from the start and endpoints, the shorter distance between these points explains the larger standard deviation.

4. Discussion

This paper presented a route planning module for an autonomous compost turner. The module automatically detects the windrows on the composting plant in an initialization phase using LiDAR. This is achieved by first generating a global point cloud of the windrows from which the individual windrow clusters are extracted. Then the ridge points of each windrow cluster are retrieved and an optimal route through them is estimated. The routes through each windrow are then incorporated into a global route planning algorithm. Via user input, the windrows that the robot should turn can be selected and the optimal route to, through, and between the selected windrows can be calculated. The algorithm was tested in a simulation environment.

Tests have shown that the algorithm reliably manages to detect all windrows after the initialization phase. The start and endpoints of each route through a windrow were detected with a standard deviation of less than 0.15 m, while the orientation of each route through the windrow was estimated with a standard deviation of less than 0.5°. Therefore, this algorithm is suitable for the estimation of the windrow start and endpoints.

Future research will focus on the evaluation of the method using real-world data, including tests at a real composting plant. Currently, an obstacle detection and avoidance module as well as a precise local approach strategy to the windrows are being developed.

On the path toward smart compost plant management, autonomous compost turners play an important role. When equipped with additional sensors, they can be used to georeference and monitor temperatures and greenhouse gas emissions. This can help to ensure quality control. Furthermore, autonomous compost turners allow increasing efficiency at the composting site.

Author Contributions: Conceptualization, F.T., C.S. and E.R.; methodology, F.T.; software, F.T., C.S. and E.R.; validation, F.T.; formal analysis, F.T.; investigation, F.T.; resources, F.T., C.S. and E.R.; data curation, F.T.; writing—original draft preparation, F.T. and E.R.; writing—review and editing, E.R. and C.S.; visualization, F.T. and C.S.; supervision, M.W.; project administration, E.R.; funding acquisition, E.R. and M.W. All authors have read and agreed to the published version of the manuscript.

Funding: This research was funded by the Austrian Federal Ministry for Climate Action, Environment, Energy, Mobility, Innovation, and Technology via Österreichische Forschungsförderungsgesellschaft (Austrian Research Promotion Agency) (885368).

Institutional Review Board Statement: Not applicable.

Informed Consent Statement: Not applicable.

Data Availability Statement: The data presented in this study are available upon request from the corresponding author.

Conflicts of Interest: The authors declare no conflict of interest. The funders had no role in the design of the study; in the collection, analyses, or interpretation of data; in the writing of the manuscript; or in the decision to publish the results.

Abbreviations

The following abbreviations are used in this manuscript:

ESKF	Error State Kalman Filter
GNSS	Global Navigation and Satellite System
IMU	Inertial Measurement Unit
LiDAR	Light Detection and Ranging
PCL	Point Cloud Library
RANSAC	Random Sample Consensus
ROS	Robot Operating System
RTK	Real-Time Kinematic
TEB	Timed Elastic Band
UAV	Unmanned Aerial Vehicle

References

1. European Union. Directorate-General for Environment, Communication from the Commission to the European Parliament, the Council, the European Economic and Social Committee and the Committee of the Regions. A New Circular Economy Action Plan For a Cleaner and More Competitive Europe. Available online: https://eur-lex.europa.eu/legal-content/EN/ALL/?uri=COM:2020:98:FIN (accessed on 30 April 2023) .
2. European Parliament, Council of the European Union. Directive (EU) 2018/851 of the European Parliament and of the Council of 30 May 2018 Amending Directive 2008/98/EC on waste. Available online: http://data.europa.eu/eli/dir/2018/851/oj (accessed on 30 April 2023) .
3. European Environment Agency. *Bio-Waste in Europe: Turning Challenges into Opportunities*; Publications Office of the European Union: Luxembourg, 2020; ISSN 1977-8449. [CrossRef]
4. Razza, F.; D'Avino, L.; L'Abate, G.; Lazzeri, L. The Role of Compost in Bio-waste Management and Circular Economy. In *Designing Sustainable Technologies, Products and Policies: From Science to Innovation*; Benetto, E., Gericke, K., Guiton, M., Eds.; Springer International Publishing: Cham, Switzerland , 2018; pp. 133–143. [CrossRef]
5. Reitbauer, E.; Schmied, C.; Wieser, M. Autonomous Navigation Module for Tracked Compost Turners. In Proceedings of the 2020 European Navigation Conference (ENC), Dresden, Germany, 23–24 November 2020; pp. 1–10. [CrossRef]
6. Reitbauer, E.; Schmied, C.; Schedler, M. Integrated Navigation for Tracked Compost Turners Using GNSS, INS, Odometers, Stereo Camera and 3D Map. In Proceedings of the 33rd International Technical Meeting of the Satellite Division of The Institute of Navigation (ION GNSS+ 2020), Virtual, 21–25 September 2020; pp. 2518–2532, ISSN 2331-5954. [CrossRef]
7. Reitbauer, E.; Schmied, C. Performance Analysis of GNSS/INS/VO/Odometry Sensor Fusion Algorithms for Tracked Agricultural Vehicles. In Proceedings of the 34th International Technical Meeting of the Satellite Division of The Institute of Navigation (ION GNSS+ 2021), St.Louis, USA, 20–24 September 2021; pp. 3250–3262, ISSN 2331-5954. [CrossRef]
8. Reitbauer, E.; Schmied, C. Bridging GNSS Outages with IMU and Odometry: A Case Study for Agricultural Vehicles. *Sensors* **2021**, *21*, 4467. [CrossRef]
9. Reitbauer, E. Multi-Sensor Positioning for the Automatic Steering of Tracked Agricultural Vehicles. Ph.D. Thesis, Graz University of Technology, Graz, Austria, 2022.
10. Cichocki, M.; Reitbauer, E.; Theurl, F.; Schmied, C. Composting 4.0: From the Automatic Steering of Compost Turners Towards an Autonomous Plant Management. In Proceedings of the 7th Central European Biomass Conference (CEBC 2023), Graz, Austria, 18–20 January 2023; p. 189.
11. Hart, P.E.; Nilsson, N.J.; Raphael, B. A Formal Basis for the Heuristic Determination of Minimum Cost Paths. *IEEE Trans. Syst. Sci. Cybern.* **1968**, *4*, 100–107. [CrossRef]
12. Duchoň, F.; Babinec, A.; Kajan, M.; Beňo, P.; Florek, M.; Fico, T.; Jurišica, L. Path Planning with Modified a Star Algorithm for a Mobile Robot. *Procedia Eng.* **2014**, *96*, 59–69. [CrossRef]
13. Erke, S.; Bin, D.; Yiming, N.; Qi, Z.; Liang, X.; Dawei, Z. An improved A-Star based path planning algorithm for autonomous land vehicles. *Int. J. Adv. Robot. Syst.* **2020**, *17*, 1729881420962263. [CrossRef]
14. Tang, G.; Tang, C.; Claramunt, C.; Hu, X.; Zhou, P. Geometric A-Star Algorithm: An Improved A-Star Algorithm for AGV Path Planning in a Port Environment. *IEEE Access* **2021**, *9*, 59196–59210. [CrossRef]
15. Dijkstra, E.W. A note on two problems in connexion with graphs. *Numer. Math.* **1959**, *1*, 269–271. [CrossRef]
16. Fox, D.; Burgard, W.; Thrun, S. The dynamic window approach to collision avoidance. *IEEE Robot. Autom. Mag.* **1997**, *4*, 23–33. [CrossRef]
17. Roesmann, C.; Feiten, W.; Woesch, T.; Hoffmann, F.; Bertram, T. Trajectory modification considering dynamic constraints of autonomous robots. In Proceedings of the ROBOTIK 2012; 7th German Conference on Robotics, Munich, Germany, 21–22 May 2012; pp. 1–6.
18. Rusu, R.B.; Cousins, S. 3D is here: Point Cloud Library (PCL). In Proceedings of the IEEE International Conference on Robotics and Automation (ICRA), Shanghai, China, 9–13 May 2011; IEEE: Shanghai, China, 2011.

19. Quigley, M.; Gerkey, B.; Conley, K.; Faust, J.; Foote, T.; Leibs, J.; Berger, E.; Wheeler, R.; Ng, A. ROS: an open-source Robot Operating System. In Proceedings of the IEEE International Conference on Robotics and Automation (ICRA) Workshop on Open Source Robotics, Kobe, Japan, 12–17 May 2009.
20. Rösmann, C.; Hoffmann, F.; Bertram, T. Integrated online trajectory planning and optimization in distinctive topologies. *Robot. Auton. Syst.* **2017**, *88*, 142–153. [CrossRef]

engineering proceedings

MDPI

Proceeding Paper

Ice Navigation in Arctic Offshore Wind Parks: Traffic Coordination Using Route Exchange and Moving Havens †

Thomas Porathe

Department of Design, Norwegian University of Science and Technology (NTNU), NO-7491 Trondheim, Norway; thomas.porathe@ntnu.no; Tel.: +46-706681819

† Presented at the European Navigation Conference 2023, Noordwijk, The Netherlands, 31 May–2 June 2023.

Abstract: New areas for the production of fossil-free renewable offshore energy production are planned in Arctic waters in the Bay of Bothnia, in the very north of the Baltic Sea between Sweden and Finland. These are waters normally covered with 30–60 cm of ice every winter and shipping is carried out with icebreaker assistance. This article suggests the need for research on the wind turbine's effect on ice behavior and ice navigation. Some concepts from previous research are suggested as a solution.

Keywords: ice navigation; Arctic; offshore wind farms; ship traffic management; route exchange; e-Navigation; Moving Havens

1. Introduction

The global climate situation and UN's zero emission goal for 2050 is, in Sweden, like elsewhere, leading to an increased pressure for new fossil-free energy production. Public resistance to wind turbines on land has increased interest in offshore wind farms. In the north of Sweden, the steel industry is today using large quantities of coal, but new fossil-free methods will use clean, electro-generated hydrogen. Future demand for electric power will by far overrun the present capacity. So, there is a huge pressure for new, close-range, offshore wind production, and several new wind farms are proposed for the Bay of Bothnia, which is located furthest north in the Baltic Sea (see Figure 1).

Citation: Porathe, T. Ice Navigation in Arctic Offshore Wind Parks: Traffic Coordination Using Route Exchange and Moving Havens. *Eng. Proc.* **2023**, *54*, 55. https://doi.org/10.3390/ENC2023-15462

Academic Editors: Tom Willems and Okko Bleeker

Published: 29 October 2023

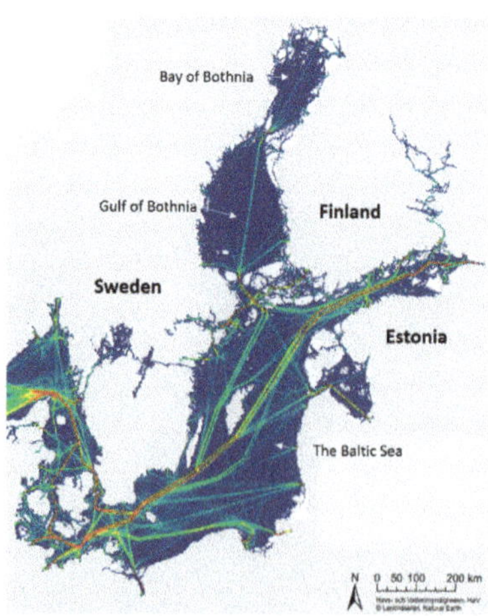

Figure 1. Map of the Nordic countries with the Baltic Sea, Gulf of Bothnia, and the Bay of Bothnia. Color-graded traffic density from shows the relative traffic flow. Reprinted with permission from Ref. [1]. 2018, The Swedish Agency for Marine and Water Management.

In December 2022, the Swedish Maritime Administration (SMA) expressed concerns regarding plans for offshore wind turbine parks in the Bay of Bothnia. Presently, the SMA is handling over 60 applications for new offshore wind parks around the Swedish coast, but the ones in the Bay of Bothnia are special in the sense that these waters are covered with ice most winters [2]. Sweden and Finland are sustaining the world's second largest icebreaker fleet to keep the Arctic ports in the area open all year around. As can be seen by the ship traffic density map in Figure 1, the traffic to the Bay of Bothnia is relatively limited in relation to traffic in the Baltic Sea. The number of vessels assisted by icebreakers in the Bay of Bothnia in the winter season 2020–2021 was 1095 ships, with 141 days of restricted traffic [3].

The SMA points to a need to avoid letting wind farms become choke points blocking areas vital to cargo transportation [2]. The direct tracks used during the summer season cannot always be used in the winter. When the sea is covered with ice, the routes must change to accommodate the ice situation. "Icebreaking is mostly about avoiding thick ice and if it is a severe winter the icebreaker must take the possible route. The ice is not static but dynamic and moves depending on meteorological circumstances," Johan Wahlstöm, at the SMA, says [4]. Figure 2 shows the northern part of the Bay of Bothnia with the big steel port Luleå at the top left in both maps. The left-hand map shows the summer tracks used as well as plans for new wind farms merged from Swedish and Finnish sources [2,5]. The right-hand map overlays the tracks of icebreakers during the 2022–2023 season [2]. As can be seen, the tracks of icebreaker-assisted operations do not correspond to the open water ship traffic patterns in the left-hand map. It is an open question how wind farms will affect the icebreaker-assisted winter shipping in the Bay of Bothnia. Experience and research are lacking.

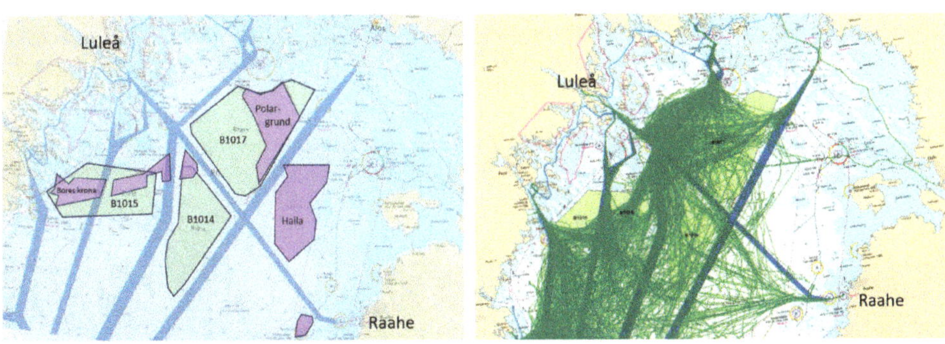

Figure 2. To the (**left**) are the ordered tracks of summer shipping in blue. Overlayed are suggested wind parks in green from [2] and in purple from [5]. To the (**right**) are the green tracks from Swedish icebreakers during the winter season 2022–2023. Adapted with permission from Ref. [2]. 2023, The Swedish Maritime Administration.

2. Present Winter Navigation in the Bay of Bothnia

Winter conditions in the Bay of Bothnia generally start in December and last until May. During this time about nine or ten Swedish and Finish icebreakers assist ice-classed cargo ships to ports in this area. In April 2011, this author did a field study onboard the Swedish icebreaker Frej. The icebreaker service consisted of opening new ice channels in 30–55-centimeter-thick ice (see Figure 3, left), breaking loose the ships that were stuck in the ice, assisting convoys of ships though ice channels (Figure 3, right), and occasionally assisting single ships in heavy ice by notch towing.

Figure 3. (**Left**) the icebreaker Frej, at 5 knots, is opening a new channel through the 30–55 cm thick ice. A minute after this photo is taken, by the red arrow, Frej is brought to a complete stop by this seemingly insignificant ice ridge that shows very little of its true size above the ice sheet. (**Right**) convoy service. Often the ships in the convoy become stuck and the icebreaker must go back and break them free. During this maneuver, the other ships become stuck. It is tedious work. Photo by the author.

All these situations were experienced during the field study which was undertaken in the EU project MONALISA Ice (MICE). The goal of the study was to collect user needs for e-Navigation in Arctic conditions [6].

The Swedish and Finnish icebreakers in the Bay of Bothnia are coordinated and help each other. Each icebreaker or group of icebreakers has a limited area in which they are responsible for escorting ships. This gives them an opportunity to learn the local ice conditions. North- or southbound convoys of ships are then handed over from one icebreaker to the next as they transit through the ice. In Figure 4, the ice situation in the Bay of Bothnia on the 15 April 2023 can be seen [7]. The red color denotes severe ice (in this case, 30–60 centimeters thick in the central part of the Bay of Bothnia). The map also shows the areas where the Swedish icebreakers (Atle, Frej, Ymer, Ale) and the Finnish icebreakers (Konto, Sisu, Polaris, Otso, Zeus) are assisting. Note the blue "lead" of open water created on the Finnish side due to easterly winds pressing the ice sheet westwards. Though such a lead, ships can move without assistance. If the wind becomes westerly, the ice sheet will move back over to the Finnish side and a lead may form along the Swedish coast. Such an ice drift could be prevented by wind farms.

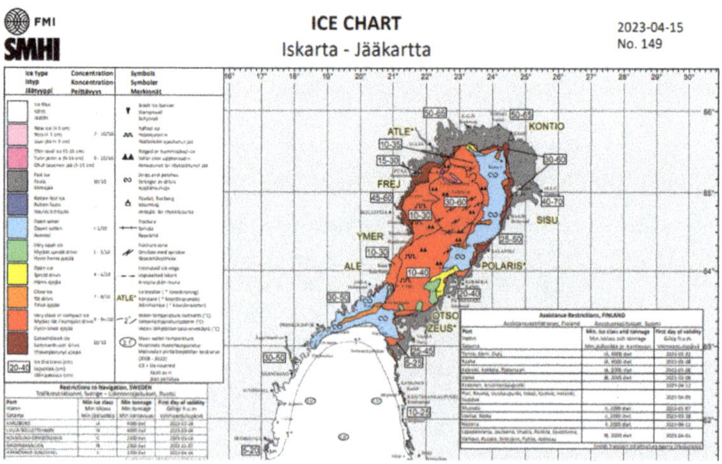

Figure 4. Example of a color-coded ice chart for 15 April 2023 [7]. These charts are published daily during the ice season. Reprinted with permission from the Swedish Meteorological and Hydrological Institute.

One important finding during the study in 2011 was the problem of finding the right, open, ice channel in the maze of old, abandoned and frozen-solid channels in the ice sheet.

Once the ships have passed through the channel, the crushed ice rubble floats back into the channel and fills the passage, and if the temperature is below the freezing point, immediately starts to freeze solid again. In Figure 5, on the right, such an ice channel is shown. It is almost impossible to judge whether the channel is open and negotiable for a cargo vessel, or if it is an old frozen channel. Frozen channels are more difficult to break open and, in such cases, icebreakers often prefer to make a new channel by its side.

Figure 5. (**Left**) the ice sheet seen from the air. The maze of ice channels, some old and abandoned, with the open, central ice channel and a ship visible. (**Right**) a photo of a new, open channel filled with ice rubble after the icebreaker had just passed. An old, frozen-solid channel would look exactly the same. The question is, how do you find the right channel? Photo by the author.

Even for the experienced eye, it is difficult to visually determine which channels are open and which are frozen solid. But the icebreakers had a trick. During the study, I noticed that they, besides their ordinary Furuno ECDIS, used an Adveto system which could record their own or other ships' AIS tracks in high resolution in a local database. By uploading a part of such a historic track, they could use it as a "footpath" through the ice. Often, the ice, with the track, could have drifted since their last passage, but it would still give a general clue as to the latest, still open, channel (and with the added value of developing a hunch for the movement of the ice sheet).

In Figure 6, on the left, a screenshot from the Adveto system can be seen. The time is a little before nine in the morning; Frej has been freeing two ships (the Fembria and the Helena) from the ice (the historic track from his—icebreakers are always "he"—maneuvers loosening the ship can be seen) and the icebreaker is now leading the convoy through the ice using his own track from about 2300 h the night before. In this case, the ice has drifted some 50 m to the NE. In Figure 6, on the right, the situation (a little earlier) can be seen on the icebreaker's regular Furuno ECDIS screen. This system had then no possibility to display historic tracks in the same way as the Adveto.

A wish expressed by the bridge officers onboard Frej in 2011 was an ability to cut out a relevant snippet of their high-resolution historic track and send it addressed to a particular ship so that they could amend it to their active route in any ECDIS, allowing them to follow the open ice channel by themselves. In that way, the bridge officers said, the icebreaker could park in the middle of their area and function as a traffic center and only need to fetch the most vulnerable ships [6].

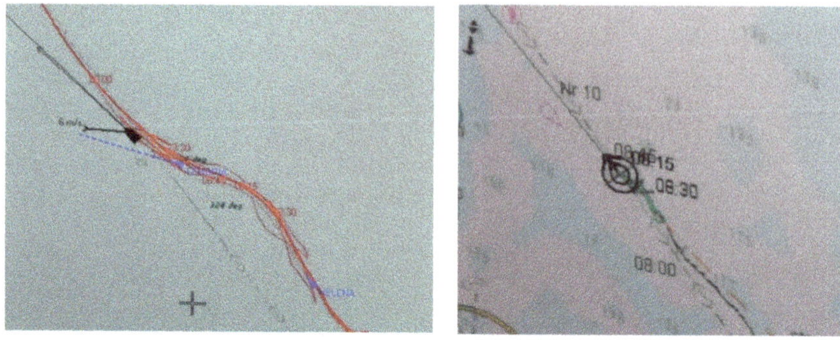

Figure 6. (**Left**) a screenshot from the Adveto ECDIS; the icebreaker is escorting a convoy of two ships at 8.45 in the morning. He is using the historic track of himself from about 23.00 the previous night, showing that the ice sheet has been relatively stable. To the (**right**) is a screenshot from the ship's regular Furuno ECDIS a little earlier. The historic track cannot be called up here. Photo by the author.

MICE and The RTZ Route Format

This need expressed by the bridge officers of the icebreaker Frej coincided with work on "route exchange" ("intended routes" and "route suggestions") concurrently investigated in the e-Navigation projects EfficenSea and MONALISA [8]. During the subsequent work in the MICE project, Andres Rydlinger at Transas Marin took part. Anders was a member of an IEC working group rewriting the route standard used in conjunction with ECDIS. The new standard was to become IEC 61174:2015 (RTZ version 1.0) and was published in 2015. The S-100-compliant successor standard of the RTZ format was developed by the IEC TC80 working group 17 as IEC 63173 S-421, officially published in 2021 [9].

3. Planned Wind Farms' Potential Influence on Ice Navigation

To prevent offshore wind power becoming an obstacle to shipping, it must be designed with regard to shipping conditions. A full expansion will have major consequences for shipping, the SMA says [10]. Shipping and offshore wind farms could coexist given the right conditions and considerations. The ships that transport goods daily between the harbors in the north normally sail along predefined tracks, but when the sea freezes, the conditions become different. Ice breaking is mostly about avoiding thick ice, and during a severe winter, the icebreakers must use the route that is possible. "Work with the ice, not against it" Captain Johan Buysse recommends in the Nautical Institutes practical guide to the handling of ships in ice [11].

The effects of wind farms are twofold: the ice pressure will affect the turbine foundations (an engineering problem) and the turbine foundations will affect the movement, structure, and behavior of the ice sheet (an ice navigation problem).

3.1. Effects of Ice on Solid Structures

A good understanding of ice conditions and ice loads is essential for the proper construction of the foundations for offshore wind power [12]. The applied practices behind the design of bridges, lighthouses, and other offshore constructions were originally created without much knowledge of ice mechanics. Now, experience has been gathered for a long time for ice-exposed constructions close to land. The design of bridges has, for instance, not caused any essential setback, whereas a few small lighthouses have been damaged [12] (see Figure 7).

Figure 7. (**Left**) in 1969, the Swedish caisson lighthouse Nygrån was broken down by the ice pressure (photo G. Kjellgren). (**Right**) in 1986, another Swedish caisson lighthouse, Björnklack, was tilted by the ice (photo by unknown). Both lighthouses were in the Bay of Bothnia.

Some research on solid offshore deep water structures in ice has been conducted, but there is a large gap of knowledge. The recent plans for wind farms in the Bay of Bothnia have sparked research projects in Finland and in Sweden. The greatest dangers seem to be attributed to ice ridges. The thickest ice ridge that has been recorded in the Bay of Bothnia was 28 m deep, but is normally 10–20 m. [13].

Ridges are usually present in all ice-covered sea areas if the ice concentration is high enough. Ice ridges are very common ice features in the Baltic Sea area [14].

First-year ice ridges are elongated accumulations of broken ice floes. Ridges form when sea ice deforms and fractures due to driving forces from winds and sea currents. In that process, large sea ice flows are compressed or sheared against each other, causing the ice to fail into smaller pieces, forming a ridge. Ice rubble is a result of the crushing and flexural failure of ice during the ridge formation process. A compression ridge is often highly irregular, both in direction and in vertical height. A shear ridge is normally very straight, because separate ice fields move laterally in opposite directions. A ridge contains a large number of ice pieces of varying sizes and shapes that are piled arbitrarily, and they float in the seawater [15].

After ridges have been formed, they drift due to wind and sea current. When a ridge collides with an offshore structure, it introduces a significant load on the structure.

3.2. *Effects by Solid Structures on Ice Drift and Structure*

From what I understand, the formation of ice and ridges are very complex and difficult to model. How constructions of many hundreds of turbines—present-day plans include more than 600 wind turbines for the Bay of Bothnia [5]—will affect the ice formation is not yet known. Two scenarios seem obvious: The ice floes or thin newly formed sea ice, propelled by the wind and current, will continue to drift past the turbine foundations which will act as icebreakers and open up channels in the ice. On the leeward side of the turbine foundation, new patterns of ice formations will appear, forming ridges or ice channels. Different interference patterns may also form. There is no research on this yet, to my knowledge. When freezing continues, the ice sheet will grow thicker and, at one point, it may be too thick for the turbine foundations to break. Then, we come to the other likely scenario.

At some point, the ice will become so strong that the prevailing wind or current can no longer exceed the force needed for the turbine foundation to keep breaking the ice. The whole ice sheet might then become fixed, "nailed" in position by the turbine foundations, and may resemble fast ice.

Or, it may be a combination of these two scenarios depending on sea and air temperature, along with wind and current direction and speed.

In the first case it is possible that the creation of many more ice channels may increase the formation of ice ridges, making the ship channels between the wind farms more difficult to negotiate. But, it may also be possible that, as in the second scenario, a fast ice sheet, held in place by the wind turbines, may cause a more stable situation and keep negotiable ice channels in place for longer periods of time, thus facilitating ice navigation.

The leads of open water which today are formed along the Swedish or Finnish coasts on the windward side, depending on the wind direction, may cease to exist in the same way if the ice sheet becomes fixed by the wind farms.

4. e-Navigation and Route Exchange for Ship Traffic in Wind Farms

In the ACCSEAS project (2013–2015), we saw plans for offshore wind farms appear in the English Channel, much to the surprise of the shipping community. In Figure 8, on the left, is an image from an article published 2013 in the ACCSEAS project [16]. At that time, very few wind turbines were yet built; but, the plans were amazing.

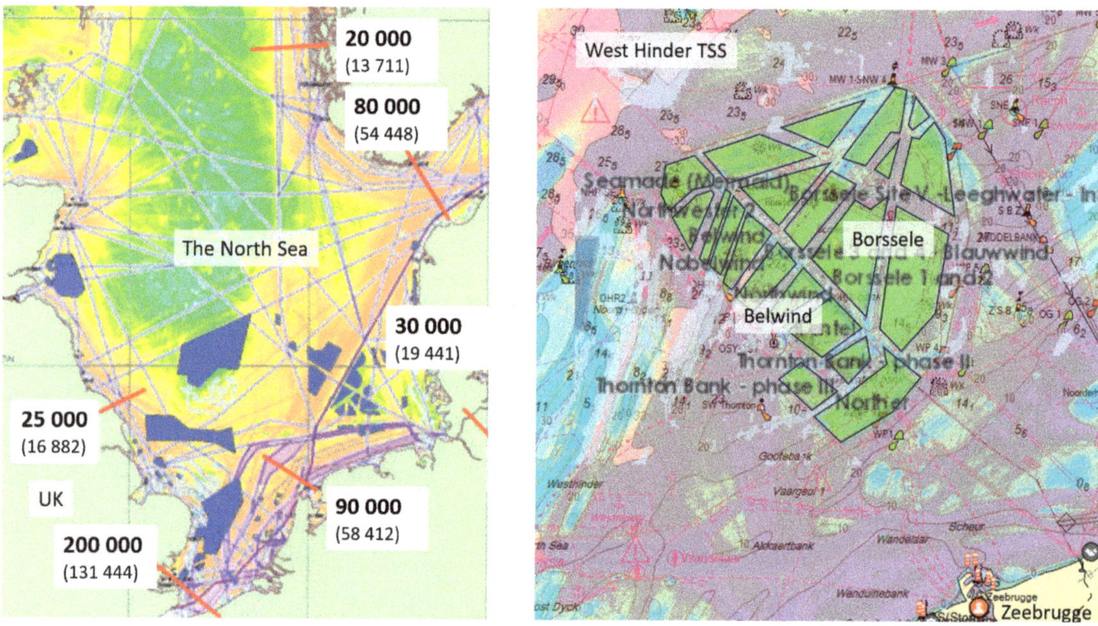

Figure 8. To the (**Left**), a 2013 prognosis for the future state of ship traffic in the North Sea. The figures are the number of ships expected to pass the respective red lines in 2025 (the numbers in brackets are the actual numbers in 2012). The blue areas where the suggested wind farms then [16]. To the (**Right**), the present-day situation in the English Channel outside Zeebrugge in Belgium. The green areas are the Belgian and Dutch Belwind and Borssele wind farms causing complex street-like conditions for shipping in lanes of varying widths [5,17]. Left figure by the author, right figure adapted with permission from Refs. [5,17]. 2023, Marine Traffic.

In Figure 8, on the right, the present situation off the Belgian and Dutch coast today is depicted. We see the Belgian Belwind wind farm, and the Dutch Borssele farm.

With a standard safety zone around each wind turbine of 500 m and each turbine in a grid 1000 m apart, an offshore wind farm creates a completely closed NoGo zone for ships (the green areas in Figure 8, right). Instead, shipping is confined to street-like conditions in corridors of varying widths in between the wind farm areas. This is good enough for normal weather, but what happens when extreme weather forces ships to head into wind and waves from whatever direction? Most likely, traffic will need to be organized and coordinated in such a way that "traffic lights" at each junction will not be necessary.

Is it possible that the situation in the Bay of Bothnia might end up like the sketch in Figure 9? Where the white areas are traffic-separated shipping routes and the blue areas are NoGo areas for wind power or fish farming? In such a street-like traffic environment, ship traffic must be coordinated to a much higher degree that before. And what happens during winter conditions? How will the wind farms affect winter traffic? Some tools are already available today in the form of route exchange, investigated in projects previously mentioned here, and presently in the STM and the Balt Safe projects.

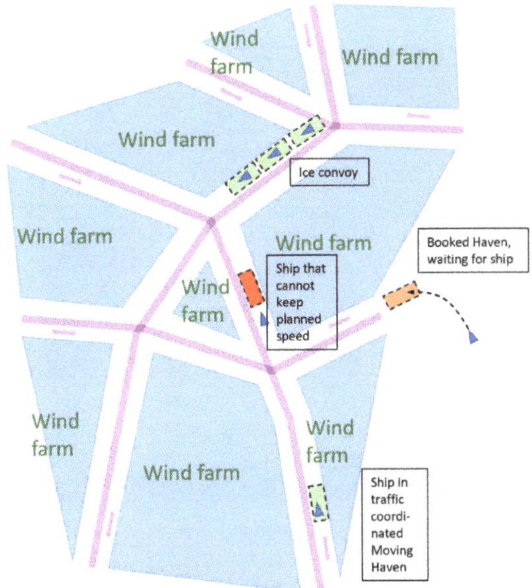

Figure 9. A possible future view of wind farms in the Bay of Bothnia. The blue areas are closed NoGo zones and ship traffic is confined to only the white corridors. Traffic coordination is executed through "moving havens", traffic separated moving boxes which each ship should stay within, one for each ship. Concept sketch by the author.

Route Exchange and Moving Havens

All ships sail in accordance with a voyage plan. IMO regulations mandate ships to make a voyage plan "from berth to berth" before leaving the port of departure. This voyage plan also contains an estimated time of arrival (ETA) at the destination. The voyage plan must be programmed into the electronic chart system (ECDIS) and can thus be visualized as a dotted line on the chart display which the watch officer should sail along. The ETA at the destination is used for the booking of pilots, locks, and linesmen. To keep your final ETA, you have an implicit ETA at each waypoint along the route. This can include speed changes. Given this, you can visualize the position where you need to be at each moment during the voyage with a box, a "Moving Haven", that keeps moving in you chart system. By keeping your ship inside the box, you are on track, and on time. A theoretical example of how a Moving Haven with a time precision of 1 min is shown in Figure 10, on the top [18]. What it can look like in an experimental ECDIS is shown in Figure 10, on the bottom. Moving Havens can be used as an organization tool for ship traffic in wind farm areas.

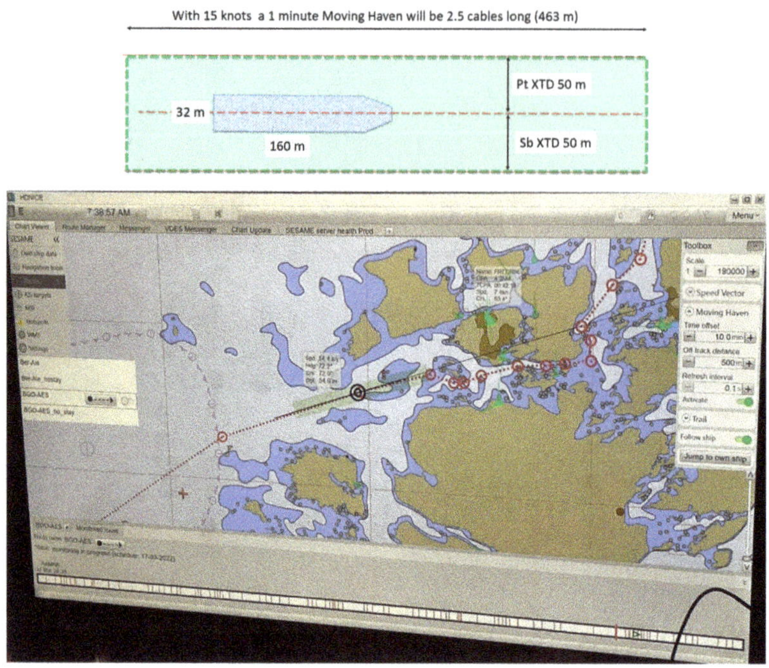

Figure 10. (**Top**) a theoretical visualization of a Moving Haven for a ship with a cross track distance set to 50 m port and starboard and a time precision of 1 min in 15 knots. (**Bottom**) a screenshot from a Kongsberg HDNICE experimental ECDIS-like display with a ship in a Moving Haven (green box, center screen). Note the timeline with all waypoints at the bottom of the screen. Illustration and photo by the author.

5. Discussion

What will the future look like and how can we prepare for it? These are questions that research is constantly trying to answer. We have half a century of experience with navigating in ice with icebreakers, but how wind farms with fixed foundations will influence ice behavior and ice navigation we have next to no knowledge of. We know for certain that climate change will come, and with it a huge need for fossil-free renewable energy production. Maybe ice in the Bay of Bothnia will not be a problem in the future? We do not know. But we know that extreme weather will come more often. Research must in any case continue to gather knowledge, identify potential problems, and suggest solutions.

Traffic coordination using slot-thinking where the bridge officer is reduced to an actuator whose job is to keep the ship in a box on a screen will be annoying to many. But, tomorrow, some ships might sail automatically, and then it will be highly interesting to be able to exactly predict their future movements.

6. Conclusions

This paper has presented new research needs for icebreaker-assisted navigation in wind farms in Arctic conditions. The effects of bottom-fixed wind turbines on ice behavior and on icebreaker-assisted shipping needs to be investigated. Problems as well as potential benefits are possible. Earlier e-Navigation research regarding route exchange and Moving Havens are suggested as potential solutions to traffic organization in such wind farm areas.

Funding: This research has been conducted within the LOAS project funded by the Norwegian Research Council (No. 296527), which is hereby gratefully acknowledged.

Institutional Review Board Statement: Not applicable.

Informed Consent Statement: Not applicable.

Data Availability Statement: Data are available in this manuscript.

Conflicts of Interest: The author declares no conflict of interest.

References

1. Swedish Agency for Marine and Water Management. Havsplan Bottniska viken: Förslag till samrådshandling. 15 February 2018. Available online: http://styrelsemote.se/PublicTimra/files/Kommunstyrelsen/2018-05-29%200829%20%20KS%202018-05-29/11%20Remissyttrande%20havsplan%20f%C3%B6r%20Bottniska%20viken/11f%20v1.1%20Samr%C3%A5dshandling%20f%C3%B6rslag%20t...ka%20viken.pdf (accessed on 13 May 2023).
2. Sjörapporten. Skydda Farlederna när Vindkraften till Havs Växer. Swedish Maritime Administration. 2023. Available online: https://www.sjofartsverket.se/sv/om-oss/sjorapporten/ (accessed on 13 May 2023). (In Swedish).
3. Baltic Icebreaking Management. Baltic Sea Icebreaking Report 2020–2021. Available online: https://baltice.org/app/static/pdf/BIM%20Report%202020-2021.pdf (accessed on 15 May 2023).
4. Tidningen Näringslivet. Sjöfartsverket: Vindkraft Till Havs Stoppar Kritiska Sjötransporter. 20 December 2022. Available online: https://www.tn.se/naringsliv/24363/sjofartsverket-vindkraft-till-havs-stoppar-kritiska-sjotransporter/ (accessed on 15 May 2023).
5. 4C Offshore.com. Global Offshore Map. Available online: https://www.4coffshore.com/ (accessed on 13 May 2023).
6. Porathe, T. e-Navigation in Arctic Conditions: Transmitting High Resolution Ice Routes in First-year Ice. In Proceedings of the Scandinavian Maritime Conference 2012, Horten, Norway, 28–29 November 2012; Available online: https://publications.lib.chalmers.se/records/fulltext/local_166919.pdf (accessed on 13 May 2023).
7. Swedish Meteorological and Hydrological Institute. Swedish Ice Service. Available online: https://www.smhi.se/en/services/scandinavian-waters/swedish-ice-service-1.8715 (accessed on 13 May 2023).
8. Porathe, T.; Lützhöft, M.; Praetorius, G. What is your intention? Communicating routes in electronic nautical charts. *Procedia—Soc. Behav. Sci.* **2012**, *48*, 3266–3273. [CrossRef]
9. Bergmann, M. Route Exchange in IEC 61174 ed 4 and S-421. Available online: http://www.marine-fields.com/wp-content/uploads/2021/08/IEC-61174-ed-4-and-S-421.pdf (accessed on 13 May 2023).
10. Vindkraftsnyheter.se. Sjöfarten Behöver Skyddas när Vindkraften Till Havs Byggs ut. Available online: https://www.vindkraftsnyheter.se/20230331/7646/sjofarten-behover-skyddas-nar-vindkraften-till-havs-byggs-ut (accessed on 13 May 2023).
11. Buysse, J. *Handling Ships in Ice: A Practical Guide to Handling Class 1A and 1AS Ships*; Nautical Institute: London, UK, 2007.
12. Fransson, L.; Bergdah, L. *Recommendations for Design of Offshore Foundations Exposed to Ice Loads*; Elforsk Rapport 09:55; Elforsk: Stockholm, Sweden, 2009.
13. Isaksson, A. Konflikter Väntar Till Havs om Vindkraften NSD, Luleå. 2023. Available online: https://nsd.se/nyheter/lulea/artikel/vindkraftsparker-till-havs-problematiska-for-sjofarten/lz3dwv2j (accessed on 13 May 2023).
14. Leppäranta, M.; Hakala, M. The structure and strength of first-year ice ridges in the Baltic Sea. *Cold Reg. Sci. Technol.* **1992**, *20*, 295–311. [CrossRef]
15. Tikanmäki, M.; Heinonen, J. Estimating extreme level ice and ridge thickness for offshore wind turbine design: Case study Kriegers Flak. *Wind. Energy* **2022**, *25*, 639–659. [CrossRef]
16. Porathe, T.; de Vries, L.; Prison, J. Ship voyage plan coordination in the MONALISA project: User tests of a prototype ship traffic management system. In *Proceedings of the Human Factors and Ergonomics Society Europe Chapter 2013 Annual Conference*; de Waard, D., Brookhuis, K., Wiczorek, R., di Nocera, F., Brouwer, R., Barham, P., Weikert, C., Kluge, A., Gerbino, W., Toffetti, A., Eds.; Human Factors and Ergonomics Society: Santa Monica, CA, USA, 2014; ISSN 2333-4959. Available online: http://hfeseurope.org (accessed on 13 May 2023).
17. Marine Traffic.com. Available online: https://www.marinetraffic.com (accessed on 13 May 2023).
18. Porathe, T. Moving Havens: An application of the e-Navigation service route exchange. In Proceedings of the International Conference on Human Factors Royal Institute of Nautical Architects Human Factors Conference, London, UK, 19–20 February 2020. Available online: https://www.sintef.no/globalassets/project/hfc/documents/2020-moving-havens.pdf (accessed on 15 May 2023).

engineering proceedings

MDPI

Proceeding Paper

Risk-Based UAV Flight Path Optimization in Accordance with SORA †

Jannik Heinze and Maarten Uijt de Haag *

Institute of Aeronautics and Astronautics, Technical University of Berlin (TU Berlin), 10587 Berlin, Germany; j.heinze@campus.tu-berlin.de
* Correspondence: maarten.uijtdehaag@tu-berlin.de

† Presented at the European Navigation Conference 2023, Noordwijk, The Netherlands, 31 May–2 June 2023.

Abstract: With the EU regulation of drone operations varying based on the specific type of drone, path planning can be done to consider risk mitigation. This paper proposes a transition-based rapidly exploring random tree star (T-RRT*) path planning algorithm for fixed-wing drone operations over rural areas. Risk is decoded in the cost function, which mainly considers population density and special infrastructure types. It was found that the algorithm is capable of finding paths that minimize exposure of the general population and infrastructure. However, path and computational inefficiencies were found. Usage of another data structure or algorithm might improve performance.

Keywords: optimal path planning; complex cost-space planning; sampling-based path planning; T-RRT*; rural drone operations; (EU) VO 2019/947

1. Introduction

On 31 December 2020 a new regulation with regard to the operation of Unmanned Aircraft Systems (UASs) in the European Union (EU) was enacted [1]. This regulation categorizes drone operations into three categories (open, specific and certified) and imposes different restrictions [1]. Of particular interest is the specific category, since it covers operation types relevant to the current industry. One peculiar use case of UASs seen in many places is the delivery of rather small and light objects over distances up to tens of kilometers. These operations are necessarily executed beyond visual line of sight (BVLOS), which puts them into a specific category [1]. Examples of these kinds of operations are those of Zipline and Wingcopter. According to the EU regulation, the risk of the operation depends on the route that is actually flown. This leads to the problem of finding a route that minimizes the risks according to the EU regulations while also taking into account the flight performance of the used Unmanned Aerial Vehicle (UAV) and economic parameters.

Ortlieb and Adolf used the Special Operation Risk Assessment (SORA) and Acceptable Means of Compliance (AMC) as well as additional data to create risk maps for minimal risk planning [2]. Work regarding path planning in higher dimensions and complex cost spaces as well as easy access rules for the mentioned regulations are available [1]. Examples of path planning for aerial vehicles are [3,4]. Schopferer and Benders also worked on risk-based path planning using a sampling-based algorithm [5]. Sampling-based path planning algorithms have been proven useful for large configuration spaces [4]. One of those is the Rapidly exploring Random Tree (RRT) algorithm, which has been altered into many different sub-variants with different strengths [4,6–8].

This contribution introduces a way to decode the Special Operation Risk Assessment (SORA) for a Specific Assurance and Integrity Level (SAIL) three together with other operational parameters into a cost function. Additionally, practical considerations for the application of the T-RRT* algorithm are introduced. This type of a priori fight path planning also fits into the efforts of the U-spaces outlined in the EU regulation, where proposed flight path planning could be carried out with information about expected UAV traffic.

Citation: Heinze, J.; Uijt de Haag, M. Risk-Based UAV Flight Path Optimization in Accordance with SORA. *Eng. Proc.* **2023**, *54*, 56. https://doi.org/10.3390/ENC2023-15473

Academic Editors: Tom Willems and Okko Bleeker

Published: 29 October 2023

The target vehicles for the algorithm have a characteristic dimension of one to three meters or an expected kinetic energy between 700 joules and 34 kilo-joules. The algorithm will not take into account most dynamic behaviors and assumes a static ground speed with no wind effects or other disturbances.

2. Materials and Methods

The path planner proposed in this work is based on the T-RRT* algorithm by Devaurs, Simeon and Cortés [4]. The pseudocode of this algorithm can be found in Algorithm 1. This algorithm combines the well-known RRT* algorithm with filtering of the randomly sampled configuration by the relative cost with respect to the nearest configuration. It was chosen due to the stated benefits in complex cost spaces in [4].

The configuration space is modeled as $\mathbb{R}^6$ with three spacial coordinates in the UTM frame and three Euler angles. The UTM frame is chosen to have a Cartesian coordinate system, which also supports the data sources described below. The roll and pitch angles are limited to account for the flight performance of the UAV. The course angle is defined in the mathematical positive direction from the x-axis, against the standard definition in aviation. It is also defined in the interval between $-\pi$ and π.

For path planning, the UAV is considered to be a point of mass. Therefore, all obstacles have to be expanded in size to account for the Flight Geography (FG), Contingency Volume (CV) and Ground Risk Buffer (GRB). This is discussed in more detail during the data pre-processing.

Additionally, the UAV is modeled to be quasi-stationary. In order to abstract the algorithm proposed from any platform, the algorithm does not account for most dynamic behaviors. For example, it is assumed that the UAV has the same true airspeed, no matter the maneuver. In curved flights, it is assumed that the UAV instantly acquires the necessary bank angle. For changes to the path angle, a constant angular velocity $\dot{\gamma}$ with an instantaneous increase or decrease in the load factor n_z as a cause is assumed. These assumptions may not reflect the nature of the UAV perfectly but are considered useful for simple and vehicle-independent path planning.

According to the SORA process, SAIL 3 operations are only possible for a Ground-Risk Class (GRC) below or equal to 4 and residual Air-Risk Class (ARC) below or equal to b. Since ARC-a is not reasonable, the operation has to take place below 500 ft above ground level (AGL) in uncontrolled airspace over rural airspace. Depending on the encountered population density, ground risk mitigations (e.g., M1 and usage of a parachute for M2) need to be applied. The threshold for that is 300 inhabitants per square kilometer or a cluster of residential buildings. This, however, is just an interpretation of the regulation. Actual decisions of the competent authority may differ.

Another aspect considered to assess the risk imposed to the ground area is the ground infrastructure. While it may be considered relatively risk-free to fly over agricultural land, flights over power plants, prisons or biosafety level 4 laboratories have to be avoided. Aided by a list of different infrastructure types relevant to risk mitigation by the Federal Ministry for Digital and Transport (BMDV) of Germany [9], a set of areas deemed too risky for operations is set up. Collisions are determined based on the horizontal coordinates alone. Flying beneath or above airspaces is not supported in this work.

The cost function is set up as an integral of the cost, as in [4]. Four cost factors will be considered: height, number of overflown people, load on the propulsion system and flight over specific ground areas. This encourages paths at reasonable heights over sparsely populated regions with the least possible power draw, while also minimizing the flight over certain ground infrastructure types. The path cost function $c_p : \Pi \rightarrow \mathbb{R}^+$ and configuration cost function $c : \mathcal{C} \rightarrow \mathbb{R}^+$ are defined as follows (see [4]). Here, L represents the length of path π.

$$c_p(\pi) = \int_0^1 c(\pi(k))dk \approx \frac{L}{n}\sum_{k=1}^{n} c\left(\pi\left(\frac{k}{n}\right)\right) \tag{1}$$

$$c(q) = \lambda_h \cdot c_h(q) + \lambda_{pop} \cdot c_{pop}(q) + \lambda_{ENG} \cdot c_{ENG}(q) + \lambda_{infra} \cdot c_{infra}(q) \quad (2)$$

In order for the T-RRT* algorithm to work, the path cost function has to be linear on the path. Otherwise, adding the costs of two path segments would not represent the cost of the joined path. Furthermore, to effectively and comprehensibly weigh the different path factors, they should scale similarly with the path length and should be normalizable. The following configuration cost functions are chosen.

$$c_h(h(k)) = L(h(k) - \bar{h})^2 \quad (3)$$
$$c_{pop}(x(k), y(k)) = L \cdot b_{OV} \cdot \rho_{pop}(x(k), y(k)) \quad (4)$$
$$c_{ENG}(\gamma(k)) = L \cdot (sin(\gamma(k)) + \epsilon \cdot cos(\gamma(k))) \quad (5)$$
$$c_{infra}(x(k), y(k)) = L \cdot infra(x(k), y(k)) \quad (6)$$

Here, $\bar{h}$ is the proposed flight height, b_{OV} is the width of the operational volume, ρ_{pop} is the population density, γ is the flight path angle, ϵ is the aerodynamic efficiency, and $infra(.,.)$ is 1 inside of penalized areas (e.g., motorways, railways, power lines, nature reserves) and otherwise 0. The propulsion system cost function is derived from linearized power consumption, yielding proportionality to the thrust required.

Normalization is achieved with the following values, restricting the normalized cost functions to $[0, L]$.

$$\lambda_{n_h} = \max_{h=\{h_{min}, h_{max}\}} (h - \bar{h})^2 \qquad \lambda_{n_{pop}} = b_{OV} \cdot 300 \left[\frac{1}{km^2}\right]$$
$$\lambda_{n_{ENG}} = 1 \qquad \lambda_{n_{infra}} = 1$$

Without hyper-parameter optimization the following weights are chosen:

$$\lambda_{w_h} = \frac{1.5}{\lambda_{n_h}} \qquad \lambda_{w_{pop}} = \frac{1}{\lambda_{n_{pop}}}$$
$$\lambda_{w_{ENG}} = \frac{1}{\lambda_{n_{ENG}}} \qquad \lambda_{w_{infra}} = \frac{10}{\lambda_{n_{infra}}}$$

The definition of the configuration cost function is analogous to [5], especially with the normalization carried out.

As shown above, three different types of data are required: terrestrial surface data, population density data and infrastructure data. The data were gathered specifically for the study case explained in Section 3.

For the surface data as ground elevation measurements, the digital surface model ('Digitales Oberflächenmodell' (DOM)) is used. It is provided by each state separately [10,11]. The DOM data for Thuringia and Saxony–Anhalt gives grid-based information of surface elevation (elevation of vegetation, buildings, terrain, etc.) for a certain area in UTM coordinates (zone UTM32U). Since the UAV may deviate from the planned path to the extent of operational volume, the ground elevation of that area has to be considered for ground collision avoidance. In order to do so with reasonable computation time, a pre-processing step with respect to the width of the operational volume b_{OV} is executed. For a given set of coordinates, the processed elevation is equal to the maximal elevation in a circle with radius of $\frac{b_{OV}}{2}$ around that coordinate. This ensures staying well-clear of the ground even when deviating from the path inside of the operational volume. In order to reduce the size of the elevation data, the grid resolution of processed data is set to 20 m.

For the population density data, the 2019 European Commission global human settlement layer is used [12]. Here, data from 2015 in a grid of 250 m are used. For each square, the number of inhabitants in that 0.125 square kilometer area is given and has to be converted to inhabitants per square kilometer. Since the georeference is given in latitude and longitude, it has to be converted into UTM32U coordinates. To account for the width of the FG, CV and GRB in the obstacle calculation, the processed population number for each 250 m by 250 m square is set to be the highest of the neighboring squares reachable with half of the width of the FG, CV and GRB, as shown in Figure 1. The orange squares are used together with the green square to calculate the processed population density of the green square. This results in an over-estimate of the overflown population and thus is conservative.

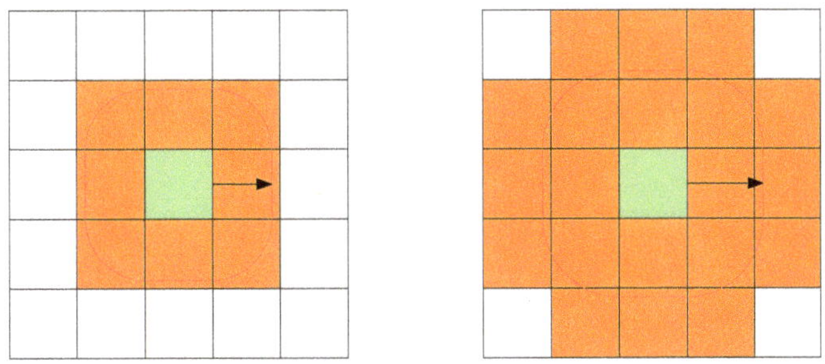

a) b)

Figure 1. Selection of used grid cells for GHSL processing depending on the width of the FG, CV and GRB. (**a**) FG, CV and GRB values are causing the selection of a square grid pattern for population density calculation. (**b**) FG, CV and GRB values lead to a non-square pattern since the outer most corners cannot be reached with those values

The ground infrastructure and type of ground area is derived from the Web Map Service (WMS) of the 'Digitale Plattform Unbemannte Luftfahrt (dipul)' by the Federal Ministry of Digital and Transport of Germany. The WMS enables generation of the image of a map showing the areas covered by the relevant infrastructure and ground areas. For both sets of infrastructure, an image is generated and is interpreted as grid data referenced in latitude and longitude. These images are converted to 2D Boolean arrays indicating the presence of an infrastructure element at the given grid point based on a color comparison with the background color. After converting the georeference to UTM32U, similar processing like the height processing is executed. Here, a grid point is considered covered if any grid point in the vicinity is covered. The vicinity for the obstacles includes both the operational volume and the GRB; for the risk calculation, only the operational volume is considered.

The connector function is set up such that the path between two configurations consists of three parts: an arc with constant bank angle and constant change in the path angle, a straight line in 3D space, and another arc like the first. The boundary conditions are the path and the bank and course angles of the start and end configurations. Since the initial and goal configuration of the planning problem are set to have zero bank, the first arc is not possible (either just not defined or because the problem would become underdefined). Figure 2 shows the horizontal path components of both cases.

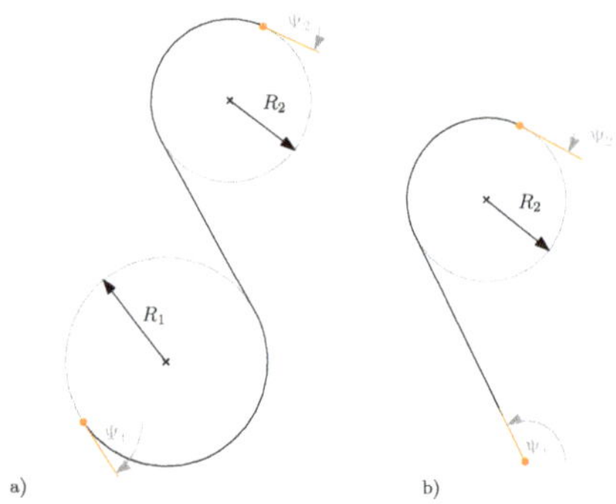

Figure 2. Distinction of two cases in the horizontal path calculation. (**a**) arc-straight-arc path segment (**b**) straight-arc path segment

In contrast to other RRT or RRT* algorithms, the goal configuration is a specific configuration rather than an arbitrary configuration inside of a goal region. Since the configuration space may be rather big, this ensures that a goal configuration at a given location is found. Relying on chance may result in no configuration being sampled in the vicinity (horizontal position error smaller than 50 m) of the specified goal configuration.

Collision detection relies on checking configurations on a path piece-by-piece since the underlying data are given in a grid format rather than a shape file. Cost calculation is done in the same fashion. To reduce the computation time, both cost calculation of a path segment and collision detection are executed at once. Therefore, in contrast to the suggestion in [4], no additional collision check function is needed.

To accelerate the collision check and path cost calculation, grid data are reduced to the segment needed based on a rectangular boundary around the query points. Therefore, the nearest neighbor interpolation only uses data points relevant to the calculation at hand.

The path planning problem is considered to be a mostly horizontal problem. This is due to the size of each dimension of the configuration space. Because of the stated problem of long-distance flights of multiple kilometers and the small altitude spectrum, horizontal position and course are the main variables of the path planning problem, and all other variables either describe dynamics of the path or are used to optimize the horizontal path in the vertical domain. Therefore, the nearest neighbor (see Algorithm 1) is calculated regarding the horizontal position and the course angle. For calculation of the set of near configurations (see Algorithm 1), the three-dimensional position and the course angle are used.

As explained by Devaurs, Simeon and Cortès, a spatial search radius in the form of a ball is assumed for determining the set of near nodes. Due to its shrinking, problems for the expansion of the tree arise. From a certain number of generated nodes, the ball used has a smaller radius than the distance d_{max} used for the expansion. Therefore, a new node at a distance of d_{max} to the nearest node cannot be connected to the tree. To address this issue, the parameter d_{max} is adjusted over the run time as follows, with the initial maximal step size $d_{max,0}$, a parameter γ as defined in [4], the number of generated nodes n, and the number of dimensions used for the ball d (here: 3).

$$d_{max} = min\{d_{max,0}, r\}$$

$$r = \gamma \cdot \left(\frac{log(n)}{n}\right)^{\frac{1}{d}}$$

Another difference from the original algorithm is the reduction of inspected nearby nodes for the rewiring. If no feasible path from a near configuration $q_i \in Q_{near}$ to a new configuration q_{new} exists, it is assumed that the path from q_{new} to q_i is also not feasible. This is meant to improve the execution time of the algorithm.

Algorithm 1 Transition-based RRT* (T-RRT*) [4]

$\mathcal{G} \leftarrow initGraph(q_{init})$

while not $stoppingCriteria(\mathcal{G})$ **do**
  $q_{rand} \leftarrow sampleRandomConfiguration(\mathcal{C})$
  $q_{near} \leftarrow findNearestNeighbour(\mathcal{G}, q_{rand})$
  $q_{new} \leftarrow extend(q_{near}, q_{rand})$

  if $q_{new} \neq null$ **and** $transitionTest(\mathcal{G}, c(q_{near}), c(q_{new})$ **then**
    $addNewNode(\mathcal{G}, q_{new})$
    $n \leftarrow numberOfNodes(\mathcal{G})$
    $Q_{near} \leftarrow nodesInBall(\mathcal{G}, q_{new}, \gamma(log(n)/n)^{1/d}$
    $q_{par} \leftarrow node_minCost_{init}(q_{new}, q_{near}, Q_{near}, c_p)$
    $addNewNode(\mathcal{G}, q_{par}, q_{new})$

    for each $q_n \in Q_{near}$ **do**
      $\pi \leftarrow pathInSpace(q_{new}, q_n)$
      if $cost_{init}(q_{new}) + c_p(\pi) < cost_{init}(q_n)$ **and** $isCollosionFree(\pi)$ **then**
        $removeEdge(\mathcal{G}, parent(q_n), q_n)$
        $addNewEdge(\mathcal{G}, q_{new}, q_n)$
      end if
    end for
  end if
end while
return $\mathcal{G}$

Due to the random nature of the algorithm, three inefficiencies emerge from the algorithm, which we try to remove. The first two are depicted in Figure 3 and deal with the combination of the sampled course and bank angle. In Figure 3a, one can see that the bank angle is in the wrong direction for the most efficient turn. In Figure 3b, the turn can be made without flying a full circle first. Both inefficiencies are solved by placing a new node at the red point and using this instead if the new path is valid. Invalidity can arise from violations of the maximum path angle or changes to the path angle.

The third inefficiency can be seen in Figure 4, where the path angle is suboptimal. This optimization is only applied as post-processing of the final path.

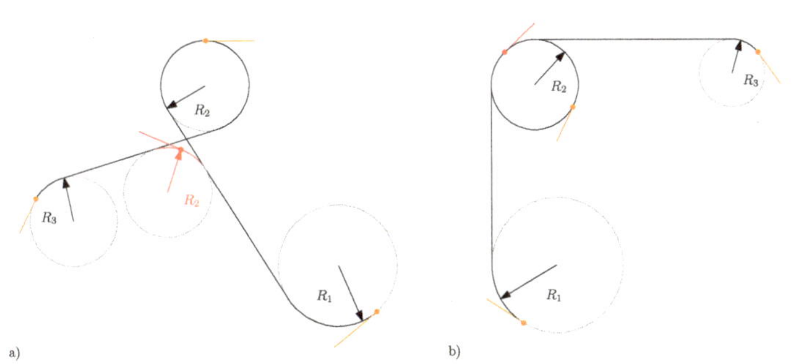

Figure 3. Horizontal path inefficiencies due to random path buildup. (**a**) wrong bank angle direction (**b**) flight includes an additional full circle

Figure 4. Vertical path inefficiency due to random path buildup.

3. Results

The test case of this work is to connect three hospitals in eastern Germany by fixed-wing UAV routes. The area is mostly rural and lies at the border region between Saxony, Saxony–Anhalt and Thuringia. For evaluation, we try to find paths from Altenburg to Zeitz and Naumburg.

The implementation of the algorithm results in two types of information: the final path (if found) together with all nodes of the tree, and information regarding the performance of the algorithm itself. In the following, both will be assessed, starting with the path information.

3.1. Analysis of the Path-Related Results

Shown in Figure 5 is the base map used for the evaluation of the horizontal flight path. Areas in black indicate no-fly zones due to high populations or restricted areas drawn from the dipul data set. Areas outlined in grey are areas associated with high risks (e.g., power lines). The green and orange areas are part of the collision-free configuration space. The color decodes population density as shown with the color bar at the right.

As one can see, the region of interest for the case study contains many small and a few large obstacles. Therefore, the path planning algorithm has to find paths around many obstacles.

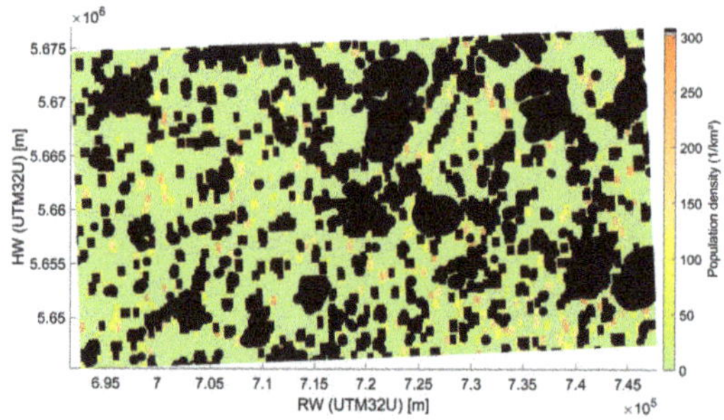

Figure 5. Horizontal obstacle map and cost function representation for $h_{max} = 100$ m and $V = 80\ \frac{\text{km}}{\text{h}}$.

However, changing the parameters of the operation (e.g., the velocity) will increase the CV and GRB, resulting in larger obstacles.

Shown in Figure 6 is the resulting horizontal path after sampling 1200 nodes for a velocity of 80 km/h and a maximum altitude of 100 m. Figure 6 only shows the area between Altenburg and Zeitz.

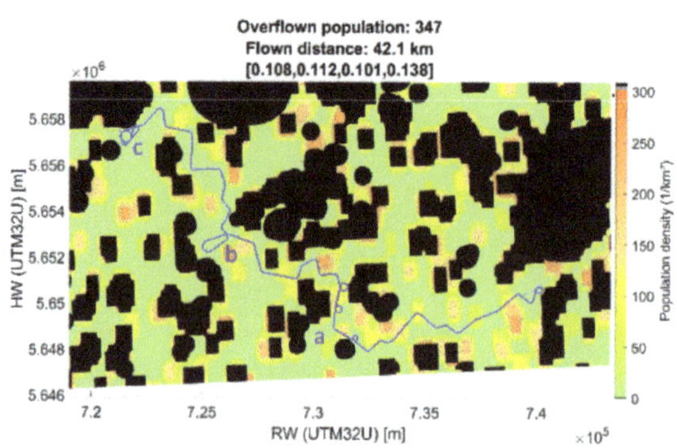

Figure 6. Found path between Altenburg and Zeitz for 1200 nodes and $d_{max,0}$ of 2.5 km.

The path found mitigates highly populated areas as well as infrastructure areas. This includes crossing, e.g., highways and power lines mostly perpendicularly. However, one can see a few inefficiencies that reveal issues with the proposed algorithm.

3.1.1. Residual Loops in the Flight Path

As one can see in Figure 6, at (a), there are still loops remaining in the path. The optimization introduced above is, therefore, not able to remove all inefficiencies. This is assumed to be due to the feasibility constraint of the optimized path. By reducing the distance flown in a horizontal arc, the change of path angle has to higher than before to match the boundary constraints of the connector function. This, then, increases the necessary load factor into an unfeasible regime. Multi-node loops such as the one at (b) of Figure 6 will be considered in Section 3.1.2.

3.1.2. Graph Structure and Re-Wiring

The issues resulting from the graph structure and re-wiring process can be seen in the unnecessary loops or jiggly lines as well as detours: for example, around the end in the northeast of Figure 6. The loop is constructed from tree nodes (two intermediate path segments), whereas a direct connection to the subsequent path would be better suited. Even though the true reason for this behavior cannot be stated with certainty, a contributing circumstance is that re-wiring only considers a certain node but nothing downstream. To connect the nodes after the loop at (b) to a better-suited parent node, this node would need to be newly sampled.

A related issue arises at the goal configuration (seen at (c)): Since the goal node is a single node, it can only have a single parent. Once a path is found and a node is sampled, connecting to the goal in a more efficient way, the first connection is removed and lost. Together with the fact that sampling a node able to connect to the goal configuration is hard, few chances arise to optimize the last section of the path.

The issue of inefficient optimization of the path due to only local re-wiring and forgetting former possible connections is further intensified by shrinking of the search radius for re-wiring. With that, newer configurations have to be even closer to an optimizable region to connect to the relevant nodes. This can be seen best at the goal configuration, as newer samples have to be closer and closer to it to connect.

Another problem arising from the spatial search volume and step size is the trade-off between fast expansion and the ability to explore more restricted areas.

3.1.3. Difficulties with the Cost Function

Figure 7 shows the path profile (blue), processed ground level (brown) and minimum and maximum heights (red). Having set $\bar{h}$ of the cost function to the mean of the extreme heights above the ground, the algorithm is capable of finding a path reasonably close to the chosen desired height. However, one can also see that the height changes quite a lot without a comprehensive reason (see, for example, around 19 km). The propulsion cost parameter was intended to minimize climb and decent maneuvers. But the linearized definition introduces quasi path-independence in the sense that this parameter is only dependent on the height difference between a given configuration and the initial configuration. As a result, the algorithm prefers nodes with low altitudes with regard to the propulsion cost, and for the path between the initial and goal configuration, this cost parameter does not contribute in an meaningful way. This demonstrates that finding a good cost function is rather hard.

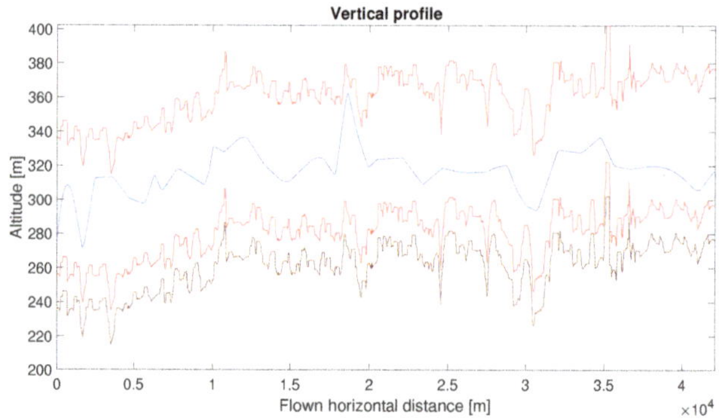

Figure 7. Vertical profile (blue) of the path shown in Figure 6. Brown shows the ground (elevation), both red lines are the limits of the specified height band.

3.2. Analysis of the Performance-Related Outputs

Figure 8a shows the percentage distribution of the sampled configurations regarding their acceptance. One can see that the distribution converges after around 4000 sampled nodes. The majority of the samples are discarded upon sampling (55%, yellow), followed by non-reachable nodes from known ones (25%, red), valid and accepted samples (15%, blue) and samples failing the transition test (5%, purple). Discarding samples sampled in obstacles does not waste too much computational expense, but the 25% of samples not able to be connected to any existing node due to blocking obstacles does. This is the issue of the described trade-off between exploration speed and the ability to navigate obstacles. Not only does the inability to navigate obstacles reduce exploration, it also wastes computational effort. Overall, the conversion rate of 15% is rather small.

Figure 8b shows the percentage distribution of the origins of the nodes in the tree. Due to the optimization, nodes that are not sampled directly are also added into the tree. After convergence, 60% of the nodes in the tree are actually sampled, and 20% each are additional nodes from optimizations in path finding and re-wiring. The number of nodes in the tree is higher than the number of sampled ones by a factor of 1.66, which lets the search radius of the T-RRT* algorithm shrink even faster without contributing too much to the exploration of the area.

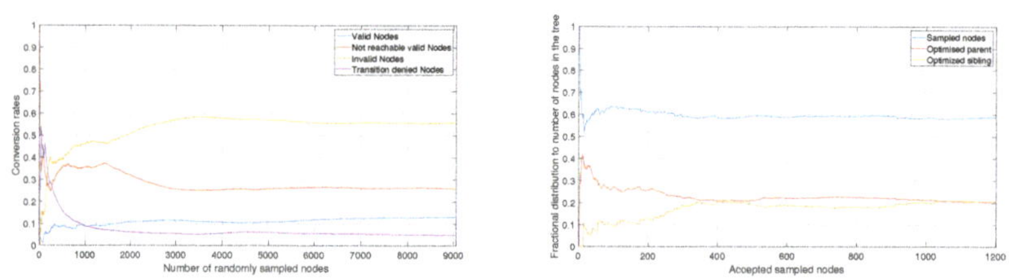

(**a**) (**b**)

Figure 8. (**a**) Percentage distribution of the validity and reasons for discarding sampled nodes. (**b**) Percentage distribution of the origins of nodes in the tree.

4. Discussion

In conclusion, the presented algorithm is capable of finding an optimized path for at least a goal configuration in the middle of the configuration space. However, it was shown that the implementation still has deficiencies regarding its performance. The computation of path costs is expensive, while only one calculated path segment at a time really contributes to the solution. Additionally, the computational expense of the cost calculation barely depends on the number of query points. With more efficient data storage and access as well as another data structure for the nodes, this efficiency could be increased. The performance of the algorithm might be improved by usage of another programming language and a more efficient cost calculation. More nodes in the tree would reduce the randomness of the solutions found, which then results in more reliable and repeatable results. Sampling more configurations would also result in finer paths able to follow straighter lines, which otherwise are only producible by logic-based optimization.

Furthermore, the algorithm proposed has trouble navigating an environment filled with obstacles. More-urban spaces with more no-fly zones or operations with a higher velocity and/or height might not be compatible with this algorithm. Finding narrow passages between obstacles is not a strength of the proposed algorithm. It might be worth exploring the usage of RRV [7].

However, this work together with Ortlieb and Adolf can be seen as a first step towards the usage of path planning algorithms in the risk management procedure for the application process of operation permits.

Author Contributions: Conceptualization, J.H.; methodology, J.H.; software, J.H.; validation, J.H.; formal analysis, J.H.; investigation, J.H.; writing—original draft preparation, J.H.; writing—review and editing, M.U.d.H.; visualization, J.H.; supervision, M.U.d.H. All authors have read and agreed to the published version of the manuscript.

Funding: This research received no external funding.

Institutional Review Board Statement: Not applicable.

Informed Consent Statement: Not applicable.

Data Availability Statement: Data sharing is not applicable to this article.

Acknowledgments: Special thanks to Globe UAV GmbH for information regarding the application process of operation permits and possible mission goals of UAV operations.

Conflicts of Interest: The authors declare no conflict of interest.

References

1. EASA, Easy Access Rules for Unmanned Aircraft Systems. September 2021. Available online: https://www.easa.europa.eu/downloads/110913/en (accessed on 12 July 2022).
2. Ortlieb, M.; Adolf, F.M. Modular Modelling of Ground and Air Risks for Unmanned Aircraft Operations Over Congested Areas. In Proceedings of the Digital Avionics Systems Conference, San Antonio, TX, USA, 11–15 October 2020 . [CrossRef]
3. Benders, S.; Schopferer, S. A Line-Graph Path Planner for Performance Constrained Fixed-Wing UAVs in Wind Fields. In Proceedings of the 2017 International Conference on Unmanned Aircraft Systems (ICUAS), Miami, FL, USA, 13–16 June 2017; pp. 79–86. [CrossRef]
4. Devaurs, D.; Simeon, T.; Cortés, J. Optimal Path Planning in Complex Cost Spaces With Sampling-Based Algorithms. *IEEE Trans. Autom. Sci. Eng. Inst. Electr. Electron. Eng.* **2016**, *13*, 415–424. [CrossRef]
5. Schopferer, S.; Benders, S. Minimum-Risk Path Planning for Long-Range and Low-Altitude Flights of Autonomous Unmanned Aircraft. In Proceedings of the AIAA Scitech 2020 Forum, Orlando, FL, USA, 6–10 January 2020. [CrossRef]
6. Gammell, J.D.; Srinivasa, S.S.; Barfoot, T.D. Informed RRT*: Optimal Sampling-based Path Planning Focused via Direct Sampling of an Admissible Ellipsoidal Heuristic. In Proceedings of the IEEE/RSJ International Conference on Intelligent Robots and Systems, Chicago, IL, USA, 14–18 September 2014. [CrossRef]
7. Tahirovic, A.; Ferizbegovic, M. Rapidly-Exploring Random Vines (RRV) for Motion Planning in Configuration Spaces with Narrow Passages. In Proceedings of the IEEE International Conference on Robotics and Automation (ICRA), Brisbane, Australia, 21–25 May 2018. [CrossRef]
8. Luders, B.D.; Kothari, M.; How, J.P. Chance Constrained RRT for Probabilistic Robustness to Environmental Uncertainty. In Proceedings of the AIAA Guidance, Navigation, and Control Conference, Toronto, ON, Canada, 2–5 August 2010.
9. Federal Ministry for Digital and Transport of Germany. Anleitung für den Web Map Service (WMS). Available online: https://www.dipul.de/homepage/de/hilfe/anleitung-web-map-service-wms/ (accessed on 13 July 2022).
10. Thüringen, F. Download Höhendaten. Available online: https://www.geoportal-th.de/de-de/Downloadbereiche/Download-Offene-Geodaten-Th%C3%BCringen/Download-H%C3%B6hendaten (accessed on 1 June 2022).
11. Anhalt, S. Kostenfreies Digitales Oberflächenmodell mit einer Gitterweite von 2 m (DOM2)—Landesweit. Available online: https://www.lvermgeo.sachsen-anhalt.de/de/dom2-landesweit.html (accessed on 1 June 2022).

12. Schiavina, M.; Freire, S.; MacManus, K. GHS Population Grid Multitemporal (1975, 1990, 2000, 2015) R2019A. European Commission, Joint Research Centre (JRC). 2019. Available online: http://data.europa.eu/89h/0c6b9751-a71f-4062-830b-43c9f432370f (accessed on 1 June 2022).

engineering proceedings

Proceeding Paper

Realism-Oriented Design, Verification, and Validation of Novel Robust Navigation Solutions †

Sorin Andrei Negru [1], Patrick Geragersian [1], Ivan Petrunin [1,*], Raphael Grech [2] and Guy Buesnel [2]

1 School of Aerospace, Transport, and Manufacturing, Cranfield University, Bedfordshire MK43 0AL, UK; sorin.negru@cranfield.ac.uk (S.A.N.); patrick.geragersian@cranfield.ac.uk (P.G.)
2 Spirent Communications PLC, West Sussex RH10 1BD, UK; raphael.grech@spirent.com (R.G.); guy.buesnel@spirent.com (G.B.)
* Correspondence: i.petrunin@cranfield.ac.uk

† Presented at the European Navigation Conference 2023, Noordwijk, The Netherlands, 31 May–2 June 2023.

Abstract: Urban environments are characterized by a set of conditions underpinning degradation Position, Navigation and Timing (PNT) signals, such as multipath and non-line of sight (NLOS) effects, negatively affecting the position and the navigation integrity during the Uncrewed Aerial Vehicles (UAVs) operations. Before the deployment of such uncrewed aerial platforms, a realistic simulation set-up is required, which should facilitate the identification and mitigation of the performance degradation that may appear during the actual mission. This paper presents the case study of the development of a robust Artificial Intelligence (AI)-based multi-sensor fusion framework using a federated architecture. The dataset for this development, comprising the outputs of a Global Navigation Satellite System (GNSS) receiver, an Inertial Measurement Unit (IMU) and a monocular camera is generated in a high-fidelity simulation framework. The simulation framework is built around Spirent's GSS7000 simulator, software tools from Spirent (SimGEN and SimSENSOR) and OKTAL-SE (Sim3D), where the realism for the vision sensor data generation is provided by a photorealistic environment generated using the AirSim software with the Unreal Engine aid. To verify and validate the fusion framework a hardware in the loop (HIL) set-up has been implemented using the Pixhawk controller. The results obtained demonstrate that the presented HIL set-up is the essential component of a more robust navigation solution development framework, providing resilience under conditions of GNSS outages.

Keywords: realistic simulation; sensor fusion; machine learning; navigation integrity

Citation: Negru, S.A.; Geragersian, P.; Petrunin, I.; Grech, R.; Buesnel, G. Realism-Oriented Design, Verification, and Validation of Novel Robust Navigation Solutions. *Eng. Proc.* **2023**, *54*, 57. https://doi.org/10.3390/ENC2023-15424

Academic Editors: Tom Willems and Okko Bleeker

Published: 29 October 2023

1. Introduction

The world of uncrewed and autonomous systems is quickly evolving in all sectors, bringing a new set of challenges for platform positioning and navigation solutions from both performance and certification points of view. Considering tighter requirements for positioning accuracy and integrity from the emerging use cases, such as drone operations in urban environments and BVLOS (beyond-visual line-of-sight conditions), extensive use of AI in positioning and fusion engines, repeatable and verifiable testing, and validation of the navigation systems gains significant importance.

As can be seen from [1,2], AI models can be used to increase the navigation robustness of UAVs under multipath and NLOS conditions [3]. Moreover, given the complexity that such navigation solutions must be performing in, a realistic yet repeatable development, testing, and validation environment is required. Different realistic environments were developed previously as can be seen in [4,5] with the aid of Unreal Engine, a photorealistic simulation software, and AirSim (Aerial Informatics and Robotics Simulation) [6], being then deployed directly on an embedded computer, to test, in real-time, the required computational load. This enables the possibility of introducing into the photorealistic environment optical sensors, such as monocular or stereo cameras, or even Lidars, leading to additional sources of positioning information. AirSim, a plug-in compatible with Unreal Engine, also allows the possibility to model the intrinsic and distortion parameters of the optical camera, mounted on the desired aerial or ground platforms. Moreover, lighting effects based on predefined daytime hours can be added, increasing the realism of the environment. In this way, it is possible to deploy visual algorithms more easily, as described in [7], where AirSim is used for the fast prototyping of algorithms used by racing drones, in a photorealistic environment.

Although it is possible to validate algorithms directly on embedded Companion Computers (CCs) before actual deployment in the field, external factors such as multipath, NLOS, and IMU stochastic and deterministic errors are not considered in the simulation environment. Multipath in urban canyons plays a disruptive force when acquiring GNSS location. Due to the obscurations from buildings, NLOS signals reduce positioning accuracy and increase uncertainty in the receiver location. Furthermore, IMU biases and noises lead to the divergence of estimated position and ground truth. This effect is known as random walk (RW). Therefore, the simulation of such physical effects is crucial when evaluating potential algorithms that can be deployed in such environments. Considering the challenges in modeling IMU and GNSS sensors in virtual urban environments, more realistic sensor-based modeling is required.

The considered case study is testing the performances of an AI-based multi-sensor positioning solution, using a federated fusion architecture, with GNSS, INS (Inertial Navigation System), and a monocular camera, in a photorealist environment. The paper is organized as follows: in Section 2, the federated fusion framework is presented along with all the sensors used; in Section 3, the hardware in the loop (HIL) configuration with all the specifications of the set-up is presented; in Section 4, it is demonstrated that the proposed testing and validation system enables realism in simulation and repeatability analysis concerning variability in noise and satellite occlusion and comprehensively covers needs in the verification of AI-based navigation algorithms; and, finally, in Section 5, the conclusions are presented.

2. Fusion Framework

Global Navigation Satellite Systems (GNSS) is a constellation of satellites that transmit signals to ground-based receivers. These signals are used to determine the receiver's position, velocity, and time, which are critical components for navigation and timing applications. The most widely used GNSS system is the Global Positioning System (GPS), which was developed by the United States Department of Defense. Other GNSS systems include the Russian GLONASS, the European Galileo, and the Chinese BeiDou.

One of the major disadvantages of GNSS is its performance in urban canyons. Urban canyons refer to areas with tall buildings and narrow streets, which can create obstructions that block or reflect GNSS signals. This can result in degraded signal quality and increased errors in position determination. In urban canyons, the signals can be reflected off buildings or blocked entirely, leading to multipath errors, where the receiver receives multiple signals of the same satellite at different times and angles, causing confusion and inaccuracies in positioning. Additionally, the signal can be absorbed or weakened by atmospheric conditions or other sources of interference, such as radio waves or electromagnetic fields.

To mitigate the effects of urban canyon environments, GNSS receivers can use a variety of techniques such as multi-constellation and multi-frequency signal reception, as well as inertial measurement sensors to estimate position and velocity when GNSS signals are blocked or unavailable. However, these techniques may not always be sufficient in highly obstructed urban environments, highlighting the need for alternative positioning technologies in some scenarios. Considering that IMU sensors based on MEMS (Micro-Electro-Mechanical System) technology are the source of large time-dependent navigation errors under long GNSS outages, visual sensors can be considered for integration as a more resilient solution. VO (Visual Odometry) algorithms can be used to estimate the camera ego-motion, providing alternative relative positioning information. Considering that urban and sub-urban environments are characterized by many details, for this framework, an ORB (Oriented FAST and rotated BRIEF) feature detector algorithm has been implemented because of its performance discussed in [8].

To fuse all the data gathered from the sensors described above, a centralized sensor fusion framework can be implemented potentially. However, the required computational load decreases the overall efficiency, leading to the so-called 'dimensional disaster effect' [9]. Instead, a decentralized fusion framework requires less computational power, allowing an easier integration on robotic platforms, for real-time applications. In these terms, the paper implements a decentralized federated sensor fusion framework as described in [2], using local filters in the form of two EKFs (Extended Kalman Filters) and a GRU (Gated Recurrent Unit) as the master filter.

GRUs are a variation of RNNs that address the vanishing gradient problem by using gating mechanisms to selectively pass information forward and backward through time. The gating mechanisms in GRUs include an update gate and a reset gate, which control how much of the current input is added to the current state and how much of the previous state is forgotten, respectively. The mechanics of which are shown in Figure 1b.

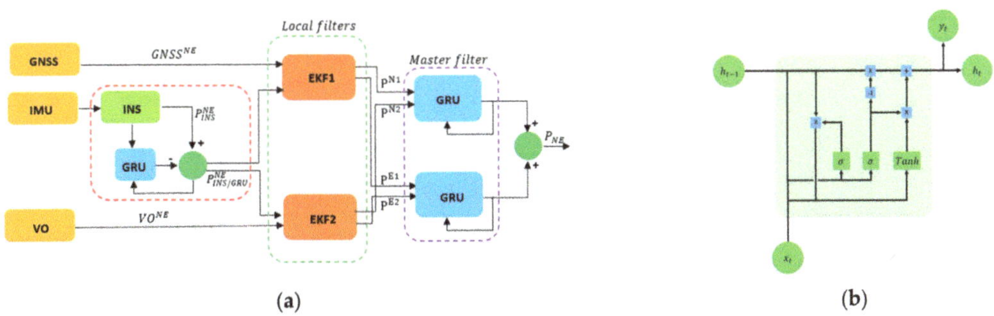

Figure 1. Federated fusion architecture (**a**) [2]; GRU unit (**b**).

GRUs have become popular for a variety of applications, including natural language processing, speech recognition, and time series prediction. They have been shown to be effective at capturing long-term dependencies in sequential data and are often used as baseline architecture in comparison to more complex models. Overall, GRUs offer a flexible and effective tool for modeling sequential data and are an important component of modern deep-learning architectures.

For reducing Inertial Navigation System (INS) position and velocity errors, the paper additionally proposes the utilization of Inertial Measurement Unit (IMU) reading prediction by using a deep neural network, utilizing GRU, to compensate for the aforementioned errors, as it can be seen in the first left part of Figure 1. Specifically, the GRU-based error correction model is trained on a dataset composed of IMU readings, where raw accelerometer and gyroscope measurements serve as input features, and the distance error between the INS-estimated position and ground truth data constitutes the corresponding output.

To obtain position data from the accelerometer and gyroscope sensors, Direction Cosine Matrix (DCM) transformation is employed to convert these sensor readings from a body frame to the navigational NED frame. Velocity and position data are subsequently derived through integration, as presented in [10]. However, it should be noted that the integration process of noise and bias from the sensors generates a random walk effect that diminishes the accuracy of INS information over extended time intervals.

To mitigate this issue, a GRU network is proposed to estimate the error engendered by the mathematical process of position information generation. By doing so, the proposed approach enhances the accuracy of INS-derived information over extended periods of time, without requiring frequent updates from Global Navigation Satellite Systems (GNSS).

Thereby, the first EKF is fusing GNSS and IMU positioning data, with the aid of a GRU block, in a loosely coupled scheme, while the second EKF is fusing the same corrected IMU data, but with the VO algorithm aid, leading to a VIO (Visual Inertial Odometry) framework. Further, the outputs from the two EKFs are processed by a trained master GRU block as presented in [2]. To decrease the computational load required by the master GRU, the main block, is divided into two sub-blocks, each processing only one coordinate, the N (North) and, respectively the E (East) coordinate.

3. Hardware in the Loop Configuration

Most of the drones are equipped with a CC (Companion Computer) usually used to process data, from external sensors, such as monocular cameras, or to deploy AI-based models to support the Flight Control Unit (FCU) computational cost, which is responsible mainly for the navigation, control, and stability part of the UAV. An HIL simulation is performed to validate the sensor fusion framework presented in the previous section. Thus, the first step to bring realism into the simulation set-up is to implement a Pixhawk 2.4 board as an FCU that is directly connected to a local computer and used as a CC. With the aid of Unreal Engine and AirSim, it is possible to link the FCU to a photorealistic environment. Using photogrammetry data in Unreal Engine, it is possible to mimic the nature of urban environments, increasing the level of detail in terms of light intensity, realistic weather effects, and material properties. The main role of the photorealistic environment is to support validation, and tests of optical sensors that the UAV is equipped with, being able to specify camera intrinsic and distortion parameters. For the presented scenario, photogrammetry data from Toulouse and San Francisco are used, as can be seen in Figure 2.

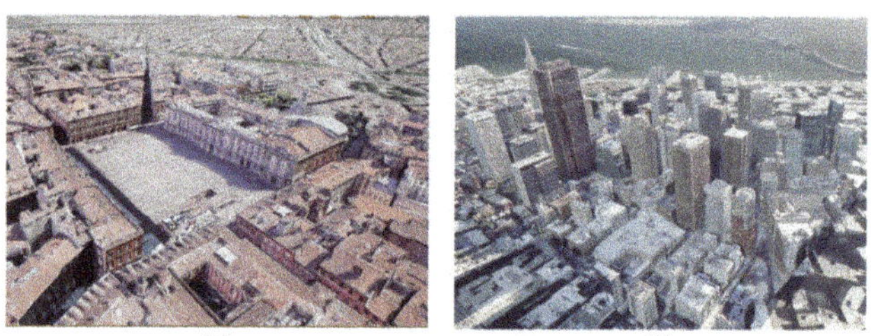

(a) (b)

Figure 2. The 3D environment model for Toulouse (**a**); 3D environment model for San Francisco (**b**).

Once the link between the FCU, AirSim, and, respectively, Unreal Engine, is established, the trajectory commands generated from a Python file that is linked to the AirSim plug-in are transmitted further to the Spirent GSS7000 simulator, as can be seen in Figure 3. This allows performing HIL tests in an iterative way, under the same trajectory assumption. At the same time, the estimated trajectory from the monocular camera is stored in an Excel file.

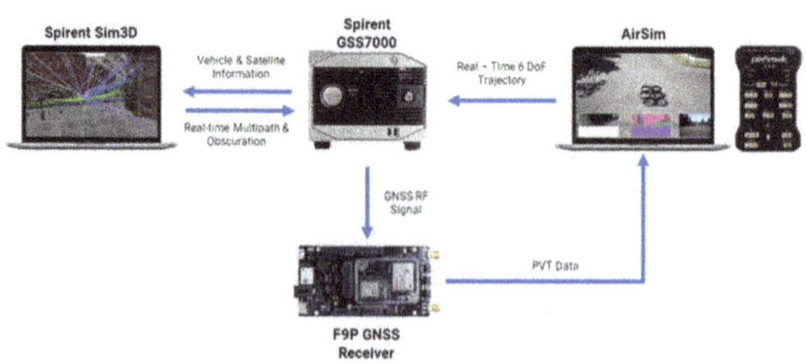

Figure 3. HIL set-up.

Thus, to assess the effectiveness of the proposed architecture, it is necessary to conduct training, utilizing realistic scenarios obtained through the use of the Spirent GSS7000 simulator and SimSENSOR obtained from Spirent PLC in Paignton, United Kingdom. Spirent GSS7000 and SimGEN simulate GNSS Radio Frequencies (RF) that are generated by satellites in each constellation. Simulation time and constellation settings can be changed to improve realism by conducting time-accurate comparisons. Furthermore, due to the repeatability of the simulation set-up, this allows for direct comparisons of different algorithm techniques to be evaluated, reducing independent variables for testing. These signals are then fed into a GNSS receiver for capturing positioning data. This enables the testing of Hardware-in-the-Loop (HIL) simulation which improves scenario realism. To produce IMU data required for training, SimSENSOR is used to generate accelerometer and gyroscope measurements that the vehicle is exhibiting. SimSENSOR allows for the simulation of realistic sensor errors by changing the SimSENSOR deterministic and stochastic error parameters such as random walk rate and bias offset. This adds an additional layer of realism to the simulation scenario. For the simulation in this paper, IMU stochastic and deterministic errors were tuned to mimic the characteristics of an Advanced Navigation Orientus IMU sensor, the specifications of which are listed in Table 1. These data are then transmitted via the UDP communication protocol to the AirSim simulator.

Table 1. INS sensor and GNSS receiver specifications.

Sensor Specifications					
Accelerometer		Gyroscope		U-Blox F9P GNSS receiver	
Scaling factor (ppm)	500	Scaling factor (ppm)	500	Pseudo-range accuracy (m)	3
Bias (mg)	0.1	Bias (deg/h)	0.001	Pseudo-range rate accuracy (m/s)	0.5
ARW (m/s/sqrt(h))	0.003	GRW (deg/sqrt(h))	0.003	Update rate (Hz)	1
Update rate (Hz)	100			Constellations	GPS L2C, GLO L2OF, GAL E5b, BDS B2I, QZSS L2C

Data collected from the sensors and simulation systems were stored in a CSV file format, with the time and date of each timestep recorded. The data were subsequently processed in MATLAB to generate the ground truth data that served as the training target. The input data consisted of the raw output from the IMU sensors (accelerometer and gyroscope in XYZ) and the GNSS latitude, longitude, and height information recorded at a rate of 1 Hz. Using interpolation, these data were synchronized to the same timestep as the accelerometer and gyroscope data, which were recorded at 100 Hz.

To further improve realism regarding urban canyon scenarios, OKTAL-SE Sim3D software is used to introduce the multipath effect on GNSS signals using the ray-tracing method. The program takes into consideration the specific geometry in the desired urban environments that are simulated in this paper. Furthermore, the different material properties underpinning the signal reflection characteristics are also considered for an additional layer of realism. This will generate line-of-sight, multipath, and non-line-of-sight conditions for the GNSS signals that are processed by the receiver. No additional multipath mitigation techniques other than those already provided by the receiver manufacturer are used to aid in these conditions.

4. Evaluation

To evaluate the proposed sensor fusion framework, Toulouse, and San Francisco, were chosen to validate the HIL configuration by adding diversity in the evaluation process.

The fusion framework performances for San Francisco can be observed in the table below. Two cases were considered, one with the corrected INS data with the aid of a GRU block, and another one, without. It can be observed that the output of the first extended Kalman filter, fusing data from the GNSS and the IMU without the GRU aid, does have inferior positioning performances compared to the one with the GRU aid leading to a horizontal position RMSE (root mean square error) of 13.22 m (95th percentile). In comparison to the second EKF, slightly better results were obtained in the case without the GRU aid for IMU, and wort results with the GRU aid, with an equivalent of 3.33 m (95th percentile). The master filter (based on GRU) and fusing data from the EKFs demonstrate the best performance, with a horizontal RMSE of 0.28 m (95th percentile). Slightly worst performances were obtained for Toulouse, as can be seen in Table 2 without the INS GRU aid, due to the VO degradation in the urban environment. It can also be observed that using the Spirent GSS7000 simulator along with SimSENSOR and OKTAL-SE Sim3D software, higher errors are introduced into the simulation. Figure 4 shows a comparison between ground truth and GNSS position in NED coordinates gathered from AirSim and, respectively, OKTAL-SE Sim3D aid, affected by multipath. Although GNSS data from AirSim are not affected much by the urban canyon or the urban environment effects, data from OKTAL-SE Sim3D are more degraded, representing a more realistic positioning output.

Table 2. Positioning performance for San Francisco and Toulouse.

San Francisco			
Position source	RMSE N, (m)	RMSE E, (m)	Horizontal RMSE (95th percentile) (m)
EKF1 IMU/GNSS (no GRU aid)	5.47	2.45	13.22
EKF2 IMU/VO (no GRU aid)	3.88	2.24	10.09
Master filter (no GRU aid)	5.12	2.40	11.10
EKF1 IMU/GNSS (with GRU aid)	0.16	0.11	0.31
EKF2 IMU/ VO (with GRU aid)	1.60	1.23	3.33
Master filter	0.15	0.10	0.28
Number of VO features (95th percentile)		22,726	
Toulouse			
Position source	RMSE N (m)	RMSE E (m)	Horizontal RMSE (95th percentile) (m)
EKF1 IMU/GNSS (no GRU aid)	4.39	2.69	12.43
EKF2 IMU/VO (no GRU aid)	4.37	2.68	12.40
Master filter (no GRU aid)	0.84	3.05	7.03
EKF1 IMU /GNSS (with GRU aid)	0.12	0.09	0.31
EKF2 IMU/VO (with GRU aid)	1.15	1.10	2.42
Master filter	0.12	0.09	0.28
Number of VO features (95th percentile)		17,941	

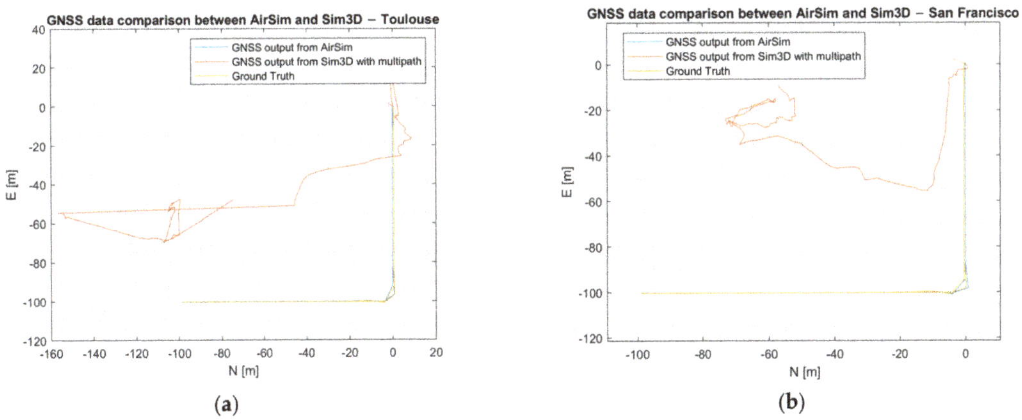

Figure 4. GNSS data with multipath in Toulouse (**a**) and in San Francisco (**b**) scenarios.

Comparing the Horizontal Dilution of Precision (HDOP) and Position Dilution of Precision (PDOP) between the Toulouse scenario and the San Francisco scenario (Figure 5), an increase in both HDOP and PDOP measurements can be observed. PDOP considers the geometry of the satellite in view and the position of the GNSS receiver. Therefore, a higher PDOP value indicates poor geometry, where the satellites are clustered in a small area of the sky, making it more difficult to determine a precise position. This is to be expected in an urban canyon scenario, where large buildings obscure direct LOS signals and reduce the area signals that can be received by the receiver. This is also visually correlated in Figure 6, where large urban canyons are present in San Francisco compared to Toulouse.

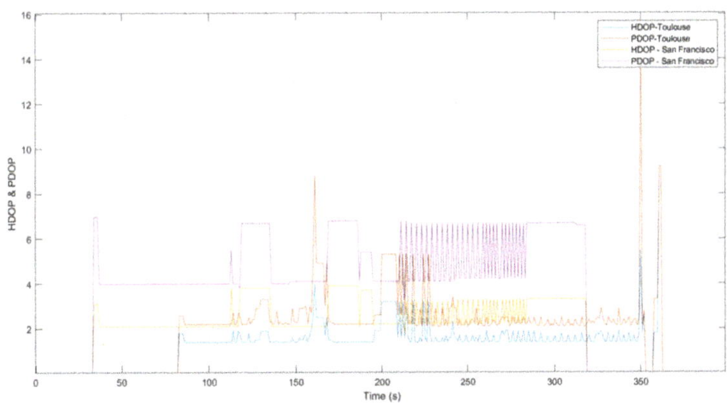

Figure 5. HDOP and PDOP for Toulouse and San Francisco scenario.

HDOP, on the other hand, specifically measures the accuracy of the horizontal position. A higher HDOP value indicates that the GPS receiver's horizontal position calculation is less accurate, meaning there is more uncertainty in the latitude and longitude coordinates. A reduced position performance is observed in San Francisco, which represents the urban canyon scenario. This reduced performance is highlighted by the HDOP in Figure 5, which shows a higher value than seen in Toulouse. Hence, taking into consideration both HDOP and PDOP evaluation, the simulations are sensitive with regard to the change in environment and therefore aid in providing realism regarding the effects of urban canyon obscuration of LOS signals and multipath effects.

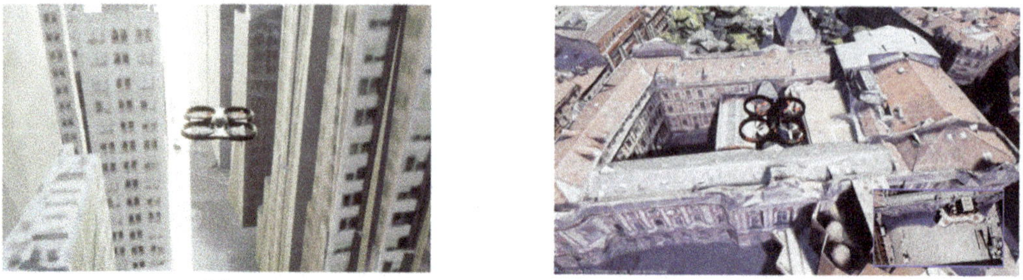

(a) (b)

Figure 6. Feature variation of the VO algorithm in San Francisco (**a**) and Toulouse (**b**).

Considering that the UAV's monocular camera is pointing downwards during the entire mission, the number of features of the environment on the ground directly affects the VO performance. The altitude of the UAV and the light intensity are two other factors that contribute to a good VO positioning output, increasing the realism of the simulation. It can be seen clearly from Table 2 that more features were extracted when the drone was flying in an urban canyon, in San Francisco, compared to the flight in the urban environment, in Toulouse. The number of features directly affects the positioning performance of the UAV, as can be seen, for the second EKF without the GRU aid for INS, leading the San Francisco flight to a horizontal RMSE of 10.09 (95th percentile) and a horizontal RMSE of 12.40 (95th percentile) for Toulouse. As a result, it can be seen that the VO algorithm is more stable in an urban canyon, rather than an urban environment, due to building height, which leads to shades induced by the lighting effects generated by the photorealistic environment. Thus, the VO algorithm does have better performances in San Francisco, rather than in Toulouse, leading to a more accurate and stable output.

5. Conclusions

The discussed multi-sensor PNT testing, and simulation framework aims to support the development of navigation solutions for aerial and terrestrial vehicles operating in diverse and complex environments, such as urban environments. The case study based on the robust federated fusion architecture illustrates how the enhanced accuracy, resilience, and integrity of the techniques, relying on the large amounts of diverse training and testing data, can be developed, and tested ensuring repeatability and realism. Using Spirent's GSS7000, SimSENSOR, and OKTAL-SE Sim 3D in combination with AirSim, it was possible to increase the HIL simulation realism, taking into consideration stochastic and deterministic errors for the IMU, multipath, and signal obstruction for the GNSS receiver. Furthermore, the photorealistic environment, with the aid of photogrammetry data, added to the realism of the vision data simulation, being able to test and validate the accuracy of the VO algorithm under realistic conditions. With the HIL simulation set-up presented in this study, it was possible to develop, optimize and evaluate more accurately the performance of a novel sensor fusion framework, designed to be used by UAVs in urban environments.

Author Contributions: Conceptualization, R.G., G.B. and I.P.; methodology, S.A.N. and P.G.; software, S.A.N. and P.G.; validation, S.A.N. and P.G.; formal analysis, S.A.N. and P.G.; investigation, S.A.N. and P.G.; resources, S.A.N., P.G. and R.G.; data curation, S.A.N. and P.G.; writing—original draft preparation, S.A.N. and P.G.; writing—review and editing, I.P., R.G. and G.B.; visualization, S.A.N. and P.G.; supervision, I.P.; project administration, I.P. and R.G.; All authors have read and agreed to the published version of the manuscript.

Funding: This research received no external funding.

Institutional Review Board Statement: Not applicable.

Informed Consent Statement: Not applicable.

Data Availability Statement: The data presented in this study are available on request from the corresponding author.

Conflicts of Interest: The authors from both Cranfield University and Spirent Communications PLC declare no conflict of interest.

References

1. Geragersian, P.; Petrunin, I.; Guo, W.; Grech, R. Multipath Detection from GNSS Observables Using Gated Recurrent Unit. In Proceedings of the 2022 IEEE/AIAA 41st Digital Avionics Systems Conference (DASC), Portsmouth, VA, USA, 18–22 September 2022. [CrossRef]
2. Negru, S.A.; Geragersian, P.; Petrunin, I.; Zolotas, A.; Grech, R. GNSS/INS/VO fusion using Gated Recurrent Unit in GNSS denied environments. In Proceedings of the AIAA Science and Technology Forum and Exposition, AIAA SciTech Forum 2023, National Harbor, MD, USA, Online, 23–27 January 2023. [CrossRef]
3. Groves, P.D.; Jiang, Z.; Rudi, M.; Strode, P. A Portfolio Approach to NLOS and Multipath Mitigation in Dense Urban Areas. In Proceedings of the 26th International Technical Meeting of the Satellite Division of The Institute of Navigation (ION GNSS+ 2013), Nashville, TN, USA, 16–20 September 2013; pp. 3231–3247.
4. Boroujerdian, B.; Genc, H.; Krishnan, S.; Duisterhof, B.; Plancher, B.; Mansoorshahi, K.; Almeida, M.; Cui, W.; Faust, A.; Reddi, V. The Role of Compute in Autonomous Aerial Vehicles. *arXiv* **2019**, arXiv:1906.10513.
5. Giovagnola, J.; Megias, J.B.M.; Fernandez, M.M.; Cuellar, M.P.; Santos, D.P.M. AirLoop: A Simulation Framework for Testing of UAV Services. *IEEE Access* **2023**, *11*, 23309–23325. [CrossRef]
6. Shah, S.; Dey, D.; Lovett, C.; Kapoor, A. AirSim: High-Fidelity Visual and Physical Simulation for Autonomous Vehicles. *arXiv* **2017**, arXiv:1705.05065.
7. Madaan, R.; Gyde, N.; Vemprala, S.; Brown, M.; Nagami, K.; Cristofalo, E.; Scaramuzza, D.; Schwager, M.; Kapoor, A. AirSim Drone Racing Lab. *arXiv* **2020**, arXiv:2003.05654.
8. Rublee, E.; Rabaud, V.; Konolige, K.; Bradski, G. ORB: An efficient alternative to SIFT or SURF. In Proceedings of the 2011 International Conference on Computer Vision, Barcelona, Spain, 6–13 November 2011; pp. 564–571. [CrossRef]
9. Lawrence, P.J.; Berarducci, M.P. Comparison of Federated and Centralized Kalman Filters with Fault Detection Considerations. In Proceedings of the 1994 IEEE Position, Location and Navigation Symposium—PLANS'94, Las Vegas, NV, USA, 11–15 April 1994; pp. 703–710. [CrossRef]
10. Geragersian, P.; Petrunin, I.; Guo, W.; Grech, R. An INS/GNSS fusion architecture in GNSS denied environments using gated recurrent units. In Proceedings of the AIAA Science and Technology Forum and Exposition, AIAA SciTech Forum 2022, San Diego, CA, USA, Virtual, 3–7 January 2022. [CrossRef]

engineering proceedings

MDPI

Proceeding Paper

Performance Improvement Provided by Global Navigation Satellite System Foresight Geospatial Augmentation in Deep Urban Environments †

Esther Anyaegbu, Paul Hansen and Bo Peng *

Spirent Communications plc, Paignton TQ4 7QR, UK; esther.anyaegbu@spirent.com (E.A.); paul.hansen@spirent.com (P.H.)
* Correspondence: bo.peng@spirent.com

† Presented at the European Navigation Conference 2023, Noordwijk, The Netherlands, 31 May–2 June 2023.

Abstract: Global navigation satellite systems (GNSSs) are an integral part of global positioning. However, because GNSS performance is impacted by signal obscuration and the presence of multipath in urban and deep urban environments, it is not accurate, reliable, and widely available enough to be a standalone system in all environments. This creates two problems: (1) the GNSS user does not know when or where GNSS performance may be degraded and (2) the GNSS user has limited ability to mitigate these issues. No mitigation strategy exists to improve the availability of GNSSs themselves. Inertial measurement units (IMUs) and sensor fusion provide other costly methods to improve positioning performance, but most systems still rely on GNSSs for absolute position. Spirent's GNSS Foresight service aims to solve both issues. As a cloud-based solution, GNSS Foresight provides satellite and signal information, and this can be employed to support the decision-making strategy and calculations in the GNSS receiver to improve its positioning solution performance, integrity, and reliability. In this paper, GNSS Foresight is introduced, and a performance evaluation of GNSS Foresight in dense urban areas is presented. Using the data collected from two urban areas in North America, we evaluated GNSS Foresight and compared the performance of GNSS positioning solutions with and without Foresight-aided data. The comparison results show the observed improvements in GNSS receiver operation. Foresight can also be used to develop measurement engine performance enhancements in the acquisition of new satellites and the tracking/re-acquisition of current satellites using line-of-sight (LOS) satellite information. In the positional computation process, Foresight enables receivers to prioritize LOS signals over degraded non-line-of-sight (NLOS) signals, hence significantly reducing positioning errors and outperforming conventional GNSS positioning, particularly in difficult urban environments.

Keywords: GNSS augmentation; multi-constellation; performance improvement; urban environments; line of sight; geospatial augmentation

Citation: Anyaegbu, E.; Hansen, P.; Peng, B. Performance Improvement Provided by Global Navigation Satellite System Foresight Geospatial Augmentation in Deep Urban Environments. *Eng. Proc.* **2023**, *54*, 58. https://doi.org/10.3390/ENC2023-15444

Academic Editors: Tom Willems and Okko Bleeker

Published: 29 October 2023

1. Introduction

1.1. The Problem—GNSS Navigation in Urban Areas

GPSs were developed over 40 years ago and are used in many different applications. Many of these applications are pushing the requirements of GNSS technology further than ever before. The concept of GNSS technology is that the GNSS receiver measures its distance from the satellite by calculating the time it takes for the GNSS signal to travel from the satellite to the receiver. The receiver needs four or more satellites to compute its position, velocity, and time [1].

GNSS signal reception is challenging in urban environments such as downtown areas and city centers due to frequent blockage of line-of-signal GNSS signals by obstructions such as tall buildings, overpasses, and bridges. The unpredictability of GNSS availability and its accuracy in urban environments has been a persistent problem in GNSSs. GNSS performance is unpredictable due to signal obscuration and the presence of multipath in deep urban environments. As shown on the left in Figure 1, signal obscuration occurs when GNSS satellite signals are blocked by buildings and only satellites that are high in the sky are visible to the GNSS receiver. Constellations formed by those few high-elevation satellites, however, have poor dilution of precision (DOP) values.

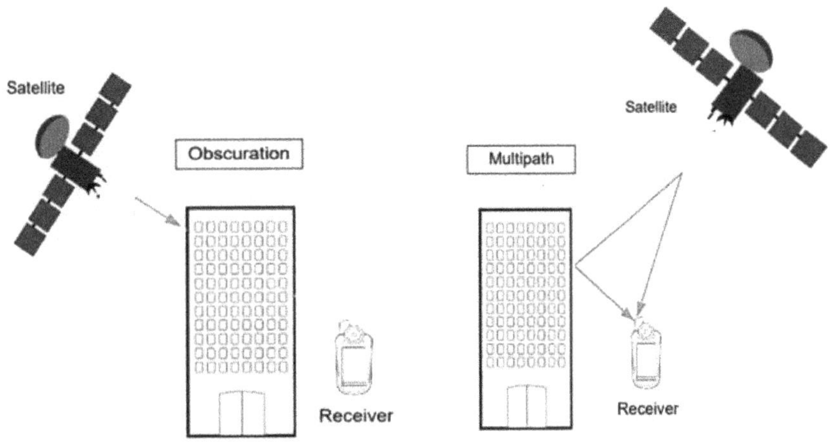

Figure 1. Signal obscuration/non-line of sight and multipath.

GNSS multipath refers to the phenomenon of when satellite signals are reflected by buildings before reaching the GNSS receiver, as shown on the right in Figure 1. Such reflections can cause significant errors in user navigation solutions. The accuracy of the computed position is dependent on the quality of the signals received by the receiver and on the geometry of the satellites in view [2].

1.2. The Problem—Which Satellites Do You Use?

Currently, there are four major GNSS constellations, and in the last ten years, the number of GNSS satellites has increased by over 50% to over 120 GNSS satellites. The increase in the number of GNSS satellites improves satellite availability. In a typical urban environment, the GNSS receiver will acquire many GNSS satellites; however, only a few of these measurements will be line of sight (LOS), while the others will be non-line of sight (NLOS).

As an example, in a drive test in downtown Indianapolis, the drive trial started in a fairly open sky area (the orange arrow in Figure 2 below), and from the start of the SNR plot in Figure 3, it is easy to see which GNSS measurements are LOS signals and which ones are NLOS signals. LOS signals tend to have a higher SNR than NLOS signals. As the car drives into the built-up area, there is then a considerable overlap between the LOS signals and the NLOS signals, and it becomes difficult to distinguish between the LOS and NLOS signals. In this situation, it is difficult for the receiver to decide on which satellites to use in computing its position.

With the truth data and some post-processing tools, the LOS and NLOS signals were worked out, as shown in Figure 4. This makes it difficult to use SNR to decide which signals are LOS signals and which signals are NLOS signals.

Figure 2. Drive test in downtown Indianapolis.

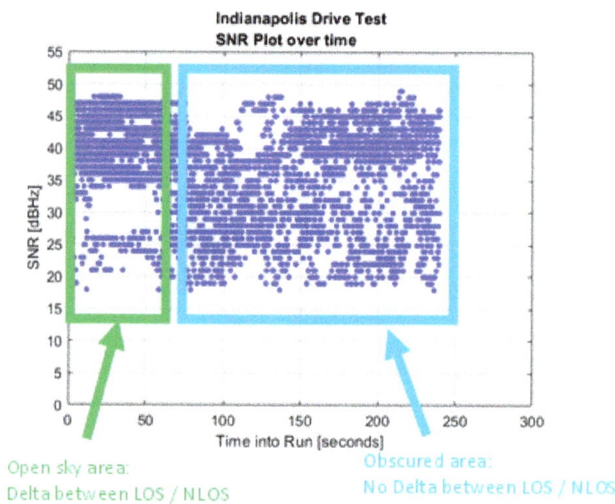

Figure 3. SNR measurements of all GNSS satellites acquired during the Indianapolis drive test.

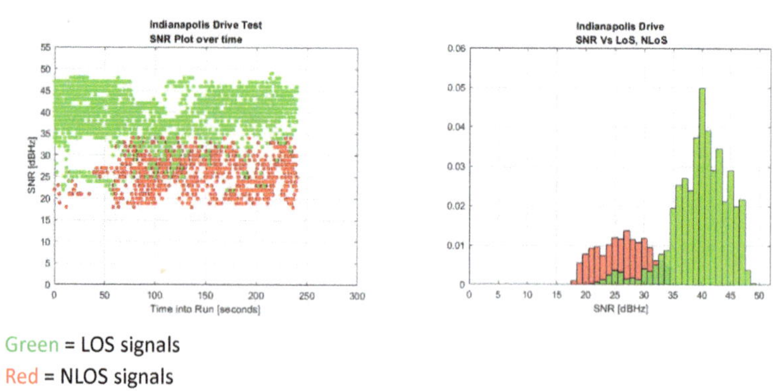

Figure 4. SNR measurement data from the downtown Indianapolis drive test.

Table 1 below highlights this issue: there is an 11 dBHz overlap between LOS and NLOS signals, so a simple cut-off cannot be used to define LOS signals, as it would result in either a significant number of NLOS signals being included or a significant number of LOS signals being excluded.

Table 1. Summary of the minimum and maximum SNR values for the LOS and NLOS signals.

SNR (dBHz)	Max SNR	Min SNR
LOS	45	22
NLOS	33	17

1.3. *Review of Approaches to Improve GNSS Position Accuracy*

Several techniques have been developed to tackle the problem of navigating GNSSs in urban areas such as:

- Improvements to the GNSS receiver hardware and signal processing algorithms
- Outlier rejection techniques
- The use of sensors
- GNSS shadow matching

1.3.1. Improvements to the GNSS Receiver Hardware and Signal Processing Algorithms

Several changes have been made to the GNSS receiver hardware and algorithms to improve GNSS position accuracy in urban environments. One of the most popular approaches is the use of narrow correlator spacing [3], which reduces the chip spacing between the tracking correlators in the tracking channels in order to mitigate the impact of multipath. Other changes include improvements to the antenna design, the use of wideband filters, and improvements to multipath detection and mitigation algorithms. More recently, the use of super correlation techniques [4] to coherently integrate the GNSS signals for longer in order to make the GNSS receiver more sensitive to the LOS signals than the NLOS signals has been employed. However, these techniques do not provide any external information about the environment and when an NLOS signal is about to become visible or when an LOS signal is about to be obscured.

1.3.2. Outlier Rejection Techniques

Another approach is the use of the outlier rejection technique to improve GNSS position accuracy. These techniques work well when the position engine has lots of measurements. The resulting position solution improves when the error in the measurements is greater than the loss in geometry by excluding satellites. However, when the error is not present in the measurements, this technique could still exclude an LOS signal in error, and this will result in worse geometry and therefore worse position accuracy. This approach does not work well in dense urban environments due to the limited number of satellites available. Also, this approach does not provide any external information about the environment and when an LOS signal is about to become visible or when an LOS signal is about to be obscured.

1.3.3. The Use of Sensors

Sensors can be used to aid GNSS receivers in order to improve their computed position. Sensors work well when they are properly calibrated and if they are started in an open sky area before moving into a dense urban environment. However, sensors do not provide information on how badly contaminated the received GNSS signals are and when an LOS signal is about to become visible or when an LOS signal is about to be obscured.

1.3.4. GNSS Shadow Matching Techniques

Shadow matching is a technique that determines position by comparing the measured signal availability and the SNR with predictions made using a 3D model [5]. However, these techniques do not provide any external information about the environment and when an LOS signal is about to become visible or when an LOS signal is about to be obscured.

A combination of these techniques can be used to improve GNSS position accuracy depending on the GNSS application:

- What equipment is available to the positioning system?
- Does the application require an instantaneous position fix or is it being used for continuous navigation?
- Does the application require future-looking route and mission planning?

2. Spirent's GNSS Foresight

Spirent's GNSS Foresight service provides GNSS users with a forecast of what the GNSS performance is going to be—just like a weather forecast but for GNSSs. This service enables users to know in advance where and when the GNSS is reliable. The Foresight service is a cloud-based service that uses a high-resolution 3D map, combined with precise orbital predictions of where the satellites are to provide LOS visibility for each satellite from current major GNSS constellations. There are two GNSS Foresight services:

(1) The Foresight Live service provides real-time GNSS performance forecast meter by meter every second over the area requested.
(2) The Foresight risk analysis provides a predictive performance analysis of the best- and worst-case GNSS performance for operations and planning.

There are two building blocks for the GNSS Foresight system, as shown in Figure 5:

- The forecast engine is where the forecast calculation is carried out using high-resolution 3D maps and precise satellite orbit information.
- A content delivery network (CDN) is used for fast data delivery. There are three interfaces in the CDN:
    - The ingress API is for receiving the forecast data generated by the forecast engine.
    - The device API is for user devices to request and receive forecast data.
    - The web console is for user configuration and management.

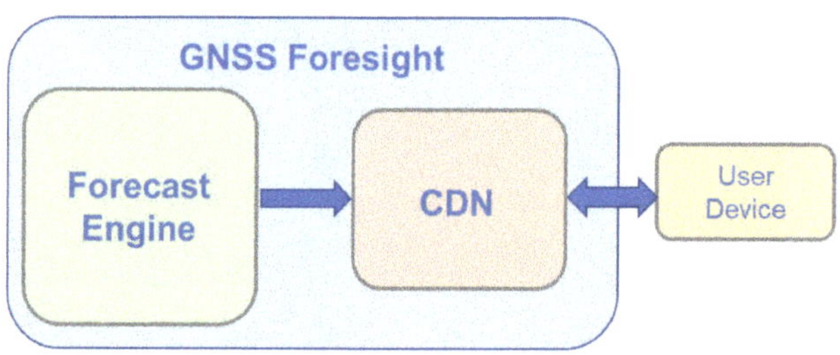

Figure 5. Foresight system architecture.

Following the publication of preliminary results at ION GNSS+ 2022 [6], progress has been made in integrating the data from Foresight with conventional GNSS position solutions.

3. Experimental Data Collection

To understand the performance of the Foresight system under various environmental conditions, we conducted a number of drive tests in two cities in North America. One represents a dense urban environment characterized by tall buildings, skyscrapers, narrow streets, and other obstacles. The other is a light urban environment with fewer buildings, lower building heights, and wider streets. GNSS measurements from multiple constellations (including GPS, GLONASS, BeiDou, and Galileo) were collected in these environments.

The selected dense urban environment is a typical downtown area in North America. The high density of buildings and other structures can cause multipath interference, where signals bounce off buildings and other surfaces before reaching the GNSS receiver, causing errors in the position measurement. Additionally, narrow streets and tall buildings can create "urban canyons" that limit the visibility of GNSS satellites, making it more difficult for the GNSS receiver to obtain a reliable position fix.

In contrast to the dense urban environment, GNSS signals are less likely to encounter multipath interference, and the GNSS receiver can more easily access a sufficient number of GNSS satellites for more accurate positioning in a light urban area. As a result, GNSS performance in a light urban environment is generally better than in a dense urban environment. However, some level of signal obstruction and multipath errors can still occur in the light urban environment due to buildings or other obstructions, especially if the receiver is located in an area with a limited view of the sky.

A mix of survey and consumer grades of GNSS receivers were used in the drive tests. At the beginning of each test, the vehicle was parked at a static open sky location for 15 min to allow the receivers to download ephemeris data and acquire a position fix. The duration of each drive test was 1 h in each environment.

4. Performance Evaluation

4.1. Methodology

With the collected GNSS receiver measurements, matching Foresight LOS satellite information was generated covering the area where the two drive tests were conducted over the testing period. The Foresight LOS information was then used to aid and support the decision-making process, where pseudo-range measurements calculated by NLOS satellites were removed. Therefore, only LOS measurements were passed on to the position and fusion engine to calculate a position fix. Note that in this experimental data set, a location is defined as one where an LOS-only position fix can be produced when there are a minimum of six LOS satellites measured at the location.

Evaluations were performed by comparing the performance aggregated over all locations for which there are positions generated using the following solutions:

- LOS-only with Foresight: The position is computed using LOS measurements only.
- LOS and NLOS measurements: This is a "conventional" GNSS position solution used by receivers that uses both LOS and NLOS measurements to calculate a position fix.

This enables a direct comparison of the benefits of using LOS information provided by the GNSS Foresight service.

4.2. Performance Evaluation

In this section, comparison results are presented for both dense and light urban environments. In particular, cumulative distribution function (CDF) plots are calculated to show the probability distribution of the position errors for the solution with and without Foresight aiding. Figure 6 shows the CDF of the horizontal position error of solutions with and without GNSS Foresight aiding in a dense urban environment. The CDF curves for both solutions are close together at low levels of the position error range, but as the position error increases, the solution without GNSS foresight aiding begins to shift to the right, indicating greater position errors at the percentile of the error distribution. In contrast, the solution with GNSS Foresight aiding remains consistently lower, indicating lower horizontal position error. This difference in position accuracy is particularly noticeable when the position error is less than 10 m, as shown in Figure 7. Overall, both figures provide clear evidence that GNSS Foresight can significantly improve the accuracy of horizontal position estimation.

Table 2 shows the position errors for specific percentiles, computed with and without GNSS Foresight aiding in a dense urban environment. The results show that position errors are significantly lower with GNSS Foresight aiding compared to the LOS + NLOS scenario without GNSS aiding across all computed percentile levels. The table also shows the error reduction in both meters and percentage when using GNSS Foresight data, which indicates that using GNSS Foresight data can lead to a 20–77.9% reduction in position error, depending on the percentile. Overall, these results confirm that incorporating

GNSS Foresight can significantly improve the accuracy of position estimation in a dense urban environment.

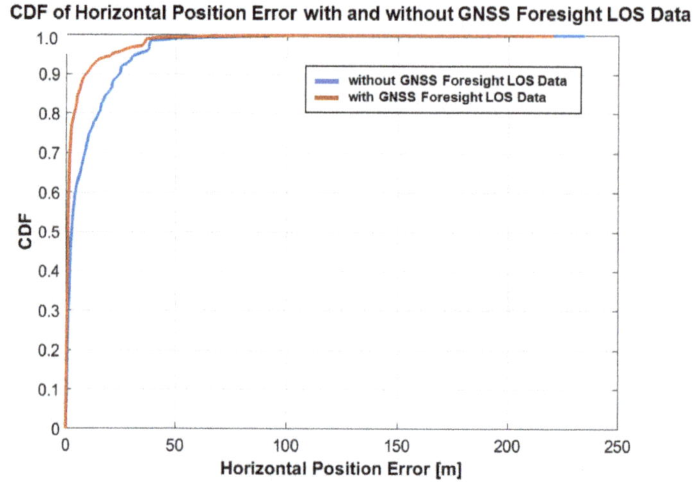

Figure 6. CDF of the horizontal position error of both solutions with and without GNSS Foresight aiding.

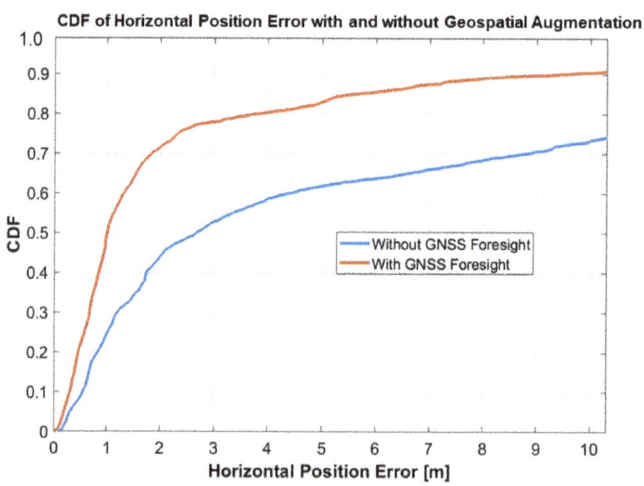

Figure 7. CDF of the horizontal position error of both solutions with and without GNSS Foresight aiding (zoomed to the 10 m position error level).

Figure 8 shows the CDF of the horizontal position error of solutions with and without GNSS Foresight aiding in a light urban environment. It can be observed that the solution with Foresight aiding still outperforms the solution without Foresight in terms of position accuracy. However, the difference in performance between the two solutions is reduced compared to what is observed in dense urban environments in Figure 6. The reason is that, in a less challenging environment where the majority of signals from satellites are LOS signals, the benefit that Foresight can provide in terms of LOS information is limited. This can also be observed in Table 3, which shows the position errors for specific percentiles, computed with and without GNSS Foresight aiding in a light urban environment.

Table 2. Position errors for specific percentiles, computed when using GNSS Foresight LOS data and when using both LOS and NLOS data, and the reduction in error determined when using GNSS Foresight LOS data (in meters and % reduction in error).

Position Solutions and Error by Percentile	25th Percentile	50th Percentile	67th Percentile	90th Percentile	95th Percentile	99th Percentile
LOS + NLOS (without GNSS Foresight aiding) [m]	1.0	2.7	7.4	24.4	32.2	47.7
LOS only (with GNSS Foresight aiding) [m]	0.6	1.0	1.6	9.3	20.7	38.2
Reduction in position error when using GNSS Foresight [m]	0.4	1.7	5.8	15.1	11.5	9.6
Reduction in position error (%) when using GNSS Foresight	42.2	62.4	77.9	62.0	35.6	20.0

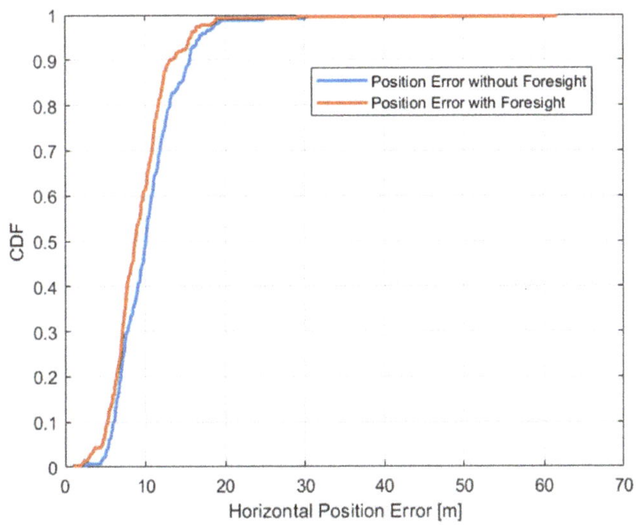

Figure 8. CDF of the horizontal position error of both solutions with and without GNSS Foresight aiding in a light urban environment.

Table 3. Position errors for specific percentiles, computed when using GNSS Foresight LOS data and when using both LOS and NLOS data, and the reduction in error determined when using GNSS Foresight LOS data (in meters and % reduction in error).

Position Solution Type and Error by Percentile	25th Percentile	50th Percentile	67th Percentile	90th Percentile	95th Percentile	99th Percentile
LOS + NLOS (without GNSS Foresight aiding) [m]	7.3	10.1	11.6	15.6	16.8	21.0
LOS only (with GNSS Foresight aiding) [m]	6.9	8.7	10.5	13.1	15.6	18.8
Reduction in position error when using GNSS Foresight [m]	0.4	1.4	1.1	2.5	1.2	2.2
Reduction in position error (%) when using GNSS Foresight data	5.4	15.6	10.6	19.5	7.7	11.4

5. Conclusions

The performance of conventional GNSS positioning in the urban environment can be degraded due to building obstructions and reflections. GNSS Foresight was developed to improve GNSS performance, integrity, and reliability in such a challenging environment by aiding the receivers with predicted LOS information. The performance of the Foresight-aided solution was evaluated by comparing it with conventional GNSS position solutions in both light and dense urban environments. The comparison results show that the Foresight aiding solution significantly reduces positioning errors and outperforms conventional GNSS positioning solutions, particularly in a dense urban environment.

Author Contributions: All authors equally contributed to all parts of the published version of the manuscript. All authors have read and agreed to the published version of the manuscript.

Funding: This research received no external funding.

Institutional Review Board Statement: Not applicable.

Informed Consent Statement: Not applicable.

Data Availability Statement: Restrictions apply to the availability of the data presented in this study.

Conflicts of Interest: The authors declare no conflict of interest.

References

1. Kaplan, E.D.; Hegarty, C. *Understanding GPS/GNSS: Principles and Applications*, 3rd ed.; Artech House Inc.: Norfolk County, MA, USA, 2017.
2. Misra, P.; Enge, P. *Global Positioning System: Signals, Measurements, and Performance*; Ganga-Jamuna Press: Lincoln, UK, 2010.
3. Van Dierendonck, A.J.; Fenton, P.; Ford, T. Theory and performance of narrow correlator spacing in a GPS receiver. *Navig. J. Inst. Navig.* **1992**, *39*, 265–283. [CrossRef]
4. Faragher, R.; Powe, M.; Esteves, P.; Couronneau, N.; Crockett, M.; Martin, H.; Ziglioli, E.; Higgins, C. Supercorrelation as a Service: S-GNSS Upgrades for Smartdevices. In Proceedings of the 32nd International Technical Meeting of the Satellite Division of The Institute of Navigation (ION GNSS+ 2019), Miami, Florida, 16–20 September 2019.
5. Groves, P.D. Shadow matching: A new GNSS positioning technique for urban canyons. *J. Navig.* **2011**, *64*, 417–430. [CrossRef]
6. Anyaegbu, E.; Paul, H. GNSS Performance Evaluation for Deep Urban Environments using GNSS Foresight. In Proceedings of the 35th International Technical Meeting of the Satellite Division of The Institute of Navigation (ION GNSS+ 2022), Denver, CO, USA, 19–23 September 2022; pp. 1127–1136.

engineering proceedings

MDPI

Proceeding Paper

Hardware in the Loop Laboratory Test Systems for Medium Frequency R-Mode Receivers †

Lars Grundhöfer *, Filippo Giacomo Rizzi, Niklas Hehenkamp and Stefan Gewies

Institute of Communications and Navigation, German Aerospace Center (DLR), 17235 Neustrelitz, Germany; filippo.rizzi@dlr.de (F.G.R.); niklas.hehenkamp@dlr.de (N.H.); stefan.gewies@dlr.de (S.G.)
* Correspondence: lars.grundhoefer@dlr.de
† Presented at the European Navigation Conference 2023, Noordwijk, The Netherlands, 31 May–2 June 2023.

Abstract: Terrestrial radio navigation systems are an important data source for increasing the integrity of position, navigation, and timing information and for strengthening the immunity to the spoofing and jamming of satellite-based systems. A possible solution for maritime use is the medium frequency R-Mode. Here, the first real results were presented in the Baltic Sea within the R-Mode Baltic project. For the development of receivers and future signal evolution of R-Mode, there is a great need to provide simulated signals to the receiver hardware. For our laboratory work, we developed different hardware in the loop test systems which enable the simulation of each of the three components and the entire medium frequency R-Mode signal. With this setup, we are able to conduct reproducible tests of the R-Mode receiver's ranging and positioning performance without the necessity of field tests. Furthermore, the impact of R-Mode signal modifications can be initially analyzed without the need of an implementation in the real-world test bed. Firstly, we describe the usage of arbitrary wave generators that can be used to replay received or simulated signals. Due to their wide distribution in electronic laboratories, there are cost-efficient ways to build up test capabilities. For this work, we tested the Tektronix AFG 3022 and the Rigol DG1032. For further tests, we utilize software-defined radios that are capable of streaming continuous signals. We utilize the ETTUS N210 to directly output the simulated signals. Additionally, we test the LimeSDR with an external down-converter. To generate these signals, we utilize software packages that were created to support the development of the digital signal processing. This approach allows us to test our receiver with a continuous integration from pure software to hardware in the loop test. A comprehensive summary gives an overview of the pros and cons for the different suggested systems.

Keywords: R-Mode; medium frequency; positioning; hardware in the loop

Citation: Grundhöfer, L.; Rizzi, F.G.; Hehenkamp, N.; Gewies, S. Hardware in the Loop Laboratory Test Systems for Medium Frequency R-Mode Receivers. *Eng. Proc.* **2023**, *54*, 59. https://doi.org/10.3390/ENC2023-15464

Academic Editors: Tom Willems and Okko Bleeker

Published: 29 October 2023

1. Introduction

Nowadays, high confidence and trust is placed in global navigation satellite system (GNSS) receivers to retrieve position, navigation, and timing (PNT) data. On the one hand, GNSS provides high accuracy with global coverage, which makes its usage so appealing. On the other hand, it constitutes a single point of failure due to the low power level of the satellite signal at the earth's surface, which can be easily disrupted by simple jamming techniques or more advanced and hazardous spoofing attacks. To counteract these threats and increase the resilience of maritime navigation, the idea of establishing an alternative PNT system arose under the name of R(anging)-Mode.

The general idea of R-Mode is to reuse existing infrastructure to provide positioning and timing information to mariners. Therefore, two main systems have been considered: the D(ifferential)GNSS IALA beacon, transmitting in the 300 kHz band, and the Very High Frequency Data Exchange System (VDES), which is the evolution of AIS, transmitting at around 162 MHz [1,2].

In this work, we will concentrate on the medium frequency R-Mode subsystem, which consists of several terrestrial transmitters broadcasting minimum shift key (MSK)-modulated GNSS correction. To enable ranging with sufficient accuracy, two continuous waves (CW) are included at the fourth zero crossing of the power spectra of the modulation, below (CW_1) and above (CW_2) the carrier frequency [3]. We present the resulting signal for a 500 Hz channel, which is common in Europe, in Figure 1. When these signals are synchronized over the different stations, we are able to derive ranges from the phase estimation of the CW and, consequently, positioning and timing, when we receive at least three stations [4,5].

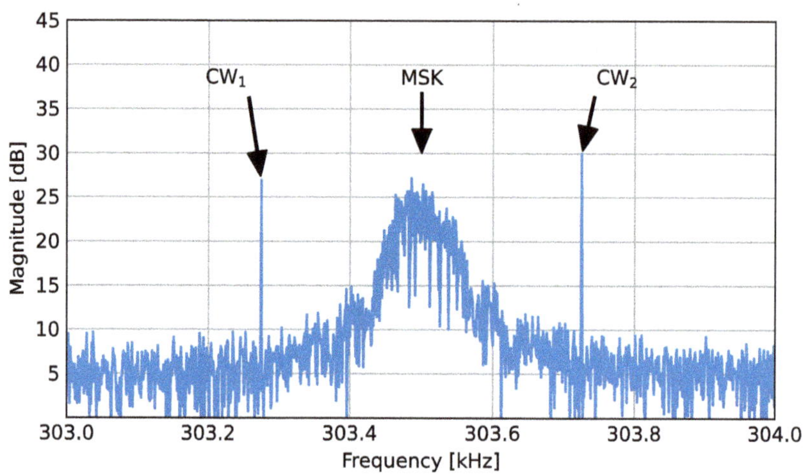

Figure 1. Example spectrum of an R-Mode MF signal.

To support our MF R-Mode receiver development, it is crucial to be able to generate signals of a single station up to a whole-system constellation. This enables us to test newly developed algorithms and verify hardware designs. The simulations described here were derived to verify our research in different publications [6–8].

In this work, we describe different forms of implementation to set up hardware in the loop simulation for different stages of development. Therefore, we introduce the general setups in Section 2. The concepts are then implemented and tested with respect to the needs of the MF R-Mode system in Section 3. Finally, we compare the results and close the paper in Section 4.

2. Simulation Setups

The simulation is oriented to the implementation of real-world R-Mode transmitters as described in [3]. In general, the report follows two possible design ideas. The first uses the existing MSK modulator and adds two continuous signals with an external signal generator. For the simulation in our laboratory, we use a software-defined radio (SDR) as a source of continuous modulated MSK signals, while the aiding carrier is provided via a signal generator that is synchronized with a precise time source. The overall setup we described is shown in Figure 2 as a block diagram. The advantage of this approach is that we can continuously stream real data. However, in the minimal configuration, only one station can be simulated, as most signal generators only provided two channels and the modulation of the messages is limited by the computational resource that drives the SDR. Moreover, not all the SDR are capable of transmitting frequencies as low as 280 kHz. When we want to utilize one of these, we need to introduce a down conversion, as shown in the block diagram in Figure 3. Additional filtering is not necessary, as the higher side band after the mixer is filtered out via the band-pass filter of the MF R-Mode receiver.

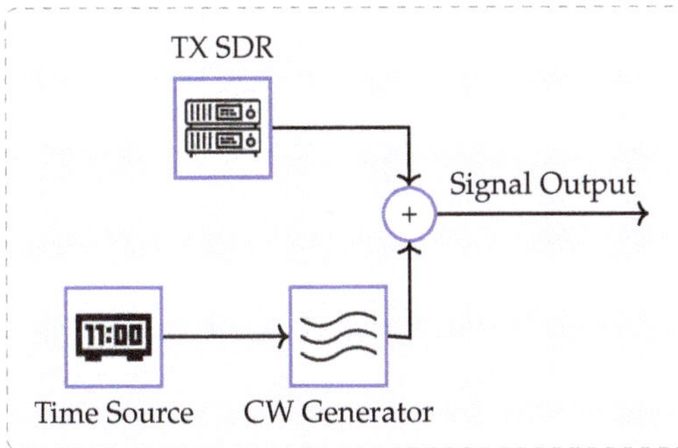

Figure 2. R-Mode signal generation with external tone generation.

This limitation can be overcome by using an arbitrary waveform generator (AWG), which replaces the aiding carrier generation with a replay of a sum of the needed carrier. Still, the number of modulated channels is limited by the computational power.

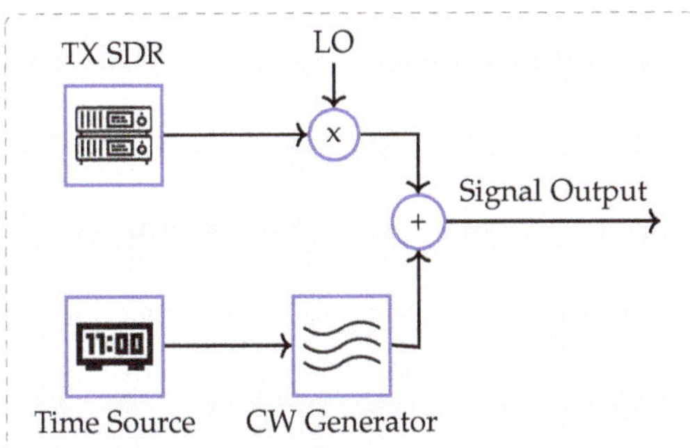

Figure 3. R-Mode signal generation with external tone generation and down conversion of the baseband.

The second approach follows the idea of utilizing an AWG to generate multiple stations by replaying the R-Mode signal for multiple channels at once. To guarantee good timing of the signal, we synchronize the AWG with an external clock source, which is described in Figure 4. The setup is similar to the transmitter station with an integrated R-Mode modulator, as described in [3]. However, with our approach, we are only able to replay recorded or pre-computed signals. These files are then repeated by continuously utilizing the sequencer of the AWG. As we simulate 16 s of the data, we assume a repeating pattern every 16 s.

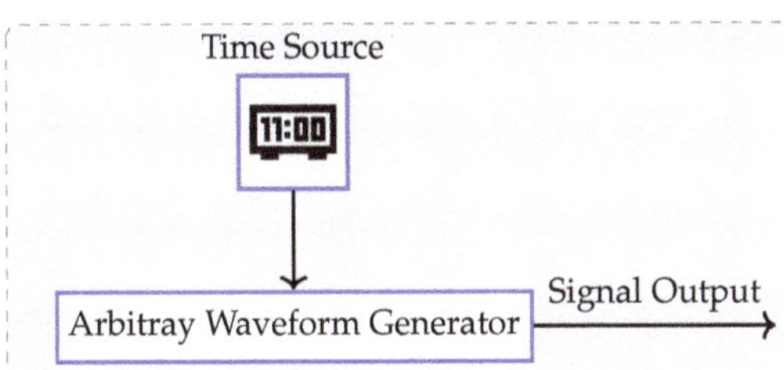

Figure 4. R-Mode signal generation with AWG.

The scenario we use for this work is obtained from the DLR simulation environment for the MF R-Mode. We have already used these methods in the past to verify different developments for the MF R-Mode [6–8].

3. Implementation

In this paper, we concentrated on the generation of a single station. In the following part, we first describe the used measurement devices. Subsequently, we discuss the result of a simulation run of half an hour, where each phase estimation considers one second of observation time. The DLR implementation of the R-Mode MF receiver based around an ETTUS N210 is used to obtain our results.

3.1. MSK Generation with ETTUS SDR

To implement the design presented in Figure 2, we utilize an ETTUS N210 with daughterboard LFTX as TX SDR. This allows us to use a baseband modulation at 300 kHz. The modulation is generated in a GNUradio flow graph, from a random byte source with a bitrate of 100 bits/s. Unfortunately, it is not feasible to generate the CW with the same SDR due to intermodulation in the generation path. As a CW generator, we use a RIGOL DG1032 two-channel generator, which is synchronized to a common oven-controlled oscillator (OCXO) shared with the receiver. The CW are generated on each channel separately with an amplitude of 100 mV and combined with the modulation signal with a Mini-Circuits ZSC-4-3-75+ signal combiner. Due to the low frequency, we neglect the mismatch between the 50 Ω signal generation and the 75 Ω of the signal combiner. Figure 5 shows the resulting setup, which is tested.

In Figure 6, we present our test run of 40 min results with zero mean. In the graph, the orange triangle indicates the results for CW_1, the blue triangle indicates CW_2 and the green points represent the beat phase, which is the phase difference between CW_1 and CW_2. We can clearly observe for the two CW a phase drift going from −0.005 rad to 0.002 rad, which corresponds to an error of 1 m. However, the phase difference is stable over the full time, with a resulting variance of 4.467×10^{-11} rad^2.

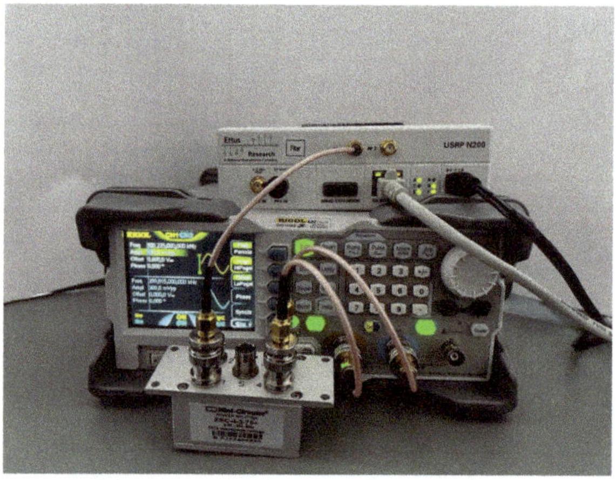

Figure 5. Test setup with ETTUS SDR.

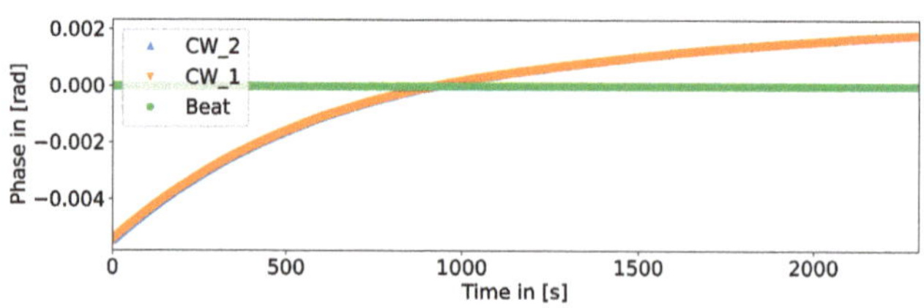

Figure 6. Phase estimation with modulation from ETTUS SDR.

3.2. MSK Generation with Lime SDR

In general, with some SDRs which do not cover the medium frequency band, a down conversion is required as described in Figure 3. As example implementation for this paper, we use the Lime SDR USB, wich is presented in Figure 7. The MSK signal is generated at a baseband frequency of 40 MHz and then down converted with the mixer Mini-Circuit ZFY-1. The local oscillator (LO) signal is generated with the Tektronix AFG31000, which is synchronized to the timing reference. The resulting intermediate frequency is then combined with two continuous signals generated using a Rigol DG1032 signal generator, corresponding to the previous setup with an amplitude of 100 mV. The result of a 40 min test run is shown in Figure 8. The orange triangles mark the estimation result with zero mean of CW_1, the blue triangle is CW_2, and the green dots are the results for the beat phase. The CWs again show a phase drift over our measurement interval, here from -0.007 rad to 0.004 rad. However, the result for CW_2 shows different outliers of the trend line after around 800 s of testing. We observe this accordingly in the result for the beat frequency, which leads to a variance of 8.528×10^{-8} rad^2. The less accurate result is potentially due to the interference with the MSK, as described in [7]. This interference occurs strongly when the carrier frequency is drifting with respect to the CWs. The problem becomes more recognizable with the mixing step, as the conversion has an influence on the stability. It is very obvious that the interference is different for the two CWs and depends on the frequency shift of the MSK. In this experiment, we can assume that the MSK has a shift towards the CW_2, producing higher interference than for the CW1.

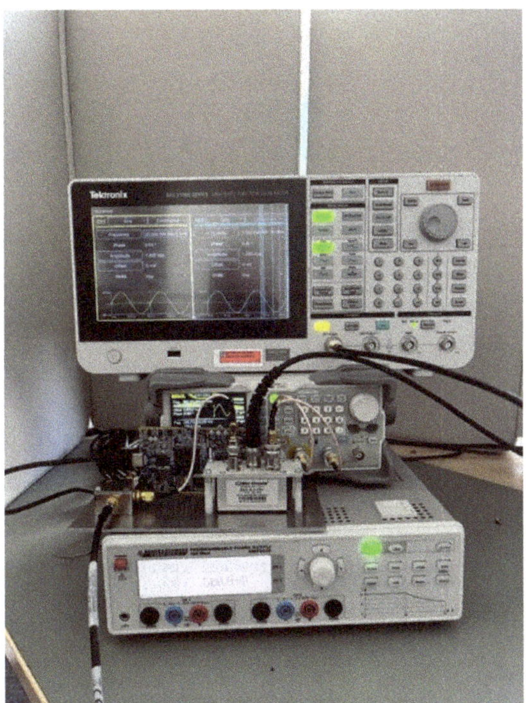

Figure 7. Test setup with Lime SDR.

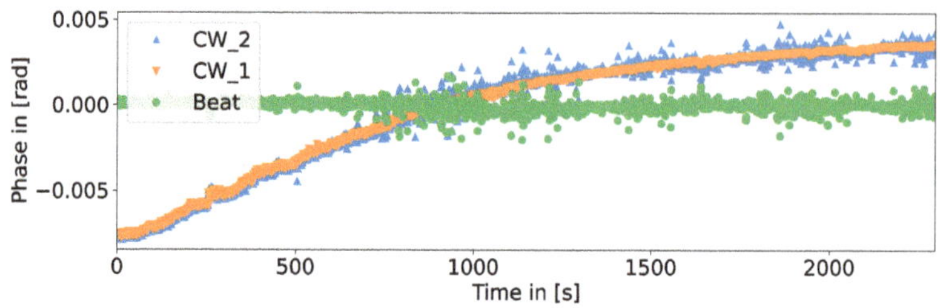

Figure 8. Phase estimation with modulation from Lime SDR.

3.3. Data Replay with AWG

The last setup is fundamentally different to those previously used. Instead of generating the signals in real time, we replay pre-simulated or recorded signals as described in Figure 4. Our setup utilizes the Tektronix AFG31000 with a common OCXO to our receiver. We obtain a single R-Mode station from the DLR python simulation environment with a sample rate of 5 Msamples/s in four comma-separated value (CSV) files. Each of the files holds 4 s of data and is loaded into the sequencer of the AWG. The corresponding CWs have an amplitude of 100 mV for CW_1 and 200 mV for CW_2. The 16 s of total data are repeated endlessly. The generated output is fed directly to the MF R-Mode receiver, as shown in Figure 9.

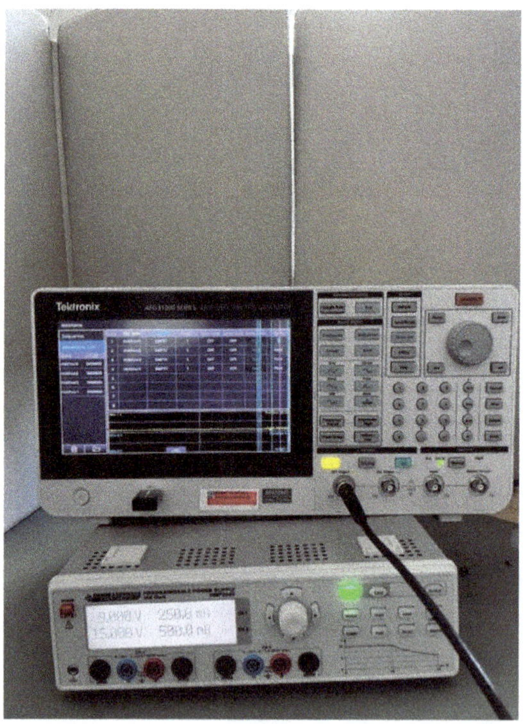

Figure 9. Test setup with AWG.

To proof the repeatability of the pattern, we compare the phase estimation of the first and second replay of the files in Figure 10. We indicate the first replay with red dots and the second replay with blue crosses. All signals have discrete values that appear repeatedly in the range of −0.2 rad to 0.2 rad. The overlay shows a good agreement.

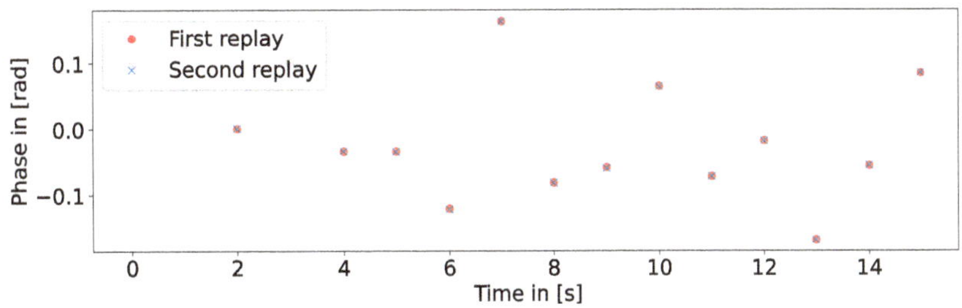

Figure 10. Phase estimation with modulation from replayed CSV files for first and second replay.

To show the attainable performance, the result of the conducted half-hour test run is shown in Figure 11 for every 16th estimation. By only evaluating at certain time steps, the repetition pattern is not visible and we only evaluate the expected performance. The estimation results are represented by orange dots for the CW_1, blue dots for the CW_2, and green dots for the beat frequency. The CW appear to drift in the range of -0.0075 rad to 0.0025 rad. Despite the issue encountered, this approach has the big advantage of generating several R-Mode stations with a predefined configuration. Therefore, it is useful to test the R-Mode receiver positioning engine, as long as we only consider snapshot solutions.

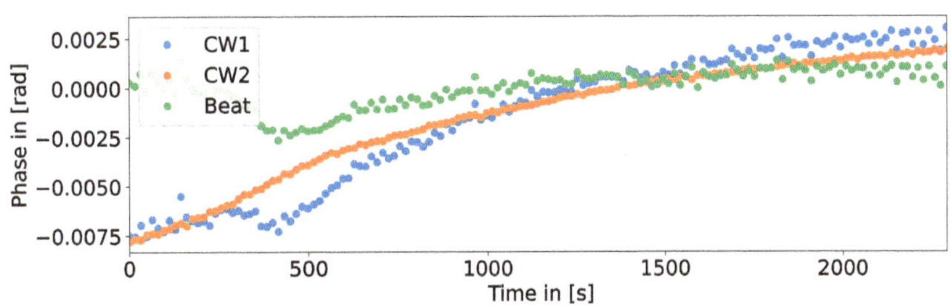

Figure 11. Phase estimation with modulation from replayed CSV files for every 16th estimation.

4. Conclusion

We introduced three different approaches for simulating R-Mode signals in a laboratory environment. We were able to generate R-Mode signals with a medium frequency baseband, a baseband conversion, and the replay of pre-processed files. Overall, the use of an external generator and an SDR that is able to transmit in the medium frequency baseband provides the highest phase stability and can therefore be used to detected even small improvements. However, this approach is limited to the simulation of a single, station, and when more stations are required, the amount of hardware grows. As an alternative, the AWG can be used to simulate multiple stations, even if the observed performance is limited.

Author Contributions: All authors contributed equally to the paper. All authors have read and agreed to the published version of the manuscript.

Funding: The study was funded by the security and defence research at DLR as part of the project SiNaKoL.

Institutional Review Board Statement: Not applicable.

Informed Consent Statement: Not applicable.

Data Availability Statement: Data sharing is not applicable to this article.

Conflicts of Interest: Te authors declare no conflict of interest.

References

1. Gewies, S.; Dammann, A.; Ziebold, R.; Bäckstedt, J.; Bronk, K.; Wereszko, B.; Rieck, C.; Gustafson, P.; Eliassen, C.G.; Hoppe, M.; et al. R-Mode Testbed in the Baltic Sea. In Proceedings of the 19th IALA Conference 2018, Incheon, Republic of Korea, 27 May–2 June 2018.
2. Wirsing, M.; Dammann, A.; Raulefs, R. Investigating R-Mode Signals for the VDE System. In Proceedings of the OCEANS 2019, Seattle, WA, USA, 27–31 October 2019.
3. Johnson, G.; Swaszek, P. *Feasibility Study of R-Mode Using MF DGPS Transmissions*; IALA: St Germain en Laye, France, 2014.
4. Johnson, G.; Dykstra, K.; Ordell, S.; Swaszek, P. R-Mode Positioning System Demonstration. In Proceedings of the ION GNSS+ 2020, St. Louis, MO, USA, 22–25 September 2020.
5. Grundhöfer, L.; Rizzi, F.G.; Gewies, S.; Hoppe, M.; Bäckstedt, J.; Dziewicki, M.; Galdo, G.D. Positioning with medium frequency R-Mode. *ION Navig.* **2021**, *68*, 829–841. [CrossRef]
6. Grundhöfer, L.; Rizzi, F.G.; Gewies, S.; Del Galdo, G. Improving medium frequency R-Mode ranging with GMSK modulation. In Proceedings of the 34th International Technical Meeting of the Satellite Division of the Institute of Navigation ION GNSS + 2021, St. Louis, MO, USA, 20–24 September 2021; the Institute of Navigation: Manassas, VA, USA, 2021.
7. Grundhöfer, L.; Wirsing, M.; Gewies, S.; Galdo, G.D. Phase Estimation of Single Tones Next to Modulated Signals in the medium frequency R-Mode system. *IEEE Access* **2022**, *10*, 73309–73316. [CrossRef]
8. Grundhöfer, L.; Meinert, N.; Rizzi, F.G.; Gewies, S.; Ziebold, R. Calculating Lower Bounds within the PyTorch Framework. In Proceedings of the 35th International Technical Meeting of the Satellite Division of the Institute of Navigation ION GNSS + 2022, Denver, CO, USA, 19–23 September 2022; the Institute of Navigation: Manassas, VA, USA, 2022.

engineering proceedings

MDPI

Proceeding Paper

Satellite Navigation Signal Interference Detection and Machine Learning-Based Classification Techniques towards Product Implementation †

Jelle Rijnsdorp [1,*,‡], Annemarie van Zwol [2,‡] and Merle Snijders [2]

1 S[&]T, 2616 LR Delft, The Netherlands
2 Netherlands Aerospace Centre (NLR), 1059 CM Amsterdam, The Netherlands; annemarie.van.zwol@nlr.nl (A.v.Z.); merle.snijders@nlr.nl (M.S.)
* Correspondence: jelle.rijnsdorp@stcorp.nl
† Presented at the European Navigation Conference 2023, Noordwijk, The Netherlands, 31 May–2 June 2023.
‡ These authors contributed equally to this work.

Abstract: Many critical applications highly depend on Global Navigation Satellite Systems (GNSS) for precise and continuously available positioning and timing information. To warn a GNSS user that the signals are compromised, real-time interference detection is required. Additionally, real-time classification of the interference signal allows the user to select the most effective mitigation methods for the encountered disturbance. A compact proof of concept has been built using commercial off-the-shelf (COTS) components to analyse the jamming detection and classification techniques. It continuously monitors GNSS frequency bands and generates warnings to the user when interference is detected and classified. Various signal spectrum analyses, consisting of kurtosis and power spectral density (PSD) calculations, as well as a machine learning model, are used to detect and classify anomalies in the incoming signals. The system has been tested by making use of a COTS GNSS signal simulator. The simulator is used to generate the upper L-band GNSS signals and different types of interferences. Successful detection and classification is demonstrated, even for interference power levels that do not degrade the performance of a commercial reference receiver.

Keywords: GNSS; jamming detection; machine learning; jamming classification; product implementation

Citation: Rijnsdorp, J.; van Zwol, A.; Snijders, M. Satellite Navigation Signal Interference Detection and Machine Learning-Based Classification Techniques towards Product Implementation. *Eng. Proc.* **2023**, *54*, 60. https://doi.org/10.3390/ENC2023-15449

Academic Editors: Tom Willems and Okko Bleeker

Published: 29 October 2023

1. Introduction

Many critical applications exhibit a high dependency on Global Navigation Satellite Systems (GNSS) for precise and continuously available positioning and timing information. Interference of the low-power GNSS signals has become an increasing threat for society [1,2]. In terms of intentional Radio Frequency Interference (RFI), jamming aims to disrupt GNSS signals by overpowering them, leading to the receiver losing the position, velocity and timing (PVT) solution. To warn a GNSS user that the GNSS signals are compromised, real-time interference detection is required. Additionally, real-time classification of the interference signal allows the user to select the most effective mitigation methods for the encountered disturbance. Classification can also be used to estimate disturbance signal properties and help point the user towards likely sources of the interference signal.

Previous publications [3–5] have shown the potential of using artificial intelligence for pre-correlation interference classification. The aim in this paper is to use various signal spectrum analysis methods as well as a machine learning model to develop a compact GNSS interference detection and classification module. The module is constructed using widely available commercial off-the-shelf (COTS) components. The goal is to continuously monitor GNSS frequency bands, generate warnings to the user when interference is detected and inform the user on how the interference signal is classified.

This paper is structured to first provide an overview of the implemented interference detection and classification techniques. Then, hardware and software designs of the module are briefly discussed. The strategy and results of an initial over-the-cable test campaign are presented, and finally, a short outlook on future product development is provided.

2. Detection

This section elaborates on the techniques that are used for jamming detection. As the monitoring system is designed to be independent of a GNSS receiver, the selected detection techniques are based on the signal spectrum that is obtained directly from the antenna.

2.1. Statistical Analysis

The first method is a statistical analysis based on the spectrum of the incoming GNSS signal. The method consists of a calculation of the Kurtosis value K of the spectrum, and is defined as:

$$K = \frac{\mu^4}{\sigma^4}, \quad (1)$$

where μ is the fourth central moment and σ is the standard deviation of the spectrum. For natural thermal emissions, the distribution is Gaussian and has a nominal Kurtosis of 3.0. Since GNSS signals are buried under the noise level, a sample data set containing GNSS signals should resemble white noise. If the signal is corrupted by RFI of sufficient power, the distribution will deviate from normality, leading to a deviation from the nominal Kurtosis value. The Kurtosis calculation can be used to detect jammer types that add non-noise-like energy to the signal.

2.2. Power Spectral Density Ratio (PSD)

For partial band and narrowband interference signals, analysis of the frequency domain provides valuable information in terms of detection. By calculating the power spectral density (PSD) ratio, the presence of additive energy in the signal spectrum can be detected. The frequency spectrum X can be computed using the discrete Fourier transform:

$$X_k = \sum_{n}^{N-1} x[n] e^{-j\frac{2\pi n}{N}}, \quad (2)$$

where x is the GNSS signal, k is the frequency bin, N is the number of samples in the frequency bin and n is an iterator. From the frequency domain, the PSD can be calculated. The algorithm that is implemented in the monitoring system uses the PSD from both the received signal, as well as the PSD of a pre-defined reference signal, which is recorded in a non-RFI scenario and provides a baseline of the received power by the system. The maximum PSD ratio r_{max} is then defined as

$$r_{max} = \max\left(\frac{\|X_{test}\|^2}{\left\|X_{ref}\right\|^2}\right), \quad (3)$$

and if increased above a certain threshold, it indicates an RFI signal in the spectrum. This detection technique is used for finding added energy to the signal spectrum and is not bound to specific jamming types.

3. Classification

As an extension to the aforementioned jamming detection methods, a classification model can be used to provide classification of the type of jamming signal, which is useful information when considering mitigation techniques. Additionally, it can be used as a detection technique.

3.1. Model Setup

The classification model is constructed using a Convolutional Neural Network (CNN) in combination with transfer learning and is trained on recognizing spectral features in images created by taking a spectrogram of the incoming GNSS signal. As the different jammer types induce different spectral features, this method can be very effective for the classification of jamming. The set of included jammer types consist of Continuous Wave (CW), chirp or Swept Continuous Wave (SCW), Additive White Gaussian Noise (AWGN), Pseudo-Random Noise (PRN), Amplitude Modulation (AM) and Frequency Modulation (FM). Figure 1 shows the spectral features on which the classification model is trained.

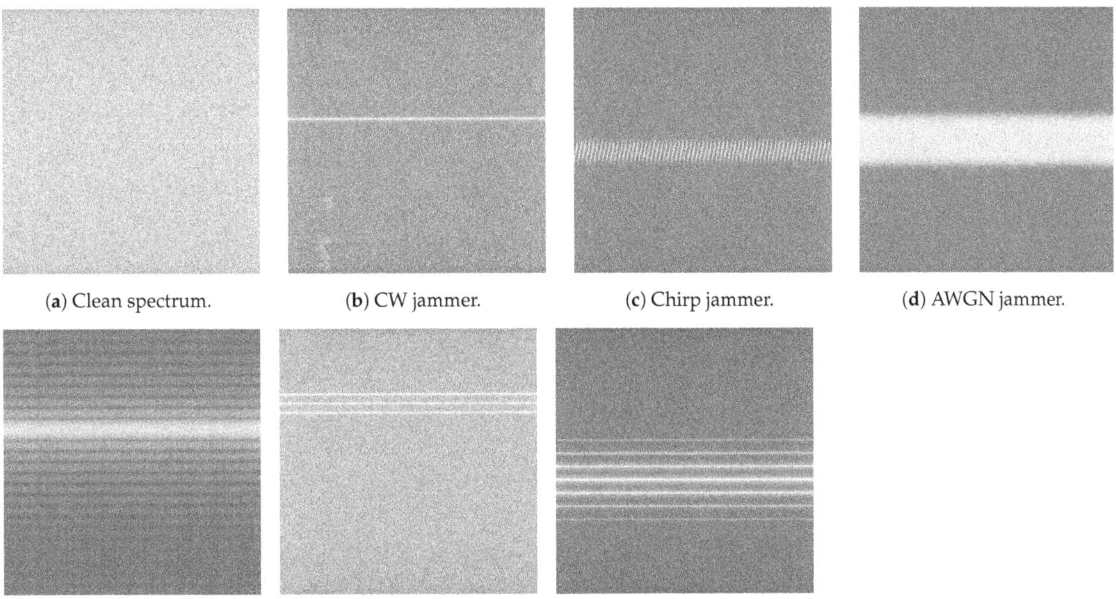

(**a**) Clean spectrum. (**b**) CW jammer. (**c**) Chirp jammer. (**d**) AWGN jammer. (**e**) PRN jammer. (**f**) AM jammer. (**g**) FM jammer.

Figure 1. Spectrograms of the clean signal without RFI and with different types of jamming signals. The horizontal axis shows the time domain and the vertical axis shows the frequency domain.

The jamming types of which the spectra are shown in Figure 1 are described in Table 1. In these models, y_J denotes the jamming signal, a_J is the (initial) amplitude, f_c is the centre frequency, t is the time and ϕ_J is the phase.

3.2. Model Training

As a baseline for the CNN, the family of EfficientNet [6] models that were specifically developed for increased accuracy and efficiency with respect to MobileNet [7] and ResNet [8] models are used. The actual model that is selected for implementation is the EfficientNet-B5 model, which has 83.6% top 1 accuracy and 96.7% top 5 accuracy, while containing more than five times fewer parameters than comparable models [6]. Training is conducted on a set of input images that contain spectrograms of the GNSS signal with and without different types of jamming signals. The Adam algorithm for gradient descent optimisation is used, which is commonly used to ensure fast training that is robust to potential noise in the input features [9]. To assess the performance of the classification model, a confusion matrix can be used which visualises the expected probability of correct classification and the expected probability for misclassification for all combinations of true and predicted classes. The confusion matrix corresponding to the model described in this section is shown in Figure 2.

Table 1. Jammer models (derived from [3,4]) that are included in the classification model.

Type	Model	Description
CW	$y_J = a_J \cdot \exp(2j\pi f_c t + \phi_J)$	Generates a single-tone jamming signal, which shows a distinct peak at a constant frequency in the signal spectrum. This translates to a horizontal line in the spectrogram (Figure 1b).
Chirp	$y_J = a_J \cdot \exp\left(2j\pi\left(f_c t \pm \frac{1}{2T}(f_{max} - f_{min})t^2\right) + \phi_J\right)$	Generates a frequency-sweeping signal that shows a moving peak over a certain range in the signal spectrum. In the spectrogram, it is characterised by a sawtooth pattern (Figure 1c) with a bandwidth enclosed by minimum and maximum frequencies f_{min} and f_{max} and width of a single tooth defined by the sweep time T.
AWGN	$y_J = a_J \cdot \int_0^t n(\tau)d\tau \cdot \exp(2j\pi f_c t + \phi_J)$	Generates a signal consisting of white Gaussian noise filtered to a specific bandwidth, which translates in a horizontal band in the spectrogram (Figure 1d). The noise n is modelled by a finite impulse response filter applied on a Gaussian distribution with zero mean and variance $\sigma^2 \sim \mathcal{N}(0, \sigma^2)$.
PRN	$y_J = a_J \cdot \int_0^t p(\tau)d\tau \cdot \exp(2j\pi f_c t + \phi_J)$	Generates one PRN-code and shows multiple peaks in the spectrogram (Figure 1e). The PRN-code p is modelled by a finite random sequence of values -1 and 1 and is repeated every millisecond.
AM	$y_J = a_J \cdot [1 + m\cos(2\pi f_m t + \phi_m)] \cdot \exp(2j\pi f_c t + \phi_J)$	Generates a carrier wave signal of which the amplitude is modulated by a message signal with frequency f_m and phase ϕ_m. This translates into a peak at the carrier frequency and two sidebands at a frequency difference of $\pm f_m$ (Figure 1f). Finally, m defines the modulation index, which is a measure for the ratio between the power of the carrier and the message signal.
FM	$y_J = a_J \cdot \exp(2j\pi f_c t + m\sin(2j\pi f_m t + \phi_m) + \phi_J)$	Generates a carrier wave signal of which the frequency is modulated by a message signal with frequency f_m and phase ϕ_m. This translates into a peak at the carrier frequency and multiple sidebands, each with a frequency shift of $n \cdot f_m$, where n is an integer (Figure 1g). Again, m denotes the modulation index, which defines the relative height of the peaks according to a set of Bessel functions.

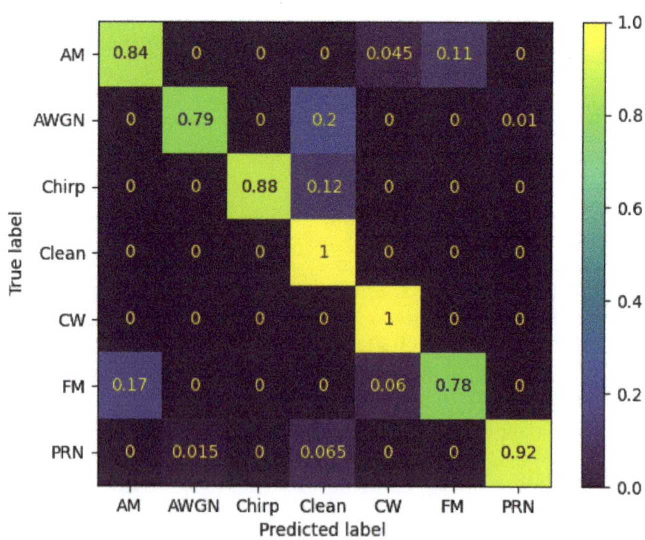

Figure 2. Confusion matrix of the classification model.

This matrix shows excellent results for the clean class and CW and good results for the other classes. The non-zero off-diagonal entries show that the spectral features of some class pairs may be similar, which is especially the case for low-power jamming scenarios. For low-power narrowband waveforms, such as AM and FM, the spectrum may not contain all sub-peaks, leading to misclassification of AM jammers as CW and FM jammers as either CW or AM, depending on the number of visible sub-peaks. The misclassification of an AWGN or PRN jammer as a non-RFI case can be explained by the fact that the energy of these waveforms is spread over a relatively large band of frequencies and the signals themselves resemble noise, which make them barely detectable for low-power scenarios.

4. Product Implementation

Moving from the theoretical methods and techniques that were described in Sections 2 and 3 towards product implementation requires a practical solution for a number of problems. The GNSS monitoring device is required to operate in real time and is to be used in a hand-held configuration. This poses limitations on the computational power, dimensions, weight and power consumption of the hardware to be used. Eventually, the hardware selection influences the software design, which should be able to fulfil the capability of (real-time) jamming detection and classification with limited available computational resources. The designed hardware and software setup are described in this section.

4.1. Hardware Setup

The aim is to keep the hardware design of the monitoring system as simple as possible by limiting the amount of hardware components. It is designed to be managed by a microprocessor that runs the software and controls the other components. To retrieve the signal, the microprocessor will be connected to a Software-Defined Radio (SDR) that is tuned to the GPS L1 frequency with an instantaneous bandwidth of 25 MHz. This SDR contains an input port for a GNSS antenna, which poses the flexibility for the user to use their own GNSS antenna. The data stream from the SDR to the microprocessor contains I/Q data, which are further handled by the software package that is described in Section 4.2.

Results from the jamming detection and classification processes will be communicated to the user via an LCD monitor that can be connected externally. When the device is used in a more mobile setting and the LCD monitor is not used, simple results and other relevant information are communicated via a status LED. To accommodate for mobile usage, a battery pack is included, which is able to power the unit for several hours; additionally, the device can be connected to an external power supply for direct power and to recharge the battery pack.

The flow diagram of the hardware design can be seen in Figure 3. All components except for the external antenna and LCD monitor will be integrated in a protective casing. As mentioned earlier in this section, the monitoring device should be usable in a hand-held configuration, which limits the size of the hardware components. Additionally, using components specifically designed for this particular system is avoided, and therefore it is chosen to use only COTS hardware.

4.2. Software Setup

The microprocessor that was mentioned in the previous paragraph contains the monitoring software, which is enabled when the device is powered on. This software package controls the SDR by starting all relevant processes that lead to the SDR providing I/Q data that describe the GNSS signal. The computational power of the microprocessor is insufficient for processing the complete 25 M samples/s I/Q data stream, and thus the data are reduced before the jamming detection and classification methods are enabled. For detection, data blocks of 100 ms are found to be sufficient, whereas for classification, blocks of only 10 ms are used.

After slicing the data into these blocks, process pools are generated, which will contain the individual processes that execute the detection and classification algorithms. This will ensure concurrency and optimization of the limited computational resources of the microprocessor. When a process is finished, it will pull new I/Q data in and again perform the detection or classification method. The results from all the algorithms are then synchronised and, with a fixed interval, a decision algorithm is executed. This algorithm compares the individual detection and classification results with specified thresholds, which will provide a single alert containing a jamming flag and corresponding jamming class, if applicable. These are then displayed by the status LED and, if connected, the LCD monitor. Additionally, more extensive results from the individual algorithms, as well as general system status, is logged to the SD card of the microprocessor.

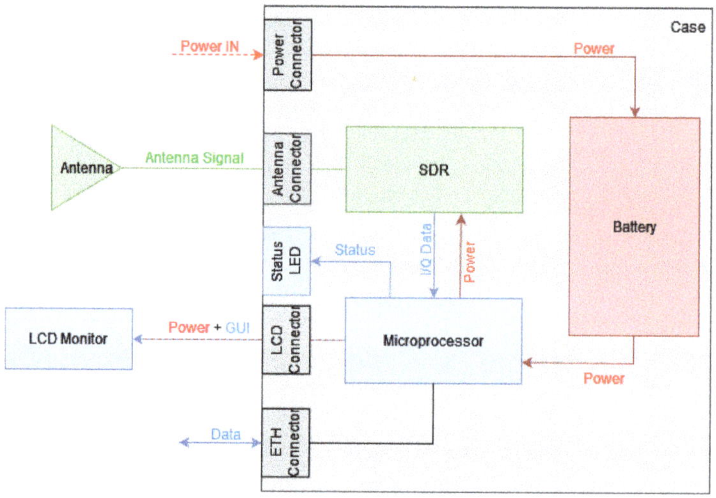

Figure 3. Schematics of the hardware setup of the GNSS monitoring device.

5. Product Tests

An initial test campaign is performed to test the performance of the proposed detection and classification module. The goal of this test campaign is threefold: validate the implemented detection strategy, validate the implemented classification strategy and obtain an initial benchmark of the jammer-to-signal ratios (J/S) for which the module can reliably detect and classify interference signals.

5.1. Experimental Setup

To ensure a clean, controlled, and repeatable test scenario and to exclude any accidental over-the-air interferences the tests are performed over-the-cable in a controlled lab environment. The device under test (DUT) is provided with a series of different input signals in which the true GNSS component is the same for every run and the characteristics of the interference are varied.

A COTS GNSS signal simulator (Orolia GSG-8) is used to simulate the clean GNSS signals, a realistic Gaussian noise floor and the interference signals. Since the DUT only operates in the L1 frequency band, only GPS L1 C/A, Galileo E1 B/C and SBAS L1 signals are included in the input signal. An overview of the settings used to generate the clean input GNSS signals for all test scenarios can be found in Table A1 in Appendix A.

The clean GNSS signals and the interference signals are simulated on two different channels and combined by an RF combiner. To be able to provide the test signal not only to the DUT, but also to a COTS reference receiver (Septentrio AsteRx3 HDC), an active four-way RF signal splitter is used to split the combined input stream to two identical output streams. In Figure 4, the schematics of the test setup are displayed.

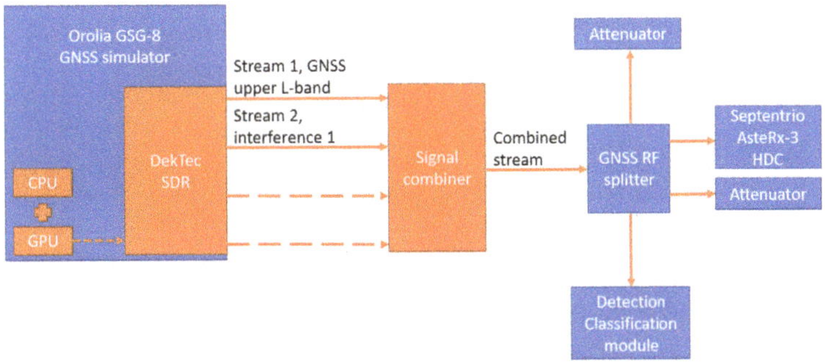

Figure 4. Schematic overview of the test setup. The GNSS simulator is used to generate the test signals. The generated signals are streamed to the DUT and reference receiver.

5.2. Experimental Results

The DUT and a reference receiver are subjected to two test scenarios in which the jamming power is increased over time. First, a wide range of J/S is used, where the ratio is increased in steps of 3 dB. For every step in this scenario, the J/S is kept constant for 60 s to allow the reference receiver to acclimatise. The second scenario, which entails a smaller range of J/S with a step size of 0.5 dB and a step duration of only 10 s, is used to improve the accuracy of the detection and classification results. The waveform characteristics of the interference signals used in this test campaign are summarised in Table A2 in Appendix A. Detection and classification are triggered when the thresholds are exceeded for three out of five subsequent measurements to reduce the number of false alarms. The reference receiver is used to monitor the effect of the interference signal on the resolving capability of a targeted GNSS receiver. The J/S for which the reference receiver experiences first loss of its PVT solution is used to put the performance of the DUT in perspective.

Both detection and correct classification take place even before the first loss of a tracked satellite is registered. The classification model proves to be the most sensitive detection method, as it is able to recognise the spectral features of an interference signal before the thresholds on the detection parameters are exceeded (highlighted values in Table 2). It will cause the system to trigger a jamming warning, albeit that no reliable classification can be made yet. The PSD method proves to be effective for narrowband waveforms, whereas the kurtosis can be used for all waveforms except for the AWGN. All detection methods are triggered at significantly lower J/S than the ratio where the reference receiver experiences loss of its PVT solution. The relatively 'late' correct classification of the FM jammer can be explained by the fact that for lower J/S, only the first sub-peaks on either side of the carrier frequency are visible, and thus this jammer is classified as an AM jammer. With the current parameters, it is expected that the module can issue a warning before a GNSS user is seriously affected by the presence of interference.

Table 2. Test results (J/S) for the different detection and classification parameters of the DUT. First loss of tracked satellite and loss of PVT are obtained from the reference receiver. The values for which the system first triggers a jamming warning are highlighted. All values are given in dB.

Interference Waveform	CW	Chirp	AWGN	PRN	AM	FM
First loss tracked satellite	−6	3	15	6	15	15
Loss of PVT	21	24	33	27	21	21
Kurtosis ($2.99 < K < 3.01$)	7	6.5	no detection	6.5	8	7.5
PSD (1 dB increase)	−7	6	11	6	−1	0
Detection with classification model	**−7**	**−1**	**1.5**	**1.5**	**−4**	**−4**
Correct classification	−6	−0.5	3.5	1.5	−4	11

6. Conclusions

The initial test results are promising, as the system is able to both detect and classify interferences before first loss of a tracked satellite by the reference receiver. It can be concluded that in terms of detection, the classification model proves to be successful as it can notice changes in the signal spectrum in low-power jamming scenarios. The PSD ratio and kurtosis are computationally inexpensive methods that can contribute to the confidence of the interference detection when the jamming power has increased a bit further. Concerning classification, the model seems to be capable of correctly classifying the jammer type before the COTS reference receiver first loses track of a satellite, although it is seen that there is significant difference in performance for different jamming waveforms.

In this paper, the implementation of real-time jamming detection strategies in an operational product are discussed. As stated, these methods will provide the user with knowledge concerning possible jamming attacks, but not with information regarding spoofing. To extend the capabilities of the product, real-time spoofing detection is considered to be implemented. In a spoofing attack, a set of counterfeit GNSS signals is created and transmitted, with the aim to misdirect a receiver into computing a false position and timing solution. Effort is already being put in the development of a spoofing detection functionality: Structural Power Content Analysis (SPCA) can be used on the pre-despreaded signal to find additional PRNs in the signal [10], thereby indicating a spoofing attack. For future development of the GNSS monitoring system, the use of SPCA as an additional detection parameter is investigated.

For the development of the operational device, it is advised to perform additional tests, which should not only account for the different waveforms, but also for variations in waveform characteristics like bandwidth, central frequency and sweeping time. The results of the additional tests can not only verify the robustness of the module against waveform variations but will also allow for fine-tuning of the implemented detection and classification thresholds. Validation should be carried out by exposing the system to real jamming and spoofing scenarios, where more fine-tuning can be performed to rule out, for example, unintentional interference sources, such as multipath.

Author Contributions: Hardware and software design and implementation, J.R.; Test design, A.v.Z and M.S.; Test execution, J.R., A.v.Z. and M.S.; Writing, review and editing, J.R., A.v.Z. and M.S. All authors have read and agreed to the published version of the manuscript.

Funding: This research received no external funding.

Institutional Review Board Statement: Not applicable.

Informed Consent Statement: Not applicable.

Data Availability Statement: Research data provided in article. For more information, please contact one of the authors.

Conflicts of Interest: The authors declare no conflicts of interest.

Abbreviations

The following abbreviations are used in this manuscript:

AM	Amplitude Modulation
AWGN	Additive White Gaussian Noise
BPSK	Binary Phase Shift Keying
CNN	Convolutional Neural Network
COTS	Commercial Off-The Shelf
CW	Continuous Wave
DUT	Device Under Test
FM	Frequency Modulation
GNSS	Global Navigation Satellite System

J/S	Jammer-to-Signal Ratio
MDPI	Multidisciplinary Digital Publishing Institute
PRN	Pseudo-Random Noise
PSD	Power Spectral Density
PVT	Position, Velocity and Time
RFI	Radio Frequency Interference
SCW	Swept Continuous Wave
SDR	Software-Defined Radio
SPCA	Structural Power Content Analysis

Appendix A. Simulator Settings and Test Signal Waveform Characteristics

Table A1. Simulator settings used to simulate the GNSS signals.

Setting	Value
Centre frequency	1575.42 MHz
Sample rate	50 Msamples/s
Simulation start time	1 April 2023, 12:00:00 (date, time)
Ionospheric model	Klobuchar
Tropospheric model	Stanag
Signal power GPS L1 C/A	−3.00 dB (with respect to −131.5 dBm at 5° elevation)
Signal power Galileo E1	−2.00 dB (with respect to −131.5 dBm at 5° elevation)
Signal power SBAS L1	0.50 dB (with respect to −131.5 dBm at 5° elevation)
Receiver position	Lat: 52.6757761°, Lon: 5.9247521°, Alt: 2.00 m
Transmitter position	Lat: 52.6803418°, Lon: 5.9119103°, Alt: 2.00 m

Table A2. Characteristics of the waveforms for which the DUT is tested.

Interference Waveform	Central Frequency	Characteristics
CW	1575.00 MHz	
Chirp	1575.42 MHz	Bandwidth: 1 MHz, sweeping time: 100 μs
AWGN	1575.42 MHz	Bandwidth: 2 MHz
PRN	1575.42 MHz	Modulation scheme: Binary Phase Shift Keying (BPSK), code rate: 1.023 MHz, code length: 1 ms
AM	1575.42 MHz	modulation index: 1, modulation frequency: 1 MHz
FM	1575.42 MHz	modulation index: 1, modulation frequency: 1 MHz

References

1. UK Government Office for Science. *Satellite-Derived Time and Position: A Study of Critical Dependencies*; Blackett review; UK Government Office for Science: London, UK, 2018.
2. EGNSS Centre of Excellence. *IKUS-II: Inventarisatie Kwetsbaarheden Uitval Satellietnavigatie*, Ministerie Infrastructuur en Waterstaat; EGNSS Centre of Excellence: Noordwijk, NL, 2022.
3. Morales Ferre, R.; de la Fuente, A.; Lohan, E.S. Jammer classification in GNSS bands via machine learning algorithms. *Sensors* **2019**, *19*, 4841. [CrossRef] [PubMed]
4. Wu, Z.; Zhao, Y.; Yin, Z.; Luo, H. Jamming signals classification using convolutional neural network. In Proceedings of the 2017 IEEE International Symposium on Signal Processing and Information Technology (ISSPIT), Bilbao, Spain, 18–20 December 2017; pp. 062–067.
5. Swinney, C.J.; Woods, J.C. GNSS jamming classification via CNN, transfer learning & the novel concatenation of signal representations. In Proceedings of the 2021 International Conference on Cyber Situational Awareness, Data Analytics and Assessment (CyberSA), Dublin, Ireland, 14–18 June 2021; pp. 1–9.

6. Tan, M.; Le, Q.V. EfficientNet: Rethinking Model Scaling for Convolutional Neural Networks. In Proceedings of the 36th International Conference on Machine Learning, Long Beach, CA, USA, 9–15 June 2019.
7. Howard, A.G.; Zhu, M.; Chen, B.; Kalenichenko, D.; Wang, W.; Wey, ; T.; Andreetto, M.; Adam, H. MobileNets: Efficient Convolutional Neural Networks for Mobile Vision Applications. *arXiv* **2017**, arXiv:1704.04861v1
8. He, K.; Zhang, X.; Ren, S.; Sun, J. Deep Residual Learning for Image Recognition. In Proceedings of the IEEE Conference on Computer Vision and Pattern Recognition, Las Vegas, NV, USA, 27–30 June 2016; pp. 770–778.
9. Kingma, D.P.; Ba, J. Adam: A Method for Stochastic Optimization. In Proceedings of the 3rd International Conference for Learning Representations, San Diego, CA, USA, 7–9 May 2015.
10. Jafarnia Jahromi, A. GNSS Signal Authenticity Verification in the Presence of Structural Interference. Ph.D. Thesis, University of Calgary, Calgary, AB, USA, 2013.

engineering proceedings

MDPI

Proceeding Paper

Antiference: New Concept for Evolutive Mitigation of RFI to GNSS †

Shahzrad Afroozeh [1,*], Vincent Bejach [2], Uros Bokan [1], André Bos [2], Bastiaan Ober [3] and Sascha Bartl [1]

1 OHB Digital Solutions GmbH, Kärntner Straße 7b/1, 8020 Graz, Austria; uros.bokan@ohb-digital.at (U.B.); sascha.bartl@ohb-digital.at (S.B.)
2 S&T—Science and Technology BV, Olof Palmestraat 14, 2616 Delft, The Netherlands; vincent.bejach@stcorp.nl (V.B.); andre.bos@stcorp.nl (A.B.)
3 Integricom, Tjalkenwerf 30, 2317 Leiden, The Netherlands
* Correspondence: shahzrad.afroozeh@ohb-digital.at
† Presented at the European Navigation Conference 2023, Noordwijk, The Netherlands, 31 May–2 June 2023.

Abstract: The past decade has shown a growing awareness of the dangers of intentional interference (especially jamming and spoofing) with GNSS signals. The Antiference project uses reconfigurable digital signal processing methods in the detection, classification, and mitigation of interference by employing machine learning techniques. The ML-based jamming classifier uses distinctive features of spectrograms for the differentiation of various jamming attacks. A residual neural net is used to map the spectrograms to the different jamming types. It relies on a fingerprinting architecture. Fingerprints summarize the characteristics of all the incoming signals, which are stored in and matched to a database of previously encountered interference types. To validate the implemented functionalities, a developed test-bed runs test scenarios and benchmarks the results against two state-of-the-art COTS receivers with interference mitigation capabilities.

Keywords: machine learning; jamming; spoofing; ResNet50; COTS receivers; interference detection and mitigation

Citation: Afroozeh, S.; Bejach, V.; Bokan, U.; Bos, A.; Ober, B.; Bartl, S. Antiference: New Concept for Evolutive Mitigation of RFI to GNSS. *Eng. Proc.* **2023**, *54*, 61. https://doi.org/10.3390/ENC2023-15451

Academic Editors: Tom Willems and Okko Bleeker

Published: 29 October 2023

1. Introduction

The past decade has shown a growing awareness in the dangers of intentional interference in the area of GNSS signals. Jamming and Spoofing (the two main intentional and harmful radio frequency interference (RFI) types) have been receiving a lot of attention in terms of the availability and affordability of hardware to induce the attacks, and the resulting costs to the victims. Jamming is the broadcasting of a radio frequency (RF) signal with the goal of blocking the reception of authentic GNSS signals and it leads to the complete blockage of the service or to the significant degradation of the service quality. Victims of jamming usually know that they are experiencing a form of interference. However, newer jamming methods, such as low-power jamming, code/spectrum-matching jamming (or PRN jammers) [1], and systemic jamming [2] lead to the degradation or blocking of the service without the victim being completely aware of what is happening.

Spoofing, on the other hand, is the targeted broadcasting of counterfeit GNSS signals so that the victim receiver will perceive them as the authentic signals and follow the counterfeit signals. In these types of attacks, the victim receivers are not aware of being attacked and can be steered to unknown locations or be prone to accidents.

The Antiference project uses reconfigurable digital signal processing (DSP) methods and tools in the detection, classification, characterization, and mitigation of harmful RFI. While some forms of RFI signal mitigation do not rely on the detection and classification of the RFI (like Notch filters or narrow correlators for multipath mitigation), effective mitigation in most cases relies on knowing the parameters of the interfering signal and can be significantly improved if accurate knowledge about the type and parameters of the attack is known. Employing data-driven artificial intelligence (AI) and machine learning (ML) techniques to facilitate the detection and classification of RFI is the main focus of this project. However, like many other defensive procedures, the recording and sharing of previously encountered attacks (RFI incidents in this case) can save time in their detection and classification and facilitate effective mitigation measures. Since ML models can suffer from explicit knowledge representation and storing that can be shared between multiple equipment (multiple instances of the same model), a fingerprinting (FP) architecture is introduced to alleviate the possible shortcomings. The FPs of each incoming signal have been generated and matched to a database that contains all the previously encountered RFI types. This results in a database of fingerprints that can be shared between devices of the possible RFIs. To test and validate the implemented functionalities, a testbed has been developed that can run test scenarios and benchmark the results with state-of-the-art COTS receivers which include RFI mitigation capabilities.

2. Background

The detection of interference has been a topic of research for a long time. Most of the classical ways to detect jamming are based on signal processing or statistical techniques, such as automatic gain control (AGC) monitoring, spectral monitoring, time-domain statistical analysis, and signal/carrier to noise ratio (CNR) monitoring. It is trickier to detect spoofing, but the most common methods use Position/Velocity/Time (PVT) estimate consistency checks, navigation data estimate consistency checks, the monitoring of AGC for anomalies, the monitoring of multiple correlation peaks and distortions of the Autocorrelation function (ACF), using additional equipment like inertial measurement units (IMU) for position check, signal polarization, or multi-antenna arrays for directing arrival detection.

Machine learning (ML)-based methods have also been explored to detect jamming and spoofing signals. In some cases, they use similar features to the classical techniques as a basis for the detection of interference. Some of the features used can be summarized as Power Law Detector (PLD), such as the monitoring of AGC-gain values; Power Spectral Density (PSD) Analysis, including methods such as the computation of Power Spectral Density (PSD), filter banks, and wavelet Transformations; Probability Density Function (PDF) Analysis, such as Kurtosis and the Gaussian White Noise test; Short Term Fourier Transform (STFT) for the detection and classification of jamming; and Cyclo-stationary analysis to detect spoofed conditions, such as pre-correlation Spectral Power Contents Analysis (SPCA) [3].

Using the pre-processed data as features is only part of the solution; the other part is the choice of model and its training and testing data and methods. Jamming detection has been implemented successfully using Support Vector Machines (SVM) in [4–6], while Neural networks (NN) such as Convolutional Neural Networks have been used in [7]. RFI classification has used SVMs [8,9], Long Short-Term Memory (LSTM) has been used in [3], and Convolutional Networks has been used in [8] with good results. Spoofing detection is carried out using SVMs in [10,11], using NNs in [11,12], and using the LASSO technique in [13]. Signal Characterization in the cognitive radio domain is carried out using different NNs [14–18] and SVMs [19].

The mitigation of jamming is performed using multiple classical techniques depending on the types of attack. The most used techniques can be summarized as the Bandpass filter (for out-of-band interfering signals), the Notch Filter (NF) [20], the Adaptive Notch Filter (ANF) [21,22], Frequency Domain Adaptive Filtering (FDAF), Pulse Blanking [20], Wavelet-based methods (such as [23]), and Karhunen–Loeve-based methods [23].

3. Models

The Antiference system uses two NN models for the detection and classification of different RFI types. One model is concerned with the detection and classification of different jammer types, and the other detects the existence of spoofing signals. Figure 1 shows the two models used for jamming detection and classification (left) and spoofing detection (right).

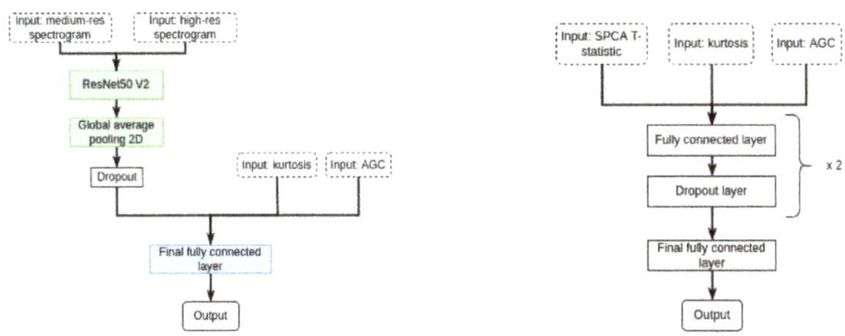

Figure 1. Detection and classification models used in Antiference for jamming (**left**) and spoofing (**right**).

Different features were used for the detection of jamming and spoofing. This implied that separate models would need to be used (otherwise the models would either not be efficient or their size would increase dramatically), which, in turn, also led to having two different FPs and FP databases. The FP of the incoming signals will be created and matched against the FP database for each model separately. If no matches are found, the needed features are extracted from the signal and sent for detection and classification through the NN. The results will be both inserted in the FP database and also sent to the mitigation unit for parameter extraction and mitigation of the signal.

The jamming detection/classification NN is built using transfer learning. It is illustrated in Figure 1, and is composed of two parts: (i) In green: this part extracts features from the spectrograms. Its main component is a pre-trained 50 layers ResNet architecture (see [24] for details about this type of architecture), available through the Keras toolbox (see [25]). The final pooling layer is used to reduce the dimension of the feature map between this step and the decision layer. This reduces the required computations down the network while acting as a "summary" of the most prominent features. In addition, a dropout layer is inserted between the extracted features and the rest of the network to ensure the robustness of the training. (ii) In blue: features are processed by the decision layer, using a fully connected (FC) decision layer. This layer combines the features extracted by the ResNet50 with additional information (the kurtosis and AGC value of the sample).

The ResNet50 V2 architecture was selected by comparing the training speed and classification performance of several architectures commonly used for transfer learning on a subset of the available training data. It is the one that showed the best compromise between the complexity of training and a good prediction of all the classes.

In addition, the parameters of this architecture and of the training process were tuned to enable the highest performance, using a process of Bayesian optimization (see [26,27] for an overview of the tool used to perform the optimization).

4. Data

Since ML-based models require a large number of samples for proper training and assessments, and based on the automotive standards that were used as the basis for assessments of the results, it was decided that the data used for training the model and final validation will use both recordings and simulations to fulfil the requirements.

To do so, three types of data were used: (i) Purely simulated data used for the simulation of all spoofing and PRN jamming datasets; (ii) Recorded base GNSS signal + simulated RFI used for all jamming types other than code/spectrum-matched jammers; and (iii) Recorded signals containing both GNSS + RFI used only for our final validation tests.

It should be noted that the broadcasting of jamming and/or spoofing signals is illegal. All the broadcasting and subsequent recordings of RFI in the real field was carried out on a military training ground in Austria by OHB Digital Solutions GmbH under a special permit. The difficulty in recording these datasets and their subsequent scarcity is the reason they were used for our final validation. To record the required signals, two data-collection campaigns were performed: firstly, Real-field RFI recordings, collected on the military training ground in Austria within a period of two days. The recordings consist of a collection of jamming and spoofing datasets in static and dynamic conditions. The recorded jammers were five COTS jammers, and broadcasted simulated jamming signals using an IZT S1000 signal generator with a connected power amplifier and transmit antenna. Secondly, Base (clean) GNSS signal recordings collected in the city of Graz and its surrounding areas in Austria. Twelve datasets containing around 8 h of recording in different urban and suburban environments from a car were collected.

Training and Test Data Preparation

To prepare the training data, the first eleven routes from Graz recordings were combined with simulated jamming signals. A few hundred jamming signals were created using different parameters, and combined with chunks of 10 ms of the recorded (base) GNSS signals. The signal combinations were mix-and-match to remove bias in the learning results of the model.

All the real-field RFI recordings, route 12 of the drive-throughs in the city of Graz, and a selection of purely simulated data were used for final validation.

5. Antiference Testbed

The testbed is designed and configured to facilitate the verification and validation of the Antiference system as well as for the benchmarking of the results in relation to the two COTS receivers which have state-of-the-art RFI mitigation techniques implemented in them. Figure 2 shows the Antiference testbed architecture. As can be seen in the image, the COTS receivers can be used to process the incoming signals with or without their internal mitigation mechanisms tuned on, or to process the mitigated signal using the Antiference signal. The testbed has the following sections:

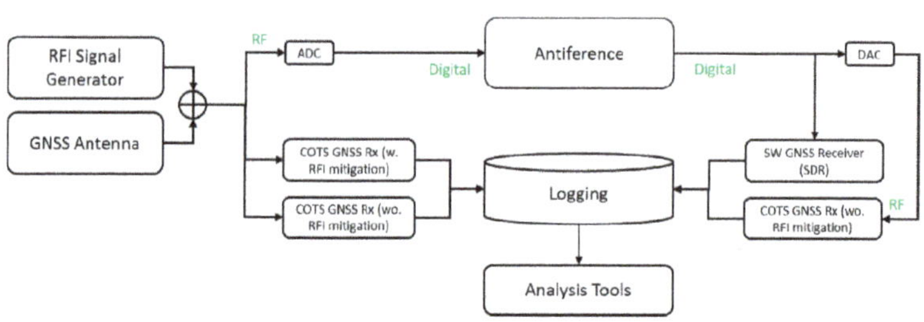

Figure 2. Antiference testbed architecture.

Signal simulator: The signal simulator can be used for the simulation of clean, clean and RFI, or only RFI signals in the GNSS frequency bands. In this project, OHB's GNSS simulator XPLORA was used for the simulation of all simulated signals. Analogue signals recorded through the GNSS antenna can be directly fed into the COTS receivers (to be directly processed by them) or be converted to digital signals to be fed into the Antiference system. Antiference only works in the digital domain. Analogue to Digital Convertor (ADC): in this project, a GNSS frontend based on a 8-bit ADC was used (TeleOrbit MGSE). Two COTS receivers are used for the benchmarking of the results: (1) a low-cost receiver (Rec01) and (2) a mid-range receiver (Rec02) suitable for the automotive domain. SDR (software defined receiver): OHB Digital Solutions' GNSS SDR was used in this project, to process the results of the Antiference, and verify if any residual RFI could be detected using the built-in RFI detectors. Digital to Analogue Convertor (DAC): For the conversion of digital to analogue signals, the built-in convertor in TeleOrbits' MGSE (Multi-GNSS

Simulation and Test Environment) was used for the broadcasting of digital signals for processing using the COTS receivers. The logging manager (or the collection of logs created by each of the elements): to be stored for analysis, signal by signal or in batches. Analysis tools: developed in MATLAB, these are used to analyze the results for the final validation and assessment of the performance

6. Test and Validation Results

We have seen good detection capabilities for both jamming and spoofing. The classification of individual signal chunks has yielded good results, and has proven to provide useful information for the mitigation stage. Table 1 summarizes some of the results.

Table 1. Detection accuracy of the Antiference system.

Datasets	AM	FM	SCW	CW	AGW
False Alarm Rate	0.001	0.002	0.002	0.0	0.002
Detection Accuracy	1.0	1.0	0.9	1.0	0.952

Having knowledge of the specifics of a jamming attack can lead to better mitigation, as was demonstrated in the validation phases. The effectiveness of the mitigation method has not been fully demonstrated yet in the user domain; although Antiference does filter jamming signals effectively and shows significantly improved C/N0 values for medium- to high-power jammers, the resulting improvement in the PVT-domain has not been convincingly demonstrated. The main reason for the shortcomings comes from the limitations in the digital domain, as the mitigation/filtering was performed only on the digitized signals, where the degradations of the RFI cannot be fully recovered. In a future evolution, the overall mitigation performance in the user domain can be improved by applying the mitigation directly at the receiver level.

Antiference has been used in combination with two COTS receivers and has been compared to Rec02′s internal RFI suppression. As the Rec01′s internal RFI mitigation cannot be turned off, Antiference can only be used in combination with it and not compared with it in isolation. The relative performances differ for the different interference types. Antiference performance is typically comparable to that of Rec02, while, in some cases, it is clearly somewhat better or somewhat worse than Rec02. Further optimization is expected to further improve the Antiference performance, but, for example, for FM interference, it already performs remarkably well, as shown in Figure 3 It is, however, hard to distinguish FM and AM interferers; however, due to the identical mitigation strategy, this does not impact performance negatively (see Figure 4).

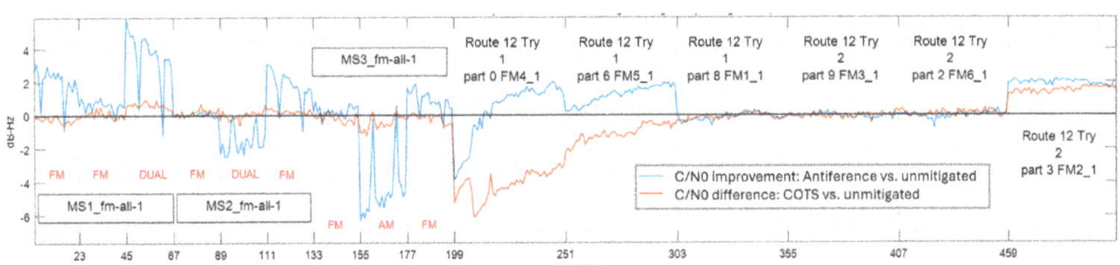

Figure 3. The improvement in observed C/N0 in the presence of FM interference for nine of our recordings (that are depicted head to tail). Antiference shows a better interference suppression than the Rec02 COTS receiver.

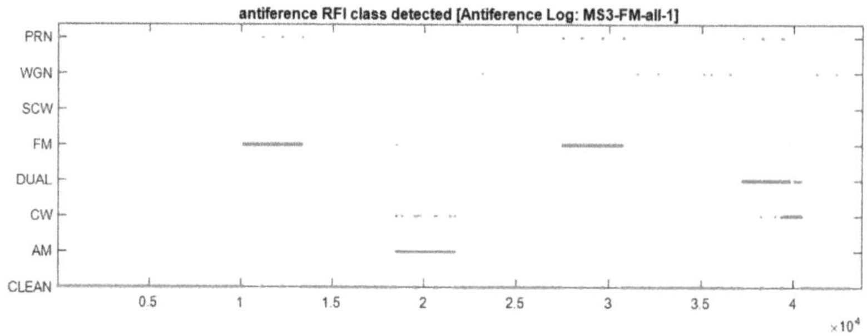

Figure 4. Antiference RFI classification outputs for the first recording from Figure 4. FM interference is hard to distinguish from AM interference and is sometimes mistaken for CW interference. As the mitigation of these types of interference is the same, this does not impact mitigation performance.

In addition, we have looked at the outcomes of the Antiference Logs that provide data on the internal workings of the module, particularly in relation to the false-detection rates. The Antiference algorithms turned out to be highly sensitive, as they effectively detect low-power CW and SCW (chirp) interferences close to 100% of the time. This is not necessarily desirable, as the detection of such low-impact interferences is not needed and comes at the expense of higher-than-needed false-detection rates and lower quality signal due to the removal of some of the signal because of filtering. A simple remedy is the use of a simple majority voting scheme over multiple subsequent RFI classifications. This has been shown to reduce the false detection rates effectively, but at the expense of an increased time-to-detection.

7. Discussion

The Antiference project has been a challenging project. Below, we summarize some of the most important lessons learned.

The detection and classification model has been successful and performs well. However, the system is very sensitive, and picks up very low-power RFI and small perturbations from the environment as well. This has led to a lack of distinguishing between harmless and dangerous RFI. The detection of RFI is made based on small, 10 ms chunks of data. While this solves the problem of detecting pulsed RFI, for the most part, and assures short detection delays, it also leads to some false negatives/positives as 100 detections and classification decisions are made every second. There are many ways this problem can be addressed, such as looking at a window of detections and employing majority voting, or having a secondary model in place that decides on the detection/classification results based on the detection–result streams.

Antiference focused on ML-based methods for the detection and classification model and the fingerprint database. However, the estimation of filter parameters can also be delegated to ML models and might lead to improvements in performance.

There is no standardized way to inform the user of the existence or detected type of RFI. Further developments are needed for meaningful interfacing with GNSS receivers and their operators.

The proper training of ML models requires a massive amount of data. Unfortunately, currently there are not enough free or publicly available recorded datasets that can be used for the training of models. This led to recording campaigns internally, in addition to processing and labelling the data samples manually. However, the amount of data was insufficient for training the models to perform the interference mitigation. In our view, a significantly increased number of GNSS datasets (both clean and containing RFI) is needed for the future widespread development of ML-based RFI detection, classification, and mitigation.

Digital data were used for the training and testing of models. While this was appropriate for detection and classification, the integration of mitigation could not be carried out in this domain. It should be noted that the effective mitigation (that can lead to increased CN/0 in all cases) requires the real-time mitigation of signals in the analogue domain and the inclusion of the effects in the loop. It was suspected that the Antiference system would not be functional in real-time, so the decision was made to keep all processing (barring the COTS receivers) in the digital domain to reduce complexity and keep tests realistic. Until more performant detection and classification models are developed, this will remain a shortcoming of the system.

8. Conclusions

The Antiference project succeeded in the implementation of a prototype, showing the high potential of the application of ML methods to RFI detection and classification. Significant improvements are still needed for the commercialization of such a system. It was convincingly demonstrated that the usage of signal fingerprints and a fingerprint database is feasible and beneficial in documenting the history and streamlining of the performance for subsequent encounters with the same source of RFI.

While the system model developed during the project was a local model, the extended of the general architecture to a cloud-based system can be envisioned, where the detection, classification, and RFI characterization can be carried out remotely on the equipment. The design of very small fingerprints and detection/classification features that can be communicated easily via networks facilitates this approach.

On the other hand, having fingerprints that are explicitly readable by humans facilitates the combination and sharing of databases between equipment. This can result in a better understanding of the growing threats that affect all GNSS receivers in terms of their daily performance and a better cataloguing of the types and frequency of occurrence of different types of RFI.

Author Contributions: Conceptualization, S.A., V.B., A.B. and B.O.; software, V.B. and U.B.; validation, V.B. and U.B.; formal analysis, B.O.; data curation, S.A.; writing—original draft preparation, S.A.; writing—review and editing, A.B., B.O. and S.B.; supervision, S.B.; project administration, S.A.; funding acquisition, S.A., A.B. and B.O. All authors have read and agreed to the published version of the manuscript.

Funding: This research was funded by the European Space Agency, under the program NAVISP Element 1, grant number 4000133535/20/NL/GP.

Institutional Review Board Statement: Not applicable

Informed Consent Statement: Not applicable.

Data Availability Statement: No datasets are publicly available.

Acknowledgments: The reported work was carried out under ESA NAVISP Element 1, which is devoted to the development of innovative PNT, systems, technologies, algorithms, and techniques.

Conflicts of Interest: The authors declare no conflicts of interest. All involved parties including OHB Digital Solutions, S&T, as well as Integricom declare no conflicts of interest.

References

1. Caparra, G.; Ceccato, P.; Formaggio, F.; Laurenti, N.; Tomasin, S. Low Power Selective Denial of Service Attacks Against GNSS. In Proceedings of the 31st International Technical Meeting of the Satellite Division of The Institute of Navigation (ION GNSS+ 2018), Miami, FL, USA, 24–28 September 2018; pp. 3028–3041.
2. Curren, J.; Closas, P.; Navarro, M. A look at the Threat of Systematic Jamming of GNSS. *Inside GNSS* **2017**, *12*, 46–53.
3. Lee, G.; Jo, J.; Park, C. Jamming Prediction for Radar Signals Using Machine Learning Methods. *Secur. Commun. Netw.* **2020**, *2020*, 2151570. [CrossRef]
4. Akos, D. Who's Afraid of the Spoofer? GPS/GNSS Spoofing Detection via Automatic Gain Control (AGC). *Navigation* **2012**, *59*, 281–290. [CrossRef]

5. Arjoune, Y.; Salahdine, F.; Islam, S.; Ghribi, E.; Kaabouch, N. A Novel Jamming Attacks Detection Approach Based on Machine Learning for Wireless Communication. In Proceedings of the 2020 International Conference on Information Networking (ICOIN), Barcelona, Spain, 7–10 January 2020; pp. 459–464.
6. Li, W.; Huang, Z.; Lang, R.; Qin, H.; Zhou, K.; Cao, Y. A Real-Time Interference Monitoring Technique for GNSS Based on a Twin Support Vector Machine Method. *Sensors* **2016**, *16*, 329. [CrossRef]
7. Akeret, J.; Chang, C.; Lucchi, A.; Refregier, A. Radio frequency interference mitigation using deep convolutional neural networks. *Astron. Comput.* **2017**, *18*, 35–39. [CrossRef]
8. Carley, E. Using supervised machine learning to automatically detect type II and III solar radio bursts. In Proceedings of the Machine Learning in Heliophysics, Amsterdam, The Netherlands, 16–20 September 2019; p. 11.
9. Ferre, M.; de la Fuente, R.; Lohan, A.; Simona, E. Jammer classification in GNSS bands via machine learning algorithms. *Sensors* **2019**, *19*, 4841. [CrossRef] [PubMed]
10. Semanjski, S.; Semanjski, I.; De Wilde, W.; Muls, A. Use of supervised machine learning for gnss signal spoofing detection with validation on real-world meaconing and spoofing data—Part I. *Sensors* **2020**, *20*, 1171. [CrossRef] [PubMed]
11. Sun, M.; Qin, Y.; Bao, J.; Yu, X. GPS Spoofing Detection Based on Decision Fusion with a K-out-of-N Rule. *Int. J. Netw. Secur.* **2017**, *19*, 670–674.
12. Shafiee, E.; Mosavi, M.; Moazedi, M. Detection of spoofing attack using machine learning based on multi-layer neural network in single-frequency GPS receivers. *J. Navig.* **2018**, *71*, 169–188. [CrossRef]
13. Schmidt, E.; Gatsis, N.; Akopian, D. A GPS spoofing detection and classification correlator-based technique using the LASSO. *IEEE Trans. Aerosp. Electron. Syst.* **2020**, *56*, 4224–4237. [CrossRef]
14. Kokalj-Filipovic, S.; Miller, R.; Morman, J. Autoencoders for training compact deep learning rf classifiers for wireless protocols. In Proceedings of the 2019 IEEE 20th International Workshop on Signal Processing Advances in Wireless Communications (SPAWC), Cannes, France, 2–5 July 2019; pp. 1–5.
15. O'Shea, T.; Corgan, J.; Clancy, T. Convolutional radio modulation recognition networks. In *Engineering Applications of Neural Networks: Proceedings of the 17th International Conference (EANN 2016), Aberdeen, UK, 2–5 September 2016*; Springer International Publishing: Berlin/Heidelberg, Germany, 2016; Volume 17, pp. 213–226.
16. Shi, Y.; Davaslioglu, K.; Sagduyu, Y.; Headley, W.; Fowler, M.; Green, G. Deep learning for RF signal classification in unknown and dynamic spectrum environments. In Proceedings of the 2019 IEEE International Symposium on Dynamic Spectrum Access Networks (DySPAN), Newark, NJ, USA, 11–14 November 2019; pp. 1–10.
17. Tang, Z.; Li, S.; Yu, L. Implementation of deep learning-based automatic modulation classifier on FPGA SDR platform. *Electronics* **2018**, *7*, 122. [CrossRef]
18. Shah, H.; Koo, I. Reliable machine learning based spectrum sensing in cognitive radio networks. *Wirel. Commun. Mob. Comput.* **2018**, *2018*, 5906097. [CrossRef]
19. de Vriez, C.; Simic, L.; Mahonen, P. The importance of being earnest: Performance of modulation classification for real RF signals. In Proceedings of the 2018 IEEE International Symposium on Dynamic Spectrum Access Networks (DySPAN), Seoul, Republic of Korea, 22–25 October 2018; pp. 1–5.
20. Dovis, F. *GNSS Interference Threats and Countermeasures*; Artech House: Norwood, MA, USA, 2015.
21. Regalia, P. An improved lattice-based adaptive IIR notch filter. *IEEE Trans. Signal Process.* **1991**, *39*, 2124–2128. [CrossRef] [PubMed]
22. Regalia, P. A complex adaptive notch filter. *IEEE Signal Process. Lett.* **2010**, *17*, 937–940. [CrossRef]
23. Chen, Y.; Chien, Y.; Tsao, H. Chirp-like jamming mitigation for GPS receivers using wavelet-packet-transform-assisted adaptive filters. In Proceedings of the 2016 International Computer Symposium (ICS), Chiayi, Taiwan, 15–17 December 2016; pp. 458–461.
24. He, K.; Zhang, X.; Ren, S.; Sun, J. Deep residual learning for image recognition. In Proceedings of the IEEE Conference on Computer Vision and Pattern Recognition, Las Vegas, NV, USA, 27–30 June 2016; pp. 770–778.
25. Keras Applications. Available online: https://keras.io/api/applications/ (accessed on 5 April 2023).
26. KerasTuner API. Available online: https://keras.io/api/keras_tuner/ (accessed on 5 April 2023).
27. Keras BayesianOptimization Tuner. Available online: https://keras.io/api/keras_tuner/tuners/bayesian/ (accessed on 5 April 2023).

MDPI
St. Alban-Anlage 66
4052 Basel
Switzerland
www.mdpi.com

Engineering Proceedings Editorial Office
E-mail: engproc@mdpi.com
www.mdpi.com/journal/engproc

www.ingramcontent.com/pod-product-compliance
Lightning Source LLC
Chambersburg PA
CBHW041015270926
51850CB00024B/119
* 9 7 8 3 7 2 5 8 0 5 9 1 4 *